Fluorine-Containing Synthons

ACS SYMPOSIUM SERIES **911**

Fluorine-Containing Synthons

Vadim A. Soloshonok, Editor
The University of Oklahoma

**Sponsored by the
ACS Divisions of Fluorine Chemistry, Medicinal Chemistry,
and Organic Chemistry**

American Chemical Society, Washington, DC

Library of Congress Cataloging-in-Publication Data

Fluorine-Containing Synthons / Vadim A. Soloshonok, editor.

p. cm.—(ACS symposium series ; 911)

Includes bibliographical references and index.

ISBN 0-8412-3911-8 (alk. paper)

1. Organic compounds—Synthesis—Congresses. 2. Fluorine compounds—Congresses.

I. Soloshonok, V. A. II. American Chemical Society. Divisions of Fluorine Chemistry, Medicinal Chemistry, and Organic Chemistry. III. Series.

QD262. F584 2005
547′.2—dc22 2005041194

The paper used in this publication meets the minimum requirements of American National Standard for Information Sciences—Permanence of Paper for Printed Library Materials, ANSI Z39.48–1984.

Distributed by Oxford University Press

The cover art is reproduced from Scheme 3.2 in Chapter 31.

PRINTED IN THE UNITED STATES OF AMERICA

Foreword

The ACS Symposium Series was first published in 1974 to provide a mechanism for publishing symposia quickly in book form. The purpose of the series is to publish timely, comprehensive books developed from ACS sponsored symposia based on current scientific research. Occasionally, books are developed from symposia sponsored by other organizations when the topic is of keen interest to the chemistry audience.

Before agreeing to publish a book, the proposed table of contents is reviewed for appropriate and comprehensive coverage and for interest to the audience. Some papers may be excluded to better focus the book; others may be added to provide comprehensiveness. When appropriate, overview or introductory chapters are added. Drafts of chapters are peer-reviewed prior to final acceptance or rejection, and manuscripts are prepared in camera-ready format.

As a rule, only original research papers and original review papers are included in the volumes. Verbatim reproductions of previously published papers are not accepted.

ACS Books Department

Contents

Fluorine-Containing Aromatic and Heteroaromatic Synthons

Trifluoromethyl-Containing Synthons

Difluoromethylene-Containing Synthons

Application of Fluorinated Synthons in Crystal Engineering

Fluorine-Containing Amino Acids and Peptides: Fluorinated Synthons for Life Sciences

Indexes

Preface

Recent years have witnessed a spectacular growth of interest in selectively fluorinated chemicals and their utilization in the rational design of new compounds with desired properties. Thus, it becomes customary to see at least one fluorine atom or trifluoromethyl group incorporated into the structure of many newly designed agrochemicals, drugs, and materials (liquid crystals, for example). It is interesting to note that fluorine-containing organic molecules, found extremely rarely in nature, have now gained the status of privileged structural features in the de novo design of biologically relevant compounds. The growing demand for fluorinated compounds has fueled new developments in synthetic methodology, in particular, for the chemo-, regio- and stereoselective introduction of fluorine into organic compounds. One of the most promising developments in this area is the effective exploitation of a building-block, or synthon, approach. In this strategy, incorporation of low molecular weight polyfunctional fluorine-containing synthons becomes a key part of the synthetic scheme. Due to the synthetic versatility associated with the synthon approach, this area of research has been enjoying extraordinary success in compound development. The rapid progress and the importance of this research prompted the ACS Division of Fluorine Chemistry to ask Vadim Soloshonok and Vyacheslav Petrov to organize a symposium entitled "Fluorinated Synthons" at the 226th ACS Meeting in New York City (September 7–11, 2003). Vadim Soloshonok and Vyacheslav Petrov are very grateful for the support of this symposium from the ACS Division of Medicinal Chemistry, Asahi Glass Company, Central Glass Company, IM&T Research Company and Dupont Company.

This ACS Symposium Series book, *Fluorine-Containing Synthons*, is based largely on the topics presented at the symposium. However, to give the readers more complete overview of the current activity in the field, have included in the book several invited chapters written by eminent scientists who recently contributed important new results in synthetic methodology of fluoro-organic chemistry. Therefore, the book is designed to be a comprehensive "snap-shot" of the current state-of-the-art of fluoro-organic methodology and to be useful to practitioners, in both academic and industrial laboratories, working in various areas of chemistry, such as fluorine, organic, bio-organic, medicinal, biochemistry, crystal engineering and material science.

As an organizer of the symposium and the editor of this book, I take this opportunity to gratefully acknowledge the truly outstanding contributions from 30 speakers at the symposium and 35 authors of the book. Their efforts have made this project a very enjoyable task, as well as professionally highly rewarding.

Vadim A. Soloshonok
Associate Professor
Department of Chemistry and Biochemistry
The University of Oklahoma
620 Parrington Oval, Room 208
Norman, OK 73019-3051
Phone: (405)-325-8279
Fax: (405)-325-6111
e-mail: vadim@ou.edu
http://cheminfo.chem.ou.edu/faculty/vas.html

Direct Introduction of Fluorine and Fluorine-Containing Groups into Organic Molecules

Chapter 1

Recent Advances in Perfluoroalkylation Methodology

Teruo Umemoto

IM&T Research, Inc., 6860 North Broadway, Suite B, Denver, CO 80221

Recent developments in perfluoroalkylations are reviewed. They are classified and disscused in the following three categories: electrophilic, nucleophilic, and free radical perfluoroalkylations. Great advances have been made in electrophilic and nucleophilic trifluoromethylation areas involving the generation and synthetic application of CF_3^+ and of CF_3^- which had both long been desired because of their expected broad application.

Introduction

The unique behavior of fluorinated compounds is attributed to the high electronegativity of fluorine (4.0), which is the highest of all elements, combined with its size, close to that of hydrogen (van der Waal's radii; F, 1.47 Å; H, 1.2 Å). Perfluoroalkyl groups, $R_f=C_nF_{2n+1}$, regarded as a gathering of fluorine atoms, have strong effects such as high electronegativity, stability, and lipophilicity (*1,2*). In addtion, long perfluoroalkyl groups of six or more carbons have high water- and oil-resistance. Therefore, the introduction of a perfluoroalkyl group into an organic compound can bring about remarkable changes in the physical, chemical, and biological properties that may result in new chemicals or materials for various applications (*1~4*). Among the perfluoroalkyl groups, the introduction of the trifluoromethyl group has received much interest in medicinal and agricultural chemistry because it is expected to bring positive changes in biochemical activity (*3,4*).

This review covers the development of perfluoroalkylation methodology in the last one or two decades. The perfluoroalkylations are classified and discussed in the following three categories: electrophilic (R_f^+), nucleophilic (R_f^-), and free radical ($R_f^\cdot$) perfluoroalkylations. Some of this content may overlap with the chapters written by other authors in this book and readers are directed to the chapters written by Prof. Prakash and Prof. Langlois, with regards to CF-$_3$SiMe$_3$, CF_3H, and other nucleophilic trifluoromethylating agents.

1. Electrophilic Perfluoroalkylation

In 1984, Yagupolskii *et al.* reported the synthesis and reactivity of (*p*-chloropheyl)(2,4-dimethyl- and *p*-methoxyphenyl)(trifluoromethyl)sulfonium salts **2** and **3** (Scheme 1) (*5*). They synthesized **2** and **3** by treating **1** with *m*-xylene and anisole and reported that these trifluoromethyl sulfonium salts reacted with sodium *p*-nitrothiophenolate to give trifluoromethyl *p*-nitrophenyl sulfide, but did not react with *N,N*-dimethylaniline. This report had not been noticed for some time because of the low reactivity of sulfonium salts.

Scheme 1

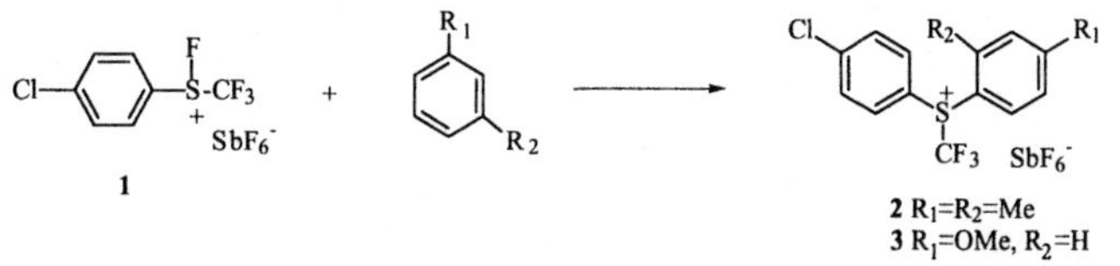

1 + (R1, R2-substituted benzene) → 2, 3 (SbF_6^-)

2 R_1=R_2=Me
3 R_1=OMe, R_2=H

Scheme 2

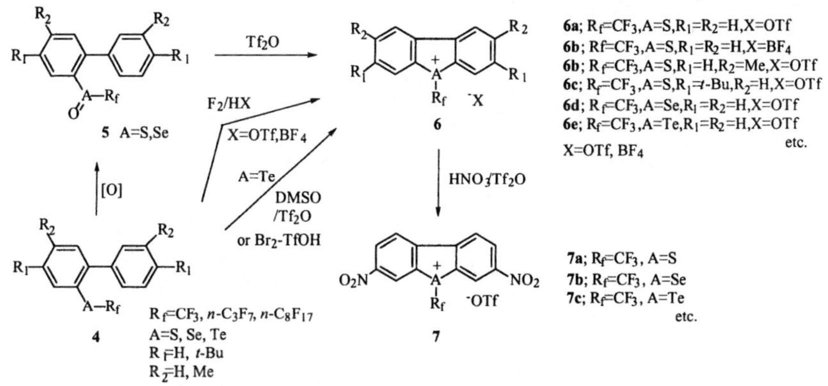

5 A=S,Se; **4**: R_f=CF_3, *n*-C_3F_7, *n*-C_8F_{17}; A=S, Se, Te; R_1=H, *t*-Bu; R_2=H, Me

Reagents: Tf_2O; F_2/HX (X=OTf,BF$_4$); A=Te: DMSO/Tf_2O or Br_2-TfOH; [O]; HNO_3/Tf_2O

6a; R_f=CF_3,A=S,R_1=R_2=H,X=OTf
6b; R_f=CF_3,A=S,R_1=R_2=H,X=BF_4
6b; R_f=CF_3,A=S,R_1=H,R_2=Me,X=OTf
6c; R_f=CF_3,A=S,R_1=*t*-Bu,R_2=H,X=OTf
6d; R_f=CF_3,A=Se,R_1=R_2=H,X=OTf
6e; R_f=CF_3,A=Te,R_1=R_2=H,X=OTf
etc.
X=OTf, BF_4

7a; R_f=CF_3, A=S
7b; R_f=CF_3, A=Se
7c; R_f=CF_3, A=Te
etc.

The author recognized the importance of Yagupolskii's report. In 1990, he and his coworkers reported a new system of heterocyclic chalcogenium salts, *S*,

Se, and *Te*-(perfluoroalkyl)dibenzothio-, -seleno-, and -tellurophenium salts **6** and **7** as power-variable electrophilic perfluoroalkylating agents (Scheme 2) *(6-8)*. 2-R_fS(O)- or R_fSe(O)-biphenyl underwent easy intramolecular cyclization to give the *S* and *Se*-R_f-dibenzothio- and -selenophenium salts **6** in high yields. **6** were also prepared by direct fluorination in the presence of triflic acid (TfOH) or HBF_4 (or BF_3). 2-R_fTe-biphenyl was treated with DMSO and Tf_2O to give *Te*-R_f-dibenzotellurophenium salts **6** (A=Te) in high yields. These triflates were nitrated with HNO_3/Tf_2O to give **7**. **6a** and **6b** were prepared on a large scale by counteranion-exchange reaction of **6** (X=HSO_4) with NaOTf and $NaBF_4$, respectively (*8*).

The reactivity increases in the order of Te<Se<S and the ring substituents alkyl<H<NO_2. More powerful perfluoroalkylating agents react with less reactive nucleophiles with high yields, and less powerful reagents react with more reactive nucleophiles with high yields. Moderately powerful reagents react with moderately reactive nucleophiles also with high yields. Three salts **6d**, **6a**, and **7a** were chosen as useful power-variable trifluoromethylating reagents. The power order is **6d**<**6a**<**7a**, where **7a** is the most powerful reagent.

Scheme 3

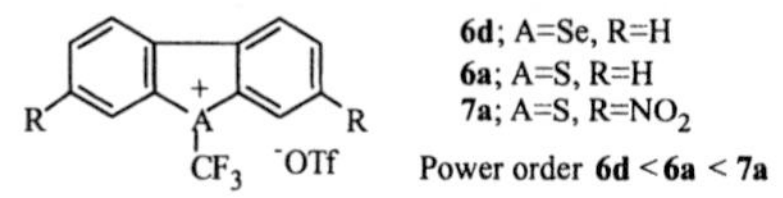

6d; A=Se, R=H
6a; A=S, R=H
7a; A=S, R=NO_2
Power order **6d** < **6a** < **7a**

Scheme 4

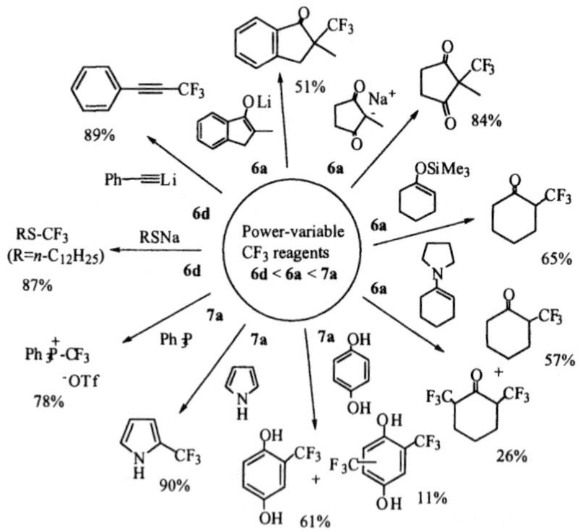

By means of the combination rule, the power-variable trifluoromethylating agents, **6d**, **6a**, and **7a**, made it possible to trifluoromethylate a wide range of nucleophiles such as carbanions, silyl enol ethers, phenols, anilines, phosphines, thiolates, and halides as shown in Scheme 4.

Another new method for the perfluoroalkylation of metal enolates was developed by the combination of *S*-(perfluoroalkyl)dibenzothiophenium triflate **6a** and the boron Lewis acid **8** (Scheme 5) (*9*). **8** was chosen among Lewis acids, which have different Lewis acidities, as the most suitable compound to balance metal enolates with the power of **6a**. By this method, various potassium enolates were trifluoromethylated with high yields. This led to 45%ee enantioselective trifluoromethylation of potassiun enolate of propiophenone using an optical active boron Lewis acid.

Scheme 5

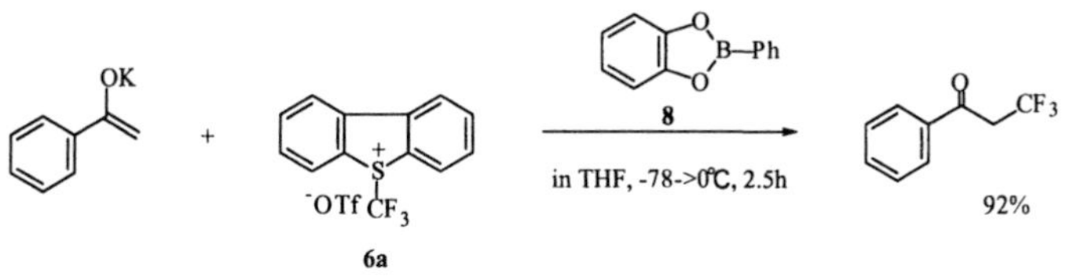

OK + $^{-}$OTf CF$_3$ **6a** — **8** (B–Ph), in THF, -78->0°C, 2.5h → CF$_3$ 92%

Furthermore, the corresponding counteranion-bound salts, *S*, *Se*, and *Te*-(perfluoroalkyl)dibenzo-thio-, -seleno-, and -telluro-phenium-3-sulfonates and their nitro derivatives **9**, were developed (Scheme 6) (*8,10*). **9a** was prepared on a large scale by treatment of 2-$CF_3S(O)$-biphenyl with fuming sulfuric acid (*8*). These salts are useful because water-soluble dibenzoheterocyclic sulfonic acids are easily removed by washing with water after the perfluoroalkylation.

Scheme 6

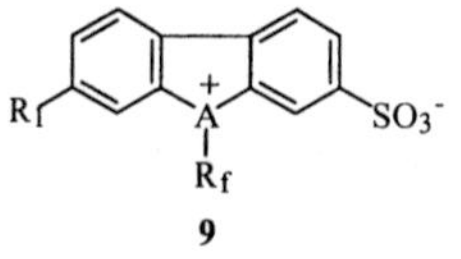

R_1, A$^+$, SO_3^-, R_f **9**

9a; R_f=CF_3,A=S, R_1=H
9b; R_f=CF_3,A=S, R_1=NO_2
9c; R_f=CF_3,A=Se, R_1=H
9d; R_f=CF_3,A=Te, R_1=H
etc.

Direct trifluoromethylation of porphyrins and chlorins was carried out with the most powerful dinitro reagent, **7a**, by Tamiaki *et al.* (*11*). Cahard *et al.* reported mild electrophilic trifluoromethylations of β-keto esters and silyl enol ethers with *S*-(trifluoromethyl)dibenzothiophenium tetrafluoroborate **6b** (*12*). β-Keto esters were treated with **6b** in the presence of excess potassium carbonate and a catalytic amount of tetrabutylammonium iodide in DMF at room temperature for 3 h to give α-trifluoromethyl-β-keto esters with fair to high yields. The diastereomeric excess obtained by the reaction of a (-)-menthyl β-keto ester was 4% (chemical yield 97%). Silyl enol ethers were treated with **6b** in DMF in the presence of $Bu_4N^+[Ph_3SnF_2]^-$ to give α-trifluoromethyl ketones with fair to high yields.

The corresponding trifluoromethyloxonium salts, *O*-(trifluoromethyl)-dibenzofuranium salts, were reported by the author and his coworkers (*13*). For the first time, thermally unstable *O*-trifluoromethyloxonium salts **11** were synthesized by photolysis of the diazonium salts **10** at very low temperature (Scheme 7). NMR analysis at low temperature demonstrated the formation of the *O*-CF_3 salts **11**. As the temperature rises, the salts begin to decompose gradually at -70°C or more and then rapidly at -30°C. The decomposition product was CF_4. This indicates easy generation of CF_3^+ from the CF_3 oxonium salts. The stability increases in the order of the ring-substituents, F<H<alkyl, and the counteranions, $BF_4^- < PF_6^- < SbF_6^- < Sb_2F_{11}^-$. More electrodonating ring-substituents and less nucleophilicity of the counteranions increase the stability of the CF_3-oxonium salts.

Scheme 7

R CF3O
N2+ X-
10
hv
-99~-90℃
in CH2Cl2
R
O+
CF3 -X
11
X=BF4, PF6, SbF6, Sb2F11
R=H, *t*-Bu, F

The *O*-CF_3 salts made possible direct *O*-and *N*-triflurometylations as shown in Scheme 8. Alcohols and phenols were treated with **11a** in the presence of pyridine or a bulky trialkyl amine base at low temperature in methylene chloride to give trifluoromethyl ethers in good yields. **11a** reacted with primary and secondary amines to give *N*-CF_3-amines in good yields. The reaction with *tert.*-amines and pyridines to give *N*-CF_3-ammonium and -pyridinium salts also produced good yields.

Scheme 8

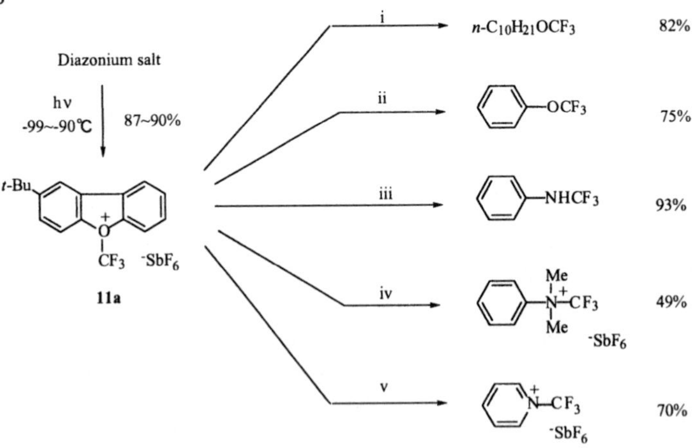

i. *n*-$C_{10}H_{21}OH$(1 eq)/(*i*-Pr)$_2$NEt(1eq). ii. phenol(1 eq)/(*i*-Pr)$_2$NEt(1 eq).
iii. aniline(2 eq). iv. *N,N*-dimethylaniline(1 eq). v. pyridine(1 eq).
Reaction conditions; -90 -> -10℃, 3~5 h

The reactivity between *O*- and *S*-CF_3 onium salts was compared. With aniline, the *O*-CF_3 salts produced *N*-CF_3 aniline, while *S*-CF_3 produced *C*-CF_3 products, a mixture of *o*- and *p*-CF_3 anilines. With an alkyl amine, the *O*-CF_3 salt produced the *N*-CF_3 amine, while the *S*-CF_3 salts decomposed with the formation of CF_3H. These remarkable differences are more likely due to the difference between the reaction mechanism. The *O*-CF_3 salts should undertake S_N1 or S_N2 mechanism including real CF_3^+ or $CF_3^{\delta+}$, while the *S*-CF_3 salts should undertake an electron-transfer type mechanism through the complex. The differences in the mechanisms are likely due to the electron-withdrawing effect of the hetero atoms bonding to the CF_3 carbon. The O-atom of the CF_3-O salt should have a strong electron-withdrawing effect, enough to generate the real CF_3^+, but the S-atom of the CF_3-S salt should not have the effect strong enough to generate the real CF_3^+.

Shreeve *et al.* reported new electrophilic trifluoromethylating agents (Scheme 9) (*14*). A non-heterocyclic system, fluoro- and nitro-substituted trifluoromethyldiarylsulfonium triflates **13** and **14**, were developed. A mixture of **12** and triflic anhydride was treated with fluorobenzene or *m*-difluorobenzene to give salts **13** in good yields. **13** were nitrated with HNO_3/fuming sulfuric acid to give nitro derivatives **14** in good yields. The reactivity of the sulfonium salts depends on the electron-withdrawing effect of the substituents, and **14b** has the highest reactivity in this series. Phenols, anilines, and pyrrole were trifluoromethylated in good yields with **14b**.

Scheme 9

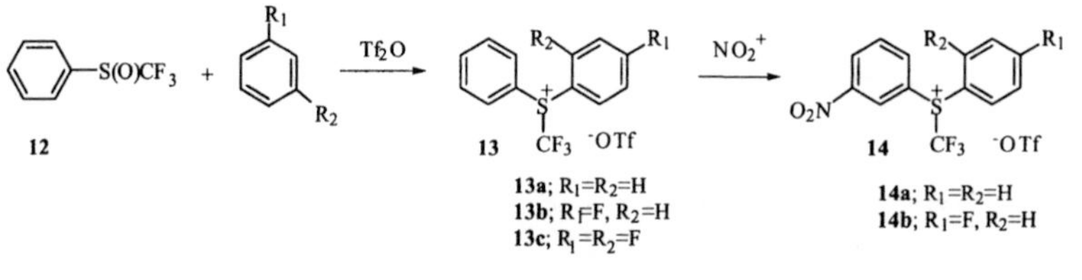

A kinetic study of the trifluoromethylation with *S*-CF_3 sulfonium salts **6a**, **7a**, and **13a** was carried out (*15*). It was clarified that the high reactivity of the heterocyclic salts **6a** and **7a**, compared to a non-heterocylic salt **13a**, is due to the greatly enhanced activation entropy, steric factors, and the high reactivity of salt **7a** is due to the great electron deficiency caused by two nitro groups in addition to the heterocyclic system.

Most recently, regioselective trifluoromethylation of anilines, indolines, and indoles by treatment with CCl_4/HF/SbF_5 followed by HF/pyridine was reported (*16,17*). For example, *p*-toluidine was regioselectively trifluoromethylated to give 3-CF_3-*p*-toluidine in 85% yield, while *m*-toluidine gave 4-CF_3-*m*-toluidine in 81% yield. The real reactive species of these reactions is CCl_3^+, which reacts with aromatics to form CCl_3-aromatics, which undertake halogen-exchange reaction with fluorine to give the CF_3-products. This reaction was successfully applied to the trifluoromethylation of a vasodilator (*17*).

2. Nucleophilic Perfluoroalkylation

In this area, two reports in 1989 and 1991 were noticed. One was Prakash's report that R_fSiMe_3 reacted with carbonyl compounds in the presence of a catalytic amount of fluoride anion to give Rf carbinols **15** in high yields (Scheme 10) (*18*). Stahly *et al.* also reported the similar trifluoromethylation of benzoquinones (*19*). The other was Shono's report that CF_3H reacted with benzaldehyde in DMF in the presence of base **16** (R=n-C_8H_{17}) generated by electrolysis to give a CF_3 carbinol in high yields (Scheme 11) (*20*). He also reported that CF_3H reacted with benzaldehyde in DMF in the presence of LiH, NaH, and *t*-BuOK to give the CF_3 carbinol in 0, 28, and 40% yields, respectively.

Scheme 10

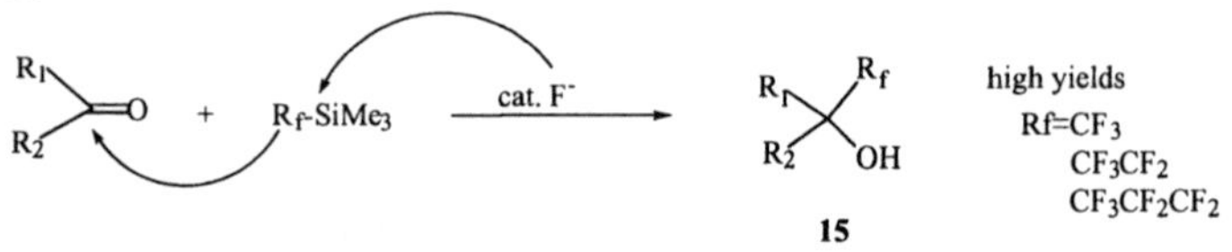

$R_1R_2C{=}O$ + R_f-$SiMe_3$ —(cat. F^-)→ $R_1R_2C(R_f)OH$ **15**

high yields
Rf=CF_3
CF_3CF_2
$CF_3CF_2CF_2$

Scheme 11

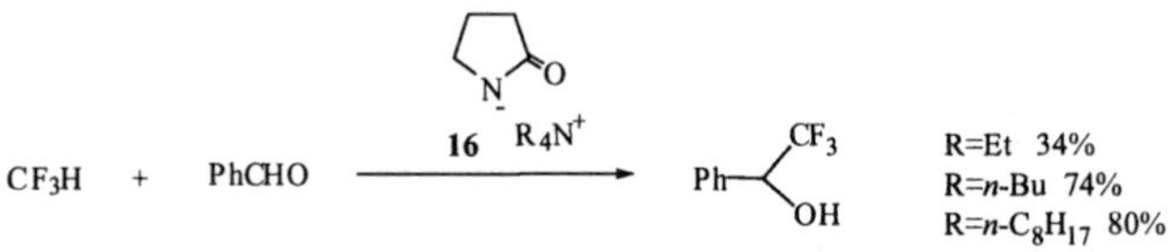

CF_3H + PhCHO —(**16** R_4N^+ [2-pyrrolidinone anion])→ $PhCH(CF_3)OH$

R=Et 34%
R=n-Bu 74%
R=n-C_8H_{17} 80%

Prakash's report has had a great impact upon synthetic chemists. Since then, nucleophilic perfluoroalkylation using R_fSiMe_3 has been extensively studied by Prakash and many other chemists. Readers can refer to some reviews which have been published (*21~23*). The advances of these nucleophilic perfluoroalkylations are also described in the chapter by Prof. Prakash in this book.

On the other hand, Shono's report came as a surprise to many organofluorine chemists, because CF_3^- has long been believed to be very unstable and to immediately decompose to reactive CF_2 carbene and fluoride anion ($CF_3^- \rightarrow CF_2{:} + F^-$). After a while, Shono's reaction was examined in detail by many organofluorine chemists (*24~29*). It was elucidated that the solvent, DMF, has a crucial role in the reactions. The real trifluoromethylating agent is an adduct of CF_3^- to a DMF molecule, trifluoroacetaldehyde hemiacetal **17** (Scheme 12). The adduct has been demonstrated to act as a useful trifluoromethylating agent for various carbonyl compounds. Readers can refer to a review which was recently written by Prof. Langlois *et al.* (*30*). The findings led to recent advances for the use of the adduct and its derivatives and these are described in the chapter by Prof. Langlois in this book.

Scheme 12

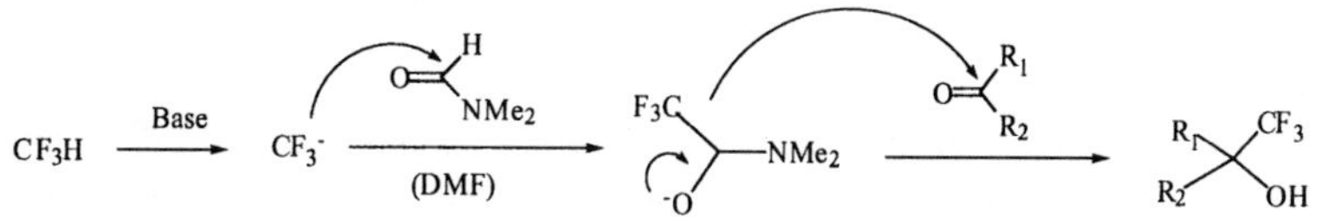

CF_3H —Base→ CF_3^- —(DMF)→ **17** → $R_1R_2C(CF_3)OH$

In 1996, another remarkable report appeared. Yokoyama *et al.* succeeded in the activation of $PhSCF_3$ by a germyl anion which has strong affinity to the sulfur atom. They reported that $PhSCF_3$ reacted with aldehydes in the presence of Et_3GeNa in THF/HMPA at -60°C to give CF_3-carbinols **18** in 92-96% yield (Scheme 13) (*31*). They suggested the formation of CF_3Na in the reactions. Et_3GeLi and Et_3GeK did not give the products and the reaction temperature was an important factor: 12% yield at -20°C and only trace amounts at -78°C. $PhSCF_3$ is another type of nucleophilic trifluoromethylating agent in place of CF_3SiMe_3. This method was also used for trifluoromethylation of imines **19** (Scheme 13) (*32*) and for chemoselective trifluoromethylation of methyl esters (*33*).

Scheme 13

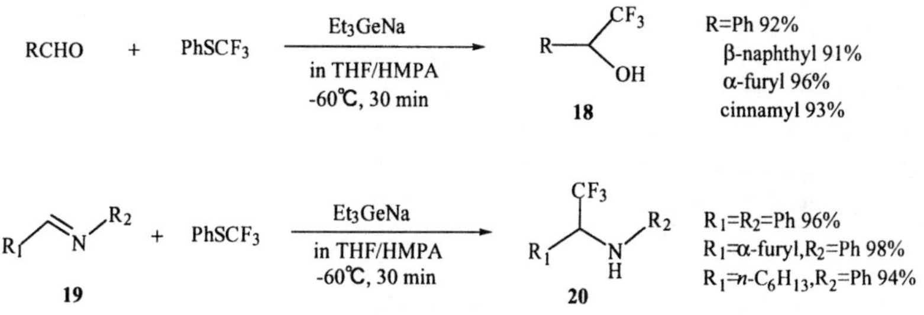

RCHO + $PhSCF_3$ —Et_3GeNa, in THF/HMPA, -60°C, 30 min→ **18**

R=Ph 92%
β-naphthyl 91%
α-furyl 96%
cinnamyl 93%

19 + $PhSCF_3$ —Et_3GeNa, in THF/HMPA, -60°C, 30 min→ **20**

$R_1=R_2=Ph$ 96%
R_1=α-furyl, R_2=Ph 98%
R_1=*n*-C_6H_{13}, R_2=Ph 94%

Most recently, Prakash *et al.* reported that $PhSO_2CF_3$ or $PhSOCF_3$ reacted with carbonyl compounds in DMF at -50°C to room temperature in the presence of *t*-BuOK to give CF_3 carbinols in high yields (Scheme 14) (*34*). Since this reaction in DMSO solvent gave a similar yield of the product as in DMF solvent, the adduct **17** of CF_3^- and DMF is not a necessary intermediate.

Scheme 14

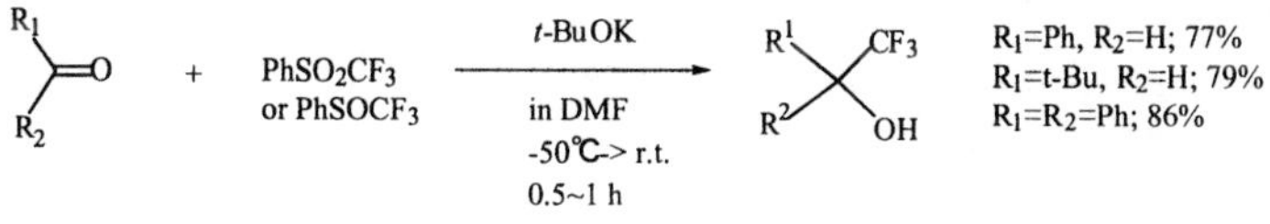

$R_1R_2C=O$ + $PhSO_2CF_3$ or $PhSOCF_3$ —*t*-BuOK, in DMF, -50°C-> r.t., 0.5~1 h→ $R^1R^2C(CF_3)OH$

R_1=Ph, R_2=H; 77%
R_1=t-Bu, R_2=H; 79%
$R_1=R_2$=Ph; 86%

Dolbier *et al.* reported nucleophilic trifluoromethylation of carbonyl compounds based on the photoinduced reduction of CF_3I by tetrakis(dimethylamino)ethylene (TDAE) (*35*). They found that the reaction of CF_3I, carbonyl compounds, and TDAE in DMF was significantly accelerated by photoirradiation (Scheme 15). This method afforded high yields of CF_3 carbinols. The adduct **17** of CF_3^- and DMF is not a necessary intermediate because DME solvent and a mixture of DME/HMPA gave similar results. They suggested the formation of a complex between CF_3^- anion and $TDAE^{2+}$ dication as an active trifluoromethylating agent and this method was applied to trifluoromethylation of vicinal diol cyclic sulfates (*36*).

Scheme 15

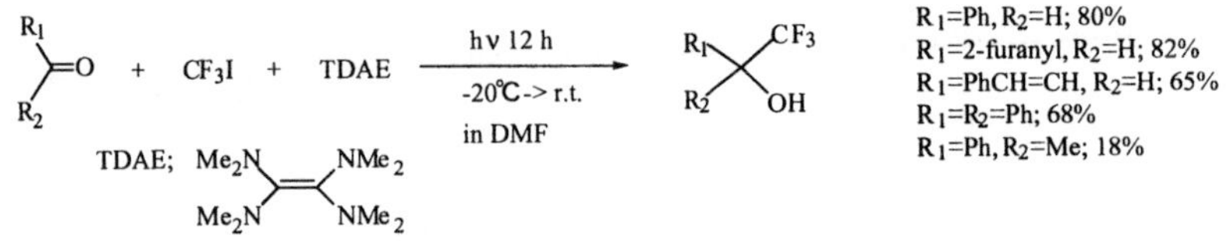

$R_1R_2C{=}O$ + CF_3I + TDAE —(hν 12 h, -20°C -> r.t., in DMF)→ $R_1R_2C(CF_3)OH$

TDAE; $(Me_2N)_2C{=}C(NMe_2)_2$

R_1=Ph, R_2=H; 80%
R_1=2-furanyl, R_2=H; 82%
R_1=PhCH=CH, R_2=H; 65%
R_1=R_2=Ph; 68%
R_1=Ph, R_2=Me; 18%

Scheme 16

CF_2X_2 (X=Br, Cl) —(Cd or Zn, in DMF, r.t.)→ $CF_3MX + (CF_3)_2M$ (M=Cd or Zn) 80-95% —(Cu(I) salt, -80°C -> r.t.)→ CF_3Cu 90-100% —(Ar-I)→ Ar-CF_3

Ar; Ph (100%), *o*-NO_2Ph (95%)
o-MePh (95%), *o*-MeOPh (78%)
p-BrPh (95%)

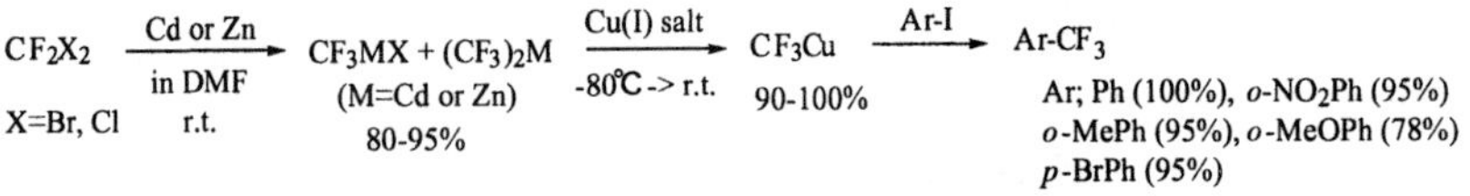

As mentioned above, difluorocarbene is a reactive species which results from decomposition of CF_3^-. If its reverse reaction (CF_2: + F^- → CF_3^-) is synthetically usable, it would be another method of generating CF_3^-. CF_3^- is stabilized by the vacant orbitals of transition metals (Cu, Zn, Cd, Hg cations etc.). In 1986, Burton *et al.* first succeeded in using the reverse reaction for the preparation of a useful trifluoromethylating agent, CF_3Cu, through CF_3MX and $(CF_3)_2M$ (M=Cd or Zn) (Scheme 16). CF_2Br_2 reacted with Cd or Zn and then with CuI to give CF_3Cu, which was used for the halogen exchange reaction of aryl iodides, giving trifluoromethylarenes in high yields (*37*).

Chen *et al.* reported a new method for the direct generation of CF_3Cu via CF_2: using ICF_2SO_2F/Cu (*38*) or FSO_2CF_2COOMe/CuI (*39*). In the last decade, Chen and his coworkers found other fluorinated compounds to be sources of CF_3^- for the CF_3Cu reagent. CF_3Cu was generated by using $ClCF_2COOMe$/KF/CuI (*40*), XCF_2COOMe(X=Cl,Br)/KF/CuI/CdI_2 (*41*), $FSO_2CF_2CF_2OCF_2COOMe$/CuI (*42*), FSO_2CF_2COOMe/CuI (*43~49*), or $FSO_2CF_2CF_2COCF_2COOK$/CuI (*50*). For example, phenyl iodide reacted with $ClCF_2COOMe$, KF, and CuI in DMF at 120°C for 8 h to give $PhCF_3$ with an 88% yield (Scheme 17). The mechanism for the generation of CF_3Cu from XCF_2COOMe/KF/CuI is also shown in Scheme 17.

As an alternative method for generating CF_3Cu, using CF_3SiMe_3/CuI/KF was reported (*51*) as an extension of Fuchikami's method (*52*).

Scheme 17

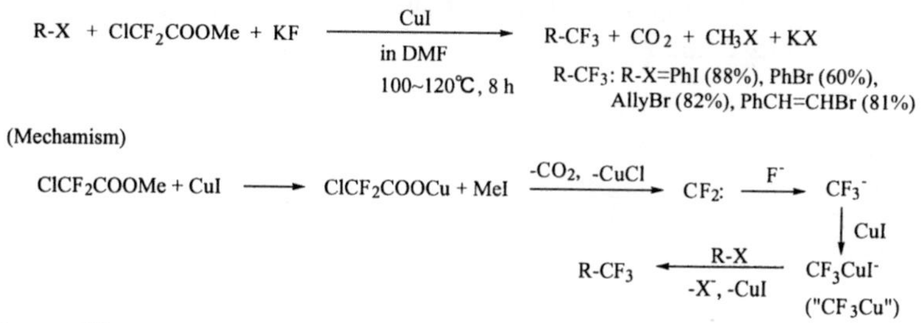

$$\text{R-X} + \text{ClCF}_2\text{COOMe} + \text{KF} \xrightarrow[\text{in DMF, } 100\sim120^{\circ}\text{C, 8 h}]{\text{CuI}} \text{R-CF}_3 + \text{CO}_2 + \text{CH}_3\text{X} + \text{KX}$$

R-CF_3: R-X=PhI (88%), PhBr (60%), AllyBr (82%), PhCH=CHBr (81%)

(Mechamism)

$$\text{ClCF}_2\text{COOMe} + \text{CuI} \longrightarrow \text{ClCF}_2\text{COOCu} + \text{MeI} \xrightarrow{-\text{CO}_2,\ -\text{CuCl}} \text{CF}_2\text{:} \xrightarrow{\text{F}^-} \text{CF}_3^- \xrightarrow{\text{CuI}} \text{CF}_3\text{CuI}^-\ (\text{"CF}_3\text{Cu"}) \xrightarrow[-\text{X}^-,\ -\text{CuI}]{\text{R-X}} \text{R-CF}_3$$

Scheme 18

$$\text{R-X} + \text{FSO}_2\text{COOMe} + \text{S}_8 \xrightarrow[\text{in HMPA, } 100\sim110^{\circ}\text{C, 8h}]{\text{CuI}} \text{R-SCF}_3 + \text{SO}_2 + \text{CO}_2 + \text{CH}_3\text{X}$$

R-SCF_3: R-X=PhI (60%), *p*-ClC_6H_4I (63%), *p*-$NO_2C_6H_4I$ (74%)

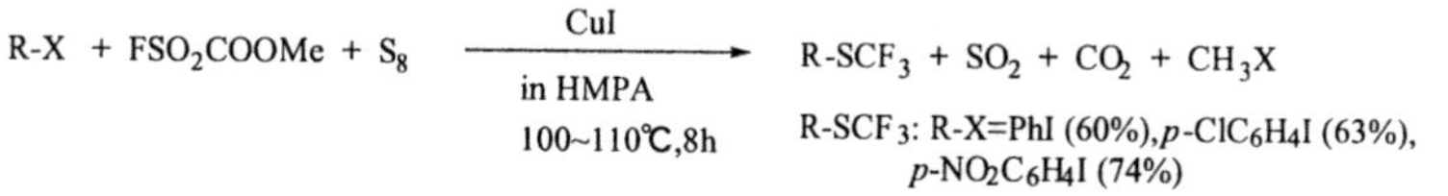

$FSO_2CF_2COOMe/S_8/CuI$ gave CF_3SCu and this method provided CF_3S-containing compounds from halo compounds (Scheme 18) (*53*).

3. Free Radical Perfluoroalkylation

In the last decade, many free radical perfluoroalkylations have been reported. A new approach to the photochemical trifluoromethylation of aromatic compounds using CF_3COOAg as a source of CF_3 free radicals and powdered TiO_2 as a photocatalyst was described by Mallouk *et al.* (*54*). By this method, various aromatics were trifluoromethylated in low to moderate yields (Scheme 19).

Another new system for radical trifluoromethylation, using $Bi(CF_3)_3$/$Cu(OAc)_2$, was reported by Naumann (*55*). The Cu salt was essential for this reaction. By this method, enamines were trifluoromethylated to give CF_3-enamines (Scheme 20). *N,N*-Diethylaniline gave a 1.7:1 mixture of *o*- and *p*-CF_3-*N,N*-diethylaniline in 48% yield. Thiophenolate produced a mixture of CF_3SPh and PhSSPh.

Scheme 19

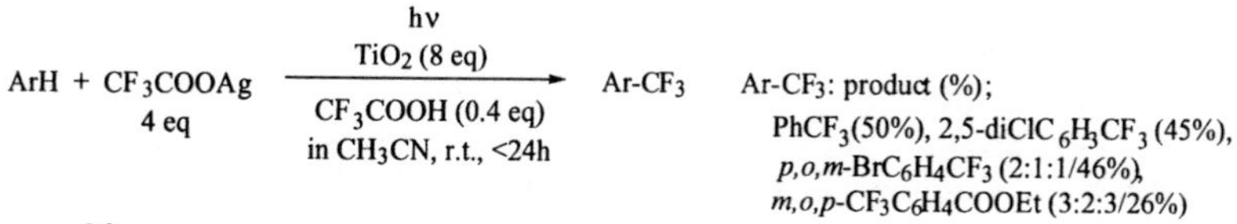

$$\text{ArH} + \underset{4\text{ eq}}{\text{CF}_3\text{COOAg}} \xrightarrow[\text{CF}_3\text{COOH (0.4 eq), in CH}_3\text{CN, r.t., <24h}]{h\nu,\ \text{TiO}_2\text{ (8 eq)}} \text{Ar-CF}_3$$

Ar-CF_3: product (%);
$PhCF_3$(50%), 2,5-di$ClC_6H_3CF_3$ (45%),
p,o,m-$BrC_6H_4CF_3$ (2:1:1/46%),
m,o,p-$CF_3C_6H_4COOEt$ (3:2:3/26%)

Scheme 20

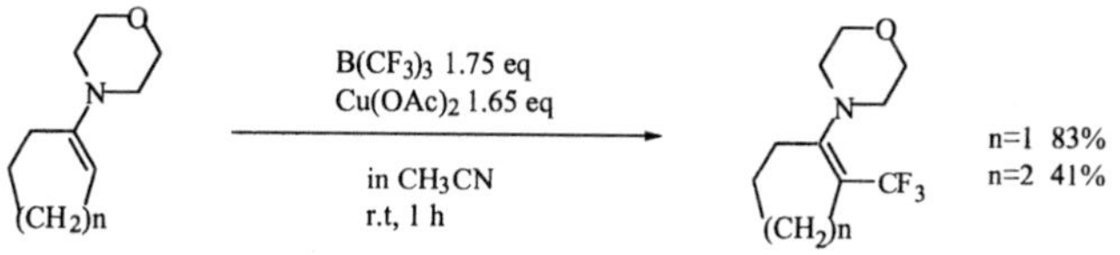

Morpholino enamine (ring $(CH_2)n$) → [$B(CF_3)_3$ 1.75 eq, $Cu(OAc)_2$ 1.65 eq, in CH_3CN, r.t, 1 h] → CF_3-substituted morpholino enamine (ring $(CH_2)n$)

n=1 83%
n=2 41%

Diastereoselective trifluoromethylation of chiral imide enolates with excess CF_3I mediated by Et_3B was described by Iseki *et al.* (*56*). A maximum of 86% de was achieved (Scheme 21).

Scheme 21

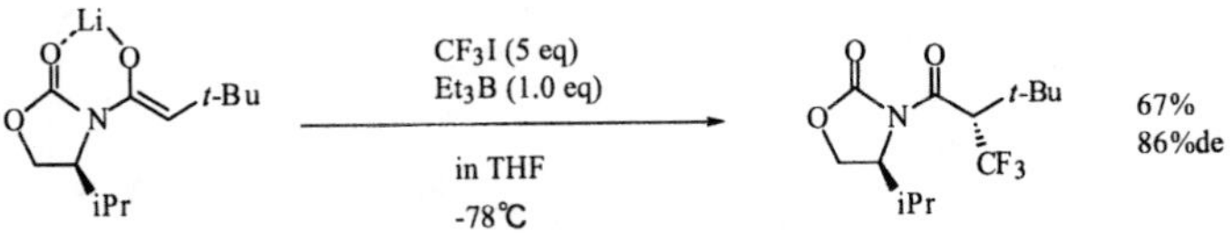

Li
O
O
O
N
t-Bu
iPr
CF3I (5 eq)
Et3B (1.0 eq)
in THF
-78℃
CF3
67%
86%de

Nagano *et al.* reported hydroxytrifluoromethylation of α,β-unsaturated esters with CF_3I(excess)/Et_3B/KF/H_2O in a THF solvent at room temperature (Scheme 22) (*57*).

Scheme 22

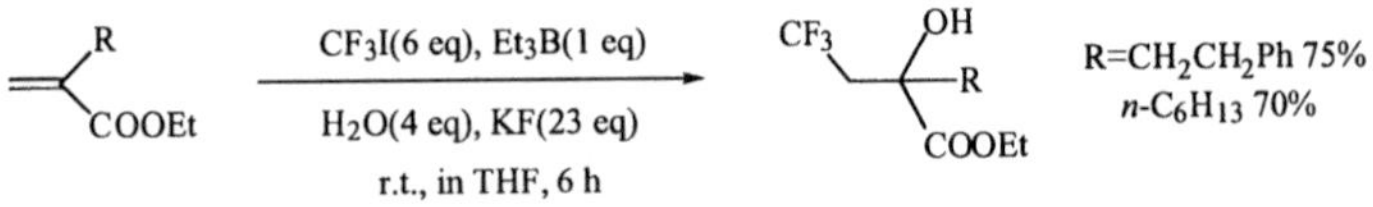

R
COOEt
CF3I(6 eq), Et3B(1 eq)
H2O(4 eq), KF(23 eq)
r.t., in THF, 6 h
CF3
OH
-R
COOEt
R=CH2CH2Ph 75%
n-C6H13 70%

CF_3S-containing α-amino acids were synthesized from dithio-amino acids with CF_3SO_2Na/*t*-BuOOH (*58*). Perfluoroalkylation of aliphatic thiols with R_fI or R_fBr in the presence of NaO_2SCH_2OH in DMF/H_2O at room temperature was described (*59*). Trifluoromethylations of aromatic amines using CF_3I/Zn/SO_2 (*60*), a D-mannose derived ketene acetal using CF_3I or CF_3Br/HCO_2Na/SO_2 (*61*), and [60]- and [70]fullerenes using CF_3COOAg at 300°C (*62*) were also described. Photolysis of CF_3SO_2SR or CF_3COSR in the presence of alkenes provided CF_3-alkenes and CF_3SR (*63*) (Scheme 23).

Scheme 23

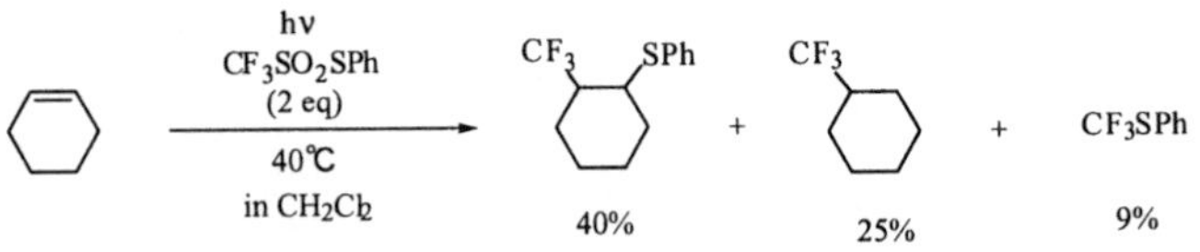

hν
CF3SO2SPh
(2 eq)
40℃
in CH2Cl2
CF3
SPh
40%
+
CF3
25%
+
CF3SPh
9%

In the last decade, many electrochemical perfluoroalkylations were reported. Electrochemical trifluoromethylations of butyl acrylate (*64*), dialkyl fumarates (Scheme 24) (*65*), dimethyl acetylenedicarboxylate (*66*), α,β-unsaturated nitriles (*67*), aromatic compounds (*68*), and benzonitrile (*69*) using trifluoroacetic acid as a free radical CF_3 source were described. Electrochemical trifluoromethylations of aromatic compounds and alkenes using CF_3SO_2K (*70*) and thiophenol using CF_3Br/SO_2 (*71*) were also reported. It was reported that SO_2 may serve as an efficient catalysis for the electrochemical trifluoromethylation of thiophenol with CF_3Br (*71*).

Scheme 24

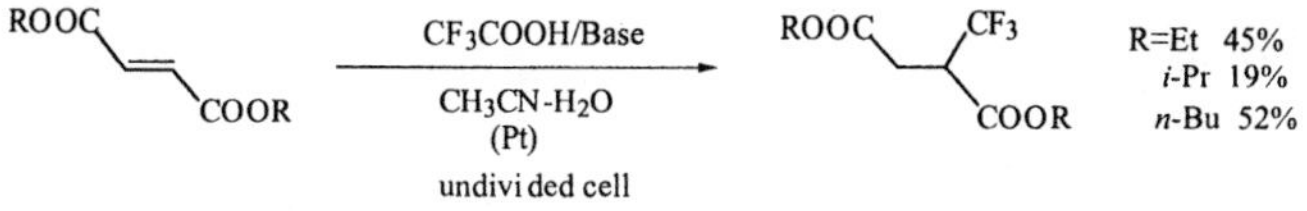

$$\text{ROOC-CH=CH-COOR} \xrightarrow[\text{CH}_3\text{CN-H}_2\text{O}\ (\text{Pt})\ \text{undivided cell}]{\text{CF}_3\text{COOH/Base}} \text{ROOC-CH}_2\text{-CH(CF}_3\text{)-COOR}$$

R=Et 45%
i-Pr 19%
n-Bu 52%

Conclusion

As mentioned above, perfluoroalkylation has been extensively studied and developed in the previous two decades. In particular, there were two breakthroughs in the field, the successful generation and synthetic applications of CF_3^+ and of CF_3^-. Until now it had been believed that both were difficult to generate. CF_3^+ was difficult to generate due to three strongly electronegative fluorine atoms bonding to the cationic carbon of CF_3^+, and CF_3^- is very unstable due to repulsion between the lone paired electrons of three fluorine atoms and the electrons of the carbon atom of CF_3^-. These obstacles have been overcome by means of the use of other elements such as O, S, Se, N, Si, Ge, Zn, Cd, and Cu.

Literature cited

1. *Organofluorine Chemistry: Principles and Commercial Applications*; Banks, R. E.; Smart, B. E.; Tatlow, J. C., Eds.; Plenum: New York, 1994, and references cited therein.
2. *Chemistry of Organic Fluorine Chemistry II*; Hudlicky, M.; Pavlath, A. E., Eds.; ACS Monograph 187, American Chemical Society: Washington, DC, 1995, and references cited therein.
3. *Biomedical Frontiers of Fluorine Chemistry*; Ojima, I., McCarthy, J. R.; Welch, J. T., Eds.; ACS Symposium Series 639; American Chemical Society: Washington, DC, 1996, and references cited therein.
4. *Organofluorine Compounds in Medicinal Chemistry and Biomedical Applications*; Filler, R.; Kobayashi, Y.; Yagupolskii, L. M., Eds.; Elsevier: Amsterdam, 1993, and references cited therein.
5. Yagupolskii, L. N.; Kondratenko, N. V.; Timofeeva, G. N. *J. Org. Chem. USSR* **1984**, *20*, 103-105.
6. Umemoto, T.; Ishihara, S. *Tetrahedron Lett.* **1990**, *31*, 3579-3582.
7. Umemoto, T.; Ishihara, S. *J. Am. Chem. Soc.* **1993**, *115*, 2156-2164.
8. Umemoto, T.; Ishihara, S. *J. Fluorine Chem.* **1999**, *98*, 75-81.
9. Umemoto, T.; Adachi, K. *J. Org. Chem.* **1994**, *59*, 5692-5699.
10. Umemoto, T.; Ishihara, S.; Adachi, K. *J. Fluorine Chem.* **1995**, *74*, 77-82.
11. Tamiaki, H.; Nagat, Y.; Tsudzuki, S. *Eur. J. Org. Chem.* **1999**, 2471-2473.
12. Ma, J. -A.; Cahard, D. *J. Org. Chem.* **2003**, *68*, 8726-8729.
13. Umemoto, T. *Chem. Rev.* **1996**, *96*, 1757-1777: Adachi, K.; Ishihara, S.; Umemoto, T. *15th International Symposium on Fluorine Chemistry*,

Vancouver, Canada, August **1997**, Abstract O(2)C-6: Umemoto, T.; Adachi, K.; Ishihara, S. *219th American Chemical Society National Meeting*, March **2000**, San Francisco, U.S.A., Abstract FLUO 31: Jpn. Kokai Tokkyo Koho JP H7(1995)-330703: U. S. Patent 6,239,289 B1: full paper in preparation.

14. Yang, J. -J.; Kirchmeier, R. L.; Shreeve, J. M. *J. Org. Chem.* **1998**, *63*, 2656-2660.
15. Ono, T.; Umemoto, T. *J. Fluorine Chem.* **1996**, *80*, 163-166.
16. Debarge, S.; Violeau, B.; Bendaoud, N.; Jouannetaud, M. -P.; Jacquesy, J. -C. *Tetrahedron Lett.* **2003**, *44*, 1747-1750.
17. Debarge, S.; Kassou, K.; Carreyre, H.; Violeau, B.; Jouannetaud, M. -P.; Jacquesy, J. -C. *Tetrahedron Lett.* **2004**, *45*, 21-23
18. Prakash, G. K. S.; Krishnamurti, R.; Olah, G. A. *J. Am. Chem. Soc.* **1989**, *111*, 393-395; also see Krishnamurti, R.; Bellew, D. R.; Prakash, G. K. S. *J. Org. Chem.* **1991**, *56*, 984-989.
19. Stahly, G. P.; Bell, D. R. *J. Org. Chem.* **1989**, *54*, 2873-2877.
20. Shono, T.; Ishifune, M.; Okada, T.; Kashimura, S. *J. Org. Chem.* **1991**, *56*, 2-4.
21. Prakash, G. K. S.; Yudin, A. K. *Chem. Rev.* **1997**, *97*, 757-786.
22. Singh, P. R.; Shreeve, J. M. *Tetrahedron* **2000**, *56*, 7613-7632.
23. Prakash, G. K. S.; Mandal, M. *J. Fluorine Chem.* **2001**, *112*, 123-131.
24. Barhdadi, R.; Troupel, M.; Perichon, J. *J. Chem. Soc., Chem. Commun.* **1998**, 1251-1252.
25. Folleas, B.; Marek, I.; Normant, J. F.; Saint-Jalmes, L. *Tetrahedron Lett.* **1998**, *39*, 2973-2976.
26. Folleas, B.; Marek, I.; Normant, J. F.; Saint-Jalmes, L. *Tetrahedron* **2000**, *56*, 275-283.
27. Russell, J.; Roques, N. *Tetrahedron* **1998**, *54*, 13771-13782.
28. Large, S.; Roques, N.; Langlois, B. R. *J. Org. Chem.* **2000**, *65*, 8848-8856.
29. Roques, N.; Russell, J.; Langlois, B.; Saint-Jalmes, L.; Large, S. *Patent* WO 98/22435, **1998**.
30. Langlois, B. R.; Billard, T. *Synthesis* **2003**, 185-194.
31. Yokoyama, Y.; Mochida, K. *Synlett* **1996**, 1191-1192.
32. Yokoyama, Y.; Mochida, K. *Tetrahedron Lett.* **1997**, *38*, 3443-3446.
33. Yokoyama, Y.; Mochida, K. *Synlett* **1997**, 907-908.
34. Prakash, G. K. S.; Hu, J.; Olah, G. A. *Org. Lett.* **2003**, *5*, 3253-3256.
35. Ait-Mohand, S.; Takechi, N.; Medebielle, M.; Dolbier Jr, W. R. *Org. Lett.* **2001**, *3*, 4271-4273.
36. Takechi, N.; Ait-Mohand, S.; Medebielle, M.; Dolbier Jr, W. R. *Org. Lett.* **2002**, *4*, 4671-4672.
37. Wiemers, D. M.; Burton, D. J. *J. Am. Chem. Soc.* **1986**, *108*, 832-834.
38. Chen, Q. -Y.; Wu, S. -W. *J. Chem. Soc. Perkin Trans.1*, **1989**, 2385-2387.
39. Chen, Q. -Y.; Wu, S. -W. *J. Chem. Soc. Chem. Commun.* **1989**, 705-706.
40. Duan, J. -X.; Su, D. -B.; Chen, Q. -Y. *J. Fluorine Chem.* **1993**, *61*, 279-284.
41. Chen, Q. -Y.; Duan, J. -X. *Tetrahedron Lett.* **1993**, *34*, 4241-4244.

42. Chen, Q. -Y.; Duan, J. -Y. *J. Chem. Soc., Chem. Commun.* **1993**, 1389-1391.
43. Duan, J. -X.; Su, D. -B.; We, J. -P.; Chen, Q. -Y. *J. Fluorine Chem.* **1994**, *66*, 167-169.
44. Duan, J.; Dolbier Jr, W. R.; Chen, Q. -Y. *J. Org. Chem.* **1998**, *63*, 9486-9489.
45. Zhang, X.; Qing, F. -L.; Yu, Y. *J. Org. Chem.* **2000**, *65*, 7075-7082.
46. Zhang, X.; Qing, F. -L.; Peng, Y. *J. Fluorine Chem.* **2001**, *108*, 79-82.
47. Qing, F. -L.; Zhang, X. *Tetrahedron Lett.* **2001**, *42*, 5929-5931.
48. Qing, F. -L.; Zhang, X.; Peng, Y. *J. Fluorine Chem.***2001**, *111*, 185-187.
49. Qiu, X. -L.; Qing, F. -L. *J. Chem. Soc., Perkin Trans. 1*, **2002**, 2052-2057.
50. Long, Z. -Y.; Duan, J. -X.; Lin, Y. -B.; Guo, C. -Y.; Chen, Q. -Y. *J. Fluorine Chem.* **1996**, *78*, 177-181.
51. Cottet, F.; Schlosser, M. *Eur. J. Org. Chem.* **2002**, 327-330.
52. Urata, H.; Fuchikami, T. *Tetrahedron Lett.* **1991**, *32*, 91-94.
53. Chen, Q. -Y.; Duan, J. -X. *J. Chem. Soc. Chem. Commun.* **1993**, 918-919.
54. Lai, C.; Malloak, T. E. *J. Chem. Soc. Chem Commun.* **1993**, 1359-1361.
55. Kiriji, N. V.; Pasenok, S. V.; Yagupolskii, Y. L.; Tyrra, W.; Naumann, D. *J. Fluorine Chem.* **2000**, *106*, 217-221.
56. Iseki, K.; Nagai, T.; Kobayashi, Y. *Tetrahedron: Asymmetry* **1994**, *5*, 961-974.
57. Yajima, T.; Nagano, H.; Saito, C. *Tetrahedron Lett.* **2003**, *44*, 7027-7029.
58. Langlois, B.; Montegre, D.; Roidot, N. *J. Fluorine Chem.* **1994**, *68*, 63-66.
59. Anselmi, E.; Blazejewski, J. C.; Tordeux, M.; Wakselman, C. *J. Fluorine Chem.* **2000**, *105*, 41-44.
60. Strekowski, L.; Hojjat, M.; Patterson, S. E.; Kiselyov, A. S. *J. Heterocyclic Chem.* **1994**, *31*, 1413-1416.
61. Fouland, G.; Brigaud, T.; Portella, C. *J. Org. Chem.* **1997**, *62*, 9107-9113.
62. Darwish, A. D. *Chem. Commun.* **2003**, 1374-1375.
63. Billard, T.; Roques, N.; Langlois, B. R. *Tetrahedron Lett.* **2000**, *41*, 3069-3072.
64. Sato, Y.; Matanabe, S.; Uneyama, K. *Bull. Chem. Soc. Jpn.* **1993**, *66*, 1840-1843.
65. Dan-oh, Y.; Uneyama, K. *Bull. Chem. Soc. Jpn.* **1995**, *68*, 2993-2996.
66. Dmowski, W.; Biernacki, A. *J. Fuorine Chem.* **1996**, *78*, 193-194.
67. Dmowski, W. *Tetrahedron,* **1997**, *53*, 4437-4440.
68. Grinberg, V. A. *Russian Chem. Bull.* **1997**, *46*, 1131-1135.
69. Depecker, C.; Marzouk, H.; Trvin, S.; Devynck, J. *New J. Chem.* **1993**, *23*, 739-742.
70. Tommasino, J. -B.; Brondex, A.; Medebielle, M.; Thomalla, M.; Langlois, B. R.; Billard, T. *Synlett* **2002**, 1697-1699.
71. Koshechko, V. G.; Kiprianova, L. A.; Fileleeva, L. I.; Rozhkova, Z. Z. *J. Fluorine Chem.* **1995**, *70*, 277-278.

Chapter 2

New Nucleophilic Fluoroalkylation Chemistry

G. K. Surya Prakash and Jinbo Hu

Loker Hydrocarbon Research Institute and Department of Chemistry, University of Southern California, Los Angeles, CA 90089–1661

This chapter summarizes the new nucleophilic fluoroalkylation chemistry developed in our laboratory over the past several years. These include trifluoromethylation with $TMSCF_3$, trifluoromethylation and difluoromethylation with tri- and difluoromethyl sulfones and sulfoxides, the introduction of difluoromethylene group via fluorinated enol silyl ethers, etc. The related work by others are also covered in this review.

1. Introduction

Selective introduction of fluorine atoms or fluorine-containing building blocks into organic molecules has continued to be a blooming filed since 1980s. Given the fact that over 10 % of newly registered pharmaceutical drugs and some 40 % of newly registered agrochemicals contain one or more fluorine atoms, fluorine is highlighted as the second favorite hetero-element (after nitrogen) in life science-oriented research (*1*). Nucleophilic fluoroalkylation, especially nucleophilic trifluoromethylation, is one of most intensively studied and widely used methods in this field (*2,3*). Trifluoromethyl organometallic compounds, unlike other long-chain perfluoroalkyl analogs, are readily decomposed (via α-elimination of fluoride) and thus not suitable as efficient nucleophilic trifluoromethylating agents (*2*). In 1989, we discovered the first straightforward and reliable nucleophilic trifluoromethylating method (*4*), using (trifluoromethyl)trimethylsilane ($TMSCF_3$), a compound first prepared by Ruppert (*5*). A comprehensive review on the work before 1997 is available (*4c*). Recently, two review papers have covered the chemistry of $TMSCF_3$ from 1997 to early 2001 (*6,7*). The reviews for other nucleophilic perfluoroalkylation chemistry with organosilicon reagents (*4c*) or organometallic reagents (*8,9*) are also available.

In the late 1990s, nucleophilic trifluoromethylation and perfluoroalkylation using trifluoromethane (CF_3H) and other hydroperfluoroalkanes (R_fH) have been developed by several research groups (*10*). *N,N*-Dimethylformamide (DMF) was found to be the best solvent since it can form CF_3^-/DMF hemiaminolate adduct as the real trifluoromethylating species. (*10c-g*). Based on this rationale, a variety of new nucleophilic trifluoromethylation methods have been demonstrated in the recent several years, such as trifluoromethylation using stable hemiaminals of fluoral (*11*), 2,2,2-trifluoroacetophenone and its adduct (*12*), trifluoroacetic acid and trifluorosulfinic acid derivatives (*13*), among others. A recent review paper on the nucleophilic trifluoromethylation using trifluoromethane and stable hemiaminals of fluorals is available (*11*). Langlois and co-workers (*14a*) have also reported enantiopure trifluoroacetamide derived from ephedrine, which can act as a trifluoromethylating agent in the presence of a catalytic amount of fluoride. However, they did not achieve the expected enantioselectivity with this reagent system (*14a*). Trifluoromethyl iodide has also been successfully used as a nucleophilic trifluoromethylating agent under the activation of electron-transfering agent tetrakis(dimethylamino)ethylene (TDAE) (*14b*). However, these recently developed nucleophilic trifluoromethylating methods have a common drawback: most of them do not work effectively with enolizable electrophilic systems. With the enolizable substrates, $TMSCF_3$ is still the best choice as a nucleophilic trifluoromethylating agent.

This chapter summarizes the recent progress of new nucleophilic fluoroalkylation chemistry in our laboratory, including nucleophilic trifluoromethylation, difluoromethylation, difluoromethylenation, and others.

This chapter also covers the recent studies on this topic from 2001 and includes some that have been omitted in previous reviews.

2. Trifluoromethylation and Perfluoroalkylation with (Trifluoromethyl)-trimethylsilane ($TMSCF_3$) and (Perfluoroalkyl)trimethylsialnes ($TMSR_f$)

2.1. Diastereoselective trifluoromethylation. Asymmetric introduction of trifluoromethyl group into organic molecules is of great importance for many applications. The first diastereoselective trifluoromethylation with $TMSCF_3$ into the ketonic function of steroids to give corresponding trifluoromethylated carbinols as single stereomers, was reported by us in 1989 (*4*). Thereafter, diastereoselective trifluoromethylation of carbonyl groups with $TMSCF_3$ (with fluoride activation) has been successfully applied to various systems such as carbohydrates and inositol derivatives (*15*), steroides (*16*), azirines (*17*) and amino acid derivatives (*18*). Diastereoselective trifluoromethylation of chiral *tert*-butanesulfinimines to give chiral trifluoromethyl amines has been achieved in our laboratory in 2001 (*19*). Following this success, we extended the similar chemistry to *α,β-unsaturated imines* (**3**) as outlined in Scheme 1 (*20*).

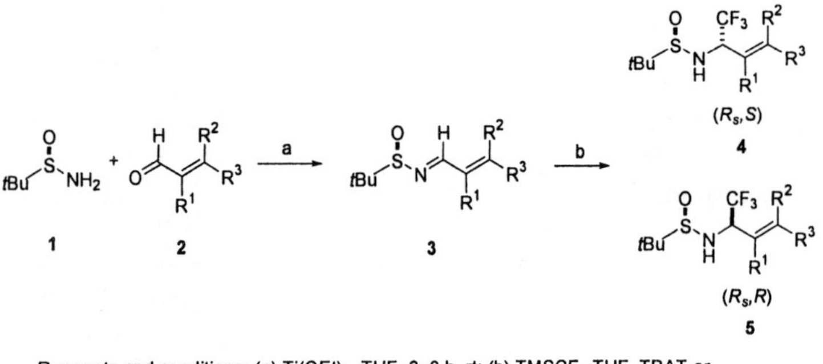

Reagents and conditions: (a) $Ti(OEt)_4$, THF, 2~6 h, rt; (b) $TMSCF_3$,THF, TBAT or TMAF, - 25 °C.

Scheme 1. Diastereoselective trifluoromethylation of α,β-unsaturated chiral sulfinimines.

Chiral α,β-unsaturated sufinimines (**3**) were prepared from corresponding α,β-unsaturated aldehydes (**2**, with known stereochemistry) and (*R*)-*N*-*tert*-butanesulfinamide (**1**) in the presence of 3 equiv of $Ti(OEt)_4$. This preparation method tolerates all the substituents at α- and β-positions and provides the products (**3**) in 70~90 % yields (*20*). Since α,β-unsaturated sufinimines are less reactive toward nucleophilic addition reactions compared to nonconjugated imines, the trifluoromethylation reactions were carried out at relatively higher temperature (– 25 °C). The scope of this reaction is summaried in Table 1. Similar to the trifluoromethylation of usual α,β-unsaturated carbonyl compounds,

Table 1. Diasteroselective trifluoromethylation of sufinimines (**3**).

sufinimines (**3**)	Sulfinamides	
	4 (R_S,S) : **5** (R_S, R)[a]	yields (%)[b]
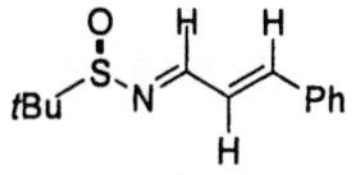 **3a**	90:10	55
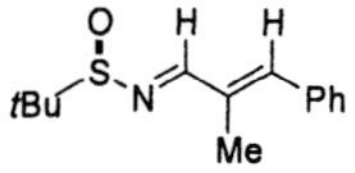 **3b**	>99	73
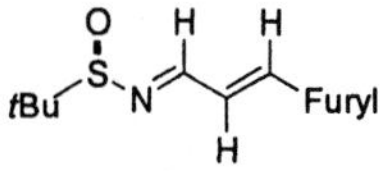 **3c**	92:08	76
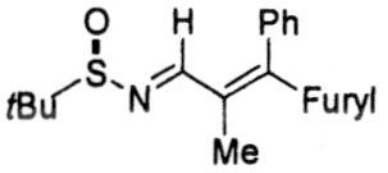 **3d**	98:02	50
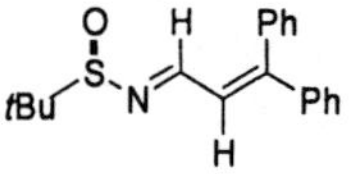 **3e**	>99	62
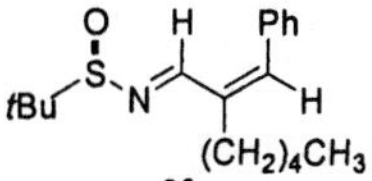 **3f**	>99 (90:10)[c]	25 (82)[c]
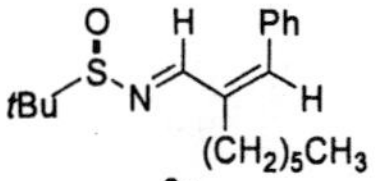 **3g**	>99 (92:08)[c]	20 (75)[c]
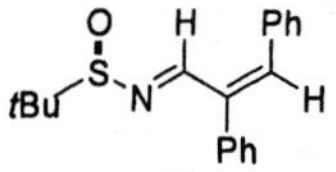 **3h**	(93:07)[c]	(62)[c]

[a]Diastereomeric ratios were determined by ^{19}F NMR from the crude products. Unless otherwise mentioned, yields and diastereomeric ratios are for TBAT as the fluoride source. [b]Isolated yields of the major diastereomer. [c]Diastereomeric ratios and yields when TMAF (tetramethylammonium fluoride) was used as a fluoride source.

no 1,4-additon products were observed for the reactions of **3**. Substitution at the β-position (imine **3e**) has no influence on the reaction yield. The reaction tolerat--es the heterocycle substitutions at β-position (imines **3c** and **3d**), while a long-chain substitution at the α-position (**3f** and **3g**) gave lower yields of the addition products as a single diastereomer. This problem was overcome by reducing the steric volume of the counterion of the fluoride source. When TMAF (tetramethylammonium fluoride) was used as the fluoride activator instead of tetrabutylammonium triphenyldifluorosilicate (TBAT). The rationale for the high *anti*-Ellman products **4** in these reactions could be explained by Cram-Davis' open transition state model (*20*). The sulfinamide products **4** are highly useful precursors for the chiral trifluoromethylated allylic amines.

We have further extended this robust diastereoselective trifluoromethylation methodology to *α-amino N-tert-butanesufinimines* (**7**) (*21*). α-Amino *N-tert*-butanesufinimines were prepared from the smooth condensations of Reetz's α-amino aldehydes (**6**) and Ellman's *N-tert*-butanesulfinimines (**1**) using $Ti(OEt)_4$ as the dehydrating agent (Scheme 2). As shown in Scheme 2, by using less bulky TMAF (rather than TBAT) as the fluoride source, the imines **7a–f** were trifluoromethylated to the corresponding vicinal ethylenediamine adducts **8a–f** in good to excellent yields with high diastereoselectivities (>99:1). However, the imine derived from D-amino aldehyde **7g** gave the corresponding product **8g** in 80:20 diastereomeric ratio with a 60 % yield of the major diastereomer. This suggests that both the chiral centers presented in the molecules of **7a–f** direct the incoming nucleophile to the *re* face of the imines (*21*).

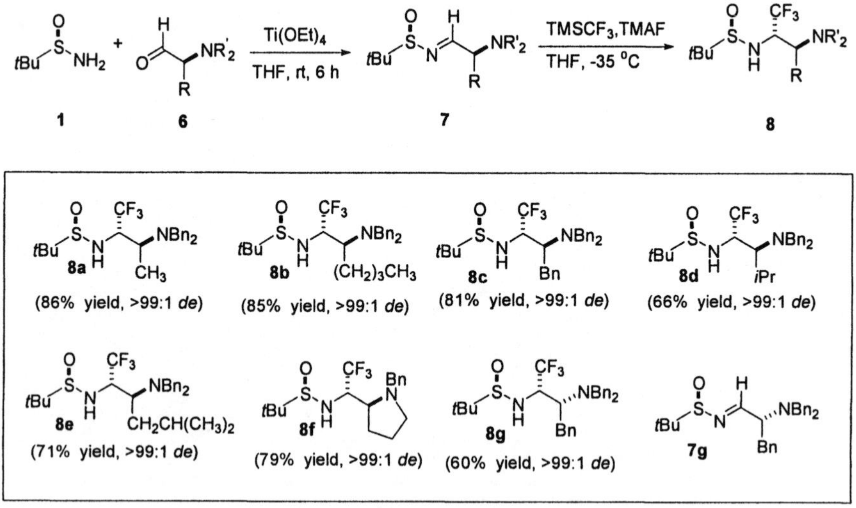

1 + 6 → [$Ti(OEt)_4$, THF, rt, 6 h] → 7 → [$TMSCF_3$, TMAF, THF, -35 °C] → 8

8a ($R = CH_3$): (86% yield, >99:1 *de*)
8b ($R = (CH_2)_3CH_3$): (85% yield, >99:1 *de*)
8c (R = Bn): (81% yield, >99:1 *de*)
8d (R = *i*Pr): (66% yield, >99:1 *de*)
8e ($R = CH_2CH(CH_3)_2$): (71% yield, >99:1 *de*)
8f: (79% yield, >99:1 *de*)
8g: (60% yield, >99:1 *de*)
7g

Scheme 2. Diastereoselective trifluoromethylation of α-amino sulfinimines (**7**).

The above synthesized sulfimamides **8** can be hydrolyzed into trifluoromethyl vicinal ethylenediamines (**9**). Scheme 3 demonstrates the further elaborations of **9**. Protection of the free amino group with ethyl chloroformate gave the corresponding carbamate (**10**), and **10** was debenzylated and treated with triphosgene to give the 2-oxo-1-imidazolidinyl derivatives **11**. Catalytic hydrogenation of **9** gave the free diamine **12** in good yield (*21*).

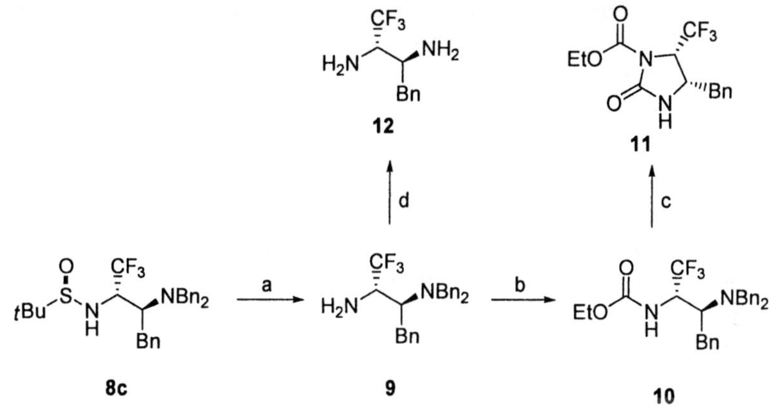

Reaction conditions: (a) 1. 10 equiv of HCl in dioxane, MeOH, 70 °C, 3h. 2. Saturated $NaHCO_3$ washing; 92 % yield. (b) EtOCOCl, 50 % K_2CO_3, dioxane; 95 % yield; (c) 1. Pd/H_2, MeOH/DCM, 24 h. 2. $(Cl_3CO)_2CO$, Et_3N, THF; 60 % yield; (d) Pd/C, H_2, MeOH/DCM, 24 h.

Scheme 3. Synthetic applications of **8**.

2.2. Enantioselective Trifluoromethylation. Enantioselective trifluoromethylation is a practically important method for the preparation of enantiopure CF_3-containing compounds, which are highly useful in agrochemistry and pharmaceutical industry. Compared with diasteroselective trifluoromethylation, enantioselective trifluoromethylation is more challenging and thus less developed. The enantioselective trifluoromethylation with $TMSCF_3$ are usually based on the use of a chiral tetraalkylammonium fluoride as the chiral inducer. During the reaction, the chiral ammonium cation is closely associated with the silyloxy anionic adduct (tight ion pair) and thus induces enantioselectivity. In 1993, we found that the use of *N*-benzylquinidinium fluoride in dichloromethane at – 78 °C allows trifluoromethylation of 9-anthraldehyde in 95 % ee (*22*). *N*-[4-(Trifluoromethyl)benzyl]cinchonium fluoride (10~20 mol %) has been used by Kobayashi and co-workers (*23*) for the enantioselective introduction of trifluoromethyl group into carbonyl compounds in high yields with moderate enantiomeric excesses. The same group also found that quinine itself was capable of enantioselective trifluoromethylation of aldehydes using Et_3SiCF_3, although with low enantioselectivities and yields (*24*). Chiral triaminosulfonium triphenyldifluorostannate (10 mol %) was similarly attempted as fluoride source, and only moderate enantioselectivities (*ee* = 10–52 %) were observed (*25*).

Very recently, Pfizer research group reported an improved enantioselective trifluoromethylation by carefully modifying the cinchonine-derived fluoride catalysts' structure and the reaction conditions (*26*). It was observed that the introduction of a bulky subunit at the quinuclidine nitrogen atom of cinchona alkaloids led to an remarkable enhancement of the enantioselectivity. Finally, they found that the chiral fluoride catalyst **13** can be used as low as 4 mol % in the trifluoromethylation to give the desired product **14** in up to 92 % ee (Scheme 4). However, this method was only shown to work with aryl ketones and aldehydes.

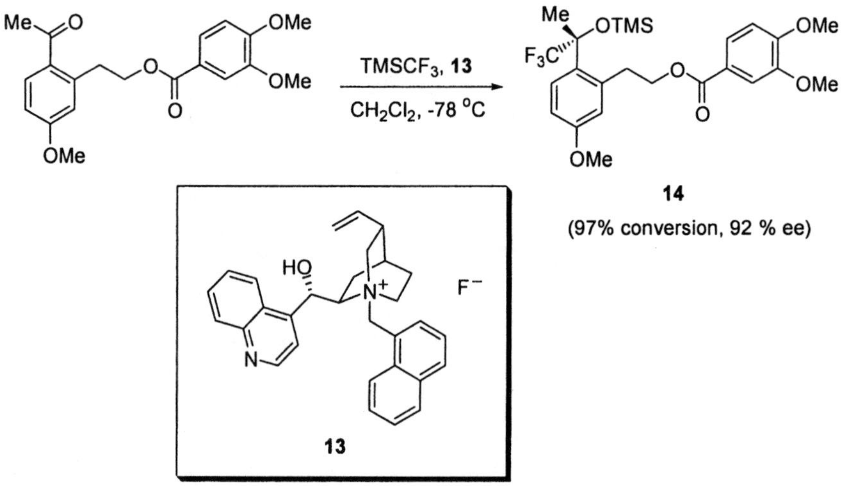

Me O OMe OMe O O OMe

TMSCF$_3$, **13**

CH$_2$Cl$_2$, -78 °C

Me OTMS F$_3$C OMe OMe O O OMe

14

(97% conversion, 92 % ee)

HO N N+ F$^-$

13

Scheme 4. Enantioselelctive trifluoromethylation of aryl carbonyl compounds.

2.3. Trifluoromethylation Using Trimethylamine *N*-Oxide as a Catalyst. Since our initial discovery in 1989 (*4*), trifluoromethylation using $TMSCF_3$ have been commonly accompanied by an activation with catalytic amount of fluoride (such as TBAF, CsF, TBAT, TMAF, etc.) or metal alkoxide. Although the yields of the trifluoromethylated adducts with a variety of electrophiles were high, the reactions were not catalytic in initiators. Furthermore, due to the basic nature of these inintiators we sought a milder process. Recently, advances in Lewis base catalyzed organic transformation for the silylated species that take the advantage of the high strength of the Si–O bond as well as judicious selection of kenetic reactivity can also be applied to $TMSCF_3$ chemistry. We envisioned that we would be able to activate $TMSCF_3$ for CF_3 transfer towards the carbonyl compounds in a similar fashion using oxygen nucleophiles that can act as true catalysts (*27*).

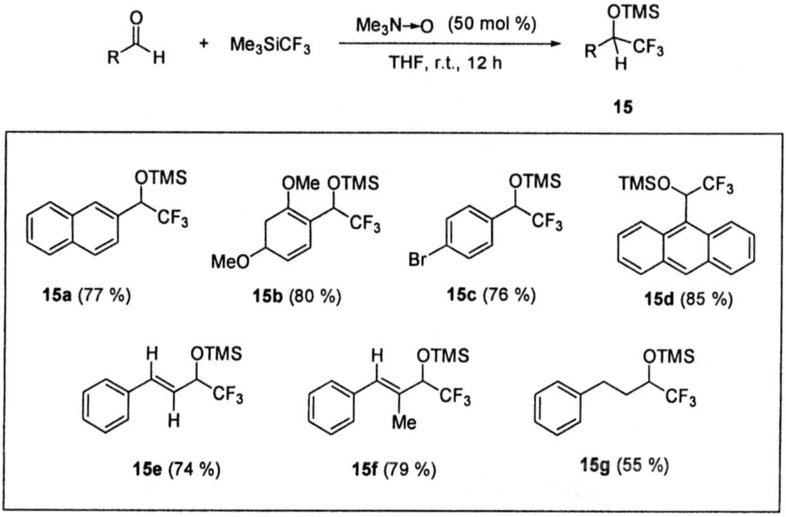

Scheme 5. Trifluoromethylation using trimethylamine oxide.

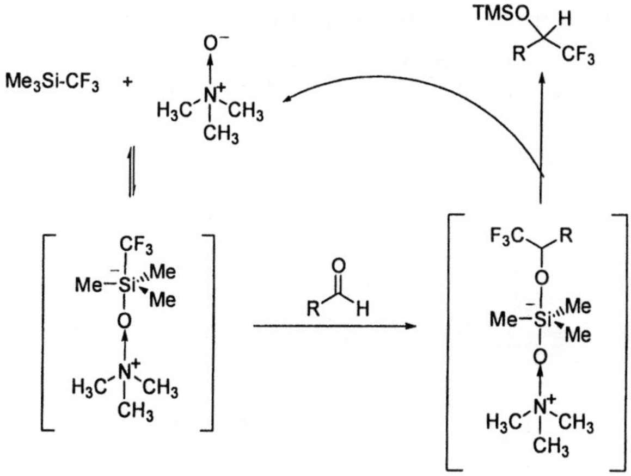

Scheme 6. Catalytic cycle of trimethylamine N-oxide catalyzed trifluoromethylation.

We have screened the reactivity of commercially available amine oxides such as pyridine *N*-oxide, *p*-chloropyridine *N*-oxide, *N,N*-dimethylpyridine *N*-oxide as well as hexamethylphosphoramide (HMPA). The use of stoichiometric amount of HMPA gave only < 10 % conversion after 48 h with the reaction of $TMSCF_3$ and 2-naphthaldehyde. Trialkylamine *N*-oxides are more nucleophilic compared to aromatic *N*-oxide and therefore, we chose trimethylamine *N*-oxide as the catalyst for the trifluoromethylation with $TMSCF_3$. It was found that 50 mol % of trimethylamine *N*-oxide was able to catalyze the CF_3 transfer at room temperature over 12 h to afford good yields of TMS-protected products (Scheme 5). It is important to note that no free alcohol was observed, indicating true catalysis (vide supra). The scope and the yields of the reaction are shown in Scheme 5. The probable path for the observed catalysis by trimethylamine *N*-oxide is depicted in Scheme 6 *(27)*.

2.4. *1,4*-Trifluoromethylation and Perfluoromethylation. Nucleophilic trifluoromethylation and perfluoroalkylation using $TMSCF_3$ or other $TMSR_f$ reagents (with an initiator) usually proceed as a 1,2-addition of the CF_3 or other R_f groups to the carbonyl functionality of the α,β-unsaturated carbonyl compounds (*4*). Recently, Sosnovskikh and co-workers (*28*) have found that when the β-position of the α,β-unsaturated carbonyl compounds is substituted with an strong electron-withdrawing group such as CF_3 group, the trifluoromethylation and other perfluoroalkyaltion reactions proceeded mainly in 1,4-addition mode with high regioselectivity and good yields (Scheme 7). Obviously, this unusual regioselectivity is directed by the high electron-withdrawing substituent at the β-carbon. This 1,4-trifluoromethylation chemistry has been applied to the synthesis of fluorinated analogs of natural 2,2-dimethylchroman-4-ones and 2,2-dimethylchromenes (*28*).

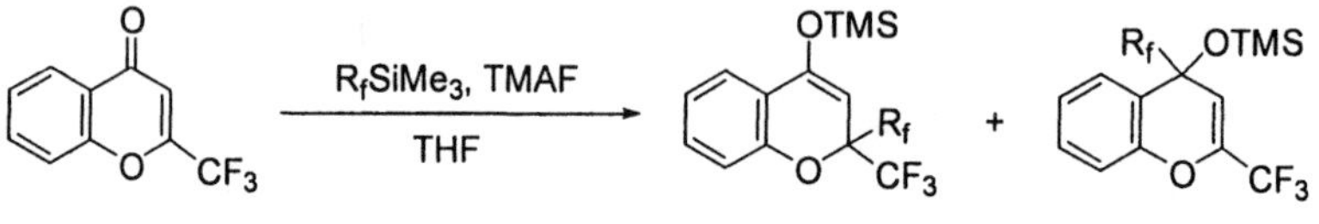

R_fSiMe_3, TMAF
THF
OTMS
R_f
CF_3
16a
R_f OTMS
CF_3
16b

	$R_f = CF_3$ **16a:16b**	$R_f = C_2F_5$ **16a:16b**
- 30 °C:	92 : 8	82 : 18
0 °C:	90 : 10	77 : 23
25 °C:	87 : 13	72 : 28

Scheme 7. 1,4-Trifluoromethylation and perfluoroalkylation.

The same group (*29*) have demonstrated another methodology for the 1,4-trifluoromethylation of α,β-enones by using a bulky Lewis acid (methyl aluminum bis(2,5-Di-*tert*-butyl-4-methylphenoxide, **MAD**) to complex with the carbonyl group, preventing the addition of the trifluoromethide at this position (Scheme 8). However, the reaction yields of this method is only moderate *albeit* its high regioselectivity (*29*).

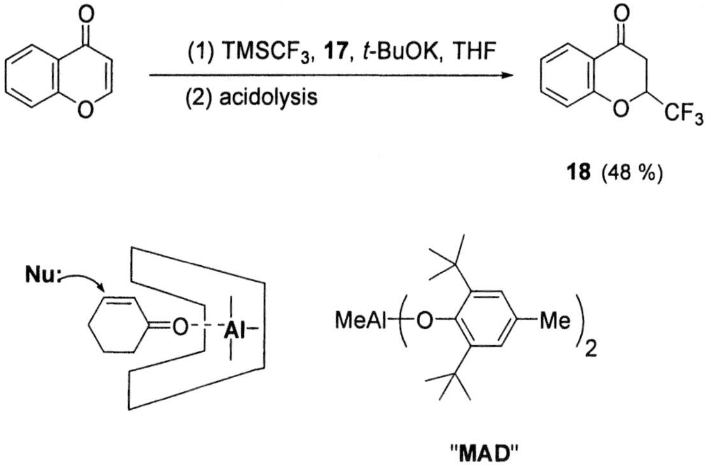

(1) $TMSCF_3$, **17**, *t*-BuOK, THF

(2) acidolysis

CF_3

18 (48 %)

Nu:

O- - Al-

MeAl-(O-...-Me)$_2$

"**MAD**"

Scheme 8. 1,4-Trifluoromethylation using '*protect-in-situ*' methodology.

2.5. Trifluoromethylations and perfluoroalkyaltions of Simple Inorganic Molecules. Fluoride-induced nucleophilic trifluoromethylation and perfluoroalkylation with $TMSCF_3$ or related reagents has continued to expand its synthetic application in both organic and inorganic fields in last several years. In 2002, Shreeve and co-workers (*30*) reported the stoichiometric amount of cesium fluoride-induced trifluoromethylation and pentafluoroethylation of a variety of inorganic substrates using $TMSCF_3$ and $TMSCF_2CF_3$ in glyme as a solvent. By this method, they demonstrated that NOCl (or NO/NO_2), SO_2, and CO_2 could be transformed into perfluoronitrosoalkanes (R_fNO, R_f = CF_3, CF_3CF_2), perfluoroalkanesulfinic acids, and perfluoroalkanoic acids, respectively (Scheme 9). Thereafter, Tyrra and co-workers (*31a*) reported a new synthesis of tetramethylammonium, cesium and di(benzo-15-crown-5)cesium trifluoromethanethiolates from $TMSCF_3$, S_8 and corresponding fluorides (Scheme 9). Very recently, Yagupolskii and co-workers (*31b*) have reported trifluoromethylation and perfluoroalkylation of CS_2, CO_2 and CS_2 using R_fSiMe_3 reagents in the presence of TMAF (Scheme 9).

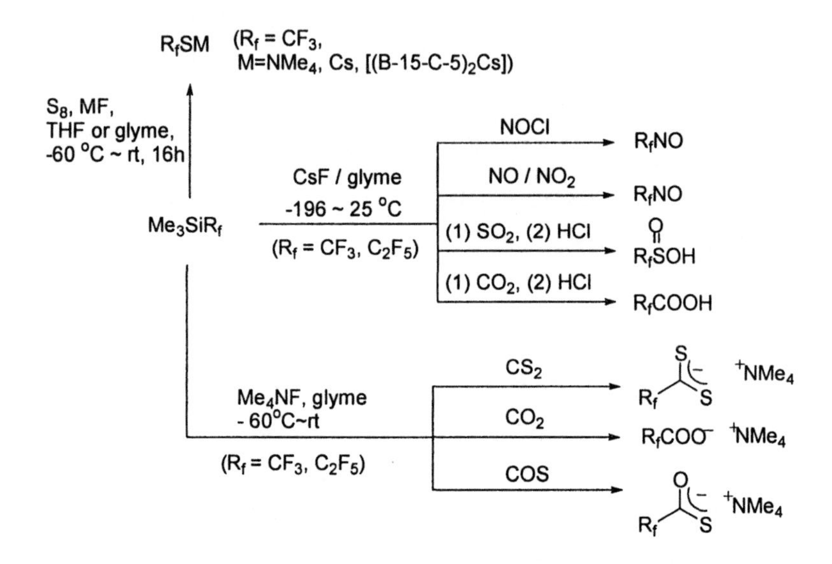

Scheme 9. Perfluoroalkylation of simple inorganic molecules.

2.6. Trifluoromethylation of Thioketones. The trifluoromethylation of thioketones with $TMSCF_3$ is rather complicated compared to the reaction with normal ketones (*32a*). Treatment of 2,2,4,4-tetramethylcyclobutane-1,3-dione (**19**) in THF with $TMSCF_3$ in the presence of TBAF yielded the corresponding 3-(trifluoromethyl)-3-[(trimethylsilyl)oxy]-cyclobutanone (**20**) (Scheme 10, eq. 1) via nucleophilic addition of a trifluoromethide at the C=O group and subsequent silylation of the alcoholate (*32a*). Under similar conditions, the 'monothione' (**21**) reacted to give thietane derivatives (**22**), whereas in the case of 'dithione' (**23**) only the dispirodithietane (**24**), the dimer of **23**, was formed (Scheme 10, eqs. 2,3). A conceivable mechanism for the formation of **22** is the ring opening of the primary formed CF_3 adduct **25** followed by ring closure via the S-atom (Scheme 10, eq. 2).

Langlois and co-workers (*32b*) reported that by using a stoichiometric amount of $Bu_4NF{\cdot}3H_2O$, $TMSCF_3$ reacted with aromatic thiones in both thiophilic and carbophilic ways to give, in medium yields, a mixture of (trifluoromethylthio)diarylmethane (major) and 1,1-diaryl-2,2,2-trifluoroethanethiol (minor) (see Scheme 10, eq. 4).

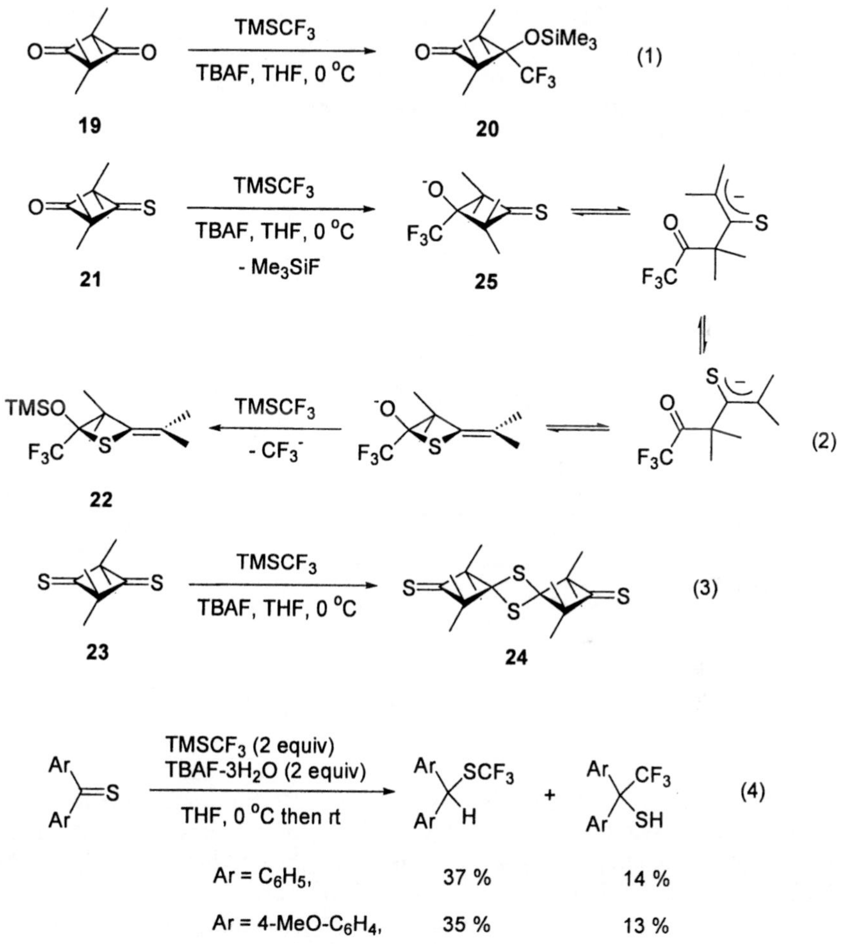

TMSCF$_3$, TBAF, THF, 0 °C (1) 19 → 20 (OSiMe$_3$, CF$_3$)

TMSCF$_3$, TBAF, THF, 0 °C, - Me$_3$SiF; 21 → 25; TMSCF$_3$, - CF$_3^-$ → 22 (TMSO, F$_3$C, S) (2)

TMSCF$_3$, TBAF, THF, 0 °C (3) 23 → 24

TMSCF$_3$ (2 equiv), TBAF-3H$_2$O (2 equiv), THF, 0 °C then rt (4): $Ar_2C{=}S$ → $Ar_2CH(SCF_3)$ + $Ar_2C(CF_3)SH$

Ar = C_6H_5,	37 %	14 %
Ar = 4-MeO-C_6H_4,	35 %	13 %

Scheme 10. Trifluoromethylation of ketones and thioketones.

2.7. Trifluoromethylation and Perfluoroalkylation of Trialkoxyboranes. Recently, Molander and co-workers (*33*) reported an improved synthesis of potassium (trifluoromethyl)trifluoroborate [K(CF_3BF_3] by the reaction of $TMSCF_3$ and trimethoxyborane in the presence of potassium fluoride, followed by aqueous HF treatment (Scheme 11, eq. 1). This procedure gave the final product in 85 % overall yield. Soon after, similar trifluoromethylation and perfluoroalkylation (Scheme 11, eq. 2) was described by Kolomeitsev, Roschenthaler and co-workers (*34*). The further deoxylation of the the prepared fluorinated borates by methyl triflate, methyl tosylate, methanesulfonyl chloride and chlorotrimehtylsilane gave the free boronic esters. Such perfluoroalkyl borates and boronic esters are claimed to be potential partners for Suzuki and Petasis reactions, although no such success has been demonstrated (*34*).

$$TMSCF_3 + B(OMe)_3 \xrightarrow[\text{rt, 18 h}]{\text{KF (1 equiv), THF}} K[CF_3B(OCH_3)_3]$$

$$\xrightarrow[\text{rt, 16 h}]{\text{48 \% HF (6 equiv)}} K[CF_3BF_3] \quad (1)$$

85 % overall yield

$$B(OR)_3 + R_fSiMe_3 \xrightarrow[Q^+F^-]{\text{THF or glyme}} R_fB(OR)_3^- Q^+ \xrightarrow{E} R_fB(OR)_2 \quad (2)$$

R= Me, Et
R_f= CF_3, C_2CF_5, $(EtO)_2P(O)CF_2$
Q^+= K^+, Me_4N^+

Scheme 11. Trifluoromethylation and perfluoroalkylation of trialkoxyboranes.

2.8. Trifluoromethylation of Phenyl Trifluoromethanesulfonate. Sevenard and co-workers have recently disclosed some interesting results on the trifluoromethylation of phenyl trifluoromethanesulfonate (*35*). The bis(trifluoromethyl)sulfone intermediate (**26**) was unexpectedly further trifluoromethylated to the hypervalent [10-*S*-5] sulfuranide dioxide salts (**27**) with three S-C bonds (Scheme 12). The compounds **27** are stable enough to be characterized by single-crystal X-ray diffraction (*35*).

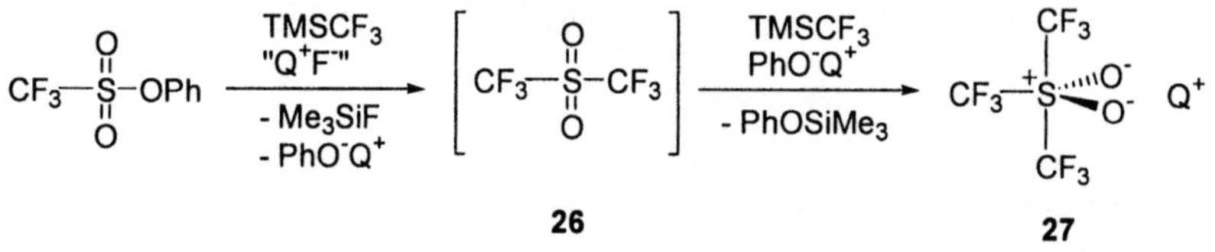

Entry	"Q^+F^-"	Product / Q^+	Yield, %
1	KF / 18-crown-6	**27a** / [K(18-crown-6)]$^+$	50
2	Me_4NF	**27b** / Me_4N^+	79
3	$(Me_2N)_3S^+$ $Me_3SiF_2^-$	**27c**/ $(Me_2N)_3S^+$	65

Scheme 12. Synthesis of hypervalent [10-*S*-5] sulfuranide dioxide salts.

2.9. Other Trifluoromethylations. Fluoride-induced trifluoromethylation with $TMSCF_3$ has also been successful in introducing a CF_3 group into *N*-(sulfinyl)trifluoromethanesulfonamide (*36*). Schlosser and co-workers (*1*) used $TMSCF_3$/KF/CuI system for the preparation of trifluromethyl-substituted pyridine- and quinolinecarboxylic acid derivates. In 2003, Kiplinger and co-workers (*37a*) reported the synthesis and characterization of the first early transition-metal perfluoroalkyl complexes, $Cp_2Ti(CF_3)(F)$ (**28**) and $Cp_2Ti(CF_3)(OTf)$ (**29**) by using the trifluoromethylation of Cp_2TiF_2 with $TMSCF_3$ (Scheme 13). More recently, Diederich and co-workers reported a nucleophilic trifluoromethylation of cyclic imides using $TMSCF_3$ (*37b*).

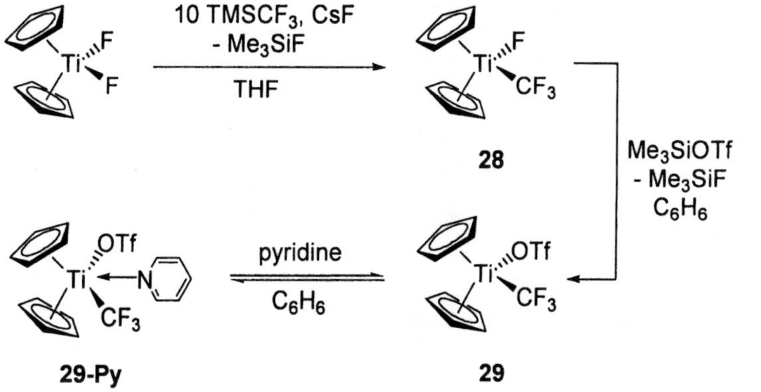

Scheme 13. Preparation of trifluoromethyl titanium complexes.

3. Trifluoromethylation with Trifluoromethyl Phenyl Sulfone, Sulfoxide and Sulfide

3.1. Reductive Trifluoromethylation. We recently developed a new efficient magnesium metal-mediated reductive nucleophilic trifluoromethylation method using trifluoromethyl phenyl sulfone (**30a**), sulfoxide (**30b**) and sulfide (**30c**) as the trifluoromethylating agents (*38*). Compounds **30a-c** can be conveniently prepared from non-ozone-depleting precursors such as trifluoromethane (CF_3H) or trifluoroacetate (*39*). Our new reductive trifluoromethylation method is proved to be highly useful for the preparation of structurally diverse trifluoromethylsilanes from chlorosilanes. This method can be used as a catalytic process for the $TMSCF_3$ preparation from non-ozone-depleting trifluoromethane (CF_3H), which has many advantages over the previously known methods that commonly use ozone-depleting trifluoromethyl bromide as precursor.

In the trifluoromethyl sulfones or sulfoxides, due to the strong electron withdrawing effect of CF_3 group, the bond between the pseudohalide and the sulfur atom is sufficiently polarized with the pseudohalide group bearing substantial negative charge. Thus, when the electrons are transferred from magnesium metal to the sulfones and sulfoxides, reductive cleavage of the C–S bond to generate anionic "CF_3^-" species is prefered over the C-F bond fission that occurs for trifluoromethyl ketones (Scheme 14). It is important to note that, since trifluoromethyl anion ("CF_3^-") is a kinetically very unstable species and prone to α-elimination, "CF_3^-" in this whole text only represents a solvated short-lived trifluoromethyl anion species or simply as a *de facto* trifluoromethyl anion.

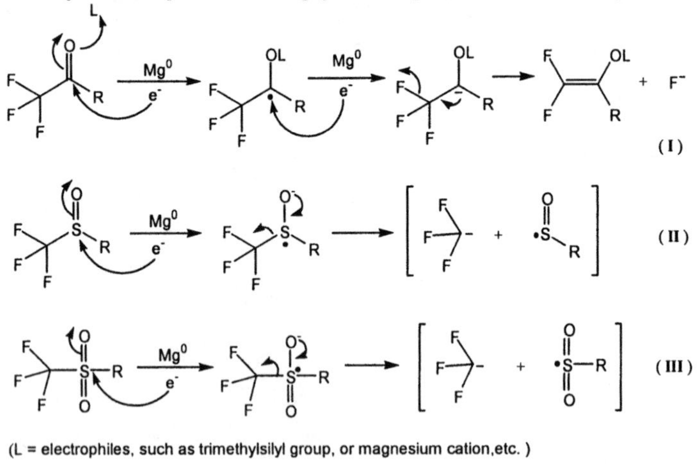

Scheme 14. Mechanistic explanation of magnesium-mediated reductive defluorination and trifluoromethylation.

Reaction of sulfone **30a** with 3 equivalents of maganesium metal and the chlorotriethylsilane in DMF solution at 0 °C gave exclusively (trifluoromethyl)triethylsilane, the only product detected by ^{19}F NMR. After work-up and purification, (trifluoromethyl)triethylsilane was isolated in 95% yield. Diphenyl disulfide (PhSSPh) was also collected as a byproduct. TMS-CF_3 was also prepared similarly in quantitative conversions as identified by ^{19}F NMR. The reaction works equally well for diverse type of chlorosilanes with trifluoromethyl sulfone (**30a**) and sulfoxide (**30b**) (see Table 2).

Table 2. Preparation of trifluoromethylsilanes through Mg^0-mediated reductive cleavage of C-S bonds.

$$PhS(O)(O)CF_2X \ (\mathbf{30}) + R^1R^2R^3Si{-}Cl \ (\mathbf{31}) \xrightarrow[DMF]{Mg^0} R^1R^2R^3Si{-}CF_2X \ (\mathbf{32}) + PhSSPh$$

entry	sulfur compound **30**	chlorosilane **31**	temperature[a]	time (h)[b]	product **32**	yields (%)[c]
a	$PhSO_2CF_3$	Me_3SiCl	0°C ~ r t.	0.5 ~ 2	Me_3SiCF_3	100 (83)
b	$PhS(O)CF_3$	Me_3SiCl	0°C ~ r.t.	0.5 ~ 2	Me_3SiCF_3	100 (81)
c	$PhSCF_3$	Me_3SiCl	r. t.	4	Me_3SiCF_3	45
d	$PhSO_2CF_3$	Et_3SiCl	0°C ~ r.t.	1.3	Et_3SiCF_3	100 (95)
e	$PhS(O)CF_3$	Et_3SiCl	r. t.	0.5	Et_3SiCF_3	98
f	$PhSO_2CF_3$	$t\text{-}BuMe_2SiCl$	-30°C ~ r. t.	3	$t\text{-}BuMe_2SiCF_3$	75 (57)
g	$PhS(O)CF_3$	$t\text{-}BuMe_2SiCl$	r. t.	0.5	$t\text{-}BuMe_2SiCF_3$	73
h	$PhSO_2CF_3$	$(Me_3Si)_3SiCl$	-40 °C ~ r.t.	0.5	$(Me_3Si)_3SiCF_3$	85 (62)
i	$PhSCF_2Br$	Me_3SiCl	r.t.	1.0	$PhSCF_2Si(CH_3)_3$	86 (85)
j	$PhSO_2CF_2Br$	Me_3SiCl	0°C ~ r.t.	0.5	$Me_3SiCF_2CF_2SiMe_3$	76 (55)
					$Me_3SiCF_2SiMe_3$	18

[a]The reaction temperature control is crucial due to the exothermic nature of the reaction. Larger scale reactions normally need lower temperature. [b]The reaction time may vary according to the different reaction scales. [c]The yields are determined by ^{19}F NMR, and the data in parenthese are isolated yields.

In the case of phenyl trifluoromethyl sulfide **30c** (entry **c**), the reaction was sluggish indicating that the fluoroalkyl carbon-sulfur bond is not efficient in accepting the electron from the Mg metal. This was also confirmed by the fact that for the bromodifluoromethyl phenyl sulfide ($PhSCF_2Br$), the Barbier product ($PhSCF_2SiMe_3$, **32i**) was produced in high yield without the C-S bond cleavage (see entry i). In the case of bromodifluoromethyl phenyl sulfone (entry j), 1,2-bis(trimethylsilyl)-1,1,2,2-tetrafluoroethane **(32j)** was generated as the major product. This indicates that a Barbier-type coupling intermediate [$PhS(O)_2CF_2CF_2S(O)_2Ph$] **(33)** is presumably formed, which is subsequently transformed into **32j** via a similar reductive fluoroalkylation process (see Scheme 15, Path A).

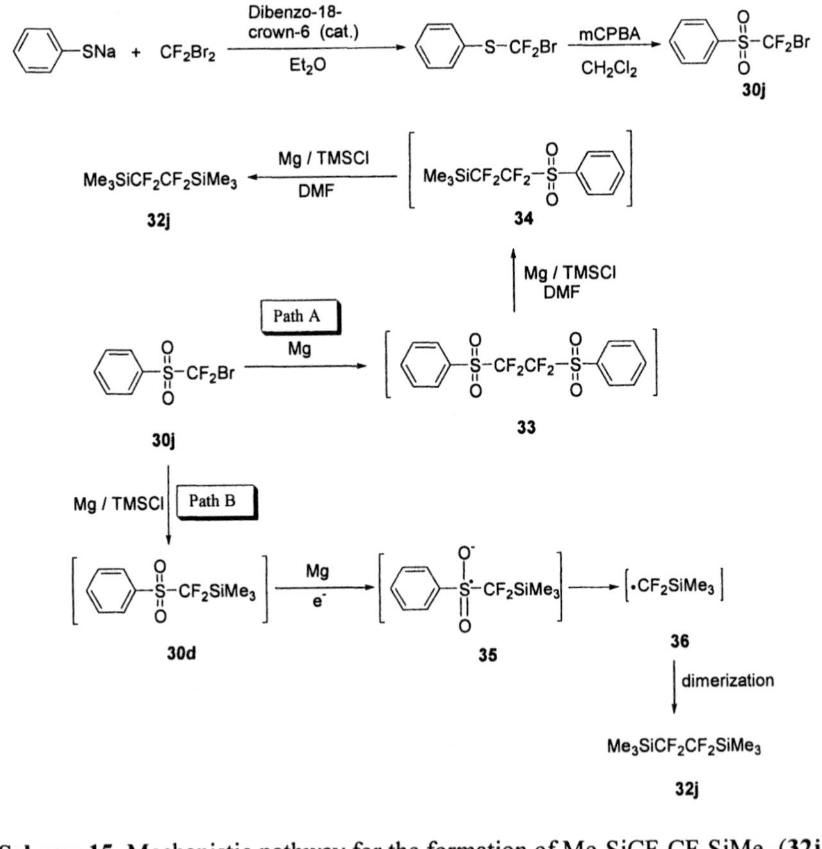

Scheme 15. Mechanistic pathway for the formation of $Me_3SiCF_2CF_2SiMe_3$ **(32j)**.

It is also possible that an alternative Barbier type coupled intermediate **30d** is formed, which can generate the trimethylsilyldifluoromethyl radical species **36** that homo-couples to produce **32j**. Likelihood of path B was supported by the experimental result that, under similar reaction conditions using Mg and TMSCl in DMF, both sulfone **32d** ($PhSO_2CF_2SiMe_3$) and bissulfone **32e** ($PhSO_2CF_2O_2SPh$) readily produce compound **32j** in good yields (Scheme 16).

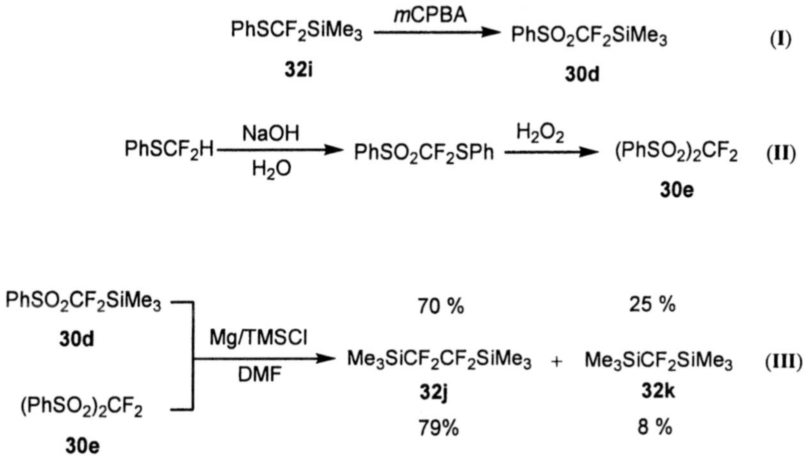

$$PhSCF_2SiMe_3 \ (\mathbf{32i}) \xrightarrow{mCPBA} PhSO_2CF_2SiMe_3 \ (\mathbf{30d}) \qquad (I)$$

$$PhSCF_2H \xrightarrow[H_2O]{NaOH} PhSO_2CF_2SPh \xrightarrow{H_2O_2} (PhSO_2)_2CF_2 \ (\mathbf{30e}) \qquad (II)$$

$$PhSO_2CF_2SiMe_3 \ (\mathbf{30d}) \text{ or } (PhSO_2)_2CF_2 \ (\mathbf{30e}) \xrightarrow[DMF]{Mg/TMSCl} Me_3SiCF_2CF_2SiMe_3 \ (\mathbf{32j}) + Me_3SiCF_2SiMe_3 \ (\mathbf{32k}) \qquad (III)$$

From **30d**: 70 % (**32j**), 25 % (**32k**); from **30e**: 79% (**32j**), 8 % (**32k**).

Scheme 16. Preparation of $Me_3SiCF_2CF_2SiMe_3$ (**32j**).

It should also be mentioned that the use of several reducing metals such as zinc, aluminum, indium, sodium and lithium were explored to replace magnesium as the reducing agent, among which only zinc worked but only with low yield of the products (~30 %). Other reducing or electron-donating reagents, such as samarium iodide (SmI_2) and tetrakis(dimethylamono)ethylene, were also investigated to no avail. Attempts to improve the reactivity of phenyl trifluoromethyl sulfide (**30f**) via electrochemistry using magnesium or zinc rod as the sacrificial anode, and platinum as the cathode in DMF, were also not successful. DMF is not the only solvent required for this reaction. Other solvents such as THF can also be used, although it needs prolonged reaction times. This indicates that there is no need to invoke CF_3^-/DMF adduct, as proposed earlier (*39a*), acting as the intermediate for these reactions. On the other hand, methyl phenyl sulfone in the presence of magnesium and TMSCl under similar conditions did not produce any tetramethylsilane.

Concerning the mechanism, we propose that a single electron transfer from magnesium metal to sulfones or sulfoxides facilitate a reductive cleavage of the C-S bond to form anionic tri- and difluoromethyl species and a sulfur-containg radical species (see Scheme 17). This mechanism provides a working model for this novel type of trifluoromethylation and difluoromethylation. The isolation of PhSSPh as a byproduct further confirms the possibility of the sulfur radical species. This mechanism is also supported by the fact that, when we used 2,2,2-trifluoroethyl phenyl sulfone [$PhS(O)_2CH_2CF_3$] **30g** as the reactant with Mg and TMSCl under similar reaction conditions, 1,1-difluoroethene was produced readily. Obviously, 1,1-difluoroethene was obtained through the β-elimination of the fluoride (F^-) from the in situ generated anionic species ($CF_3CH_2^-$) from **30g** via a smilar mechanism as described above (Scheme 18). Scheme 18 also shows that in the case of 2,2,2-trifluoroethyl phenyl sulfone, sulfoxide and sulfide, the order of reactivity is sulfone > sufoxide > sulfide.

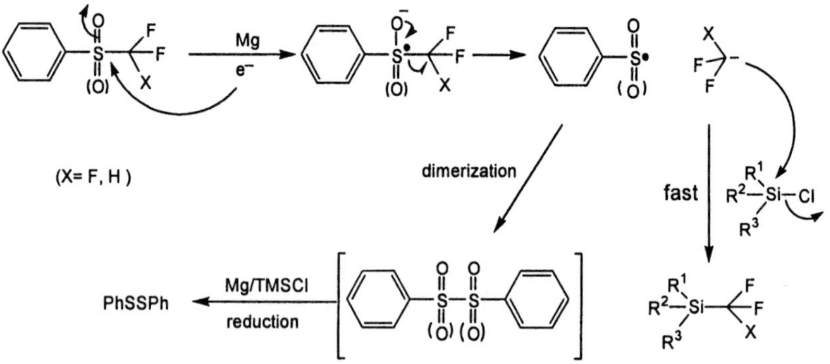

Scheme 17. Proposed mechanism of magnesium-metal mediated trifluoromethylation.

It should be also mentioned that, methyl trifluoromethyl sulfone ($CH_3SO_2CF_3$) also reacts with magnesium metal and TMSCl in DMF to produce TMS-CF_3 in moderate yields (~ 40 % over a period of 20 hours at room temperature). However, the reaction appears to be sluggish. This indicates that the aromatic ring conjugation in **30a** is important to facilitate the initial electron transfer process. Furthermore, the reductive fluoroalkylation chemistry was also attempted with other electrophiles such as aldehydes, ketones, allyl bromide, benzyl chloride, or tributyltin chloride with no success. Even tributyltin hydride and allyltrimethylsilane showed no reactivity. The reason for such a behavior is not currently clear.

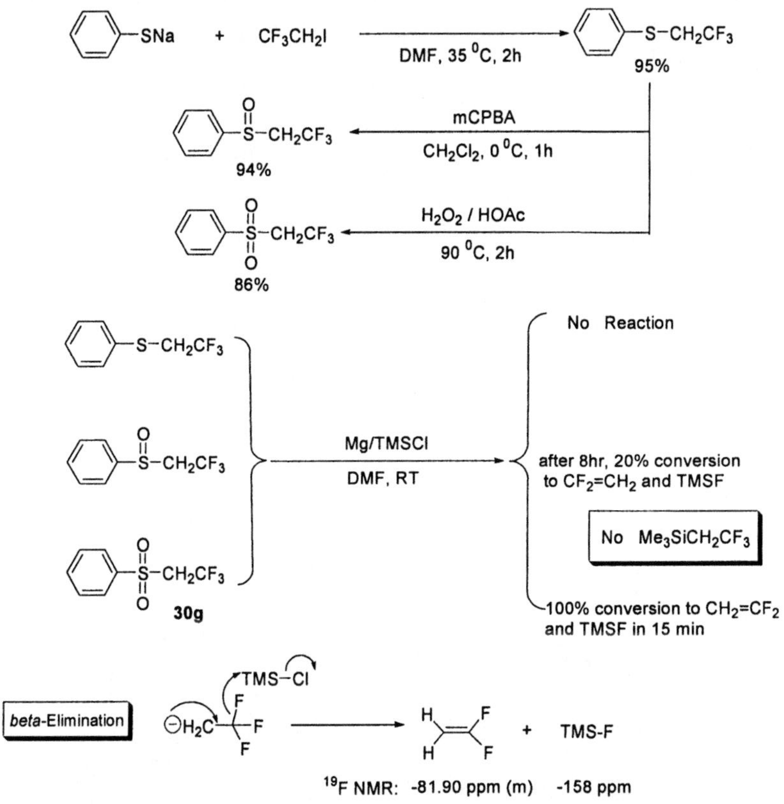

PhSNa + CF_3CH_2I → (DMF, 35 °C, 2h) → $PhS-CH_2CF_3$ 95%

→ (mCPBA, CH_2Cl_2, 0 °C, 1h) → $PhS(O)-CH_2CF_3$ 94%

→ (H_2O_2 / HOAc, 90 °C, 2h) → $PhS(O)_2-CH_2CF_3$ 86%

$PhS-CH_2CF_3$, $PhS(O)-CH_2CF_3$, $PhS(O)_2-CH_2CF_3$ **30g** → (Mg/TMSCl, DMF, RT) →

No Reaction

after 8hr, 20% conversion to $CF_2=CH_2$ and TMSF

No $Me_3SiCH_2CF_3$

100% conversion to $CH_2=CF_2$ and TMSF in 15 min

beta-Elimination: $^{\ominus}H_2C-CF_3$ + TMS-Cl → $H_2C=CF_2$ + TMS-F

^{19}F NMR: -81.90 ppm (m) -158 ppm

Scheme 18. The reactions of 2,2,2-trifluoroethyl phenyl sulfide, sulfoxide, and sulfone.

It is well known that the phenyl trifluoromethyl sulfone **30a** and sulfoxide **30b** can be readily prepared from trifluoromethane (manufactured from methane) and diphenyl disulfide. Since in our fluoroalkylation process, diphenyl disulfide is produced as a reductive byproduct, the presently developed method provides a novel and useful catalytic pathway (in diphenyl disulfide) for the production of (trifluoromethyl)silanes from trifluoromethane and chlorosilanes (see Scheme 19).

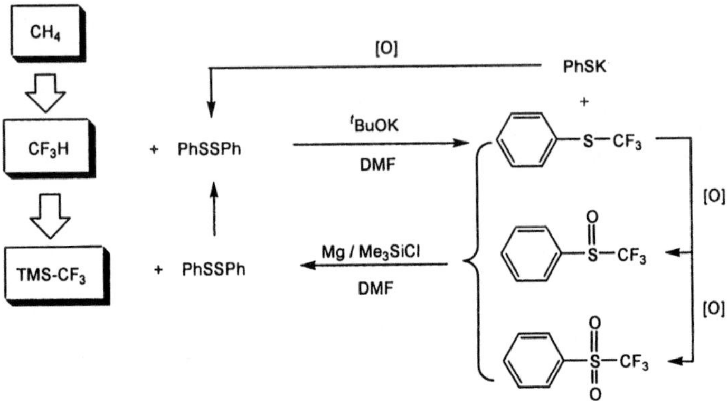

Scheme 19. A catalytic process for the preparation of $TMSCF_3$ from CF_3H.

3.2. Alkoxide- and Hydroxide-Induced Trifluoromethylation Using Trifluoromethyl Phenyl Sulfones or Sufoxide. In section **3.1** we described the reductive trifluoromethylation using trifluoromethyl sulfides, sulfoxides and sulfones as trifluoromethyl (CF_3) group precursors (*38*). However, under the reductive condition, where magnesium metal was used, the reaction only worked with chlorosilanes as electrophiles, while attempts to react with carbonyl compounds failed. We anticipated that by using a nucleophilic base such as alkoxides, the carbon-sulfur bond of trifluoromethyl phenyl sulfone (**30a**) or sulfoxide (**30b**) can be cleaved to give a trifluoromethyl anion (CF_3^-) synthon that can undergo addition to carbonyl compounds (Scheme 20). The driving force of this substitution is the formation of strong S-O bond (348 ~ 551 kJ/mol) (*40*) and the high polarity of C-S bond of sulfone **30a** or sulfoxide **30b**. The generation of pseudohalide CF_3^- species is somewhat similar to the reaction between benzenesulfonyl halides with alkoxides.

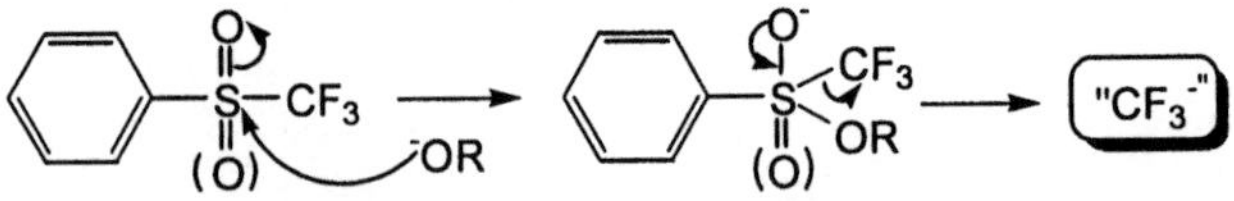

Scheme 20. Mechanistic explanation of alkoxide and hydroxide induced trifluoromethylation.

Nucleophilic displacement of the trifluoromethyl group has been reported for trifluoromethyl aryl sulfones with sodium methoxide (*41*). Similar reaction between trifluoromethyl aryl sulfone and Grignard reagents has been reported for the preparation of sulfones (*42*). More recently, Cheburkov *et al.* reported that perfluoroalkyl sulfones react with metal hydroxides in water or alcohol solution and with ammonia to form fluorinated sulfonic acid derivatives (*43*). Recently, we reported the first alkoxide induced nucleophilic trifluoromethylation of carbonyl compounds, disulfides and other electrophiles, using trifluoromethyl phenyl sulfone **30a** or sulfoxide **30b** (*44*). The trifluoromethyl sulfone **30a** or sulfoxide **30b** can be used as "CF_3^-" synthon. Both phenyl trifluoromethyl sulfone **30a** and sulfoxide **30b** are commercially available (b.p. 203 °C/760 mmHg for **30a**, b.p. 85~87°C/10 mmHg for **30b**) and can also be conveniently prepared from trifluoromethane in high yields. Thus the new methodology provides a convenient route for efficient nucleophilic trifluoromethylation.

Potassium *tert*-butoxide (*t*BuOK) was first used as nucleophile to attack the sulfur center of phenyl trifluoromethyl sulfone **30a** generating trifluoromethyl anion (Scheme 20). Into an equimolar mixture of sulfone **30a** and benzaldehyde in DMF at – 50 °C, a DMF solution of ^{t}BuOK (two molar equivalents) was slowly added. The reaction mixture was strried at – 50 °C for 1 h, and then warmed to room temperature over period of 2 h. 1-Phenyl-2-trifluoromethylethanol **38** was produced in 71 % yield (Scheme 21).

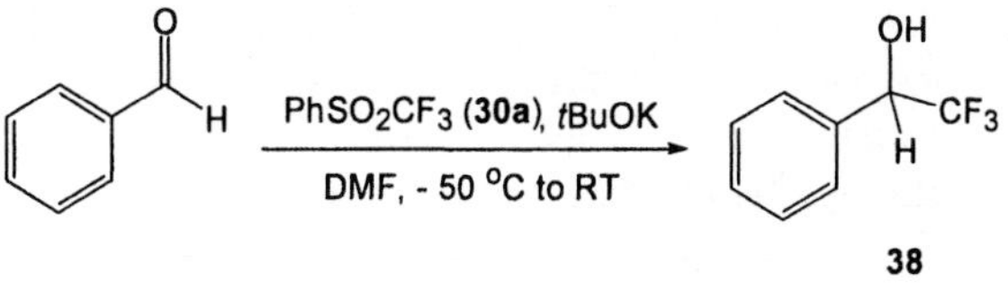

$PhSO_2CF_3$ (**30a**), *t*BuOK

DMF, - 50 °C to RT

OH H CF_3

38

Scheme 21. Trifluoromethylation using **30a**.

Shono and co-workers (*10a*) found that when they used $CF_3H/^tBuOK/DMF$ to react with benzaldehyde at – 50 °C, benzyl alcohol and benzoic acid were formed by the competing Cannizzaro reaction. Russell and Roques (*10e*) repeated the reaction using excess CF_3H (9.5 eq.) and *t*BuOK (2.2 eq.) at – 50 °C, and 67 % yield of trifluoromethylated product **38** was formed and no benzyl alcohol was detected. They mentioned that the high reaction temperature with excess base could lead to Cannizzaro reaction and jeopardize the nucleophilic trifluoromethylation. In our reaction, as shown in Scheme 21, only traces of benzoic acid and benzyl alcohol were detected by NMR. This implies that at low temperature the Cannizzaro reaction rate is much slower than the *tert*-butoxide induced trifluoromethylation process. The reaction conditions have been optimized (Table 3). When excess benzaldehyde was introduced, the yield of the product **38** can be improved based on the amount of sulfone **30a** used (Table 3, entry b).

Table 3. Trifluoromethylation of benzaldehyde by $PhSO_2CF_3$ induced by alkoxide or hydroxide

	Benzaldehyde	Sulphone **30a**	Alkoxide	Solvent	Temperature	Time	Yield (**38**,%)
a	1 eq.	1 eq.	t-BuOK (2 eq.)	DMF	- 50 °C	1 h	71
b	3 eq.	1 eq.	t-BuOK (2 eq.)	DMF	- 30 °C~r.t.	30 min	91
c	1 eq.	2 eq.	t-BuOK (2 eq.)	DMF	-50 °C~r.t.	3 h	84
d	3 eq.	1 eq.	MeONa (3 eq.)	DMF	- 30 °C~r.t.	3 h	30
e	3 eq.	1 eq.	KOH (8 eq.)	DMF	r.t.	20 h	45
f	3 eq.	1 eq.	t-BuOK (2 eq.)	DMSO	r.t..	1 h	76

[a]Yields were determined by ^{19}F NMR based on the amount of **30a** used.

Besides ^tBuOK, sodium methoxide (CH_3ONa) and potassium hydroxide (KOH) were tried as nucleophiles (Table 3, entries d and e). Both of them worked but gave lower yields. There are several possible reasons: first of all, both sodium methoxide and KOH can not readily dissolve in DMF, which affects the reaction rate. Second, unlike potassium *t*BuOK, sodium methoxide may react with benzaldehyde via a Meerwein-Ponndorf-Verley type reduction pathway. Third, KOH has lower nucleophilicity than *t*BuOK, so the reaction rate can be slow. Here, Cannizzaro reaction may still happen as a competing side-reaction, but it should not be a dominating factor to affect the yield since excess benzaldehyde is present.

It is worth noting that DMF is not the only convenient solvent for the reaction. Dimethyl sulfoxide (DMSO) was also used, and the reaction worked well (Table 3, entry f). This again indicates that the CF_3^-/DMF adduct[6,10,11a,39a] is not a necessary intermediate for this new type of nucleophilic trifluoromethylation. However, the intermediate species **37** (in Scheme 20) may play an important role in the CF_3 group transfer process.

We also attempted to carry out the reaction by using catalytic amount of ^tBuOK. However, when small amount of *t*BuOK was introduced, only low yield of product **38** was obtained (after hydrolysis) which can be monitored by ^{19}F NMR. Introduction of additional *t*BuOK increased the product yield, and excellent yield of **38** was achieved when excess *t*BuOK was used.

Excess amount of *t*BuOK (2 eq.) was found to be helpful to achieve high yields in the trifluoromethylation reactions. There are several reasons and advantages. *t*BuOK reacts with water readily and it may be partially hydrolyzed

Table 4. Reaction of trifluoromethyl phenyl sulfone (**30a**) or sulfoxide (**30b**) (2 equiv) with non-enolizable carbonyl compounds (1 equiv) and *t*BuOK (2.5 equiv) in DMF at – 50 °C to room temperature.

Entry	Carbonyl compound	Trifluoromethylating agent	Product	Yield[a] (%)	^{19}F NMR (ppm)[b]
1	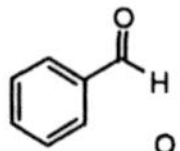 O, H	**30a**	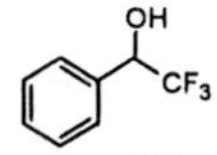 OH, CF_3	77	-78.5 (d)
2	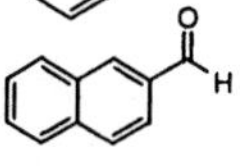 O, H	**30a**	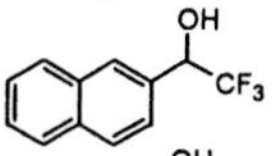 OH, CF_3	62	-78.4 (d)
3	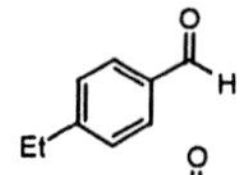 O, H, Et	**30a**	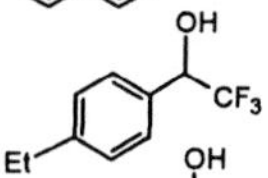 OH, CF_3, Et	83	-78.9 (d)
4	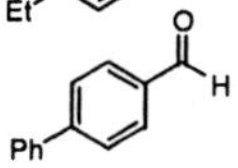 O, H, Ph	**30a**	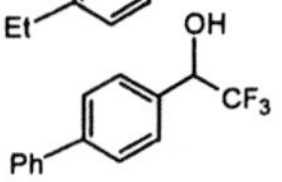 OH, CF_3, Ph	76	-78.7 (d)
5	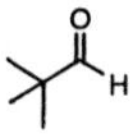 O, H	**30a**	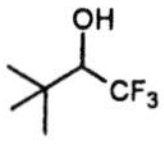 OH, CF_3	79[c]	-72.3 (d)
6	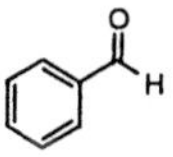 O, H	**30b**	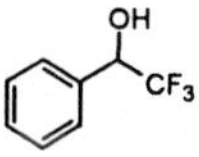 OH, CF_3	68	-78.5 (d)
7	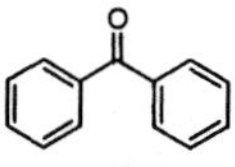 O	**30a**	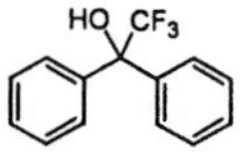 HO, CF_3	86	-74.5
8	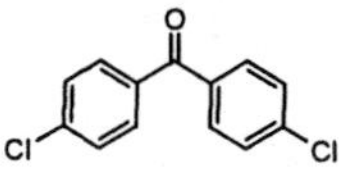 O, Cl, Cl	**30a**	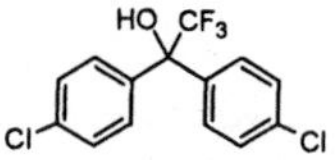 HO, CF_3, Cl, Cl	74	-75.1
9	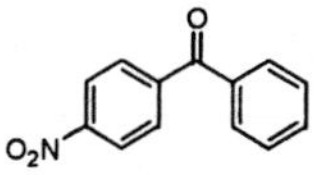 O, O_2N	**30a**	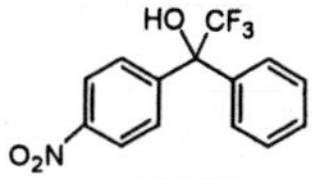 HO, CF_3, O_2N	83	-74.7
10	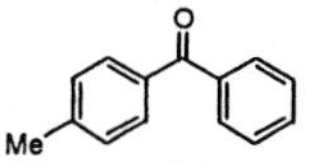 O, Me	**30a**	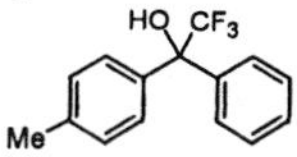 HO, CF_3, Me	85	-74.4
11	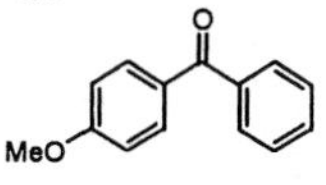 O, MeO	**30a**	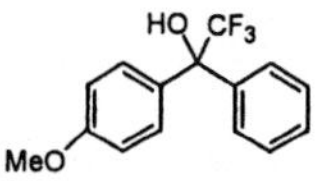 HO, CF_3, MeO	73	-75.0
12	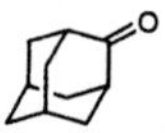 O	**30a**	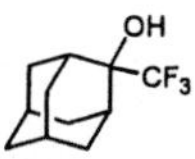 OH, CF_3	82	-76.1
13	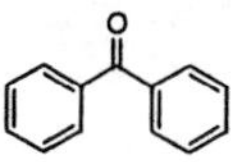 O	**30b**	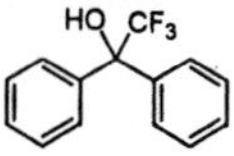 HO, CF_3	83	-74.5

[a]Isolated yields. [b]Use $CFCl_3$ as internal reference. [c]Determined by ^{19}F NMR using $PhOCF_3$ as internal standard.

during storage and handling. Furthermore, excess *t*BuOK also removes the moisture from the solvent and reagents. More importantly, excess *t*BuOK in the reaction mixture eliminates the possibility of hydrolysis of CF_3^- to form CF_3H. It is known that CF_3H can be deprotonated by *t*BuOK and undergo trifluoromethylation of carbonyl compounds. Thus, the present methodology allows to produce trifluoromethylated products in high yields in the case of non-enolizable aldehydes (Table 4, entries 1~5). Similarly, non-enolizable ketones can also be easily trifluoromethylated using this methodology (Table 4, entries 7~12). Since there is no Cannizarro reaction between benzophenone and *t*BuOK, this reaction can be carried out even at higher temperatures (such as 25 °C). Due to lower reactivity of ketone compared with aldehydes, the ketone reactions need a little bit longer time (2 ~ 3 h) to complete. However, with enolizable aldehydes and ketones, only low yield (10~30 %) of trifluoromethylated products were observed, because of competing and facile cross-adol reactions.

It is noteworthy that phenyl trifluoromethyl sulfoxide **30b** worked equally well as **30a**, the similar trifluoromethylations were observed with aldehyde and ketone (Table 4, entries 6 and 13).

Another advantage of the reaction is the simple work-up procedure. Since the by-product of the reaction is *tert*-butyl benzenesulfonate, it can be readily hydrolyzed into *t*-butyl alcohol and benzenesulfonic acid derivatives. Aqueous washing thus can remove most of the by-products and simplifies the purification process.

The novel trifluoromethylation method was also found to work with disulfides. As shown in Scheme 22, $PhSO_2CF_3$ (1 eq.) and *t*BuOK (2 eq.) reacted with diphenyl disulfide (**39**, 1.2 eq.) at – 30 °C to room temperature in 30 minutes to give quantitative conversion (87 % isolated) of phenyl trifluoromethyl sulfide **40** (^{19}F NMR: - 43.3 ppm). This reaction was even more facile than that of carbonyl compounds.

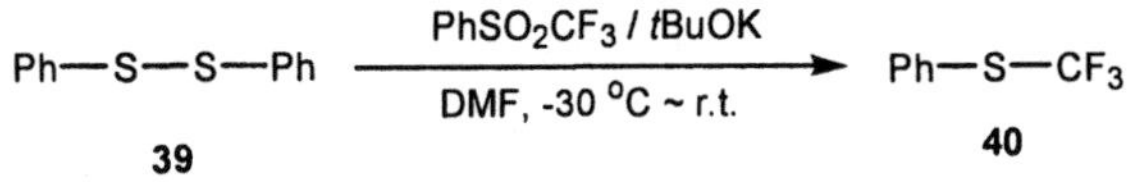

Ph—S—S—Ph (**39**) —[$PhSO_2CF_3$ / *t*BuOK; DMF, -30 °C ~ r.t.]→ Ph—S—CF_3 (**40**)

Scheme 22. Trifluoromethylation of PhSSPh with **1a**.

The **30a**/*t*BuOK trifluoromethylation method was also applied to other systems. For instance, methyl benzoate can be trifluoromethylated to generate 2,2,2-trifluoroacetophenone in 30 % yield at –50 ~ –20 °C . CF_3Cu can be *in situ* generated with **30a**/*t*BuOK and copper iodide (CuI), and then further react with iodobenzene at 80 °C for 20 h to give α,α,α-trifluorotoluene in 26 % yield.

We also attempted the use of other type of sulfones such as methyl trifluoromethyl sulfone as the trifluoromethylating agent. When diphenyl disulfide was used as the model substrate, and reaction, however, only gave minimal yield of product **40** (~ 2 %). This is probably due to the facile deprotonation of the methyl group by *t*BuOK, leading to other products.

Thus, potasium *tert*-butoxide induced trifluoromethylation using phenyl trifluoromethyl sulfone ($PhSO_2CF_3$, **30a**) or sulfoxide ($PhSOCF_3$, **30b**) enables us to transfer CF_3 group into non-enolizable carbonyl compounds and disulfides in high yields. Since both compounds **30a** and **30b** are stable high-boiling liquids that can be readily produced from non-ozone-depleting trifluoromethane, the reported new trifluoromethylating method offers an inexpensive and convenient synthetic methodologies.

5. Difluoromethylation and Difluoromethylenation with Difluoromethyl Phenyl Sulfone

5.1. Difluoromethylation Using Difluoromethyl Phenyl Sulfone. Difluoromethyl group is known to be isosteric and isopolar to hydroxy (OH) group, and it has attracted much attention in medicinal chemistry and drug discovery (*3*). However, in contrast to trifluoromethylation chemistry, the methods for direct introduction of difluoromethyl group (CF_2H) into organic molecules are rare. In 1995, Fuchikami and co-workers (*45*) reported the difluoromethylation of carbonyl compounds using (difluoromethyl)dimethylphenylsilane (**41**) in the presence of fluoride or alkoxide. However, the reactions took place only at high temperature (100 °C) due to the less polarity of the Si—CF_2H bond of **41**, which limits the scope and yields of the reaction. In 1997, we disclosed the difluoromethylation of carbonyl compounds at room temperature in good yields using bis(trimethylsilyl)difluoromethane (**42**) (Scheme 23) (*46*).

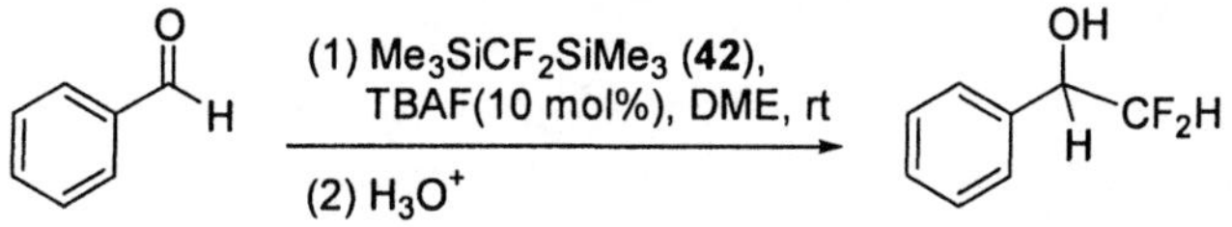

(1) $Me_3SiCF_2SiMe_3$ (**42**), TBAF(10 mol%), DME, rt
(2) H_3O^+

43 (70 %)

Scheme 23. Difluoromethylation with **42**.

Similar to the reductive trifluoromethylation with trifluoromethyl phenyl sulfone (**30a**), we discovered that under the induction of reducing metals, difluorometh l phenyl sulfone (**44**) can act as a difluoromethyl anion ("CF_2H^-") synthon (*38*). For instance, difluoromethylsilanes (R_3SiCF_2H) can be prepared from the magnesium metal-mediated reductive difluoromethylation of chlorosilanes using difluoromethyl phenyl sulfone (Scheme 24). The magnesium metal acts as a single-electron transfer (SET) source for difluoromethyl phenyl sulfone to generate the difluoromethyl anion species, which reacts with chlorosilanes to produce difluoromethylsilanes. (Difluoromethyl)trimethylsilane and (difluoromethyl)triethylsilane have been prepared by this method. These (difluoromethyl)trialkylsilanes can be further used as difluoromethylating agent for other electrophiles such as carbonyl compounds.

$$PhSO_2CF_2H + R_3SiCl \xrightarrow{Mg,\ DMF,\ 0\ ^{\circ}C,\ 1.5\ h} R_3SiCF_2H$$

44 **45**

(R = Me, 76 %
= Et, 51%)

Scheme 24. Difluoromethylation with difluoromethyl phenyl sulfone.

5.2. Difluoromethylenation Using Difluoromethyl Phenyl Sulfone (*47*). Due to the unique properties of fluorine atom, more and more of organofluorine compounds have been found to have biological effects like mimic, block, polar and lipophilic effects (*3*). For instance, C-F bond is known to mimic C-H bond because of its similar bond length, and difluoromethylene group is known to be isosteric and isopolar to an ethereal oxygen (*46*). Thus, the synthesis of fluorine-containing analogs of bioactive natural products are of great interest for their potential applications in pharmaceutical industry (*48*). Since *anti*-1,3-diol functionality is a fundamental unit in many naturally occurring compounds, its stereoselective preparation is always attractive to synthetic organic chemists (*49*). *anti*-2,2-Difluoropropan-1,3-diols (**46**) are a group of interesting compounds, however, not much is known about their synthesis. To our best knowledge, the only reported method to synthesize these compounds is via the diasteroselective Meerwein-Pondorff-Verley reduction of α,α-difluoro-β-hydroxy ketones (*50*). The disadvantage of this approach is the requirement of preparation of α,α-difluoro-β-hydroxy ketones as the precursors.

In 1997, we reported the preparation of difluorobis(trimethylsilyl)methane (TMSCF$_2$TMS) as a potential difluoromethylene dianion ("$^-CF_2^-$") equivalent (*46*). However, TMSCF$_2$TMS was found only to couple with one molecule of aldehyde such as benzaldehyde to give 2,2-difluoro-1-phenylethanol (after acidic hydrolysis) (*46*).

In section **3.2** we have described the alkoxide and hydroxide induced nucleophilic trifluoromethylation of non-enolizable carbonyl compounds and disulfides using trifluoromethyl sulfone or sulfoxide. The chemistry is based on the nucleophilic attack by alkoxide (commonly potassium *tert*-butoxide) or hydroxide on the sulfur center of trifluoromethyl phenyl sulfone **30a** (or sulfoxide **30b**) to release a trifluoromethyl anion (Scheme 25, Eq. i). Similarly, we assumed that a similar type of S-C bond cleavage could occur with difluoromethyl phenyl sulfone **44** ($PhSO_2CF_2H$). Compared to trifluoromethyl sulfone **30a**, the chemistry of difluoromethyl sulfone **44** is even more interesting (Scheme 25, Eq. ii). It is known that the CF_2H hydrogen of compound **44** is rather acidic, and a common base such as sodium methoxide or even aqueous sodium hydroxide can deprotonate it in an equilibrium mode to generate $PhSO_2CF_2^-$ anion **47** (*51,52*). In 1989, Stahly has shown that the *in situ* generated anion **6** can react with aldehydes to give difluoromethylated carbinols in aqueous NaOH media in the presence of a phase-transfer agent (*52*). However, in Stahly's study, he did not observe any S-C bond cleavage under the aqueous NaOH conditions (at room temperature for 4 h). Obviously, aqueous NaOH is not nucleophilic enough to activate the S-C bond scission. It also indicates that with hydroxide or alkoxide, the deprotonation on difluoromethyl sulfone **44** is much faster than the S-C bond cleavage. Thus, by use of a proper alkoxide such as *t*BuOK working both as a base and a nucleophile, sulfone **44** may react stepwise with two electrophiles to give new difluoromethylene-containing products (Scheme 25, Eq. ii). Thus, difluoromethyl phenyl sulfone **44** can be regarded as a selective difluoromethylene dianion ("$^-CF_2^-$") synthon.

With above considerations in mind, we first reacted $PhSO_2CF_2H$/*t*BuOK system with diphenyl disulfide (PhSSPh) as an electrophile. The results are shown in Table 5. With the different reactant ratio, both mono-substitution product **52** and double-substitution product **53** can be obtained at room temperature in high selectivity (Table 5, entries b and g). This result confirms our previous assumption that the reactivities of deprotonation and S-C bond cleavage are different and these two steps can be controlled selectively (see Scheme 26). Excess *t*BuOK facilitates the completion of S-C bond cleavage process, which is similar to our previous observations with trifluoromethyl sulfone system (*44*). Furthermore, the formation and consumption of $PhSCF_2H$ (**55**) with time (Table 5, entries e and f) indicate that there is an equilibrium between anionic species **58** and **55** under the protonation/deprotonation with *t*BuOH/*t*BuOK (Scheme 26).

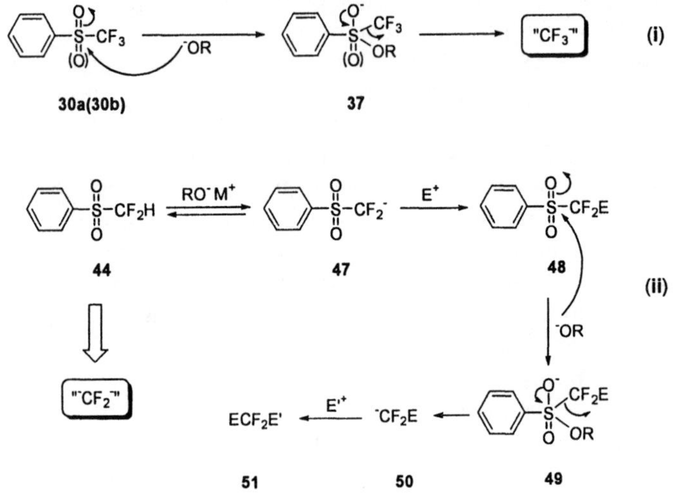

(i)

30a(30b) 37 "CF_3^-"

(ii)

44 47 48 49 50 51 "$^-CF_2^-$"

(R = H, alkyl group. E, E' = electrophiles, such as disulfides, aldehydes)

Scheme 25. Mechanistic considerations of alkoxide-mediated trifluoromethylation and difluoromethylenation.

Table 5. Difluoromethylenation of PhSSPh with **44**.

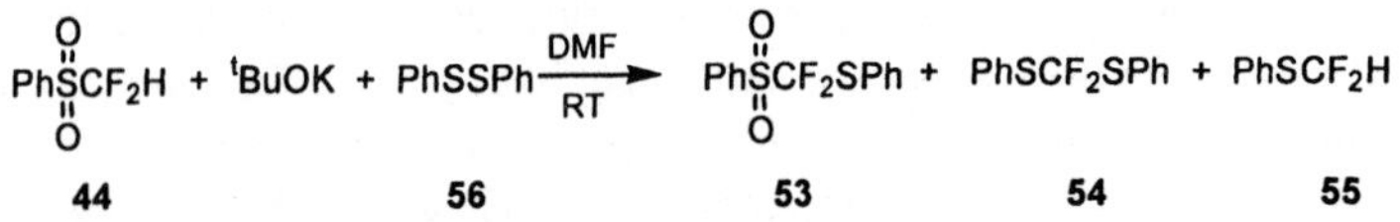

$PhSO_2CF_2H$ (**44**) + ^tBuOK + PhSSPh (**56**) →(DMF, RT) $PhSO_2CF_2SPh$ (**53**) + $PhSCF_2SPh$ (**54**) + $PhSCF_2H$ (**55**)

Entry	Reactant ratio			Reaction	Product yields		
	44	^tBuOK	56	time	53	54	55
a	1 eq.	1.0 eq.	1.0 eq.	30 min	76 %	0 %	5 %
b	1 eq.	1.5 eq.	1.0 eq.	50 min	91 %	3 %	6 %
c	1 eq.	2.5 eq.	2.0 eq.	14 h	64 %	22 %	14 %
d	1 eq.	3.0 eq.	2.0 eq.	4 h	41 %	44 %	14 %
e	1 eq.	3.5 eq.	2.0 eq.	4 h	0 %	84 %	16 %
f	1 eq.	3.5 eq.	2.0eq.	15 h	0 %	97 %	3 %
g	1 eq.	4.0 eq.	2.0 eq.	4 h	0 %	99 %	0 %

The yields were determined by ^{19}F NMR with $PhOCF_3$ as the internal standard.

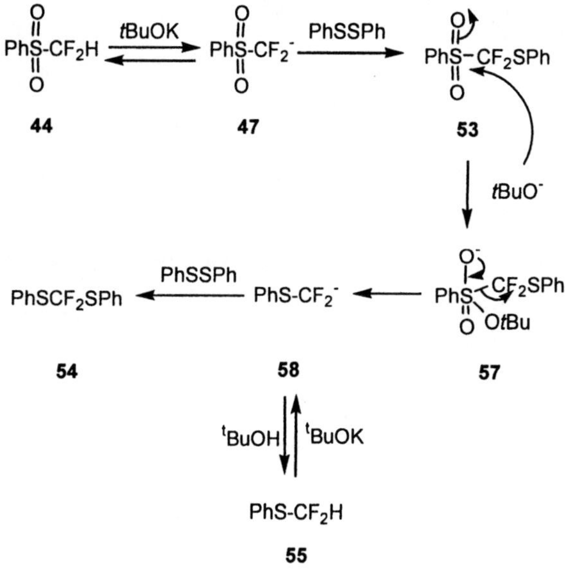

Scheme 26. Stepwise formation of **53** and **54**.

Reaction of benzaldehyde **59a** and sulfone **44**/*t*BuOK in DMF is much more intriguing and rewarding (see Scheme 27). Similar to the reaction with diphenyl disulfide, $PhSO_2CF_2H$ (1.0 equiv.)/*t*BuOK (3.0 equiv.) reacts with benzaldehyde **59a** (2.0 equiv.) at – 50 °C ~ RT for 90 minutes to generate mono-substituted product **60** (41% ^{19}F NMR yield) as earlier shown by Stahly in the aqueous NaOH medium (*52*), and di-substituted product **61a** (58 % ^{19}F NMR yield, *anti*/*syn* = 98:1). When **44** (1.0 equiv.)/*t*BuOK (4.0 equiv.) reacted with 3.0 equiv. of PhCHO at – 50 °C ~ RT for 8 h, under the activation of *t*BuOK, the *in situ* formed alkoxide **61a** undergoes S-C bond fission to generate a dianionic intermediate **63** (see Scheme 28), which can further react with another molecule of benzaldehyde to form di-substituted *anti-* diol product **61a** in excellent yield (92 % ^{19}F NMR yield, 82 % isolated) and high diastereoselectivity (*anti*/*syn* = 97:3, *de* = 94 %). The observed high diastereoselectivity can be interpreted by the charge-charge repulsion effect during the second addition (Scheme 28). To our knowledge, this may be the first example in which high diastereoselectivity has been achieved in a reaction where a dianionic species reacts with another neutral electrophile involving an intramolecular charge-charge repulsion effect (during the product formation) rather than the traditional steric control (based on the Cram's rule) (*53*). Furthermore, using DFT theory (B3LYP/6-31G**//B3LYP/6-31G* + ZPE level) on 3,3-difluoro-2,4-pentane diolate dianion as a model, the *anti* structure was found to be 5.5 kcal/mol more stable

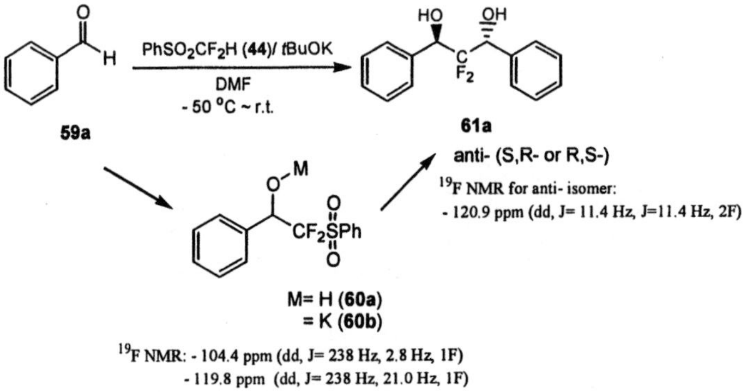

$PhSO_2CF_2H$ (**44**)/ *t*BuOK, DMF, - 50 °C ~ r.t.

59a → **61a** anti- (S,R- or R,S-)

^{19}F NMR for anti- isomer:
- 120.9 ppm (dd, J= 11.4 Hz, J=11.4 Hz, 2F)

M= H (**60a**)
= K (**60b**)

^{19}F NMR: - 104.4 ppm (dd, J= 238 Hz, 2.8 Hz, 1F)
- 119.8 ppm (dd, J= 238 Hz, 21.0 Hz, 1F)

Entry	Reactant ratio			Reaction	Product yields	
	44	^tBuOK	**59a**	time	**61a** (anti-/syn-)	**60a**
i	1.0 eq.	3.0 eq.	2.0 eq.	90 min	58 % (98:1)	41 %
ii	1.0 eq.	4.0 eq.	3.0 eq.	8 h	92 % (97:3)	0 %

The yields were determined by ^{19}F NMR with internal satndard.

Scheme 27. Reaction of **44**, benzaldehyde and *t*BuOK.

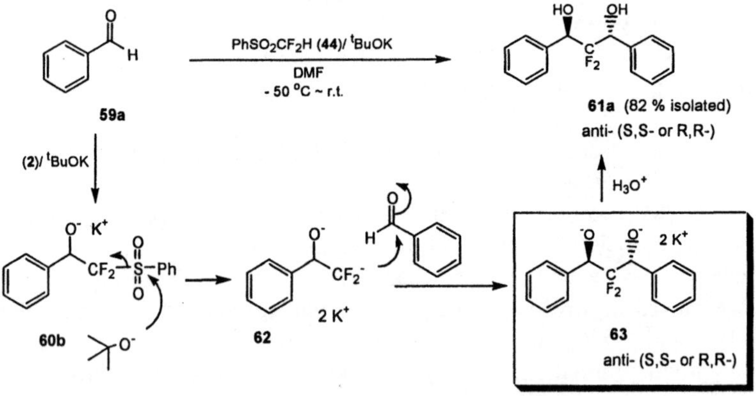

$PhSO_2CF_2H$ (**44**)/ ^tBuOK, DMF, - 50 °C ~ r.t.

59a → **61a** (82 % isolated) anti- (S,S- or R,R-)

(**2**)/ ^tBuOK

60b → **62** (2 K^+) → **63** anti- (S,S- or R,R-) (2 K^+) → H_3O^+ → **61a**

Scheme 28. Proposed mechanism of diastereoselective formation of **61a** from PhCHO and **44**/*t*BuOK.

than the corresponding *syn* structure. We also found no change in the *anti*/*syn* diol ratios upon prolonged base treatment indicating lack of product reversibility under the reaction conditions.

Table 6 demonstrates the application of this methodology to synthesize various 2,2-difluoropropan-1,3-diols (**61a–g**) with high stereoselectivity from non-enolizable aldehydes. The yields of diols are little bit lower for electron-rich aldehydes (entries d and g), probably due to the relative instability of the corresponding dianion intermediates.

Table 6. Preparation of 2,2-difluoropropan-1,3-diols (**61**) from aldehydes (3 equiv.) and difluoromethyl phenyl sulfone (2, 1 equiv.) with *t*BuOK (4 equiv.) in DMF at –50 °C to room temperature.

Entry	Substrates **1**	Products **3**	Yield (%)[a]	anti-/syn-ratio[b]	de (%)
a	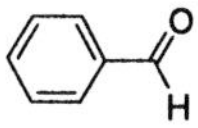 O H	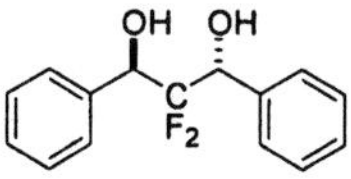 OH OH C F_2	82	97:3	94
b	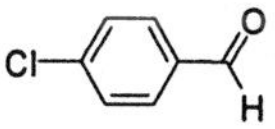 Cl O H	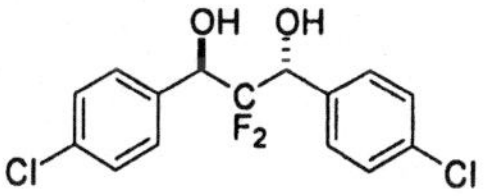 OH OH C F_2 Cl Cl	78	94:6	88
c	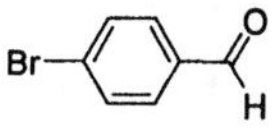 Br O H	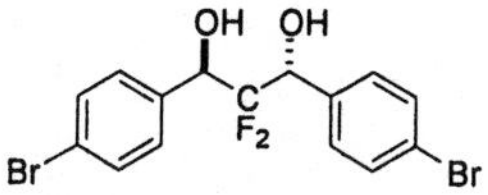 OH OH C F_2 Br Br	70	96:4	92
d	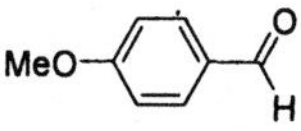 MeO O H	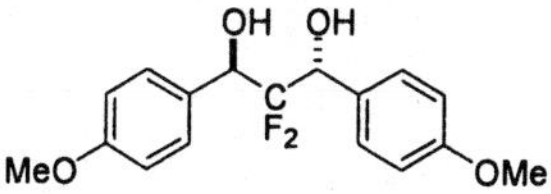 OH OH C F_2 MeO OMe	52	94:6	88
e	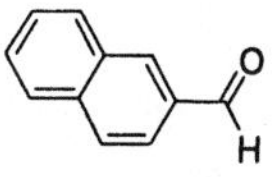 O H	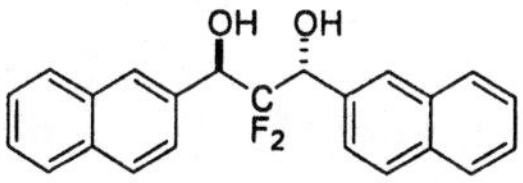 OH OH C F_2	69	97:3	94
f	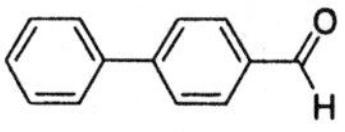 O H	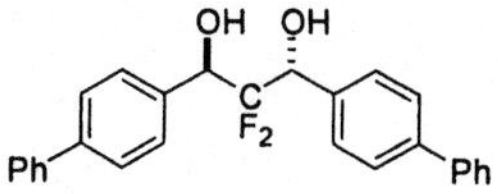 OH OH C F_2 Ph Ph	75	96:4	92
g	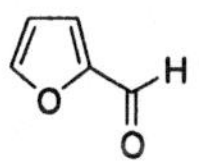 O H O	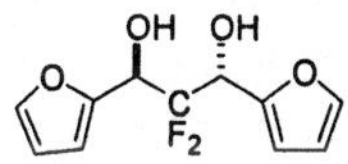 OH OH O C O F_2	63	93:7	86

[a]Isolated yields. [b]Anti-/syn- ratio were determined by ^{19}F NMR.

Besides the symmetrical *anti*-2,2-difluoropropan-1,3-diols, this new methodology can also be used to synthesize unsymmetrical *anti*-2,2-difluoropropan-1,3-diols. Scheme 29 shows one example of this type of synthesis. Difluoro(phenylsulfonyl)methyl substituted alcohol **64** can be easily obtained and isolated through Stahly's approach in high yield (*52*). The activation of **64** with *t*BuOK generates the dianion intermediate **65**, which further reacts with *p*-chlorobenzaldehyde **1b** to give unsymmetrical *anti*-2,2-difluoropropan-1,3-diol **66** (after hydrolysis) with high diastereoselectivity.

$PhSO_2CF_2H$ (**44**), $NaOH/H_2O/CH_2Cl_2$, 90 %[8] → **64**

64 (1 eq.) + Cl–C$_6$H$_4$–CHO **1b** (2 eq.), *t*BuOK (3 eq.), DMF, - 50 °C to RT → **66** (S,S- or R,R-), 76 % yield, de = 92 %

^{t}BuOK; H_3O^+; 2K$^+$; CF_2^-; **65** (S,S- or R,R-)

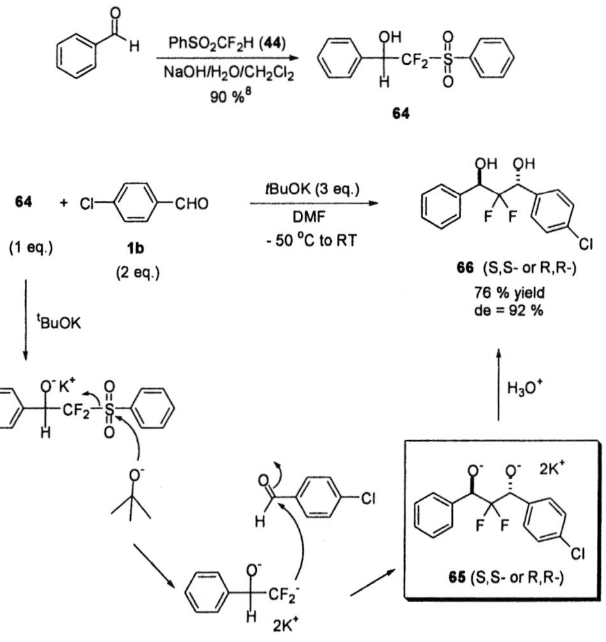

Scheme 29. Synthesis of unsymmetrical *anti*-2.2-difluoropropan-1,3-diol.

Potassium *tert*-butoxide induced difluoromethylenation using difluoromethyl phenyl sulfone enables us to synthesize both symmetrical and unsymmetrical *anti*-2,2-difluoropropan-1,3-diols with high diastereoselectivity (up to 94 % de). This unusual type of high diastereoselectivity was obtained via an intramolecular charge-charge repulsion effect rather than the traditional steric control (based on the Cram's rule). Difluoromethyl phenyl sulfone can be used as a selective difluoromethylene dianion synthon ($^-CF_2^-$), which can couple two electrophiles (such as diphenyl disulfide or non-enolizable aldehydes) to give new difluorinated products. Since difluoromethyl phenyl sulfone can be readily prepared from inexpensive chemicals such as CF_2ClH or CF_2Br_2, this new methodology provides a convenient and efficient synthetic tool for many potential applications.

6. Introduction of Difluoromethylene and Monofluoromethylene Groups via Fluorinated Silyl Enol Ethers

As mentioned earlier, difluoromethylene is isosteric and isopolar to ethereal oxygen, and thus it is frequently desired to be introduced into bioactive organic molecules (*3*). Although there are many ways to achieve this, 2,2-difluoro enol silyl ethers (**66**) and 2-monofluoro enol silyl ethers (**67**) are easily accessible precursors for difluoro- and monofluoromethylene compounds. Silyl enol ethers **66** are readily prepared from chlorodifluoromethyl ketones (*54*) or acylsilanes (*55*). In 1999, Uneyama and co-workers developed the preparation of **66** by magnesium metal promoted selective C-F bond cleavage of trifluoromethyl ketones (*56*). Thereafter, we prepared monofluoro enol silyl ethers **67** from difluoromethyl ketones using similar procedures (*57*).

OSiR'3, R, F, F — **66**; OSiR'3, R, H, F — **67**

2,2-Difluoro enol silyl ethers **66** are mainly used for the Mukaiyama-type of aldol condensenations to prepare the new difluoromethylene compounds (Scheme 30) (*55,56*).

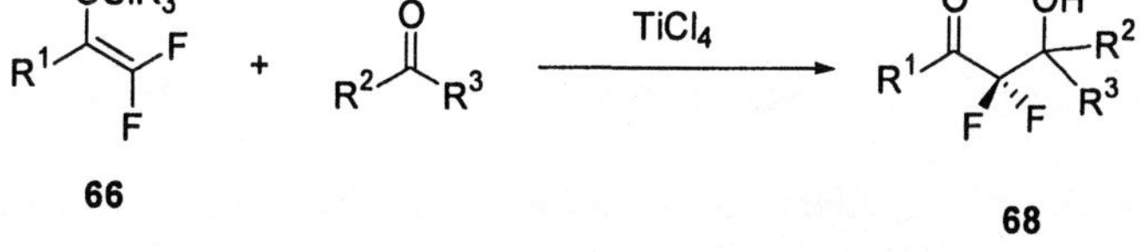

66 (OSiR3, R1, F, F) + R2C(O)R3 —TiCl4→ **68** (R1, O, OH, R2, R3, F, F)

Scheme 30. Mukaiyama reaction with 2,2-difluoro enol silyl ethers.

Recently, by adopting Uneyama's procedure, we developed a new method for the preparation of difluoromethyl and monofluoromethyl ketones from trifluoromethyl ketones via fluorinated enol silyl ethers (Scheme 31) (*57*). These fluorinated methyl ketones are useful as potential protease enhibitors.

Furthermore, we developed a general method halogenation for the synthesis of α-halodifluoromethyl ketones using 2,2-difluoro enol silyl ethers **66** (Scheme 32) (*58*). The results are summarized in Table 7.

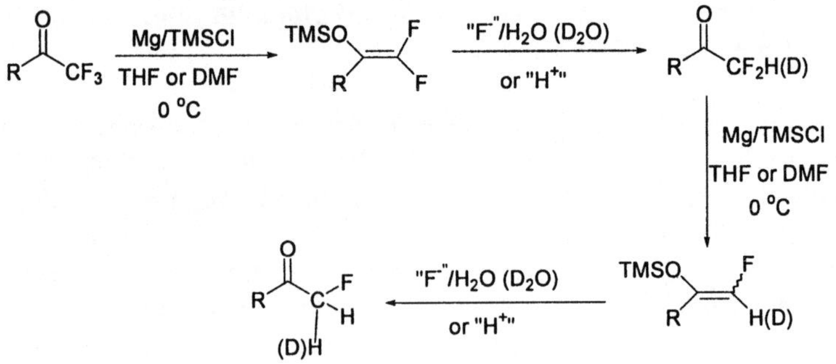

Scheme 31. Prepapration of di- and monofluoromethyl ketones from trifluoromethyl ketones.

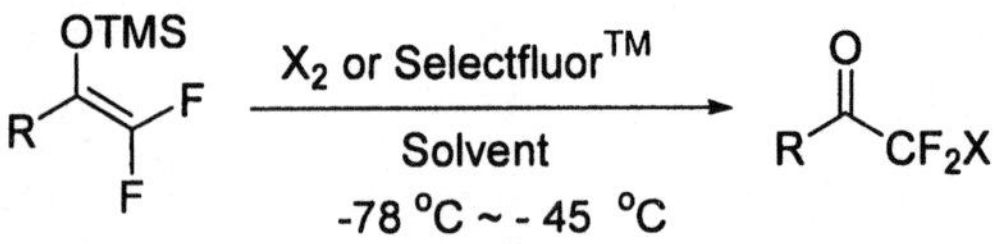

(X = F, Br, I; Solvent = CH_3CN, $CFCl_3$, CH_2Cl_2)

Scheme 32. Halogenation of 2,2-difluoro enol silyl ethers.

This facile halogenation method has been extended by us for the preparation of [^{18}F]-labeled α-trifluoromethyl ketones (*59*). Reactions of 2,2-difluoro-1-aryl-1-trimethylsiloxyethenes (**70a-d**) with [^{18}F]-F_2 at low temperature produced [^{18}F]-labeled α-trifluoromethyl ketones (**71a-d**) (Scheme 33). Radiolabeled products were isolated by purification with column chromatography. The radiochemical yields of these compounds were 45-55%, decay corrected (d. c.) in 3 runs per compound. Radiochemical purity was > 99% with specific activities 15-20 GBq/mmol at the end of synthesis (EOS). The synthesis time was 35-40 min from the end of bombardment (EOB). This one step simple method is highly useful for radiochemical synthesis of potential biologically active [^{18}F]-labeled α-trifluoromethyl ketones for PET imaging.

Bromodifluoromethyl phenyl ketone (Table 7, **69f**) has been used by us in the preparation of 2-phenyl-1,1,2,2-tetrafluoroethanesulfonic acid (**72**) (see Scheme 34) (*60*).

Table 7. Halogenation of 2,2-difluoro enol silyl ethers.

Entry	Substrate **66**	Halogenating agent	Product **69**	Yield (%) [a]
1	R= Ph (**66a**)	F_2	$PhC(O)CF_3$ (**69a**)	69
2	2-thienyl (**66b**)	F_2	2-thienyl-$C(O)CF_3$ (**69b**)	78[b]
3	Ph (**66a**)	Selectfluor™	$PhC(O)CF_3$ (**69a**)	89
4	$4-F_3C-C_6H_4$ (**66d**)	Selectfluor™	$4-F_3C-C_6H_4C(O)CF_3$ (**69c**)	88
5	$4-H_3C-C_6H_4$ (**66e**)	Selectfluor™	$4-H_3C-C_6H_4C(O)CF_3$ (**69d**)	87
6	$4-Cl-C_6H_4$ (**66c**)	Selectfluor™	$4-Cl-C_6H_4C(O)CF_3$ (**69e**)	90
7	Ph (**66a**)	Br_2	$PhC(O)CF_2Br$ (**69f**)	85
8	$4-F_3C-C_6H_4$ (**66d**)	Br_2	$4-F_3C-C_6H_4C(O)CF_2Br$ (**69g**)	87
9	$4-H_3C-C_6H_4$ (**66e**)	Br_2	$4-H_3C-C_6H_4C(O)CF_2Br$ (**69h**)	88
10	$4-Cl-C_6H_4$ (**66c**)	I_2	$4-Cl-C_6H_4C(O)CF_2I$ (**69i**)	60
11	$4-H_3C-C_6H_4$ (**66e**)	I_2	$4-H_3C-C_6H_4C(O)CF_2I$ (**69j**)	39

[a] Isolated yield; [b] determined by ^{19}F NMR.

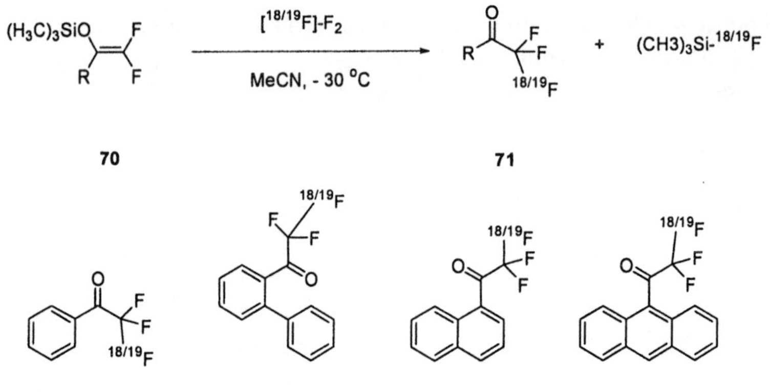

Scheme 33. [^{18}F]-Labeled synthesis of trifluoromethyl ketones.

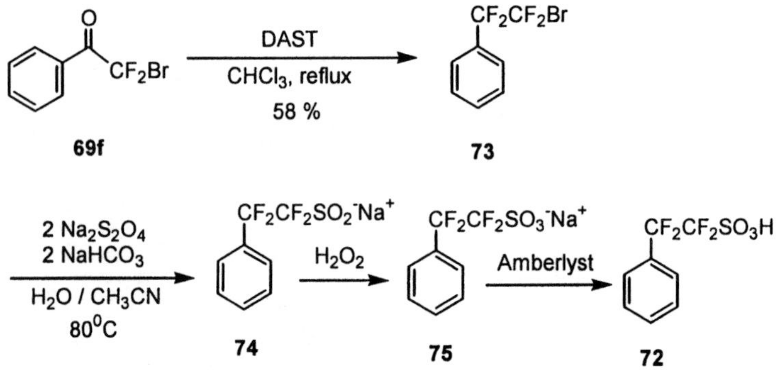

Scheme 34. Synthesis of 2-phenyl-1,1,2,2-tetrafluoroethanesulfonic acid (72) from bromodifluoromethyl phenyl ketone (69f).

7. Conclusion

In this chaper, we have summarized the new nucleophilic alkylation chemistry developed in our laboratory in the past four years. These include (1) stereoselective trifluoromethylation with imines using $TMSCF_3$, (2) amine *N*-oxide catalyzed trifluoromethylation with $TMSCF_3$, (3) reductive trifluoromethylation and difluoromethylation using tri- and difluoromethyl sulfones, sufoxides and sulfides, (4) alkoxide and hydroxide induced trifluoromethylation using trifluoromethyl phenyl sulfone and sulfoxide, (5) one-pot synthesis of *anti*-2,2-difluoro-propane-1,3-diols using difluoromethyl phenyl sulfones as difluoromethylene dianion synthon, (6) preparation of di- and monofluoromethyl ketones from trifluoromethyl ketones via fluorinated silyl enol ethers, (7) a general method for halogenation of 2,2-difluoro enol silyl ether to prepare α-halodifluoromethyl ketones, (8) [F-18] labeled synthesis of trifluoromethyl ketones, (9) preparation of partially fluorinated alkanesulfonic acid from α-halodifluoromethyl ketone, among others. We have also included recent work by others in this area to make this review more comprehensive. Given the fascinating properties and reactivities of fluorine, we believe more nucleophilic fluoroalkylation chemistry (chemo-, regio- and stereoselective) will be discovered and find practical applications in the coming years.

Acknowledgments: This review is respectfully dedicated to Professor George A. Olah to commemorate the 10th anniversary of his receipt of Nobel Prize in Chemistry. Support of our work by the Loker Hydrocarbon Research Institute is greatly acknowledged.

Literatures cited

1. Cottet, F.; Marull, M.; Lefebvre, O.; Schlosser, M. *Eur. J. Org. Chem.* **2003**, 1559-1568; and the references therein.
2. *Synthetic Fluorine Chemistry*; Olah, G. A.; Chamber, R. D.; Prakash, G. K. S., Eds., Wiley: New York, 1992.
3. *Organofluorine Compounds: Chemistry and Applications*; Hiyama, T., Ed., Springer: New York, 2000.
4. (a) Prakash, G. K. S.; Krishnamuti, R.; Olah, G. A. *J. Am. Chem. Soc.* **1989**, *111*, 393-395. (b) Krishnamurti, R.; Bellew D. R.; Prakash G. K. S., *J. Org. Chem.* **1991**, *56*, 984-989. (c) Prakash, G. K. S.; Yudin, A. K. *Chem. Rev.* **1997**, *97*, 757-786.
5. Ruppert, I.; Schlich, K.; Volbach, W. *Tetrahedron Lett.* **1984**, *25*, 2195-2198.
6. Singh, R. P.; Shreeve, J. M. *Tetrahedron* **2000**, *56*, 7613-7632.
7. Prakash, G. K. S.; Mandal, M. *J. Fluorine Chem.* **2001**, *112*, 123-131.
8. McClinton, M. A.; McClinton, D. A. *Tetrahedron* **1992**, *48*, 6555-6666.
9. Burton, D. J.; Yang, Z. Y. *Tetrahedron* **1992**, *48*, 189-275.

10. (a) Shono, T.; Ishifume, M.; Okada, T.; Kashimura, S. *J. Org. Chem.* **1991**, *56*, 2-4. (b) Barhdadi, R.; Troupel, M.; Perichon, J. *Chem. Comm.* **1998**, 1251-1252. (c) Folleas, B.; Marek, I.; Normant, J.-F.; Saint-Jalmes, L. *Tetrahedron Lett.* **1998**, *39*, 2973-2976. (d) Folleas, B, Marek, I.; Normant, J.-F.; Saint-Jalmes, L., *Tetrahedron* **2000**, *56*, 275-283. (e) Russell, J.; Roques, N. *Tetrahedron* **1998**, *54*, 13771-13782. (f) Large, S.; Roques, N.; Langlois, B. R. *J. Org. Chem.* **2000**, *65*, 8848-8856. (g) Roques, N.; Russel, J.; Langlois, B. R.; Saint-Jalmes, L.; Large, S., *PCT Int. Appl.,*1998, WO 9822435.
11. See review: Langlois, B. R.; Billard, T. *Synthesis* **2003**, 185-194.
12. (a) Motherwell, W. B.; Storey, L. J., *Synlett* **2002**, 646-648. (b) Jablonski, L.; Billard, T.; Langlois, B. R. *Tetrahedron Lett.* **2003**, *44*, 1055-1057.
13. (a) Jablonski, L.; Joubert, J.; Billard, T.; Langlois, B. R. *Synlett*, **2003**, 230-232. (b) Inschauspe, D.; Sortais, J.-P.; Billard, T.; Langlois, B. R. *Synlett*, **2003**, 233-235.
14. (a) Joubert, J.; Roussel, S.; Christophe, C.; Billard, T.; Langlois, B. R.; Vidal, T., *Angew. Chem. Int. Ed.* **2003**, *42*, 3133-3136. (b)Ait-Mohand, S.; Takechi, N.; Medebielle, M.; Dolbier, W., Jr. *Org. Lett.* **2001**, *3*, 4271-4273.
15. (a) Schmit, C. *Synlett* **1994**, 241-242. (b) Johnson, C. R.; Bhumralkar, D. R.; De Clercq, E. *Nucleosides & Nucleotides* **1995**, *14*, 185-194. (c) Lavaire, S.; Plantier-Royon; Portella, C. *J. Carbohydr. Chem.* **1996**, *15*, 361-370. (d) Lavaire, S.; Plantier-Royon, R.; Portella, C. *Tetrahedron: Asymmetry* **1998**, *9*, 213-226. (e) Eilitz, U.; Bottcher, C.; Hennig, L.; Burger, K.; Haas, A.; Gockel, S.; Sieler, J. *J. Het. Chem.* **2003**, *40*, 329-335.
16. Wang, Z.; Ruan, B. *J. Fluorine Chem.* **1994**, *69*, 1-3.
17. Felix, C. P.; Khatimi, N.; Laureat, A. J. *Tetrahedron Lett.* **1994**, *35*, 3303-3304.
18. (a) Walter, M. W.; Adlington, R. M.; Baldwin, J. E.; Chuhan, J.; Schofield, C. J. *Tetrahedron Lett.* **1995**, *36*, 7761-7764. (b) Walter, M. W.; Adlington, R. M.; Baldwin, J. E.; Schofield, C. J. *Tetrahedron Lett.* **1997**, *53*, 7275-7290. (c) Walter, M. W.; Adlington, R. M.; Baldwin, J. E.; Schofield, C. J. *J. Org. Chem.* **1998**, *63*, 5179-5192. (d) Qiu, X. L.; Qing, F. L. *J. Org. Chem.* **2002**, *67*, 7162-7164.
19. Prakash, G. K. S.; Mandal, M.; Olah, G. A. *Angew. Chem. Int. Ed.* **2001**, *40*, 589-590.
20. Prakash, G. K. S.; Mandal, M.; Olah, G. A. *Org. Lett.* **2001**, *3*, 2847-2850.
21. Prakash, G. K. S.; Mandal, M. *J. Am. Chem. Soc.* **2002**, *124*, 6538-6539.
22. (a) Prakash, G. K. S., Presented at the 29th Western Regional Meeting of American Chemical Society and 32nd Annual Meeting of the Southern California Section of the Society for Applied Spectroscopy, Pasadena, CA,

October 19–23, Paper No. 123. (b) Yudin, A. K., Ph.D. dissertation, University of Southern California, 1996.

23. Iseki, K.; Nagai, T.; Kobayashi, Y. *Tetrahedron Lett.* **1994**, *35*, 3137-3138.
24. Iseki, K.; Kobayashi, Y. *Rev. Hetereoat. Chem.* **1995**, *12*, 211-237.
25. Kuroki, Y.; Iseki, K. *Tetrahedron Lett.* **1999**, *40*, 8231-8234.
26. Caron, S.; Do. N. M.; Arpin, P.; Larivee, A. *Synthesis* **2003**, *11*, 1693-1698.
27. Prakash, G. K. S.; Mandal. M.; Panja, C.; Mathew, T.; Olah, G. *J. Fluorine Chem.* **2003**, *123*, 61-63; and the references therein.
28. (a) Sosnovskikh, V. Y.; Sevenard, D. V.; Usachev, B. I.; Roschenthaler, G.-V. *Tetrahedron Lett.* **2003**, *44*, 2097-2099. (b) Sosnovskikh, V. Y.; Usachev, B. I.; Sevenard, D. V.; Roschenthaler, G.-V. *J. Org. Chem.* **2003**, *68*, 7747-7754.
29. Sevenard, D. V.; Sosnovskikh, V. Y.; Kolomeitsev, A. A.; Konigsmann, M. H.; Roschenthaler, G.-V., *Tetrahedron Lett.* **2003**, *44*, 7623-7627.
30. Singh, R. P.; Shreeve, J. M. *Chem. Comm.* **2002**, 1818-1819.
31. (a) Tyrra, W.; Naumann, D.; Hoge, B.; Yagupolskii, Y. L. *J. Fluorine Chem.* **2003**, *119*, 101-107. (b) Babadzhanova, L. A.; Kirij, N. V.; Yagupolskii, Y. L. *J. Fluorine Chem.* **2004**, *125*, 1095-1098.
32. (a) Mloston, G.; Prakash, G. K. S.; Olah, G. A.; Heimgartner, H. *Helvetica Chemica Acta* **2002**, *85*, 1644-1658. (b) Large-Radix, S.; Billard, T; Langlois, B. R. *J. Fluorine Chem.* **2003**, *124*, 147-149.
33. Molander, G. A.; Hoag, B. P. *Organometallics* **2003**, *22*, 3313-3315.
34. Kolomeitsev, A. A.; Kadyrov, A. A.; Szczepkowska-Sztolcman, J.; Milewska, M.; Koroniak, H.; Bissky, G.; Barten, J. A.; Roschenthaler, G.-V. *Tetrahedron Lett.* **2003**, *44*, 8273-8277.
35. Sevenard, D. V.; Kolomeitsev, A. A.; Hoge, B.; Lork, E.; Roschenthaler, G.-V. *J. Am. Chem. Soc.* **2003**, *125*, 12366-12367.
36. Garlyaukayte, R. Y.; Bezdudny, A. V.; Michot, C.; Armand, M.; Yagupolskii, Y. L.; Yagupolskii, L. M. *J. Chem. Soc., Perkin Trans. 1*, **2002**, 1887-1889.
37. (a) Taw, F. L.; Scott, B. L.; Kiplinger, J. L. *J. Am. Chem. Soc.* **2003**, *125*, 14712-14713. (b) Hoffmann-Roder, A.; Seiler, P.; Diederich, F. *Org. Biomol. Chem.* **2004**, *2*, 2267.
38. Prakash, G. K. S.; Hu, J.; Olah, G. A. *J. Org. Chem.*, **2003**, *68*, 4457-4463.
39. (a) Russell, J.; Roques, N. *Tetrahedron* **1998**, *54*, 13771-13782. (b) Gerard, F.; Jean-Mannel, M.; Laurent, S.-J. *Eur. Pat. Appl.* **1996**, EP 733614, 13 pp. (c) Chen, Q.-Y.; Duan, J.-X. *Chem. Comm.* **1993**, 918-919. (d) Yang, J.-J.; Kirchmeier, R. L.; Shreeve, J. M. *J. Org. Chem.* **1998**, *63*, 2656-2660.
40. Dean, J. A. *Lange's Handbook of Chemistry (14th Ed.)*, McGraw Hill: New York, **1992**, p 4.34.
41. Shein, S. M.; Krasnopol'skaya, M. I.; Boiko, V. N., *Zh. Obshei Khim.* **1966**, *36*, 2141-2147.

42. Steensma, R. W.; Galabi, S.; Tagat, J. R.; McCombie, S. W., *Tetrahedron Lett.* **2001**, *42*, 2281-2283.
43. Barrera, M. D.; Cheburkov, Y.; Lamanna, W. M. *J. Fluorine Chem.* **2002**, *117*, 13-16.
44. Prakash, G. K. S.; Hu, J.; Olah, G. A. *Org. Lett.* **2003**, *5*, 3253-3256.
45. Higiwara, T.; Fuchikami, T. *Synlett* **1995**, 717-718.
46. Yudin, A. K.; Prakash, G. K. S.; Deffieux, D.; Bradley, M.; Bau, R.; Olah, G. A. *J. Am. Chem. Soc.* **1997**, *119*, 1572-1581.
47. Prakash, G. K. S.; Hu, J.; Mathew, T.; Olah, G. A. *Angew. Chem. Int. Ed.* **2003**, *42*, 5216-5219.
48. McCarthy, J., *Utility of Fluorine in Biologically Active Molecules*, ACS Fluorine Division Tutorial, 219th National ACS Meeting, San Francisco, March 26, 2000.
49. Masamune, S.; Choy, W. *Aldrichimica Acta* **1982**, *15*, 47-64.
50. Kuroboshi, M.; Ishihara, T. *Bull. Chem. Soc. Jpn.* **1990**, *63*, 1185-1190.
51. Hine, J.; Porter, J. J. *J. Am. Chem. Soc.* **1960**, *82*, 6178-6181.
52. Stahly, P. G. *J. Fluorine Chem.* **1989**, *43*, 53-66.
53. Recently, two groups reported the electrostatic-repulsion controlled stereoselectivity. (a) In 1997, Mall and Stamm reported that the electrostatic repulsion by the charged tail of a radical controls the stereochemistry of coupling with the anthracenide radical anion: T. Mall, H. Stamm, *J. Org. Soc., Perkin Trans. 2*, **1997**, 2135-2140. (b) In 2001, Uneyama et al. reported a diastereoselectivity controlled by electrostatic repulsion between the negative charge density on a trifluoromethyl group and that of electron-poor aromatic rings: T. Katagiri, S. Yamaji, M. Handa, M. Irie, K. Uneyama, *Chem. Commun.* **2001**, 2054-2055; T. Katagiri, K. Uneyama, *Chirality*, **2003**, *15*, 4-9.
54. Yamana, M.; Ishihara, T.; Ando, T. *Tetrahedron Lett.* **1983**, *24*, 507-510.
55. Brigaud, T.; Doussot, P.; Portella, C. *Chem. Comm.* **1994**, 2117-2118.
56. Amii, H.; Kobayashi, T.; Hatamoto, Y.; Uneyama, K. *Chem. Comm.* **1999**, 1323-1324.
57. Prakash, G. K. S.; Hu, J.; Olah, G. A. *J. Fluorine. Chem.* **2001**, *112*, 357-362.
58. Prakash, G. K. S.; Hu, J.; Alauddin, M. M.; Conti, P. S.; Olah, G. A. *J. Fluorine Chem.* **2003**, *121*, 239-243.
59. Prakash, G. K. S.; Alauddin, M. M.; Hu, J.; Conti, P. S.; Olah, G. A. *J. Labelled Compd. & Radiopharm.* **2003**, *46*, 1087-1092.
60. Prakash, G. K. S.; Hu, J.; Simon, J.; Bellew, R.; Olah, G. A. *J. Fluorine Chem.* **2004,** *125,* 595-601.

Chapter 3

Fluoroform, Fluoral, Trifluoroacetic, and Trifluoromethanesulfinic Acids Derivatives as New Reagents for the Introduction of Polyfluorinated Moieties

Bernard R. Langlois and Thierry Billard

Laboratoire SERCOF (UMR CNRS 5181), Université Claude Bernard-Lyon I, 69622 Villeurbanne, France

Several new classes of stable reagents have been designed for nucleophilic trifluoromethylation, starting from fluoroform, fluoral, trifluoroacetamides or trifluoromethanesulfinamides. Several are stable enough to be considered as laboratory reagents. Trifluoroacetamides or trifluoromethanesulfinamides derived from O-TMS-1,2-aminoalcools constitute the most attractive family. Some of them have delivered promising results for enantioselective trifluoromethylation. Hemiaminals of fluoral are also potent precursors of trifluoromethylated iminiums and difluoroketene aminals, which are wide-scope key-intermediates for the expedient preparation of elaborated compounds, especially nitrogen-containing cycles.

Because of the unique properties of the fluorine atom (small size, low polarizability, high electronegativity, ability to form strong bonds with carbon) (1), fluorinated molecules take a more and more prominent place in materials (2) and biologically active compounds (3,4) as well as in very diverse areas (2). Among fluorinated products, trifluoromethylated ones constitute an important family because of the hydrophobicity (and often lipophilicity) attached to the trifluoromethyl substituent, that enhances their bioavailability (1,5).

Consequently arises the question : how to prepare trifluoromethyl-containing compounds ? In fact, two main strategies can be considered. The first one is the building block strategy, starting from a simple trifluoromethylated material, which is functionnalized and sophisticated until the target is reached. However, this linear and stepwise strategy needs to design a specific route for each target and, moreover, the starting synthon is not always easily available. The second approach, which emerged more recently, is the direct introduction of a trifluoromethyl moiety on an organic substrate. In principle, it seems more powerful, since the expensive CF_3 substituent could be attached on an already complex structure, close from the final target. For this purpose, three possibilities are available :

- an electrophilic approach, using an equivalent of the putative cation $^+CF_3$; nevertheless, generators of this species, such as (trifluoromethyl)iodonium or sulfonium salts [Umemoto's reagents (6) and Shreeve's reagent (7)], are still expensive and not available on a large scale at the moment.

- a radical approach using the electrophilic radical $^{\cdot}CF_3$ (8), but this technique is rarely compatible with stereoselectivity.

- a nucleophilic approach, using an equivalent of the unstable anion $^-CF_3$, which, being more adapted to stereoselective reactions, is at present the most attractive method, all the more so since new classes of chiral bioactive compounds, such as Efavirenz (9a,b) or Befloxatone (9c,d), are emerging (Scheme 1). This is the reason why a sustained activity is focused on nucleophilic trifluoromethylation for some years.

Scheme 1

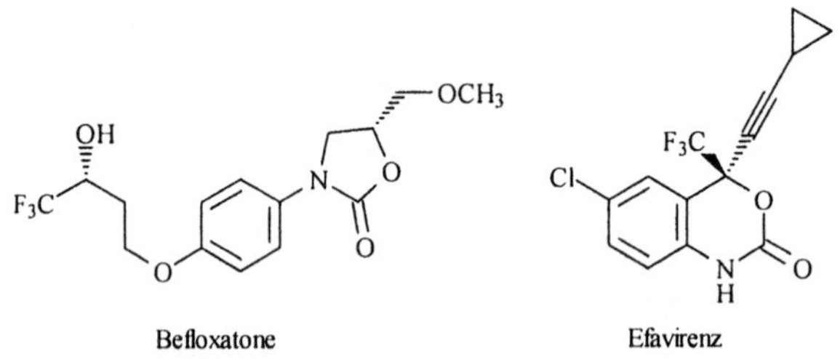

Befloxatone

Efavirenz

These different methodologies have been already reviewed in 1992 (10) but, since this time, numerous new methodologies appeared, especially for nucleophilic trifluoromethylation.

These different methodologies have been already reviewed in 1992 (10) but, since this time, numerous new methodologies appeared, especially for nucleophilic trifluoromethylation.

For a long time, the development of this latter technique has been slowed down by the instability of the trifluoromethyl anion : as carbon-fluorine bonds are short (and shorten with the accumulation of fluorine atoms on the same carbon) (1) and fluorine is the most electronegative element, $^{-}CF_3$ is the seat of huge coulombian destabilizing interactions between the anionic charge and the p-electrons of the three fluorine atoms. Consequently, it constitutes a transient species which, in the absence of any stabilization, collapses very rapidly into a fluoride anion and a difluorocarbene. In contrast, the latter is efficiently stabilized by back-donation of fluorine p-electrons into the vacant orbital of the carbon since difluorocarbene exhibits a singlet ground-state (11) (Scheme 2).

Scheme 2

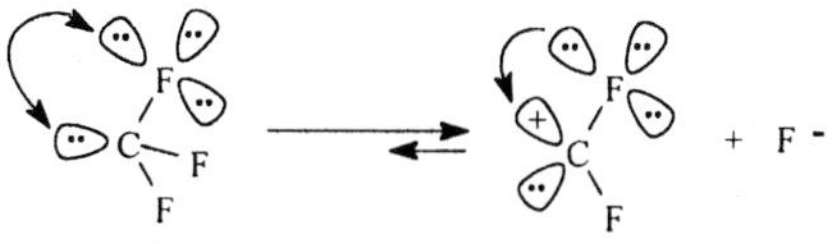

destabilized stabilized + F -

In order to stabilize the trifluoromethyl anion, three strategies can be considered. All of them tend to decrease the electronic density on the carbon or to take the carbanionic charge away from fluorine atoms.

In the first strategy, the carbanionic charge is dispersed into the vacant orbitals of main-group (Hg, Zn) or transition metals (Cd, Cu) (12). Practically, trifluoromethyl Cu(I) species are the most convenient for synthetic purposes since they are easily obtained through thermal decarboxylations, either from alkaline trifluoroacetate in the presence of CuI (13a,b) or $ClCF_2CO_2Me/KF/CuI$ (13c-f), as well as $CF_3CO_2Me/CsCl/CuI$ (13g,h) (Scheme 3). Under these conditions, a mixture of CF_3Cu, $[CF_3CuI]^-$ and $(CF_3)_2Cu^-$ is usually generated, all these species acting as trifluoromethylating agents. However, they are sluggish and usually reserved for the substitution of aromatic iodides or bromides, under thermal activation (up to 160-180 °C).

The second possibility is to engage the anionic charge into a labile σ-bond, for example with tin or silicon, the trifluoromethyl anion being regenerated by fluoride activation when required. In this respect, (trifluoromethyl)trimethylsilane (Ruppert-Prakash reagent, CF_3SiMe_3) is the most popular tool which has been extensively used, under fluoride activation, for the trifluoromethylation of carbonyl compounds, activated imines, disulfides (or diselenides) (14), as well as thiocyanates (or selenocyanates) (15-17). The scope of this reagent has been reviewed by Prakash *et al.* (18) and Shreeve *et al.* (19) and is developed in another chapter of this book. It must be noticed that, for a long time, scaling-up the preparation of CF_3SiMe_3 has been slowed down because of the deleterious effect of bromotrifluoromethane, the starting material, towards stratospheric ozone. Fortunately, a new and environmentally begnin synthesis of the Ruppert-Prakash reagent, starting from phenyl trifluoromethyl sulfone, appeared recently (20).

Scheme 3

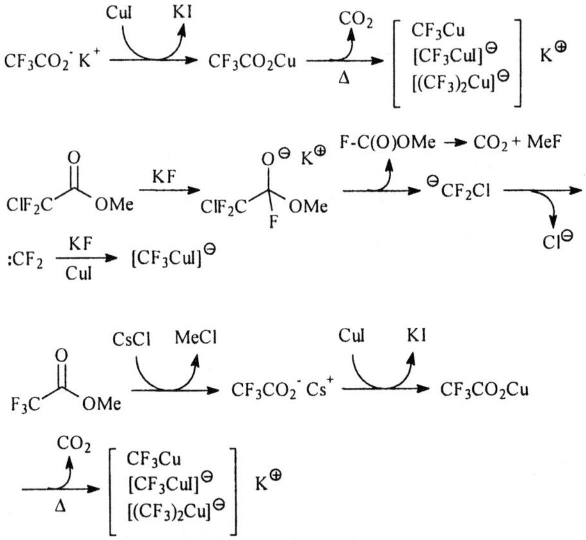

$CF_3CO_2^-K^+$ —(CuI, KI)→ CF_3CO_2Cu —(CO_2, Δ)→ [CF_3Cu, $[CF_3CuI]^\ominus$, $[(CF_3)_2Cu]^\ominus$] $K^\oplus$

$ClF_2C(O)OMe$ —(KF)→ $ClF_2C(F)(O^\ominus K^\oplus)OMe$ —(F-C(O)OMe → CO_2 + MeF)→ $^\ominus CF_2Cl$ —($Cl^\ominus$)→

$:CF_2$ —(KF, CuI)→ $[CF_3CuI]^\ominus$

$F_3C(O)OMe$ —(CsCl, MeCl)→ $CF_3CO_2^-Cs^+$ —(CuI, KI)→ CF_3CO_2Cu

—(CO_2, Δ)→ [CF_3Cu, $[CF_3CuI]^\ominus$, $[(CF_3)_2Cu]^\ominus$] $K^\oplus$

The third way to stabilize the trifluoromethyl anion is to trap it, as soon as formed, by a very powerful electrophile, present in large excess, in order that the charge is taken away from the fluorine atoms. Formamides, and especially DMF used as solvent, are able to play this role.

These three possibilities are summarized in Scheme 4.

Scheme 4

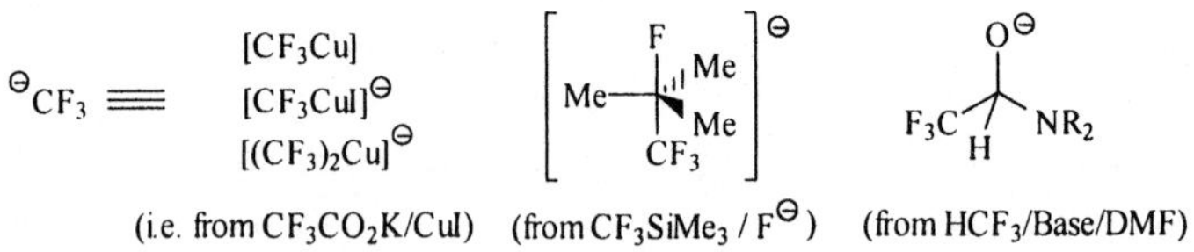

$^\ominus CF_3$ ≡ [CF_3Cu], $[CF_3CuI]^\ominus$, $[(CF_3)_2Cu]^\ominus$ (i.e. from CF_3CO_2K/CuI) ; $[Me_3Si(F)(CF_3)]^\ominus$ (from CF_3SiMe_3 / $F^\ominus$) ; $F_3C\text{-}CH(O^\ominus)NR_2$ (from HCF_3/Base/DMF)

Trifluoromethylation with trifluoromethane (fluoroform).

The first evidences of the latter strategy arose when, simultaneously with other groups (21), we tried to trifluoromethylate benzaldehyde with fluoroform in the presence of a strong base such as potassium *tert*-butoxide. This attempt was motivated by the fact that fluoroform is a very attracting candidate for trifluoromethylation because it is the smallest trifluoromethyl-containing molecule and, moreover, does not interact with stratospheric ozone. For a long time, such a reaction was claimed to fail, except in one case reported by Shono *et al.* but without mechanistic explanation (22). However, provided that DMF was used as solvent, it worked smoothly at -10 °C and delivered the expected (trifluoromethyl)carbinol in a good yield. It was demonstrated that $^{-}CF_3$ was trapped by DMF, prior transfer to the carbonyl substrate, in the form of an hemiaminolate **1** which constituted a "CF_3 reservoir" and was the real trifluoromethylating species (Scheme 5) (21b-d,23).

Scheme 5

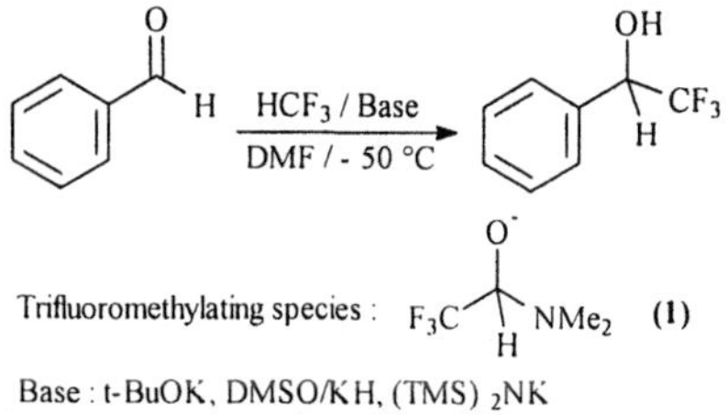

HCF_3 / Base, DMF / - 50 °C

Trifluoromethylating species : F_3C–CH(O^-)–NMe_2 (1)

Base : t-BuOK, DMSO/KH, $(TMS)_2NK$

Of course, as a strong base was used in a stoichiometric amount, this reaction could not be applied to enolizable carbonyl substrates. Under such conditions, aldolization extensively prevailed. However, it was found that the strong base $(Me_3Si)_2N^-$ can be generated very slowly from *tris*-(trimethylsilyl)amine and anhydrous tetramethylammonium fluoride in DMF (one hour was needed to complete this desilylation in the absence of any other substrate) (23). Thus, the possibility to trifluoromethylate benzophenone, used as model substrate, with fluoroform in DMF was examined in the presence of low steady-state concentrations of tetramethylammonium *bis*-(trimethylsilyl)amide, generated with such a system. Indeed, 1,1-diphenyl-2,2,2-trifluoroethanol was formed in interesting yields and, moreover, both tetramethylammonium fluoride and DMF could be used in catalytic amounts. Consequently, the reaction could be carried out in THF, which was easier to eliminate than DMF, and the final (trifluoromethyl)carbinol was obtained as a silylated derivative. In such a way, globally speaking, the system HCF_3 / $N(TMS)_3$ behaved, in THF, like CF_3SiMe_3 when activated with a catalytic amount of fluoride and DMF (Scheme 6) (23).

These reactions could be rationalized by the mechanism proposed in Scheme 7. It must be noticed that, in order to take into account that unstable $^{-}CF_3$ cannot be generated in the free form without extensive decomposition, trifluoromethyl transfer was assumed to occur in a concerted way, through a six-membered transition state.

Scheme 6

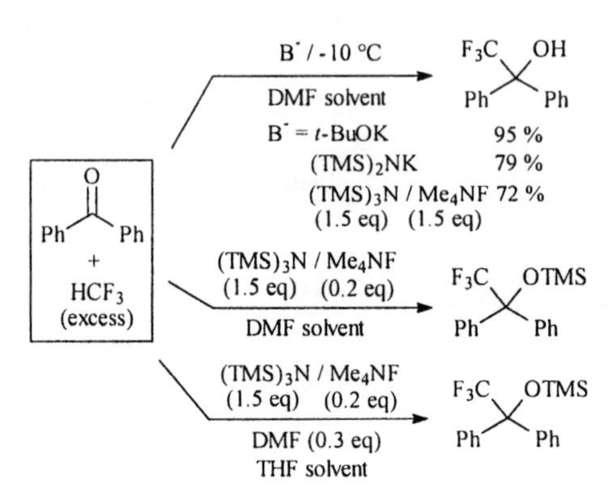

Scheme 7

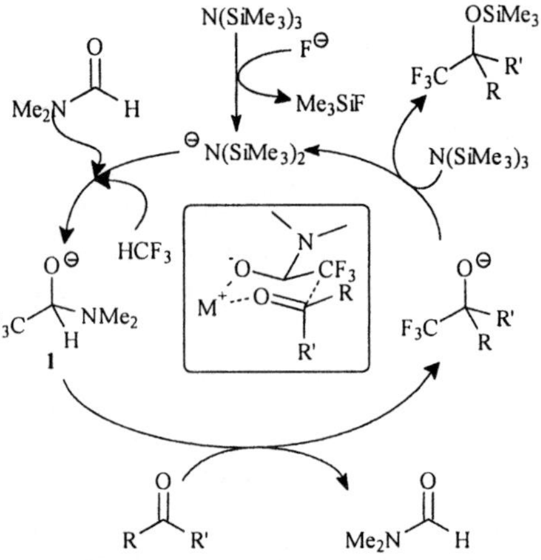

As expected, such a multi-component system (HCF_3/$N(TMS)_3$/Me_4NF cat./DMF cat.) trifluoromethylated enolizable ketones, though in a modest yield (acetophenone : 26 %, propiophenone : 28 %), except when enolization was slowed down for steric reasons (2,6-dimethylcyclohexanone : 50 %) (23). Nevertheless, these results were far better than those resulting from *t*-BuOK with which aldolization exclusively occurred. In the same way, transformation of *n*-octyl disulfide into *n*-octyl trifluoromethyl sulfide was improved with such a system (yield : 73 %, compared to 54 % with *tert*-BuOK/DMF solvent) (23). Again, a cyclic transition state could be assumed (Scheme 8). This synthesis of trifluoromethyl sulfides has been invoked by Prakash *et al.* to reinforce the interest of their new preparation of CF_3SiMe_3 (20).

Scheme 8

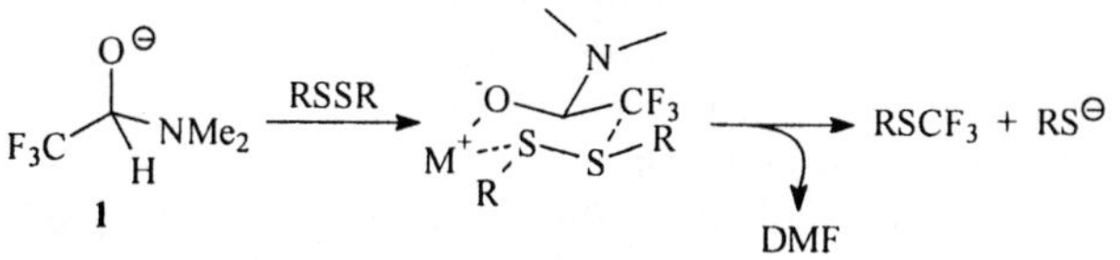

$$F_3C-C(O^{\ominus})(H)-NMe_2 \; (\mathbf{1}) \xrightarrow{RSSR} [\text{cyclic intermediate } M^+ \cdots O, S\text{–}S, CF_3, NMe_2] \xrightarrow{-DMF} RSCF_3 + RS^{\ominus}$$

Though trifluoromethylating species **1** has been characterized by NMR, attempts to isolate it failed. However, it was suspected that, in the absence of any other electrophile, the reaction of DMF with fluoroform, $N(TMS)_3$ and tetramethylammonium fluoride in THF could deliver the silylated derivative of **1** (Scheme 9, NR_2 = NMe_2). Indeed, **2a** was obtained but was very difficult to purify and, moreover, exhibited a very limited shelf-life (few hours). Fortunately, treatment of N-formylmorpholine under the same conditions led to a stable product **2b** (Scheme 9, NR_2 = morpholino) which was easily purified by flash chromatography and could be stored quite indefinitely without any alteration (24). Thus, **2b** constitutes a new reagent which was able to trifluoromethylate non enolizable ketones and aldehydes, at 80 °C under activation with sub-stoichiometric quantities of cesium fluoride, provided that two equivalents of **2b** were used (Scheme 9) (24). As O-silylated trifluoromethyl carbinols were obtained and N-formylmorpholine was quantitatively recovered as by-product, $HCF_3/N(TMS)_3$ can be considered again, in this process, as an equivalent of the Ruppert-Prakash reagent.

Scheme 9

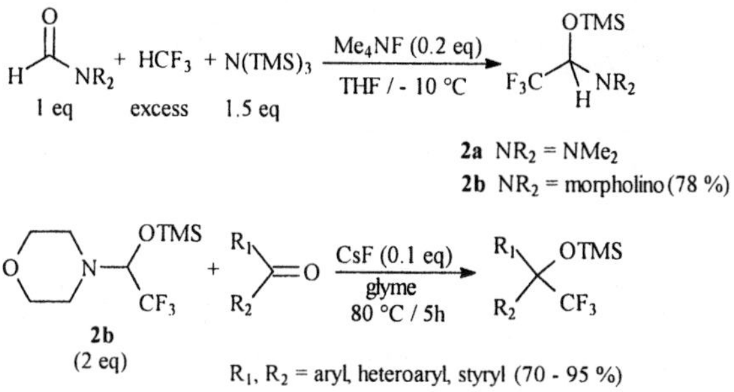

$$H-C(=O)-NR_2 \; (1\text{ eq}) + HCF_3 \; (\text{excess}) + N(TMS)_3 \; (1.5\text{ eq}) \xrightarrow[\text{THF / - 10 °C}]{Me_4NF\ (0.2\text{ eq})} F_3C-CH(OTMS)-NR_2$$

2a NR_2 = NMe_2
2b NR_2 = morpholino (78 %)

$$\mathbf{2b}\ (2\text{ eq}) + R_1R_2C=O \xrightarrow[\text{glyme, 80 °C / 5h}]{CsF\ (0.1\text{ eq})} R_1R_2C(OTMS)CF_3$$

R_1, R_2 = aryl, heteroaryl, styryl (70 - 95 %)

Trifluoromethylation with Stable Hemiaminals of Fluoral

Though gaseous fluoroform (Eb = -82.2 °C) is obviously the best trifluoromethylating reagent in terms of atom economy, its manipulation is not very easy on the lab scale. Consequently, another strategy for preparing **2b** had to be considered. In fact, **2b** cannot be seen only as an adduct between fluoroform and N-formylmorpholine but also as an adduct between morpholine and trifluoroacetaldehyde (fluoral). Indeed, fluoral is also a gas but is commercially available in the form of its stable and liquid methyl hemiketal, easy to handle. When opposed to morpholine in the presence of a methanol trap (Na_2SO_4, $MgSO_4$ or 4 Å molecular sieves), this latter compound led, under mild conditions, to an hemiaminal of fluoral which, after silylation under neutral conditions with N-(trimethylsilyl)imidazole, delivered **2b** in a good yield (25). In the same way, another O-silylated hemiaminal of fluoral **2c** has been obtained from N-benzylpiperazine (Scheme 10) (25). Like **2b**, **2c** was able to trifluoromethylate non enolizable carbonyl substrates but under milder conditions : the same results were obtained with one equivalent of **2c** at 60 °C instead of two equivalents of **2b** at 80 °C (Scheme 10) (25). By analogy with CF_3SiMe_3 (15), **2b** and **2c** were also able to transform disulfides and diselenides into trifluoromethyl sulfides and selenides, respectively, but, in contrast with the reaction of carbonyl substrates, the best results were obtained, whatever the reagent (**2b** or **2c**), at 80 °C with two equivalents of **2b,c** and $Bu_4N^+ (Ph_3SiF_2)^-$ (TBAT) (Scheme 10) (26).

Scheme 10

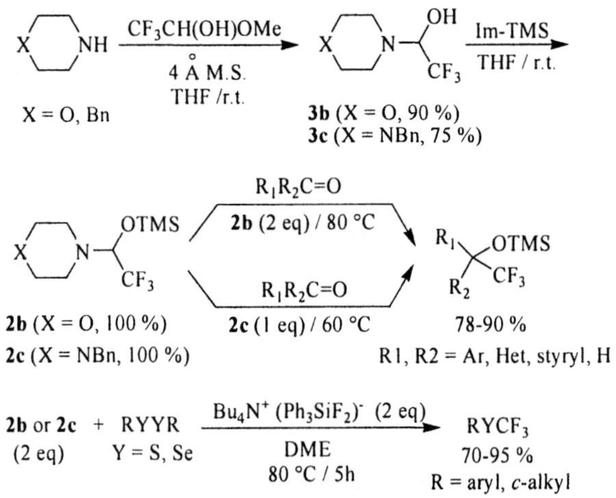

X = O, Bn; NH + $CF_3CH(OH)OMe$ (4 Å M.S., THF /r.t.) → N–CH(OH)CF_3; Im-TMS, THF / r.t.

3b (X = O, 90 %)
3c (X = NBn, 75 %)

$R_1R_2C{=}O$, **2b** (2 eq) / 80 °C; $R_1R_2C{=}O$, **2c** (1 eq) / 60 °C → $R_1R_2C(OTMS)CF_3$

2b (X = O, 100 %)
2c (X = NBn, 100 %)

78-90 %
R1, R2 = Ar, Het, styryl, H

2b or **2c** (2 eq) + RYYR (Y = S, Se) —[$Bu_4N^+ (Ph_3SiF_2)^-$ (2 eq), DME, 80 °C / 5h]→ $RYCF_3$ 70-95 %, R = aryl, *c*-alkyl

As indicated in Scheme 10, reagents **2b** and **2c** were respectively obtained from hemiaminals **3b** and **3c** which were also stable compounds. Though **3b** was an oil which could not be purified from contaminating morpholine, **3c** was a solid which could be recrystallized and remained indefinitely stable on storage. As expected, **3c** could be quantitatively deprotonated by strong bases, such as *tert*-butoxide, hydride or *bis*-(trimethylsilyl)amide, to deliver an hemiaminolate **3c**$^-$ that, when associated to a potassium cation (sodium or lithium bases failed), efficiently trifluoromethylated non enolizable ketones and aldehydes in THF

(Scheme 11) (27). Again, two equivalents of **3c**$^-$ K^+ were needed to reach the best results in THF. However, the addition of DMF (at least one equivalent) or 18-C-6 crown ether (1 eq) allowed the use of only one equivalent of **3c**$^-$, indifferently associated to a potassium or a sodium cation. **3c**$^-$ was also able to deliver aromatic trifluoromethyl sulfides or selenides from the corresponding aromatic dichalcogenides but aliphatic ones failed, probably because of side-deprotonations in α-position to the heteroatom (Scheme 11) (26).

Scheme 11

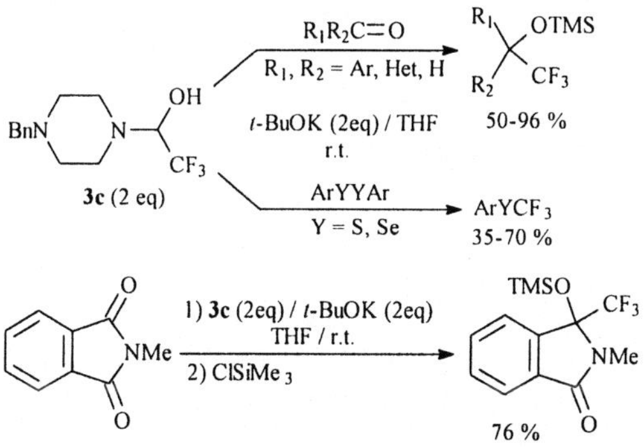

As previously mentioned, trifluoromethylation of non enolizable carbonyl substrates in THF delivered the best yields with two equivalents of **2b** or **3c**, but only one equivalent of **2c** was sufficient. Thus, it could be assumed that, in this solvent, **2c** was a monomer whereas **2b** and **3c** were present in a head-to-tail dimeric form, in the first case because of the strong interaction between oxygen and silicon (25) and, in the second case, because of the chelation of potassium cations by piperazinyl moieties (Scheme 12) (27). In this latter case, such a complexation could self-activate the reagent. Moreover, it can be easily understood that, when stronger potassium solvating or chelating compounds were added (DMF, 18-C-6), dimeric **3c**$^-$ was dissociated.

Concerning the mechanism of trifluoromethylation with these three species, it must be taken into account, again, that the trifluoromethyl moiety could not be transfered as a naked trifluoromethyl anion which should be too unstable to survive during this step. Thus, a concerted process, involving six-membered transition states, could be proposed, as illustrated in Scheme 13 (which has been simplified by considering monomeric species only).

Scheme 12

Scheme 13

Generation and Uses of Fluoral Iminiums

All the preceding examples concern the trifluoromethylation of non-enolizable carbonyl compounds. As expected, treatment of enolizable ketones, such as acetophenone, under the conditions employed to trifluoromethylate benzophenone or benzaldehyde (2 eq **3c**, 2 eq *tert*-BuOK, r.t.), did not result in any trifluoromethylation but in extensive self-aldolization. However, when catalytic amounts of potassium *tert*-butoxide were used and the medium heated at 80 °C for 24 h, 3-trifluoromethyl-3-[1-(N'-benzyl)piperazinyl]-propiophenone was obtained in an excellent yield from acetophenone and **3c** (28). **2b** or **2c**, opposed to the same substrate and catalytic amounts of cesium fluoride at 80 °C for 5h, led to the same result. This reaction have been successfully extended to other ketones, including cyclohexanone which could be functionnalized twice with four equivalents of **3c**, and to diethyl malonate (Scheme 14) (28). In all cases, two equivalents of hemiaminal must be employed to get good yields but, after reaction, one equivalent remained unaffected, as indicated by NMR. The resulting β-trifluoromethyl-β-(dialkylamino)ketones were usually formed in good crude yields (up to 82 %) but exhibited a limited stability under chromatographic conditions. Thus, they were transformed, by acid treatment, into more stable β-trifluoromethyl-α,β-enones of exclusive *E*-configuration, (28) which are valuable dienophiles or Michaël acceptors (29). These enones could be prepared in one pot from the starting ketones (Scheme 14).

Scheme 14

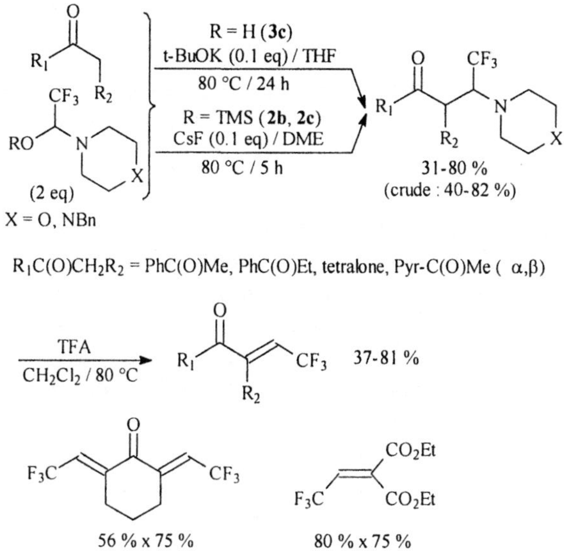

R = H (**3c**) t-BuOK (0.1 eq) / THF 80 °C / 24 h

R = TMS (**2b**, **2c**) CsF (0.1 eq) / DME 80 °C / 5 h

(2 eq) X = O, NBn

31-80 % (crude : 40-82 %)

$R_1C(O)CH_2R_2$ = PhC(O)Me, PhC(O)Et, tetralone, Pyr-C(O)Me (α,β)

TFA, CH_2Cl_2 / 80 °C → 37-81 %

56 % x 75 %

80 % x 75 %

Mechanistically speaking, the above-mentioned β-trifluoromethyl-β-dialkylamino-ketones look like condensation products between an enol and an iminium species. Thus, we first verified that **3c** was not in equilibrium, at least to a small extend, with an iminium hydroxide : when **3c** was opposed to ^{18}O-labelled water or methanol, no ^{18}O or methoxy incorporation was detected in **3c**. It has been also verified that no reaction occurred between **3c** and acetophenone in the absence of base. In order to rationalize these observations and the fact that two equivalents of **3c** must be employed, a mechanism starting from a complex between dimeric **3c** and a potassium enolate has been proposed (Scheme 15) (28). Because of the known acidity of hydroxyl functions in α-position to a trifluoromethyl group, this system was probably in equilibrium with a complex between an enol, **3c** and $\mathbf{K^+}$ $\mathbf{3c^-}$. This equilibrium could be then shifted by a push-pull loss of hydroxide, due to the interaction between **3c** and $\mathbf{K^+}$ $\mathbf{3c^-}$ within the K^+ coordination sphere, resulting in the subsequent formation of an iminium.

Scheme 15

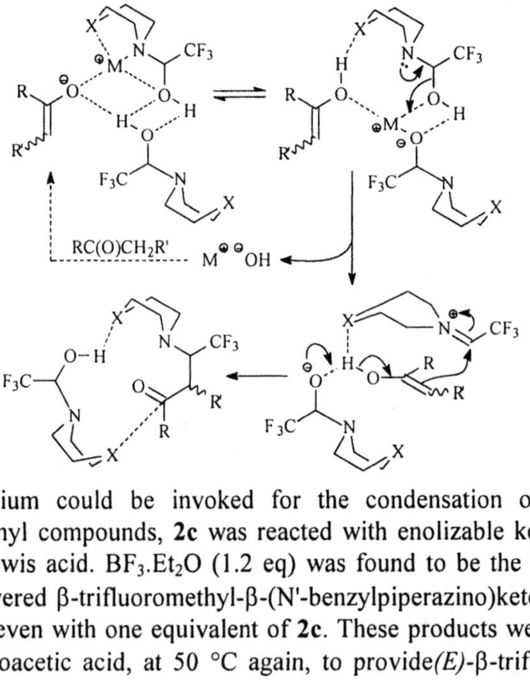

As an iminium could be invoked for the condensation of **2b,c** with enolizable carbonyl compounds, **2c** was reacted with enolizable ketones in the presence of a Lewis acid. $BF_3.Et_2O$ (1.2 eq) was found to be the best one. As expected, it delivered β-trifluoromethyl-β-(N'-benzylpiperazino)ketones in good yields at 50 °C, even with one equivalent of **2c**. These products were treated *in situ* with trifluoroacetic acid, at 50 °C again, to provide*(E)*-β-trifluoromethyl-α,β-enones. As analogues of **2c** could be prepared from chlorodifluoroacetaldehyde or pentafluoropropionaldehyde hydrates, β-chlorodifluoromethyl- and β-pentafluoroethyl- α,β-enones were also produced in the same way (Scheme 16) (28). This one-pot two-steps synthesis of *(E)*-β-trifluoromethyl-α,β-enones was very efficient and simpler than other routes to these products (30), all the more so since it could be extended to longer fluoroalkyl chains.

Scheme 16

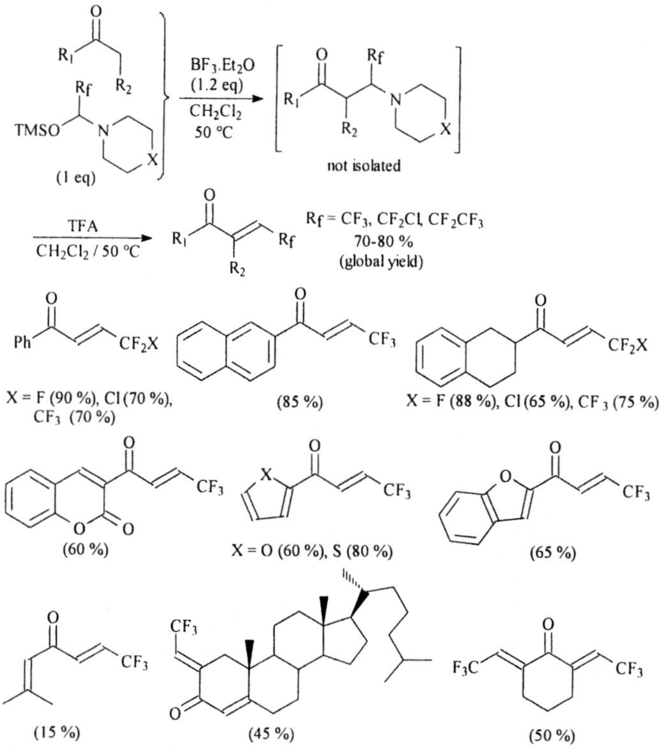

α-(Perfluoroalkyl)iminiums, generated in such an efficient way from perfluoroaldehyde hydrates or hemiketals, have been opposed to other nucleophiles, such as alcohols and amines (even chiral ones), silyl hydrides, silyl azides, allylsilanes, potassium vinyl trifluoroborates and electron-rich aromatic compounds (Scheme 17) (31).

Several trifluoromethylated analogues of bioactive compounds have been prepared from these expedient methodologies. For example, starting from an hemiaminal of fluoral arising from N'-(4,4'-dihalobenzhydryl)piperazine, analogues of Cinnarizine (vasodilatator) and Flunarizine (used for migraine prophylaxis) were obtained in three steps, the two first ones carried out in one pot (Scheme 18) (32).

Scheme 17

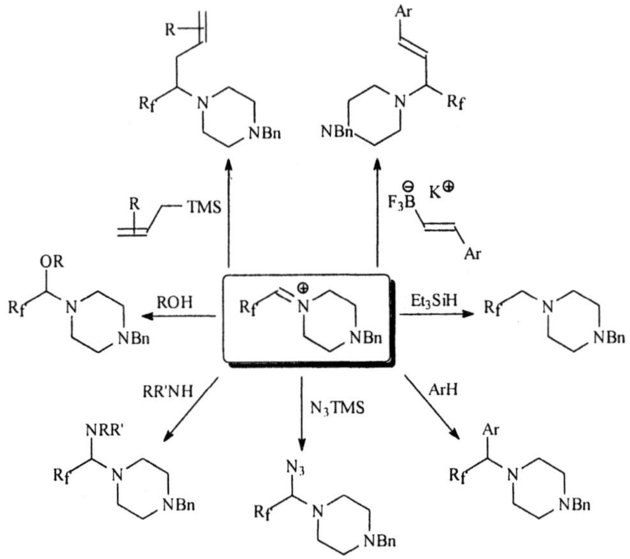

Scheme 18

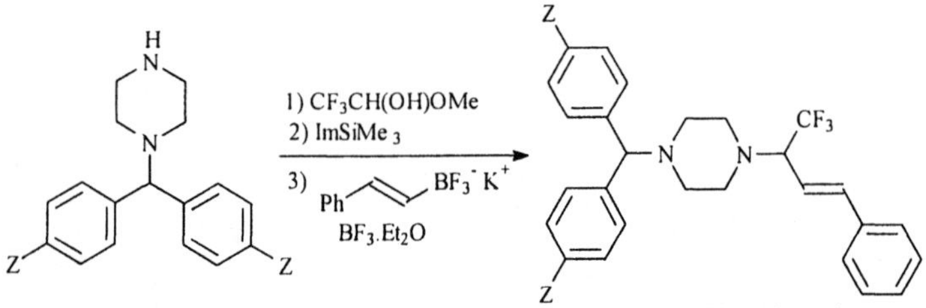

Z = H : 57 % (Cinnarizine analogue)
Z = F : 70 % (Flunarizine analogue)

On the other hand, **2c** and its chlorodifluoro analogue were reacted with indole, in the presence of $BF_3.Et_2O$, and delivered *bis*-indolic derivatives **4a,b**. These compounds are trifluoro- and chlorodifluoro- analogues of Vibrindole, a natural product exhibiting antibiotic properties (Scheme 19). Preliminary studies show that **4a** is 2-fold and **4b** 4-fold more efficient than Vibrindole against *Staphyloccocus Aureus* (32).

N-[1-(trifluoromethyl)but-3-enyl]-N'-benzylpiperazine, resulting from **2c** and allyltrimethylsilane (*cf.* Scheme 17)**,** is a derivative of α-(trifluoromethyl)-homoallylamine bearing the useful piperazine pharmacophore, but it cannot be easily converted into α-(trifluoromethyl)homoallylamine itself. To prepare this amine, another O-silylated hemiaminal of fluoral has been prepared from benzophenone imine. Under treatment with $BF_3.Et_2O$ at 50 °C, the corresponding strained iminium **5** was formed and reacted *in situ* with allylsilane. After hydrolysis, α-(trifluoromethyl)-homoallylamine **6** was obtained (Scheme 20) (31,32). Unfortunately, **5** did not react cleanly with potassium styryltrifluoroborate.

Scheme 19

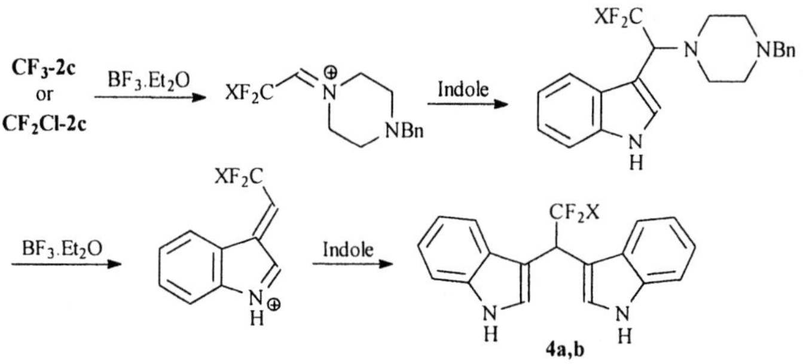

4a X = F (80 %) CMI = 6 μg.mL^{-1}
4b X = Cl (88 %) CMI = 3 μg.mL^{-1}
Vibrindole (CH_3 instead of CF_2X) CMI = 12.5 μg.mL^{-1}

Scheme 20

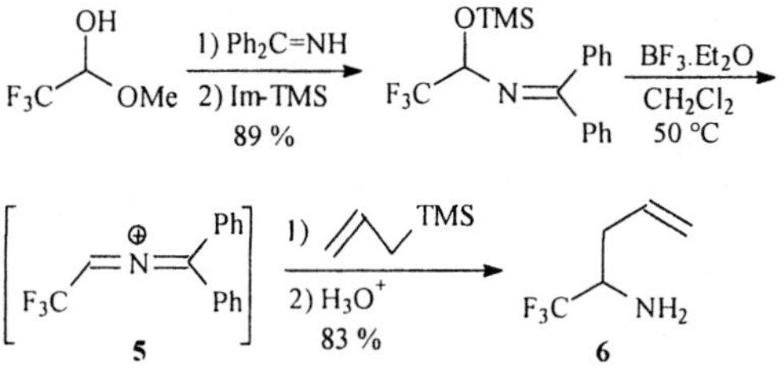

Through acylation with unsaturated acyl chlorides, or nitrogen protection followed by allylation or propargylation, **6** was the key-starting material for polyunsaturated compounds which, after ruthenium catalyzed ring closure metathesis, offered medium to large nitrogen-containing cycles (Scheme 21) (32,33).

Such polyunsaturated substrates can be prepared in a shorter way from the hemiaminal of fluoral (**8**) arising from benzylamine, which was synthesized by a modified procedure using pyridinium tosylate (PPTS). Formation of the corresponding iminium and subsequent allylation also needs peculiar conditions, since $BF_3.Et_2O$ must be replaced by a milder Lewis acid such as trimethylsilyl triflate (Scheme 22) (32). The resulting N-benzyl-α-(trifluoromethyl)-homoallylamine **9** is, however, more difficult to allylate or to acylate than **7** and ring closure metathesis of the resulting products needs larger amounts of Grubb's catalyst (32).

Scheme 21

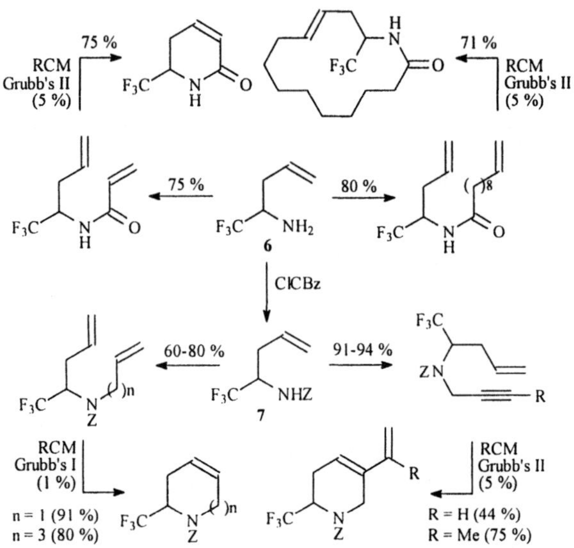

RCM Grubb's II (5 %) 75 % · 71 % RCM Grubb's II (5 %) · 75 % · 80 % · 6 · ClCBz · 60-80 % · 91-94 % · 7 · RCM Grubb's I (1 %) · n = 1 (91 %) n = 3 (80 %) · RCM Grubb's II (5 %) · R = H (44 %) R = Me (75 %)

Scheme 22

$BnNH_2$ / PPTS, MS 4Å / r.t. / 5 h → **8** (> 95 %)

7 Allyl-TMS / TfOTMS (1 eq), CH_2Cl_2 / r.t. / 4h → **9** (65 %)

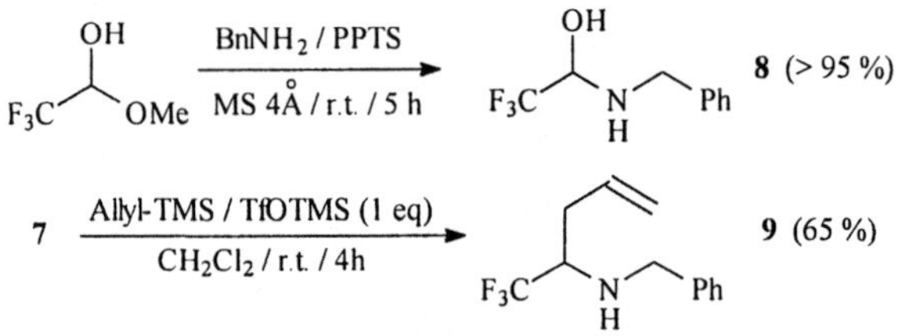

Generation and Uses of Difluorocarboxylate Anions

Methyl aminal of fluoral **10** have been easily prepared by direct methylation of **3c** with methyl iodide or from **2c** and methanol, in the presence of $BF_3.Et_2O$ (*cf.* Scheme 17) (31). **10** underwent an easy dehydrofluorination with BuLi and provided the aminal of difluoroketene **11** which could not be isolated but was characterized *in situ* by NMR. Provided that the by-produced fluoride was first eliminated with trimethylsilyl triflate, this ketene aminal was nucleophilic enough to react with aldehydes [even enolizable ones, in contrast with Dolbier's reagent CF_2=C(NMe_2)$_2$ (34)] and acyl chlorides, but not with less electrophilic ketones (35). When fluoride was not expelled in a volatile form, Li^+ and F^- formed a too tight ion pair which prevented the entrance of the carbonyl function into the coordination sphere of the lithium cation (Scheme 23). In such a way, intermediate 3-alkoxide-2,2-difluoro-1-methoxyiminiums were formed from aldehydes. After acidic hydrolysis, methyl β-hydroxy-α,α-difluoroacetates were exclusively obtained from the most reactive aldehydes but were contaminated by some corresponding amide from the less reactive ones. However, replacement of acidic hydrolysis by work-up with aqueous lithium chloride led to the exclusive formation of amides : such a result came from a Krapcho-type elimination of methyl chloride (Scheme 23). For the same reason, intermediate 3-keto-2,2-difluoro-1-methoxyiminiums arising from acyl chlorides provided directly and exclusively 3-keto-2,2-difluoroacetamides prior hydrolysis since chloride anions were generated during the condensation step (Scheme 23).

This methodology has been illustrated by an expedient synthesis, from 3-formyl-N-Boc-indole, of N-benzyl-N'-(3-indolyl-3-hydroxy-2,2-difluoro-propionyl)piperazine which is a ligand for 5-HT_{1D} brain receptors (36).

Thus, **11** behaved as an equivalent of $^-CF_2C(O)X$ (X = OMe, NR_2). This technique constitutes a good alternative to the Reformatsky reaction (37,38) using halodifluoroacetates and zinc, or to Iseki's method with CF_2=C(OEt)OTMS (also prepared from $BrCF_2CO_2Et$ and zinc) (40). It can be also advantageously compared to Dolbier's methodology using CF_2=C(NMe_2)$_2$ (prepared by copper mediated substitution of ozone-depleting CF_3CHCl_2) (34) or Uneyama's methodology using TMS-CF_2CO_2R (arising from reduction of CF_3CO_2R with Mg) (39). All these methods, useful to graft equivalents of $^-CF_2C(O)OR$, complete other processes designed to introduce $CF_2C(O)$ moieties with CF_2=C(R)OTMS [prepared from halodifluoromethyl ketones and metals (39a-b,41,42) or from acylsilanes (43,44)], with CF_2=C(R)OMe (45) or with CF_2=CHOMEM (prepared from trifluoroethanol) (46). Difluoroenolates have been also synthesized by metal-free reduction of chlorodifluoroketones ; so formed, these enolates can add on carbonyl electrophiles or substitute cyanide from thiocyanates (47).

As far as readily available hemiaminals of fluoral and their derivatives are concerned, it can be concluded that they constitute very versatile reagents (Scheme 24) since :

- they behave as stable equivalents of the unstable trifluoromethyl anion, and can trifluoromethylate non-enolizable ketones and aldehydes, as well as disulfides,
- they can generate trifluoromethylated iminiums which allow the synthesis of a variety of nitrogen- and fluorine-containing products,
- they can generate difluoroketene aminals and thus can be considered as equivalents of difluoroacetates ans difluoroacetamides anions.

Scheme 23

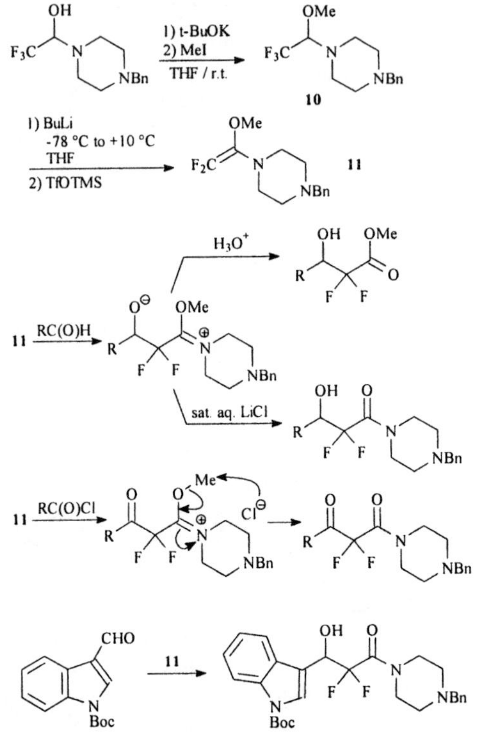

OH
F_3C N NBn
1) t-BuOK
2) MeI
THF / r.t.
OMe
F_3C N NBn
10

1) BuLi
-78 °C to +10 °C
THF
2) TfOTMS
OMe
F_2C N NBn
11

H_3O^+
OH OMe
R O
F F

11 RC(O)H
$O^{\ominus}$ OMe
R ⊕N NBn
F F

sat. aq. LiCl
OH O
R N NBn
F F

11 RC(O)Cl
O O Me $Cl^{\ominus}$
R ⊕N NBn
F F
O O
R N NBn
F F

CHO
N
Boc
11
OH O
N NBn
F F
N
Boc

Scheme 24

OH
F_3C OMe
(commercially available)
2 steps
1 or 2 steps

OTMS
F_3C N Ph Ph

OR
F_3C N X (X = O, NBn)

F_3C ⊕NR_2

$^{\ominus}CF_3$

O
$F_2C^{\ominus}$ Y
Y = OMe
piperazino

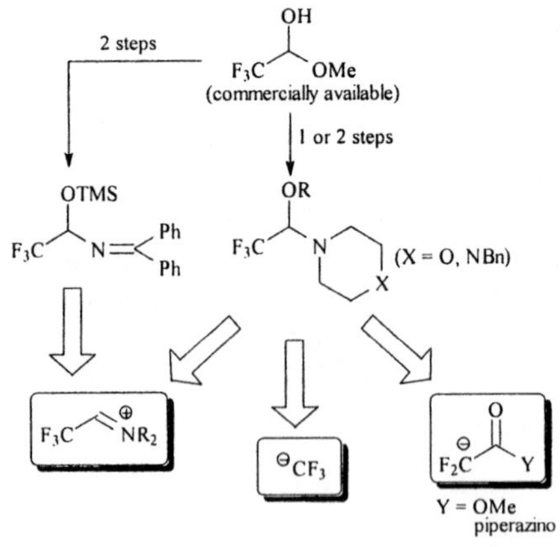

Trifluoromethylation with trifluoroacetic and trifluoromethanesulfinic derivatives.

Intermolecular process.

As fluoral is usually produced from cheaper trifluoroacetic acid, the trifluoromethylating ability of trifluoroacetic derivatives has been examined. Methyl trifluoroacetate (48), trifluoroacetamides (48), and α,α,α-trifluoroacetophenone (49) were studied first. Indeed, these compounds behave, under the action of potassium *tert*-butoxide, as efficient trifluoromethylating reagents towards benzophenone, chosen as model for non enolizable carbonyl substrates (Scheme 25). Such a result was not completely surprising since, a long time ago, formation of fluoroform has been noticed during alkaline hydrolysis of trifluoroacetates (50) and trifluoroacetophenone (51). Nevertheless, no synthetic use of these reactions has been reported until now.

Scheme 25

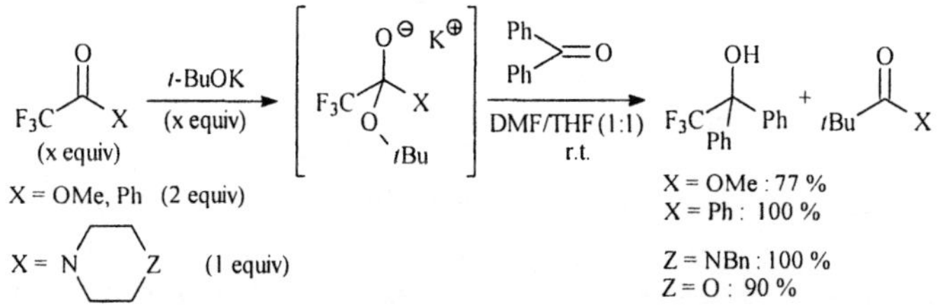

F_3C–C(O)–X (x equiv) —[t-BuOK (x equiv)]→ [F_3C–C($O^{\ominus}$ $K^{\oplus}$)(X)–O–tBu] —[$Ph_2C{=}O$, DMF/THF (1:1), r.t.]→ F_3C–C(OH)(Ph)–Ph + tBu–C(O)–X

X = OMe, Ph (2 equiv)

X = N(piperazine/morpholine ring)Z (1 equiv)

X = OMe : 77 %
X = Ph : 100 %

Z = NBn : 100 %
Z = O : 90 %

Though efficient towards benzophenone, methyl trifluoroacetate exhibited a limited scope in terms of substrates. Trifluoroacetophenone was very active towards other aromatic or heteroaromatic substrates than benzophenone, but two equivalents of reagent and base were needed to ensure reproducible yields. Thus, the trifluoroacetamide derived from N-benzylpiperazine were preferred since, in this case, generally good and reproducible yields were obtained with only one equivalent of *tert*-BuOK. However, it was essentially adapted to the trifluoromethylation of simple non enolizable ketones but not to that of chalcone nor that of aldehydes, which essentially underwent Cannizaro processes under these conditions. In contrast, N-trifluoromethanesulfinyl-N'-benzylpiperazine was active both towards non enolizable ketones and aldehydes since it is very electrophilic and trapped *tert*-BuOK faster than aldehydes, so that Cannizaro reaction was overcome (Scheme 26) (52). This reagent also trifluoromethylated aromatic and aliphatic disulfides (52). It must be noticed that trifluoromethanesulfinamides are readily available from potassium trifluoromethanesulfinate (53) which, itself, is prepared in one step from potassium trifluoroacetate (54).

Scheme 26

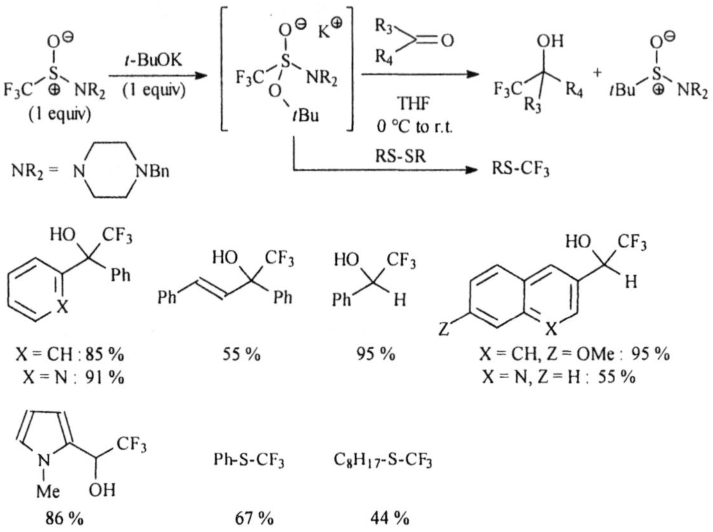

NR_2 = N-benzylpiperazino (N NBn)

F_3C-S(O⁻)($\oplus$)-NR_2 (1 equiv) → t-BuOK (1 equiv) → [F_3C-S(O⁻ K^+)(NR_2)-O-tBu] → R_3R_4C=O, THF, 0 °C to r.t. → F_3C-C(OH)(R_3)R_4 + tBu-S(O⁻)($\oplus$)-NR_2

RS-SR → RS-CF_3

X = CH : 85 %
X = N : 91 %

55 %

95 %

X = CH, Z = OMe : 95 %
X = N, Z = H : 55 %

86 %

Ph-S-CF_3 67 %

C_8H_{17}-S-CF_3 44 %

Intramolecular process.

Although such trifluoroacetic and triflinic derivatives are easily available and ecologically friendly, the strong base needed for their activation precluded the use of any enolizable substrate. This drawback should be circumvented by a strategy in which the alcoholate would be delivered slowly enough to favor, over deprotonation of the substrate, its attack on the trifluoroacetamido site of the reagent. Moreover, an intramolecular process would also favor nucleophilic attack and bring a beneficial entropic factor. This led to design a fluoride-induced reaction of trifluoroacetamides and trifluoromethanesulfinamides, derived from O-silylated *vic*-aminoalcohols, with carbonyl substrates. By analogy with trifluoromethylation using hemiaminals of fluoral or trifluoroacetamides, the following process was expected (Scheme 27).

For this purpose, a panel of trifluoroacetamides derived from O-TMS-1,2-aminoalcools (**14a-f**) has been prepared from ethanolamine, *rac*-1-phenyl-2-aminoethanol, (*R*)- and (*S*)-phenylglycinol as well as ephedrine, by reductive alkylation followed by silylation and trifluoroacetylation (Scheme 28) (55).

14a-f were evaluated, in the presence of a catalytic amount of cesium fluoride, towards different carbonylated substrates : benzophenone (as a model for non enolizable ketones), benzaldehyde (as a model for aldehydes) and acetophenone (as a model for enolizable ketones) (Scheme 29) (55).

Scheme 27

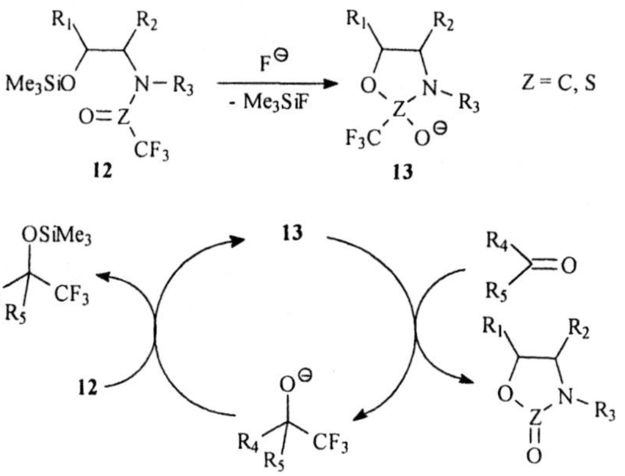

R_1 R_2 Me_3SiO N − R_3 O=Z CF_3 **12**

$F^{\ominus}$

$- Me_3SiF$

R_1 R_2 O N R_3 Z F_3C $O^{\ominus}$ **13**

Z = C, S

$OSiMe_3$ R_4 R_5 CF_3

13

R_4 R_5 =O

R_1 R_2 O N R_3 Z O

12

$O^{\ominus}$ R_4 R_5 CF_3

Scheme 28

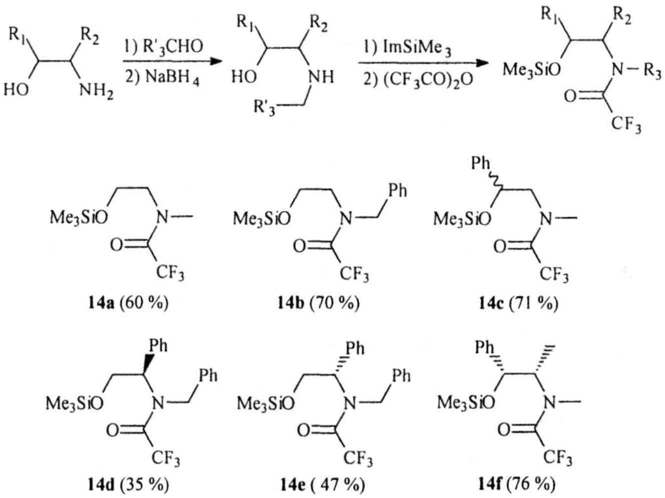

R_1 R_2 HO NH_2

1) R'_3CHO
2) $NaBH_4$

R_1 R_2 HO NH R'_3

1) $ImSiMe_3$
2) $(CF_3CO)_2O$

R_1 R_2 Me_3SiO N − R_3 O= CF_3

Me_3SiO N— O= CF_3 **14a** (60 %)

Me_3SiO N Ph O= CF_3 **14b** (70 %)

Ph Me_3SiO N— O= CF_3 **14c** (71 %)

Ph Me_3SiO N Ph O= CF_3 **14d** (35 %)

Ph Me_3SiO N Ph O= CF_3 **14e** (47 %)

Ph Me_3SiO N— O= CF_3 **14f** (76 %)

Scheme 29

R_1 R_2 TMSO N − R_3 O= CF_3 **14a-c,e,f** + O R_4 R_5

1) CsF (0.1 equiv)/DME
2) Bu_4NF (1 equiv)

OH R_4 R_5 CF_3 + R_1 R_2 O N − R_3 O **15**

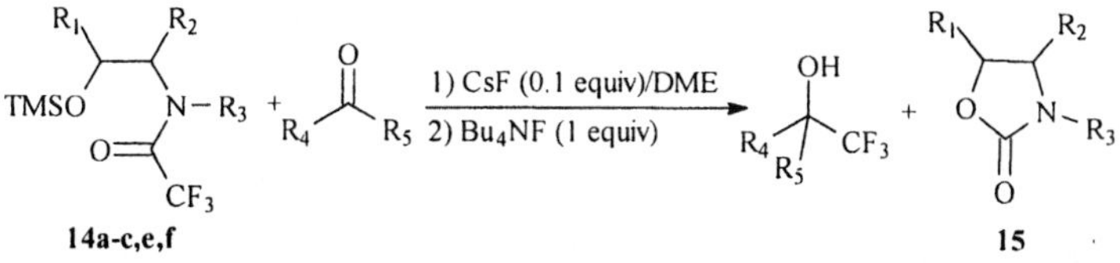

The expected trifluoromethylation occured with reagents **14a-f** and isolation of oxazolidinone **15** brought an evidence of the proposed mechanism. As expected also, these reagents were able to trifluoromethylate both enolizable and non enolizable ketones, as well as benzaldehyde. From the parametric study, it resulted that unsubstituted O-trimethylsilyl-N-trifluoroacetyl-aminoethanols (**14a,b**), or those without any substituent in α position to nitrogen (**14c**), needed a thermal activation, up to 80 °C, to transfer efficiently their trifluoromethyl group. At this temperature, the nitrogen substituent had a minor importance, so that **14a** and **14b** exhibited similar reactivities. In contrast, the presence of a substituent in α position to nitrogen dramatically changed the reactivity since **14d** and **14e** were able to provide, from benzophenone, benzaldehyde and acetophenone, good to excellent yields at 40 °C only, whereas **14f** reacted even at room temperature. Acetophenone was the most sluggish substrate but its trifluoromethylation could be accelerated by replacing cesium fluoride and DME with $Bu_4N[Ph_3SiF_2]$ (TBAT) and THF : under these conditions, a 60 % yield was reached from **14f** within seven hours instead of five days.

14f, derived from ephedrine, appeared as the most interesting reagent, all the more so since it is a very stable white solid whereas **14a-c** exhibited a limited shelf-life ($t_{1/2}$ < 48 h). Its reactivity towards diverse carbonyl compounds has been examined (Scheme 30) (55).

Scheme 30

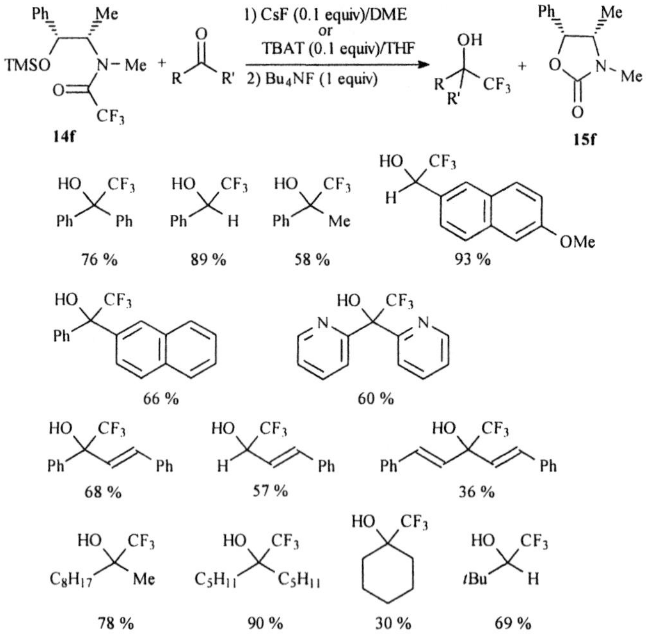

1) CsF (0.1 equiv)/DME or TBAT (0.1 equiv)/THF
2) Bu_4NF (1 equiv)

14f + R(C=O)R' → R(R')C(OH)CF_3 + **15f**

76 % · 89 % · 58 % · 93 %

66 % · 60 %

68 % · 57 % · 36 %

78 % · 90 % · 30 % · 69 %

Though **14f** was able to trifluoromethylate a wide range of non-enolizable or enolizable substrates with good to excellent yields, it could not trifluoromethylate very acidic carbonyl substrates such as valeraldehyde and dibenzyl ketone : in these cases, enolization was only occuring. Thus, the trifluoromethylating power of N-trifluoromethanesulfinyl-O-TMS-ephedrine **16f** has been examined (Scheme 31). Engaged in trifluoromethylations leading to the α-(trifluoromethyl)carbinols listed in Scheme 30, **16f** delivered results very similar to those obtained from **14f**, but, in addition, fair yields were reached from valeraldehyde (30 % with CsF/DME) and dibenzyl ketone (45 % with TBAT/THF) (56).

Scheme 31

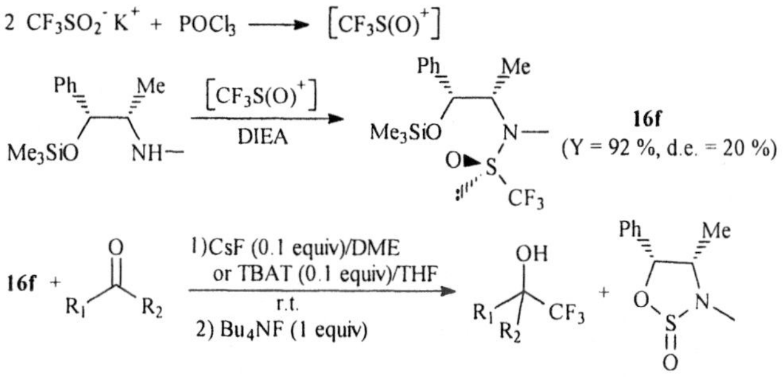

$2\ CF_3SO_2^-K^+ + POCl_3 \longrightarrow [CF_3S(O)^+]$

Ph Me Me$_3$SiO NH— $\xrightarrow[\text{DIEA}]{[CF_3S(O)^+]}$ Ph Me Me$_3$SiO N— O S CF$_3$ **16f** (Y = 92 %, d.e. = 20 %)

16f + R$_1$C(O)R$_2$ — 1)CsF (0.1 equiv)/DME or TBAT (0.1 equiv)/THF, r.t. 2) Bu_4NF (1 equiv) → R$_1$R$_2$C(OH)CF$_3$ + Ph Me O N S O

Enantioselective trifluorometylation.

14d-f are asymmetric compounds, prepared from enantiopure precursors [(*R*)- and (*S*)-phenylglycinols, ephedrine]. **16f**, in which S(IV) is an extra stereogenic center, has been also obtained from enantiopure ephedrine but with a moderate diastereomeric excess (d.e. = 20 %). Nevertheless, one of the diastereoisomer has been isolated in an enantiopure form. Thus, induction of chirality has been evaluated during the trifluoromethylation of benzaldehyde with **14d-f** or diastereo- and enantio-pure **16f**. Unfortunately, the resulting 1-phenyl-2,2,2-trifluoroethanol was obtained as a racemic mixture from any of these reagents, even from **16f**, in which chirality is brought by the reactive center.

Enantioselective trifluoromethylation remains a real challenge since, at present, few tools are available for this purpose. Indeed, if high diastereoselectivity can be reached from trifluoromethylation of chiral substrates, such as chiral sulfinimides, with Ruppert-Prakash reagent (18g,h), trifluoromethylation of prochiral substrates with chiral trifluoromethyl-containing systems usually leads to medium enantiomeric excesses. Until now, the best results have been obtained with CF_3SiMe_3, activated by chiral fluorides (57) or chiral Lewis bases (58) based on quinquina alcaloids, but enantiomeric excesses exceeding 50 % are rarely reached, except when the substrate is very hindered or when its conformation is such that any nucleophile would add in a stereoselective way (59).

Consequently, several other O-TMS-N-trifluoroacetyl-aminoalcools **14g-n** were prepared, in a pure enantiomeric form, from *nor*-ephedrine, *pseudo*-ephedrine, (*1R,2S*)- and (*1S,2R*)-diphenylaminoethanol, prolinol and *(1R,2S)*-1-hydroxymethyl-2-aminocyclohexane (Scheme 32).

Scheme 32

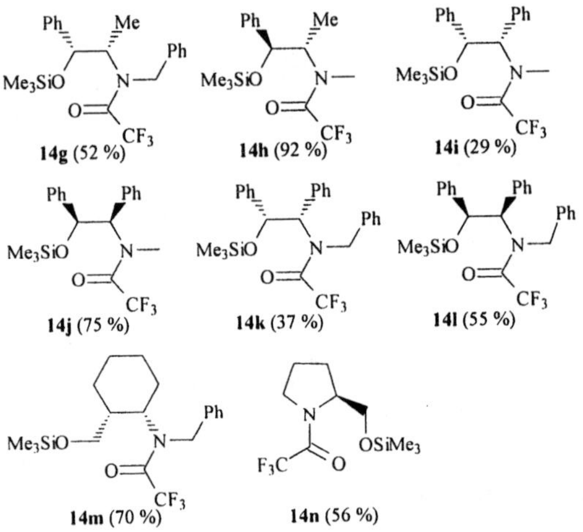

14g (52 %) **14h** (92 %) **14i** (29 %)

14j (75 %) **14k** (37 %) **14l** (55 %)

14m (70 %) **14n** (56 %)

14g-n were opposed to benzaldehyde, a model of prochiral substrates, in the presence of cesium fluoride [CsF (0.1 equiv), DME, r.t., 24h] or TBAT [TBAT (0.1 equiv), THF, r.t., 7h]. Good trifluoromethylation yields were obtained with all these reagents, except with **14n** that did not react, even under heating, and was recovered in its desilylated form. May be **14n** could exhibit a rigid conformation in which the silyoxy group is, for coulombian reasons, far enough from the trifluoroacetyl moiety to prevent nucleophilic ring closure.

Racemic phenyl(trifluoromethyl)carbinol was obtained from **14i-m** and a tiny enantiomeric excess (3-5 %) was observed from pseudoephedrine derivative **14h**. **14g** was the only reagent to induce a noticeable transfer of chirality since it delivered 1-phenyl-2,2,2-trifluoroethanol, from benzaldehyde at room temperature, with a 10 to 12 % enantiomeric excess. The reaction could not be carried out at a lower temperature since reaction became too slow.

Such poor results could arise from the fact that, in **14g-n**, the chiral centers were too far from the reaction site. Thus, analogous trifluoromethanesulfinamides were prepared from enantiopure (*1S,2R*)-1,2-diphenyl-2-aminoethanol, (*R*)- and (*S*)- phenylglycinol, prolinol, α,αdiphenylprolinol and (*1R,2S*)-1-hydroxymethyl-2-aminocyclohexane (Scheme 33). Because of the chirality of sulfur(IV), these compounds were obtained as a mixture of two diastereoisomers (d.e. = 10-52 %), from which at least one of them has been isolated in a pure form or as an enriched fraction.

Scheme 33

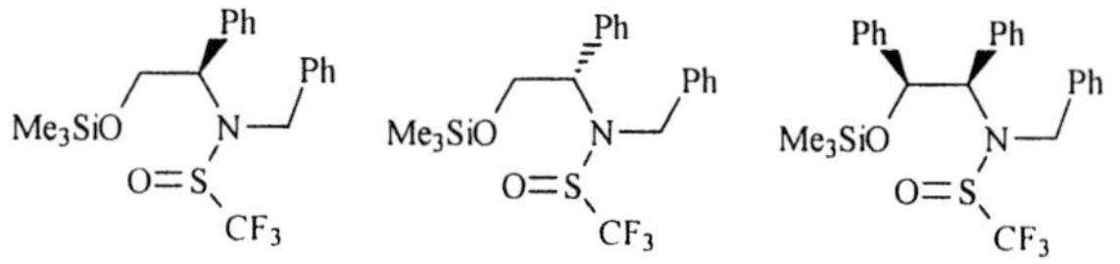

16d (75 %, de : 52 %) **16e** (32 %, de : 50 %) **16l** (68 %, de : 46%)

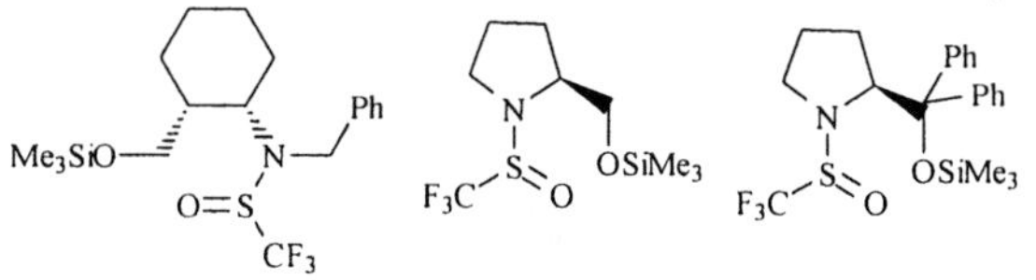

16m (41 %, de : 44 %) **16n** (75 %, de : 10 %) **16p** (75 %, de : 10 %)

Sulfinamides **16d,e,l-p** were reacted, at room temperature, with benzaldehyde in the presence of CsF. Except from **16n**, which was unreactive as its acetamido analog, medium to good trifluoromethylation yields were obtained. On that point, it must be noticed that, if **14d,e** had to be warmed to 40 °C to trifluoromethylate, their sulfinyl analogs **16d,e** could be successfully used at room temperature. However, though reactive, **16e,f,l,p** led to an almost racemic (trifluoromethyl)carbinol. In contrast, diastereomerically enriched **16d** (d.e. = 84 %) delivered 1-phenyl-2,2,2-trifluoroethanol with a 73 % yield and a 20 % enantiomeric excess. Enriched **16m** (d.e. = 84 %), though less reactive (yield = 33 %) led to a comparable enantiomeric excess (e.e. = 17 %). Once more, these results demonstrated the higher interest of trifluoromethanesulfinamides over trifluoroacetamides as trifluoromethylating reagents.

Finally, **16d** has been reacted with benzaldehyde in the presence of N-benzylcinchonium fluoride, the chiral ammonium fluoride that has been already used by Kobayashi *et al.* to induce chirality during trifluoromethylation with CF_3SiMe_3 (56). Indeed, the enantiomeric excess of the resulting 1-phenyl-2,2,2-trifluoroethanol jumped to 30 % (Scheme 34). As already mentioned for the homologous trifluoroacetamides, no improvement could be found by decreasing temperature since, at 0 °C, no conversion of benzaldehyde was observed after 72h (60).

Other chiral trifluoromethylating reagents derived from aminoalcohols, associated to other chiral catalysts, are presently under investigation.

Scheme 34

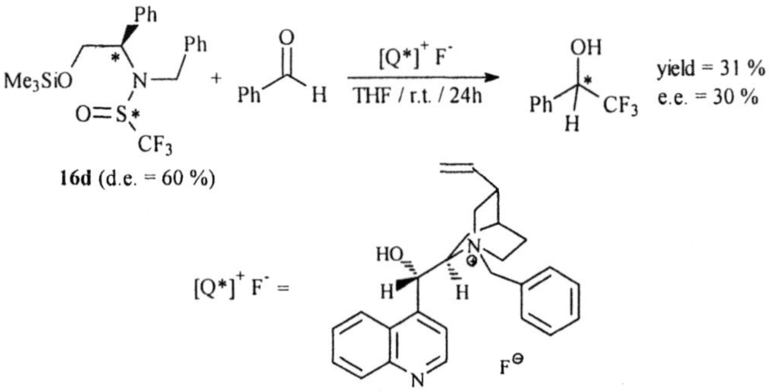

Me₃SiO / Ph / Ph / N / O=S* / CF₃ + PhCHO → [Q*]⁺ F⁻, THF / r.t. / 24h → PhCH(OH)CF₃ yield = 31 %, e.e. = 30 %

16d (d.e. = 60 %)

$[Q^*]^+ F^-$ =

Conclusion

In conclusion, several new reagents have been designed for the nucleophilic trifluoromethylation. All of them are environmently friendly, since they arise from fluoroform, fluoral, trifluoroacetamides or trifluoromethanesulfinamides. All of them trifluoromethylate efficiently non enolizable carbonyl compounds or disulfides. It must be added that some are so stable (**2b**, **2c**, **3c**, **14f**) that they can be considered as real laboratory reagents. Trifluoroacetamides or trifluoromethanesulfinamides derived from O-silylated-1,2-aminoalcools constitute the most attractive family since they exhibit the larger scope and are also able to trifluoromethylate enolizable carbonyl substrates, even at room temperature with a suited substituent pattern. Promising results have been obtained, with some members of this family, concerning enantioselective trifluoromethylation of prochiral substrates.

On the other hand, hemiaminals of fluoral are also potent precursors of trifluoromethylated iminiums and difluoroketene aminals, which constitute wide-scope key-intermediates for the expedient preparation of elaborated compounds, especially nitrogen-containing cycles.

References

1. (a) Smart, B.E. in *Organofluorine Chemistry, Principles and Commercial Applications*; Banks, R.E.; Smart, B.E.; Tatlow, J.C., Eds.; Plenum Press: New York, **1994**, Chapt. 3. (b) Hiyama, T. *Organofluorine Compounds, Chemistry and Applications*; Springer: Berlin, **2000**. (c) Hudlicky, M.; Pavlath, A.E. *Chemistry of Organic Fluorine Compounds II, a Critical Review*; ACS Monograph 187;

American Chemical Society: Washington DC, **1995**. (d) Mikami, K.; Itoh, Y.; Yamanaka, M. *Chem. Rev.* **2004**, *104*, 1.

2. (a) Banks, R.E.; Smart, B.E.; Tatlow, J.C. *Organofluorine Chemistry, Principles and Commercial Applications*; Plenum Press: New York, **1994**. (b) Zhang, W. *Chem. Rev.* **2004**, *104*, 2531.
3. a) Filler, R.; Kobayashi, Y.; Yagupolskii, L.M. Organofluorine Compounds in Medicinal Chemistry and Biomedical Applications; Elsevier: Amsterdam, (**1993**. (b) Ojima, I.; McCarthy, J.R.; Welch, J.T. *Biomedical Frontiers of Fluorine Chemistry*; American Chemical Society: Washington, DC, **1996**.
4. Becker, A. *Inventory of Industrial Fluoro-Biochemicals*; Eyrolles: Paris, **1996**.
5. Hansch, C.; Leo, A. *Substituent Constants for Correlation Analysis in Chemistry and Biology*; Wiley: New York, **1979**.
6. (a) Umemoto, T.; Kuriu, Y.; Shuyama, H.; Miyano, O. *J. Fluorine Chem.* **1982**, *20*, 695. (b) Umemoto, T.; Kuriu, Y.; Nakayama, S.I. *Tetrahedron Lett.* **1982**, *23*, 1169. (c) Umemoto, T.; Kuriu, Y.; Nakayama, S.I.; Miyano, O. *Tetrahedron Lett.* **1982**, *23*, 1471. (d) Umemoto, T.; Kuriu, Y.; Nakayama, *Tetrahedron Lett.* **1982**, *23*, 4101. (e) Umemoto, T.; Ishihara, S. *J. Am. Chem. Soc.* **1993**, *115*, 2156. (f) Umemoto, T.; Ishihara, S.; Adachi, K. *J. Fluorine Chem.* **1995**, *74*, 77. (g) Umemoto, T.; Adachi, K. *J. Org. Chem.* **1994**, *59*, 5692.
7. Yang, J.J.; Kirchmeier, R.L.; Shreeve, J.M. *J. Org. Chem.* **1998**, *63*, 2656.
8. (a) Dolbier, W.R., Jr in *Organofluorine Chemistry, Fluorinated Alkenes and Reactive Intermediates*; Chambers, R.D., Ed.; Springer: Berlin, **1997**, pp. 97-163. (b) Clavel, J.L.; Langlois, B.; Laurent, E.; Roidot, N. *Phosphorus, Sulfur & Silicon* **1991**, *59*, 169. (c) Langlois, B.; Laurent, E.; Roidot, N. *Tetrahedron Lett.* **1991**, *32*, 7525. (d) Langlois, B.; Laurent, E.; Roidot, N. *Tetrahedron Lett.* **1992**, *33*, 1291. (e) Langlois, B.; Montègre, D.; Roidot, N. *J. Fluorine Chem.* **1994**, *68*, 63. (f) Langlois, B.; Billard, T.; Guérin, S.; Large, S.; Roidot-Perol, N. *Phosphorus, Sulfur & Silicon* **1999**, *153-154*, 323. (g) Tommasino, J.B.; Brondex, A. Médebielle, M.; Thomalla, M.; Langlois, B.R.; Billard, T. *Synlett* **2002**, 1697. (h) Billard, T.; Langlois, B.; Large, S.; Anker, D.; Roidot, N.; Roure, P. *J. Org. Chem.* **1996**, *61*, 7545. (i) Billard, T.; Roques, N.; Langlois, B.R. *J. Org. Chem.* **1999**, *64*, 3813. (j) Roques, N.; Langlois, B.R. *Tetrahedron Lett.* **2000**, *41*, 3069.
9. (a) Ren, J.; Milton, J.; Weaver, K.L.; Short, S.A.; Stuart, D.I.; Stammers, D.K. *Structure* **2000**, *8*, 1089. (b) Pedersen, O.S.; Pedersen, E.B. *Synthesis* **2000**, 479. (c) Barasseda, X.; Sorbera, L.A.; Castaner, J. *Drugs Fut.* **1999**, *24*, 1057. (d) Wouters, J.; Moureau, F.; Evrard, G.; Koenig, J.J.; Jegham, S.; George, P.; Durant, F. *Bioorg. Med. Chem.* **1999**, *7*, 1683.
10. McClinton, M.A.; McClinton, D.A. *Tetrahedron* **1992**, *48*, 6555.
11. Burton, D.J.; Hahnfeld, J.L. The Preparation and Reactions of Fluoromethylenes; in *Chemistry Reviews*, Vol. 8; Tarrant, P., Ed.; Marcel Dekker Inc.: New York, **1977**, 153.
12. (a) Burton, D.J.; Yang, Z.Y. *Tetrahedron* **1992**, *48*, 189. (b) Burton, D.J.; Yang, Z.Y.; Morken, P.A. *Tetrahedron* **1994**, *50*, 2993. (c) Burton, D.J.; Lu, L. in *Organofluorine Chemistry, Techniques and Synthons*; Chambers, R.D., Ed.; Springer: Berlin, **1997**, pp. 46-89.

13. (a) Matsui, K.; Tobita, E.; Ando, M.; Kondo, K. *Chem. Lett.* **1981**, 1719. (b) Quiclet-Sire, B.; Saicic, R.N.; Zard, S.Z. *Tetrahedron Lett.* **1996**, *37*, 9057. (c) McNeil, J.G. Jr; Burton, D.J. *J. Fluorine Chem.* **1991**, *55*, 225. (d) Su, D.B.; Duan, J.X.; Chen, Q.Y. *Tetrahedron Lett.* **1991**, *32*, 7689. (e) Chen, Q.Y.; Duan, J.X. *Tetrahedron Lett.* **1993**, *34*, 4241. (f) Duan, J.X.; Chen, Q.Y. *J. Chem. Soc., Perkin Trans. 1*, **1994**, 725. (g) Roques, N. PhD dissertation, University Claude Bernard-Lyon I (France), **1996**. (h) Langlois, B.; Roques, N. WO Patent 94/01383, **1994**.
14. Billard, T.; Langlois, B. *Tetrahedron Lett.* **1996**, *37*, 6865.
15. Billard, T.; Large, S.; Langlois, B. *Tetrahedron Lett.* **1997**, *38*, 65.
16. Bouchu, M.N.; Large, S.; Steng, M.; Langlois, B.; Praly, J.P. *J. Carbohydrate Res.* **1998**, *314 (1-2)*, 37.
17. Granger, C.E.; Felix, C.P.; Parrot-Lopez, H.; Langlois, B.R. *Tetrahedron Lett.* **2000**, *41*, 9257.
18. (a) Prakash, G.K.S.; Yudin, A.K. *Chem. Rev.* **1997**, *97*, 757. (b) Wiedemann, J.; Heiner, T.; Mloston, G.; Prakash, G.K.S.; Olah, G.A. *Angew. Chem. Int. Ed.* **1998**, *37*, 820. (c) Prakash, G.K.S.; Mandal, M. *J. Fluorine Chem.* **2001**, *112*, 123. (d) Blazejewski, J.C.; Anselmi, E.; Wilmshurst M.P. *Tetrahedron Lett.* **1999**, *40*, 5475. (e) Prakash, G.K.S.; Mandal, M.; Olah, G.A. *Synlett* **2001**, 77. (f) Prakash, G.K.S.; Mandal, M.; Olah, G.A. *Org. Lett.* **2001**, *3*, 2847. (g) Prakash, G.K.S.; Mandal, M.; Olah, G.A. *Angew. Chem. Int. Ed.* **2001**, *40*, 589. (h) Prakash, G.K.S.; Mandal, M. *J. Am. Chem. Soc.* **2002**, *124*, 6538.
19. (a) Singh, R.P.; Shreeve, J.M. *Tetrahedron* **2000**, *56*, 7613. (b) Singh, R.P.; Cao, G.; Kirchmeier, R.L.; Shreeve, J.M. *J. Org. Chem.* **1999**, *64*, 2873. (c) Singh, R.P.; Shreeve, J.M. *J. Org. Chem.* **2000**, *65*, 3241.
20. Prakash, G.K.S.; Hu, J.; Olah, G.A. *J. Org. Chem.* **2003**, *68*, 4457.
21. (a) Barhdadi, R.; Troupel, M.; Périchon, J. *J. Chem. Soc., Chem. Commun.* **1998**, 1251. (b) Folléas, B.; Marek, I.; Normant, J.F.; Saint-Jalmes, L. *Tetrahedron Lett.* **1998**, *39*, 2973. (c) Folléas, B.; Marek, I.; Normant, J.F.; Saint-Jalmes, L. *Tetrahedron* **2000**, *56*, 275. (d) Russell, J.; Roques, N. *Tetrahedron* **1998**, *54*, 13771.
22. Shono, T.; Ishifune, M.; Okada, T.; Kashimura, S. *J. Org. Chem.* **1991**, *56*, 2.
23. (a) Large, S.; Roques, N.; Langlois, B.R. *J. Org. Chem.* **2000**, *65*, 8848. (b) Roques, N.; Russell, J.; Langlois, B.; Saint-Jalmes, L.; Large, S. Patent WO 98/22435, **1998**.
24. Billard, T.; Bruns, S.; Langlois, B.R. *Org. Lett.* **2000**, *2*, 2101.
25. Billard, T.; Langlois, B.R.; Blond, G. *Tetrahedron Lett.* **2000**, *41*, 8777.
26. Blond, G.; Billard, T.; Langlois, B.R. *Tetrahedron Lett.* **2001**, *42*, 2473.
27. Billard, T.; Langlois, B.R.; Blond G. *Eur. J. Org. Chem.* **2001**, 1467.
28. Blond, G.; Billard, T.; Langlois, B.R. *J. Org. Chem.* **2001**, *66*, 4826.
29. Work in progress, in collaboration with Pr. G. Haufe (University of Münster, Germany).
30. (a) Latypov, R.R.; Belogai, V.D.; Pashkevich, K.I. *Izv. Akad. Nauk SSSR, Ser. Khim.* **1986**, 123. (b) Takaya, J.; Kagoshima, H.; Akiyama, T. *Org. Lett.* **2000**, *2*, 1577. (c) Fuchigami, T.; Nakagawa, Y.; Nonaka, T. *J. Org. Chem.* **1987**, *52*, 5491. (d) Xu, Y.; Dolbier, W.R. Jr *Tetrahedron Lett.* **1998**, *39*, 9151. (e) Xu, Y.; Dolbier, W.R. Jr *J. Org. Chem.* **2000**, *65*, 2134. (f) Kubota, T.; Ijima, M.; Tanaka, T.

Tetrahedron Lett. **1992**, *32*, 1351. (g) Ates, C.; Janousek, Z.; Viehe, H.G. *Tetrahedron Lett.* **1993**, *34*, 5711.
31. Billard, T.; Langlois, B.R. *J. Org. Chem.* **2002**, *67*, 997.
32. Billard, T.; Gille, S.; Ferry, A.; Barthelemy, A.; Christophe, C.; Langlois, B.R. *J. Fluorine Chem.*, in press.
33. Gille, S.; Ferry, A.; Billard, T.; Langlois, B.R. *J. Org. Chem.* **2003**, *68*, 8932.
34. (a) Xu, Y.; Dolbier, W.R. Jr; Rong, X.X. *J. Org. Chem.* **1997**, *62*, 1576. (b) Ding, Y.; Wang, J.; Abboud, K.A.; Xu, Y.; Dolbier W.R. Jr; Richards N.G.J. *J. Org. Chem.* **2001**, *66*, 6381.
35. Blond, G.; Billard, T.; Langlois, B.R. *Chem. Eur. J.* **2002**, *8*, 2917.
36. Van Niel, M.B. et al. *J. Med. Chem.* **1999**, *42*, 2087.
37. (a) Hallinan, E.A.; Fried, J. *Tetrahedron Lett.* **1984**, *25*, 2301. (b) Shen, Y.; Qi, M. *J. Fluorine Chem.* **1994**, *67*, 229. (c) Braun, M.; Vonderhagen, A.; Waldmüller, D. *Liebigs Ann.* **1995**, 1447. Lang, R.W.; Schaub, B. *Tetrahedron Lett.* **1988**, *29*, 2943. (d) Kitagawa, O.; Taguchi, T.; Kobayashi, Y. *Tetrahedron Lett.* **1988**, *29*, 1803. (e) Tozer, M.J.; Herpin, T.F. *Tetrahedron*, **1996**, 8619.
38. (a) Mcharek, S.; Sibille, S.; Nédélec, J.Y.; Périchon, J. *J. Organomet. Chem.* **1991**, *401*, 211. (b) Shono, T.; Kise, N.; Oka, H. *Tetrahedron Lett.* **1991**, *32*, 6567.
39. (a) Uneyama, K.; Mizutani, G.; Maeda, K.; Kato, T. *J. Org. Chem.* **1999**, *64*, 6717. (b) Uneyama, K.; Amii, H. *J. Fluorine Chem.* **2002**, *114*, 127. (c) Amii, H.; Kobayashi, T.; Uneyama, K. *Synthesis*, **2000**, *14*, 2001. (d) Uneyama, K.; Mizutani, G. *Chem. Lett.* **1999**, 613.
40. (a) Iseki, K.; Kuroki, Y.; Asada, D.; Kobayashi, Y. *Tetrahedron Lett.* **1997**, *38*, 1447. (b) Iseki, K.; Kuroki, Y.; Asada, D.; Takahashi, M.; Kishimo, S.; Kobayashi, Y. *Tetrahedron* **1997**, *53*, 10271.
41. (a) Ishihara, T.; Yamanaka, T.; Ando, T. *Chem. Lett.* **1984**, 1165. (b) Kuroboshi, M.; Ishihara, T. *Tetrahedron Lett.* **1987**, *28*, 6481. (c) Kuroboshi, M.; Ishihara, T. *Bull. Chem. Soc. Jpn.* **1990**, *63*, 428. (d) Yamana, M.; Ishihara, T.; Ando, T. *Tetrahedron Lett.* **1983**, *24*, 507.
42. Amii, H.; Kobayashi, T.; Hatamoto, Y.; Uneyama, K. *J. Chem. Soc., Chem. Commun.* **1999**, 1323.
43. Jin, F.; Jiang, B.; Xu, Y. *Tetrahedron Lett.* **1992**, *33*, 1221.
44. (a) Brigaud, T.; Doussot, P.; Portella, C. *J. Chem. Soc., Chem. Commun.* **1994**, 2117. (b) Lefebvre, O.; Brigaud, T.; Portella, C. *Tetrahedron* **1998**, *54*, 5939. (c) Lefebvre, O.; Brigaud, T.; Portella, C. *J. Org. Chem.* **2001**, *66*, 1941. (d) Saleur, D.; Brigaud, T.; Bouillon, J.P.; Portella, C. *Synlett* **1999**, *4*, 432. (e) Saleur, D.; Bouillon, J.P.; Portella, C. *J. Org. Chem.* **2001**, *66*, 4543. (f) Saleur, D.; Brigaud, T.; Bouillon, J.P.; Portella, C. *Tetrahedron* **1999**, *55*, 7233. (g) Lefebvre, O.; Brigaud, T.; Portella, C. *J. Org. Chem.* **2001**, *66*, 4348. (h) Brigaud, T.; Lefebvre, O.; Plantier-Royon, R.; Portella, C. *Tetrahedron Lett.* **1996**, *37*, 6115. (i) Berber, H.; Brigaud, T.; Lefebvre, O.; Plantier-Royon, R.; Portella, C. *Chem. Eur. J.* **2001**, *7*, 903.
45. Kodama, Y.; Yamane, H.; Okumura, M.; Shiro, M. *Tetrahedron* **1995**, *51*, 12217.
46. (a) Broadhurst, M.J.; Percy, J.M.; Prime, M.E. *Tetrahedron Lett.* **1997**, *38*, 5903. (b) Broadhurst, M.J.; Brown, S.J.; Percy, J.M.; Prime, M.E. *J. Chem. Soc. , Perkin Trans. 1* **2000**, 3217. (c) Kariuki, B.M.; Owton, W.M.; Percy, J.M.; Pintat, S.;

Smith, C.A.; Spencer, S.C.; Thomas, A.C.; Watson, M. *J. Chem. Soc., Chem. Commun.* **2002**, 228. (d) Howarth, J.A.; Owton, W.M.; Percy, J.M.; Rock, M.H. *Tetrahedron* **1995**, 37, 10289. Howarth, J.A.; Owton, W.M.; Percy, J.M. *J. Chem. Soc., Chem. Commun.* **1995**, 757.

47. (a) Médebielle, M.; Keyrouz, R.; Langlois, B.; Billard, T.; Dolbier, W.R. Jr; Burkholder, C.; Ait-Mohand, S.; Okada, E.; Ashida, T. "Electron Transfer Methodologies to the Synthesis of Organo-Fluorine Compounds" in *Recent Research Development in Electron Transfer Reactions in Organic Synthesis*; Vanelle, P., Ed.; Research Signpost: Trivandrum, **2002**, pp. 89-97. (b) Billard, T.; Langlois, B.R.; Médebielle, M. *Tetrahedron Lett.* **2001**, *42*, 3463.
48. Jablonski, L.; Joubert, J.; Billard, T.; Langlois, B.R. *Synlett* **2002**, 230.
49. Jablonski, L.; Billard, T.; Langlois, B.R. *Tetrahedron Lett.* **2002**, *44*, 1055.
50. Bergman, E. *J. Org. Chem.* **1958**, *23*, 476.
51. (a) Simons, J.H.; Ramler, E.O. *J. Am. Chem. Soc.* **1943**, *65*, 389. (b) Guthrie, J.P.; Cossar, J. *Can. J. Chem.* **1990**, *68*, 1640. (c) Delgado, A.; Clardy, J. *Tetrahedron Lett.* **1992**, *33*, 2789.
52. Inschauspe, D.; Sortais, J.-B.; Billard, T.; Langlois, B.R. *Synlett* **2002**, 233.
53. Billard, T.; Greiner, A.; Langlois, B.R. *Tetrahedron* **1999**, *55*, 7243.
54. Forat, G. et al. Eur. Pat. 733,614 (1995) (to Rhodia Co.).
55. Joubert, J.; Roussel, S.; Christophe, C.; Billard, T.; Langlois, B.R.; Vidal, T. *Angew. Chem. Int. Ed.* **2003**, *42*, 3133.
56. Roussel, S.; Billard, T.; Langlois, B.R.; SaintJalmes, L. *Synlett* **2004** (12), 2119.
57. Iseki, K.; Nagai, T.; Kobayashi, Y. *Tetrahedron Lett.* **1994**, *35*, 3137.
58. Hagiwara, T.; Kobayashi, T.; Fuchikami, T. *Main Group Chem.* **1997**, *2*, 13.
59. Caron, S.; Do, N.M.; Arpin, P.; Larivée, A. *Synthesis*, **2003**, 1693.
60. Roussel, S.; Billard, T.; Langlois, B.R.; SaintJalmes, L. *Chem. Eur. J.*, in press.

Chapter 4

SF_5-Synthons: Pathways to Organic Derivatives of SF_6

R. W. Winter, R. A. Dodean, and G. L. Gard

Department of Chemistry, Portland State University, Portland, OR 97207

In this review, the synthesis and use of the primary synthons SF_5X (X = F, Cl, Br, SF_5) in forming organic derivatives of SF_6 will be described. We have found that using primary SF_5X synthons, convenient pathways exist for preparing secondary synthons that include SF_5 – alkanes, alkenes, alkynes, iodides and aromatics. The primary synthons also offer pathways for preparing and studying the reactive SF_5^+, SF_6^-, and $SF_5\cdot$ intermediates.

Introduction – Overview of SF_5X

The present review will include the more recent developments in the last several years of the chemistry of SF_5-C systems; as is appropriate SF_5-N and SF_5-O compounds will be briefly mentioned. It is worthwhile to point out that a number of significant reviews have been published dealing with SF_5 – chemistry *(1-5)*. The interest in SF_5 chemistry was first started by Moissan who, in 1891, prepared the first high valent sulfur fluoride (SF_6) by the direct fluorination of sulfur *(1)*.

Primary and Secondary Synthons

Halogens / Pseudohalogens

There are a number of simple SF_5X (X = halogen/pseudohalogen) compounds that have been reported and they are listed in Table I.

Table I. Halogen / Pseudohalogen Compounds

SF_6	SF_5Cl	SF_5Br	S_2F_{10} (F_5S-SF_5)
SF_5OSO_2F	SF_5OF	SF_5OCl	SF_5OCN
SF_5CN	SF_5NC	SF_5NCO	SF_5NSO
SF_5NCS	SF_5NF_2	SF_5NCl_2	

The outstanding chemical stability of sulfur hexafluoride (SF_6) is carried over to a large number of its organic derivatives RSF_5. While there are various pathways for preparing RSF_5 *(2-4)*, this review will limit itself to the more recent use of SF_5X (X = F, Cl, Br, SF_5) systems and the use of secondary systems prepared from SF_5X.

In general, the parent SF_5X, SF_6, is not useful for preparing RSF_5 compounds. It will react with hot alkali metals, accelerate the pyrolysis of paraffin hydrocarbons, remove silicon from a platinum catalyst and catalyze the reaction of ammonia with a ketone and an aldehyde to give a pyridine*(1)*. Sulfur hexafluoride will form a solid hydrate and other inclusion compounds; the hydrate decomposes above 0° C *(1)*. It will react with electophilic agents such as $AlCl_3$ and SO_3 to give AlF_3, Cl_2, S_2Cl_2 and SO_2F_2, respectively *(6)*. It has also been shown that methyl radicals can abstract fluorine from SF_6 above 140° C *(7)*.

In a low temperature plasma, SF_6, SF_5CF_3 and SF_5Cl reacts with C_6H_6 and gives mainly halogenated aromatic compounds and minor quantities of $SF_5C_6H_5$. With C_6H_5Cl, $SF_5C_6H_5$ and $SF_5C_6H_4Cl$ are found *(8)*. In a microwave study of SF_6 with chlorine it was found that the percentage conversion into SF_5Cl was as high as 25% when the SF_6:Cl_2 molar ratio was 0.17 to 1.0 *(9)*. Recently, it was found that SF_6 may be converted to the SF_5^+ anion; this will be discussed in the section dealing with SF_5^- anions. More recently, SF_6 had found use as a plasma etching gas; SF_6 with O_2 in plasma will etch carbon or silicon based materials *(10, 11)*.

After the report of SF_6, the next SF_5X compound to be prepared and studied was S_2F_{10} (1933), which was obtained as a by-product from the fluorination of sulfur *(12)*. Several useful preparative procedures have been reported; in one method, SF_5Cl undergoes reduction with H_2 in the presence of light *(13)* and another method involves the photodecomposition of SF_5Br into S_2F_{10} and Br_2 *(14)*.

The compound, SF_5Cl (1960) was formed as a minor product in the fluorination of SCl_2 *(15)*. Other useful preparative methods include chlorination of S_2F_{10} with Cl_2 *(16)*, reaction of ClF with SF_4 *(17)* and reaction of MF (CsF, KF) with SF_4 and Cl_2 *(18, 19)*.

SF_5Br (1962) was prepared by the reaction of SF_4 with Br_2 and BrF_5 at 100° C under pressure *(20)*. In 1965, a method using S_2F_{10} and Br_2 was reported *(21)*; this method was further studied in order to improve both yield and purity *(22, 23)*. Additional methods have used SF_4, Br_2, BrF_3 or BrF with CsF *(24, 25)*. Our laboratory found that SF_5Br can be prepared in situ, in large quantities, and in high yields by the following steps (26):

$BrF_3 + Br_2 \rightarrow 3BrF$ 6 - 11 days

$SF_4 + BrF \rightarrow SF_5Br$ 36 days @ r.t. (yield 99%)

20 days with heating (yield 88%)

SF_5^+, SF_6^-, SF_5^- and $SF_5\cdot$ Intermediates

SF_6 is a colorless, odorless, non-toxic and inert gas at room temperature. There is a resonance capture of electrons of about 2 eV energy to form the SF_6^- and SF_5^- in about equal amounts *(1)*. The appearance potentials (volts) for the following positive ions are observed in a mass spectrometer are: SF_5^+ (15.9). SF_4^+ (18.9), SF_3^+ (20.1), SF_2^+ (26.8) and SF^+ (31.3) *(1)*. While it has not been possible to prepare a compound containing the SF_5^+ ion, derivatives like RSF_4^+ are known; these ions are generated via reaction of RSF_5 with SbF_5 where R = CH_3, $CH{=}CH_2$, $C{\equiv}CH$. These compounds are not stable at room temperature *(24)*.

Compounds containing the SF_5^- anion are conveniently generated via the reaction of SF_4 with MF *(18, 19, 27, 28)*:

$$MF + SF_4 \rightarrow M^+SF_5^- \quad [M = K, Rb, Cs, Ag, (CH_3)_4N^+, Cs(18\text{-crown-}6)_2^+]$$

The salt, $(Cs^+)_6(SF_5^-)_4(HF_2^-)_2$ is also known *(29)*. In a recent patent report, salts of the form $Q^+SF_5^+$, with Q^+ = certain NR_4^+, PR_4^+, R_3S^+, phosphazenium and amidinium, and of the form $Q^{2+}X^-SF_5^-$, with Q^{2+} = dications of tetraaminoethylenes, tetraalkylthioethylenes, or of *o*- and *p*- phenylenediamines.

In particular, it is found that S_2F_{10} and SF_6 will react with tetrakis-(dimethylamino)-ethylene to give $[((CH_3)_2N)_2C{=}C(N(CH_3)_2)_2]^{2+}(SF_5^-)_2$ (which was isolated) and $[((CH_3)_2N)_2C{=}C(N(CH_3)_2)_2]^{2+}F^-SF_5^-$ (which was not isolated) *(30)*. These pentafluorosulfuranides may be used to generate SF_4 (the process $SF_6 \rightarrow SF_4$), as fluorinating agents (e.g. R-OH → R-F), and apparently also to directly introduce the SF_5 group into aliphatic ($RCH_2OH \rightarrow RCH_2SF_5$) and aromatic (Aryl-X) systems. It is also claimed that SF_5-aryl, SF_5-acetylene, and SF_5-olefin systems can be prepared. This would be highly significant with respect to the use of SF_6 (which is very easily made) and the possibility to utilize the nucleophile "SF_5^-", which has not been possible up to now.

The $SF_5\cdot$ radical can be generated from S_2F_{10} (heat, 125-140° C) *(21)* from SF_5Cl (light or heat with peroxide catalyst) *(2)*, from SF_5Br (heat or light) *(31)* and from SF_6 (microwave or electrical discharge) *(1)*. $SF_5\cdot$ is the chain carrying agent in reaction with SF_5X and unsaturated compounds *(32)*. The first matrix isolation and infrared identification of the $SF_5\cdot$ radical was achieved by the vacuum ultraviolet photolysis of SF_5X (X = F, Cl, Br). In addition to the simple SF_5-X bond rupture, there is a significant competing process that involves the molecular photoelimination of XF and SF_4 *(31)*.

Secondary Synthons

A number of additional noteworthy SF_5-systems that can be used as secondary synthons. These include the following compounds: SF_5NH_2* $SF_5N{=}CX_2$ (X = F, Cl), SF_5NSX_2 (X = F, Cl)*, SF_5NSF_4*, $SF_5NHC(O)F$, $SF_5(O)_nSF_5$ (n = 2, 3), SF_5OOH, $SF_5CX_2C(O)Cl$ (X = H, F) $SF_5CX_2SO_3H$, $SF_5\overline{NCF_2O}$ and $SF_5\overline{CFXCF_2OSO_2}$ (X = F, H). The asterix denotes compounds that are not prepared from SF_5X.

In particular, the following secondary synthons prepared from SF_5X (X = Cl, Br, SF_5) include: $SF_5CF{=}CF_2$, $SF_5C{\equiv}CH$, $SF_5C{\equiv}CCF_3$, $SF_5CZ_2CY_2X$ (Z, Y = H, F, Cl, Br and X = Br, I), $SF_5(CF_2)_nI$ and $SF_5(CF_2)_nC_6H_5$; their reaction pathways are given below:

$SF_5X + F_2C{=}CHF \rightarrow SF_5CF_2CHFX$ which with base gives $SF_5CF{=}CF_2$

$SF_5X + HC{\equiv}CH \rightarrow SF_5CH{=}CHX$ which with base gives $SF_5C{\equiv}CH$

$SF_5Br + CF_3C{\equiv}CH \rightarrow CF_3C(Br){=}CHSF_5$ which with base gives $SF_5C{\equiv}CCF_3$

$SF_5X + CZ_2{=}CY_2 \rightarrow SF_5CZ_2CY_2X$ which with base gives $SF_5CZ{=}CY_2$

$$S_2F_{10} + ICF_2CF_2I + 1/2nCF_2{=}CF_2 \xrightarrow{\Delta,P} SF_5(CF_2)_nI$$

$$SF_5(CF_2)_nI + C_6H_6 \rightarrow SF_5(CF_2)_nC_6H_5$$

The chemistry of these secondary synthons (SF_5-Carbon synthons) along with the primary synthons will be discussed in the following pages.

SF_5-Organic Derivatives from Primary and Secondary Synthons

There are a number of ways to prepare SF_5-organic derivatives *(1-5)*:

(1) treatment of disulfides, mercaptans or sulfides with fluorine or via electrochemical fluorination
(2) oxidation of aromatic disulfides with AgF_2 or with F_2
(3) addition of SF_5X ($X = Cl, Br, SF_5$) across multiple bonds
(4) reaction of SF_5-olefins/alkynes with various reagents
(5) use of $SF_5(CF_2)_nI$ with $n = 2,4,6,8,10$

SF_5-Perfluoroalkyl Iodides

Perfluoroalkyl iodides are an important class of compounds used in the preparation of many useful fluoroorganic and fluoroorganometallic derivatives. By comparison, there are only a few studies dealing with the chemistry of SF_5-perfluoroalkyl iodides. The known and characterized SF_5-perfluoroalkyl iodides are: SF_5CF_2I *(33)*, $SF_5(C_2F_4)_nI$ where n = 1,2,3 *(34, 35)*, $SF_5(C_2F_4)_nI$ where n = 4,5 *(36)*, $SF_5(C_2F_4)_2I$ *(37)*, $SF_5CF(I)CF_3$ and isomer $SF_5(CF_2)_2I$ *(35, 38, 39)*, $SF_5CF{=}CFI$ *(40, 41)*, $SF_5C{\equiv}CI$ *(42)*, $SF_5CF(I)CF_2CF_3$ *(43)*. Since SF_5I is unknown, the chemistry of the readily accessible $SF_5CF_2CF_2I$ has been studied.

A homologous series of $SF_5(CF_2)_nI$ can be conveniently prepared from S_2F_{10} *(34, 35)*:

$$S_2F_{10} + ICF_2CF_2I + nCF_2{=}CF_2 \xrightarrow{\Delta,P} SF_5(CF_2)_{2n}I$$

Mass spectral analysis report values of n up to 9 *(34, 35)*.

The chemistry of the $SF_5(CF_2)_nI$ with alkene and alkyne systems has been carried out; particular emphasis was given to $SF_5CF_2CF_2I$ *(34)*. The reactions of $SF_5CF_2CF_2I$ with alkenes/alkynes are carried out in a Carius tube containing mercury; irradiation was achieved with a halogen lamp. In this manner a number of adducts were formed: $SF_5(CF_2)_n(CH_2)_yI$ where n = 2, 4, 6 with y = 2, and where n = 2 with y = 2,4,6 *(34, 44)*; $SF_5(CF_2)_4(CH_2)_2I$ *(36, 37)*, $SF_5CF(CF_3)(CH_2)_2I$ *(38)*, $SF_5(CF_2)_2(CHFCF_2)_nI$ and isomers with n = 1,2 *(34)*, $SF_5(CF_2)_2CH{=}CHI$ *(34)*, $SF_5(CF_2)_2CH_2CHICH_2CF{=}CF_2$ *(45)*.

The SF_5-polyfluoroalkyl iodides serve as starting points for preparing a number of useful derivatives: $SF_5(CF_2)_nCH{=}CH_2$ with n = 2,4 *(34, 45)*, $SF_5(CF_2)_n(CH_2)_2OH$ with n = 2,4 *(45, 46)*, $SF_5(CF_2)_2(CH_2)_2OC(O)CH{=}CH_2$ and polymer *(47, 48)*, $SF_5(CF_2)_2(CH_2)_2OCH_2C(C(O)OCH_2CH_3){=}CH_2$ and polymer *(47, 48)*, $SF_5(CF_2)_n(CH_2)_2SiCl_3$ with n = 2,4 *(47, 49)*, $[SF_5(CF_2)_n(CH_2)_2SiO_{1.5}]_x$ with n = 2,4 *(47, 49)*, $SF_5(CF_2)_2\overline{CHCH_2O}$ and polymer *(46, 50)*, $SF_5(CF_2)_2(CH_2)_2OCH_2\overline{CHCH_2O}$ and polymer *(46, 50)*, $SF_5(CF_2)_6(CH_2)_2SH$ *(51)*, $SF_5(CF_2)_6(CH_2)_2OC(O)CH{=}CH_2$ and polymer *(52)*. The reaction of $SF_5(CF_2)_nI$ with ClF_3 or with $(CF_3)_3COCl$ gave $SF_5(CF_2)_2IF_2$, $SF_5(CF_2)_2IF_4$, $SF_5(CF_2)_4IF_4$ or $SF_5(CF_2)_2I[OC(CF_3)_3]_2$, respectively *(53, 54)*.

The first SF_5-substituted iodoperfluoroalkene, $SF_5CF{=}CFI$, was prepared by reacting $SF_5CF{=}CF_2$ with trimethylphosphine; the intermediate, $[SF_5CF{=}CFP(F)(CH_3)_3]$, when treated with the etherate of BF_3 gave the salt, $[SF_5CF{=}CFP(CH_3)_3]^+[BF_4]^-$. This salt forms $SF_5CF{=}CFI$ by reaction with I_2, Na_2CO_3, and water*(40)*. An improved method has also been found using tri-n-butylphosphine in place of the trimethylphosphine; the crystal structure of $[SF_5CF{=}CFP(CH_3)_3]^+[BF_4]^-$ was also determined *(41)*. In an interesting development, SF_5-Grignard reagent , (E)-$SF_5CF{=}CFMgI$, was made from $SF_5CF{=}CFI$ and magnesium; when reacted with various phosphorus (III) chlorides the following derivatives were produced: $SF_5CF{=}CFPR_2$ where R can be OEt, OMe, NEt_2, $OCH_2CH_2N(Me)$, i-Pr. For the case with R = OMe, reaction with hexafluoroacetone gives the corresponding SF_5-dioxaphospholane, $SF_5\overline{CF{=}CFPOC(CF_3)_2C(CF_3)_2O}$ *(55)*.

SF_5-Alkyl/Fluoroalkyl Halides

A large number of SF_5-alkyl/fluoroalkyl halides have been prepared by the direct addition of SF_5X to alkenes and alkynes and much of this work has already been reported *(2-5)*. It has been determined that S_2F_{10} adds poorly to double bonds *(56)*. New SF_5 adducts are prepared by reaction of SF_5X (X = Cl, Br) with alkenes and alkynes *(2-5)*. For example, reaction of the appropriate SF_5X with $CH_2{=}CH_2$, $CH_2{=}CF_2$, $CHF{=}CF_2$, $CH_2{=}CCl_2$, $CH_2{=}CHBr$, $CF_2{=}CFBr$,

CH_2=CHOAc, CH_2=CHCOOR, CH_2=C=O, and CH_2=$CHCH_2CF$=CF_2 gave the expected addition products and derivatives:

$SF_5CH_2CH_2Cl$, $SF_5CH_2CH_2Br$, $SF_5CH_2CF_2Br$, SF_5CHFCF_2X, $SF_5CH_2CCl_2Br$, $SF_5CH_2CHBr_2$, and $SF_5CF_2CFBr_2$. These addition products give rise to other useful materials such as SF_5CH=CH_2, $SF_5CH_2CH_2Y$ (Y= SO_3H, OTos, OH, H), SF_5CH=CF_2, $SF_5CH_2SO_3H$, $SF_5CH_2CF_2SO_3H$, SF_5CF=CF_2, $SF_5CHFCF_2SO_3H$, SF_5CHFSO_3H, $SF_5CF_2SO_3H$, SF_5CH=CCl_2, SF_5CH_2CHO, $SF_5CH_2COOCH_3$, $SF_5CH_2CH(OR)_2$, (R = CH_3, C_2H_5, CH_3CO) $SF_5CH_2CH_2COOH$, $SF_5(CH_2)_3Z$ (Z = OH, Br, I, OTos), SF_5CH_2COCl, SF_5CH_3, SF_5CH_2COOH, SF_5CH_2COCl, SF_5CH_2Br, SF_4=CH_2, $SF_5CH_2CHBrCH_2CF$=CF_2 and $SF_5(CH_2)_3CF$=CF_2. In a similar fashion, the addition of SF_5X to alkynes can lead to a number of significant products; with HC≡CH, CH_3C≡$COCH_3$, SF_5C≡CH, CF_3C≡CH, HC≡CSi(R,R'R"), and HC≡$CCOOCH_3$ the following products and derivatives are found: SF_5CH=CHX, SF_5C≡CH, SF_5CH=$CHOCH_3$, SF_5CH_2CHO, $SF_5C_6H_5$, $SF_5(CH_3)C$=$CClOCH_3$, $SF_5CHBrCH_3$ SF_4=$CHCH_3$, SF_5CBr=$CHSF_5$, CF_3CBr=$CHSF_5$, SF_5C≡CSF_5, CF_3C≡CSF_5, SF_5CH=CBrSi(R,R',R") (e.g. R,R',R"=isopropyl, or R,R'=isopropyl, R"=t-butyl) SF_5C≡$CSi(CH_3)_3$, SF_5C≡CH, and SF_5CH=$CBrCOOCH_3$, SF_5CH=$CHCOOCH_3$.

One can see that a vast number of functional groups are compatible with SF_5X addition to the unsaturated systems. Many of the above compounds will be discussed further in this text. In general, SF_5Br is more reactive than SF_5Cl *(57, 58)*. It is also clear that there is a certain reactivity range of the olefinic bond which allows for successful SF_5X addition. This addition is limited by low reactivity on one side and high reactivity (electron-rich olefins) on the other side; high reactivity results in XF addition and SF_4 expulsion *(57)*.

Recently, several noteworthy results in this area have been reported. In particular, SF_5Cl addition to olefins was found possible, even for very reactive double bonds, by using trialkyl boron at low temperatures *(59)*. This method is generally applicable to a free-radical addition process. A noteworthy example of this process shows that SF_5Cl can be added to styrene; without the trialkyl boron present rapid polymerization of styrene occurs *(60)*.

Another interesting development involves chain-extension and was found by reacting SF_5Br with CH_2=$CHC(O)OCH_2CH_3$; reduction of the adduct with (n-butyl)$_3$SnH produced the $SF_5CH_2CH_2C(O)OCH_2CH_3$ which when treated with $BH_3{\cdot}(CH_3)_2S$ gave $SF_5(CH_2)_3OH$ *(57)*; the alcohol can be used to prepare the corresponding three carbon chain iodide, tosylate and bromide *(46)*.

In our laboratories, we have found that a Diels-Alder adduct with butadiene is formed using SF_5CH=CHCOOR *(46)*.

SF_5-Aliphatic Hydrocarbons (Alkanes, Alkenes, Alkynes)

Only several SF_5-alkanes are known: SF_5CH_3 is produced by reduction of SF_5CH_2Br, prepared from $SF_5CH_2C(O)OAg$ and Br_2 *(61)*. $SF_5CH_2CH_3$ is prepared by a series of reactions: reduction of $SF_5CH_2C(O)Cl$ with $LiAlH_4$ gives $SF_5CH_2CH_2OH$ which then forms the corresponding tosylate, $SF_5CH_2CH_2OSO_2C_6H_4CH_3$; the last step is the reduction of the tosylate with $LiAlH_4$ *(24)*. It is possible to form $SF_5CH_2CH_3$ (with about 70% yield) from the reduction of $SF_5CH_2CH_2Br$ with R_3SnH in diphenyl ether *(46, 62)*. It is claimed that SF_5 - containing fluoroalkyl iodides may be reduced with H_2 at temperatures up to 400° C forming SF_5 - containing fluoroalkanes *(63)*.

Since the last review of SF_5-alkenes/alkynes *(4)*, the following systems have been reported: $SF_5(CF_2)_nCH{=}CH_2$, (n= 2,4), $SF_5(CF_2)_2(CH_2)_3CF{=}CF_2$ *(34, 45)*.

The use of $SF_5CF{=}CF_2$ as a secondary synthon has led to a number of interesting and important developments. In the 1994 review of SF_5-alkenes/alkynes, the chemistry of $SF_5CF{=}CF_2$ is discussed *(4)*; in particular, the formation of SF_5 oxetanes, sultones, perfluoroalkyl iodides, an exocyclic alkene, a N-chlorosulfenyl-aziridine, and several copolymers with $CF_2{=}CF_2$ or $CFH{=}CF_2$. Also listed are the very interesting phosphonate derivatives: $SF_5CF{=}CFP(O)(OR)_2$ where R = Et, $SiMe_3$. By comparison, $SF_5CH{=}CF_2$ when reacted with silylphosphites, $(RO)_2POSiMe_3$ with R = Et, $SiMe_3$, produced the corresponding phosphonates, $SF_5CH{=}CFP(O)(OR)_2$; for the case where R = $SiMe_3$, hydrolysis gave the phosphonic acid $SF_5CH{=}CFP(O)(OH)_2$ *(64)*. Using this same method, the fully fluorinated acid $SF_5CF{=}CFP(O)(OH)_2$ was made *(64)*. The novel nitroso compound, $SF_5(CF_3)CFNO$ has been reported *(65)*.

Only a handful of bis and tris-SF_5 carbon compounds exist. The first compound containing two SF_5 groups was $(SF_5)_2CF_2$; it was prepared either by direct fluorination of CS_2 *(66)* or electrochemical fluorination *(67)*. The addition of S_2F_{10} to $CF_3CF{=}CF_2$ produced small quantities of $SF_5CF(CF_3)CF_2SF_5$ *(13)*;with vinyl chloride, $SF_5CHClCH_2SF_5$ is produced *(13)*. The compound, bis(pentafluorosulfur) ethane, $(SF_5)_2C_2H_3Br$, was prepared by adding SF_5Br to $SF_5CH{=}CH_2$ *(68)*. When $AgCF(SF_5)CF_3$ is reacted with Br_2, $BrCF(SF_5)CF_3$ and $[CF(SF_5)CF_3]_2$ are formed *(69)*; the dimer is unstable at 100° C producing S_2F_{10} and $CF_3CF{=}CFCF_3$. Another dimer, $(SF_5CFCF_2)_2$ prepared from $SF_5CF{=}CF_2$ and CsF has been reported; its structure is unknown *(70)*. When $SF_5CF{=}CF_2$ is treated with TASF, the dimer, $SF_5CF{=}CFCF(CF_3)SF_5$ is formed *(71)*. Surprisingly, CF_3OCl with $SF_5CCl{=}CH_2$ gives the following butane, $SF_5CCl_2CH_2CH_2CCl_2SF_5$ *(72)*. The following interesting compounds have also been reported: $(SF_5)_2C{=}CF_2$, $(SF_5)_2CHCOOH$, $(SF_5)_2CH_2$ *(73, 74)*; it is interesting to note that treatment of $(SF_5)_2CH_2$ with base and heat gives $SF_5C{\equiv}SF_3$ *(73)*. The diester, $[SF_5CF(SO_2F)C(O)OCH_2CF_2]_2CF_2$, is formed from $HOCH_2(CF_2)_3CH_2OH$ and $SF_5\overline{CFCF_2OSO_2}$ *(75)*.

Several aromatic di/tri substituted SF_5 derivatives are known; $(SF_5)_2C_6H_3NO_2$ *(76)* and $(SF_5)_3C_6H_3$ *(42, 77)*. The compounds $SF_5(CF_3)NN(CF_3)SF_5$, $(SF_5)_2NR_f$ (R_f=CF_3, C_2F_5) and $(SF_5)_3N$ are known *(18, 78, 79)*. Recently, the following materials have been produced: $SF_5CH_2CHBrCH_2SF_5$, trans-SF_5CH_2CH=$CHCH_2SF_5$, and $(SF_5CH_2)_2C$=CH_2 *(46)*.

Additional examples of bis-SF_5 compounds will be found at the end of the next section.

SF_5-Alcohols, Alkoxides, Aldehydes, Ketones, Thioketones, Ketenes Acetals, Enol Ethers

The known SF_5-alcohols include: $SF_5(CH_2)_nOH$ (n = 2, 3) *(24, 80)*: $SF_5(CF_2)_2(CH_2)_2OH$ *(45)*. It was found that a transesterification reaction of $SF_5(CH_2)_2OC(O)CH_3$ or $SF_5(CF_2)_nCH_2CH_2OC(O)CH_3$ (n = 2,4) gave the corresponding alcohols in high yields *(81)*. The esters were prepared by reacting $CH_3C(O)OAg$ with either $SF_5(CH_2)_2Br$ or $SF_5(CF_2)_nCH_2CH_2I$ (n = 2, 4) *(81)*.

The first SF_5-containing ene-ol, SF_5CH=$CHCH_2OH$, was prepared by the base treatment of $SF_5CH_2CHClCH_2OAc$ *(58)*.

The only known aldehyde, SF_5CH_2CHO, was prepared by first forming $SF_5CH_2CHCl_2$ from SF_5Cl; dehydrohalogenation and treatment with $NaOCH_3$ gives SF_5CH=$CHOCH_3$. Acid hydrolysis of the methyl ester produces SF_5CH_2CHO *(82)*. In another procedure, the hydrate of the aldehyde, $SF_5CH_2CH(OH)_2$ was prepared by base treatment of $SF_5CH_2CHClOC(O)CH_3$; the pure aldehyde was obtained by dehydration with P_4O_{10} *(83)*. A later study of this reaction found that the trimer $(SF_5CH_2CHO)_3$ was a by-product; the crystal structure of the trimer was also determined *(84)*.

The following reactions illustrate the relationship between acetals, acylals, enol ethers and vinyl halides:

The first SF_5-ketone, $SF_5CH_2C(O)CH_3$, was prepared by heating SF_5Cl and isopropenyl acetate together at 100° C *(83)*. The fully fluorinated ketone, $SF_5C(O)CF_3$ has also been reported; bromine is added to SF_5CH=CF_2 giving $SF_5CH(Br)CF_3$ which when treated with $S_2O_6F_2$ gives $SF_5C(O)CF_3$ and $S_2O_5F_2$ *(85)*. The thioketone, $SF_5(CF_3)C$=S, is formed from $SF_5(CF_3)C$=SF_2 and BI_3; the intense blue-violet compound will dimerize to the 1,3-dithietane, $SF_5(CF_3)CS_2C(CF_3)SF_5$ *(86)*.

There are two interesting ketenes, $SF_5C(SO_2F)$=C=O and SF_5CH=C=O. The first ketene is prepared from the ester $SF_5CH(SO_2F)C(O)OCH(CH_3)_2$ which was formed from the corresponding $SF_5CHCF_2OSO_2$ and $(CH_3)_2CHOH$; treatment of this ester with P_4O_{10} or treatment of the acid fluoride with $BF_3{\cdot}N(C_2H_5)_3$ gave $SF_5C(SO_2F)$=C=O *(87, 88)*. The second ketene was prepared by dehydration of $SF_5CH_2C(O)OH$ with P_4O_{10}; this ketene dimerizes with heat or in glass and at 270-290° C is isomerized to SF_4=$CHC(O)F$. In the presence of metal fluorides, SF_5CH=C=O will polymerize to a rubbery solid *(89)*.

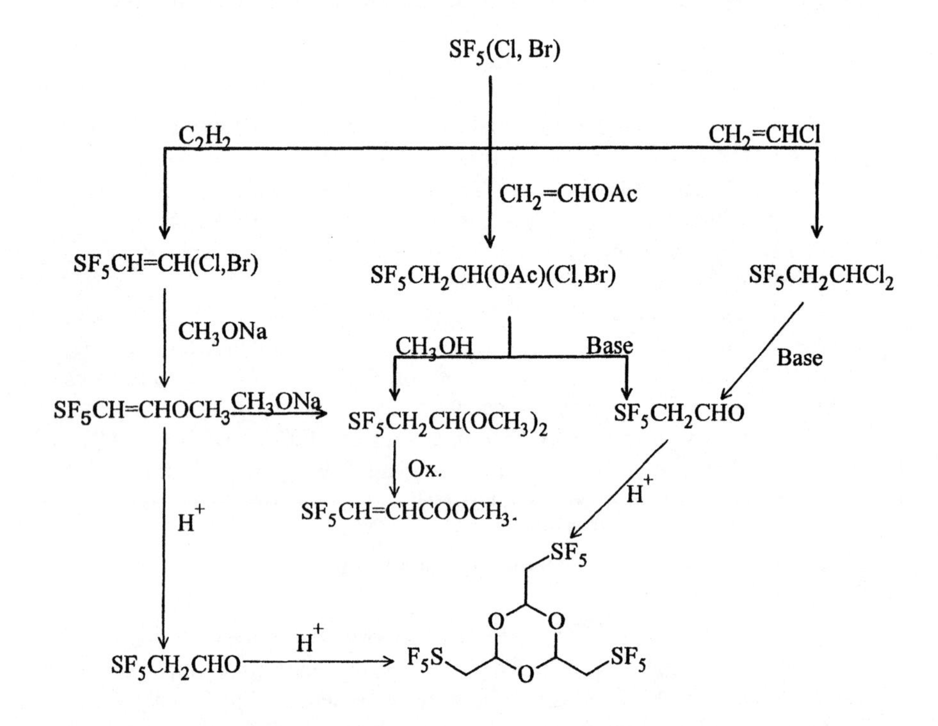

SF5(Cl, Br)
C2H2
CH2=CHCl
CH2=CHOAc
SF5CH=CH(Cl,Br)
SF5CH2CH(OAc)(Cl,Br)
SF5CH2CHCl2
CH3ONa
CH3OH
Base
Base
SF5CH=CHOCH3
CH3ONa
SF5CH2CH(OCH3)2
SF5CH2CHO
Ox.
H+
SF5CH=CHCOOCH3.
H+
SF5
O
O
O
H+
SF5CH2CHO
F5S
SF5

SF_5-Acids, Acid Halides, Esters, Salts

One area of particular interest deals with the synthesis and reactions of a number of SF_5-carboxylic acids and salts. The first SF_5-carboxylic acid to be reported, $SF_5CF_2C(O)OH$, was prepared via electrochemical fluorination of $HSCH_2CO_2H$ *(90)*. In 1970, an alternate method was developed by adding SF_5Cl to alkyl trifluorovinyl ethers. The products, $SF_5CF_2CFClOR$ (R=CH_3,CH_3CH_2), when treated with H_2SO_4 at 75-80° C give the corresponding esters; treatment of the ethyl ester with base and H_2SO_4 gave $SF_5CF_2C(O)OH$ *(91)*.

The addition of SF_5Cl to ketene gives $SF_5CH_2C(O)Cl$ which upon hydrolysis yields $SF_5CH_2C(O)OH$ *(83)*; also the oxidation of the SF_5CH_2CHO with $KMnO_4$ gives this acid, too *(92)*. Addition of $C(O)F_2$ to $SF_5CF{=}CF_2$ with CsF affords the preparation of $SF_5CF(CF_3)C(O)X$ (X = F, OH) *(70)*. The interesting acid, $(SF_5)_2CHCO_2H$, was prepared by addition of SF_5Cl to $SF_5CH{=}CF_2$; the addition product when treated with K_2CO_3 produces $(SF_5)_2CF{=}CF_2$ which undergoes hydrolysis to $(SF_5)_2CHCO_2H$ *(74)*. A number of additional acids have been prepared; in Table II is a list of the SF_5-containing carbonyl halides, carboxylic acids and some salts that can be prepared from SF_5X.

The longer chain acid, $SF_5(CH_2)_2COOH$, was prepared by a novel chain-extension method: $SF_5CH_2CHBrC(O)OC(CH_3)_3$ prepared from SF_5Br and $CH_2{=}CHC(O)OC(CH_3)_3$, is reduced with tri(n-butyl)tin hydride producing $SF_5(CH_2)_2C(O)OC(CH_3)_3$ which undergoes hydrolysis to $SF_5(CH_2)_2C(O)OH$ *(57)*; the crystal structure of this acid has been reported *(93)*. Also, a number of other SF_5-perfluoroalkyl carbonyl fluorides, SF_5-perfluoroalkyl carboxylic acids/salts are known and have been prepared by the electrochemical fluorination of sulfur containing compounds; for example, electrochemical fluorination of 2-thiophenecarbonyl chloride gives, $SF_5(CF_2)_4C(O)F$, $SF_5(CF_2)_4C(O)OH$, $SF_5(CF_2)_4C(O)OM$ (M = Li, NH_4^+). Also, by the same method using di-(thiohexanoyl chloride) the following compounds were obtained: $SF_5(CF_2)_3CF(CF_3)C(O)F$, $SF_5CF(CF_3)(CF_2)_3C(O)F$, and $SF_5(CF_2)_5C(O)F$; these compounds produced the corresponding carboxylic acids and $SF_5(CF_2)_5CO_2NH_4$ *(94)*; additional work in this area for $SF_5(CF_2)_nC(O)F$ (n= 3,4) and derivatives are given in reference *(95)*.

In addition to the SF_5-carboxylic acids discussed above, a number of SF_5-containing sulfonic acids, their halides, esters and salts have been made (see table III). There are several pathways for preparing these materials. The first method involves reacting SF_5-alkyl bromide (SF_5CHXCY_2Br where X,Y = H, F) with sodium sulfite; the sodium salt is isolated and then acidified giving the corresponding SF_5-fluoroalkyl or alkyl acids *(96)*. In the second method, the SF_5-sultones ($SF_5CXCF_2OSO_2$ with X = H, F), formed from the reaction of $SF_5CX{=}CF_2$ with monomeric SO_3, are hydrolyzed producing $SF_5CXCF_2SO_2F$; the sulfonyl fluorides are reacted with base and acidified in order to give the corresponding SF_5-sulfonic acids *(75, 97)*. The perfluorinated acid, $SF_5CF_2SO_3H$

(SF_5-triflic acid), is prepared by fluorinating the precursor SF_5CFHSO_2F with F_2; treatment of the $SF_5CF_2SO_2F$ with base and acid gives the corresponding $SF_5CF_2SO_3H$ *(98)*. Electrochemical fluorination of 2-thiophenesulfonyl fluoride and dimethyl sulfide produced $SF_5(CF_4)_4SO_2F$ *(95)*; base treatment gave $SF_5(CF_4)_4SO_3Li$. The m-$SF_5(CF_2)_n$-aromatic sulfonic acids (n = 2, 4, 6) are prepared by reacting $SF_5(CF_2)_nC_6H_5$ with HSO_3Cl; the intermediates, *m*-$SF_5(CF_2)_nC_6H_4SO_2Cl$, are hydrolyzed to the corresponding sulfonic acids *(46, 99)*. A number of interesting salts have been formed from either the SF_5-alkylsulfonyl fluoride by treatment with base or from the SF_5-alkylsulfonic acids *(100, 101)*.

In a recent report, the following ester derivatives of β-SF_5-acrylic acid were reported: $SF_5CH{=}CBrC(O)OC_2H_5$, $SF_5CH{=}CHC(O)OC_2H_5$ and $SF_5CH{=}C(CH_3)C(O)OCH_3$ *(102)*.

In addition, a large number of SF_5-esters have been prepared from SF_5-acids, acid halides, and SF_5-alcohols. They are discussed in other parts of this review.

SF_5-Organic Metal Complexes

The first SF_5-transition metal complex, $PtCl(SF_5)(PPh_3)_2$, was prepared in 1969 from SF_5Cl *(103)*. In the reaction of $SF_5C{\equiv}CCF_3$ with $Ni(CO)_4$, a yellow liquid, $Ni(CO)_2(SF_5C{=}CCF_3)$ is produced; this material upon further reaction with triphenyl phosphine gives a bright yellow solid, $Ni[P(Ph)_3]_2(SF_5C{=}CCF_3)$ *(104)*. In both compounds, a 3-membered cyclic structure with Ni coordinated to the carbon atoms of C=C bond is found. A number of interesting cobalt compounds are formed from $SF_5C{\equiv}CH$ and $Co_2(CO)_8$; $Co_2(CO)_n(SF_5C{\equiv}CH)$ with n = 4, 5, 6; the complex with n = 4 when treated with Br_2 produces the aromatic compound, $(SF_5)_3C_6H_3$ *(42)*. The crystal structure of the cobalt complex with n = 4 is also reported. Further work of $Co_2(CO)_8$ with $RC{\equiv}CSF_5$ (R = H, CF_3) gave 1:1, 1:2 and 1:3 type products; for R = H, all three types of products were isolated *(105)*. Several additional SF_5-metal complexes were prepared using SF_5-cyclopentadiene, $SF_5C_5H_4$; reaction with thallium ethanolate gave $Tl^+[SF_5C_5H_4]^{-1}$ which when reacted with [(cod)$RhCl_2Rh$(cod)] produced $SF_5C_5H_4Rh$(cod) or when reacted with $[(CO)_2RhCl_2(CO)_2]$ produced $SF_5C_5H_4Rh(CO)_2$ *(106)*. It is expected that using $Tl^+[SF_5C_5H_4]^-$, many organometallic cyclopentadienyl complexes should now be accessible *(106)*; the crystal structure of the $Tl[SF_5C_5H_4]{\cdot}THF$ complex revealed a bent chain of rings linked through the Tl^+cation.

A number of SF_5 salts have been prepared as discussed above (see table III). Also, a number of SF_5 salts formed from SF_4 and MF. In general, intermediate addition products, such as $M^+[SF_5C(CF_3)_2]$ (M = Rb, Cs), $M^+[SF_5CFCF_3]^-$ (M = Cs, TASF), have not been isolated but are used in preparing SF_5 products *(18, 29, 70)*. A number of additional salts have been isolated.;

Table II. SF_5-Carbonyl Halides, Carboxylic Acids and Salts

Acid Halides	*Acids*	*Salts*
$SF_5C(O)F$	$SF_5CF_2C(O)OH$	$SF_5CF_2C(O)OAg$
$SF_5CF_2C(O)F$	$SF_5CF(CF_3)C(O)OH$	$SF_5CH_2C(O)OM$ (Ba,Ag)
$SF_5CF(CF_3)C(O)F$	$SF_5(CH_2)_nC(O)OH$	
$SF_5CH_2C(O)X$		

NOTE: X = F, Cl, n = 1,2,4

Table III. SF_5-Sulfonyl Fluorides, Sulfonic Acids, and Salts

SF_5-SO_2X	*SF_5-SO_3H*	*SF_5-SO_3M*
$SF_5CH_2SO_2F$	$SF_5CH_2SO_3H$	$SF_5CH_2SO_3M$ (M= Ca, Li)
SF_5CHFSO_2F	SF_5CHFSO_3H	SF_5CHFSO_3M (M=Ca, Na, Li)
$SF_5CF_2SO_2F$	$SF_5CF_2SO_3H$	$SF_5CF_2SO_3M$ (M= Ca, Ba, Li)
$SF_5CH(COF)SO_2F$	$SF_5CH_2CH_2SO_3H$	$SF_5CH_2CH_2SO_3M$ (Na, Li)
$SF_5CF(COF)SO_2F$	$SF_5CH_2CF_2SO_3H$	$SF_5CH_2CF_2SO_3M$ (Na, Li)
$SF_5CH(Cl)SO_2F$	$SF_5CHFCF_2SO_3H$	$SF_5CHFCF_2SO_3M$ (Na, Li)
$SF_5CH(Br)SO_2F$		$SF_5(CF_4)_4SO_3Li$
$SF_5CBr_2SO_2F$		
$SF_5(CF_4)_4SO_2F$		
m-$SF_5(CF_2)_nC_6H_4SO_2Cl$	*m*-$SF_5(CF_2)_nC_6H_4SO_3H$	*m*-$SF_5(CF_2)_nC_6H_4SO_3K$ (n=2,4,6)

$Ag[SF_5CFCF_3]\cdot CH_3CN$ [69], $M^+[SF_5C(SO_2F)COX]^-$ where (X = F, OCH_3, M = Ag, Na, Cs), and $[(CH_3CH_2)_3NH]^+[SF_5C(SO_2F)C(O)OCH_3]^-$ *(107)*.

The Grignard reagent in both *m-,p*-$SF_5C_6H_4Br$ leads to the formation of the corresponding *m-,p*-$SF_5C_6H_4CO_2H$ acids *(77)*.

SF_5-Polymer Chemistry and Surface Studies

The synthons, SF_5X, [X = Cl, Br, SF_5, $CF{=}CF_2$, $(CF_2)_nCH_2OC(O)CH{=}CH_2$] are useful in preparing novel SF_5-polymers (see Table IV). In our laboratory, we have found that by using SF_5Br and fluoroolefins not only monomeric liquid products but also polymeric solids are formed. In this study, a number of trends coincide with the increasing number of fluorine atoms in the structure:

- increase in melting point
- increase in stability and
- decrease in solubility

It is remarkable that partially fluorinated polymers are soluble in acetone and that the $SF_5(CF_2CF_2)_{16}Br_{0.3}$ polymer is more stable than poly(tetrafluoroethylene) *(108)*. In later work, it was found that $CF_2{=}CFBr$ and SF_5Br formed a waxy polymer *(109)*. Polymer studies that include SF_5-polyesters, SF_5-epoxides, SF_5-acrylates and SF_5-silicones have been reported; extensive studies on the surface properties of the latter two polymeric systems have been carried out .

In order to prepare SF_5-polyesters, several SF_5- fluorosultones, $SF_5\overline{CFXCF_2OS}O_2$, with X = H, F, were prepared by reacting monomeric SO_3 with the corresponding olefin, $SF_5CX{=}CF_2$ *(110)*. The SF_5-fluorosultones react with allyl alcohol giving the following allyl esters- $SF_5CFX(SO_2F)C(O)OCH_2CH{=}CH_2$ (X = H, F). The corresponding polyester derivatives were prepared by ultraviolet irradiation in $CFCl_3$ *(110)*. The properties and characteristics of these polyesters with SF_5 groups were determined. The decomposition temperatures are 170°C and 150°C, respectively; both polyesters are soluble in acetone but insoluble in water and $CFCl_3$.

A number of new SF_5 monomers that contain a polymerizable functionality such as an epoxide group, an acrylate group or a chlorosilane group have also been prepared and studied. Several polymer systems based on $SF_5(CF_2)_2\overline{CHCH_2O}$ and $SF_5(CF_2)_2(CH_2)_2OCH_2\overline{CHCH_2O}$ have been prepared and characterized *(46)*. In addition to preparing epoxide systems, successful methods have been developed in order to prepare a number of terminal SF_5-fluorinated acrylates *(48, 111)*. The following acrylate monomers, bearing side chains terminated in SF_5 groups, have been synthesized; $SF_5(CF_2)_2(CH_2)_2OCH_2C(C(O)OCH_2CH_3)=CH_2$ and $SF_5(CF_2)_n(CH_2)_2OC(O)CH=CH_2$ with n = 2, 6 *(48, 52, 111)*. The first monomer was polymerized by heating with AIBN; the polymer had a melting range of 126-150° C, and a contact angle (H_2O) of 96°. The second SF_5-acrylate monomer was photo- copolymerized with HEMA; contact angles up to 100° were observed. All copolymer films containing the SF_5-acrylate had a surface composition corresponding to an enriched overlayer of pure SF_5-acrylate. The thickness of the surface-enriched overlayer decreases as the bulk concentration of SF_5-acrylate monomer decreases. At the lowest bulk copolymer concentration examined, 1 wt % loading of SF_5-monomer, the overlayer thickness was approximately 20 Å. It is now possible to prepare surfaces entirely composed of SF_5-groups using long chain acrylates *(44)*; aqueous contact angles for non-annealed photopolymerized SF_5-longchain acrylate were 106±2.4°. After annealing under a nitrogen environment for 24 h, the angles were 112±3.6°. Additional studies with SF_5-groups using long chain thiol have also been carried out *(51)*.

In an alternate route, new apolar SF_5-containing surfaces utilizing SF_5-terminated perfluoroalkylsilanes and resulting SF_5-polysiloxanes as components for films and coatings were formed *(49, 111)*. In order to prepare SF_5-polysiloxane films and coatings, the new SF_5-terminated perfluoroalkylsilanes, $SF_5(CF_2)_nCH_2CH_2SiCl_3$ (n = 2, 4) were prepared and characterized. Thin films of these new SF_5-silanes on clean silicon oxide substrates were characterized by aqueous contact angle and XPS; the longer SF_5-silane exhibits a high aqueous static contact angle of 106°. XPS analysis of the SF_5-six carbon silane film strongly supports formation of organized SF_5-silane monolayers despite probable steric and structural problems from the terminal SF_5-group. Bulk SF_5-polysiloxane materials with SF_5-containing perfluoroalkyl groups are prepared by bulk hydrolysis of the new SF_5-silanes *(49)*. These materials are expected to show new, interesting chemical and physical properties including high chemical and thermal stability, high dielectric strength, low refractive index and low surface energy. These unique chemical and physical properties are derived in part from the similarities of the SF_5-perfluoroalkyl groups to the parent compound, SF_6.

Applications of all of the SF_5-polymers include high-performance lubricants and optical materials, water repellent and oil-resistant sealants, antifoaming agents and protective surface coatings.

Table IV. SF_5-Fluoropolymers / Telomers

1. SF_5Cl with $CF_2{=}CF_2$, *(112, 113)*; $SF_5(CF_2)_nCl$, MW = 2500, m.p.>400° C
2. S_2F_{10} with $CF_3CF{=}CF_2$ *(114)*
3. $SF_5CF{=}CF_2$ with $CF_2{=}CF_2$, *(115)*; Polymer has m.p. that decreases with % $SF_5CF{=}CF_2$ (0.1 molar % $SF_5CF{=}CF_2$ gave product with m.p. 326° C; 50 molar %, m.p. 280° C)
4. $SF_5CF{=}CF_2$ (I) with $CF_2{=}CH_2$ (II) and $CF_3CF{=}CF_2$ (III), *(116)*; tough elastomers with I & II, and I, II ,III
5. $SF_5CF{=}CF_2$ with $CF_2{=}CH_2$ *(43)*; white copolymer
6. SF_5Br with $CF_2{=}CF_2$,$CF_2{=}CFH$, $CF_2{=}CH_2$, $CFH{=}CH_2$ *(117)*
7. $SF_5\overline{CXCF_2OSO_2}$ (X = H, F) with polyvinyl alcohol *(118)*
8. SF_5Br with $CF_2{=}CFBr$, *(109)*; $SF_5(CF_2CFBr)_{10}Br$; waxy solid
9. $SF_5(CF_2)_4CH_2OH$ with acrylic acid *(94, 95)*; $[F_5S(CF_2)_4CH_2OC(O)CH\text{-}CH_2\text{-}]_n$ Calculated surface energy (based on contact angles) was 13.1 dynes/cm; $[CF_3(CF_2)_6CH_2\text{-}CO_2C(CH_3)CH_2\text{-}]_n$ surface energy was 10.7 dynes/cm
10. Fluorination of poly(S-vinyl-O-t-butylthiocarbonate) will give both hydrocarbon and perfluorocarbon vinyl polymers containing the SF_5 group *(119)*
11. SF_5-Fluoroalkyl Acrylate polymers that include *(47, 48)*: $[SF_5(CF_2)_n(CH_2)_2OC(O)CHCH_2]_x$ (n = 2, 6) and $[SF_5(CF_2)_n(CH_2)_2OCH_2C(C(O)OCH_2CH_3)CH_2]_x$
12. SF_5-Polysiloxanes *(47, 49)*: $[SF_5(CF_2)_n(CH_2)_2SiO_{1.5}]_x$ and n = 2, 4
13. SF_5-Fluoroalkyl Epoxide polymers that include *(46)*: $[SF_5(CF_2)_2CHCH_2O]_x$ and $[SF_5(CF_2)_2(CH_2)_2OCH_2CHCH_2O]_x$
14. Polymides containing the $SF_5C_6H_3$- and $SF_5(CF_2)_2C_6H_3$- groups have been prepared from $SF_5C_6H_3(NH_2)_2$ and $SF_5(CF_2)_2C_6H_3(NH_2)_2$ *(76, 46)*
15. SF_5-Aromatic Acrylate and Styrene polymers *(120)*: $[SF_5(CF_2)_2C_6H_4OC(O)CHCH_2\text{-}]_n$ and $[SF_5(CF_2)_2C_6H_4CHCH_2\text{-}]_n$

SF_5-Heterocycle Chemistry

While it is possible to prepare many aliphatic and aromatic SF_5-compounds, only a relatively few SF_5-heterocycle compounds have been prepared. Heterocycles containing a metal site are presented above. In this section we will review only the more recently reported SF_5-heterocycles. Some of the known SF_5-heterocycles have already been covered in references *(4)* and include: pyrazoles, $SF_5\overline{C{=}CHNHN{=}C}H$ with isomer and carbamoylated derivatives, $SF_5\overline{C{=}CRNHN{=}C}H$ where R = C_6H_5, p-ClC_6H_4, $CH_3OC_6H_4$, *m*-$HO_2C_6H_4$, $(CH_3O)_3C_6H_2$ and derivatives; oxetanes, $SF_5\overline{CFCF_2OC}(CF_3)_2$ and isomers;

aziridines, such as $SF_5NCH_2C(CF_3)X$ where X = Cl, N_3 and $SF_5NCH_2C(CF_3)CH_2X$; oxyaziridine, SF_5NCF_2O; succinimide, $SF_5NC(O)CF_2CF_2C(O)$; epoxides, SF_5CXCF_2O where X = H, F; sultones, $SF_5CXCF_2OSO_2$ where X = H, F, Cl, OMe.

The more recent SF_5-heterocycle systems include: SF_5CFCF_2NSCl prepared by heating $SF_5CF{=}CF_2$ and $(NSCl)_3$ together at 50° C *(121)*. A number of 1,3,4-oxadiazoles have been prepared that include for $R_1C{=}NN{=}CR_2O$ where R_1, R_2 = $SF_5(CF_2)_n$ (n = 1, 2, 3); $=SF_5CH_2$; $=C(NO_2)_3CH_2CH_2$; $=C(NO_2)_2FCH_2CH_2$; $=CF_3(CF_2)_n$ and n = 0, 1, 2, 6 *(122)*. These materials were prepared from the corresponding diacyl hydrazines, $R_1C(O)NHNHC(O)R_2$, by treatment with pure PCl_5, $POCl_3$ or a mixture of PCl_5, $POCl_3$. The diacyl hydrazines were obtained by reaction of $R_1C(O)NHNH_2$ with $R_2C(O)Cl$. By reacting a number of SF_5-polyfluoroalkyl iodides, Y = $SF_5(CF_2)_n(CH_2)_mI$ (n = 2, 4 and m = 2, 4) with R = pyridine, N-methyl imidazole, N-propyl triazole and pyridazine new quaternary salts having the general formula, $[RY]^+I^-$. These quaternary salts were metathesized with $LiN(SO_2CF_3)_2$ giving the following derivatives $[RY]^+[N(SO_2CF_3)_2^-]$; almost all of these salts melted or showed a phase transition below 25° C *(123)*. Two new SF_5-epoxides, $SF_5(CF_2)_2CHCH_2O$ and $SF_5(CF_2)_2(CH_2)_2OCH_2CHCH_2O$, have been prepared *(50)*. In a recent study, the following SF_5-sulfolane/sulfolene compounds were prepared; $SF_5CHCH(Br)CH_2SO_2CH_2$ and $SF_5CHCH{=}CHSO_2CH_2$ *(81)*.

SF_5 Aromatic Chemistry

Review and New Pathways to $SF_5C_6H_5$

The introduction of the SF_5 group into an aromatic system was first reported by Sheppard in 1962 *(77)*.

$$X\text{-}C_6H_4\text{-}S\text{-}S\text{-}C_6H_4\text{-}X \; + \; \xrightarrow{AgF_2} \; X\text{-}C_6H_4\text{-}SF_3 \; \xrightarrow[120°C]{AgF_2} \; X\text{-}C_6H_4\text{-}SF_5$$

Solvent: $FCCl_2CF_2Cl$
Copper reaction vessel
X = H, *m*-NO_2, *p*-NO_2

From $SF_5C_6H_4NO_2$ a number of *m, p* -derivatives were prepared.
X = NH_2, NHC(O)R, $N_2^+X^-$, OH, Br, CO_2H, $CH(OH)CH_3$, $CH{=}CH_2$
Since the SF_5 group is meta directing, nitration gives meta-$SF_5C_6H_4NO_2$.

More recently (2000) a useful procedure using elemental fluorine has been developed *(124)*:

$$O_2N\text{-}C_6H_4\text{-}S\text{-}S\text{-}C_6H_4\text{-}NO_2 + F_2/N_2 \xrightarrow{-5^\circ C} SF_5C_6H_4NO_2$$

(yield ~ 40%)

Solvent: CH_3CN

New SF_5 aromatic compounds prepared using, in particular, cross-coupling reactions:

m,p-$SF_5C_6H_4NHC(O)CH_3$, *m,p*-$SF_5C_6H_4I$, *m,p*-$SF_5C_6H_4C{\equiv}CPh$,
m,p-$SF_5C_6H_4CH{=}CHCO_2Me$, *m,p*-$SF_5C_6H_4Ph$

Also, the crystal structure of m,p-$SF_5C_6H_4NHC(O)CH_3$ was reported.

The first aromatic compound containing two SF_5 groups was also prepared by Sheppard *(77)*:

$$3,5\text{-}(SO_2Cl)_2C_6H_3NO_2 \xrightarrow{HI} [\text{-}S\text{-}C_6H_3(NO_2)\text{-}S\text{-}] \xrightarrow{AgF_2} 3,5\text{-}(SF_5)_2C_6H_3NO_2$$

The first SF_5 trisubstituted benzene derivatives were made by the photolysis of $SF_5C{\equiv}CH$ with SF_5Cl or the reaction of $SF_5C{\equiv}CH$ with $Co_2(CO)_8$ and bromine *(42)*:

$$1,3,5\text{-}(SF_5)_3C_6H_3 \qquad 1,2,4\text{-}(SF_5)_3C_6H_3$$

The preparation of the first ortho-$SF_5C_6H_4F$ was recently announced *(125)*:

$$(O_2N\text{-}C_6H_3F\text{-}S)_2 \xrightarrow{AgF_2} O_2N\text{-}C_6H_3(F)\text{-}SF_5$$

Solvent: $FCCl_2CF_2Cl$

o-$SF_5C_6H_4F$ was converted into a number of ortho-substituted derivatives:

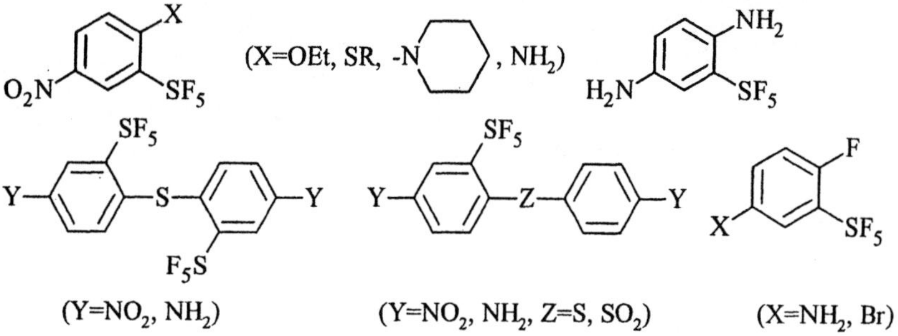

(X=OEt, SR, -N(piperidino), NH_2)

($Y=NO_2$, NH_2) ($Y=NO_2$, NH_2, Z=S, SO_2) ($X=NH_2$, Br)

The parent compound, $SF_5C_6H_5$ can also be prepared in low yield from S_2F_{10} and benzene *(13)*. Also, a multistep method with $SF_5C{\equiv}CH$ has produced $SF_5C_6H_5$ *(126)*. A new and efficient method for preparing $SF_5C_6H_5$ is summarized *(127)*:

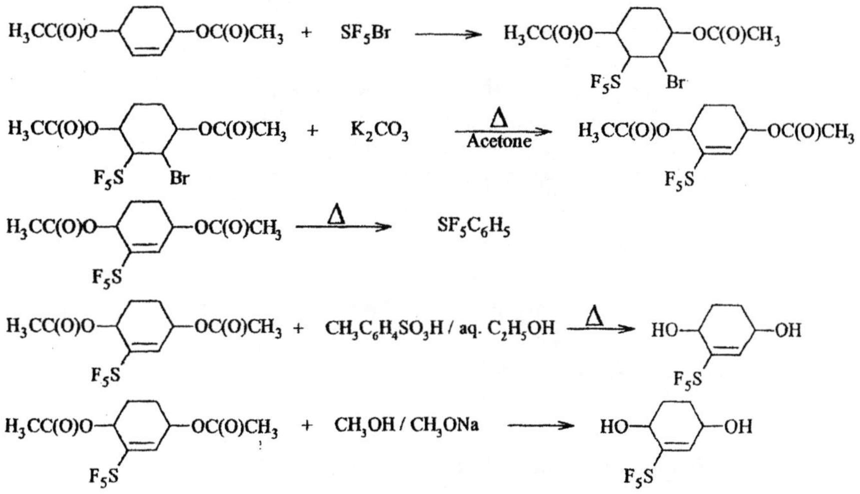

$H_3CC(O)O$–C_6H_8–$OC(O)CH_3$ + SF_5Br ⟶ $H_3CC(O)O$–$C_6H_8(SF_5)(Br)$–$OC(O)CH_3$

$H_3CC(O)O$–$C_6H_8(SF_5)(Br)$–$OC(O)CH_3$ + K_2CO_3 $\xrightarrow[\text{Acetone}]{\Delta}$ $H_3CC(O)O$–$C_6H_7(SF_5)$–$OC(O)CH_3$

$H_3CC(O)O$–$C_6H_7(SF_5)$–$OC(O)CH_3$ $\xrightarrow{\Delta}$ $SF_5C_6H_5$

$H_3CC(O)O$–$C_6H_7(SF_5)$–$OC(O)CH_3$ + $CH_3C_6H_4SO_3H$ / aq. C_2H_5OH $\xrightarrow{\Delta}$ HO–$C_6H_7(SF_5)$–OH

$H_3CC(O)O$–$C_6H_7(SF_5)$–$OC(O)CH_3$ + CH_3OH / CH_3ONa ⟶ HO–$C_6H_7(SF_5)$–OH

HO–(cyclohexene, F_5S)–OH + P_4O_{10} $\xrightarrow{\Delta}$ $SF_5C_6H_5$

SF_5Cl + (cyclohexene) ⟶ (cyclohexane, F_5S, Cl)

(cyclohexane, F_5S, Cl) + $KOC(CH_3)_3$ / Ether $\xrightarrow{r.t.}$ (cyclohexene, F_5S)

(cyclohexene, F_5S) + (N–Br, O, O) ⟶ Br–(cyclohexene, F_5S)–Br (cis / trans)

Br–(cyclohexene, F_5S)–Br + $KOC(CH_3)_3$ / Ether ⟶ $SF_5C_6H_5$

A similar method using 4,5-dichlorocyclohexene and SF_5Cl has been announced; the adduct when treated with sodium ethoxide in ethanol gave $SF_5C_6H_5$ in high yield (an overall yield of 71 %) *(128)*. In an attempt to prepare SF_5-benzenes, SF_6, SF_5Cl and CF_3SF_5 were reacted with C_6H_6, C_6H_5X (X = Cl, Br) in a low temperature radio-frequency plasma. In the reaction products, $SF_5C_6H_5$ was present in only minor quantities; with C_6H_5X (X = Cl, Br), $SF_5C_6H_4X$ were present along with numerous halogenated benzenes *(8)*.

Prior to 2001, the only known benzene synthons containing an SF_5-group were $SF_5C_6H_5$, $SF_5OC_6H_5$, *p*-$SF_5OC_6H_4Cl$ and derivatives. The $SF_5OC_6H_4X$ compounds are easily prepared from SF_5OOSF_5 and benzene, toluene and chlorobenzenes and serve as synthons for several derivatives. The nitration and sulfonation of $SF_5OC_6H_5$ gives *p*-$SF_5OC_6H_4NO_2$ and *p*-$SF_5OC_6H_4SO_3H$, respectively. The oxidation of *p*-$SF_5OC_6H_4CH_3$ with CrO_3 produces $SF_5OC_6H_4C(O)OH$ in high yield *(129)*.

New SF_5 Containing Aromatic Synthons

In 2001, a new homologous series of SF_5-aromatic compounds was reported. The method of preparation was *(130)*:

$$SF_5(CF_2)_nI + xs.\ C_6H_6 \xrightarrow{\Delta} SF_5(CF_2)_nC_6H_5 \quad (n=2,4,6,8)$$

A large number of $SF_5(CF_2)_2C_6H_4X$ derivatives have been prepared *(99, 120)*:

$$SF_5(CF_2)_2C_6H_5 \longrightarrow m\text{-}SF_5(CF_2)_2C_6H_4X$$

X = NO_2, NH_2, $N_2^+X^-$, N_3, OH, OC(O)CH=CH_2, I, Br, CH=CH_2, NHC(O)CH_3, SO_3(H,K)

The polymers for X = OC(O)CH=CH_2 and X = CH=CH_2 have also been prepared.

Another method for preparing $SF_5CF_2CF_2C_6H_5$ was found *(131)*:

$$SF_5Br + F_2C{=}CFC_6H_5 \xrightarrow{sunlamp / CH_2Cl_2} SF_5CF_2CFBrC_6H_5$$

(71% yield, 2½ h)

$$SF_5CF_2CFBrC_6H_5 + AgBF_4 \xrightarrow{CH_2Cl_2} SF_5CF_2CF_2C_6H_5 + AgBr + BF_3$$

(92% yield, ¾ h)

The above method was modified in order to prepare new *o,m,p*-$SF_5CF_2CF_2C_6H_4X$ derivatives. In the first step of this procedure, *o,m,p*-$CF_2{=}CFC_6H_4X$ were made. The second step involved the addition of SF_5Br and was followed by fluorination with $AgBF_4$ *(132)*:

$$F_2C{=}CFC_6H_4X + SF_5Br \xrightarrow{sunlamp / CH_2Cl_2 / 0°C} o,m,p\text{-}SF_5CF_2CFBrC_6H_4X$$

$$\text{o,m,p-}SF_5CF_2CFBr\text{-}C_6H_4\text{-}X + AgBF_4 \xrightarrow{\Delta / CH_2Cl_2} \text{o,m,p-}SF_5CF_2CF_2\text{-}C_6H_4\text{-}X$$

X = *m*-Br, *p*-Br, *p*-Cl, *p*-CH_3, *p*-CF_3, *p*-NO_2, *o*-F, *o*-CF_3, *o*-$CH(CH_3)_2$

The reaction of $SF_5CF_2CF_2C_6H_5$ with $BF_3 \cdot 2CH_3C(O)OH$ gave $SF_5CF_2C(O)C_6H_5$ and $CF_3C(O)C_6H_5$. A $SF_5CF_2CF_2$-dye was prepared by condensing m- $SF_5CF_2CF_2C_6H_4NH_2$ with Fischer's aldehyde *(46)*.

Properties and Applications of SF_5 compounds-Overview and Recent Results

For many years, the premiere SF_5 containing material was SF_6. This material is a colorless, odorless, nontoxic and inert gaseous substance. It is used as an electrical insulator in coaxial cables and in high voltage equipment; G.H. Cady has published an excellent review of the properties of SF_6 and other properties of fluorine-containing compounds of sulfur *(1)*.

The outstanding chemical and thermal stability of SF_6 is carried over to many of its organic derivatives. It is known that carbon compounds containing a SF_5 group have interesting and useful properties which include electronic properties, selective biological activity, high thermal and chemical stability, surface activity for SF_5 monomers/polymers *(1-5)*.

Electronic Properties

Gaseous / Liquid Insulating Medium.

For many years, SF_6 has been used as a gaseous insulator with a breakdown strength (ball to plane) of 16-17 KV. Under similar conditions, $SF_5CF{=}CF_2$ has a breakdown strength of 21.5 KV *(133)*. It is reported that a 52% (SF_6) - 48% ($SF_5CF{=}CF_2$) gas mixture has approximately the same breakdown value as does 100% $SF_5CF_2{=}CF_2$ *(133)*. Also, a 67.4% (N_2) - 32.6% ($SF_5CF{=}CF_2$) gas mixture has the same breakdown strength as does 100% SF_6 *(133)*. SF_5CF_3 appears to be comparable to $SF_5CF{=}CF_2$ *(134)*. Other dielectric gases are C_2F_6, C_3F_8, c-C_4F_8 and C_4F_{10}. These have breakdown strengths less than $SF_5CF{=}CF_2$ and CF_3SF_5 *(134)*. SF_5 - oxetanes also have excellent dielectric properties *(135)*.

SF_5C1/TFE telomers: $SF_5(CF_2)_nC1$, where n = 2, 4, 6 and polymer. These telomers have high thermal stability (above 250° C), are chemically inert, are non-toxic, have very low surface tensions (19-22 dynes/cm), and have high dielectric strengths and low electrical loss factors. They are chemically inert to acids and alkali at 200° C *(136)*.The use of RSF_5 (R = CH_3, C_6H_5) as a solvent for a high-output double layer capacitor has been patented *(137)*.

SF_5 fluoroalkyl sulfonate salts give rise to outstanding electronic properties. For example, bis(ethylenedithio)tetrathiafulvalene (ET) forms the following salts: $(ET)_2SF_5CH_2CF_2SO_3$ (superconductor at 5.2°K); $(ET)_2SF_5CHFCF_2SO_3$ (semiconductor); $(ET)_2SF_5CH_2SO_3$ (semiconductor), $(ET)_2SF_5CHFSO_3$ (metallic) and $(ET)_2SF_5CF_2SO_3$ (insulator) *(138)*. The lithium salt of $SF_5CF_2SO_3$ when complexed with polyethylene oxide gives a material with outstanding conductivity *(101)*.

Biological Activity

The influence on biological activity of molecules brought about by substitution with fluorine or with the trifluoromethyl-group is well known and documented *(139)*.

It is expected, based on the Hansch Lipophilicity Parameter π^*, and the Hammet substituent constant, σ; that SF_5 compounds will show an interesting range of biological activity. The Hansch Lipophilicity Parameter, π^*, is taken as a measure of the lipophilic character and the larger its value the greater is the lipophilicity; the relative values for F, CF_3 and SF_5 are $\pi = 0.14$, 0.88 and 1.51, respectively *(140)*. For biological activity, the Hammett substituent constant σ, can be used-a positive σ contributes to high biological activity; the σ values for CF_3 and SF_5 are ~0.5 and ~0.7, respectivitely *(140)*.

It is known that $SF_5CF{=}CF_2$ *(141)* and $SF_5CF(I)CF_3$ *(39)* are effective fumigants. $SF_5CF{=}CF_2$ showed some activity as a vapor-phase bactericide. $SF_5N{=}CF_2$ *(142)* is useful as a fumigant for roaches, flies, etc. Compounds derived from SF_5Cl with ethylene and propylene have insecticidal properties *(143)*. SF_5Cl is a powerful lung irritant *(136)*.

More recently, a number of patents have shown that by incorporation of the SF_5-group, derivatives are produced that show behavior as fungicides, herbicides and insecticides *(144-152)*.

Nitrogen mustard compounds and prodrugs have been prepared for use in therapy and treatment of cancer. Thus {3,5-difluoro-4-[bis(2-iodoethyl)amino]benzoyl}-L-glutamic acid showed antitumor activity against breast carcinoma in mice. The inclusion of SF_5, CF_3, SCF_3, OCF_3, CF_3CF_2 groups in these compounds have been patented *(153)*.

Chemical and Thermal Stability

Many saturated SF_5-carbon systems possess excellent chemical and thermal stability. The perfluoroderivatives of SF_6, R_fSF_5, rival the chemical and thermal stability of SF_6; for example, CF_3SF_5 is not attacked by 6N NaOH and reacts with alkali metals only at dull red heat *(154)*. At high temperatures, CF_3SF_5 will act as a fluorinating agent; at 500° C, it will fluorinate P_3N_5 and give a mixture of $(PNF_2)_3$ and $(PNF_2)_4$ *(155)*. The oxetane, $SF_5\overline{CFCF_2OC}(CF_3)_2$ and its isomer survived prolonged exposure to 6N KOH and concentrated H_2SO_4 even at high temperatures (3 months at r.t.; 1 day at temps > 100° C). The above oxetane was thermally destroyed at 314-350° C for 2 hours but survived temperatures of 250° C *(156)*.

Phenylsulfur pentafluoride ($C_6H_5SF_5$) heated at 400-450° C for 7 hours results in only minor decomposition; heating for 4 hours at 100° C with aqueous NaOH produced no decomposition *(157)*. $SF_5CF{=}CF_2$ is unaffected when heated to 300° C in a Pyrex glass vessel for 6 hours and is destroyed only slowly when subjected to U.V. or γ radiation in a silica container *(43)*.

Therefore, many of these compounds are useful as "stable liquid" materials, e.g. fluids for high temperature power transmissions, hydraulic systems or liquid - coupled mechanical drives. They should also be useful as liquid dielectric materials.

Surface Activity & Surface Tension

$SF_5(CF_2)_nCl$ (n = 2, 4, 6, + polymer) have low surface tensions (19-22 dynes/cm) *(136)*. It has been found that polymer surfaces composed of SF_5 groups have a very low critical surface tension. For the polymer $[F_5S(CF_2)_4CH_2OC(O)CHCH_2]_n$, the calculated surface energy (based on contact angles) was 13.1 dynes/cm *(94, 95)*.

The surface tension of aqueous solutions of $SF_5(CF_2)_4CO_2M$ (M = Li, NH_4) and $SF_5(CF_2)_5CO_2NH_4$ have been measured. The latter salt solution has a surface tension value of 22 dynes/cm when its concentration was 5000 ppm; for $CF_3(CF_2)_6CO_2NH_4$ the value was 29.6 at similar concentration *(94, 95)*. New SF_5-aromatic sulfonate salts, *m*-$SF_5(CF_2)_nC_6H_4SO_3K$ with n = 2, 4, 6 have been prepared and characterized. The minimal surface tensions of an aqueous solution are 34.3 mN/m, 32.4 mN/m and 32.3 mN/m for n = 2, 4, 6, respectively. The corresponding bulk concentrations were 42.8 mM, 17.3 mM and 1.67 mM *(46)*.

Solvents for Highly Fluorinated Polymers

A 10% by weight solution of a low mw, low melting TFE polymer (melting range 83-150° C) is prepared by warming the polymer in $SF_5C_6H_5$ until the polymer dissolves. Strips of cellulose filter paper are partly immersed in the warm solution for 0.5-1.0 minutes. The treated area of the filter paper is washed with acetone and dried in air. The treated area is not wetted by drops of H_2O and does not support combustion. It is also possible to use *m*- hydroxyphenylsulfur pentafluoride and 3-biphenylsulfur pentafluoride *(157)*.

$SF_5N{=}CCl_2$, $SF_5N{=}CClR_f$ and $SF_5N{=}CFR_f$ are useful as solvents for low molecular weight polytetrafluoroethylenes *(142)*. SF_5NHCF_3, and SF_5NHR_f dissolves high MW polyoxymethylene at room temperature (2g dissolves 0.1g) while polyamide resins in contact with these materials swell and on moderate warming become soft enough to be molded readily *(142)*.

The compound $SF_5N(CF_3)N(CF_3)SF_5$ is an excellent solvent for very low molecular weight polytetrafluoroethylene at room temperature *(142)*. SF_5CH_2COOH above its m.p. is an active solvent for the polyamide from hexamethylenadipamide, forming solutions which do not crystallize on cooling. These solutions are useful for forming films, and fibers of polyamides *(158)*.

Perfluorinated Blood Substitute

The blood substituted perfluoroalkylsulfur hexafluoride and preferably $(R_f)_nSF_{6-n}$ can be emulsified with a nonionic surfactant and serves as a blood substitute medium *(159)*.

Explosives

There are few reports of energetic materials containing the SF_5-group. In 1978, the compound $SF_5CH_2CO_2CH_2C(NO_2)_2F$ was prepared and characterized as a very dense, thermally stable and insensitive liquid *(160)*.

Also, polynitroaliphatic explosives containing the SF_5 group have been prepared *(161)*:$SF_5CH_2C(O)OR$ with R = $CH_2C(NO_2)_2CH_3$, $CH_2CH_2CF(NO_2)_2$, $CH_2CH_2C(NO_2)_3$. Unfortunately, these compounds had low melting points and were not further evaluated. However, the system $SF_5CH_2CH_2OR$ with R = $(NO_2)_3CCH_2NHC(O)$-, $[(NO_2)_2\text{-}FCCH_2]_2NC(O)$-, $(NO_2)_3FCCH_2CH_2C(O)$-, $[(NO_2)_2\ FCCH_2O]_3C$- had higher melting points and good densities. The tris nitro-derivative had a high melting point (96° C), good thermal stability and low impact sensitivity *(161)*. This work indicates that the SF_5-group is effective in reducing the impact sensitivity of nitro explosives.

Additionally, a large number of SF_5-imine polynitro compounds have been prepared by reacting $SF_5N{=}CCl_2$ with a polynitroalcohol in the presence of a base *(162)*: $SF_5N{=}CR_1R_2$ where R_1 and R_2 represent polynitro groups such as –$OCH_2CH_2C(NO_2)_3$, $-OCH_2C(NO_2)_2CH_3$, $-OCH_2C(NO_2)_3$, $-OCH_2CFC(NO_2)_2$, -$OCH_2CF_2C(NO_2)_2$, $-OCH_2CH_2C(NO_3)_3$ *(162)*. A more recent study suggested that pentafluorosulfanyl-nitramide salts may be useful as an oxidizer ingredients; these salts include $Z^+SF_5N(NO_2)^-$ where Z = NH_4, K, Na, $NH_2C(=NH_2)NH_2$, $NH_2C(=NH_2)NHNH_2$ *(163)*. The compound, $NH_4^+N(NO_2)^-$, is of interest as an oxidizer for rockets.

Refrigerant Fluids

CF_3SF_5 is a thermally stable, non-toxic and non-combustible gas and useful as a working fluid for heat pumps and refrigerators *(164)*. Azeotropic compositions containing either CF_3SF_5 or HCF_2SF_5 or $CF_3CF_2SF_5$ and H_2CF_2 or HCF_2CF_2H or other fluorocarbons have a low ozone depletion potential and are useful as refrigerant fluids *(165)*.

Plasma Etching Gas

SF_6 with oxygen in a plasma will etch carbon-based or silicon-based materials *(10, 11)*. There are a considerable number of patents that use SF_6 as a etching gas.

Polymers

Incorporation of the SF_5 group into a polymeric system can produce high stable polymers with low surface energies.

Liquid Crystals

There have been relatively few studies of liquid crystals dealing with the effect of the polar terminal SF_5-group. The first SF_5-containing liquid crystals reported in 1999 were $SF_5C_6H_4X$ systems, where X was one of the following:

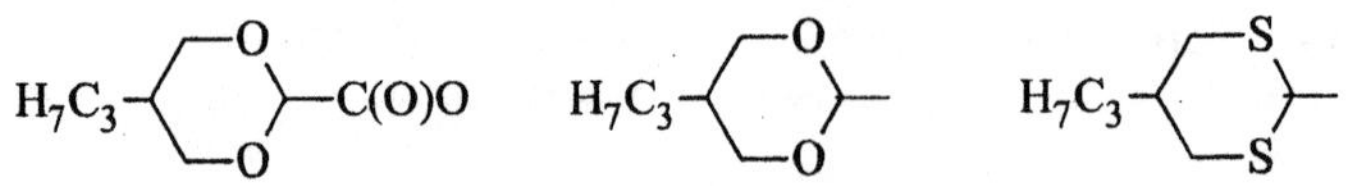

H_7C_3– (ring: O, O) –C(O)O H_7C_3– (ring: O, O) – H_7C_3– (ring: S, S) –

Some advantages of these systems were (i) most polar class of liquid crystals that are compatible with an active matrix display technology and (2i) have adequate dielectric anisotropy *(166)*. More recently, it was found that liquid crystals can be prepared in which the polar terminal $SF_5CF_2CF_2$-group is present *(167)*. Mesomorphism of parent homologue series is preserved upon derivatization with this group.

Some advantages found (i) entiotropic nematic liquid crystals are present and can be used in mixtures commonly used in liquid crystal displays, (2i) quantum calculations indicate high dipole moments for the para-derivatives.

There has been a rapid increase in the number of patents which list the SF_5 group and other groups in liquid crystals and in numerous applications *(168-175)*

Ionic Liquids

A number of SF_5-polyfluoralkyl iodides, $SF_5(CF_2)_n(CH_2)_mI$, were reacted with pyridine, n-methyl imidazole, n-propyl triazole and pyridazine in order to produce the corresponding iodide salts which when treated with $LiN(SO_2CF_3)_2$, at 25° C in water or in a water/acetone mixture, formed in general new salts that melted or showed a phase transition below 25° C *(123)*.

References

1. Cady, G.H. In *Advances in Inorganic Chemistry and Radiochemistry;* Emeléus, H. J.; Sharpe, A. G., Eds.; Academic Press, N.Y., N.Y.,USA, 1960; pp 105-157.
2. Roberts, H.L., *Quart. Revs. Chem. Soc.* **1961**, *15,* 30.
3. Verma, R.D.; Kirchmeier, R.L.; Shreeve, J.M. *Advances in Inorganic Chem.* **1994**, *41,* 125.
4. Winter, R.; Gard, G.L. In *ACS Symposium Series, 555. Inorganic Fluorine Chemistry: Toward the 21st Century;* Thrasher, J.S.; Strauss, S.H.; Eds.; ACS Press, Washington, D.C. 1994; pp 128-147.
5. Seppelt, K.; Lentz, D. *Chemistry of Hypervalent Compounds;* Wiley-VCh., 1999; p 295.
6. Case, J.R.; Nyman, F. *Nature* **1962**, *193*, 473.
7. Batt, L.; Cruickshank, F.R. *J.Phys.Chem.* **1966**, *70*, 723.
8. Klampfer, P.; Skapin, T.; Kralj, B.; Žigon, D.; Jesih, A. *Acta. Chim. Solv.* **2003,** *50,* 29.
9. Emeléus, H.J.; Tittle, B. *J. Chem. Soc.* **1963**, 1644.
10. Kopalidis, P.M.; Jacob, J. *J. Electrochem. Soc.* **1993**, *140*, 3037.
11. Katoh, M.; Izumi, Y.; Kimura, H.; Ohte, T.; Kojima, A.; Ohtani, S. *Appl. Surf. Sci.* **1995**, 226.
12. Denbigh, K.G.; Whytlaw-Gray, R. *Nature* **1933,** *133,* 763.
13. Roberts, H.L. *J. Chem. Soc.* **1962**, 3183.
14. Winter, R.; Nixon, P.G.; Gard, G.L. *J. Fluorine Chem.* **1998**, *87*, 85.

15. Roberts, H.L.; Ray, N.H. *J. Chem. Soc.* **1960**, 665.
16. George, J.; Cotton, F.A. *Proc. Chem. Soc.* **1959**, 317.
17. Nyman, F.; Roberts, H.L. *J. Chem. Soc.* **1962**, 3180.
18. Tullock, C.W.; Coffman, D.D.; Muetterties, E.L. *J. Am. Chem.* Soc. **1964**, *86*, 357.
19. Jonethal, U.; Kuschel, R.; Seppelt, K. *J. Fluorine Chem.* **1998**, *88*, 3.
20. Merrill, C. PhD Thesis, University of Washington, Seattle, WA,. 1962.
21. Cohen, B.; MacDiarmid, A. *Inorg. Chem.* **1965**, *4*, 1782.
22. Kovacina, T.A.; Berry, A.D.; Fox, W.B. *J. Fluorine Chem.* **1976**, *7*, 430.
23. Rahbarnoshi, R.; Sams, L.C. *Inorg. Chem.* **1983** *22*, 840.
24. Wessel, J.; Kleemann, G.; Seppelt, K. *Chem. Ber.* **1983**, *116*, 2399.
25. Christe, K.O.; Curtis, E.E.; Schack, C.J.; Roland, A. *Spectrochim. Acta* **1977**. *33A*,69.
26. Winter, R.; Terjeson, R.J.; Gard, G.L. *J. Fluorine Chem.* **1998** *89*, 105.
27. Tunder, R.; Siegel, B. *J. Inorg. Nucl. Chem.* **1963**, *26*, 1097.
28. Clark, M.; Kellen-Yuen, C.J.; Robinson, K.O.; Zang, H.; Yang, X.-Y.; Madappat, K.V.; Fuller, J.W.; Atwood, J.L.; Thrasher, J.S. *Eur. J. Solid State Inorg. Chem.* **1992**, *29*, 809.
29. Bittner, J.; Fuchs, J.; Seppelt, K. *Z. Anorg. Allg. Chem.* **1988**, *557*, 182.
30. Kirsch, P.; Röschenthaler, G.-V.; Sevenard, D.; Kolomeitsev, A., Offenbarunqsschrift De10321114A1, 2003 and De10220901A1, 2003.
31. Smardzewski, R.R.; Fox, W.B. *J. Fluorine Chem.* **1976**, *7*, 456.
32. Sidebottom, H.W.; Tedder, J.M.; Walton, J.C. *J. Chem. Soc.* **1969**, 2103.
33. Bekker, R.A.; Dyatkin, B.L.; Knunyants, I.L. *Izv. Akad. Nauk SSSR, Ser. Khim.* **1970**, *12*, 2738.
34. Terjeson, R.J.; Renn, J.; Winter, R.; Gard, G.L. *J. Fluorine Chem.* **1997**, *82*, 73.
35. Hutchinson, J. *J. Fluorine Chem.* **1973/74**, *3*, 429.
36. Nixon, P.G.; Mohtasham, J.; Winter, R.; Gard, G.L.; Twamley, B.; Shreeve, J.M. *J. Fluorine Chem.* **2004**, *125*, 553.
37. Behr, F.E.; Hansen, J.C.; Savu, P.M., 203rd ACS National Meeting, San Francisco, CA, 1992.
38. Gard, G.L.; Woolf, C. *J. Fluorine Chem.* **1971/72**, *1*, 487.
39. Gard, G.L.; Bach, J.L.; Woolf, C. Brit. Patent 1,167,112, 1969.
40. Wessolowski, H.; Röschenthaler, G.-V.; Winter, R.; Gard, G.L. *Z. Naturforsch* **1991**, *46b*, 123.
41. Wessolowski, H.; Röschenthaler, G.-V.; Winter, R.; Gard, G.L.; Pon, G.; Willet, R. *Eur. J. Solid State Inorg. Chem.* **1992**, *29*, 1173.
42. Wessel, J.; Hartl, H.; Seppelt, K. *Chem. Ber.* **1986**, *119*, 453.
43. Banks, R.E.; Barlow, M.G.; Haszeldine, R.N.; Morton, W.D. *J.C.S. Perkin I.* **1974**, 1266.
44. Winter, R.W.; Nixon, P.G.; Terjeson, R.J.; Mohtasham, J.; Holcomb, N.R.; Grainger, D.W.; Graham, D.; Castner, D.G.; Gard, G.L. *J. Fluorine Chem.* **2002**, *115*, 107.
45. Nixon, P.G.; Renn, J.; Terjeson, R.J.; Choi, Y.S.; Winter, R. *J. Fluorine Chem.* **1998**, *91*, 13.
46. In-house work at Portland State University.
47. Gard, G.L.; Winter, R.; Nixon, P.G.; Hu, Y-H.; Holcomb, N.R.; Grainger, D.W.; Castner, D.G. *Polymer Preprint* **1998**, *39*, 962.
48. Winter, R.; Nixon, P.G.; Gard, G.L.; Castner, D.G.; Holcomb, N.R.; Hu, Y.-H.; Grainger, D.W. *Chem. Mater.* **1999**, *11*, 3044.

49. Nixon, P.G.; Winter, R.; Castner, D.G.; Holcomb, N.R.; Grainger, D.W.; Gard, G.L. *Chem. Mater.* **2000,** *12,* 3108.
50. Gard, G.L.; Winter, R.W., 226th ACS National Meeting, NY, NY, 2003.
51. Winter, R.; Nixon, P.G.; Gard, G.L.; Radford, D.H.; Holcomb, N.R.; Grainger, D.W. *J. Fluorine Chem.* **2001**, *107,* 23.
52. Winter, R.; Nixon, P.G.; Terjeson, R.J.; Mohtasham, J.; Holcomb, N.R.; Grainger, D.W.; Graham, D.; Castner, D.G.; Gard, G.L. *J. Fluorine Chem.* **2002**, *115,* 107.
53. Oates, G.; Winfield, J.M. *J. Fluorine Chem.* **1974**, *4,* 235.
54. Canich, J.M.;.Lerchen, M.E; Gard, G.L.; Shreeve, J.M. *Inorg. Chem.* **1986**, *25,* 3030.
55. Wessolowski, H.; Gentzsch, A.; Röschenthaler, G.-V.; Gard, G.L. *Heteroatom Chem.* **1997**, *8,* 467.
56. Tremblay, M. *Can. J. Chem.* **1965**, *43,* 219.
57. Winter, R.; Gard, G.L. *J. Fluorine Chem.* **2000**, *102,* 79.
58. Winter, R.; Gard, G.L. *J. Fluorine Chem.* **1994**, *66,* 109.
59. Dolbier Jr., W.R.; Ait-Mohand, S. *Org. Letts.* **2002**, *4,* 2066.
60. Case, J.R.; Ray, N.H.; Roberts, H.L. *J. Chem. Soc.* **1961**, 2066.
61. Kleemann, G.; Seppelt, K. *Angew. Chem. Int. Ed. Engl.* **1978**, *17,* 516.
62. Winter, R.;.Gard, G.L., 15th International Symposium on Fluorine Chemistry, Vancouver, Canada, 1997.
63. Anton, D.R.; Krespan, C.G.; Sievert, A.C. U.S. Patent 6,525,231, 2003.
64. Wessolowski, H.; Röschenthaler, G.-V.; Winter, R.; Gard, G.L., *Phosphorus, Sulfur, and Silicon*, **1991**, *60,* 201.
65. Kinkead, S.A.; Shreeve, J.M. *Inorg. Chem.* **1984**, *23*, 4174.
66. Tyczkowski, E.A.; Bigelow, L.A. *J. Am. Chem. Soc.* **1953**, *75,* 3523.
67. Clifford, A.F.; El-Shamy, H.K.; Emeleus, H.J.; Haszeldine, R.N. *J. Chem. Soc.* **1953**, 2372.
68. Berry, A.D.; Fox, W.B. *J. Fluorine Chem.* **1976**, *7,* 449.
69. Noftle, R.E.; Fox, W.B. *J. Fluorine Chem.* **1977**, *9,* 219.
70. DeBuhr, R.; Howbert, J.; Canich, J.M.; White, H.F.; Gard, G.L. *J. Fluorine Chem.* **1982**, *20,* 515.
71. Hare, M. M.S. Thesis, Portland State University, Portland, OR, 1992.
72. Terjeson, R.J.; Gupta, K.D.; Sheets, R.M.; Gard, G.L.; Shreeve, J.M. *Revue de Chimie Minerale* **1986**, *23,* 1.
73. Gerhardt, R.; Grelbig, T.; Buschmann, J.; Luger, P.; Seppelt, K. *Angew. Chem. Int. Ed. Engl.* **1988**, *27,* 1534.
74. Gerhardt, R.; Seppelt, K. *Chem. Ber.* **1989**, *122,* 463.
75. Terjeson, R.J.; Mohtasham, J.; Gard, G.L. *Inorg. Chem.* **1988**, *27,* 2916.
76. St.Clair, A.K.; St.Clair, T.L.; Thrasher, J.S. U.S. Patent 5,220,070, 1993.
77. Sheppard, W.A. *J. Am. Chem. Soc.* **1962**, *84,* 3064.
78. Choudhry, M.R.; Harrold, J.W.; Nielsen, J.B.; Thrasher, J.S. *J. Chem. Physics* **1988**, *89,* 5353.
79. Nielsen, J.B.; Thrasher, J.S. *J. Fluorine Chem.* **1990**, *48,* 407.
80. Winter, R.; Gard, G.L. *J. Fluorine Chem.* **2002**, *102,* 79.
81. Winter, R.W.; Gard, G.L., 228th ACS National Meeting, Philadelphia, Pa, USA, 2004.
82. Ray, N.H. Brit. Patent 941,392, 1963.
83. Coffman, D.D.; Tullock, C.W. U.S. Patent 3,102,903, 1963.
84. Winter, R.; Willett, R.D.; Gard, G.L. *Inorg. Chem.* **1989**, *28,* 2499.

85. DeMarco, R.A.; Fox, W.B. *J. Org. Chem.* **1982**, *47,* 3772.
86. Kuschel, R.; Seppelt, K. *Inorg. Chem.* **1993**, *32,* 3568.
87. Winter, R.; Gard, G.L. *Inorg. Chem.* **1988**, *27,* 4329.
88. Winter, R.W.; Gard, G.L. *J. Fluorine Chem.* **1991**, *52,* 73.
89. Krügerke, T.; Buschmann, J.; Kleemann, G.; Luger, P.; Seppelt, K. *Angew. Chem. Int. Ed. Engl.* **1987**, *26,* 799.
90. Haszeldine, R.N.; Nyman, F. *J. Chem. Soc.* **1956**, 2684.
91. Bekker, R.A.;.Dyatkin, B.L; Knunyants, I.L. *Akad. Nauk SSSR, Bull. Div. of Chem. Sci.*, **1970**, 2575.
92. Ray, N.H. Brit. Patent 941,393, 1963.
93. Yokochi, A.F.T.; Winter, R.; Gard, G.L., *Acta Christalographica*, Sec. E, **2002,** *58,* 01133.
94. Hansen, J.C.; Savu, P.M. U.S. Patent 5,286,352, 1994.
95. Hansen, J.C.; Savu, P.M. U.S. Patent 5,159,105, 1992.
96. Willenbring, R.J.; Mohtasham, J.; Winter, R.; Gard, G.L. *Can. J. Chem.* **1989,** *67,* 2037.
97. Canich, J.M.; Ludvig, M.M.; Gard, G.L.; Shreeve, J.M. *Inorg. Chem.* **1984**, *26,* 4403.
98. Gard, G.L.; Waterfeld, A.; Mews, R.; Mohtasham, J. *Inorg. Chem.* **1990**, *29,* 4588.
99. Hodges, A.M.; Winter, R.W.; Winner, S.W.; Preston, D.A.; Gard, G.L. *J. Fluorine Chem.* **2002**, *114,* 3.
100. Hamel, N.N.; Nixon, P.G.; Gard, G.L.; Nafshun, R.L.; Lerner, M.M. *J. Fluorine Chem.* **1996**, *79,* 81.
101. Nafshun, R.L.; Lerner, M.M.; Nixon, P.G.; Hamel, N.H.; Gard, G.L. *J. Electrochem. Soc*, **1995**, *142,* 153.
102. Winter, R.W.; Dodean, R.A.; Holmes, L.; Gard, G.L. *J. Fluorine Chem.* **2004**, *125,* 37.
103. Kemmitt, R.D.W.; Peacock, R.D.; Stocks, J. *Chem. Comm.* **1969**, 554.
104. Berry, A.D.; DeMarco, R.A. *Inorg. Chem.* **1982**, *21,* 458.
105. Henkel, T.; Klauck, A.; Seppelt, K. *J. Organometallic Chem.* **1995**, *501,* 1.
106. Klauck, A.; Seppelt, K. *Angew. Chem. Int. Ed. Engl.* **1994**, *33,* 93.
107. Winter, R.; Mews, R.; Noltemeyer, M.; Gard, G.L. *J. Fluorine Chem.* **1993**, *60,* 189.
108. Terjeson, R.J.; Gard, G.L. *J. Fluorine Chem.* **1987**, *35,* 653.
109. Terjeson, R.J.; Willenbring, R.; Gard, G.L. *J. Fluorine Chem.* **1996**, *76,* 63.
110. Mohtasham, J.; Gard, G.L. *Coordination Chemistry Reviews* **1992**, *112,* 47.
111. Gard, G.L.; Winter, R.; Nixon, P.G.; Hu, Y-H.; Holcomb, N.R.; Grainger, D.W.; Castner, D.G. *Polymer Preprints* **1998**, *39,* 962.
112. Roberts, H.L. U.S. Patent 3,063,972, 1962.
113. Roberts, H.L. U.S. Patent 3,063,922, 1962.
114. Dale, J.W. U.S. Patent 3,126,366, 1964.
115. Sherratt, S. Brit. Patent 929,990, 1963.
116. Banks, R.E.; Haszeldine, R.N. Brit. Patent 1,145,263, 1969.
117. Terjeson, R.J.; Gard, G.L. *J. Fluorine Chem.* **1987**, *35,* 653.
118. Terjeson, R.J.; Mohtasham, J.; Sheets, R.M.; Gard, G.L. *J. Fluorine Chem.* **1988**, *38,* 3.
119. Kawa, H.; Partovi, S.N.; Ziegler, B.J.; Lagow, R.J. *J. Polymer Science:* Part C: Polymer Letters **1990**, *28,* 297.
120. Winter, R.W.; Winner, S.W.; Preston, D.A.; Mohtasham, J.A.; Smith, J.A.; Gard, G.L. *J. Fluorine Chem.* **2002**, *115,* 101.

121. Lork, A.; Gard, G.; Hare, M.; Mews, R.; Stohrer, W-D.; Winter, R. *J. Chem. Soc. Chem. Commun.*, **1992**, 898.
122. Sitzmann, M.E. *J. Fluorine Chem.* **1995**, *70,* 31.
123. Singh, R.P.; Winter, R.W.; Gard, G.L.; Gao, Y.; Shreeve, J.M. *Inorg. Chem.* **2003**, *42,* 6142.
124. Bowden, R.D.; Comina, P.J.; Greenhall, M. P.; Kariuki, B.M.; Loveday, A.; Philip, D. Tetrahed. **2000**, *56,* 3399.
125. Sipyagin, A.M.; Bateman, C.P.; Tan, Y.-T.; Thrasher, J.S. *J. Fluorine Chem.* **2001**, *112,* 287.
126. Hoover, F.W.; Coffman, D.D. *J. Org. Chem.* **1964**, *29,* 3567.
127. Winter, R.; Gard, G.L. 16th Winter Fluorine Conference, St. Pete Beach, Fl, USA, 2003 and 226th ACS Meeting, New York, N.Y. USA, 2003. Patent pending. Published in *J. Fluorine Chem.* **2004**, *125,* 549.
128. Dolbier Jr., W.R.; Ait-Mohand, S.; Sergeeva, T.; Schertz, T., 226th ACS National Meeting, NY, NY, 2003.
129. Case, J.R.; Price, R.; Ray, N.H.; Roberts, H.L.; Wright, J., *J. Chem. Soc.*, **1962**, 2107.
130. Hodges, A.M.; Winter, R.W.; Mohtasham, J.; Bailey, P.; Gard, G.L. *J. Fluorine Chem.* **2001**, *110,* 1.
131. Winter, R.W.; Gard, G.L. *J. Fluorine Chem.* **2002**, *118,* 157.
132. Winter, R.W.; Dodean, R.; Smith, J.A.; Anilkumar, R.; Burton, D.J.; Gard, G.L. *J. Fluorine Chem.* **2003**, *122,* 251.
133. Gard, G.L.; Shaw, R.M.; Woolf, C. U.S. Patent 3,506,774, 1970 and Brit. Patent1,167,228, 1969.
134. Cove, L.S.; Seffl, R.J. *Proceed. of 6th Elec. Insul. Conference, 238.* 1965.
135. Gard, G.L.; Shaw, R.M.; Woolf, C. U.S. Patent 3,519,725, 1970.
136. Data sheet published by ICI, 1964.
137. Matsuo, H. Jpn. Patent 158,145, 2002.
138. Geiser, U.; Schlueter, J.A.; Kini, A.M.; Wang, H.H.; Ward, B.H.; Whited, M.A.; Mohtasham, J.; Gard, G.L. *Synthet. Met.* **2003,** *133-134,* 401.
139. McCarthy, J. *Utility of Fluorine in Biologically Active Molecules*, presented at the 219th National ACS meeting (Division of Fluorine Chemistry), 2000.
140. Hansch, C.; Muir, R.M.; Fujita, T.; Maloney, P.P.; Geiger, F.; Streich, M. *J. Am. Chem. Soc.* **1963**, *85,* 2817.
141. Gilbert, E.E.; Gard, G.L. U.S. Patent 3,475,453, 1969.
142. Tullock, C.W. U.S. Patent 3,228,981, 1966.
143. Ray, N.H.; Roberts, H.L. U.S. Patent 3,086,048, 1963.
144. Crowley, P.J.; Bartholmew, D. Brit. Patent 2,276,380, 1994.
145. Mita, T.; Yoshihiro, M.; Mizukoshi, T.; Hotta, H.; Maeda, K.; Takii, S. WO Patent 2004018410, 2004.
146. Maier, T. Eur. Patent 1,238,586, 2002.
147. Helmke, H.; Hoffmann, M.; Michael, G.; Haaf, K.; Willms, L.; Auler, T.; Bieringer, H.; Menne, H. WO Patent 2003051846, 2003.
148. Schätzer, J.; Wenger, J.; Hall, R.G.; Nebel, K.; Hole, S.; Stoller, A. WO Patent 2003082012, 2003.
149. Nebel, K.; Schätzer, J.; Stoller, A.; Gittall, R.; Wenger, J.; Bondy, S.S.; Comer, D.D.; Penzolta, J.E.; Grootenhuis, P.D. WO Patent 2004002947, 2004.

150. Schätzer, J.; Nebel, K.; Stoller, A.; Hall, R.G.; Wenger, J.; Bondy, S.S.; Comer, D.D.; Renzotti, J.E.; Grootenhuis, P.D. WO Patent 2004002981, 2004.
151. Worhington, P.A.; Streeting, I.T. GB Patent 2276381, 1994.
152. Barton, J.E.D.; Mitchell, G. GB Patent 2276379, 1994.
153. Springer, C.J.; Davis, C.L. WO Patent 2000058271, 2000.
154. Silvey, G.A.; Cady, G.H. *J. Am. Chem. Soc.* **1950**, *72,* 3624.
155. Mao, T.J.; Dresdner, R.D.; Young, J.A. *J. Am. Chem. Soc.* **1959**, *81,* 1020.
156. Woolf, C.W.; Gard, G.L. U.S. Patent 3,448,121, 1969.
157. Sheppard, W.A. U.S. Patent 3,219,690, 1965.
158. Coffman, D.D.; Tullock, C.W. Can. Patent 728,186, 1966.
159. Michimasa, Y. Chem. Abstrs., **1975**, *82,* 175255g.
160. Witucki, E.F.; Frankel, M.B. *Energetic Aliphatic Sulfur Pentafluoride Derivatives,* UCRL-13809, 1978.
161. Sitzmann, M.E.; Gilligan, W.H.; Ornellas, D.L.; Thrasher, J.S. *J. Energ. Mat.* **1990**, *8,* 352.
162. Sitzmann, M.E. U.S. Patent 5,194,103, 1993.
163. Sitzmann, M.E.; Gilardi, R.; Butcher, R.G.; Koppes, W.M.; Stern, A.G.; Thrasher, J.S.; Trivedi, N.J.; Yang, Z-Y. *Inorg. Chem.* **2000**, *39,* 843.
164. Oomure, Y.; Noguchi, M.; Fujiwara, K. Jpn. Patent 89-231887, 1991
165. Minor, B.H.; Sheally, G.S., U.S. Patent 5,433,880, 1996.
166. Kirsch, P.; Bremer, M.; Heckmeier, M.; Tarumi, K. *Angew. Chem. Int. Ed. Eng.* **1999**, *38,* 1989.
167. Smith, J.; DiStasio, R.A.; Hannah, N.; Rananavare, S.B.; Weakley, T.; Winter, R.W.; Gard, G.L. *J. Physical Chem.* submitted, 2004.
168. Miyazawa, K.; Saito, S.; Harufuji, T. Jpn. Patent 2003238492, 2003.
169. Heckmeier, M.; Schön, S.; Kirsch, P. Dtsch. Bundes Patent. 10,344,474, 2004.
170. Ionescu, D.; Scott, C. Brit. Patent 2,393,966, 2004.
171. Koike, T.; Ikemura, M. Jpn. Patent 2003261478, 2003.
172. Ranger, R.; Kazhushkov, S.; de Mejiere, A.; Demus, D. Jpn. Patent 2003261470, 2003.
173. Manabe, A.; Heywang, U.; Pauluth, D.; Heckmeier, M.; Kirsch, P.; Montenegro, E., Dtsch. Bundes Patent 10247986, 2004.
174. Francis, M.; Goulding, M.; Patrick, J. Brit. Patent 2,394,475, 2004.
175. Kirsch, P.; Lenges, M.; Heckmeier, M.; Lüssem, G. Eur. Patent 1,411,104, 2004.

Fluorine-Containing Synthons with Unsaturated C–C Bonds: Alkenes, Alkynes, and Allenes

Chapter 5

Stereoselective Synthesis of Monofluoroalkenes Using Cross-Coupling Reactions

Shoji Hara

Division of Molecular Chemistry, Graduate School of Engineering, Hokkaido University, Sapporo 060–8628, Japan

Various methods for the stereoselective synthesis of fluorohaloalkenes and fluoroalkenyl metals are developed. The resulting fluorohaloalkenes and fluoroalkenyl metals are used for the cross-coupling reaction to synthesize monofluoroalkenes stereoselectively. The cross-coupling reactions using fluorohaloalkenes and fluoroalkenyl metals are a powerful tool for the stereoselective synthesis of polyfunctionalized fluoroalkenes and alkadienes.

Introduction

Introduction of a fluorine atom onto the double bond of natural compounds having interesting bioactivities is of great interest because the fluorine atom can enhance their bioactivities or reduce their undesired side-effects (*1-3*). The Horner-Wadsworth-Emmons reaction has been generally used for the stereoselective synthesis of fluoroalkenes (*4*). The reaction of a fluorophosphonate with aldehydes provides (*E*)-α-fluoro-α,β-unsaturated esters stereoselectively. However, in order to introduce the fluorine atom onto any position of the double bond in natural products, more methods for the selective synthesis of fluoroalkenes are desired.

On the other hand, cross-coupling reactions using transition metal catalysts have been recently used as a powerful tool for the stereoselective synthesis of polyfunctionalized alkenes and alkadienes (*5,6*). For the synthesis of various monofluoroalkenes (**1a-d**) by the cross-coupling reactions, stereoselective synthesis of fluorohaloalkenes (**2**) or (**4**), or fluoroalkenyl metals (**3**) or (**5**) is required (Scheme 1).

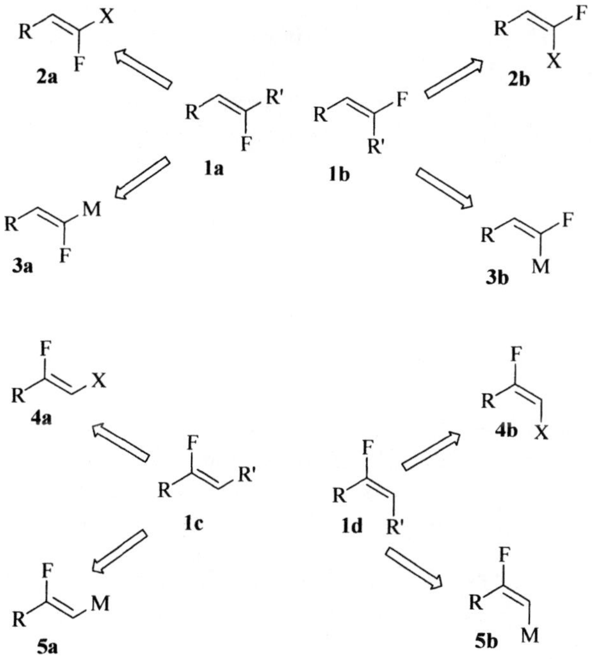

Scheme 1 R = aryl, alkyl, alkenyl
R' = aryl, alkyl, alkenyl, ester

Stereoselective Synthesis of 1-Fluoro-1-haloalkenes (2) and their Application to the Cross-coupling Reactions.

Hiyama *et al.* reported that (*E*)-1-bromo-1-fluoroalkenes **2a** (R = Ar, X = Br) can be stereoselectively prepared from aldehyde-dibromofluoromethyllithium adducts (**6**) by reductive elimination reaction (*7*) (Scheme 2).

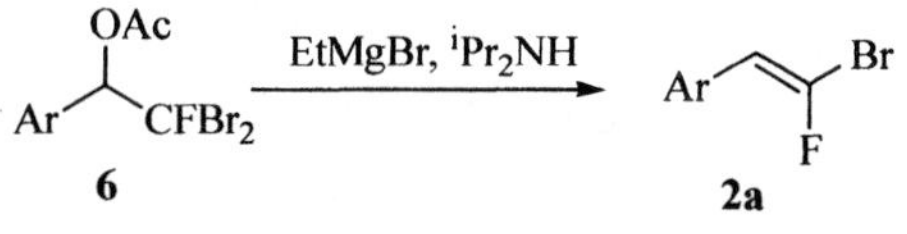

Scheme 2

Rolando *et al.* modified this methodology and used it for the synthesis of fluorinated analogs (**9**) and (**10**) of resveratrol and pterostilbene (*16*). A β-bromo-β-fluorostyrene derivative (**11**) was obtained as a mixture of stereoisomers from the 3,5-dihydroxybenzaldehyde derivative by a modified Burton method. Suzuki-Miyaura coupling reaction of **11** with phenylboronic acid derivative (**12**) gave the (*Z*)-isomer of fluororesveratrol derivative (**13**) selectively. Deprotection of **13** gave the desired fluorinated analog **9** of resveratrol. They also synthesized **10** using phenylboronic acid derivative (**14**) and bromofluorostyrene derivative (**15**) (Scheme 6).

Generally, 1-bromo-1-fluoro-1-alkenes (**2,** X = Br) have been prepared by the reaction of aldehydes with $CFBr_3$ and Ph_3P (*4,8,9*). However, a mixture of (*E*)- and (*Z*)-isomers was formed and their separation is practically difficult (Scheme 3). Several methods have been reported for the isomerization of the (*Z*)-isomer to the more stable (*E*)-isomer (*10-12*).

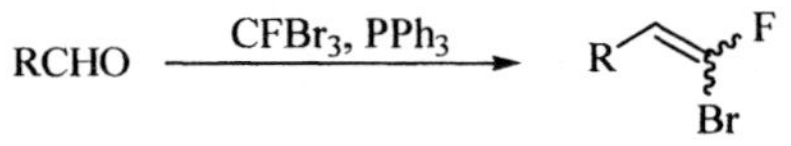

RCHO —($CFBr_3$, PPh_3)→ R–CH=C(F)(Br)

Scheme 3

McCarthy *et al.* separated both isomers of β-bromo-β-fluorostyrene (**7**) by gas chromatography and used them for the Suzuki-Miyaura coupling reaction to give (*Z*)- and (*E*)-fluorostilbenes (**8**) stereospecifically (*10,13*) (Scheme 4).

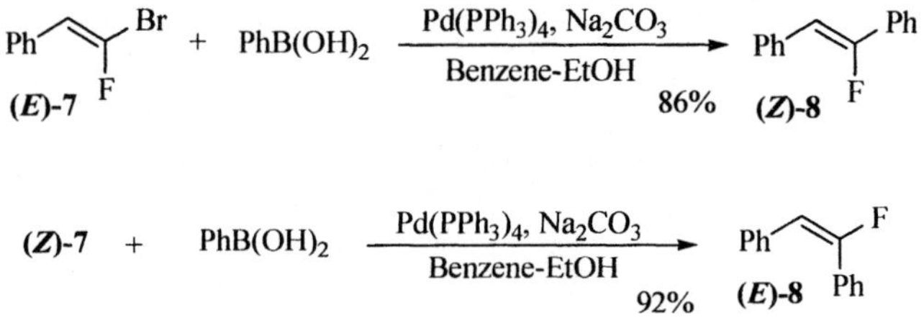

(*E*)-**7** + $PhB(OH)_2$ —($Pd(PPh_3)_4$, Na_2CO_3 / Benzene-EtOH)→ (*Z*)-**8** 86%

(*Z*)-**7** + $PhB(OH)_2$ —($Pd(PPh_3)_4$, Na_2CO_3 / Benzene-EtOH)→ (*E*)-**8** 92%

Scheme 4

In order to synthesize cross-coupling products derived from the (*E*)-isomer (**2a**, X = Br), the mixture can be used without separation, because the (*E*)-isomer reacts more rapidly than the (*Z*)-isomer (*14,15*). For instance, (*Z*)-fluorostilbene **8** could be stereoselectively synthesized from a mixture of (*E*)- and (*Z*)-**7** by the Stille coupling reaction (*12*) (Scheme 5).

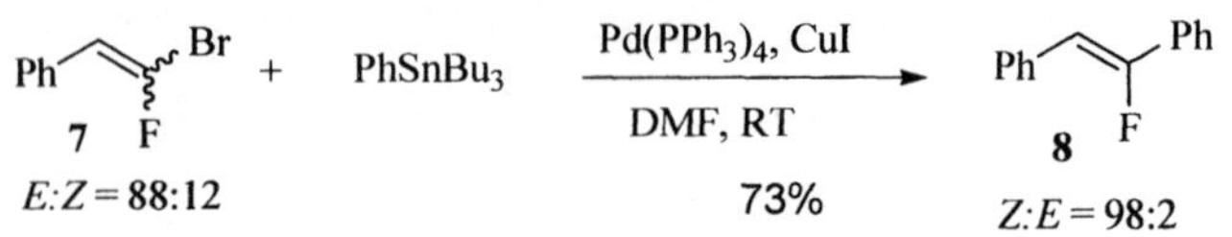

7 (*E*:*Z* = 88:12) + $PhSnBu_3$ —($Pd(PPh_3)_4$, CuI / DMF, RT)→ **8** (*Z*:*E* = 98:2) 73%

Scheme 5

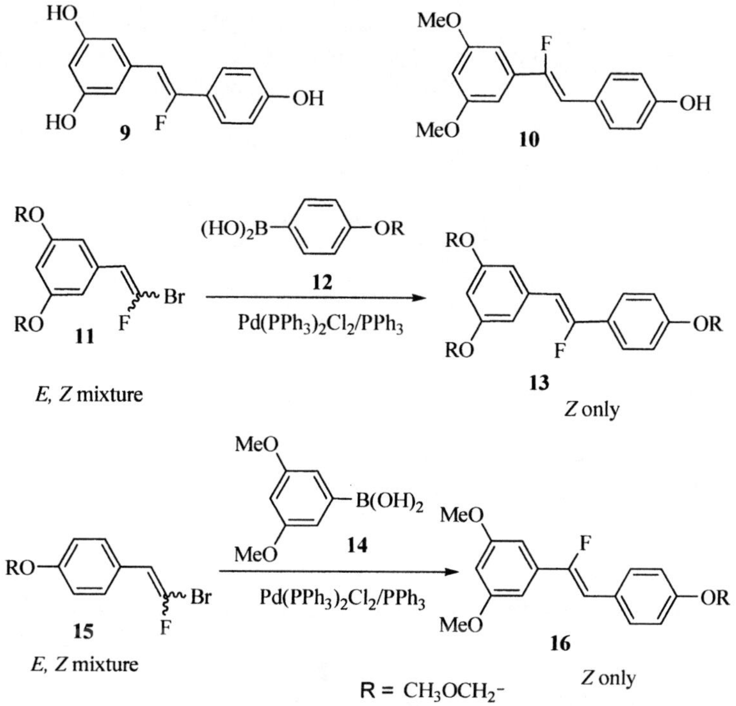

9 **10**

$(HO)_2B$–C₆H₄–OR **12**

RO **11** F Br → $Pd(PPh_3)_2Cl_2/PPh_3$ → **13**

E, Z mixture *Z* only

14 $B(OH)_2$

15 → $Pd(PPh_3)_2Cl_2/PPh_3$ → **16**

E, Z mixture *Z* only

R = CH_3OCH_2-

Scheme 6

This methodology is also applicable for other cross-coupling reactions such as the Heck reaction, Sonogashira reaction, and alkoxycarbonylation reaction. From an (*E*)- and (*Z*)-mixture of β-bromo-β-fluorostyrene **7**, (2*E*, 4*Z*)-γ-fluoro-α,β,γ,δ-unsaturated ester (**17**), (*Z*)-fluoroenyne (**18**), and (*Z*)-α-fluoro-α,β-unsaturated ester (**19**) were stereoselectively prepared (*17,15,18*) (Scheme 7).

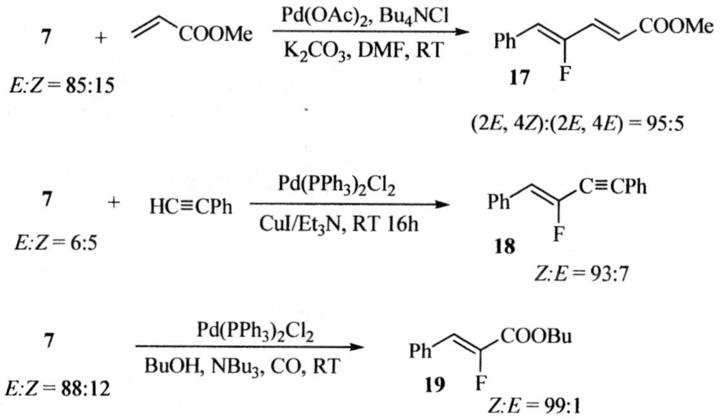

7 + COOMe → $Pd(OAc)_2$, Bu_4NCl; K_2CO_3, DMF, RT → **17**

E:*Z* = 85:15 (2*E*, 4*Z*):(2*E*, 4*E*) = 95:5

7 + HC≡CPh → $Pd(PPh_3)_2Cl_2$; CuI/Et_3N, RT 16h → **18**

E:*Z* = 6:5 *Z*:*E* = 93:7

7 → $Pd(PPh_3)_2Cl_2$; BuOH, NBu_3, CO, RT → **19**

E:*Z* = 88:12 *Z*:*E* = 99:1

Scheme 7

On the other hand, pure (*Z*)-1-fluoro-1-bromoalkenes (**2b,** X = Br) can be prepared from (*Z*)-α-fluoro-α,β-unsaturated carboxylic acids (**20**) by bromination and decarboxylative hydrogen bromide elimination sequences (*11*) (Scheme 8).

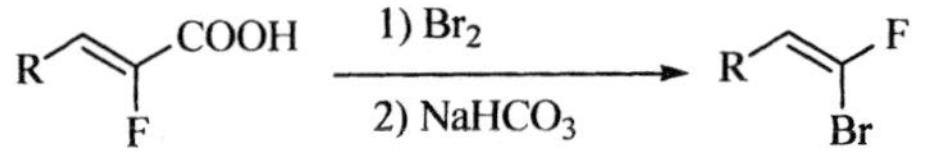

R–CH=C(F)–COOH (**20**) → 1) Br_2; 2) $NaHCO_3$ → R–CH=C(F)(Br)

Scheme 8

Rolando *et al.* used (*Z*)-**7** obtained by this method for the Sonogashira reaction, Heck reaction, and Stille reaction, and succeeded in the stereoselective synthesis of (*E*)-fluoroenyne (**21**), (2*E,* 4*E*)-γ-fluoro-α,β,γ,δ-unsaturated ester (**22**), (*E*)-fluorodiene (**23**), respectively (*11,19*) (Scheme 9).

However, this method is effective only when R in Scheme 8 is an aromatic group having no electron-donating substituent. When R is an aliphatic group or an aromatic group having the electron-donating group, good stereoselectivity cannot be expected. Burton *et al.* reported that pure **2b** (X = Br) is obtainable by a kinetic separation method. When a mixture of **7** was used for the cross-coupling reactions, (*E*)-**7** reacted selectively and (*Z*)-**7** remained unchanged (*14*). Therefore, after the coupling reaction, (*Z*)-**7** can be easily separated from the mixture, and (*E*)-fluoroenyne (**21**) was stereoselectively prepared using (*Z*)-**7** obtained by the kinetic separation method (*15*). More conveniently, (*E*)-**7** in the mixture was selectively reduced to the fluoroalkene, and the remained (*Z*)-**7** was used for the alkoxycarbonylation reaction without separation to give (*E*)-α-fluoro-α,β-unsaturated esters (**24**) stereoselectively (*18*) (Scheme 10)

Synthesis of 1-Fluoro-1-alkenyl Metals (3) and their Application to the Cross-coupling Reactions

(1-Fluoro-1-alkenyl)tributylstannane (**26**) was non-stereoselectively synthesized from aldehydes in two-steps *via* (fluoroalkenyl)sulfone (**25**). It was used for the further reactions after separation by column chromatography (*20*) (Scheme 11).

(*E*)-(1-Fluoro-2-phenylvinyl)tributylstannane (**27**) was used for the cross-coupling reaction with 5-iodo-1,3-dimethyluracil (**28**) and benzoyl chloride to give (*Z*)-5-(1-fluoro-2-phenylvinyl)-1,3-dimethyluracil (**29**) and fluorochalcone (**30**) respectively (*21,22*) (Scheme 12).

A polyfunctionalized fluorochromone derivative (**33**) was synthesized from formylchromone derivative (**31**) by this methodology. (Fluorovinyl)stannane (**32**) was synthesized from **31** via a fluorosulfone, and (*E*)-**32** was used for the cross-coupling reaction with β-iodoacrylate to give **33** stereoselectively (*22*) (Scheme 13)

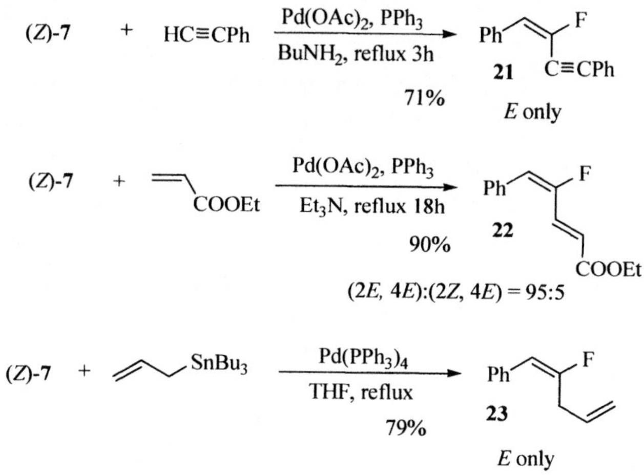

(Z)-7 + HC≡CPh $\xrightarrow[\text{BuNH}_2\text{, reflux 3h}]{\text{Pd(OAc)}_2\text{, PPh}_3}$ 21, 71%, E only

(Z)-7 + CH_2=CHCOOEt $\xrightarrow[\text{Et}_3\text{N, reflux 18h}]{\text{Pd(OAc)}_2\text{, PPh}_3}$ 22, 90%

(2E, 4E):(2Z, 4E) = 95:5

(Z)-7 + allyl-$SnBu_3$ $\xrightarrow[\text{THF, reflux}]{\text{Pd(PPh}_3)_4}$ 23, 79%, E only

Scheme 9

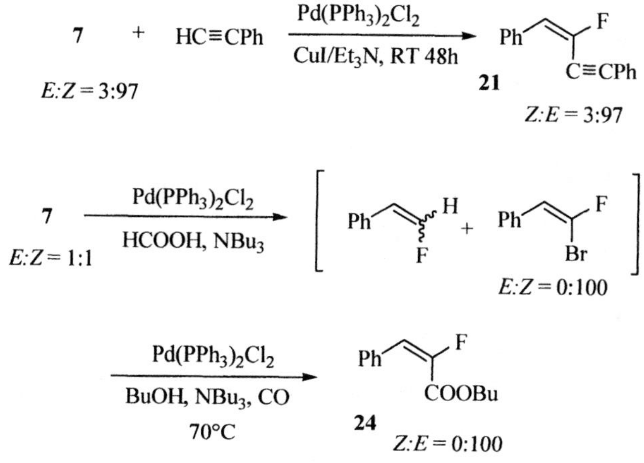

7 (E:Z = 3:97) + HC≡CPh $\xrightarrow[\text{CuI/Et}_3\text{N, RT 48h}]{\text{Pd(PPh}_3)_2\text{Cl}_2}$ 21 (Z:E = 3:97)

7 (E:Z = 1:1) $\xrightarrow[\text{HCOOH, NBu}_3]{\text{Pd(PPh}_3)_2\text{Cl}_2}$ [PhCH=CHF + PhCH=CFBr (E:Z = 0:100)]

$\xrightarrow[\text{BuOH, NBu}_3\text{, CO, 70°C}]{\text{Pd(PPh}_3)_2\text{Cl}_2}$ 24 (COOBu), Z:E = 0:100

Scheme 10

PhCH_2CH_2CHO $\xrightarrow{(\text{EtO})_2\text{P(O)CFSO}_2\text{Ph}^{\ominus}}$ 25 (SO_2Ph, F), 100% (E:Z = 66:33)

25 $\xrightarrow[\text{AIBN}]{\text{HSnBu}_3}$ (E)-26 ($SnBu_3$, F), 58% + (Z)-26 (F, $SnBu_3$), 18%

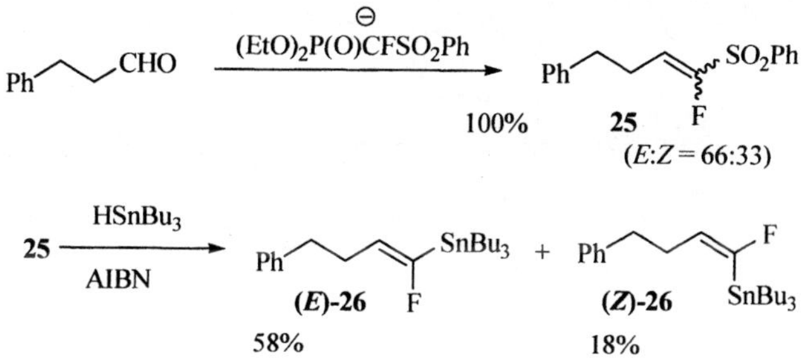

Scheme 11

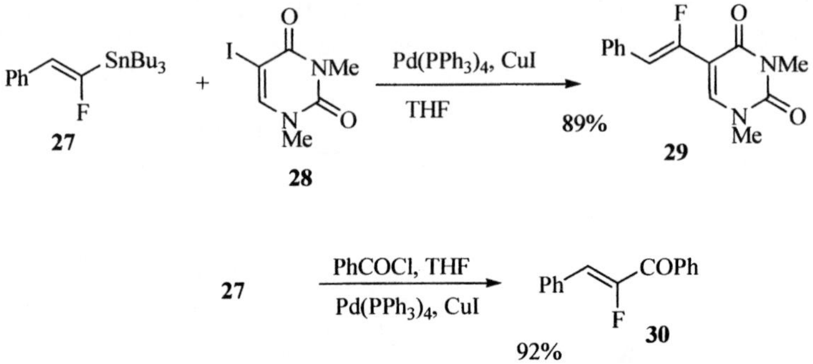

Scheme12

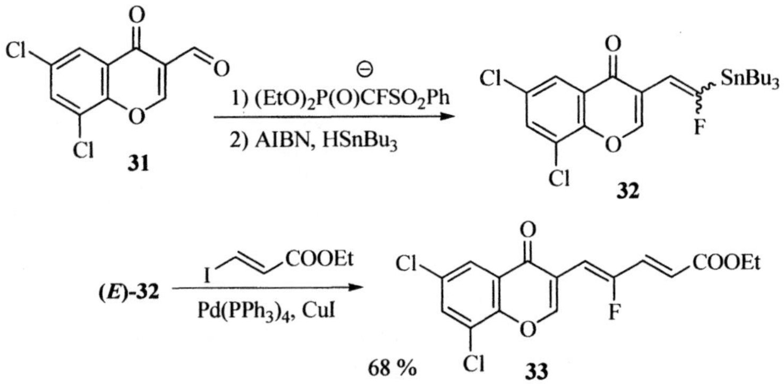

Scheme13

No other successful methods for the synthesis of 1-fluoro-1-alkenyl metals (**3**) have been reported. The reaction of β-bromo-β-fluorostyrene **7** with bis(tributyltin) or bis(pinacolato)diboron for the synthesis of **27** or (*E*)-1-fluoro-1-alkenylborane (**34**) resulted unsuccessfully (*23,24*). A homo coupling product (**35**) or (**36**) was obtained instead of the expected **27** or **34**. The cross-coupling reaction between the generated **27** or **34** with **7** must be fast and it is difficult to terminate the reaction at the formation of **27** or **34** (Scheme 14).

Stereoselective Synthesis of 2-Fluoro-1-haloalkenes (4) and their Derivatives, and their Application to Cross-coupling Reactions

(*E*)-2-Fluoro-1-bromoalkenes (**4b**, X = Br) were stereoselectively prepared by the addition of *in-situ*-generated BrF from *N*-bromoacetamide and HF to 1-alkynes (*25*). Though good stereoselectivity (*E* = 95%) was attained, a drawback was the use of hazardous anhydrous HF as the F source (Scheme 15).

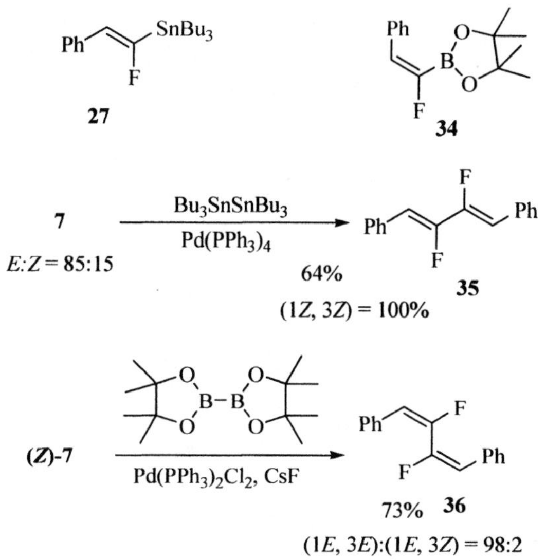

Ph SnBu$_3$ F

27

Ph O B O F

34

7 $\xrightarrow[\text{Pd(PPh}_3)_4]{\text{Bu}_3\text{SnSnBu}_3}$ Ph F F Ph **35**

E:*Z* = 85:15

64%

(1*Z*, 3*Z*) = 100%

(*Z*)-7 $\xrightarrow[\text{Pd(PPh}_3)_2\text{Cl}_2\text{, CsF}]{}$ Ph F F Ph **36**

73%

(1*E*, 3*E*):(1*E*, 3*Z*) = 98:2

Scheme 14

Bu-C≡CH $\xrightarrow[\text{HF}]{\text{BrHNCOCH}_3}$ Bu F Br

47.9%

E:*Z* = 95:5

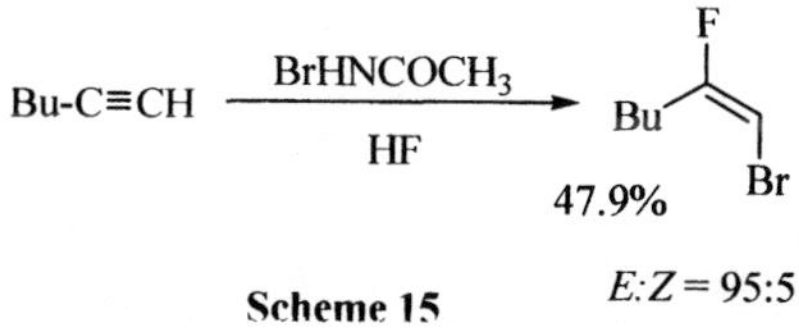

Scheme 15

Rolando *et al.* also synthesized the (*E*)-2-fluoro-1-bromoalkene (**37**) from 1-alkyne using 1,3-dibromo-5,5-dimethylhydantoin as a Br source and pyridine-HF as the F source which is easier to handle than anhydrous HF (*26*). They also synthesized (*E*)-2-fluoro-1-iodoalkene (**38**) with high stereoselectivity (*E* = 98%) from 1-alkyne using bis(pyridine)iodonium tetrafluoroborate and pyridine-HF (*19*) (Scheme 16).

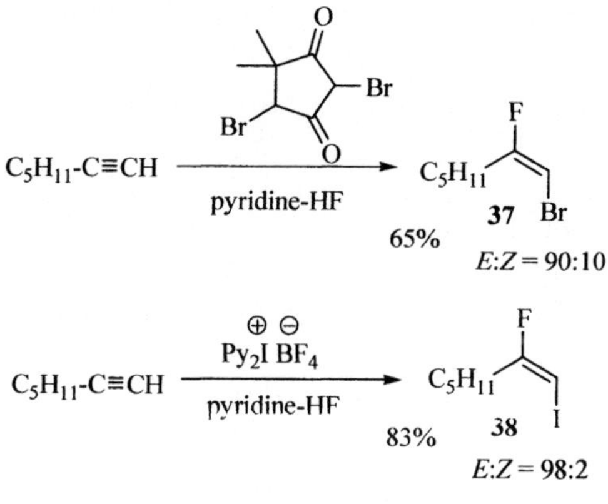

C_5H_{11}-C≡CH $\xrightarrow[\text{pyridine-HF}]{}$ C_5H_{11} F Br **37**

65%

E:*Z* = 90:10

C_5H_{11}-C≡CH $\xrightarrow[\text{pyridine-HF}]{\text{Py}_2\text{I}^{\oplus}\ \text{BF}_4^{\ominus}}$ C_5H_{11} F I **38**

83%

E:*Z* = 98:2

Scheme 16

They used **38** for the Sonogashira reaction and Heck reaction to give (*E*)-2-fluoroenyne (**39**) and (*E,E*)-fluoroalkadiene (**40**) stereoselectively (*19*) (Scheme 17).

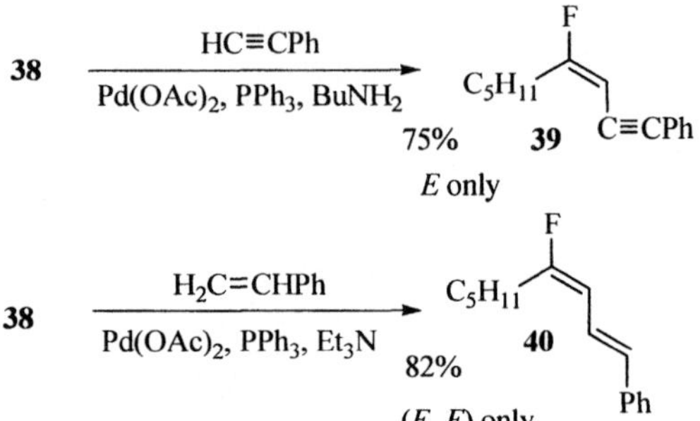

38 —[HC≡CPh; Pd(OAc)$_2$, PPh$_3$, BuNH$_2$]→ 39 (C_5H_{11}, F, C≡CPh) 75%, *E* only

38 —[H$_2$C=CHPh; Pd(OAc)$_2$, PPh$_3$, Et$_3$N]→ 40 (C_5H_{11}, F, Ph) 82%, (*E, E*) only

Scheme 17

(*E*)-2-Fluoro-1-alkenyliodonium salts (**41**) could be stereoselectively synthesized by the addition of iodotoluene difluoride (**42**) (*27,28*) to 1-alkynes in the presence of Et$_3$N-5HF (*29*) (Scheme 18).

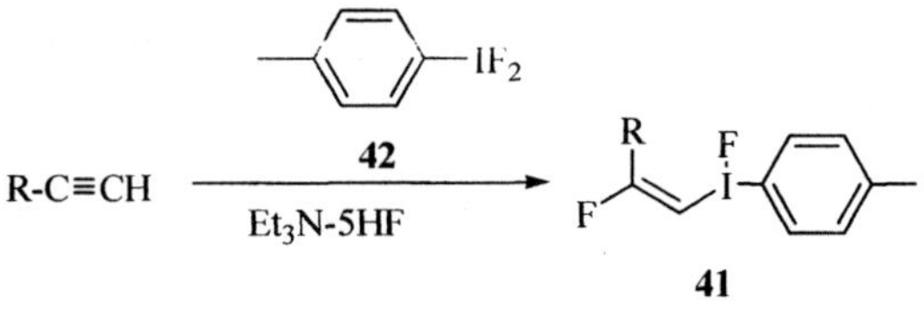

R-C≡CH —[42 (p-tolyl-IF$_2$); Et$_3$N-5HF]→ 41

Scheme 18

As the fluoroalkenyliodonium salts **41** are reactive hypervalent compounds, the cross-coupling reaction using **41** proceeds under mild conditions. Methoxycarbonylation reaction and Heck reaction of **41** proceeded at room temperature and the corresponding (*E*)-β-fluoro-α,β-unsaturated ester (**43**) and (*E, E*)-δ-fluoro-α,β,γ,δ-unsaturated carbonyl compounds (**44**) could be stereoselectively obtained, respectively (*30-32*) (Scheme 19).

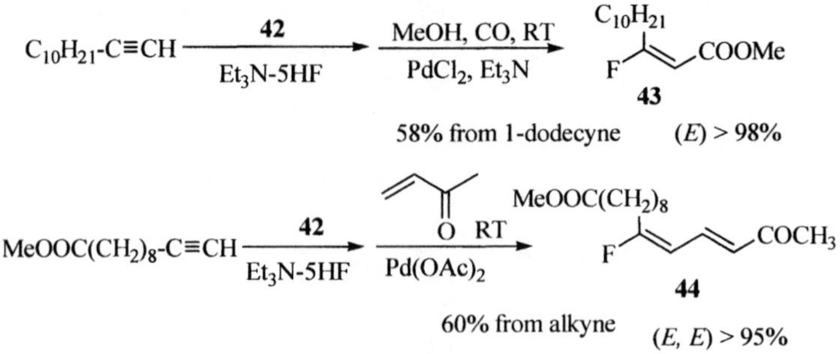

$C_{10}H_{21}$-C≡CH —[42; Et$_3$N-5HF]→ —[MeOH, CO, RT; PdCl$_2$, Et$_3$N]→ 43 ($C_{10}H_{21}$, F, COOMe)

58% from 1-dodecyne (*E*) > 98%

MeOOC(CH$_2$)$_8$-C≡CH —[42; Et$_3$N-5HF]→ —[methyl vinyl ketone, RT; Pd(OAc)$_2$]→ 44 (MeOOC(CH$_2$)$_8$, F, COCH$_3$)

60% from alkyne (*E, E*) > 95%

Scheme 19

This methodology was applied for the synthesis of a fluorinated analog (**45**) of a polyunsaturated fatty acid metabolite, (9*Z*, 11*E*)-13-hydroxy-9,11-octadecadienoic acid (coriolic acid) (*33*). Methyl 9-decynoate was converted to the corresponding fluoroalkenyliodonium salt which was used for the Heck reaction with 1-octen-3-one to give methyl (9*E*, 11*E*)-9-fluoro-13-oxooctadecadienoate (**46**). By the reduction of the keto group, the desired **45** could be obtained in racemic form (*31,32*) (Scheme 20).

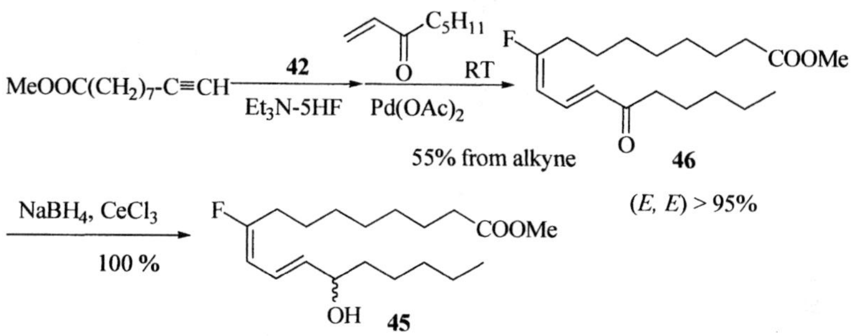

$MeOOC(CH_2)_7$-C≡CH → (**42**, Et_3N-5HF) → (1-octen-3-one (C_5H_{11} vinyl ketone), RT, $Pd(OAc)_2$) → **46** (F, COOMe; 55% from alkyne; (*E*, *E*) > 95%)

→ ($NaBH_4$, $CeCl_3$; 100 %) → **45** (F, COOMe, OH)

Scheme 20

When (*E*)-2-fluoro-1-alkenyliodonium salts (**47**) and (**48**) were used for the Sonogashira reaction and Suzuki-Miyaura reaction, a significant amount of by-products (**50**) and (**52**) was formed (*34,35*). They were generated by the coupling reaction of alkyne or arylboronic acid with the tolyl group on the iodonium salts. When the fluorine atom was not attached, the cross-coupling reaction took place selectively at the alkenyl part and the arylated by-products were not formed (*36*). Therefore, the fluorine atom on the alkenyl group retarded the oxidative addition of the Pd catalyst and decreased the selectivity (Scheme 21).

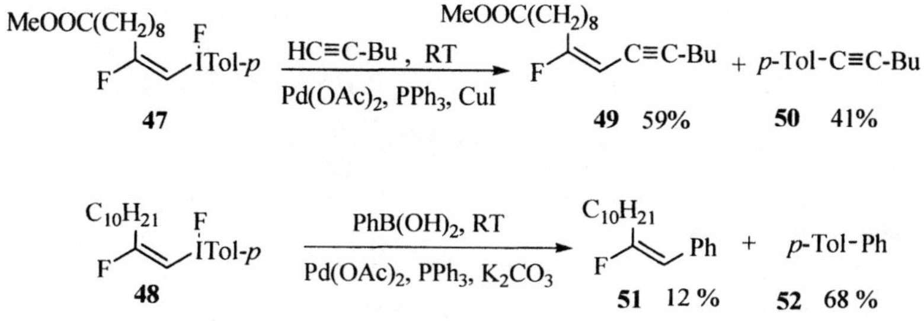

47 ($MeOOC(CH_2)_8$, F, F, ITol-*p*) → (HC≡C-Bu, RT; $Pd(OAc)_2$, PPh_3, CuI) → **49** ($MeOOC(CH_2)_8$, F, C≡C-Bu) 59% + *p*-Tol-C≡C-Bu **50** 41%

48 ($C_{10}H_{21}$, F, F, ITol-*p*) → ($PhB(OH)_2$, RT; $Pd(OAc)_2$, PPh_3, K_2CO_3) → **51** ($C_{10}H_{21}$, F, Ph) 12 % + *p*-Tol-Ph **52** 68 %

Scheme 21

(*E*)-2-Fluoro-1-alkenyliodonium salts **47** and **48** were converted to the corresponding (*E*)-2-fluoro-1-iodoalkenes (**52**) and (**53**) without isolation by Ochiai's method (*37*). Though **52** and **53** are less reactive than **47** and **48**, and higher reaction temperature was required in the Suzuki-Miyaura reactions, the desired (*E*)-fluoroenyne (**49**) and (*E*)-fluoroalkene (**51**) could be obtained selectively (*34,35*) (Scheme 22).

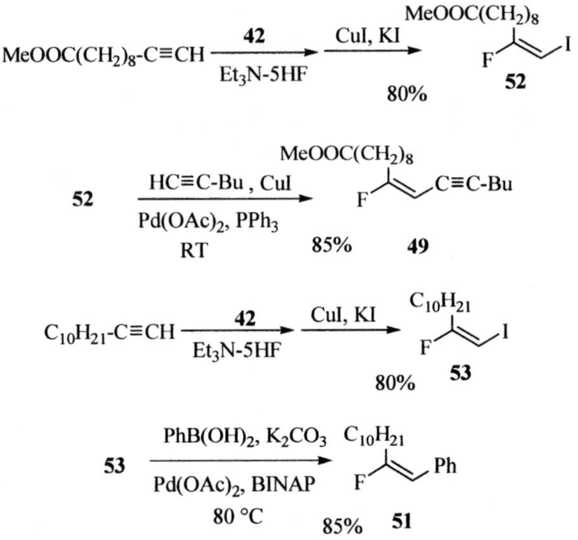

$MeOOC(CH_2)_8$-C≡CH → (**42**, Et_3N-5HF) → (CuI, KI; 80%) → **52** [$MeOOC(CH_2)_8$, F, I]

52 → (HC≡C-Bu, CuI; $Pd(OAc)_2$, PPh_3, RT; 85%) → **49** [$MeOOC(CH_2)_8$, F, C≡C-Bu]

$C_{10}H_{21}$-C≡CH → (**42**, Et_3N-5HF) → (CuI, KI; 80%) → **53** [$C_{10}H_{21}$, F, I]

53 → ($PhB(OH)_2$, K_2CO_3; $Pd(OAc)_2$, BINAP, 80 °C; 85%) → **51** [$C_{10}H_{21}$, F, Ph]

Scheme 22

The methodology was applied for the synthesis of fluorinated analogs (**54**) and (**55**) of (9*Z*, 11*E*)-1-acetoxy-9,11-tetradecadiene, a pheromone of the Egyptian cotton leaf worm, and 11,12-dehydrocoriolic acid ester (*38*). 9-Decyn-1-ol was converted to (*E*)-9-fluoro-10-iodo-9-decen-1-ol (**56**) by the reaction with **42**, followed by the treatment of CuI and KI. The cross-coupling reaction of **56** with 1-butenylborane followed by the acetylation of alcohol gave the (9*E*, 11*E*)-1-acetoxy-9-fluoro-9,11-tetradecadiene **54**, the fluorinated analog of the Egyptian cotton leaf worm pheromone (*39*). Racemic 9-fluoro-11,12-dehydrocoriolic acid methyl ester **55** was synthesized by the Sonogashira coupling reaction using fluoroiodoalkene (**57**) prepared from methyl 9-decynoate as shown in Scheme 23 (*34*).

(*Z*)-2-Fluoro-1-alkenyliodonium salts were stereoselectively synthesized by the addition of CsF to 1-alkynyliodonium salts (*40,41*). However, due to the low nucleophilicity and low solubility of the metal fluoride, the yields were low (15 – 20 %). This problem could be overcome by using aq HF as the fluoride source, and the (*Z*)-2-fluoro-1-alkenyliodonium salts could be obtained in good yield with high stereoselectivity (*42*) (Scheme 24).

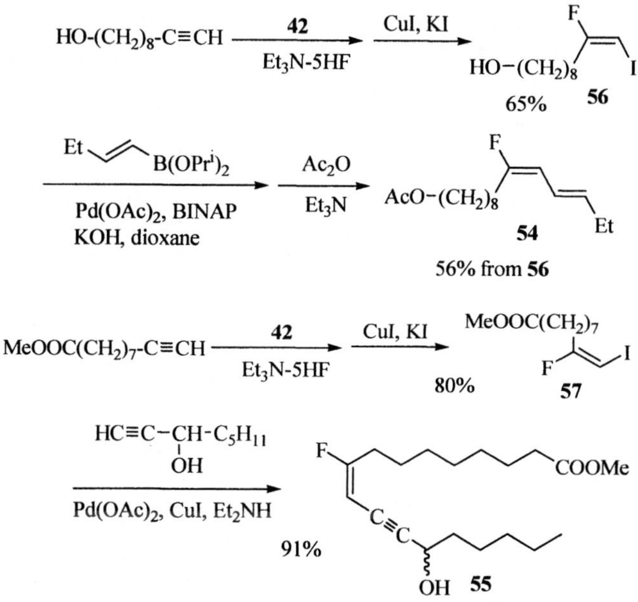

HO-$(CH_2)_8$-C≡CH —(**42**, Et_3N-5HF)→ —(CuI, KI)→ HO−$(CH_2)_8$(F)C=CH–I **56** 65%

—(Et–CH=CH–B(OPr^i)$_2$, Pd(OAc)$_2$, BINAP, KOH, dioxane)→ —(Ac_2O, Et_3N)→ AcO−$(CH_2)_8$ diene with F and Et **54** 56% from **56**

MeOOC$(CH_2)_7$-C≡CH —(**42**, Et_3N-5HF)→ —(CuI, KI)→ MeOOC$(CH_2)_7$(F)C=CH–I 80% **57**

—(HC≡C−CH(OH)−C_5H_{11}, Pd(OAc)$_2$, CuI, Et_2NH)→ 91% **55** (F, COOMe, C≡C, OH)

Scheme 23

C_8H_{17}-C≡C−$\overset{\oplus}{I}$Ph $\overset{\ominus}{BF_4}$ —(CsF, acetone-CH_2Cl_2)→ C_8H_{17}(F)C=CH−$\overset{\oplus}{I}$Ph $\overset{\ominus}{BF_4}$ 20% **58**

$C_{10}H_{21}$-C≡C−$\overset{\oplus}{I}$Ph $\overset{\ominus}{BF_4}$ —(20% aq HF, $CHCl_3$)→ $C_{10}H_{21}$(F)C=CH−$\overset{\oplus}{I}$Ph $\overset{\ominus}{BF_4}$ 84% **59**

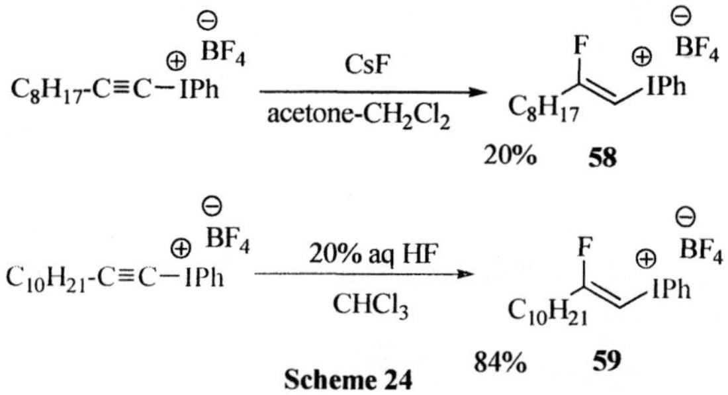

Scheme 24

Application of (*Z*)-2-fluoro-1-alkenyliodonium salt **59** for methoxycarbonylation reaction (*43*), Heck reaction (*39*), and Stille reaction (*39*) gave (*Z*)-β-fluoro-α,β-unsaturated ester (**60**), (3*E*, 5*Z*)-δ-fluoro-α,β,γ,δ-unsaturated carbonyl compound (**61**), and (*Z*)-fluoroalkadiene (**62**) stereoselectively (Scheme 25).

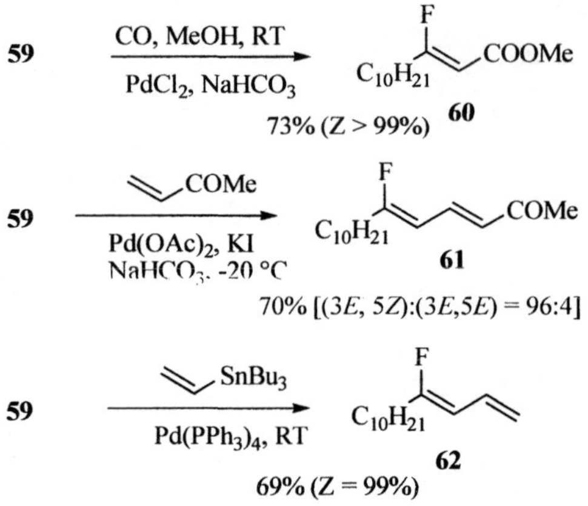

59 $\xrightarrow[\text{PdCl}_2\text{, NaHCO}_3]{\text{CO, MeOH, RT}}$ (F)($C_{10}H_{21}$)C=CH–COOMe **60** 73% (Z > 99%)

59 $\xrightarrow[\text{Pd(OAc)}_2\text{, KI, NaHCO}_3\text{, -20 °C}]{\text{CH}_2\text{=CH–COMe}}$ (F)($C_{10}H_{21}$)C=CH–CH=CH–COMe **61** 70% [(3*E*, 5*Z*):(3*E*,5*E*) = 96:4]

59 $\xrightarrow[\text{Pd(PPh}_3)_4\text{, RT}]{\text{CH}_2\text{=CH–SnBu}_3}$ (F)($C_{10}H_{21}$)C=CH–CH=CH$_2$ **62** 69% (Z = 99%)

Scheme 25

In order to avoid the formation of by-products, **59** was converted to (*Z*)-2-fluoro-1-iodo-1-alkenes (**63**) (*42*), and used for the Suzuki-Miyaura reaction and Sonogashira reaction. From **63**, (*Z*)-fluoroalkene (**64**), (*E, Z*)-fluoroalkadiene (**65**), and (*Z*)-fluoroenyne (**66**) could be obtained stereoselectively (*39*) (Scheme 26).

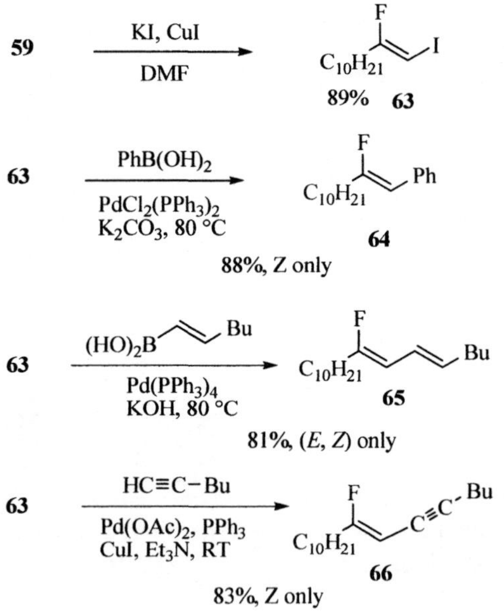

59 $\xrightarrow[\text{DMF}]{\text{KI, CuI}}$ (F)($C_{10}H_{21}$)C=CH–I 89% **63**

63 $\xrightarrow[\text{PdCl}_2\text{(PPh}_3)_2\text{, K}_2\text{CO}_3\text{, 80 °C}]{\text{PhB(OH)}_2}$ (F)($C_{10}H_{21}$)C=CH–Ph **64** 88%, Z only

63 $\xrightarrow[\text{Pd(PPh}_3)_4\text{, KOH, 80 °C}]{\text{(HO)}_2\text{B–CH=CH–Bu}}$ (F)($C_{10}H_{21}$)C=CH–CH=CH–Bu **65** 81%, (*E, Z*) only

63 $\xrightarrow[\text{Pd(OAc)}_2\text{, PPh}_3\text{, CuI, Et}_3\text{N, RT}]{\text{HC≡C–Bu}}$ (F)($C_{10}H_{21}$)C=CH–C≡C–Bu **66** 83%, Z only

Scheme 26

Synthesis of 2-Fluoro-1-alkenyl Metals (5)

Though 2-fluoro-1-alkenyl metals (**5**) could be useful precursor for various fluoroalkenes, practical methods for **5** has not been reported yet.

References

(1) Welch, J. T. *Tetrahedron,* **1987**, *43*, 3123-3197.
(2) *Fluorine in Bioorganic Chemistry*; Welch, J. T.; Eswarakrishnan, S.; Wiley, New York, 1991.
(3) *Fluorine-containing Amino Acids*; Kukhar', V. P.; Soloshonok, V. A.; Ed.; Wiley, Chichester, 1994.
(4) Burton, D. J.; Yang, Z.-Y.; Qiu, W. *Chem. Rev.* **1996**, *96*, 1641-1715.
(5) *Metal-catalyzed Cross-coupling Reactions*; Diederich, F.; Stang, P. J. Ed.; Wiley-VCH: Weinheim, 1998.
(6) *Cross-Coupling Reactions;* Miyaura, N.; Ed.; Springer: Berlin, 2002, and references cited.
(7) Shimizu, M.; Yamada, N.; Takebe, Y.; Hata, T.; Kuroboshi, M.; Hiyama, T. *Bull. Chem. Soc. Jpn.*, **1998**, *71*, 2903-2921.
(8) Vanderhaar, R. W.; Burton, D. J.; Naae, D. G. *J. Fluorine Chem.*, **1971**, *1*, 381-383.
(9) Burton, D. J. *J. Fluorine Chem.*, **1983**, *23*, 339-357.
(10) Chen, C.; Wilcoxen, K.; Strack, N.; McCarthy, J. R. *Tetrahedron Lett.*, **1999**, *40*, 827-830.
(11) Eddarir, S.; Francesch, C.; Mestdagh, H.; Rolando, C. *Tetrahedron Lett.*, **1990**, *31*, 4449-4452.
(12) Xu, J.; Burton, D. J. *Tetrahedron Lett.*, **2002**, *43*, 2877-2879.
(13) Chen, C.; Wilcoxen, K.; Huang, C. Q.; Strack, N.; McCarthy, J. R. *J. Fluorine Chem.*, **2000**, *101*, 285-290.
(14) Zhang, X.; Burton, D. J. *J. Fluorine Chem.*, **2001**, *112*, 47-54.
(15) Zhang, X.; Burton, D. J. *J. Fluorine Chem.*, **2001**, *112*, 317-324.
(16) Eddarir, S.; Abdelhadi, Z.; Rolando, C. *Tetrahedron Lett.*, **2001**, *42*, 9127-9130.
(17) Xu, J.; Burton, D. J. *J. Fluorine Chem.*, **2004**, *125*, 725-730.
(18) Xu, J.; Burton, D. J. *Org. Lett.*, **2002**, *4*, 831-833.
(19) Eddarir, S.; Francesch, C.; Mestdagh, H.; Rolando, C. *Bull. Soc. Chim. Fr.*, **1997**, *134*, 741-755.
(20) McCarthy, J. R.; Huber, E. W.; Le, T.-B.; Laskovics, F. M.; Matthews, D. P. *Tetrahedron*, **1996**, *52*, 45-58.
(21) Chen, C.; Wilcoxen, K.; Kim, K.; McCarthy, J. R. *Tetrahedron Lett.*, **1997**, *38*, 7677-7680.

(22) Chen, C.; Wilcoxen, K.; Zhu, Y.-F.; Kim, K.; McCarthy, J. R. *J. Org. Chem.*, **1999**, *64*, 3476-3482.
(23) Xu, J.; Burton, D. J. *Tetrahedron Lett.*, **2002**, *43*, 4565-4567.
(24) Eddarir, S.; Rolando, C. *J. Fluorine Chem.*, **2004**, *125*, 377-380.
(25) Dear, R. E. A. *J. Org. Chem.*, **1970**, *35*, 1703-1705.
(26) Eddarir, S.; Mestdagh, H.; Rolando, C. *Tetrahedron Lett.*, **1991**, *32*, 69-72.
(27) Carpenter, W. *J. Org. Chem.*, **1966**, *31*, 2688-2689.
(28) Sawaguchi, M.; Ayuba, S.; Hara, S. *Synthesis*, **2002**, 1802-1803.
(29) Hara, S.; Yoshida, M.; Fukuhara, T.; Yoneda, N. *Chem. Commun.*, **1998**, 965-966.
(30) Hara, S.; Yamamoto, K.; Yoshida, M.; Fukuhara, T.; Yoneda, N. *Tetrahedron Lett.*, **1999**, *40*, 7815-7818.
(31) Yoshida, M.; Hara, S.; Fukuhara, T.; Yoneda, N. *Tetrahedron Lett.*, **2000**, *41*, 3887-3890.
(32) Yoshida, M.; Nagahara, D.; Fukuhara, T.; Yoneda, N.; Hara, S. *J. Chem. Soc., Perkin Trans. 1*, **2001**, 2283-2288.
(33) Kato, T.; Yamaguchi, Y.; Hirano, T.; Yokoyama, T.; Uyehara, T.; Namai, T.; Yamanaka, S.; Harada, N. *Chem. Lett.*, **1984**, 409-412.
(34) Yoshida, M.; Yoshikawa, S.; Fukuhara, T.; Yoneda, N.; Hara, S. *Tetrahedron*, **2001**, *57*, 7143-7148.
(35) Yoshida, M.; Ota, D.; Fukuhara, T.; Yoneda, N.; Hara, S. *J. Chem. Soc., Perkin Trans. 1*, **2002**, 384-389.
(36) Kang, S.-K.; Lee, H.-W.; Jang, S.-B.; Ho, P.-S. *J. Org. Chem.*, **1996**, *61*, 4720-4724.
(37) Ochiai, M.; Sumi, K.; Takaoka, Y.; Kunishima, M.; Nagao, Y.; Shiro, M.; Fujita, E. *Tetrahedron*, **1988**, *44*, 4095-4112.
(38) Kobayashi, Y.; Okamoto, S.; Shimazaki, T.; Ochiai, Y.; Sato, F. *Tetrahedron Lett.*, **1987**, *28*, 3959-3962.
(39) Yoshida, M. Dissertation, HokkaidoUniversity, 2004.
(40) Ochiai, M.; Oshima, K.; Masaki, Y. *Chem. Lett.*, **1994**, 871-874.
(41) Ochiai, M.; Kitagawa, Y.; Toyonari, M.; Uemura, K.; Oshima, K.; Shiro, M. *J. Org. Chem.*, **1997**, *62*, 8001-8008.
(42) Yoshida, M.; Hara, S. *Org. Lett.*, **2003**, *5*, 573-574.
(43) Yoshida, M.; Komata, A.; Hara, S. *J. Fluorine Chem.*, **2004**, *125*, 527-529.

Chapter 6

A New In Situ Room Temperature Approach to Fluorinated Vinylzinc Reagents and Fluorinated Styrenes

Anilkumar Raghavanpillai[1,2] and Donald J. Burton[1,*]

[1]Department of Chemistry, The University of Iowa, Iowa City, IA 52246
[2]Current address: Department of Chemistry, Dartmouth College, Hanover, NH 03755

The reaction of LDA with a solution of $ZnCl_2$ and CF_3CH_2X (X = F, Cl, Br, I) at 15-20 °C provides a high yield *in situ* preparation of [CF_2=CXZnCl]. The resulting [F_2C=CXZnCl] intermediates are readily functionalized by Pd(0) catalyzed cross-coupling with aryl iodides to provide a one flask entry directly to the corresponding fluorinated styrenes from commercially available precursors. The stability of the intermediate vinyl lithium [F_2C=CXLi] and the nature and reactivity of these zinc reagents [F_2C=CXZnCl] are discussed.

Introduction

Introduction of fluorine into organic molecules *via* fluorinated organometallic building blocks have very high synthetic potential, as they avoid direct use of nasty fluorinating agents. Although fluorinated organometallics have been less studied compared to their hydrocarbon counterparts, much chemistry has evolved in the last two decades in which various fluorinated synthons (vinyl, alkynyl, allyl, benzyl, propargyl, aryl etc) were prepared and subsequent reactions with these reagents were explored (*1,2,3,4*). But unlike their hydrocarbon analogues, the chemistry with many of the fluorinated organometallic reagents was difficult due to their poor thermal stability. But the development of fluorinated zinc and tin reagents made this chemistry attractive because of their superior thermal stability.

Among the fluorinated organometallic reagents, fluorovinyl organometallic reagents have received special attention as they can be utilized as synthons to introduce a fluorinated vinyl group directly into an organic molecule. Fluorinated vinyllithium is the first choice of reagent to effect this transformation, but the poor thermal stability of these reagents makes them less attractive in large scale synthetic preparations (*1,5,6*). Subsequently, fluorinated vinylzinc reagents were developed that exhibited excellent thermal stability and desirable chemical reactivity. Fluorinated vinyl zinc reagents have been prepared by two different methods; i) Exchange of the corresponding alkenyllithium reagents at low temperature with zinc chloride. ii) Direct insertion of zinc into the carbon – halogen bond of a fluorinated vinyl halide. The first method involves low temperature generation of a fluorinated vinyllithium from corresponding vinyl halides or hydroalkenes (*6*). Addition of zinc halide to the fluorovinyllithium solution generates the vinylzinc halides in excellent yields (*6,7,8,9,10,11,12,13,14,15,16,17,18*). The reaction temperature for this transformation depends largely on the thermal stability of the vinyllithium reagent and is also solvent dependent. Trifluorovinyl, 2,2-difluorovinyl, 1-chloro-2-fluorovinyl and (*E*)-1,2-difluorovinylzinc chlorides were generated at temperatures below -100 °C, while some (*Z*)-1,2-difluorovinylzinc chlorides could be prepared in ether at -30 °C. Examples of various zinc reagents prepared by this methodology are illustrated in Scheme 1. In the presence of a palladium

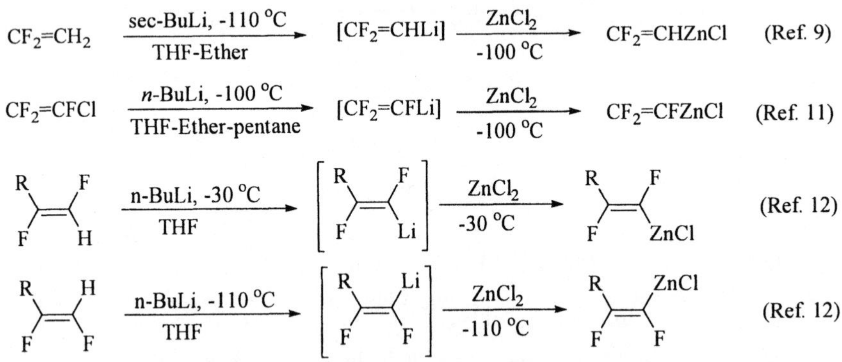

$$CF_2{=}CH_2 \xrightarrow[\text{THF-Ether}]{\text{sec-BuLi, -110 °C}} [CF_2{=}CHLi] \xrightarrow[\text{-100 °C}]{ZnCl_2} CF_2{=}CHZnCl \quad \text{(Ref. 9)}$$

$$CF_2{=}CFCl \xrightarrow[\text{THF-Ether-pentane}]{n\text{-BuLi, -100 °C}} [CF_2{=}CFLi] \xrightarrow[\text{-100 °C}]{ZnCl_2} CF_2{=}CFZnCl \quad \text{(Ref. 11)}$$

$$\text{R(F)C=C(F)H} \xrightarrow[\text{THF}]{\text{n-BuLi, -30 °C}} [\text{R(F)C=C(F)Li}] \xrightarrow[\text{-30 °C}]{ZnCl_2} \text{R(F)C=C(F)ZnCl} \quad \text{(Ref. 12)}$$

$$\text{R(F)C=C(H)F} \xrightarrow[\text{THF}]{\text{n-BuLi, -110 °C}} [\text{R(F)C=C(Li)F}] \xrightarrow[\text{-110 °C}]{ZnCl_2} \text{R(F)C=C(ZnCl)F} \quad \text{(Ref. 12)}$$

Scheme 1. Preparation of fluorovinylzinc reagents by the metallation of vinyl halides or hydroalkanes via fluorovinyllithium.

catalyst these fluorovinyl zinc reagents undergo cross-coupling reactions with aryl iodides, vinyl iodides, acid chlorides and iodoalkynes to give corresponding fluorinated alkenyl derivatives (Scheme 2).

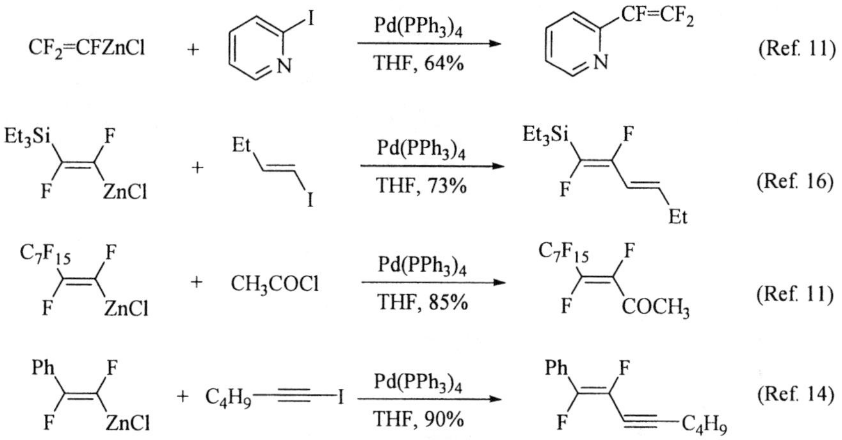

CF_2=CFZnCl + 2-iodopyridine $\xrightarrow[\text{THF, 64\%}]{Pd(PPh_3)_4}$ 2-(CF=CF_2)pyridine (Ref. 11)

Et_3Si(F)C=C(F)ZnCl + Et–CH=CH–I $\xrightarrow[\text{THF, 73\%}]{Pd(PPh_3)_4}$ Et_3Si(F)C=C(F)–CH=CH–Et (Ref. 16)

C_7F_{15}(F)C=C(F)ZnCl + CH_3COCl $\xrightarrow[\text{THF, 85\%}]{Pd(PPh_3)_4}$ C_7F_{15}(F)C=C(F)$COCH_3$ (Ref. 11)

Ph(F)C=C(F)ZnCl + C_4H_9–C≡C–I $\xrightarrow[\text{THF, 90\%}]{Pd(PPh_3)_4}$ Ph(F)C=C(F)–C≡C–C_4H_9 (Ref. 14)

Scheme 2. Pd(0) Catalyzed coupling reaction of fluorovinylzinc reagents produced by the metallation of fluorinated vinyl halides or hydroalkenes.

The second method of preparation of fluorovinylzinc reagents involves the reaction of fluorinated vinyl bromides or iodides with acid-washed zinc in a variety of solvents such as DMF, DMAC, triglyme, tetraglyme, THF and acetonitrile (*19*). The reaction is feasible at room temperature in the case of fluorinated vinyl iodides whereas vinyl bromides reacted at rt to 60 °C. ^{19}F NMR analysis of the zinc reagents showed as a *mono/bis* zinc reagent mixture with varying ratio based on the structure of the vinyl halide and the solvent. The zinc insertion reaction was stereospecific, as (*E*) or (*Z*) perfluoropropenyliodides reacted with zinc in TG to form the corresponding (*E*) or (*Z*) perfluoropropenylzinc reagents, respectively (*19*). Typical examples of zinc reagents prepared by this method are illustrated in Scheme 3 (*19,20,21,22,23,24,25,26,27*).

In the presence of a palladium catalyst, these zinc reagents undergo cross-coupling reactions with aryl or vinyl halides to produce corresponding aryl or vinyl substituted alkenes in excellent isolated yields (Scheme 4) (*20,26,28,29,30,31,32,33,34,35*). The cuprous halide catalyzed coupling reaction of these zinc reagents with allyl halides or acid chlorides afforded the corresponding alkenylated compounds (Scheme 5) (36,37).

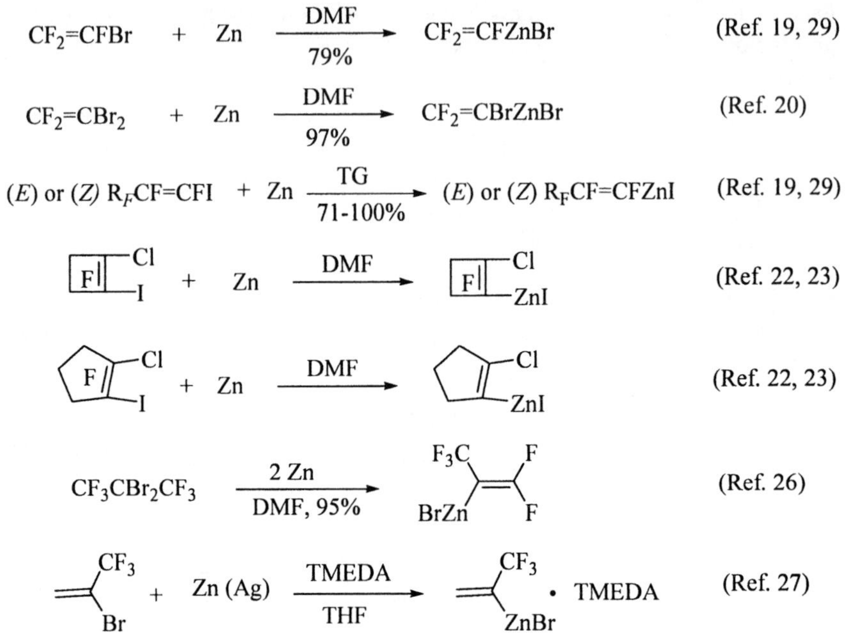

$CF_2{=}CFBr$ + Zn $\xrightarrow[79\%]{DMF}$ $CF_2{=}CFZnBr$ (Ref. 19, 29)

$CF_2{=}CBr_2$ + Zn $\xrightarrow[97\%]{DMF}$ $CF_2{=}CBrZnBr$ (Ref. 20)

(*E*) or (*Z*) $R_FCF{=}CFI$ + Zn $\xrightarrow[71\text{-}100\%]{TG}$ (*E*) or (*Z*) $R_FCF{=}CFZnI$ (Ref. 19, 29)

F(cyclobutene)(Cl)(I) + Zn $\xrightarrow{DMF}$ F(cyclobutene)(Cl)(ZnI) (Ref. 22, 23)

F(cyclopentene)(Cl)(I) + Zn $\xrightarrow{DMF}$ (cyclopentene)(Cl)(ZnI) (Ref. 22, 23)

$CF_3CBr_2CF_3$ $\xrightarrow[DMF,\ 95\%]{2\ Zn}$ $(F_3C)(BrZn)C{=}CF_2$ (Ref. 26)

$CH_2{=}C(CF_3)Br$ + Zn (Ag) $\xrightarrow[THF]{TMEDA}$ $CH_2{=}C(CF_3)ZnBr$ • TMEDA (Ref. 27)

Scheme 3. Preparation of fluorovinylzinc reagents from fluorinated vinyl bromides and iodides by zinc insertion.

Fluorovinylzinc reagents *via* a new *in situ* approach

Even though fluorovinylzinc reagents were generated in good yields by the above discussed methods, the low temperature involved and the use of environmentally hazardous vinyl halides as starting materials were major concerns. Herein we present the chemistry that we have developed recently to generate fluorovinylzinc reagents at temperatures near to room temperature starting from environmentally friendly hydrofluorocarbons. Though we began this project as part of our synthetic attempts to develop a cost-effective high-yield synthesis of α,β,β-trifluorostyrenes from readily available halocarbon HFC-134a *via* the formation of intermediate trifluorovinylzinc reagent, we explored similar possibilities for various α-halo-β,β-difluorovinylzinc reagents from the corresponding halocarbon sources. The zinc reagents thus generated could be readily functionalized with aryl or vinyl halides using Pd(0) catalyst to provide a one flask entry directly to the corresponding fluorinated alkenes under mild reaction conditions.

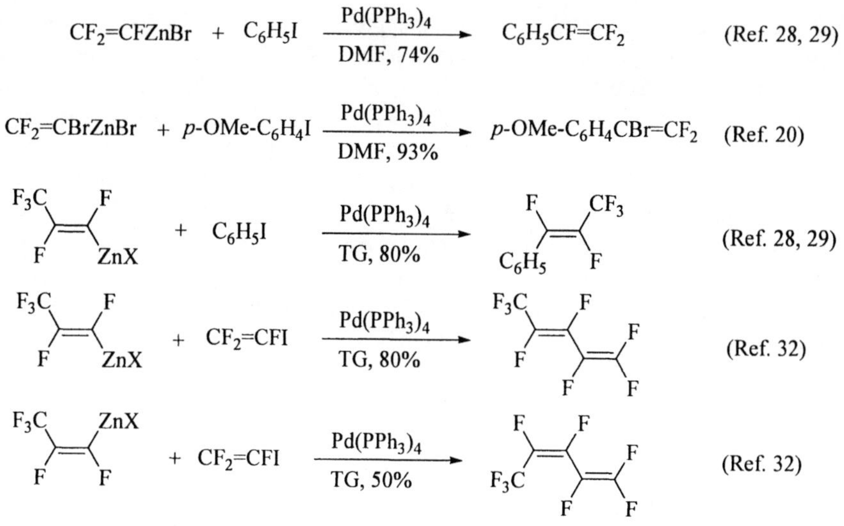

$CF_2{=}CFZnBr$ + C_6H_5I —[$Pd(PPh_3)_4$ / DMF, 74%]→ $C_6H_5CF{=}CF_2$ (Ref. 28, 29)

$CF_2{=}CBrZnBr$ + *p*-OMe-C_6H_4I —[$Pd(PPh_3)_4$ / DMF, 93%]→ *p*-OMe-$C_6H_4CBr{=}CF_2$ (Ref. 20)

F_3C, F / F, ZnX + C_6H_5I —[$Pd(PPh_3)_4$ / TG, 80%]→ F, CF_3 / C_6H_5, F (Ref. 28, 29)

F_3C, F / F, ZnX + $CF_2{=}CFI$ —[$Pd(PPh_3)_4$ / TG, 80%]→ F_3C, F / F; F / F, F (Ref. 32)

F_3C, ZnX / F, F + $CF_2{=}CFI$ —[$Pd(PPh_3)_4$ / TG, 50%]→ F, F / F_3C; F / F, F (Ref. 32)

Scheme 4. Pd(0) Catalyzed coupling reaction of fluorovinylzinc reagents (prepared by zinc insertion) with aryl or vinyl halides.

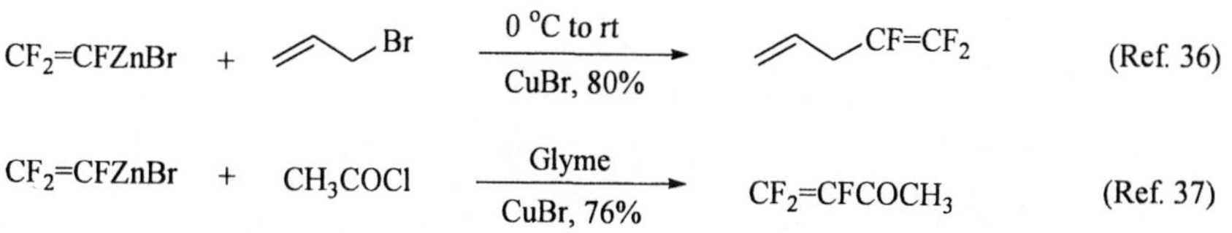

$CF_2{=}CFZnBr$ + allyl-Br —[0 °C to rt / CuBr, 80%]→ allyl-$CF{=}CF_2$ (Ref. 36)

$CF_2{=}CFZnBr$ + CH_3COCl —[Glyme / CuBr, 76%]→ $CF_2{=}CFCOCH_3$ (Ref. 37)

Scheme 5. The cuprous halide catalyzed coupling reaction of fluorovinylzinc reagents (prepared by zinc insertion) with allyl halides and acid chlorides.

Preparation of trifluorovinylzinc reagent from halocarbon HFC-134a and the synthesis of α,β,β-trifluorostyrenes.

α,β,β-Trifluorostyrene (TFS) is one of the important monomers among the fluorinated monomers capable of polymerization (*38,39,40,41,42*). Polymers of 1,2,2-trifluorostyrene, namely poly(1,2,2-trifluorostyrene) (PTFS) and its co-polymers with other alkenes, have high thermal as well as chemical stability, found extensive application as ion exchange membranes for fuel cell separators and dialysis membranes (*38,43,44,45,46,47,48,49,50,51,52,53*). The challenge in the application of TFS has been the preparation of TFS itself. Practical synthesis of TFS in high yield and purity was a difficult task. The earlier methods developed by Cohen (*54*) and by Prober (*55*) involved many steps and suffered from low overall yields (<20%). Pyrolytic methods are used for the bulk production but have the difficulty of low yield and the formation of many unwanted side products (*56*). Formation of the TFS through organometallic reagents has been the most impressive route. One of the earlier methods developed by Dixon (*57*), where reaction of phenyllithium with tetrafluoroethylene (TFE) generated the 1,2,2-trifluorostyrene in low yields (30%) along with 1,2-difluorostilbene as side product. With an excess of phenyllithium, the disubstituted product was favored (*58*). Sorokina and co-workers developed a synthetic route for TFS by palladium catalyzed cross-coupling reaction of the trifluoroethenylzinc, trifloroethenyltin reagents with aryl iodides in DMF, HMPA medium (*7,59*). Reactions of the phenyl derivatives of lanthanides (Yb, Sm, Ce) with tetrafluoroethylene produced the TFS in poor yield (*60,61*). Recently, we have developed a highly efficient synthesis of 1,2,2-trifluorostyrenes by dehydrohalogenation of the corresponding precursors using DBU or lithium heaxamethyldisilazide as the base (*62*) The best general method for the synthesis of TFS and substituted TFS derivatives was developed from this laboratory where the trifluovinylzinc reagent was generated from zinc and bromotrifluoroethylene (BTFE) in a variety of solvents (*28,29*). The palladium catalyzed cross-coupling of the zinc reagent with aryl iodides generated 1,2,2-trifluorostyrenes in very good yields. Since this reaction was performed under mild conditions the usual thermal cyclodimerization of TFS was absent. Also, the stilbene formation observed in Dixon's method was eliminated. But the cost, availability and above all the environmental concern of BTFE was a problem of concern and thus made it difficult for commercial applications. Alternatively, trifluorovinylzinc reagent could be generated *via* fluorovinyllithium reagent from a fluorocarbon of the type CF_2=CFY, where Y = H, Cl, Br, I. Here the lithium reagent was generated by the reaction of the fluoroalkene with alkyl lithium at very low temperature, which was then converted to zinc reagent by transmetallation with zinc halide (*6,11*). But unlike fluorovinylzinc reagents, the corresponding lithium reagents are thermally unstable and have to be generated at very low temperature (*1,5,6*). So the preparation of trifluorovinylzinc reagent *via* the trifluorovinyllithium reagent suffered two major drawbacks; the lithium reagent has to be generated at very low temperature and having to use CFC type starting materials. The only viable cheap, large volume, commercially available precursor for the introduction of the trifluoroethenyl group is 1,1,1,2-tetrafluoroethane (HFC-134a). In the late 1990's Coe and co-workers recognized the potential of HFC-134a as a trifluorovinyl synthon (*63,64,65*). They generated trifluorovinyllithium from HFC-134a at -78 °C by a sequence of dehydrofluorination and metallation reactions and carried out low temperature functionalization with various electrophiles (Scheme 6) (*65,66*). With zinc halides as an electrophile, trifluorovinyllithium was converted to trifluorovinylzinc halide at low temperature. Even though trifluorovinylzinc reagent was generated in good yield from HFC-134a, the low temperature involved remained as a problem for large scale reactions or industrial applications. In addition, trifluorovinyllithium readily decomposes to difluoroacetylene >-70 °C which leads to the formation of polymeric by-products.

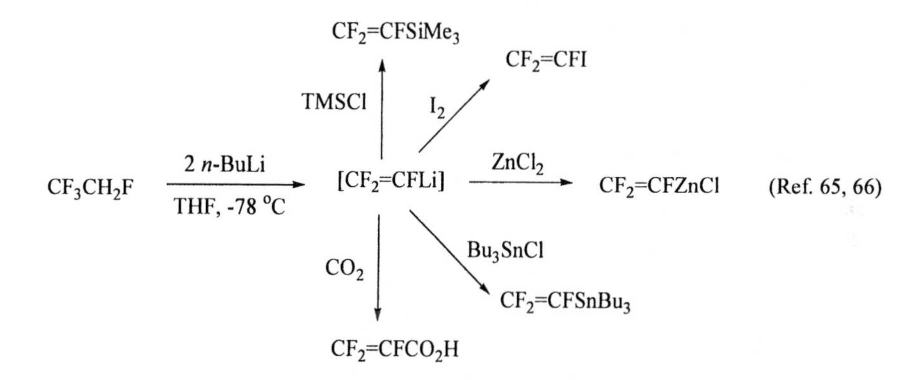

Scheme 6. Preparation of trifluorovinyllithium from HFC-134a and its reaction with various electrophiles.

In order to make this reaction feasible at ambient temperatures, we decided that the metallation-transmetallation process should be made *in situ*, so that the trifluorovinyllithium formed could be trapped faster than it decomposed. Thus *in situ* metallation reaction was attempted under a variety of reaction conditions by changing the medium (solvent and co-solvent), base, temperature and the zinc halide used for the transmetallation. Table I summarizes some of the selected trials used to optimize this *in situ* metallation reaction of HFC-134a with various alkyllithium and lithium amide bases in the presence of a zinc salt. When the *in situ* metallation reaction of HFC-134a was performed using n-BuLi in the presence of zinc iodide at -26 °C (boiling point of HFC-134a), the trifluorovinylzinc reagent was formed in poor yield (entry 1) and the major reaction was the reaction of base with the zinc salt. But with the more sterically hindered base, t-BuLi and with excess of HFC-134a resulted a fair yield (66%) of the trifluorovinylzinc reagent (entry 2). Though the presence of co-solvents like DMPU, DMI in conjunction with THF are known to accelerate the formation of fluorinated vinyllithium (67), they did not improve the yield of the trifluorovinylzinc reagent when the *in situ* reaction was performed in THF/DMPU medium (entry 3). Use of sterically hindered lithium amide bases like 4-methoxy-LTMP improved the yield of the trifluorovinylzinc reagent to 76% when the reaction was performed at -26 °C with excess of HFC-134a (entry 4), whereas lithium hexamethyldisilazide failed to metallate HFC-134a under similar conditions. Use of the non-hygroscopic $ZnCl_2$.TMEDA complex in this reaction improved the yield of the trifluorovinylzinc reagent to 82% (entry 5). The success of 4-OMe-LTMP prompted us to consider a relatively inexpensive base like lithium diisopropylamide (LDA) for effecting this *in situ* transformation. This reaction produced a clean reaction mixture with an 86% yield of the zinc reagent (entry 6). As a next step, we reduced the concentration of HFC-134a from 4 eq. to 1.5 eq., and the reaction was performed at –26 °C using LDA with the $ZnCl_2$.TMEDA salt. This reaction produced an 84% yield of the zinc reagent (entry 7). Further experiments with TMEDA as a co-solvent along with zinc chloride also produced excellent yields of the trifluorovinylzinc reagent, and pre-formation of the $ZnCl_2$.TMEDA complex was not required (entry 8). Further increase in the reaction temperature to 0 °C or room temperature did not significantly reduce the yield of the trifluorovinylzinc reagent (entry 9, 10). An interesting observation noticed in the TMEDA mediated reaction was the formation of traces of saturated zinc reagent (~10%) ($CF_3CHFZnCl$) along with the trifluorovinylzinc reagent. This species $CF_3CHFZnCl$ was gradually transformed to the vinyl zinc reagent. It is assumed that the $CF_3CHFZnCl$ is formed by the reaction of the initial mono anion (CF_3CHF-) generated from HFC-134a with zinc halide before it eliminates the fluoride ion (Scheme 7). In fact, a trace amount of such a species ($CF_3CHFSnBu_3$) was detected in the metallation reaction of HFC-134a with LDA and Bu_3SnCl (63).

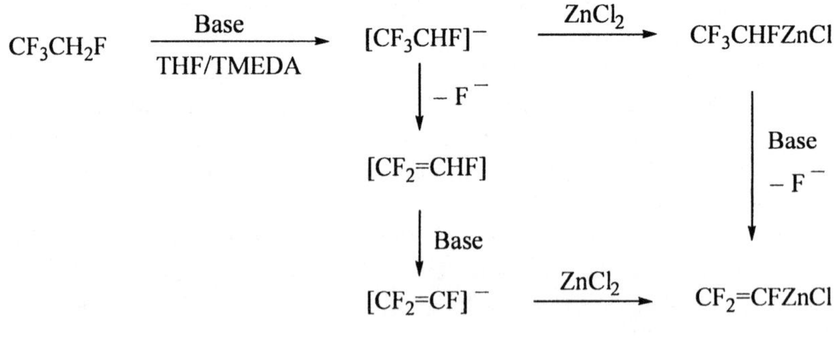

$$CF_3CH_2F \xrightarrow[\text{THF/TMEDA}]{\text{Base}} [CF_3CHF]^- \xrightarrow{ZnCl_2} CF_3CHFZnCl$$

$$[CF_3CHF]^- \xrightarrow{-F^-} [CF_2{=}CHF] \xrightarrow{\text{Base}} [CF_2{=}CF]^- \xrightarrow{ZnCl_2} CF_2{=}CFZnCl$$

$$CF_3CHFZnCl \xrightarrow[-F^-]{\text{Base}} CF_2{=}CFZnCl$$

Scheme 7. Formation of $CF_3CHFZnCl$ and CF_2=CFZnCl during the metallation of HFC-134a.

Table I. *In situ* reactions of HFC-134a with various alkyllithium and lithium amide bases in the presence of a zinc salt

$$CF_3CH_2F \xrightarrow[ZnX_2,\ \text{co-solvent}]{\text{Base, THF}} [CF_2{=}CFZnX]$$

Trail	*HFC-134a*	*Base (2 eq.)*	*Zinc salt (1 eq)*	*Temp*	*Solvent/co-solvent*	*% yield of zinc reagent*
1	1.2 eq.	*n*-BuLi	$ZnCl_2$	-26	THF	23
2	4.0 eq.	*t*-BuLi	ZnI_2	-26	THF	66
3	4.0 eq.	*t*-BuLi	ZnI_2	-26	THF/DMPU	63
4	4.0 eq.	4-OMe-LTMP	ZnI_2	-26	THF	76
5	4.0 eq.	4-OMe-LTMP	$ZnCl_2$.TMEDA	-26	THF	82
6	4.0 eq.	LDA	$ZnCl_2$.TMEDA	-26	THF	86
7	1.5 eq.	LDA	$ZnCl_2$.TMEDA	-26	THF	84
8	1.5 eq.	LDA	$ZnCl_2$	-26	THF/TMEDA	80
9	1.3 eq.	LDA	$ZnCl_2$	0	THF/TMEDA	82
10	1.2 eq.	LDA	$ZnCl_2$	22	THF/TMEDA	78
11	1.2 eq.	LDA	$ZnCl_2$	0	THF/Bipyridyl	79
12	1.2 eq.	LDA	$ZnCl_2$	0	THF/$^iPr_2NH$	77
13	1.2 eq.	LDA	$ZnCl_2$	20	THF	73

Yield of zinc $CF_2{=}CFZnX$ from ZnX_2 based on ^{19}F NMR analysis of the reaction mixture (*vs* $PhCF_3$ as internal standard).

Other ligands like 2,2'-bipyridyl (entry 11) and diisopropylamine (entry 12) offered improved yield of the trifluorovinylzinc reagent. Thus the presence of an amine co-solvent is essential in order to obtain high yield of the trifluorovinylzinc reagent in this *in situ* metallation reaction. Metallation using LDA without any added amine co-solvent at rt also produced reasonably good yield of the zinc reagent (the diisopropylamine liberated from LDA could be acting as an amine co-solvent) (entry 13). Finally, the best feasible reaction conditions for the generation of trifluoroethenylzinc reagent were chosen as follows; addition of LDA (2.0 eq) to a medium of saturated zinc chloride (1.0 eq) and HFC-134a (1.2 eq.) in THF or THF-TMEDA at 15-20 °C. The THF reaction produced a 73% yield of the trifluorovinylzinc reagent where as THF-TMEDA produced the zinc reagent in 78% yield. During our optimization experiments, it was noticed that the trifluorovinylzinc reagent was obtained as a mixture of *mono* and *bis* species in varying ratio from 60:40 to 85:15. The ^{19}F

chemical shifts of the *mono* and *bis* zinc reagent varied slightly depending on the co-solvent (THF or THF/TMEDA or THF/diisopropylamine, THF/2,2'-bipyridyl) and the counter ion (Cl, Br, I). The trifluorovinylzinc reagent generated in THF/Diisopropylamine medium showed spectral patterns corresponding to both *mono* and *bis* zinc reagents complexed to THF and diisopropylamine with slight difference in chemical shifts in the upfield direction. Addition of TMEDA to the medium produced one set of peaks resulting from the preferential complexation of *mono* and *bis* zinc reagents to TMEDA over diisopropylamine or THF. The spectral data for the *mono* and *bis* zinc reagents were assigned by a set of experimental observations (Scheme 8). When half an equivalent of zinc chloride was used for the *in situ* metallation reaction exclusive formation of the *bis* zinc reagent was observed. The ^{19}F NMR chemical shifts and the splitting pattern for this zinc reagent matched exactly for the minor component of the usual reaction, thus demonstrating that the major zinc reagent in the usual reaction was *mono*. Addition of excess of zinc chloride to the *bis* zinc reagent shifted the equilibrium towards *mono* side and obtained a 75:25 mixture of *mono* and *bis* zinc reagents. In another experiment we have increased the concentration of $ZnCl_2$ from 0.5 to 0.75 equivalent and this reaction produced a 40:60 mixture of *mono* and *bis* zinc reagent and the ratio changed to 75:25 when excess $ZnCl_2$ was added to this reaction mixture. The Schlenk equilibrium for the *mono* and *bis* zinc reagent is shown in Scheme 8. It was not possible to generate exclusively the *mono* zinc reagent even with a large excess of $ZnCl_2$.

$$CF_3CH_2F \xrightarrow[\text{THF/TMEDA } 15\ ^{\circ}C]{\text{LDA, } ZnCl_2} [CF_2{=}CFZnCl]\cdot TMEDA + [(CF_2{=}CF)_2Zn]\cdot TMEDA$$

$ZnCl_2$ (1eq)	70:30
$ZnCl_2$ (0.5 eq)	0:100
$ZnCl_2$ (0.75 eq)	40: 60

$$\underset{100}{[(CF_2{=}CF)_2Zn]\cdot TMEDA} \xrightarrow{\text{Excess } ZnCl_2} \underset{75}{[CF_2{=}CFZnCl]\cdot TMEDA} + \underset{25}{[(CF_2{=}CF)_2Zn]\cdot TMEDA}$$

$$2\ [CF_2{=}CFZnCl]\cdot TMEDA \rightleftharpoons [(CF_2{=}CF)_2Zn]\cdot TMEDA + ZnCl_2$$

Scheme 8. Schlenk equilibrium between $CF_2{=}CFZnCl$, $(CF_2{=}CF)_2Zn$ and $ZnCl_2$.

The trifluorovinylzinc reagent generated by the room temperature metallation of HFC-134a was coupled with iodobenzene using a tetrakis(triphenylphosphine)palladium catalyst at 60 °C to produce α,β,β-trifluorostyrene (TFS) in 69% isolated yield (*68,69,70*). This reaction was subsequently utilized for the synthesis of a variety of substituted α,β,β-trifluorostyrenes and representative examples are shown in Table II.

Table II. Synthesis of α,β,β-trifluorostyrenes

$$CF_3CH_2F + ZnCl_2 + LDA \xrightarrow[\text{2. ArI, Pd(PPh}_3)_4\text{, rt to 65 °C}]{\text{1. 15-20 °C/THF}} ArCF{=}CF_2$$

1.1 eq. 1.0 eq. 2.0 eq.

Entry	*Ar*	*Temp/Time*	*NMR Yield*	*Yield (%)*[a]
1	C_6H_5-	65 °C, 3 h	99	69
2	*o*-FC_6H_4-	65 °C, 2 h	98	74
3	*m*-$NO_2C_6H_4$-	60 °C, 1 h	94	82
4	*m*-$CF_3C_6H_4$-	65 °C, 1 h	95	67
5	*p*-$OMeC_6H_4$-	60 °C, 1 h	94	82
6	*o*-$(CH_3)_2CC_6H_4$-	65 °C, 2 h	96	86
7[b]	*p*-BrC_6H_4-	rt, 18 h	95	75
8[c]	2-Thienyl	60 °C, 1 h	84	59

[a] Isolated yield of pure styrene. [b] No *bis* styrene was detected after 12 h at rt; ~3 % of *bis* styrene was formed after 24 h at rt. [c] A better NMR yield (95%) was observed when stirred at RT for 24 h.

Source: Reproduced with permission from reference 68. Copyright 2002.

Preparation of the α-chloro-β,β-difluorovinyl zinc reagent from halocarbon HCFC-133a and Halothane; Synthesis of α-chloro-β,β-difluorostyrenes

Halogen containing 1,1-difluoroalkenes are building blocks in organofluorine chemistry due to the reactivity of the gem-difluoromethylene unit towards nucleophilic reagents and also due to the possible functionalization at the halogen (chlorine, bromine, iodine) site. Generally, olefins of the type $R_2C{=}CF_2$, where R = alkyl, aryl, perfluoroalkyl or hydrogen have been synthesized by Wittig type reactions (*71,72*), but this approach is not adaptable to the synthesis of $RCX{=}CF_2$, where X = Cl, Br or I due to the acylation of the Wittig reagents. Alternatively this type of olefins can be prepared by the reaction of the Grignard reagent with $CF_2{=}CFX$, but such a reaction provides an undesired isomeric olefin or an exchange reaction occurs to give the trifluorovinyl organometallic compound (*73*). Normant and co-workers generated the α-chloro-β-fluoroalkenylzinc reagents at low temperature *via* the corresponding vinyllithium starting from RCF=CHCl precursors (*8,11,74,75*). But due to the highly unstable nature of RCF=CClLi, zinc reagent formation and further reactions had to be performed at low temperature. Kumudaki and co-workers reported the generation of the α-chloro-β,β-difluorovinylzinc reagent under relatively higher temperature (-60 °C) by *in situ* metallation of CF_2=CHCl and 1-bromo-1-chloro-2,2,2-trifluoroethane (Halothane) using *s*-BuLi base. But the interpretation of this zinc reagent as a new zinc reagent from the one generated by Normant's method was quite confusing and the spectral data described were different from each other (*76,77,78*). Coe and co-workers generated chlorodifluorovinyllithium from 1-chloro-2,2,2-trifluoroethane (HCFC-133a) at -78 °C by a sequence of dehydrofluorination and metallation reactions and carried out low temperature functionalization with various electrophiles (*65,79,80*). This work established HCFC-133a as an excellent precursor for the chlorodifluorovinyl synthon and again the challenge was to generate this synthon at ambient temperature. Following a similar strategy to the previously discussed *in situ* metallation of HFC-134a, we decided that the metallation of HCFC-133a at 15-20 °C in presence of zinc halide could lead to the α-chloro-β,β-difluorovinylzinc reagent. So the metallation reaction was performed with halocarbon HFC-133a and LDA in the presence of $ZnCl_2$ at 15 °C (conditions standardized for the metallation of HFC-134a) (*81*). This reaction produced a 91% yield of the zinc reagent (Scheme 9).

$$CF_3CH_2Cl \xrightarrow[\text{THF/ 15-20 °C}]{\text{LDA, ZnCl}_2} \underset{91\%}{CF_2{=}CClZnCl} \xrightarrow[\text{PhI, rt, THF}]{\text{Pd(PPh}_3)_4} \underset{77\%}{PhCCl{=}CF_2}$$

Scheme 9. Pd(0) catalyzed coupling of α-chloro-β,β-difluorovinyl zinc reagent with iododbenzene.

The increase in the yield of the chlorodifluorovinylzinc reagent (91% by the metallation of HCFC-133a) compared to that of trifluorovinylzinc reagent (73% by the metallation of HFC-134a) under similar reaction conditions is most likely due to the increased stability of the intermediate chlorodifluorovinyllithium compared to trifluorovinyllithium. In order to confirm this rationalization, chlorodifluorovinyllithium (CF_2=CClLi) was generated at -80 °C using LDA in THF medium and the resulting brown solution was warmed to -10 °C (turned completely dark) and quenched with zinc chloride. A 60% yield of the chlorodifluorovinylzinc reagent was observed by ^{19}F NMR analysis of the reaction mixture. However, a similar sequence of experiments with [CF_2=CFLi] produced only ~5% yield of the trifluorovinylzinc reagent under similar conditions. These experiments demonstrate the enhanced stability of [CF_2=CClLi] compared to [CF_2=CFLi] at temperatures close to room temperature. Similar to the trifluorovinylzinc reagent, the chlorodifluorovinylzinc reagent generated in THF/diisopropylamine medium showed spectral patterns corresponding to both *mono* and *bis* zinc reagents complexed to THF and diisopropylamine. The zinc reagent was also formed as a mixture of *mono* and *bis* (*mono/bis* 70:30). Addition of TMEDA to the medium produced one set of peaks resulting from the preferential complexation of the *mono* and *bis* zinc reagent to TMEDA. The chemical shifts for both *mono* and *bis* chlorodifluorovinylzinc reagents were unequivocally assigned based on similar Schlenk equilibrium experiments described for the trifluorovinylzinc reagent.

We also considered 1-bromo-1-chloro-2,2,2-trifluoroethane as an alternative precursor for the chlorodifluorovinylzinc reagent owing to its cheapness and availability as a commercial anesthetic as well as user friendly (being a liquid). Thus, *via* the conditions standardized for the metallation of HFC-134a, the metallation of halothane was carried out using LDA at 15 °C in the presence of zinc chloride (Scheme 10). This reaction produced a black reaction mixture with the formation of many products along with a small amount of the desired zinc reagent. The major products formed in this reaction were identified as CF_2=CClZnCl (25%), CF_2=CClBr (9%), CF_2=CClH (4%) along with unreacted halothane (40%). Use of other bases and lowering the temperature to -78 °C changed the ratio of the different products slightly but not significantly. These results differ with the results reported by Kumudaki during the metallation of halothane in the presence of zinc chloride. We have not observed any species corresponding to his claimed new zinc reagent (*77*). The small amount of zinc reagent formed during this reaction was spectroscopically similar to the chlorodifluorovinylzinc reagent generated by the metallation of CF_3CH_2Cl (*81*).

$$CF_3CHClBr + \text{Base} \xrightarrow[\text{THF, 15 °C or -78 °C}]{ZnX_2} CF_2CCl{=}ZnCl + CF_2{=}CClBr + CF_2{=}CClH + CF_3CHClBr$$

Base : *n*-BuLi, *t*-BuLi, LDA, LTMP

ZnX_2 : $ZnCl_2$, ZnI_2, $ZnCl_2$.TMEDA

Scheme 10. Metallation of 1-bromo-1-chloro-2,2,2-trifluoroethane (halothane) in presence of zinc chloride.

The α-chloro-β,β-difluorovinylzinc reagent generated by the metallation of HCFC-133a was coupled with iodobenzene using tetrakis(triphenylphosphine)palladium catalyst at 60 °C to produce the α-chloro-β,β-difluorostyrene in excellent isolated yield (Scheme 9) (*81*). A series of substituted α-chloro-β,β-difluorostyrenes were synthesized in excellent isolated yield by utilization of this methodology (Table III).

Table III. Synthesis of α-chloro-β,β-difluorostyrenes

$$CF_3CH_2Cl\ (1.1\ eq.) + ZnCl_2\ (1.0\ eq.) + LDA\ (2.0\ eq.) \xrightarrow[\text{2. ArI, Pd(PPh}_3)_4\text{, rt to 65 °C}]{\text{1. 15-20 °C/THF}} ArCCl{=}CF_2$$

Entry	*Ar*	*Temp/Time*	*NMR Yield*	*Yield (%)*[a]
1	*p*-MeC_6H_4-	rt, 12 h	99	83
2	*p*-$EtO_2CC_6H_4$-	65 °C, 3 h	91	70
3	*m*-$NO_2C_6H_4$-	rt, 15 h	90	77
4	*o*-$(CH_3)_2CC_6H_4$-	65 °C, 12 h	95	75
5	*m*-$OMeC_6H_4$-	60 °C, 4 h	94	79
6[b]	*m*-BrC_6H_4-	rt, 18 h	96	85
7	2-Thienyl	rt, 4 h; then 60 °C, 3h	87	65

[a] Isolated yield of pure styrene. [b] 1% of the *bis* styrene was formed.

Source: Reproduced with permission from reference 81. Copyright 2002.

Preparation of the α-bromo-β,β-difluorovinylzinc reagent from CF_3CH_2Br and CF_2=CHBr; Synthesis of α-bromo-β,β-difluorostyrenes

We have further extended the room temperature *in situ* metallation method to the synthesis of the α-bromo-β,β-difluorovinylzinc reagent (CF_2=CBrZnCl) as this reagent can be utilized for the synthesis of α-bromo-β,β-difluorostyrenes. The previously reported preparation of the bromodifluorovinylzinc reagent was from this laboratory where zinc insertion into CF_2=CBr_2 in DMF at room temperature produced the bromodifluorovinylzinc reagent in excellent yield (*20*). Following the strategy similar to the metallation of HFC-134a, a THF solution of CF_3CH_2Br and anhydrous zinc chloride was treated with LDA at 15-20 °C to obtain an 89% yield of the α-bromo-β,β-difluorovinylzinc reagent (Scheme 11) as a 70:30 *mono/bis* mixture (*82*). The zinc reagent generated *in situ* was then coupled with iodobenzene using Pd(0) catalyst to give the α-bromo-β,β-difluorostyrene in 76% yield (Scheme 11).

The α-bromo-β,β-difluorovinylzinc reagent could also be generated in 95% yield from CF_2=CHBr by *in situ* treatment with one eq. of LDA and zinc chloride at 15 °C (Scheme 11). The zinc reagent prepared by this route also underwent cross-coupling reaction with iodobenzene in the presence of Pd(0) catalyst to produce a quantitative yield of the corresponding α-bromo-β,β-difluorostyrene (*82*).

The increase in the yield of the α-bromo-β,β-difluorovinylzinc reagent (>89%) by the metallation of CF_3CH_2Br or CF_2=CHBr compared to that of trifluorovinylzinc reagent (73% by the metallation of HFC-134a) under similar

$$CF_3CH_2Br \text{ Or } CF_2{=}CHBr \xrightarrow[15\text{-}20\ ^oC/THF]{LDA,\ ZnCl_2} [CF_2{=}CBrZnCl]\ (>89\%) \xrightarrow[Pd(PPh_3)_4,\ rt - 65\ ^oC]{C_6H_5I} C_6H_5CBr{=}CF_2\ (76\%)$$

Scheme 11. Preparation of α-bromo-β,β-difluorovinyl zinc reagent from CF_3CH_2Br or $CF_2{=}CHBr$ and synthesis of α-bromo-β,β-difluorostyrene.

reaction conditions is most likely due to the increased stability of the intermediate bromodifluorovinyllithium ($CF_2{=}CBrLi$) compared to trifluorovinyllithium ($CF_2{=}CFLi$). In order to confirm this rationalization, bromodifluorovinyllithium ($CF_2{=}CBrLi$) was generated at -80 °C using LDA in THF medium and the resulting brown solution was warmed to -10 °C (turned completely dark) and quenched with zinc chloride. A 63% yield of the bromodifluorovinylzinc reagent was observed by ^{19}F NMR analysis of the reaction mixture. However, a similar sequence of experiments with [$CF_2{=}CFLi$] produced only ~5% yield of the trifluorovinylzinc reagent under similar conditions. These experiments demonstrate the enhanced stability of [$CF_2{=}CBrLi$] compared to [$CF_2{=}CFLi$] at temperatures close to room temperature (*82*).

A series of substituted α-bromo-β,β-difluorostyrenes were then synthesized in excellent isolated yield by utilizing this methodology and representative examples are shown in Table IV.

Table IV. Synthesis of α-bromo-β,β-difluorostyrenes

$$\underset{1.1\ eq.}{CF_3CH_2Br} + \underset{1.0\ eq.}{ZnCl_2} + \underset{2.0\ eq.}{LDA} \xrightarrow[2.\ ArI,\ Pd(PPh_3)_4,\ rt\ to\ 65\ ^oC]{1.\ 15\text{-}20\ ^oC/THF} ArCBr{=}CF_2$$

Entry	*Ar*	*Temp/Time*	*NMR Yield*	*Yield (%)*[a]
1	*p*-ClC_6H_4-	rt, 14 h; then 60 °C 2h	95	82
2	*o*-MeC_6H_4-	60 °C, 12 h	99	86
3	*m*-$NO_2C_6H_4$-	rt, 22 h	89	75
4	*p*-$MeOC_6H_4$-	65 °C, 5 h	94	82
5	*p*-IC_6H_4-	65 °C, 5 h	88	69
6[b]	2-Thienyl	rt, 24 h; then 60 °C, 4h	84	64

[a] Isolated yield of pure styrene. [b] This reaction was sluggish and the product styrene was contaminated with ~5% 2-iodothiophene.

Source: Reproduced with permission from reference 82. Copyright 2004.

Preparation of the α-iodo-β,β-difluorovinyl zinc reagent from CF_3CH_2I and synthesis of α-iodo-β,β-difluorostyrenes

After the successful metallation of HFC-134a, HCFC-133a, CF_3CH_2Br and CF_2=CHBr we attempted the metallation of CF_3CF_2I as the resulting α-iodo-β,β-difluorovinylzinc reagent could be useful for further functionalization at the iodine site. Also, the α-iodo-β,β-difluorovinylzinc reagent could be used for the synthesis of α-iodo-β,β-difluorostyrenes. Following the strategy similar to the *in situ* metallation of HFC-134a, a THF solution of CF_3CH_2I and anhydrous $ZnCl_2$ was treated with LDA at 15-20 °C to obtain an 87% yield of the α-iodo-β,β-difluorovinylzinc reagent (Scheme 12) as a 70:30 *mono/bis* mixture. The metallation occurred only at the hydrogen site and no product was detected by the metallation at the iodine site (CF_2=CHZnCl).

$$CF_3CH_2I \xrightarrow[\text{15-20 °C/THF}]{\text{LDA, ZnCl}_2} \underset{87\%}{[CF_2{=}CIZnCl]} \xrightarrow[\text{rt 12 h, 65 °C 1h}]{\underset{\text{Pd(PPh}_3)_4}{C_6H_5I}} \underset{79\%}{C_6H_5CI{=}CF_2}$$

Scheme 12. Preparation of α-iodo-β,β-difluorovinyl zinc reagent from CF_3CH_2I and synthesis of α-iodo-β,β-difluorostyrene.

The α-iodo-β,β-difluorovinylzinc reagent was coupled with iodobenzene using Pd(0) catalyst at rt to 65 °C to afford the α-iodo-β,β-difluorostyrene in 79% isolated yield. The same sequence was repeated with a series of aromatic iodides to produce the corresponding substituted α-iodo-β,β-difluorostyrenes in good yields, and representative examples are tabulated in Table V. It was observed that the coupling reaction of the aromatic iodides with α-iodo-β,β-difluorovinylzinc reagent was not facile at rt as that of other halodifluorovinylzinc reagents. This is most likely due to the presence of bulky iodine at the α-carbon hindering the transmetallation process in the palladium cycle. But heating of the α-iodo-β,β-difluorovinylzinc reagent mixture at the beginning of the reaction resulted in decomposition of the zinc reagent with the formation of a significant amount of reduced product (CF_2=CHI) along with coupled product and unreacted iodide. The best results were obtained when the reaction mixture was stirred at rt for a prolonged time (~80-90% conversion) followed by heating at 65 °C for a short period to effect complete conversion.

Table V. Synthesis of α-iodo-β,β-difluorostyrenes

$$CF_3CH_2I \; (1.1\ eq.) + ZnCl_2 \; (1.0\ eq.) + LDA \; (2.0\ eq.) \xrightarrow[\text{2. ArI, Pd(PPh}_3)_4\text{, rt to 65 °C}]{\text{1. 15-20 °C/THF}} ArCI{=}CF_2$$

Entry	*Ar*	*Temp/Time*	*NMR Yield*	*Yield (%)*[a]
1	*p*-ClC_6H_4-	rt, 14 h; then 65 °C, 3 h	89	78
2	*o*-MeC_6H_4-	rt, 18 h; then 60 °C,5 h	85	72
3	*m*-$NO_2C_6H_4$-	rt, 15 h; then 65 °C, 4 h	86	78
4	*m*-$MeOC_6H_4$-	rt, 5h; then 65 °C, 2 h	90	80
5[b]	*p*-IC_6H_4-	rt, 24 h; then 65 °C, 4 h	---	62

[a] Isolated yield of pure styrene. [b] Reaction was sluggish; product styrene contaminated with traces of mono-styrene and 1,4-diiodobenzene.

Conclusion

In summary, a room temperature *in situ* approach was developed for the trifluorovinylzinc reagent from readily available halocarbon HFC-134a. Palladium catalyzed cross-coupling of this zinc reagent with aryl iodides afforded an economically viable environmentally friendly route for commercially important 1,2,2-trifluorostyrene monomers. An extension of this method to other halocarbons resulted an efficient synthesis of α-halo-β,β-difluorovinylzinc reagents ($CF_2{=}CXZnCl$, X = Cl, Br, I) and corresponding α-halo-β,β-difluorostyrenes.

Acknowledgements

The authors greatly acknowledge Dr. Charles Stone (Ballard Power Systems Inc. BC, Canada) for fruitful discussions.

Literature Cited

1. Burton D. J; Yang, Z.; Morken, P. *Tetrahedron* **1994**, 50, 2993-3063.
2. Burton, D. J; Lu, L. in *Organo Fluorine Chemistry Techniques and Synthones. Topics in Current Chemistry*; Chambers, RD. Ed.; 1997, 193, p 46-83.
3. Davis, C. R.; Burton, D. J. in *Organozinc Reagents*; Knochel, P; Jones, P. Eds.; 1999, p 57-76.

4. *Chemistry of Organic Fluorine Compounds II*; Hudlicky, M., Pavlath, A. E. Eds.; ACS Monograph 187; American Chemical Society: Washington, DC, 1995; p 647-728.
5 Tarrant, P.; Johncock, P.; Savory, J. *J. Org. Chem.* **1963**, 28, 839-843.
6. Normant, J. F. *J. Organomet. Chem.* **1990**, 19-34.
7. Sorokina, R.S.; Rybakova, L. F.; Kalinovskii, I. O.; Beletskaya, I. P. *Izv. Akad. Nauk. SSSR, Ser. Khim.* **1985**, 1647-1649.
8. Gillet, J. P.; Sauvetre, R.; Normant, J. F. *Synthesis* **1986**, 355-360.
9. Gillet, J. P.; Sauvetre, R.; Normant, J. F. *Synthesis* **1982**, 4, 297-301.
10. Tellier, F.; Sauvetre, R.; Normant, J. F. *J. Organomet. Chem.* **1985**, 292, 19-28
11. Tellier, F; Sauvetre, R.; Normant, J. F. *Tetrahedron Lett.* **1985**, 26, 3999-4002.
12. Tellier, F.; Sauvetre, R.; Normant, J. F. *Synthesis* **1986**, 538-543.
13. Tellier, F.; Sauvetre, R.; Normant, J. F. *J. Organomet. Chem.* **1986**, 303, 309-315.
14. Tellier, F.; Sauvetre, R.; Normant, J. F. *J. Organomet. Chem.* **1987**, 328, 1-13
15. Tellier, F.; Sauvetre, R.; Normant, J. F.; Dromzee, Y.; Jeannin, Y. *J. Organomet. Chem.* **1987**, 331, 281–298.
16. Martinet, P.; Sauvetre, R.; Normant, J. F. *J. Organomet. Chem.* **1989**, 367, 1-10.
17. Martinet, P.; Sauvetre, R.; Normant, J. F. *Bull. Soc. Chim. Fr.* **1990**, 127, 86-92
18. Dubuffet, T.; Bidon, C.; Martinet, P.; Sauvetre, R.; Normant, J. F. *J. Organomet. Chem.* **1990**, 393, 161–172.
19. Hansen, S. W.; Spawn, T. D.; Burton. D. J. *J. Fluorine Chem.* **1987**, 35, 415-420.
20. Nguyen, V. B.; Burton, D. J. *J. Org. Chem.* **1998**, 63, 1714-1715.
21. Morken, P. A.; Baenziger, N. C.; Burton, D. J.; Bachand, P. C.; Davis, C. R.; Pedersen, S, D.; Hansen, S. W. *J. Chem. Soc., Chem. Commun.* **1991**, 566-567.
22. Choi S. K.; Jeong Y. T. *J. Chem. Soc., Chem. Commun.* **1988**, 1478-1479.
23. Shin S. K.; Choi, S. K. *J. Fluorine Chem.* **1989**, 43, 439-441.
24. Jeong, Y. T.; Jung, J. H.; Shin, S. K.; Kim, Y. G.; Jeong, I. H.; Choi, S. K. *J. Chem. Soc., Perkin Trans. 1* **1991**, 1601-1606.
25. Morken, P. A.; Lu, H.; Nakamura, A.; Burton, D. J. *Tetrahedron Lett.* **1991**, 32. 4271-4274.
26. Morken, P. A.; Burton, D. J. *J. Org. Chem.* **1993**, 58, 1167-1172.
27. Jiang, B.; Xu, Y.; *J. Org. Chem.* **1991**, 56, 7336-7340.
28. Heinze, P. L.; Burton, D. J. *J. Fluorine. Chem.* **1986**, 31, 115-119.
29. Heinze, P. L.; Burton, D. J. *J. Org. Chem.* **1988**, 53, 2714-2720.

30. Sprague, L. G.; Snow, A. W.; Griffith, J. R. *J. Fluorine. Chem.* **1991**, 52, 301-306.
31. Dolbier Jr, W. R.; Palmer K; Koroniak, H.; Zhang, H. Q. *J. Am. Chem. Soc.* **1991**, 113, 1059-1060.
32. Dolbier, Jr. W. R.; Koroniak, H.; Burton, D. J.; Heinze, P. L.; Bailey, A. R.; Shaw, G. S.; Hansen, S. W. *J. Am. Chem. Soc.* **1987**, 109, 219-225.
33. Nguyen, B.V.; Burton, D. J. *J. Org. Chem.* **1997**, 62, 7758-7764.
34. Davis, C. R.; Burton, D. J.; *J. Org. Chem.* **1997**, 62, 9217-9222.
35. Liu, Qibo; Burton, Donald J. *Tetrahedron Lett.* **2000**, 41, 8045-8048.
36. Mohtasham, J.; Gard, G. L.; Yang, Z. Y.; Burton, D. J. *J. Fluorine Chem.* **1990**, 50, 31-46.
37. Spawn, T. D.; Burton, D. J. *Bull. Soc. Chim. Fr.* **1986**, 876-880.
38. Nikitina, T. S. *Usp. Khim.* **1990**, 59, 995-1020. (Engl. Transl. 575-589) and references cited their in.
39. Shuyuan, L.; Mahamoud. A.; David. G. *Tetrahedron. Lett.* **2000**, 41, 4493-4497.
40. Narita, T.; Hagiwara, T.; Hamana, H.; Shibasaki, K.; Hiruta, I. *J. Fluorine Chem.* **1995**, 71, 151-153
41. Aoki, T.; Watanabe, J.; Ishimoto, Y.; Oikawa, E.; Hayakawa, Y.; Nishida, M. *J. Fluorine Chem.* **1992**, 59, 285-288.
42. Hodgdon Jr., R. B.; Macdonald, D. I. *J. Polym. Sci., Polym. Chem.* **1968**, 6, 711-717.
43. Stone. C.; Daynard, T. S.; Hu, L-Q.; Mah, C.; Steck, A.E. *J. New Mater. Electrochem. Syst.* **2000**, 3, 43-50.
44. Stone. C.; Steck, A. E. PCT Int. Appl. WO 2001058576, 2001.
45. Steck, A.E.; Stone. C. US Patent 5,834,523, 1998.
46. Momose, T.; Kitazumi, T.; Ishigaki, I.; Okamoto, J. *J. Appl. Polym. Sci.* **1990**, 39, 1221-1230.
47. Wodzki, R.; Narebska, A.; Ceynowa, J. *Angew. Makromol. Chem.* **1982**, 106, 23-35.
48. Kim, H-K. Eur. Pat. Appl. EP 1170310, 2002.
49. Aichholzer, W; Biegert, H; Graf, V; Tober, H; Toth, G. Ger. Offen. DE 10124272, 2002.
50. Cisar, A, J.; Clarke, E. T. PCT Int. Appl. WO 2000063998, 2000.
51. Stone, C.; Summers, D, A. Eur. Pat. Appl. EP 1263066, 2002.
52. Haering, T.; Kerres, J.; Zhang, W. PCT Int. Appl. WO 2003022892, 2003.
53. Lu, L.; Hu, L.; Zhang, W.; Li, W.; He, Y.; Wang, Y. Faming Zhuanli Shenqing Gongkai Shuomingshu. CN 1346707, 2002.
54. Cohen, S. G; Wolosinski, H. T; Scheuer, P. J. *J. Am. Chem. Soc.* **1949**, 71, 3439-3440.
55. Prober M. *J. Am. Chem. Soc.* **1953**, 75, 968-973.
56. Shingu, H.; Hisazumi, M. US Patent 3,489,807, 1970.

57. Dixon, S. *J. Org. Chem.* **1956**, 21, 400-403.
58. McGrath, T. F.; Levine, R. *J. Am. Chem. Soc.* **1955**, 77, 4168-4169.
59. Sorokina, R.S.; Rybakova, L. F.; Kalinovskii, I. O.; Chernoplekova, V. A.; Beletskaya, I. P. *Zh. Org. Khim.* **1982**, 18, 2458-2459.
60. Sigalov, A. B.; Rybakova, L. F.; Beletskaya, I. P. *Izv. Akad. Nauk SSSR, Ser. Khim.* **1983**, 5, 1208.
61. Sigalov, A. B.; Beletskaya, I. P. *Izv. Akad. Nauk SSSR, Ser. Khim.* **1988**, 445-450.
62. Anilkumar, R; Burton, D. J. *Tetrahedron Lett.* **2003**, 44, 6661-6664.
63. Burdon, J.; Coe, P. L.; Haslock, I. B.; Powell, R. L. *Chem. Commun.* **1996**, 49-50.
64. Burdon, J.; Coe, P. L.; Haslock, I. B.; Powell, R. L. *J. Fluorine. Chem.* **1999**, 99, 127-131.
65. Coe, P. L. *J. Fluorine. Chem.* **1999**, 100, 45-52.
66. Brisdon, A. K.; Banger, K. K. *J. Fluorine. Chem.* **1999**, 100, 35-43.
67. Funabiki, K; Ohtsuki, T; Ishihara, T; Yamanaka, H; *J. Chem. Soc., Perkin Trans.1* **1998**, 2413-2423.
68. Anilkumar, R; Burton, D. J. *Tetrahedron Lett.* **2002**, 43, 2731-2733.
69. Stone. C; Packham T. J; Burton D. J; Anilkumar R. US Patent 6,653,515, 2003.
70. Anilkumar, R; Burton, D. J. *J. Org. Chem.* **2004**, 69, (*In Press*).
71. Herkes, F. E.; Burton, D. J. *J. Org. Chem.* **1967,** 32, 1311-1318.
72. Naae, D. G.; Burton, D. J. *J. Fluorine. Chem.* **1971**, 1, 123-125.
73. *Fluorine in Organic Chemistry*, Chambers C. D. Ed.; John Wiely and Sons, Inc., N.Y. 1973, pages 154 and 351.
74. Masure, D.; Sauvetre, R.; Normant, J. F.; Villieras, J. *Synthesis* **1976**, 761-764
75. Masure, D.; Chuit, C.; Sauvetre, R.; Normant, J. F. *Synthesis* **1978**, 458-460.
76. Shigeoka, T.; Kuwahara, Y.; Watanabe, K.; Sato, K; Omote, M.; Ando, A.; Kumadaki, I. *J. Fluorine. Chem.* **2000**, 103, 99-103.
77. Ando, A.; Kumadaki, I. *J. Fluorine. Chem.* **1999**, 100, 135-146.
78. Nishihara, M; Nakamura, Y; Maruyama, N; Sato, K; Omote, M; Ando, A.; Kumadaki, I. *J. Fluorine. Chem.* **2003**, 247-249.
79. Burdon, J.; Coe, P. L.; Haslock, I. B.; Powell, R. L. *J. Fluorine Chem.* **1997**, 85, 151-153.
80. Barnes, N. A.; Brisdon, A. K.; Cross, W. I.; Fay, J. G.; Greenall, J. A.; Pritchard, R. G.; Sherrington. J. *J. Organomet. Chem.* **2000**, 616, 96-105.
81. Anilkumar, R.; Burton, D. J. *Tetrahedron Lett* **2002**, 43, 6979-6982.
82. Anilkumar, R.; Burton, D. J. *J. Fluorine. Chem.* **2004**, 125, 561-566.

Chapter 7

Vinyl Fluorides in Cycloadditions

Günter Haufe

Organisch-Chemisches Institut, Universität Münster,
D–48149 Münster, Germany

Synthetic applications of fluoroalkenes in [4+2]- and [2+1]-cycloadditions are reviewed. Vinyl fluorides are more electron rich than the parent olefins. Thus, simple fluoroalkenes such as α-fluorostyrene are quite weak dienophiles in Diels-Alder reactions, while α-fluoro-α,β-unsaturated carbonyl compounds are useful ones. On the other hand, simple vinyl fluorides are versatile substrates for carbene additions.

Introduction

Fluorinated compounds have gathered an important position in synthesis of substances exhibiting interesting properties for new materials or for agricultural or medicinal purposes (*1*). Among such compounds fluoro alkenes (vinyl fluorides) have been shown to be much better surrogates for peptides than to parent alkenes itself. Fluoro olefins, including di- and trifluoro alkenes have been reviewed as mechanism-based enzyme inhibitors almost fifteen years ago (*2*).

General considerations

First, some general remarks on the properties of the title compounds should be summarized. Generally, there is little information concerning the effect of a single fluorine substituent on the stability, the electronic properties and the resulting reactivity of the π-system of simple vinyl fluorides (*3*). Only fluoro ethylene itself and several of its halogenated derivatives have been investigated in more detail by experimental and theoretical methods (*4*). These investigations show that a single fluorine substituent has a quite weak influence. Qualitatively, the ground state of a monofluoro alkene can be described as shown in Figure 1. The general electron withdrawing effect of the fluorine atom on the σ-system is offset by inductive and mesomeric p-π-interaction, which increases the π-electron density and polarizes the p-system (*5*).

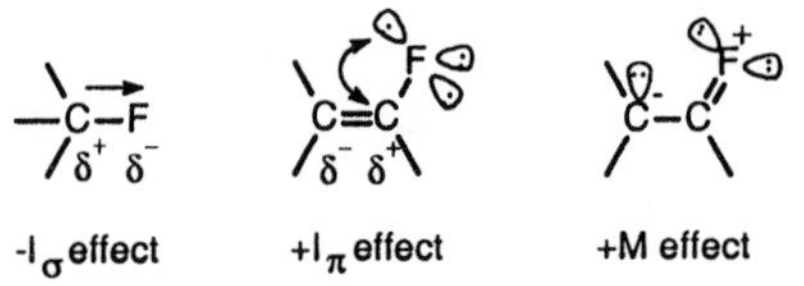

$-I_\sigma$ effect $+I_\pi$ effect   +M effect

Figure 1. Electronic effects of a single fluorine substituent

Although the atomic charges on carbon atoms are quite different for ethylene and fluoro ethylene, the HOMO and LUMO energies are only weakly affected (*4b*). The ionization potential of vinyl fluoride, determined by photoelectron spectroscopy, is only 0.14 eV higher for vinyl fluoride (*6*) and the vertical ionization potential is only 0.05 eV higher (*7*) showing the weak interaction of the π-bond with the 2p-orbital of the fluorine atom.

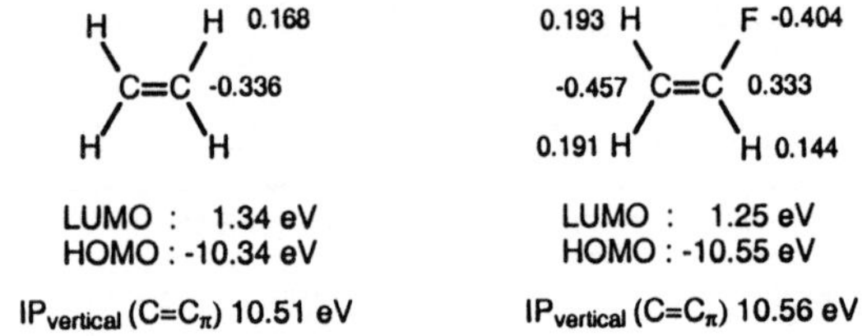

Ethylene	Fluoro ethylene
H 0.168; C -0.336	H 0.193, F -0.404; C -0.457, C 0.333; H 0.191, H 0.144
LUMO : 1.34 eV	LUMO : 1.25 eV
HOMO : -10.34 eV	HOMO : -10.55 eV
$IP_{vertical}$ ($C{=}C_\pi$) 10.51 eV	$IP_{vertical}$ ($C{=}C_\pi$) 10.56 eV

*Figure 2. NBO charges (HF/6-311++G**), HOMO and LUMO energies, and ionization potentials (PE spectroscopy) of ethylene and fluoro ethylene*

Moreover, the bond lengths and the bond angles are also influenced by a fluorine substituent to some extent (Figure 3) (*8*).

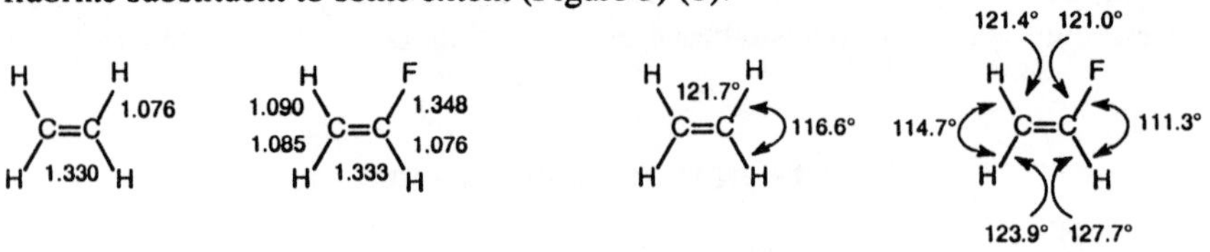

Ethylene: C–H 1.076; C=C 1.330. Fluoro ethylene: C–H 1.090, 1.085; C–F 1.348; C–H 1.076; C=C 1.333.

Ethylene: 121.7°, 116.6°. Fluoro ethylene: 121.4°, 121.0°, 114.7°, 111.3°, 123.9°, 127.7°.

Figure 3. Bond lengths and bond angles of ethylene and fluoro ethylene

Synthesis of fluorinated olefins has been summarized briefly (*9*). Principally, there are several methods for the preparation of such compounds. Most generally applicable are elimination reactions of substituted saturated fluorine compounds, such as vicinal halofluorides, fluorohydrins or its derivatives, β-fluorosulfoxides or -selenoxides. Nucleophilic substitutions of other vinylic substituents, electrophilic additions across triple bonds and Wittig-type reactions with fluorinated C-H-active compounds are also known. Moreover, several more methods for particular types of fluoro alkenes have been described in literature (*9*).

In this review we like to illustrate the applicability of fluorinated olefins in synthesis of analogs of natural products and compounds of potential biological/medicinal interest. The starting materials for our own investigations were generally produced in a two-step sequence by bromofluorination of olefins (*10*) and subsequent dehydrobromination according to ref. (*11*). In this way a selection of *p*-substituted α-fluorostyrenes was synthesized in good overall yields (*12*). Corresponding β-fluorostyrenes were obtained from the respective benzaldehydes according to a procedure described by Burton et al. (*13*).

Diels-Alder reactions of vinyl fluorides

Simple frontier orbital calculations (AM1) have shown that vinyl fluorides such as α- and β-fluorostyrene should react as dienophiles in [4+2]-cycloadditions. Surprisingly, these fluoro alkenes did not give any Diels-Alder products under thermal conditions (toluene, 120°C, sealed tube) with common dienes such as 2,3-dimethylbutadiene, cyclopentadiene, or Danishefsky's diene. Also other reactions such as [2+2]-cycloadditions or polymerizations were not observed under these conditions. Moreover, all attempts failed to catalyze the Diels-Alder reaction with Lewis acids or radical cation inducers. Furthermore, no reaction was detected under 9 kbar pressure. Only the super diene, 1,3-diphenylisobenzofuran, gave the expected diastereomeric Diels-Alder adducts after heating in toluene at 120°C (sealed tube) for several days. The reaction rate was about 30 times lower for 1-(*p*-fluorophenyl)-1-fluoroethene and about 17 times lower for (*E*)-2-(*p*-fluorophenyl)-1-fluoroethene compared to *p*-fluorostyrene. The *endo/exo* selectivity depends slightly on the position of the fluorine substituent at the double bond, but is almost independent of the ring substituent, which has also a quite weak effect on the reaction rate. From the fact that both, electron withdrawing and electron donating substituents in the aromatic ring did accelerate the reaction and from quantum chemical calculations it was deduced that these reactions are so called neutral [4+2]-cycloadditions (*14*).

X	*endo*	*exo*	Yield [%]	Relative rate*
H	60	40	63	1
p-Cl	65	35	58	3.7
p-F	60	40	52	1.3
m-Me	62	38	42	1.4

* compared to *p*-fluorostyrene, relative rate 39 (*endo*/*exo* ratio 70:30)

X	*endo*	*exo*	Yield [%]	Relative rate*
H	40	60	41	1
p-Cl	40	60	38	2.1
p-F	42	58	40	1.3
m-Me	43	57	45	1.1

* compared to *p*-fluorostyrene, relative rate 21

Thus, the *endo*-selectivity* of the reaction of α-fluorostyrene (60:40) with 1,3-diphenylisobenzofuran was slightly lower than in case of styrene itself (72:38). (*E*)-β-fluorostyrene reacted weakly *exo*-selective (40:60). The shift of selectivity towards a bigger part of *exo*-products was explained mainly by electrostatic repulsion of fluorine and oxygen in the transition state leading to the *endo*-products where these groups are in *syn*-orientation. Consistently, (*Z*)-β-fluorostyrene and (*Z*)-1-fluoronon-1-ene gave the *endo*-products exclusively. (*14*).

In contrast, α- and (*E*)-β-fluorostyrenes were shown to be useful dienophiles in reactions with polyfluorinated cyclohexa-2,4-dienones (*15*). Presuming a concerted process, these reactions proceed as [4+2]-cycloadditions with an inverse electron demand to give almost 1:1 mixtures of *endo*- and *exo*-products in case of α-fluorostyrenes. Also a *p*-substituent influencing the electronic properties of the double bond did change the ratio of diastereomers. This was

* *Endo*-selectivity means preferred *endo*-position of the phenyl group.

also true for reactions of *p*-substituted (*E*)-β-fluorostyrene derivatives, which gave the *endo*-products preferably (*16*).

benzene, reflux, 20 h

Ar	X	Y	Yield		
Ar = Ph	X = Cl	Y = F	70 %	53	47
Ar = *p*-Cl-Ph	X = Cl	Y = F	78 %	42	58
Ar = Ph	X = Ph	Y = C_6F_5	90 %	65	35

toluene, reflux, 30 h

Ar	Yield		
Ar = Ph	77 %	74	26
Ar = *p*-Cl-Ph	78 %	65	35
Ar = *p*-F-Ph	75 %	67	33
Ar = *m*-Me-Ph	79 %	63	37

Also other electron withdrawing substituents such as phenyl attached to the 2-position of the 1,3-diene moiety do support the cycloaddition, while the electron donating methoxy group did prevent this reaction, but giving rise to a 1,3-shift of a fluorine leading to the corresponding cyclohexa-2,5-diene (*16*).

toluene, reflux
70 %

It is known from reactions of non-fluorinated olefins that electron withdrawing substituents attached to the double bond accelerate the reaction rate. Thus, several α-fluoro-α,β-unsaturated carbonyl compounds have been synthesized and applied in Diels-Alder reactions. Vinyl fluorides can be hydroxylated in β-position to the fluorine substituent very selectively to give allylic alcohols (*17*), which can easily be oxidized to the corresponding ketones (*18*). These ketones as well as α-fluoro-α,β-unsaturated esters (*19,20*) or corresponding amides (*20*) have been shown to be much better dienophiles compared to simple vinyl fluorides. By way of example, 2-fluorooct-1-en-3-one with 2,3-dimethylbutadiene or *o*-quinodimethane gave the expected Diels-Alder products in 65% and 25% yield, respectively (*18*).

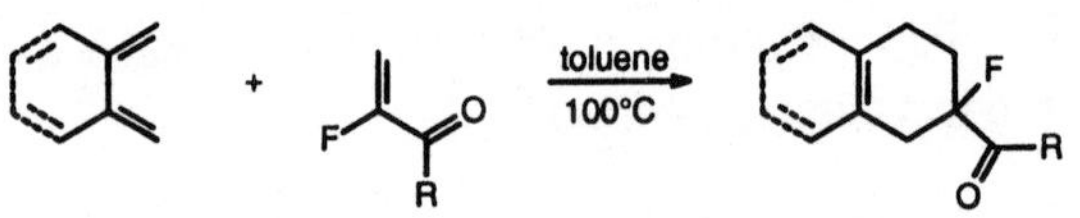

For similar cycloadditions of butadiene or Danishefsky's diene with substituted 2-fluoroacrylates quite harsh conditions were necessary (*21*,*22*).

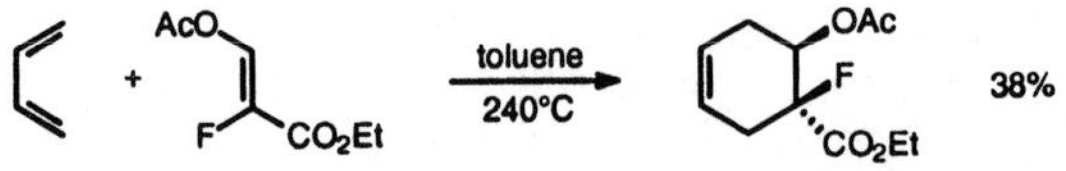

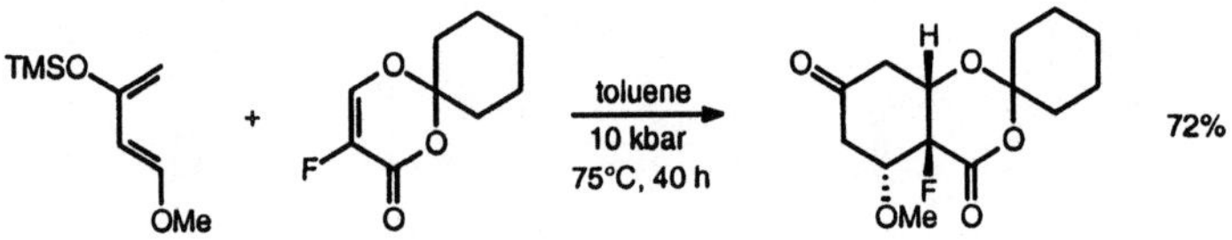

The reaction of cyclopentadiene with α-fluoroacroleine yielded 80% of a 2:1 mixture of the two diastereomers, which were not assigned (*23*). Similarly, thebaine with the same dienophile gave a mixture of the *endo*/*exo*-isomers together with 11% of a hetero Diels-Alder product (*24*). Furthermore, a fluorovinyl sulfone with cyclopentadiene, fulvene or 1-methoxy-cyclohexa-1,3-dienene, respectively, gave 1:1 mixtures of diastereomers. Without the activating phenylsulfonyl group much more drastic conditions were necessary (*25*).

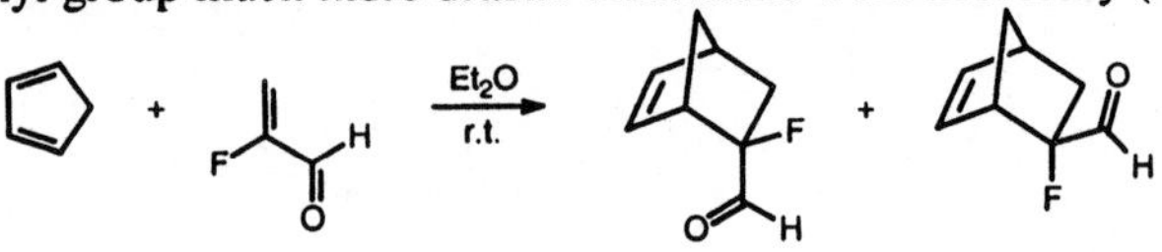

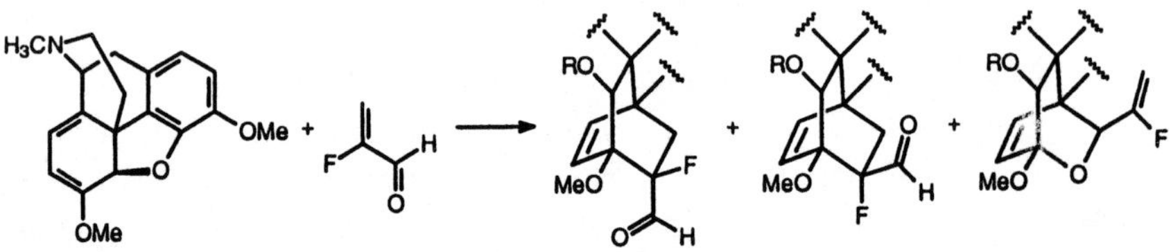

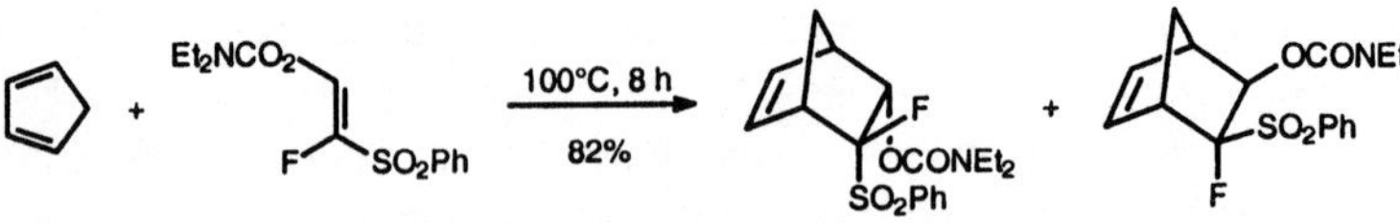

Reactions of cyclopentadiene with 2-fluorooct-1-en-3-one were shown to yield preferably the *exo*-product concerning the carbonyl moiety, both under thermal conditions (toluene, 110°C or ultra sonication) or with Lewis acid mediation at low temperature (Table). In contrast, the corresponding non-fluorinated compound yielded the *endo*-isomer as the main product (*19*).

conditions

X	Conditions	*exo*	*endo*	Yield [%]
F	toluene, 110°C, 55 min	73	27	59
H	toluene, 110°C, 30 min	25	75	74
F	ultra sound, 16 min	78	22	48
H	ultra sound, 12 min	21	79	64
F	50% $TiCl_4$, -55°C, 170 min	90	10	46
H	20% $TiCl_4$, -55°C, 15 min	4	96	73

Five years ago, Taguchi at al. used over-stoichiometric Lewis acids as mediators for the cycloaddition of cyclopentadiene with benzyl 2-fluoroacrylate to give the *exo*-product exclusively (*26*). In contrast, the thermal reaction of this diene with dimethyl fluorofumarate gave the cycloadduct bearing the fluorine in *endo*-position as the major product (*27*).

toluene reflux, 3 h 84% — 64 : 36

Recently, we became interested in optically active Diels-Alder products. Taguchi et al. already applied $TiCl_4$ and other Lewis acids as valuable mediators in cycloadditions of an optically active 2-fluoroacrylic acid amide or a 8-phenylmenthyl ester to cyclopentadiene (*20*).

Lewis acid, CH_2Cl_2, -78°C

$TiCl_4$ (1.2 eq.) 92% yield, *exo/endo* = 78:22
Et_2AlCl (1.5 eq.) 88% yield, *exo/endo* = 50:50

Lewis acid, CH_2Cl_2

$TiCl_4$ (1.2 eq.) 77% yield, 100% *exo*-, 95% de
Et_2AlCl (1.5 eq.) 79% yield, 100% *exo*-, 95% de

The first ever reported asymmetric [4+2]-cycloaddition of cyclopentadiene with 2-fluorooct-1-en-3-one in the presence of equimolar amounts of the enantiopure complexes **A**, **B**, and **C** gave mainly the *exo*-products with maximum 43% ee (Table 3), though the α-fluorine led to a significant enhancement in enantioselectivity compared to the corresponding reactions of oct-2-en-3-one, which were highly *endo*-selective.

Table 3. Results of asymmetric Diels-Alder reactions of cyclopentadiene and 2-fluorooct-1-en-3-one and oct-1-en-3-one

X	* Ti-mediator	temperature	time [h]	*exo* (ee)	*endo* (ee)	Yield [%]
F	A	-60	19	67 (6)	33 (0)	76
H	A	-60	0.5	5 (n.d.)	95 (4)	76
F	B	-40	100	79 (34)	21 (n.d.)	51
H	B	-60	15	6 (n.d.)	94 (19)	87
F	C	-20	138	72 (43)	28 (28)	62
H	C	-60	15	3 (n.d.)	97 (17)	89

Generally, a second fluorine substituent attached to the double bond seems to increase the reactivity of the corresponding dienophile (*28*). Thus, 1,1-difluoroethene derivatives with cyclopentadiene (*29*) or furan (*29*) gave the corresponding cycloadducts in reasonable yields. *Trans*-1,2-Difluorodinitro ethene and the diene reacted at ambient temperature to give the cycloadduct in moderate yield (*31*). Several perfluoroalk-1-en-1-yl phenyl or alkyl ketones having a *trans*-orientation of the vicinal fluorines reacted with cyclopentadiene quite smoothly to give mainly the products with one fluorine and the perfluoroalkyl group in *exo*-position (*32*).

R	R_f	time [min]	*endo*	*exo*	Yield [%]
Ph	C_2F_5	90	89	11	75
C_8H_{17}	C_2F_5	90	82	18	84
C_8H_{17}	CF_3	90	83	17	76
Y	C_2F_5	90	89	11	78
Z	C_2F_5	180	89	11	75

The reaction of 1,3-diphenylisobenzofuran with (*Z*)-4-fluorooct-4-en-3-one gave the *exo*-product almost exclusively (*33*), while from perfluoroalk-1-en-1-yl phenyl ketone a 1:1-mixture of *endo*- and *exo*-isomers was obtained (*32*).

toluene
2 d, reflux
55%

toluene
8 d, reflux
88%

1 : 1

Interestingly, the Diels-Alder product(s) of diphenylisobenzofuran and (1-fluorovinyl) phenyl sulfoxide is (are) not stable at the temperature of formation (100°C, 8 h), but decomposed to form 2-fluoro-1,4-diphenylnaphtalene in 80% yield (*34*).

Complete *exo*-selectivity was also observed in the synthesis of fluorine analogues of the anti-cancer active compounds cantharidin and norcantharidin. The [4+2]-cycloadditions of furan with fluoro- or difluoromaleic anhydrides led exclusively to the products bearing the anhydride function in *exo*-position (*35*).

Δ

X = H, F

The following Diels-Alder reactions of Dane's diene with different partially or perfluorinated *p*-benzoquinones shed some light on the different electronic properties of fluorinated and non-fluorinated α,β-unsaturated ketones. Fluoro-*p*-benzoquinone having only one fluorinated double bond reacted on the non-fluorinated one exclusively to give the corresponding *endo*-products and double bond isomers (shown for 2,3-difluoro-*p*-benzoquinone). From fluoro-*p*-benzoquinone regioisomers concerning the fluorine substituent were formed additionally (*36*).

Et_2O
sealed tube
50°C, 45 min
76%

73 : 27

In contrast, the reaction of 2,6-difluoro-*p*-benzoquinone with the same diene gave two regioisomeric *exo*-products in almost quantitative yield (*36*).

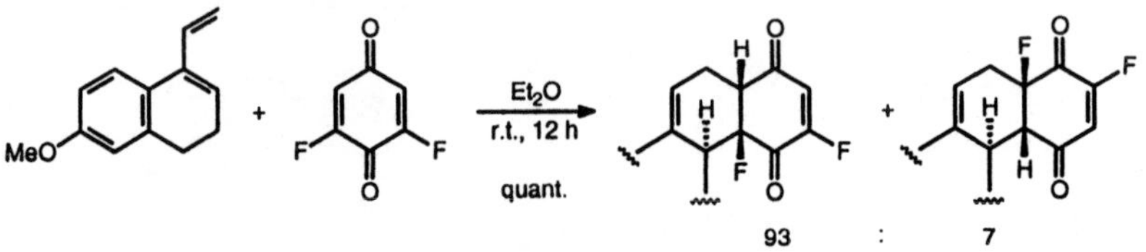

Et_2O
r.t., 12 h
quant.
93 : 7

No reaction occurred with tetrafluoro-*p*-benzoquinone (fluoranil) under similar conditions. However, refluxing in toluene for two hours gave a 89:11 mixture of the *exo*- and *endo*-isomers in moderate yield due to competing reduction of fluoranil to tetrafluorohydroquinone. The latter was isolated in 33% yield (*36*).

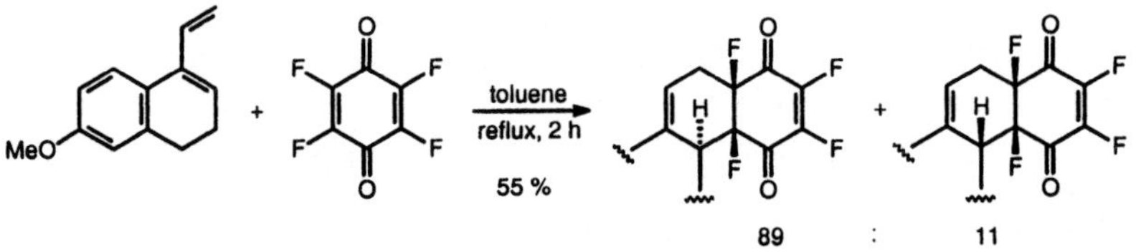

toluene
reflux, 2 h
55 %
89 : 11

Additional Diels Alder reactions of fluoranil with buta-1,3-diene (*37,38*), 1-acetoxy-1,3-butadiene (*38*), and cyclopentadiene (*38,39*) are known in literature.

In summary, simple vinylfluorides, which have to be considered as relatively electron rich species, were shown to be poor dienophiles in normal Diels-Alder reactions, but are useful in such reactions with inverse electron demand. Moreover, α-fluoro-α,β-unsaturated carbonyl compounds are versatile dienophiles, which in contrast to their non-fluorinated parent compounds do react *exo*-selectively with a variety of 1,3-dienes.

Cyclopropanations of vinyl fluorides

Earlier it has been mentioned already that the π-bond of monofluoroalkenes has to be considered as electron rich compared to the parent alkenes. Consequently, electrophilic additions should be favoured reactions. As early as in 1969 Haszeldine et al. reported on the addition of dichlorocarbene towards vinylfluoride (*40*) and we demonstrated recently that dichlorocarbene prepared from chloroform and sodium hydroxide in the presence of a phase transfer catalyst reacted slightly faster with α-fluorostyrene compared to styrene itself (*41*). In contrast, the diastereoisomeric β-fluorostyrenes did not react under these conditions.

Semiempirical MO calculations (AM1) for a solution model (in chloroform) taking into account the PTC conditions suggest a non-symmetric attack of the carbene preferably at C_2 of α-fluorostyrene. Optimization of this transition state suggests the orientation of the lone pair of the carbene away from the double bond due to frontier orbital interactions maximizing the overlap with its π^* orbital. Neither the small difference in the calculated activation enthalpies of styrene and α-fluorostyrene nor the frontier orbital energies (AM1) in chloroform ($HOMO_{styrene}$ = −9.21 eV, $HOMO_{\alpha\text{-fluorostyrene}}$ = −9.45 eV, $LUMO_{dichlorocarbene}$ = −1.02 eV) can explain the higher reactivity of α-fluorostyrene found in the experiment.

Activation Enthalpies and Gibbs Energies (AM1) for the Dichlorocarbene Addition in Chloroform [kcal/mol]

X	Approach to C-1 $\Delta H^\ddagger$	Approach to C-1 $\Delta G^\ddagger$	Approach to C-2 $\Delta H^\ddagger$	Approach to C-2 $\Delta G^\ddagger$
H	13.1	28.9	6.3	21.9
F	no t.s. found		7.4	22.2

X	Y	Approach to C-1 $\Delta H^\ddagger$	Approach to C-1 $\Delta G^\ddagger$	Approach to C-2 $\Delta H^\ddagger$	Approach to C-2 $\Delta G^\ddagger$
H	F	14.6	30.3	11.7	27.7
F	H	14.0	29.0	11.3	26.5

In contrast, the observation that (*E*)- and (*Z*)-β-fluorostyrene did not react under the reported conditions is reflected by the much higher activation barriers, which are 4-5 kcal/mol higher compared to those calculated for α-fluorostyrene (*41*).

The related reaction with dibromocarbene with α-fluorostyrene was very slow yielding 54% of the fluorinated dibromocyclopropane after bulb-to-bulb distillation. Similar to the corresponding reactions of 1-bromo-1-fluoro-cyclopropyl-allyl rearrangement with silver salts of this compound needed quite harsh conditions and was not complete after 8 days heating in acetic acid at 120°C to give 63% (GC) of the fluorobrominated allylic acetate (*41*).

$CHBr_3$, 50% NaOH, TEBAC, CH_2Cl_2, r.t., 72 h — 54 %; AgOAc, $AgNO_3$, HOAc, 120°C, 8 days — 63%, GC

Fluorinated cyclopropanes, besides addition of fluorocarbenes to olefins, were also formed from vinyl fluorides by addition of non-fluorinated carbenes. Taguchi and co-workers demonstrated the diastereoselective cyclopropanation of fluorinated allylic alcohol derivatives using Simmons-Smith reaction (*42*) and Kirk et al. published a synthesis of a racemic fluorinated ethyl cyclopropanecarboxylate from a β-fluoro-α,β-unsaturated carboxylate and diazomethane (*43*).

Et_2Zn, CH_2I_2
hexane, 0°C, 3.5 h
[50 %]
>98% de

CH_2N_2
h ν

The reaction of diazocarboxylates towards alkenes has been shown to be a very efficient method for the construction of fluorinated cyclopropanecarboxylates (*44*). In 1988 Cottens and Schlosser showed that the reaction of 2,5-dimethyl-3-fluorohexa-2,4-diene with different diazoacetates in the presence of rhodium acetate was neither regio- nor stereoselective leading both to the β- and the γ-fluorochrysanthemic acid esters each as 1:1 mixtures of the *cis/trans*-isomers, showing that there is not a big difference in reactivity of a fluorinated and a non-fluorinated double bond in a 1,3-diene system (*45*).

N_2CHCO_2R
$Rh_2(OAc)_4$
25-39% (1:1)
29-30% (1:1)

Recently, we synthesized fluorinated cyclopropanecarboxylates by this method and showed copper- or rhodium salts as efficient and selective catalysts depending on the applied diazoacetate. The cyclopropnation of α-fluorostyrene with ethyl diazoacetate was catalyzed most efficiently by $Cu(acac)_2$ to give a 1:1 mixture of *cis*- and *trans*-2-fluoro-2-phenylcyclopropanecarboxylates, which were separated chromatographically. After saponification, the carboxylic acids were transformed to the corresponding fluorinated cyclopropylamines by Curtius degradation (*44*). The *trans*-configured compound was shown to be a more active tyramine oxidase inhibitor (*46*) or human recombinant liver monoamine oxidase MAO A or MAO B inhibitor (*47*) than the non-fluorinated parent compound known as the neuroactive tranylcypromine (*48*).

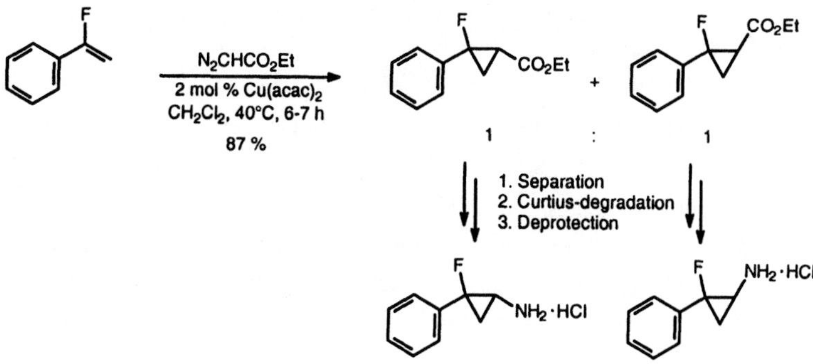

This sequence was also successful for *p*-substituted α-fluorostyrenes such as the *p*-chloro- and *p*-methyl derivatives (*49*), but was shown to be lower in yield with (*E*)-1-fluoro-2-methylstyrene or aliphatic 2-fluoroalkenes (*44*).

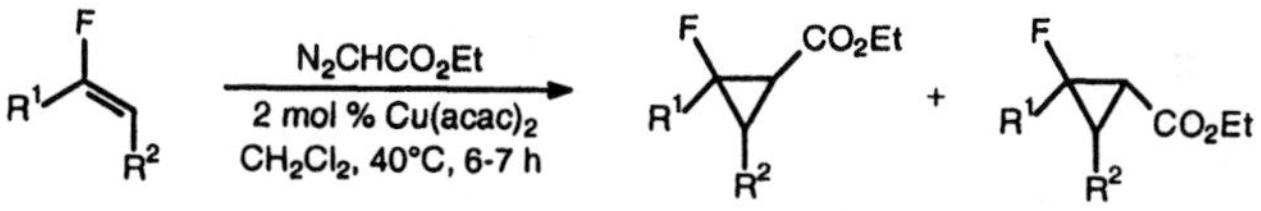

R^1	R^2	*cis*	*trans*	Yield [%]
Ph	H	50	50	87
p-Cl-Ph	H	50	50	91
p-Me-Ph	H	50	50	75
Ph	Me	33	67	63
C_4H_9	H	42	58	36

Similarly to the reactions with dihalocarbenes β-fluorostyrenes showed very low conversion under the above-mentioned conditions and even after 20 hours of reaction time only 26% conversion of the starting material was observed (*44*).

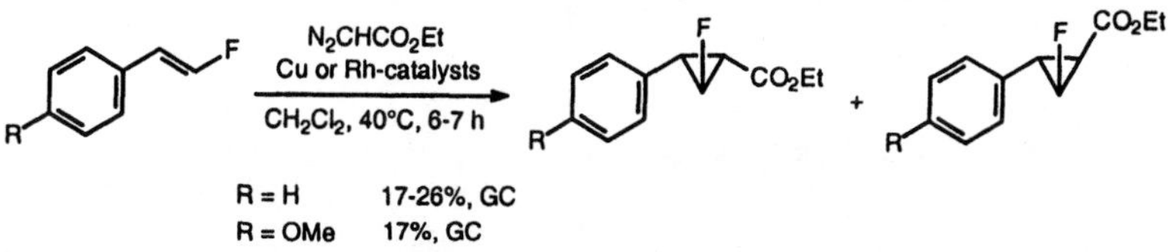

These results were attributed to the mechanism of the cyclopropanation and different intermediary stabilization of cationic centers. It is well documented that a fluorine substituent stabilizes α-carbocations, but destabilizes β-cationic structures (*50*). Provided a concerted formation of the cyclopropane carboxylates like it has been proposed by Brookhart et al. (*51*) and Doyle et al. (*52*), the carbenoid attack at the terminal position of α-fluorostyrene creates a stabilized positive partial charge in α-position to the fluorine substituent. Subsequent nucleophilic attack of the former carbenoid carbon atom and liberation of the metal catalyst leads smoothly to the corresponding cyclopropane carboxylates (shown for the *trans*-isomer).

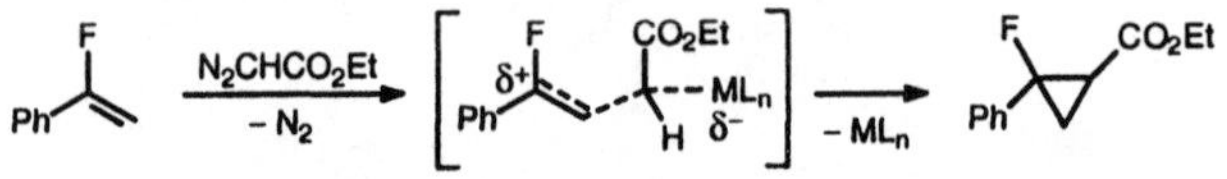

In contrast, the analogous attack of the carbenoid to (*E*)-β-fluorostyrene would lead to a cationic benzyl position bearing a β-fluorine substituent. Fluorine in such a position has a strong destabilizing effect and β-fluoroalkyl cations are not known in solution (*50*). This could be a reason for the low conversion of 1-fluoroalkenes. Also the alternative attack of the carbenoid to the benzylic carbon leading to a primary homobenzylic cationic structure which is stabilized by the fluorine seems not to be much favoured. However, the small amount of **9a** is possibly formed in this way (*44*).

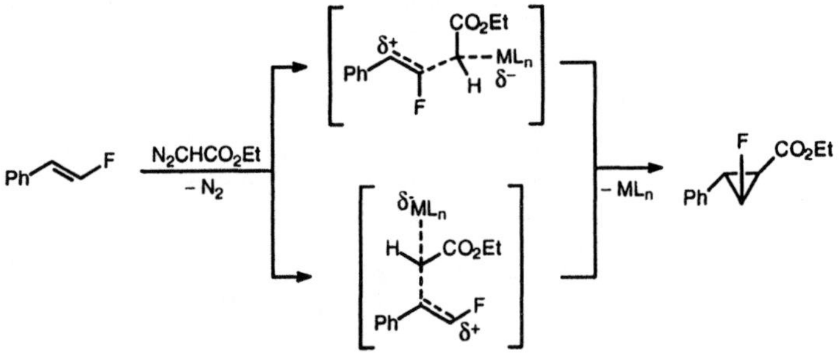

Application of copper catalysts modified by Evans' enantiopure bis(oxazolines) (*53*) was the way of choice for asymmetric cyclopropanation of α-fluorostyrene (*44*). The ligand **D** derived from (*S*)-*tert*-leucine was shown to be the most selective giving enantiomeric excesses up to 93% ee and >98% ee by double stereodifferentiation.

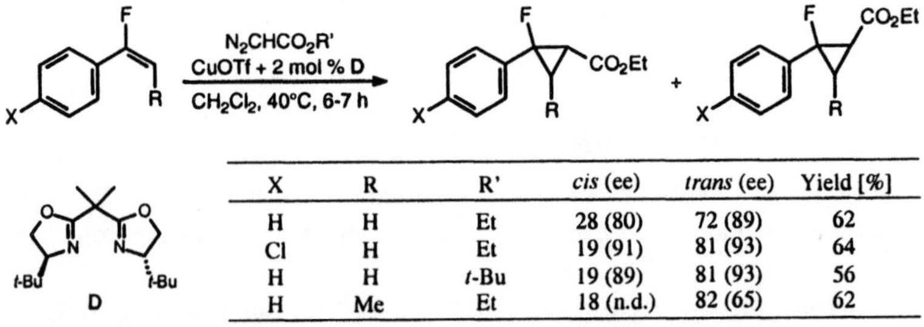

X	R	R'	*cis* (ee)	*trans* (ee)	Yield [%]
H	H	Et	28 (80)	72 (89)	62
Cl	H	Et	19 (91)	81 (93)	64
H	H	*t*-Bu	19 (89)	81 (93)	56
H	Me	Et	18 (n.d.)	82 (65)	62

The following scheme illustrates a plausible mechanism for the asymmetric cyclopropanation (shown for the enantiomers of the *trans*-products). The bis(oxazoline) ligand and the metal carbene are orthogonal to each other. In case of a *si*-attack by the olefin, a steric interaction with the bulky *tert*-butyl group would hinder the turn of the ester-group into the plane for cyclopropane formation. Since such an interaction is absent for the *re*-attack, the latter pathway is favored.

re
R
O
N
$R'O_2C$
C
Cu
N
H
O
R
favoured
F
CO_2R'
(1*S*,2*S*)-

R
O
N
$R'O_2C$
C
Cu
N
H
O
R
si
disfavoured
F
CO_2R'
(1*R*,2*R*)-

The absolute configuration of the major product was proved by X-ray analysis of the *p*-bromoanilide formed from the carboxylic acid under Schotten-Baumann conditions (*44*).

Conclusions

Readily accessable monofluoroalkenes are versatile starting materials for the synthesis of a variety of monofluorinated target compounds. While simple vinyl fluorides due to their comparably electron rich nature are quite weak dienophiles, derived α-fluoro-α,β-unsaturated carbonyl compounds are useful in Diels-Alder reactions. These reactions, in contrast to the corresponding [4+2]-cycloadditions of the non-fluorinated parent compounds, are *exo*-selective. On the other hand vinyl fluorides such as α-fluorostyrene were shown to be more reactive in dichlorocarbene additions and were useful starting materials for the synthesis of fluorinated cyclopropanecarboxylates.

References

1. *Organofluorine Chemistry: Priciples and Commercial Application*, Banks, R. E.; Smart, B. E.; Tatlow, J. C., Eds., Plenum Press: New York, 1994.

Organofluorine Compounds in Medicinal Chemistry and Biomedical Applications, Filler, R.; Kobayashi, Y.; Yagupolskii, L. M., Eds., Elsevier: Amsterdam,1993. *Biomedical Frontiers of Fluorine Chemistry*, Ojima, I.; McCarthy, J. R.; Welch, J. T., Eds., ACS Symposium Series 639, American Chemical Society: Washington, DC, 1996.

2. Bey, P.; McCarthy, J. R.; McDonald, I. A., in *Effects of Selective Fluorination on Reactivity*, Welch, J. T., Ed.; ACS Symposium Series 456, American Chemical Society: Washington, DC, 1991, pp 105-125.
3. Wilkinson, J. A. *Chem. Rev.* **1992**, 92, 505-519. T. Hiyama, *Organofluorine Compounds. Chemistry and Applications*, Springer, Berlin, 2000. Smart B. E. in *The Chemistry of Halides, Pseudohalides and Azides*, Patai, S., Ed. Wiley: Chichester, 1983, pp. 603-656. Dolbier, Jr., W. R. in *Modern Methodology in Organic Synthesis*, Shono T., Ed., VCH: Weinheim, 1992, pp. 391-408 and references cited therein. Dixon, D. A.; Smart, B. E. *J. Phys. Chem.* **1989**, 93, 7772-7780.
4. Fokin, A. V.; Landau, M. A. *Russ. Chem. Rev.* **1998**, **67**, 25-34, and references cited therein. Wiberg, K. B.; Rablen, P. R. *J. Am. Chem. Soc.* **1993**, 115, 9234-9242. Dahlke, G. D.; Kass, S. R. *J. Am. Chem. Soc.* **1991**, 113, 5566-5573.
5. Chambers, R. D.; Vaughan, J. F. S. *Top. Curr. Chem.* **1997**, *192*, 1-38.
6. Reinke, D.; Baumgärtel, H.; Cvitaš, T.; Klasinc, L.; Güsten, H. *Ber. Bunsenges. Phys. Chem.* **1974**, *78*, 1145-1147, cf. also Heaton, M. M.; El-Talbi, M. R. *J. Chem. Phys.* **1986**, *85*, 7198-7210.
7. Yamazaki, T.; Hiraoka, S.; Kitazume, T. in *Asymmetric Fluoroorganic Chemistry: Synthesis, Applications and Future Directions*, Ramachandran, P. V., Ed., ACS Symposium Series, Vol 746, American Chemical Society: Washington, DC, 2000, pp. 142-156.
8. Carlos, J. L.; Karl, R. R.; Bauer, S. H. *J. Chem. Soc., Faraday Trans. 2* **1974**, 177-187.
9. Van Steenis, J. H.; van der Gen, A. *J. Chem. Soc., Perkin Trans. 1* **2002**, 2117-2133.
10. a) Alvernhe, G.; Laurent, A.; Haufe, G. *Synthesis* **1987**, 562-564. b) Haufe, G.; Alvernhe, G.; Laurent, A.; Ernet, T.; Goj, O.; Kröger, S.; Sattler, A. *Org. Synth.* **1999**, *76*, 159-168.
11. Eckes, L.; Hanack, M. *Synthesis* **1978**, 217-219. Suga, H.; Hamatani, T.; Guggisberg, Y.; Schlosser, M. *Tetrahedron* **1990**, *46*, 4255-4260. Michel, D.; Schlosser, M. *Synthesis* **1996**, 1007-1011.
12. Ernet, T.; Haufe, G. *Tetrahedron Lett.* **1996**, *37*, 7251-7252.
13. Cox, D. G.; Gurusamy, M.; Burton, D. J. *J. Am. Chem. Soc.* **1985**, *107*, 2811-2812.
14. Ernet, T.; Maulitz, A. H.; Würthwein, E.-U.; Haufe, G. *J. Chem. Soc., Perkin Trans. 1* **2001**, 1929-1938.

15. Kobrina, L. S. *Izv. Akad. Nauk, Ser. Khim.* **2002**, 1629-1649; *Russ. Chem. Bull., Int. Ed.* **2002,** *51*, 1775-1795.
16. Bogachev, A. A.; Kobrina, L. S.; Meyer, O. G. J.; Haufe, G. *J. Fluorine Chem.* **1999,** *97*, 135-143.
17. Ernet, T.; Haufe, G. *Synthesis* **1997**, 953-956.
18. Essers, M.; Ernet, T.; Haufe G. *J. Fluorine Chem.* **2003,** *121*, 163-170.
19. Essers, M.; Mück-Lichtenfeld, C.; Haufe G. *J. Org. Chem.* **2002,** *67*, 4715-4721.
20. Ito, H.; Saito, A.; Taguchi, T. *Tetrahedron: Asymmetry* **1998,** *9*, 1979-1987 and 1989-1994.
21. a) Araki, K.; Aoki, T.; Murata, T.; Kata, T.; Tanyama, E. *14th International Symposium on Fluorine Chemistry*, Yokohama, 1994, Abstracts of Papers, 1B13. b) Tanyama, E.; Araki, K.; Sotojima, N; Murata, T.; Aoki, T. (Mitsubishi), Jpn. Pat. 0189902/1995, (*Chem. Abstr.* **1995,** *123*, 111593).
22. Iwaoka, T.; Katagari, N.; Sato, M.; Kaneko, C. *Chem. Pharm. Bull.* **1992,** *40*, 2319-2324.
23. Buddrus, J.; Nerdek, F.; Hentschel, P.; Klamann, D. *Tetrahedron Lett.* **1966,** 5379-5383.
24. Jeong, I. H.; Kim, Y. S.; Cho, K. Y. *Bull. Korean Chem. Soc.* **1990,** *11*, 178-179.
25. Crowley, P. L.; Percy, J. M.; Stansfield, K. *Tetrahedron Lett.* **1996**, *37*, 8233-8236 and 8237-8240.
26. Ito, H.; Saito, A.; Kakuuchi, A.; Taguchi, T. *Tetrahedron* **1999**, *55*, 12741-12750.
27. Essers, M. PhD Dissertation, University of Münster, 2001.
28. Percy, J. M. *Top. Curr. Chem.* **1997**, *193*, 131-195.
29. (a) Leroy, J.; Molines, H.; Wakselman, C. *J. Org. Chem.* **1987**, *52*, 290-292. (b) Percy, J. M.; Rock, M. H. *Tetrahedron Lett.* **1992**, *33*, 6177-6180. (c) Crowley, P. L.; Percy, J. M.; Stansfield, K. *Chem. Commun.* **1997**, 2033-2034.
30. Arany, A.; Crowley, P. J.; Fawcett, J.; Hursthouse, M. B.; Kariuki, B. M.; Light, M. E.; Moralee, A. C.; Percy, J. M.; Salafia, V. *Org. Biomol. Chem.* **2004**, *2*, 455-465.
31. Baum, K.; Archibald, T. G. Tzeng, D.; Gilardi, R.; Flippen-Anderson, J. L.; George, C. *J. Org. Chem.* **1991**, *56*, 537-539.
32. Chanteau, F.; Essers, M.; Plantier-Royon, R.; Haufe, G.; Portella, C. *Tetrahedron Lett.* **2002**, *43*, 1677-1680.
33. Ernet, T. PhD Dissertation, University of Münster, 1997.
34. Hanamoto, T.; Korekoda, K.; Nakata, K.; Handa, K.; Koga, Y.; Kondo, M. *J. Fluorine Chem.* **2002**, *118*, 99-101.
35. Essers, M.; Wibbeling, B.; Haufe, G. *Tetrahedron Lett.* **2001**, *42*, 5429-5433.

36. Essers, M.; Haufe, G. *J. Chem. Soc., Perkin Trans. 1* **2002**, 2719-2728.
37. Yakobson, G. G.; Shteingarts, V. D.; Kostina, N. G.; Osina, O. I.; Vorozhtsov Jr., N. N. *Zh. Obshch. Khim.* **1966**, *36*, 142-145; *Chem. Abstr.* **1966**, *64*, 17509d.
38. Hudlicky, M.; Bell, H. M. *J. Fluorine Chem.* **1975**, *5*, 189-201.
39. Wilson, R. M. *J. Org. Chem.* **1983**, *48*, 707-711.
40. Fields, R.; Haszeldine, R. N.; Peter, D. *J. Chem. Soc. C* **1969**, 165-172.
41. Haufe, G.; Meyer, O. G. J.; Mück-Lichtenfeld, C. *Coll. Czech. Chem. Commun.* **2002**, *67*, 1493-1504.
42. Morikawa, T.; Sasaki, H.; Mori, K.; Shiro, M.; Taguchi, T. *Chem. Pharm. Bull.* **1992**, *40*, 3189-3193; Morikawa, T.; Sasaki, H.; Hanai, R.; Shibuya, A.; Taguchi, T. *J. Org. Chem.* **1994**, *59*, 97-103; Taguchi, T.; Okada, M. *J. Fluorine Chem.* **2000**, *105*, 279-283.
43. Sloan, M. J.; Kirk, K. L. *Tetrahedron Lett.* **1997**, *38*, 1677-1680.
44. Meyer, O. G. J.; Fröhlich, R.; Haufe, G. *Synthesis* **2000**, 1479-1490.
45. Cottens, S.; Schlosser, M. *Tetrahedron* **1988**, *44*, 7127-7144.
46. Yoshida, S.; Meyer, O. G. J.; Rosen, T. C.; Haufe, G.; Ye, S.; Sloan, M. J.; Kirk, K. L. *J. Med. Chem.* **2004**, *47*, 1796-1806.
47. Yoshida, S.; Rosen, T. C.; Meyer, O. G. J.; Sloan, M. J.; Ye, S.; Haufe, G.; Kirk, K. L. *Bioorg. Med. Chem.* **2004**, *12*, 2645-2652.
48. a) C. G. Saysell, W. S. Tambyrajah, J. M. Murray, C. M. Wilmot, S. E. V. Phillips, M. J. McPherson, P. F. Knowels, *Biochem. J.* **2002**, *365*, 809-816; b) E. M. Shepard, H. Heather, G. A. Juda, D. M. Dooley, *Biochim. Biophys. Acta* **2003**, *1647*, 252-259. c) Rosen, T. C.; Yoshida, S.; Kirk, K. L.; Haufe, G. *ChemBioChem* **2004**, *5*, 1033-1043.
49. Rosen, T. C.; Yoshida, S.; Kirk, K. L.; Haufe, G. *J. Med. Chem.* **2004**, submitted
50. a) Smart, B. E., in *Organofluorine Chemistry*, Banks, R. E.; Smart, B. E.; Tatlow, J. C., Eds., Plenum Press, New York, 1994; pp 58-88. b) Smart, B. E., in *Chemistry of Organic Fluorine Compounds II*, Hudlický, M.; Pavlath, A. E., Eds., American Chemical Society, Monograph 187, Washington DC, 1995; pp 979-1010.
51. Brookhart, M.; Tucker, J. R.; Husk, G. R. *J. Am. Chem. Soc.* **1983**, *105*, 258-264.
52. a) Doyle, M. P.; Dorow, R. L.; Buhro, W. E.; Griffin, J. H.; Tamblyn, W. H.; Trudell, M. L. *Organometallics* **1984**, *3*, 44-52. b) Doyle, M. P.; Griffin, J. H.; Bagheri, V.; Dorow, R. L. *Organometallics* **1984**, *3*, 53-61.
53. Evans, D. A.; Woerpel, K. A.; Hinman, K. A.; Faul, M. M. *J. Am. Chem. Soc.* **1991**, *113*, 726-728.

Chapter 8

Asymmetric Synthesis of Monofluorinated Compounds Having the Fluorine Atom Vicinal to π-Systems

Michael Prakesch, Danielle Grée, and René Grée*

Université de Rennes 1, Laboratoire de Synthèse et Electrosynthèse Organiques, CNRS UMR 6510, Avenue du Général Leclerc, 35042 Rennes Cedex, France

The different methods, which have been used to date for the preparation of chiral, non racemic, derivatives having a fluorine atom vicinal to π-systems are reported. The scope and limitations of these methodologies are discussed.

Introduction

The presence of fluorine atom(s) in organic molecules induces important changes in their physical, chemical and biological properties (*1-2*). The introduction of the fluorine is often a key step in the preparation of these target molecules; this is especially the case for selective monofluorination in position vicinal to π systems: due to the delocalised structure of the intermediates, the fluorination reactions very often afford mixtures of regio- and stereoisomers (*3-4*). The purpose of this review is to describe the different methods which have been reported to date for the asymmetric synthesis of four different type of monofluorinated derivatives: the type **1** propargylic fluorides, the corresponding allylic derivatives **2** and the polyunsaturated systems **3**, as well as the type **4** benzylic fluorides (Figure1).

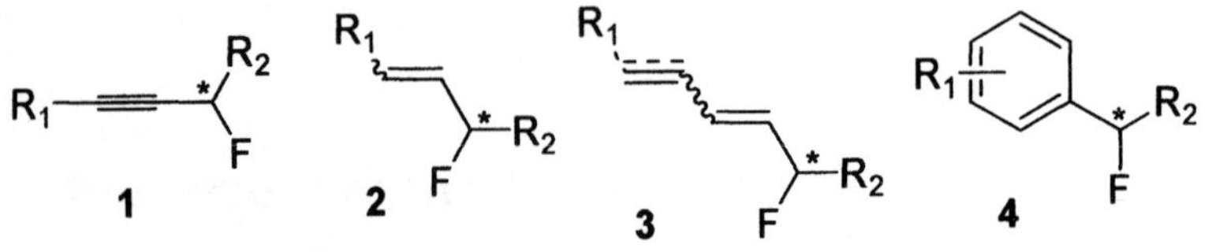

Figure 1 : monofluorinated, optically active, target molecules

For each family, we will consider the different approaches already used for the synthesis of optically active compounds and we will discuss the scope and limitations of the corresponding methodologies. In this chapter we will not consider the asymmetric synthesis and/or catalysis of derivatives having the fluorine atom vicinal to a carbonyl group, such as for instance indanone or tetralone type molecules (*5*), as well as for β-dicarbonyl derivatives (*6*) since this is discussed in another chapter of this book.

Propargylic fluorides

The synthesis of optically active propargylic fluorides has been studied only recently and the successful approach involves the dehydroxy-fluorination of type **5** propargylic alcohols (Scheme 1).

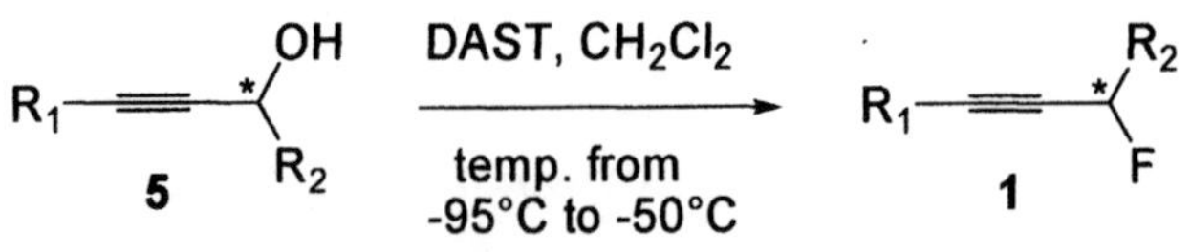

DAST, CH_2Cl_2, temp. from -95°C to -50°C

R_1 = H, alkyl, CO_2Me, $CH(OR)_2$; R_2 = alkyl

Scheme1 : synthesis of propargylic fluorides (*4*)

This reaction, which is fast even at low temperature, proved to be highly versatile. Furthermore, it presents some very important characteristics:

- The conversion of the alcohols **5** is quantitative and the desired fluorides **1** are obtained usually in high yields, the only by-products being the enynes resulting from HF elimination.
- The reaction is fully regiocontrolled and allenes could never be characterized in the crude reaction mixtures.
- The nature of substituents exerts a strong influence on the stereoselectivity of this process: when R_1 is H or an electron-withdrawing group, the reaction is highly stereoselective affording the corresponding fluorides in high ee's, with inversion of configuration. However, when R_1 is an alkyl group some loss of stereoselectivity is observed during the dehydroxy-fluorination, even at very low temperatures (*7-8*).
- The stereochemical outcome of this fluorination can be rationalized in the following way: on the contrary to allylic derivatives, the propargylic carbenium ions are not very much stabilized. Therefore, a competition between the S_N1 and S_N2 processes is possible and it will strongly depend upon the nature of substituents. With H or electron withdrawing groups in terminal position, there is no stabilization of the propargylium cation and the S_N2 reaction is occurring. On the other hand, when R_1 is an alkyl group, a small but significant extra stabilization occurs and there is a small contribution from the S_N1 pathway even at very low temperature (*9*). This is also in agreement with the results obtained using alcohols complexed to the $Co_2(CO)_6$ moiety, where the organometallic complex can stabilize the corresponding carbenium ion: this not only favors an overall retention of configuration but also slightly lowers the stereoselectivity of the fluorination, as compared to the reaction occurring with the free alcohol (*7*). As a consequence, it can be anticipated that the presence of groups stabilizing the propargylium cations will lower the stereoselectivity and therefore induce limitations in the use of this method.
- The propargylic fluorides where R_1 is a functional group are highly versatile intermediates for the total synthesis of monofluorinated analogues of bioactive natural products, as indicated later.

Allylic fluorides

The preparation of optically active allylic fluorides proved to be a very challenging problem. Several strategies were developed, as represented in Scheme 2. They use either stereoselective fluorination processes (pathways ***A-C***) or they start from already fluorinated building blocks (routes ***D-E***).

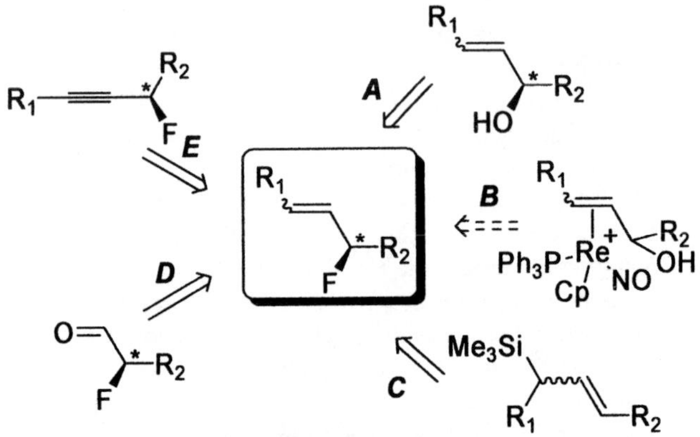

Scheme 2: Strategies developed for the synthesis of chiral, non racemic allylic fluorides

The approaches via the pathways ***A*** and ***B*** will be considered first. It is well established, from the seminal work of Middleton, that the dehydroxy-fluorination of allylic alcohols is usually neither regio- nor stereocontrolled (*10a*). Many further examples have confirmed this absence of stereocontrol (*10b*, *10c*, *11*). It is very likely related to the high stabilization of allylic cations, leading to strong preferences for S_N1 pathways (*9*). However, a recent example describes a stereospecific fluorination in the case of alcohol **6,** affording **7** with overall retention of configuration (Scheme 3) (*12*).

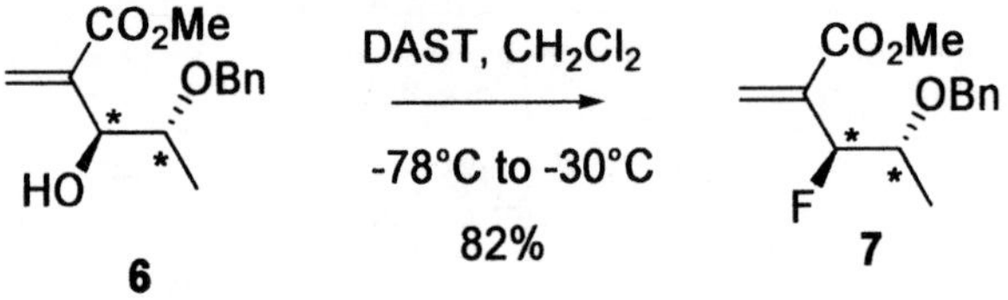

Scheme 3: stereospecific fluorination of allylic alcohol **6** (*12*)

Although no explanation was offered for this selectivity, it can be tentatively suggested that anchimeric assistance by the oxygen from the OBn group is responsible for the overall retention. This type of assistance is often observed in molecules having heteroatoms in β position (*13*).

Another way to control the regio- and stereoselectivity in such a dehydroxy-fluorination is to temporarily protect the double bond by a transition metal complex (pathway ***B***). In the case of allylic alcohols, rhenium complexes proved to be suitable and the fluorination is stereospecific (*14*). Until now, these reactions have been performed only with racemic derivatives but the starting complexes are accessible in optically pure form (*15*). However, it must be noted that no efficient decomplexation method has been reported, to date, for this type of rhenium complexes.

A novel approach has been designed recently to the synthesis of allylic fluorides via pathway *C*: it involves the electrophilic fluorodesilylation of allylsilanes (*16*). The asymmetric transformation is performed in a stoechiometric manner by using appropriate combinations of Selectfluor™ and cinchona alkaloids (Scheme 4). The latter can be either preformed chiral *N*-fluoroammonium salts or, more conveniently, in situ prepared mixtures. For this reaction, the best alkaloid is the bisdihydroquinine pyridine [$(DHQ)_2PYR$] affording excellent yields and very good enantiomeric excesses (up to 96% for R=CH_2Ph and n=1) (*17*). Lower selectivities are obtained with R=H or Me, especially for n=2. At this stage, the asymmetric version has been reported only with type **8** cyclic systems.

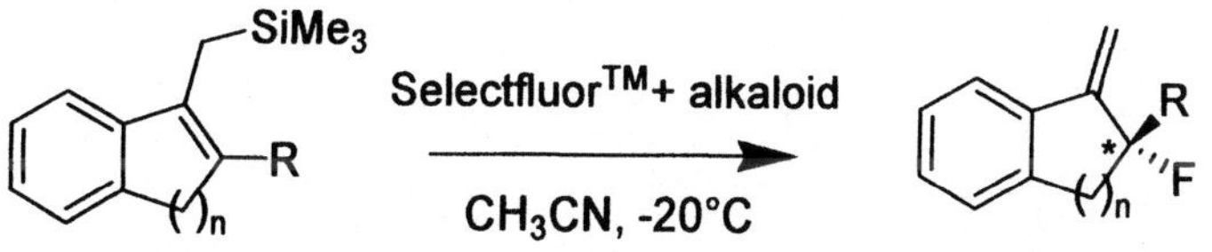

$SiMe_3$ — R — **8** — Selectfluor™+ alkaloid / CH_3CN, -20°C → **9** (R, F)

R = H, Me, CH_2Ph; n = 1, 2

Scheme 4: enantioselective fluorodesylilation of allyl silanes (*17*)

The diastereoselective electrophilic fluorination of chiral enolates is also a very useful strategy for asymmetric fluorination. Using the Evans oxazolidinone chiral auxiliary, the aldehyde **12** can be efficiently prepared. As expected, this intermediate is very prone to racemization under basic conditions but a Horner-Wadworth-Emmons reaction (pathway ***D***) can be used to obtain the allylic fluoride **13**, which is a key intermediate towards fluorosugar **14** (Scheme 5) (*18a*). It has to be further mentioned that the electrophilic fluorination of amide **15** affords the allylic fluoride **16** in a deconjugative process (*18b*).

Finally, as indicated in route ***E***, the propargylic fluorides are suitable intermediates for the preparation of both the *Z* and the *E* allylic fluorides. A representative example is given in Scheme 6: the fluoride **18** was used in the first enantiocontrolled preparation of **19** and **20**. The latter derivative is the fluorinated analogue of the 4-hydroxynonenal (4-HNE), a highly cytotoxic compound (*19*).

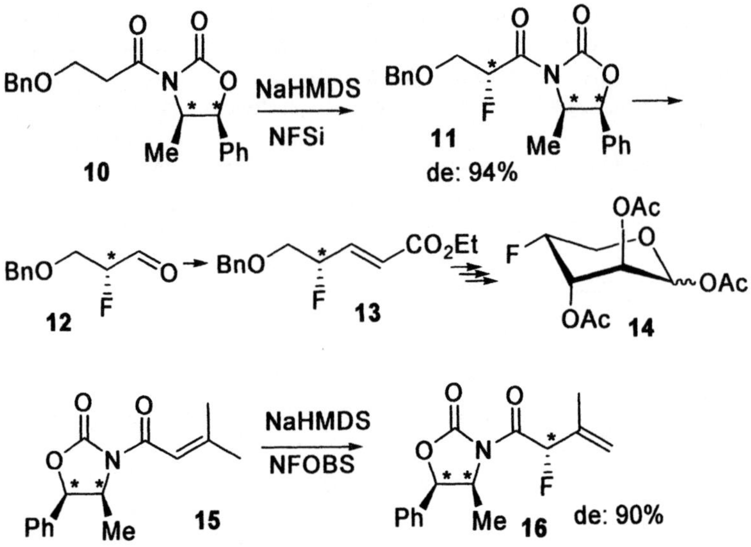

Scheme 5: Diastereoselective electrophilic fluorinations towards allylic fluorides (*18*)

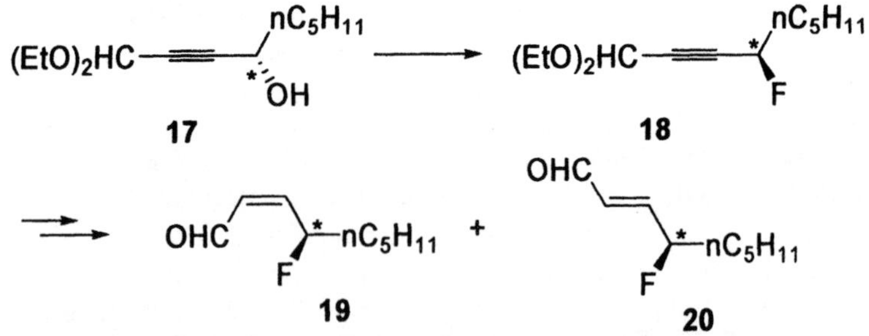

Scheme 6: Fluorinated allylic building blocks (*19*)

Dienyl and polyunsaturated allylic fluorides

Here again, the preparation of such derivatives involves either the direct fluorination of the desired molecule or the use of selected fluorinated key intermediates.

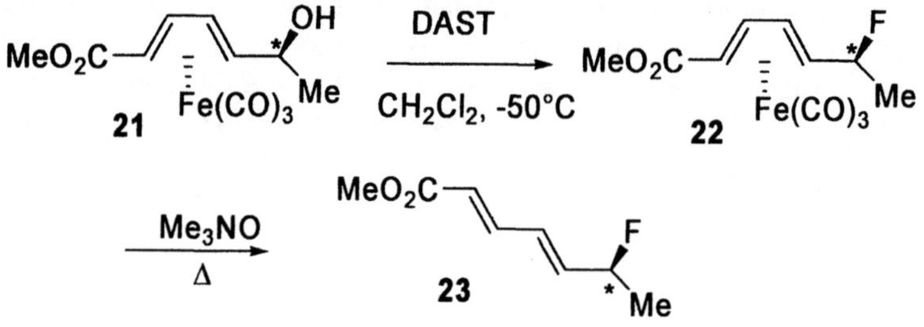

Scheme 7: Use of diene-tricarbonyl iron complexes (*20*)

As far as the first strategy is concerned, the use of chiral, non racemic, diene-tricarbonyl iron complexes is noteworthy (Scheme 7): the dehydroxy-fluorination of complex **21** is stereospecific with overall retention of configuration. The decomplexation of **22** affords then the free diene **23** in optically active form (ee >90%) (*20*). It must be noted that this method will lead only to *E,E* dienes.

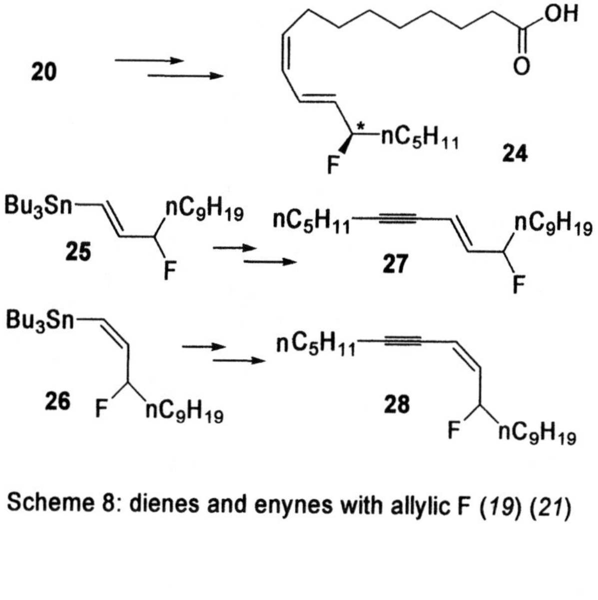

Scheme 8: dienes and enynes with allylic F (*19*) (*21*)

An alternative strategy is to start from already fluorinated building blocks (Scheme 8). A representative example is the synthesis of **24** by a Wittig reaction starting from intermediate **20**. It affords the fluorinated analogue of the 13 HODE, which is a linoleic acid metabolite with various potent biological properties (*19*).

Starting from propargylic fluorides (with R_1=H) it is also easy to prepare the vinyl stannanes **25** and **26**. After iodination, these derivatives are used in Sonogashira reactions (leading to **27** and **28**), as well as in Heck couplings affording the corresponding *E,Z* or *E,E* dienes (Scheme 8) (*21*). This strategy appears versatile since it allows to prepare either the *E* or the *Z* allylic fluorides. Although it has been described only in racemic form until now, the extension to asymmetric synthesis is possible since the propargylic fluoride, direct precursor to stannanes **25** and **26**, has been obtained in optically pure form (*8*).

Benzylic fluorides

For such derivatives, most strategies include direct fluorination of target molecules either by nucleophilic (***A*, *B*, *C***) or electrophilic (***D***) pathways.

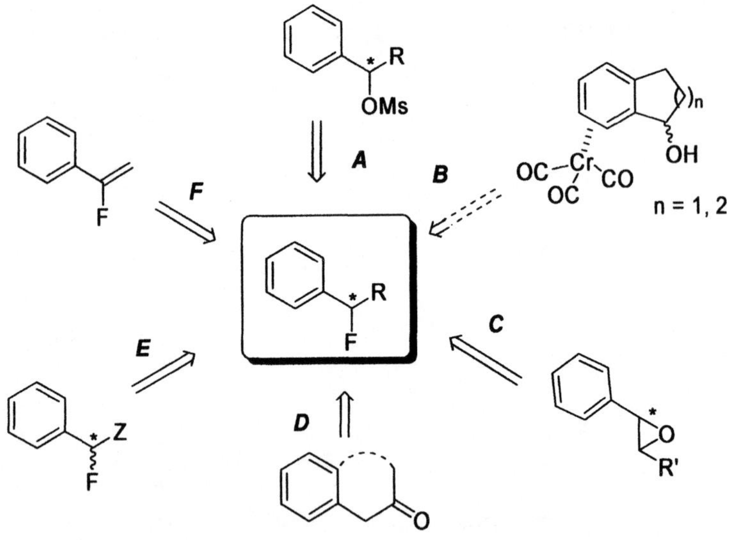

A B C D E F R OMs OC Cr CO OC OH n = 1, 2 F R F Z F O R' O

Scheme 9: strategies towards fluorination in benzylic position

Another useful possibility to be mentioned, involves the resolution of racemic benzylic fluorides (pathway ***E***); finally, it proved to be possible also to start from vinylic fluorides (pathway ***F***) (Scheme 9).

The stereoselectivity of the nucleophilic fluorination of benzyl alcohols, activated as mesylates **29**, is strongly depending upon the nature and the position of the R' group (Scheme 10): with strong electron withdrawing groups such as CN, NO_2 or CO_2Et in para position very high ee's are obtained, while in other cases only racemic derivatives are isolated (*22*). These results are probably related to the stability of the corresponding carbenium ions and the competition between S_N1 and S_N2 processes. Therefore, they give useful informations about the probable limitations of this approach.

In pathway ***B*** it was established that the dehydroxy-fluorination of indanols and tetralols **31** is completely stereocontrolled by the $Cr(CO)_3$ group: the F^- nucleophile always reacts on the face anti to the organometallic unit to give **32** (*23*). Although it has been performed only in racemic form at this stage, this strategy has excellent potentialities for the preparation of optically active benzylic fluorides since chromium complexes of this type have been described in chiral, non racemic form (*24*).

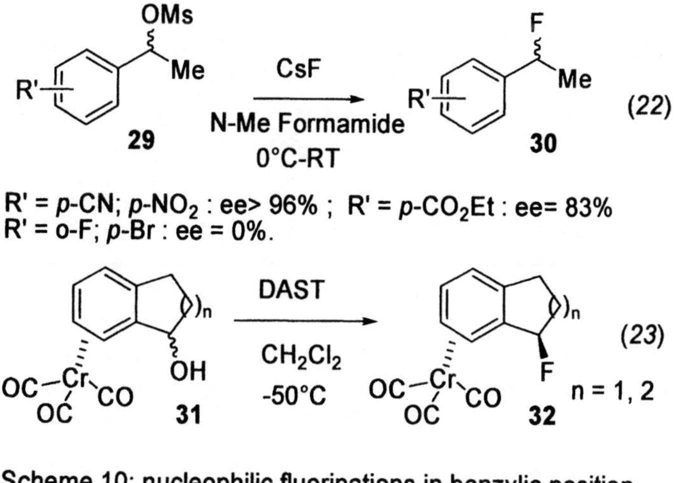

OMs Me R' **29** CsF N-Me Formamide 0°C-RT F Me R' **30** (*22*)

R' = *p*-CN; *p*-NO_2 : ee> 96% ; R' = *p*-CO_2Et : ee= 83%
R' = o-F; *p*-Br : ee = 0%.

OC–Cr(CO)(CO) OH)n **31** DAST CH_2Cl_2 -50°C OC–Cr(CO)(CO) F)n **32** n = 1, 2 (*23*)

Scheme 10: nucleophilic fluorinations in benzylic position

The ring opening hydrofluorination of enantiomerically active arylglycidic esters **33** (pathway ***C***) is another very attractive strategy, taking into account the easy access to the latter derivatives. It was discovered recently that such a fluorination could be performed under very mild reaction conditions using the BF_3-Et_2O complex (Scheme 11): it affords, stereospecifically and with a complete regiocontrol, the corresponding β-fluoro alcohols **34** in moderate to good yields (*25*).

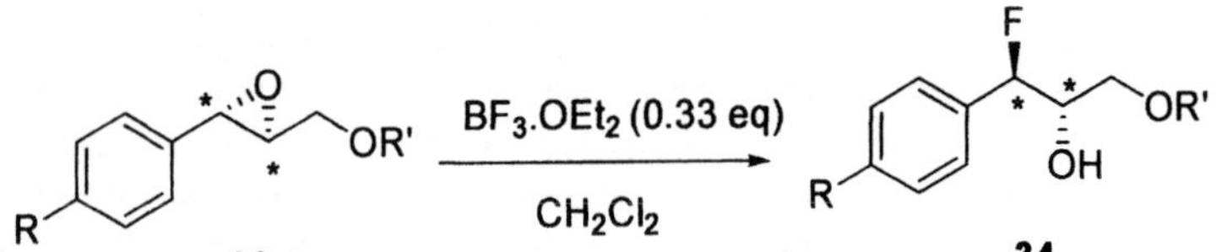

R=H, CF_3; R'=alkyl, aryl, tertiobutyldiphenylsilyl.

Scheme 11: Ring-opening hydrofluorination of phenylglycidol ethers (*25*)

The asymmetric electrophilic fluorination has been a topic of much interest during the last years, including for benzylic fluorides (pathway ***D***). The corresponding reactions have been performed using both the substrate and the reagent controls. Representative examples are given in the Schemes 12 and 13.

As part of his seminal studies in asymmetric fluorination, Differding could demonstrate that camphor derived chiral *N*-fluorosultams react with the enolate of ester **35** to afford fluoro ester **36** in moderate yields (27-34%) and ee's (< 35%) (*26*). Later on, using oxazolidinone as chiral auxiliary, Davis and Kasu obtained excellent results in asymmetric fluorination: the derivative **38** is obtained with very high ee's (> 97%), and it is a useful intermediate for the preparation of the corresponding ketones **39** (*27*).

More recently, using a combination of Selectfluor™ and the dihydroquinidine acetate it was possible to obtain the fluoro cyano benzyl esters **41** in good to excellent yields (56-87%) and ee's (68-87%) (*28*). For this synthesis, a careful optimisation of the reaction conditions was necessary. X-Ray data give good evidence for the *N*-fluoro intermediates, allowing discussions on the factors responsible for the stereocontrol. The same strategy was also very fruitful in the preparation of α-fluoro-α-phenylglycine derivatives **43**. In that case, the nitrile derivatives give systematically higher ee's (up to 94%) than corresponding esters (*29*).

The drug candidate BMS-204352 (MaxiPost) **45** is probably among the most extensively studied benzylic fluorides during the last few years. The more active (*S*) isomer can be obtained by chromatography or resolution (*30*). Asymmetric fluorination was also successful in that case: using the appropriate *N*-fluoroammonium salts of cinchona alkaloids [$(DHQ)_2$AQN], **45** is obtained in excellent yields and ee's (up to 88%) (Scheme 13) (*31, 32*).

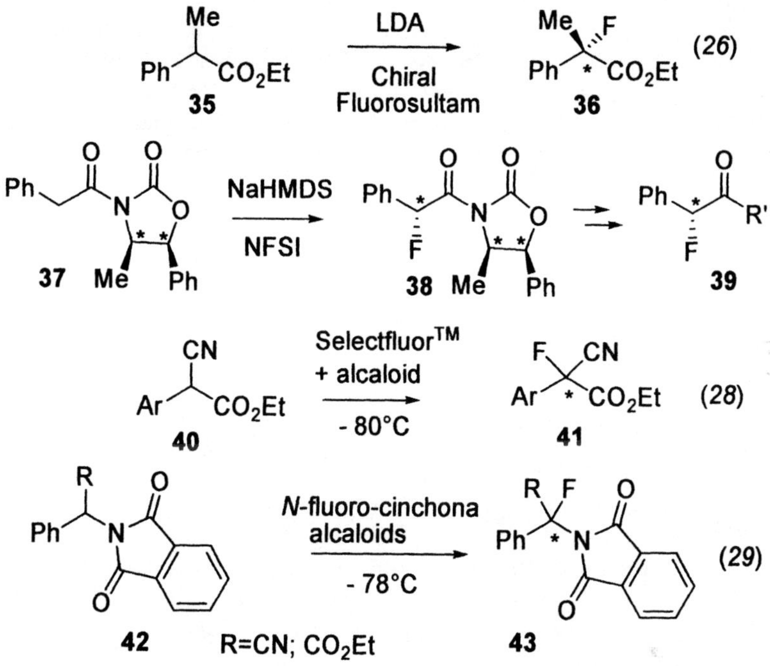

Me

Ph CO$_2$Et
35

LDA
Chiral Fluorosultam

Me F
Ph * CO$_2$Et
36

(26)

NaHMDS
NFSI

37 Me Ph

38 Me Ph

Ph * R'
F
39

CN
Ar CO$_2$Et
40

Selectfluor^{TM} + alcaloid
- 80°C

F CN
Ar * CO$_2$Et
41

(28)

N-fluoro-cinchona alcaloids
- 78°C

42 R=CN; CO$_2$Et 43

(29)

Scheme 12: electrophilic fluorinations in benzylic position

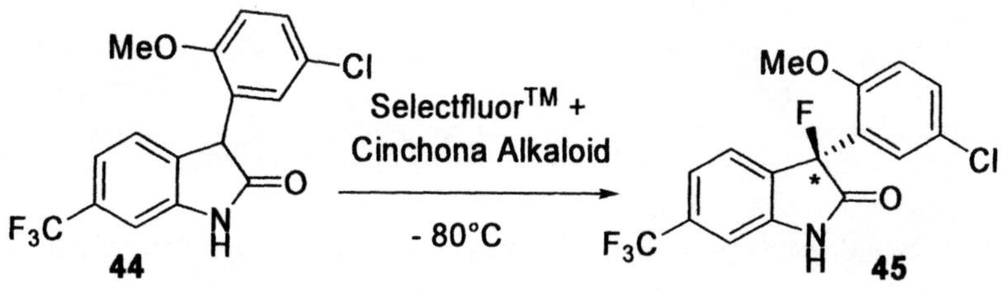

MeO Cl
F$_3$C
44

Selectfluor^{TM} + Cinchona Alkaloid
- 80°C

MeO F Cl
F$_3$C
45

Scheme 13 : Enantioselective synthesis of BMS-204352 (*31*) (*32*)

Various methods of resolution have been also successful in the preparation of benzylic fluorides (Figure 2). Lipase-catalyzed transformations were very useful for the preparation, in optically active form, of the 2-fluoro-2-aryl acetic acid derivatives **46** (*33*) and for the 2-fluoro-2-phenylalkanols **47** (*34*). The α-cyano-α-fluoro-p-tolylacetic acid **48**, which is a useful chiral derivatizing agent, has also been resolved in the same way (*35*) or by fractional crystallisation of diastereoisomeric esters (*36*). Furthermore, lipases can perform efficiently the desymmetrisation of meso derivatives: monohydrolysis of suitably functionalized malonates affords the acid **49**, while the acetate **50** is obtained by monoacetylation of 1,3 propanediols (*37*). All those derivatives are useful building blocks for the synthesis of various types of optically active benzylic fluorides.

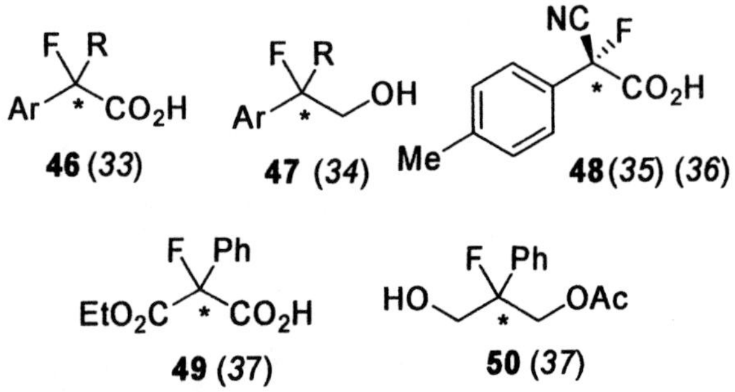

Figure 2 : Benzyl fluorides prepared by resolution

Another useful strategy is to start from vinylic fluorides and perform additions, and especially cycloaddition reactions, on the activated double bond. Although this will be reported in more detail in another chapter of this book, it is worth mentioning the excellent work of the group of G. Haufe: copper catalyzed cyclopropanation of α-fluorostyrene **51** with ethyldiazoacetate affords cyclopropanes **52** and **53**. After separation by chromatography and reduction with $LiAlH_4$, the alcohols **54** and **55** are obtained. The cis diastereoisomer (+/-)-**55** is efficiently resolved using the *Amano PS* lipase in organic solvents with various acyl donors (Scheme 14). Lower selectivities are obtained in the resolution of the trans isomer **54**. The (*1S, 2R*) absolute configuration of (+)-**55** was established by X-Ray analysis (*38*).

A complementary strategy to access such enantiopure derivatives is the asymmetric cyclopropanation (Scheme 15). The reaction of **51** with tert-butyldiazoacetate in the presence of a bis(oxazoline) derived chiral catalyst (2 mol %) and CuOTf gives a 4:1 mixture of **56** (93% ee) and **57** (89% ee) (*39*).

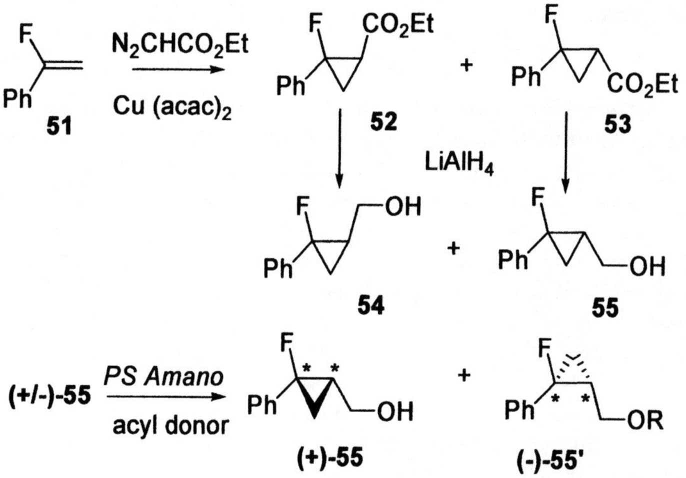

F Ph 51 —N2CHCO2Et / Cu (acac)2→ 52 + 53 —LiAlH4→ 54 + 55

(+/-)-55 —PS Amano / acyl donor→ (+)-55 + (-)-55'

Scheme 14: synthesis and resolution of fluorocyclopropanes (*38*)

After saponification and Curtius degradation, the corresponding fluorinated cyclopropylamines are obtained. The latter derivatives are potent and selective inhibitors of the monoamine oxidases. Very interesting structure-activity relationships have been obtained and, for instance, it was established that the (*1S,2S*)-2-Fluoro-2-phenylcyclopropylamine was a more potent inhibitor of both monoamine oxidases A and B than the (*1R,2R*) enantiomer (*40*).

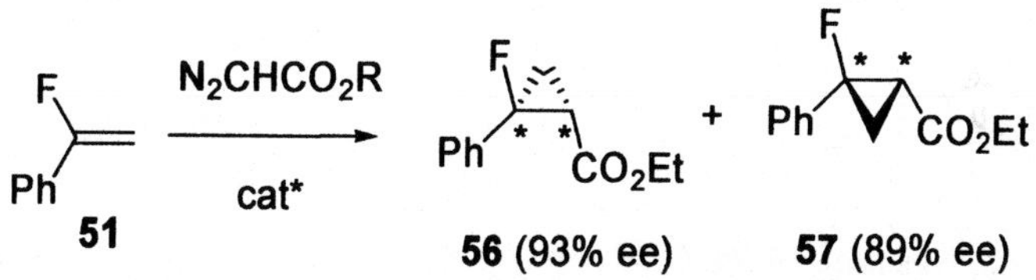

51 —N_2CHCO_2R / cat*→ 56 (93% ee) + 57 (89% ee)

Scheme 15: asymmetric synthesis of fluorocyclopropanes (*39*)

Finally it must be mentioned that Togni *et al.* observed recently that, during fluorination of 1-phenethyl bromide using thallium fluoride and a chiral ruthenium complex, a modest but significant asymmetric induction (16% ee) was observed but only at very low conversions (1-2 %) (*41*). Although very preliminary, this result gives some support to future developments in asymmetric catalytic nucleophilic fluorination (*42*).

Conclusions

The asymmetric synthesis of monofluorinated compounds is a rapidly emerging field. In the case of compounds having a fluorine atom in a position vicinal to π-systems, new strategies and novel methodologies have been developed during the last few years. Each of them has different scope and limitations and they appear highly complementary. Therefore, the choice of the method to be used will depend mainly on the target molecule. From the preceding results, it can be anticipated that asymmetric electrophilic fluorination will continue to be strongly developed, especially for catalytic versions. Although it is presently at a less advanced stage, it appears quite possible that novel catalytic nucleophilic fluorination methods will appear in a near future. These new, and more efficient, methodologies will help for the stereoselective preparation of fluorinated compounds. In that way, they will contribute to the development of fluorine chemistry in connection with life sciences as well as with material sciences.

Acknowledgments

We thank Pr. V. Gouverneur, Dr. J. C. Plaquevent, Dr. D. Cahard, Dr. B. Langlois, Dr J. P. Bégué, Dr. D. Bonnet-Delpon, Pr. G. Haufe and Pr. K. N. Houk for very fruitful discussions. We thanks CNRS and NSF for an exchange programme.

Literature Cited

1. *Experimental Methods in Organic Fluorine Chemistry*; Kitazume, T.; Yamazaki, T., Eds.; Gordon and Breach Science Publishers: Tokyo, 1998.
2. *Fluorine in Bioorganic Chemistry*; Welch, J. T.; Eswarakrishnan, S., Eds; Wiley-Interscience: New York, 1991.
3. Grée, R. L.; Lellouche, J. P. in *Enantiocontrolled Synthesis of Fluoro-Organic Compounds: Stereochemical Challenges and Biomedical Targets*; Soloshonok, V. A., Ed.; John Wiley and Sons: New York, 1999, pp 63-106.
4. Prakesch, M. ; Grée, D.; Grée, R. *Acc. Chem. Res.* **2002,** *35*, 175-181.

5. For recent representative examples see: (a) Shibata, N.; Suzuki, E.; Takeuchi, Y. *J. Am. Chem. Soc.* **2000**, *122*, 10728-10729; (b) Mohar, B.; Baudoux, J.; Plaquevent, J.C.; Cahard, D. *Org. Lett.* **2000**, *2*, 3699-3701; (c) Muniz, K. *Angew. Chem. Int. Ed.* **2001**, *40*, 1653-1656; (d) Kim, D. Y.; Park, E. J. *Org. Lett.* **2002**, *4*, 545-547 and references cited therein.
6. For recent representative examples see: (a) Hintermann, L.; Togni, A. *Angew. Chem. Int. Ed.* **2000**, *39*, 4359-4362; (b) Hamashima, Y.; Yagi, K.; Takano, H.; Tamas, L.; Sodeoka, M. *J. Am. Chem. Soc.* **2002**, *124*, 14530-14531; (c) Hamashima, Y.; Takano, H.; Hotta, D.; Sodeoka, M. *Org. Lett.* **2003**, *5*, 3225-3228; (d) Ma, J-A.; Cahard, D. *Tetrahedron: Asymmetry* **2004**, *15*, 1007-1011 and references cited therein. For the use of supported electrophilic fluorinating agents see: Thierry, B.; Audouard, C.; Plaquevent, J. C.; Cahard, D. *Synlett*, **2004**, 856-860.
7. Grée, D.; Madiot, V.; Grée, R. *Tetrahedron Lett.* **1999**, *40*, 6399-6402.
8. For the determination of ee's in the case of non functionalized propargylic fluorides, see: Madiot, V.; Lesot, P.; Grée, D.; Courtieu, J.; Grée, R. *Chem. Commun.* **2000**, 169-170.
9. Prakesch, M.; Kerouredan, E.; Grée, D.; Grée, R.; DeChancie, J.; Houk, K. N. *J. Fluorine Chem.* **2004**, *125*, 537-541.
10. (a) Middleton, W. J. *J. Org. Chem.* **1975**, *40*, 574-578; (b) Hudlicky, M. *Org. React.* **1986**, *35*, 513-637; (c) Singh, R. P.; Shreeve, J. M. *Synthesis*, **2002**, 2561- 2578.
11. For a recent example see: De Jonghe, S.; Van Overmeire, I.; Gunst, J.; De Bruyn, A.; Hendrix, C.; Van Calenbergh S.; Busson, R.; De Keukeleire, D.; Philippé, J.; Herdewijn, P. *Biorg. Med. Chem. Lett.* **1999**, *9*, 3159-3164.
12. Bernardi, L.; Bonini, B. F.; Comes-Franchini, M.; Fochi, M.; Folegatti, M.; Grilli, S.; Mazzanti, A.; Ricci, A. *Tetrahedron: Asymmetry* **2004**, *15*, 245-252.
13. For a classical example see: Gani, D.; Hitchcock, P. B.; Young, D. W. *J. Chem. Soc. Perkin Trans. I*, **1985**, 1363-1372.
14. Legoupy, S.; Crevisy, C.; Guillemin, J-C.; Grée, R. *J. Fluorine Chem.* **1999**, *93*, 171-173.
15. Agbossou, F.; O'Connor, E. J.; Garner, C. M.; Quiros Mendez, N.; Fernandez, J. M.; Patton, A. T.; Ramsden, J. A.; Gladysz, J. A. *Inorg. Synth.*, **1992**, *29*, 211-225.
16. (a) Thibaudeau S.; Gouverneur, V. *Org. Lett.* **2003**, *5*, 4891-4893; (b) Thibaudeau, S.; Fuller, R.; Gouverneur, V. *Org. Biomol. Chem.* **2004**, *2*, 1110-1112.
17. Greedy, B; Paris, J-M.; Vidal, T.; Gouverneur, V. *Angew. Chem. Int. Ed.* **2003**, *42*, 3291-3294.

18. (a) Davis, F. A.; Kasu, P. V. N.; Sundarababu, G.; Qi, H. *J. Org. Chem.* **1997**, *62*, 7546-7547; (b) Davis, F. A.; Han W. *Tetrahedron Lett.* **1992**, *33*, 1153-1156.
19. Prakesch, M.; Grée, D.; Grée, R. *J. Org. Chem.* **2001**, *66*, 3146-3151.
20. Grée, D. M.; Kermarrec, C. J. M.; Martelli, J. T.; Grée, R. L.; Lellouche, J. P.; Toupet, L. J. *J. Org. Chem.* **1996**, *61*, 1918-1919.
21. Madiot, V.; Grée, D.; Grée, R. *Tetrahedron Lett.* **1999**, *40*, 6403-6406.
22. Fritz-Langhals, E. *Tetrahedron Lett.* **1994**, *35*, 1851-1854.
23. Kermarrec, C.; Madiot, V.; Grée, D.; Meyer, A.; Grée, R. *Tetrahedron Lett.* **1996**, *37*, 5691-5694.
24. Jaouen, G.; Meyer, A. *J. Am. Chem. Soc.* **1975**, *97*, 4667-4672.
25. Islas-Gonzales, G.; Puigjaner, C.; Vidal-Ferran, A.; Moyano, A.; Riera, A. ; Pericas, M. A. *Tetrahedron Lett.* **2004**, *45*, 6337-6341.
26. Differding, E.; Lang, R. W. *Tetrahedron Lett.* **1988**, *29*, 6087-6090.
27. Davis, F. A.; Kasu, P. V. N. *Tetrahedron Lett.* **1998**, *39*, 6135-6138.
28. Shibata, N.; Suzuki, E.; Asahi, T.; Shiro, M. *J. Am. Chem. Soc.* **2001**, *123*, 7001-7009.
29. Mohar, B.; Baudoux, J.; Plaquevent, J-C.; Cahard, D. *Angew. Chem. Int. Ed.* **2001**, *40*, 4214-4216.
30. Hewasawam, P.; Gribkoff, V. K.; Pendri, Y.; Dworetzky, S. I.; Meanwell, N. A.; Martinez, E.; Boissard, C. G.; Post-Munson, D. J.; Trojnacki, J. T.; Yeleswaram, K.; Pajor, L. M.; Knipe, J.; Gao, Q.; Perronne, R.; Starrett Jr, J. E. *Biorg. Med. Chem. Lett.* **2002**, *12*, 1023-1026.
31. Shibata, N.; Ishimaru, T.; Suzuki, E.; Kirk, K. L. *J. Org. Chem.*, **2003**, *68*, 2494-2497.
32. Zoute, L.; Audouard, C.; Plaquevent, J-C.; Cahard, D. *Org. Biomol. Chem.* **2004**, *1*, 1833-1834.
33. Kometani, T.; Isobe, T.; Goto, M.; Takeuchi, Y.; Haufe, G. *J. Mol. Cat. B: Enzym.* **1998**, *5*, 171-174.
34. Goj, O.; Burchardt, A.; Haufe, G. *Tetrahedron: Asymmetry* **1997**, *8*, 399-408.
35. Takeuchi, Y.; Konishi, M.; Hori, H.; Takahashi, T.; Kometani, T.; Kirk, K. L. *Chem. Commun.* **1998**, 365-366.
36. Fujiwara, T.; Sasaki, M.; Omata, K.; Kabuto, C.; Kabuto, K.; Takeuchi, Y. *Tetrahedron: Asymmetry* **2004**, *15*, 555-563.
37. Narisano, E.; Riva, R. *Tetrahedron: Asymmetry* **1999**, *10*, 1223-1242.
38. Rosen, T. C.; Haufe, G. *Tetrahedron: Asymmetry* **2002**, *13*, 1399-1405.
39. Meyer, O. G. J.; Fröhlich, R.; Haufe, G. *Synthesis*, **2000**, 1479-1490.
40. (a) Yoshida, S.; Rosen, T. C.; Meyer, O. G. J.; Sloan, M. J.; Ye, S.; Haufe, G.; Kirk, K. L. *Biorg. Med. Chem. Lett.* **2004**, *12*, 2645-2652;

(b) Yoshida, S.; Meyer, O. G. J.; Rosen, T. C.; Haufe, G.; Ye, S.; Sloan, M. J.; Kirk, K. L. *J. Med. Chem.* **2004**, *47*, 1796-1806.

41. Togni, A.; Mezzetti, A.; Barthazy, P.; Becker, C.; Devillers, I.; Frantz, R.; Hintermann, L.; Perseghini, M.; Sanna, M. *Chimia*, **2001**, *55*, 801-805; see also: Ibrahim, H.; Togni, A. *Chem. Commun.* **2004**, 1147-1155.
42. For the first catalytic enantioselective nucleophilic fluorination by epoxide opening see: Bruns, S.; Haufe, G. *J. Fluorine Chem.* **2000**, *104*, 247-254; Haufe, G.; Bruns, S. *Adv. Synth. Cata.* **2002**, *2*, 165-171.

Chapter 9

A New Aspect of Fluoroalkylated Acetylenes: Synthesis and Applications—Hydrometallation and Carbometallation

Tsutomu Konno and Takashi Ishihara

Department of Chemistry and Materials Technology, Kyoto Institute of Technology, Matsugasaki, Sakyo-ku, Kyoto 606–8585, Japan

The facile and practical syntheses of various fluorine-containing alkynes are described. Thus-obtained alkynes undergo smooth hydrostannation, hydroboration, carbostannylation, and carbocupration reactions to afford the corresponding hydro- and carbo-metallated adducts in a highly regio- and stereoselective manner. With this synthetic methodology, a short total synthesis of estrogen-dependent antitumor agent, panomifene has been realized.

Introduction

In view of rapidly growing role of fluorine-containing substances particularly in materials and pharmaceutical science, the synthesis of fluorine-containing substances is becoming more and more important (*1*). As one of the most valuable synthetic intermediates for preparation of such organofluoro substances, fluoroalkylated alkynes **1a** have been well recognized as having potentially significant synthetic value. These alkynes have been prepared by several methods (Figure 1) (*2*). For example, the synthesis of trifluoromethylated internal alkynes has been achieved through the transition-metal catalyzed coupling reactions with the respective trifluoropropynyl lithium (*3*), Grignard (*4*) or zinc reagents (*5*). However, reliance upon trifluoropropyne as a precursor incurs the experimental difficulties associated with the handling of a gaseous reagent (b.p. –47 °C) as well as significant cost implications. In addition to limited studies on the preparation of **1a** (*6*), little attention has been paid on the synthesis of per- or poly-fluoroalkylated alkynes **1b**. To our best of knowledge, no report has appeared on the general preparation of alkynes **1c** and **1d** which possess both a per- or poly-fluoroalkyl group and an *aliphatic* chain R. Only a few examples of the synthetic applications of the alkynes have been reported thus far, such as the Diels-Alder reaction, 1,3-dipolar cycloaddition, etc., so that the development of the synthetic advantages of **1** remains a challenging task. This chapter deals with our recent studies on the practical synthesis and applications of the fluoroalkylated acetylene derivatives **1a-d**.

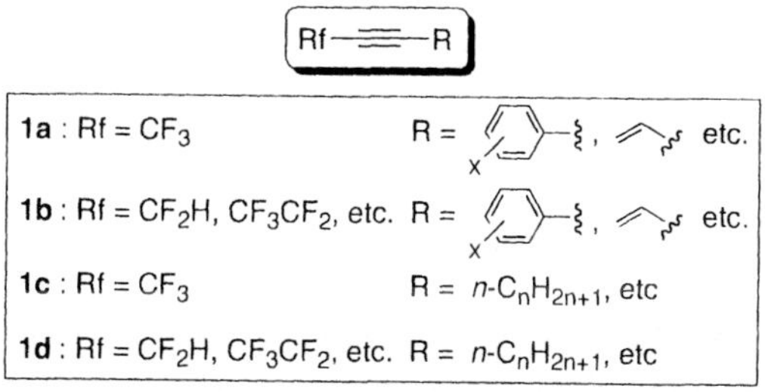

Rf—≡—R

1a : Rf = CF_3 R = X-aryl, vinyl etc.
1b : Rf = CF_2H, CF_3CF_2, etc. R = X-aryl, vinyl etc.
1c : Rf = CF_3 R = n-C_nH_{2n+1}, etc
1d : Rf = CF_2H, CF_3CF_2, etc. R = n-C_nH_{2n+1}, etc

Figure 1

The Synthesis of Various Types of Fluoroalkylated Alkynes

We have developed three synthetic methods, each of which would provide us with easy access to a variety of fluoroalkylated alkynes **1a-d** (*7*).

First of all, we have investigated the palladium-catalyzed coupling reaction of *in situ*-generated fluoroalkylated acetylides with various halides (Scheme 1). The starting vinyl iodides **2a-c** could readily be prepared from the corresponding polyfluoroalcohols **3a-c** in three steps (*8*). Thus, the commercially available **3a-c** was treated with *p*-toluenesulfonyl chloride (1.2 equiv.) and NaOH (1.2 equiv.) at room temperature to give the corresponding tosylate **4a-c** in quantitative yield. The tosylate was subjected to NaI/diethylene glycol at room temperature, followed by direct distillation, affording polyfluoroalkyl iodide **5a-c**. Finally, the HF elimination of **5a-c** by KOH in DMSO gave the desired *Z*-vinyl iodide **2a-c** in up to 56% overall yield for three steps.

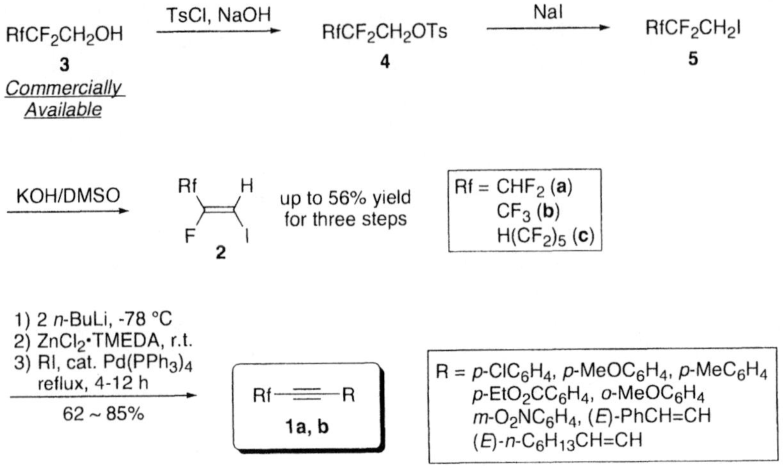

$RfCF_2CH_2OH$ (3, Commercially Available) —TsCl, NaOH→ $RfCF_2CH_2OTs$ (4) —NaI→ $RfCF_2CH_2I$ (5)

—KOH/DMSO→ Rf(H)C=C(F)(I) 2 — up to 56% yield for three steps

Rf = CHF_2 (**a**), CF_3 (**b**), $H(CF_2)_5$ (**c**)

1) 2 *n*-BuLi, -78 °C
2) $ZnCl_2$•TMEDA, r.t.
3) RI, cat. $Pd(PPh_3)_4$ reflux, 4-12 h

62 ~ 85%

→ Rf—≡—R **1a, b**

R = *p*-ClC_6H_4, *p*-$MeOC_6H_4$, *p*-MeC_6H_4, *p*-$EtO_2CC_6H_4$, *o*-$MeOC_6H_4$, *m*-$O_2NC_6H_4$, (*E*)-PhCH=CH, (*E*)-*n*-C_6H_{13}CH=CH

Scheme 1

The treatment of **2** with 2 equiv. of *n*-BuLi at –78 °C for 1 h produced the corresponding lithium acetylide *in-situ*, which was transmetallated into the zinc acetylide by addition of $ZnCl_2$•TMEDA complex to the reaction mixture. In this transformation, $ZnCl_2$•TMEDA was better than anhydrous $ZnCl_2$ as it was less hydroscopic. Then, the mixture was allowed to warm to room temperature, followed by addition of 1 equiv. of various aryl iodides and 5 mol% of $Pd(PPh_3)_4$ in this order. Heating the reaction mixture under reflux for 4-12 h resulted in the formation of fluoroalkylated alkynes **1a** or **b** in 62 ~ 85% yields. Unfortunately, the coupling reaction of fluoroalkylated acetylide with alkyl iodide in the presence of Pd(0) did not proceed, resulting in the complete recovery of the zinc acetylide.

In seeking more convenient methods for preparing fluorine-containing acetylenes, we examined the one-pot synthesis of trifluoromethylated acetylenes from commercially available, easily handled 2-bromo-3,3,3-trifluoropropene **6** (b.p. 34 ~ 35 °C) (Scheme 2). Thus, **6** was treated with 2 equiv. of LDA at –78 °C (*9*), resulting in the formation of trifluoromethylated lithium acetylide *in situ*. To this reaction mixture was added $ZnCl_2$•TMEDA, $Pd(PPh_3)_4$ and ArI, which was then heated under reflux, leading to the formation of the desired alkynes **1a**. Generally, the coupling reaction of zinc acetylide with aryl iodides gave the corresponding acetylene derivatives in excellent yields. Aryl triflates as well as aryl bromides did not give the desired coupling products. Furthermore, Stille coupling using an alkynylstannane instead of zinc acetylide afforded the desired product in very low yield

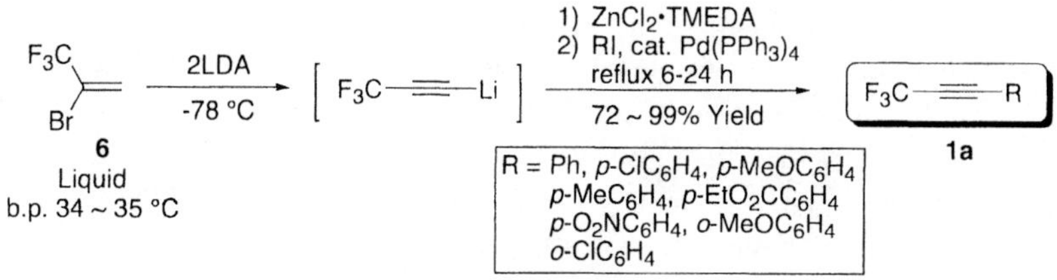

F_3C(Br)C=CH_2 **6** (Liquid, b.p. 34 ~ 35 °C) → 2LDA, -78 °C → [F_3C—≡—Li] → 1) $ZnCl_2$•TMEDA 2) RI, cat. $Pd(PPh_3)_4$ reflux 6-24 h, 72 ~ 99% Yield → F_3C—≡—R **1a**

R = Ph, *p*-ClC_6H_4, *p*-$MeOC_6H_4$, *p*-MeC_6H_4, *p*-$EtO_2CC_6H_4$, *p*-$O_2NC_6H_4$, *o*-$MeOC_6H_4$, *o*-ClC_6H_4

Scheme 2

The fluoroalkylated alkynes possessing an alkyl side chain could be synthesized *via* dehydroiodination from vinyl iodides **7** that were prepared by radical addition of perfluoroalkyl iodide to terminal alkynes in the presence of a catalytic amount of Zn/TFA according to the literature (Scheme 3) (*10*). Thus, treatment of **7** with 3 equiv. of *t*-BuOK in benzene at the reflux temperature gave the desired fluoroalkylated internal alkynes **1d** in 65-81% yield, but trifluoromethylated internal alkynes **1c** was obtained in only 24% yield. It should be noted that the use of 1.2 equiv. of *t*-BuOK at room temperature was crucial for high yields in the case of CF_3-alkyne synthesis.

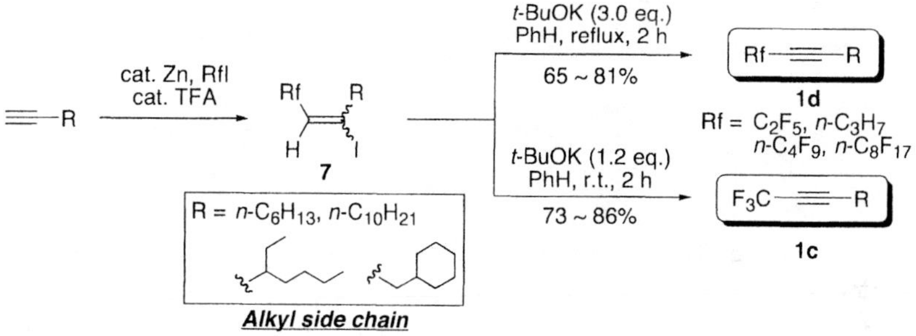

≡—R → cat. Zn, RfI, cat. TFA → Rf(H)C=C(R)(I) **7**

→ *t*-BuOK (3.0 eq.), PhH, reflux, 2 h, 65 ~ 81% → Rf—≡—R **1d** (Rf = C_2F_5, *n*-C_3H_7, *n*-C_4F_9, *n*-C_8F_{17})

→ *t*-BuOK (1.2 eq.), PhH, r.t., 2 h, 73 ~ 86% → F_3C—≡—R **1c**

R = *n*-C_6H_{13}, *n*-$C_{10}H_{21}$

Alkyl side chain

Scheme 3

The Synthetic Applications I —Hydrometallation—

Hydrostannation

As a synthetic application of the fluoroalkylated alkynes, hydrostannation was carried out to provide vinylstannanes, versatile synthesic building blocks (*11*). Thus, treatment of **1** with 1.2 equiv. of *n*-Bu_3SnH in the presence of 20 mol% of Et_3B in toluene at 0 °C for 4 h gave the corresponding vinylstannanes **8** and **9** in good to excellent yields (Table 1) (*12*). The alkynes bearing a halogen

(Cl) or an electron-donating group (Me, MeO) on an aromatic ring could participate nicely in the reaction to afford **8-*trans*** exclusively. Interestingly, switching the substituent on an aromatic ring from the electron-donating group to the electron-withdrawing group (CO_2Et, NO_2, CF_3) induced the significant change on the regio- and stereoselectivity, the opposite regioselectivity being observed in favor of **9** with high *cis* selectivity.

Rf—≡—R (**1**) → [*Sn*H (1.2 eq.), Et_3B (20 mol%), Toluene, 0 °C, 4 h; *Sn* = *n*-Bu_3] → **8-*trans*** + **8-*cis*** + **9-*trans*** + **9-*cis***

Table 1

Rf	R	Yield[a,b] /%	Isomer ratio[a] **8** (*trans* : *cis*) : **9** (*trans* : *cis*)
CF_3	*p*-ClC_6H_4	99 (95)	100 (100 : 0) : 0
CHF_2	*p*-ClC_6H_4	99 (82)	100 (100 : 0) : 0
$HCF_2CF_2CF_2$	*p*-ClC_6H_4	99 (98)	100 (100 : 0) : 0
CF_3	*p*-MeC_6H_4	99 (91)	100 (98 : 2) : 0
CF_3	*p*-$MeOC_6H_4$	99 (90)	100 (98 : 2) : 0
CF_3	*o*-ClC_6H_4	69 (55)	100 (100 : 0) : 0
CF_3	*o*-$MeOC_6H_4$	89	100 (100 : 0) : 0
CF_3	*m*-ClC_6H_4	99 (92)	100 (100 : 0) : 0
CF_3	Me_2PhSi	99 (93)	98 (82 : 18) : 2 (0 : 100)
CF_3	*n*-$C_{10}H_{21}$	32	100 (62 : 38) : 0
CF_3	*p*-$F_3CC_6H_4$	46	33 (0 : 100) : 67 (0 : 100)
CF_3	*p*-$EtO_2CC_6H_4$	78	35 (0 : 100) : 65 (0 : 100)
CF_3	*p*-$O_2NC_6H_4$	74	19 (0 : 100) : 83 (0 : 100)

a) Determined by ^{19}F NMR. b) Values in parentheses are of isolated yields.

When the hydrostannation was effected *in the absence of Et_3B*, the yield decreased significantly in the case of alkynes bearing a halogen or an electgron-donating group on an aromatic ring (Table 2). On the other hand, the reaction of alkynes having an ethoxycarbonyl or nitro group on an aromatic ring, proceeded smoothly to give the desired vinylstannanes in the almost same yield and selectivity as in the presence of Et_3B. This fact may suggest the following mechanism for the hydrostannation reaction as described in Scheme 4.

Table 2. The hydrostannation in the absence of Et_3B

R	Yield[a] of **8**+**9**/%	**8** (*trans* : *cis*) : **9** (*trans* : *cis*)[a]	Recovery[a] of **1**/%
p-ClC_6H_4	3	100 (100 : 0) : 0	97
o-ClC_6H_4	2	100 (100 : 0) : 0	98
m-ClC_6H_4	3	100 (100 : 0) : 0	97
p-$EtO_2CC_6H_4$	65	35 (0 : 100) : 65 (0 : 100)	35
p-$O_2NC_6H_4$	66	21 (0 : 100) : 79 (0 : 100)	34

a) Determined by ^{19}F NMR.

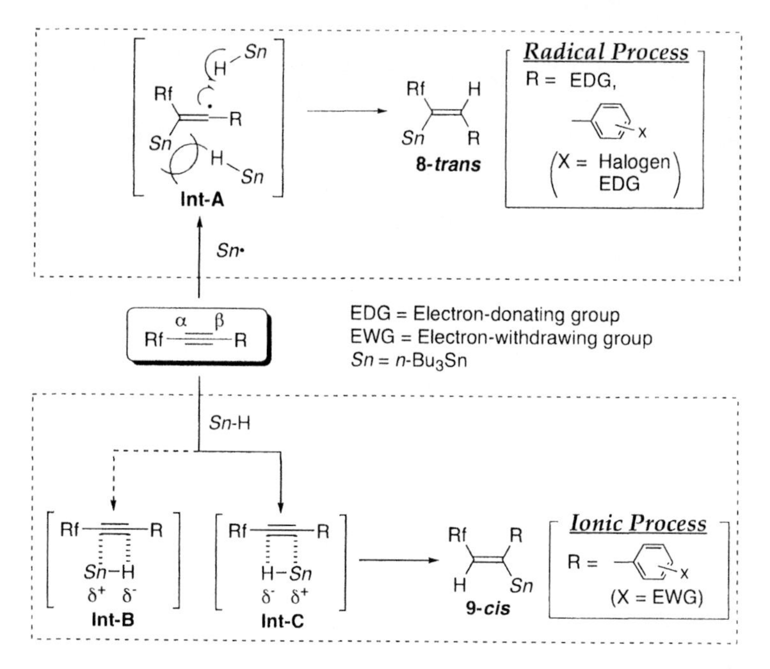

Scheme 4

In the case of acetylenes bearing a halogen or an electron-donating group on the aromatic ring, nucleophilic tributyltin radical attacks the more electrophilic α-carbon of the alkynes to give the corresponding vinyl radicals **Int-A**. Then, the tributyltin hydride approaches adjacent to the the fluoroalkyl group, avoiding a bulkier tributylstannyl group. Therefore, the vinylstannane **8-*trans*** was formed preferentially. When an electron-withdrawing group such as CO_2Et, NO_2, CF_3, was employed as a substituent X at the aromatic ring, the reaction may proceed *via* an ionic four-centered transition state (**Int-B** or **Int-C**). Presumably, tributyltin hydride may coordinate with the triple bond as in **Int-C** rather than **Int-B** because the electron density of the acetylenic carbon attached with the Rf moiety is lower than that of the other, leading to the preferential formation of **9-*cis***.

With the fluoroalkylated vinylstannane **8-*trans***, we attempted the coupling reaction with various aryl iodides through the Migita-Kosugi-Stille reaction (*13*) (Scheme 5). The vinylstannane **8a-*trans*** was treated with various aryl halides in the presence of 10 mol% each of $Pd(PPh_3)_4$ and CuI in DMF at 70 °C. After stirring for 6 h, the stereochemically defined trisubstituted alkenes **10** were obtained in excellent yields with retention of the olefinic configuration.

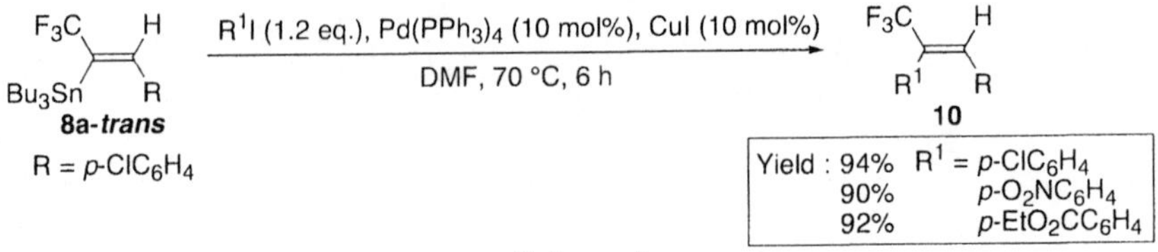

$F_3C(Bu_3Sn)C{=}CH(R)$ **8a-*trans***, R = *p*-ClC_6H_4 → [R^1I (1.2 eq.), $Pd(PPh_3)_4$ (10 mol%), CuI (10 mol%); DMF, 70 °C, 6 h] → $F_3C(R^1)C{=}CH(R)$ **10**

Yield	R^1
94%	*p*-ClC_6H_4
90%	*p*-$O_2NC_6H_4$
92%	*p*-$EtO_2CC_6H_4$

Scheme 5

Hydroboration

Hydroboration of alkynes is one of the most straightforward synthetic methods for the preparation of vinylboranes which can be transformed to variously substituted ethenes with retention of configuration through the Suzuki-Miyaura cross-coupling reaction (*14*). Indeed, the reaction of fluoroalkylated acetylenes with 1.2 equiv. of dicyclohexylborane at room temperature for 2 h proceeded smoothly to give the corresponding vinylborane **Int-D** in a highly regio- and stereoselective fashion (*15*). The *in-situ* generated vinylborane **Int-D** was treated with various aryl iodides, 10 mol% of $Pd(PPh_3)_2Cl_2$, and 3 equiv. of NaOH, at the reflux temperature of benzene for 4 h, affording trisubstituted alkenes with 83-96% regio- and stereoselectivity (Table 3).

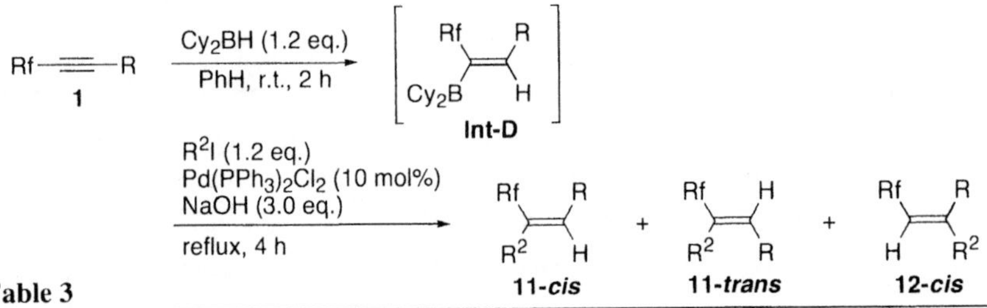

Rf—≡—R **1** → [Cy_2BH (1.2 eq.); PhH, r.t., 2 h] → [$Rf(Cy_2B)C{=}CH(R)$ **Int-D**] → [R^2I (1.2 eq.), $Pd(PPh_3)_2Cl_2$ (10 mol%), NaOH (3.0 eq.); reflux, 4 h] → $Rf(R^2)C{=}C(R)H$ **11-*cis*** + $Rf(R^2)C{=}C(H)R$ **11-*trans*** + $Rf(H)C{=}C(R)R^2$ **12-*cis***

Table 3

Rf	R	R^2	Yield[a,b] of **11+12**/%	**11** (*cis* : *trans*) : **12**[a]
CF_3	C_6H_5	C_6H_5	85 (73)	95 (100 : 0) : 5
CF_3	*p*-ClC_6H_4	C_6H_5	89 (64)	94 (98 : 2) : 6
CF_3	*p*-MeC_6H_4	C_6H_5	83 (66)	94 (100 : 0) : 6
CF_3	*p*-$MeOC_6H_4$	C_6H_5	99 (88)	96 (100 : 0) : 4
CF_3	*p*-$O_2NC_6H_4$	C_6H_5	25	88 (100 : 0) : 12
CF_3	*p*-$EtO_2CC_6H_4$	C_6H_5	99	95 (100 : 0) : 5
CF_3	*p*-$MeOC_6H_4CH_2$	C_6H_5	62	58 (100 : 0) : 42
CF_3	*p*-ClC_6H_4	*o*-ClC_6H_4	53	92 (100 : 0) : 8
CF_3	*p*-ClC_6H_4	*m*-ClC_6H_4	74	93 (100 : 0) : 7
CF_3	*p*-ClC_6H_4	*p*-ClC_6H_4	83 (74)	96 (100 : 0) : 4
CF_3	*p*-ClC_6H_4	*p*-$MeOC_6H_4$	68 (56)	98 (93 : 7) : 2
CF_3	*p*-ClC_6H_4	*p*-$O_2NC_6H_4$	54	100 (91 : 9) : 0
CF_3	*p*-ClC_6H_4	*p*-$EtO_2CC_6H_4$	99	91 (100 : 0) : 9
CHF_2	*p*-ClC_6H_4	C_6H_5	67	84 (100 : 0) : 16
$HCF_2CF_2CF_2$	*p*-ClC_6H_4	C_6H_5	78	83 (100 : 0) : 17

a) Determined by ^{19}F NMR. b) Values in parentheses are of isolated yields.

Various alkynes, carrying an electron-donating group (Me, MeO) or an electron-withdrawing group (EtO_2C) on the aromatic ring, could participate equally well in the reaction to give preferentially the corresponding trisubstituted alkenes **11-*cis*** in high yields. However, a significant decrease in yield was observed when an alkyne having a nitro group on the benzene ring was used. The regioselectivity also decreased slightly. Additionally, the fluoroalkylated alkyne having an aliphatic side chain (p-$MeOC_6H_4CH_2$-) was not a good substrate. The products were formed with low regioselectivity (**11** : **12** = 58 : 42) and **13** was produced in 26% yield. The position of the substituent on the aromatic ring in R^2I significantly influenced the coupling reaction. Thus, the dimer **14** was formed in 20% yield when *o*-chloroiodobenzene was used as R^2I, while a more satisfactory result was obtained in the case of *m*-chloroiodobenzene. The reaction using the coupling reagents with various substituents such as chloro, methyl, methoxy, and ethoxycarbonyl groups proceeded smoothly to afford the trisubstituted alkenes in a highly regio- and stereoselective manner. However, the use of *p*-nitroiodobenzene resulted in the decrease in yield. It should also be noted that changing the fluoroalkyl group from a trifluoromethyl group to a difluoromethyl or hexafluoropropyl group did not influence the course of the reaction.

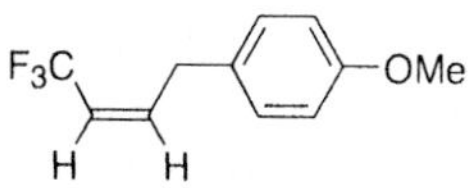

13 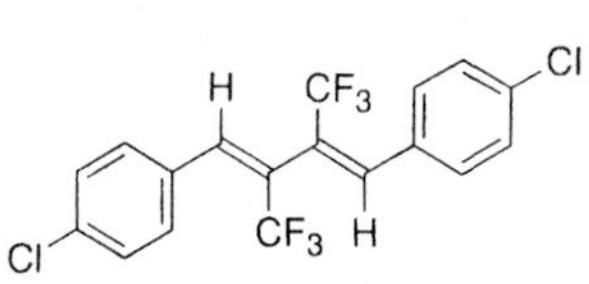 14

Figure 2

The Synthetic Applications II —Carbometallation—

Carbostannylation

Carbostannylation of alkynes has become a powerful synthetic tool in organic synthesis, because the C-C and C-Sn bonds are simultaneously introduced across a triple bond in a stereoselective fashion to give alkenylstannanes, which can be converted into variously substituted ethenes with retention of configuration through the Migita-Kosugi-Stille coupling reaction (*16*). Furthermore, the high chemoselectivity and mild reactivity of organostannanes, as compared with other organometallic reagents, make the carbostannylation and subsequent reactions extremely useful and applicable for preparing a wide variety of substances.

When the reaction of **1** with allylstannane **16a** was carried out in the presence of 5 mol% of AIBN in benzene at the reflux temperature, the reaction proceeded smoothly to give the corresponding carbostannylated products **15-*trans*** in high yields as a single isomer (Table 4) (*17*). This reaction did not occur when the allylstannanes **16b** and **16c** were employed. Alkynes having an aromatic ring with either an electron-donating group (MeO, Me) or an electron-withdrawing group (CO_2Et) were good substrates. To be noted is that the position of the substituent on the aromatic ring of the alkynes did not affect the yield. In contrast, the use of an alkyl side chain as R resulted in a significant decrease of the reaction rate (R = *p*-$MeOC_6H_4CH_2$: 44% yield, 48% recovery of the alkyne), though the reaction proceeded with high regio- and stereoselectivity. The substrates bearing various fluoroalkyl groups could participate well in the reaction forming the corresponding vinylstannanes **15-*trans*** in high yields.

It is interesting to note that the allylstannylation *without AIBN (under an atmosphere of air)* also proceeded smoothly to give the desired product **15-*trans*** in higher yields, relative to reaction in the presence of AIBN, though a longer reaction period (12 h) was needed for completion. The use of an alkyl side chain as R led to the complete recovery of the starting material. It is noteworthy that the differences in fluoroalkyl group on the alkynes significantly influenced on the reaction. Thus, the substrate with a difluoromethyl group gave the product in only 73% yield. Additionally, changing the fluoroalkyl group from a difluoromethyl group to a hexafluoropropyl group caused the significant decrease of the yield.

Rf—≡—R (**1**) → [CO_2Me-substituted allyl-$SnBu_3$ **16a** (4 eq.), AIBN (5 mol%), PhH, reflux; or without AIBN, air, PhH, reflux] → **15-*trans*** (Rf, Bu_3Sn, R, CO_2Me)

Table 4

Rf	R	In the presence of AIBN Yield[a] of **15-*trans***/%	In the absence of AIBN Yield[a] of **15-*trans***/%
CF_3	Ph	79	(90)
CF_3	*p*-ClC_6H_4	80 (60)	(92)[b]
CF_3	*m*-ClC_6H_4	81	(83)
CF_3	*o*-ClC_6H_4	85	(83)
CF_3	*p*-$MeOC_6H_4$	84	(86)
CF_3	*p*-MeC_6H_4	91	(94)
CF_3	*p*-$EtO_2CC_6H_4$	90	(86)
CF_3	*p*-$MeOC_6H_4CH_2$	44 (44)	0
CHF_2	*p*-ClC_6H_4	99 (99)	73[b]
$HCF_2CF_2CF_2$	*p*-ClC_6H_4	84 (78)	28

a) Determined by ^{19}F NMR. Valuses in parentheses are of isolated yields. b) Stirred for 2 h.

16b 16c

Figure 3

These results may allow us to postulate a reaction mechanism as follows (Scheme 6). Thus, in the initiation step, the radical generated from AIBN or the oxygen biradical reacts with the allylstannane to generate the tributylstannyl radical. As this stannyl radical is nucleophilic, the addition of the tributylstannyl radical to the alkyne would take place at more electrophilic α carbon due to the electron-withdrawing fluoroalkyl group, leading to the vinyl radical, **Int-E** exclusively. The allylstannane favors the attack from the side occupied by the fluoroalkyl group, avoiding a bulkier tributylstannyl group due to a large steric repulsion. Therefore, the vinylstannane **15-*trans*** was formed in a highly stereoselective manner.

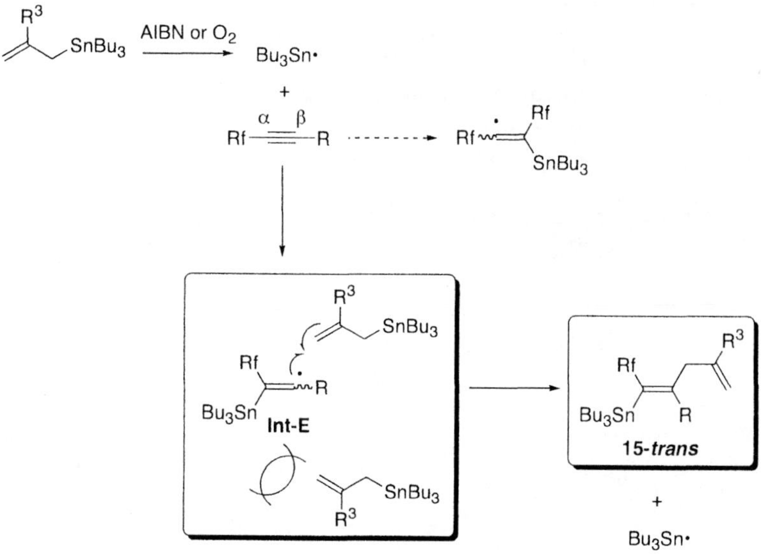

Scheme 6

Thus-obtained vinylstannane **15-*trans*** underwent the Migita-Kosugi-Stille cross-coupling reaction with iodobenzene, *p*-methoxyiodobenzene, and *p*-chloroiodobenzene successfully, the corresponding tetrasubstituted alkenes **17** being afforded with complete retention of the olefinic geometry in 96, 78, and 93% yield, respectively (Scheme 7).

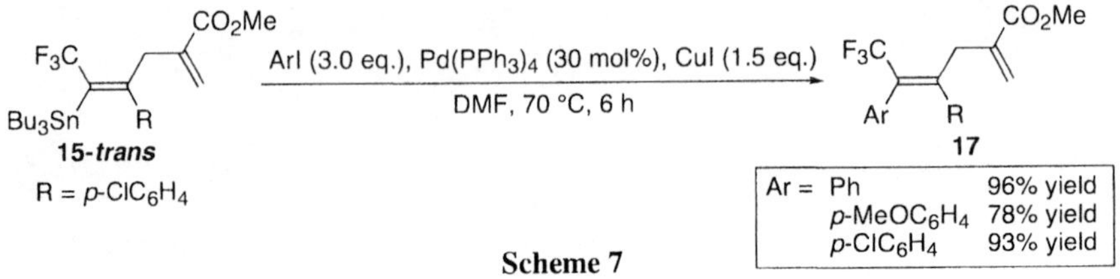

Scheme 7

Carbocupration

As shown in Scheme 8, various copper reagents prepared from organolithium reagents, Grignard reagents, and organozinc reagents, could participate nicely in the carbocupration reaction to give the corresponding vinylcoppers **Int-F** as a single isomer (*18*). The regioselectivity may be derived from the exclusive R^1-transfer from copper atom to β carbon due to the stronger Cα-Cu bond influenced by the electron-withdrawing Rf group (Figure 4). This electron-withdrawing effect also reduces the reactivity of **Int-F**, so that reaction only occurs with H_2O, iodine, substituted allyl bromide, and propargyl bromide. Such carbon electrophiles can coordinate with copper atom *via* d-π* complexation (*19*) and add to the copper metal oxidatively to form the corresponding π-allylcopper intermediate. Accordingly, methyl iodide, benzyl bromide, benzoyl chloride, ethyl chloroformate, etc. which can not coordinate with the copper, are all inactive as electrophiles.

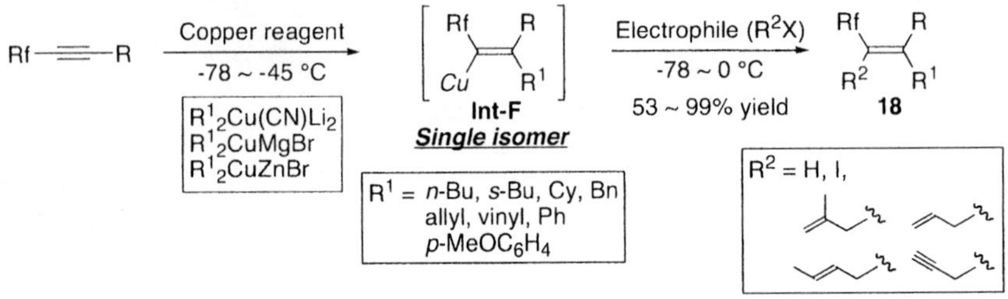

Scheme 8

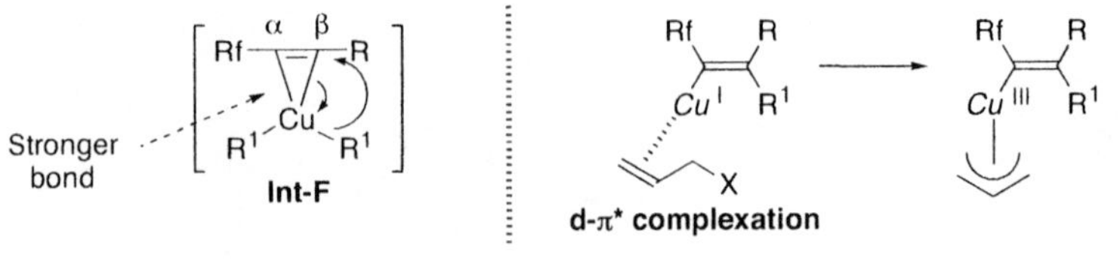

Figure 4

The vinyl iodide **18a** was subjected to the Suzuki-Miyaura or Sonogashira coupling reaction conditions (*20*) to give the corresponding alkene **19** or enyne **20** in excellent yields (Scheme 9).

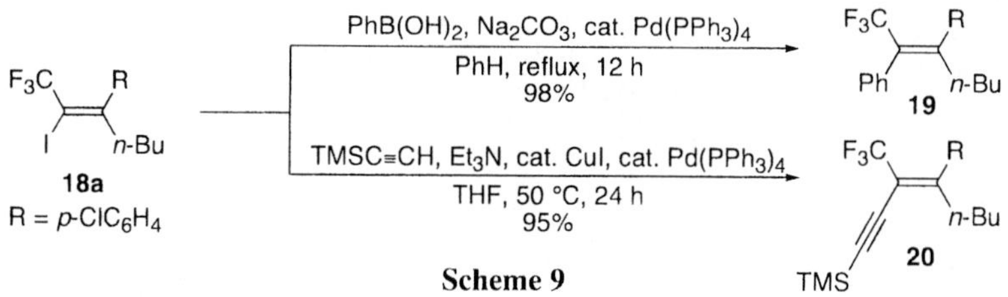

F_3C R / I *n*-Bu
18a
R = *p*-ClC_6H_4

$PhB(OH)_2$, Na_2CO_3, cat. $Pd(PPh_3)_4$
PhH, reflux, 12 h
98%

F_3C R / Ph *n*-Bu
19

TMSC≡CH, Et_3N, cat. CuI, cat. $Pd(PPh_3)_4$
THF, 50 °C, 24 h
95%

F_3C R / *n*-Bu / TMS
20

Scheme 9

With the carbocupration, the synthesis of antiestrogenic drug, panomifene **21** (*21*) was executed as follows (Scheme 10). Thus, alkyne **1** was exposed to the carbocupration reaction with $(p\text{-MeOC}_6H_4)_2Cu(CN)(MgBr)_2$ (1.2 equiv.) at −45 °C for 2 h, followed by addition of 2.4 equiv. of iodine at −78 °C, to afford vinyl iodide **18b** in 51% yield as a single isomer. The stereochemically pure **18b** was treated with 4.0 equiv. of phenylboronic acid under the Suzuki-Miyaura cross-coupling reaction conditions, producing the triarylethylene derivative **22** almost quantitatively with complete retention of the stereochemistry. Surprisingly, treatment of **22** with BBr_3 gave **23** in low yield. All attempts for improving this demethylation were unsuccessful. On the other hand, the demethylation of **18b** with BBr_3 proceeded readily to give the corresponding phenol derivative. The subsequent nucleophilic substitution reaction between phenoxide and 2-chloroethyl tosylate in DMF at the reflux temperature gave rise to the desired ether **24** in 67% yield. The Suzuki-Miyaura cross-coupling reaction of **24** with phenylboronic acid afforded the corresponding alkene **25** quantitatively. Finally, on treatment of **25** with ethanolamine in 2-methoxyethylene glycol, the desired panomifene **21** was obtained in 83% yield (28% overall yield from **1**).

Conclusion

In this review we have disclosed three types of synthetic methods for the preparation of the fluoroalkylated alkynes, enabling us to easily access various alkynes. Various alkynes were subjected to the hydrostannation, hydroboraton, carbostannylation, and carbocupration conditions, where the corresponding vinylmetals were afforded in a highly regio- and stereoselective manner. The resultant vinylmetals can be applied to the cross-coupling reactions such as Migita-Kosugi-Stille and Suzuki-Miyaura reactions, leading to the corresponding tri- or tetrasubstituted alkenes in high to excellent yields. A demonstration of this methodology resulted in a short total synthesis of anti estrogenic panomifene.

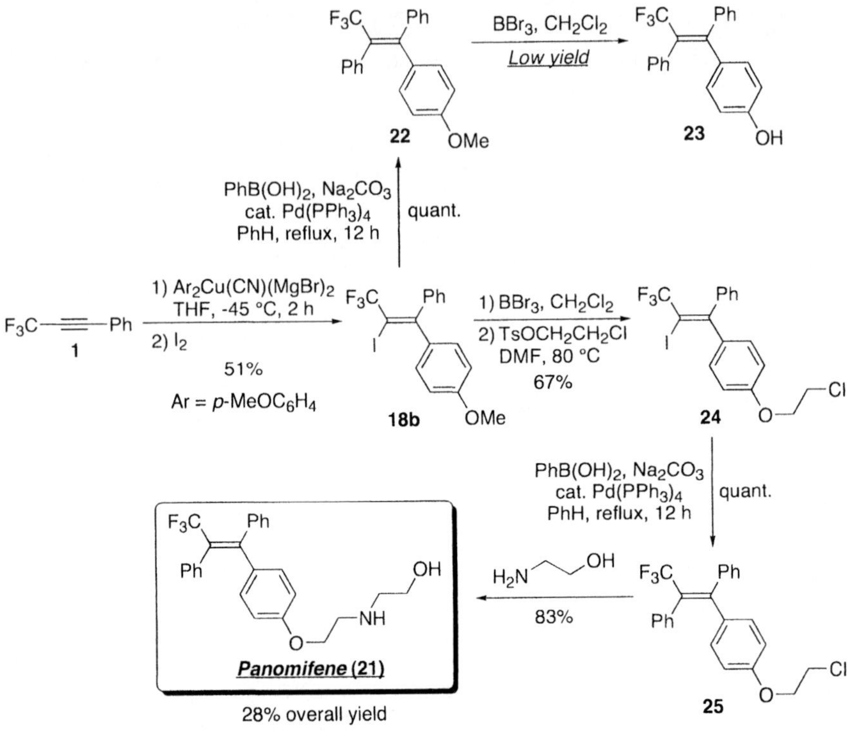

Scheme 10

References

1. (a) Banks, R. E. *Organofluorine Chemistry: Principles and Commercial Applications*; Plenum: New York, 1994. (b) Banks, R. E. *Fluorine Chemistry at the Millenium: Fascinated by Fluorine*; Elsevier: Amsterdam, 2000; and references therein.
2. (a) Henne, A. L.; Nager, M. *J. Am. Chem, Soc.* **1951**, *73*, 1042-1043. (b) Baum, K.; Bedford, C. D.; Hunadi, R. J. *J. Org. Chem.* **1982**, *47*, 2251-2257.
3. Drakesmith, F. G.; Stewart, O. J.; Tarrant, P. *J. Org. Chem.* **1968**, *33*, 280-285.
4. (a) Henne, A. L.; Nager, M. *J. Am. Chem. Soc.* **1952**, *74*, 650-652. (b) Bruce, M. I.; Harbourne, A. L.; Waugh, F.; Stone, F. G. A. *J. Chem. Soc.* (A) **1968**, 356-359.
5. Yoneda, N.; Matsuoka, S.; Miyaura, N.; Fukuhara, T.; Suzuki, A. *Bull. Chem. Soc. Jpn.* **1990**, *63*, 2124-2126.

6. (a) Porta, P. L.; Capuzzi, L.; Bettarini, F. *Synthesis* **1994**, 287-290. (b) Hiyama, T.; Sato, K.; Fujita, M. *Bull. Chem, Soc. Jpn.* **1989**, *62*, 1352-1354. (c) Brisdon, A. K.; Crossley, I. R. *Chem. Commun.* **2002**, 2420-2421.
7. Konno, T.; Chae, J.; Kanda, M.; Nagai, G.; Tamura, K.; Ishihara, T.; Yamanaka, H. *Tetrahedron* **2003**, *59*, 7571-7580.
8. Yamanaka, H.; Araki, T.; Kuwabara, M.; Fukunishi, K.; Nomura, M. *Nippon Kagaku Kaishi* **1986**, 1321-1328.
9. (a) Yamazaki, T.; Muzutani, K.; Kitazume, T. *J. Org. Chem.* **1995**, *60*, 6046-6056. (b) Mizutani, K.; Yamazaki, T.; Kitazume, T. *J. Chem. Soc., Chem. Commun.* **1995**, 51-52. (c) Katriky, A. R.; Qi, M.; Wells, A. P. *J. Fluorine Chem.* **1996**, *80*, 145-147.
10. Jennings, M. P.; Cork, E. A.; Ramachandran, P. V. *J. Org. Chem.* **2000**, *65*, 8763-8766.
11. (a) Rice, M. B.; Whitehead, S. L.; Horvath, C. M.; Muchniji, J. A.; Maleczka, Jr. R. E. *Synthesis* **2001**, 1495-1504. (b) Asao, N.; Yamamoto, Y. *Bull. Chem. Soc. Jpn.* **2000**, *73*, 1071-1087.
12. Chae, J.; Konno, T.; Kanda, M.; Ishihara, T.; Yamanaka, H. *J. Fluorine Chem.* **2003**, *120*, 185-193.
13. Milstein, D.; Stille, J. K. *J. Org. Chem.* **1979**, *44*, 1613-1618.
14. For the review, see. Miyaura, N.; Suzuki, A. *Chem. Rev.* **1995**, *95*, 2457-2483.
15. Konno, T.; Chae, J.; Tanaka, T.; Ishihara, T.; Yamanaka, H. *Chem. Commun.* **2004**, 690-691.
16. (a) Yoshida, H.; Shirakawa, E.; Nakao, Y.; Honda, Y.; Hiyama, T. *Bull. Chem. Soc. Jpn.* **2001**, *74*, 637-647. (b) Miura, K.; Saito, H.; Itoh, D.; Matsuda, T.; Fujisawa, N.; Wang, D.; Hosomi, A. *J. Org. Chem.* **2001**, *66*, 3348-3355.
17. Konno, T.; Takehana, T.; Chae, J.; Ishihara, T.; Yamanaka, H. *J. Org. Chem.* **2004**, *69*, 2188-2190.
18. Konno, T.; Daitoh, T.; Noiri, A.; Chae, J.; Ishihara, T.; Yamanaka, H. *Org. Lett.* **2004**, *6*, 933-936. For the review of the carbocupration, see (a) Nakamura, E.; Mori, S. *Angew. Chem. Int. Ed. Engl.* **2000**, *39*, 3750-3771. (b) Nakamura, E.; Yoshikai, N. *Bull. Chem. Soc. Jpn.* **2004**, *77*, 1-12.
19. Tseng, C. C.; Paisly, S. D.; Goering, H. L. *J. Org. Chem.* **1986**, *51*, 2884-2891.
20. Sonogashira, K. *Comprehensive Organic Synthesis*; Trost, B. M.; Fleming, I., Eds.; Pergamon Press: Oxford, 1991; Vol.3, Chapter 2.4, pp521-529.
21. (a) Ábrahám, G.; Horváth, T.; Toldy, L.; Borvendég, J.; Csányi, E.; Kiss, É.; Hermann-Szente, I.; Tory, K. (Gyógyszerkuntató Intézet). Hungarian Patent 17853, 1979. (b) Németh, G.; Kapiller-Dezsofi, R.; Lax, G.; Simig, G. *Tetrahedron* **1996**, *52*, 12821-12830. (c) Liu, X.; Shimizu, M.; Hiyama, T. *Angew. Chem. Int. Ed. Engl.* **2004**, *43*, 879-882.

Chapter 10

Functionalized Fluorinated Allenes

Gerald B. Hammond

Department of Chemistry, University of Louisville, Louisville, KY 40292

An overview of the literature of mono- and di-fluorinated allenes is presented with an emphasis on the synthesis and chemistry of silyl substituted mono- and di-fluorinated allenes and fluoroallenylphosphonates.

Introduction

Functionalized allenes are of paramount importance in organic chemistry, both as building blocks, and synthetic targets. In the last two years alone, over twenty reviews have been published on the subject.(*1*) The effect of fluorine substitution on the chemistry of functionalized allenylic species would be expected to impart profound changes on the biological and chemical properties of allenes. However, to date, only a minuscule amount of research in this area has been reported. The effect of one or two fluorine atoms bonded to an sp^2 allenic carbon on the known chemistry of allenes has not been fully addressed, in part because these compounds are either not easily accessible, unreactive toward nucleophilic substitution, less prone to form stable metal intermediates, or more susceptible to yield regioisomers (through propargyl-allene isomerization). This review will address advances in the synthesis of functionalized fluoroallenes containing $>C=C=CF_2$, $>C=C=CFH$ and $>C=C=CFP(O)(OEt)_2$ moieties.

Background

In general, fluoroallenes are less stable than those without fluorine substituents but they still possess reasonable stability at room temperature. Incorporation of increasing number of fluorine susbtituents on allene makes these species increasingly reactive with the result that 1,1-difluoroallene (DFA) and 1-monofluoroallene (MFA) must be stored in vacuo at –78°C to prevent oligomerization. 1,1,3-Trichloropropadiene is an unstable material that dimerizes in solution at room temperature, and in the pure state decomposes explosively. 1,1,3,3-Tetrafluoropropadiene, though still prone to dimerization, is a known material that hydrohalogenates readily with HF, HCl, and HBr. Partially fluorinated propadienes are more stable and amenable to further reaction. Phenyl subsitution renders unstable difluoroallenes. This observation is in agreement with the fact that some phenyl substituted haloallenes are quite prone to oligomerization.

Simple fluoropropadienes have been synthesized by metal reduction of dihalofluoro-olefins. The reaction of 1-trifluoromethyl-1-alkenyl bromides with n-BuLi has been employed for the synthesis of DFA (eq 1) and 1,1-difluoro-3,3-dimethyl allene (eq 2).(*2*)

F_3C–CH=CH_2 → 1) Br_2, 2) KOH → $F_3C(Br)C{=}CH_2$ → *n*-BuLi → $F_3C(Li)C{=}CH_2$ → warm → $F_2C{=}C{=}CH_2$ 55% (eq 1)

CF_3CO_2Et → 1) i-PrMgBr (59%), 2) AcCl (93%), 3) 500 °C (75%) → F_3C–CH=C(Me)Me → 4) Br_2 (79%), 5) KOH (74%), 6) BuLi, -85 °C (28%) → $F_2C{=}C{=}C(CH_3)_2$ (7% overall) (eq 2)

Although MFA could be synthesized in similar manner using 1,3-dibromo-1-fluoropropane, the yield was improved by conversion of 1,1-dichloro-2-fluorocyclopropane into fluoroallene with activated zinc dust in refluxing ethanol (eq 3).(*2*)

$CH_2{=}CH_2$ → $CHFBr_2$, ROOR, Δ → BrH_2C–CH_2–CHBrF → 1) KOH, 2) Br_2 → BrH_2C–CHBr–CHBrF → 1) KOH, 2) Zn → $H_2C{=}C{=}CHF$ 8% overall (eq 3)

Xu and Shi (*3*) used the Shapiro reaction of trifluoromethylketone 2,4,6-triisopropylbenzenesulfonylhydrazones (trisylhydrazones) as a convenient method for the preparation of monoalkyl substituted 1,1-difluoroallenes (eq 4). ^{19}F NMR spectroscopy confirmed the existence of an *anti* relationship between the trifluoromethyl and trisylamido groups. This methodology failed for the synthesis of disubstituted 1,1-difluoroallenes.

F_3C–C(=N–NHTrs)–CH_2R → *n*-BuLi (2-3eq), DME, -78°C → dianion → 0°C → [F_3C–C(⊖)=CH–R] → R(H)C=C=CF_2 (eq 4)

R=*n*-Pr, *n*-Pentyl, Bn, Ph

The first synthesis of a functionalized monofluoroallene, published by Krantz and Castelhano,(*4*) featured a fluorochloropropargyl synthon--produced by the reaction of $CHFCl_2$ with the requisite acetylide--which is then converted to fluoroallene **1** upon reduction with aluminum hydride (eq 5). This approach was used to build allenyl aminoacid **2**, a potential suicide inhibitor of GABA transaminases.

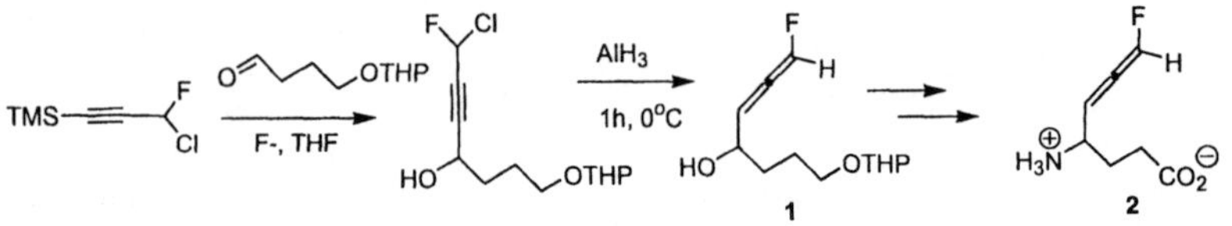

TMS, F, Cl, OTHP, F-, THF, F, Cl, HO, OTHP, AlH_3, 1h, 0°C, F, H, HO, OTHP, **1**, F, H, H_3N, CO_2, **2** (eq 5)

Cyclic or -bicyclic fluorinated compounds constitute a class of compounds, which have appeared scarcely in the literature, the synthesis of which is usually done on a case-by-case basis. An area of synthesis in which fluorinated allenes are poised to make an important contribution is in the development of ring construction methodologies. Dolbier pioneered work in this area with his investigation on the cycloaddition chemistry of DFA and MFA.(*5*) Dixon and Smart calculated geometries, energies, and several electronic properties of fluoroallenes and its isomeric acetylenes to gain qualitative insights into the bonding, in order to predict thermodynamic reactivities.(*6*) DFA was much more reactive towards [4+2] cycloadditions than MFA which in turn was more reactive than 1,1-difluoro-3,3-dimethyl allene (Scheme 1).

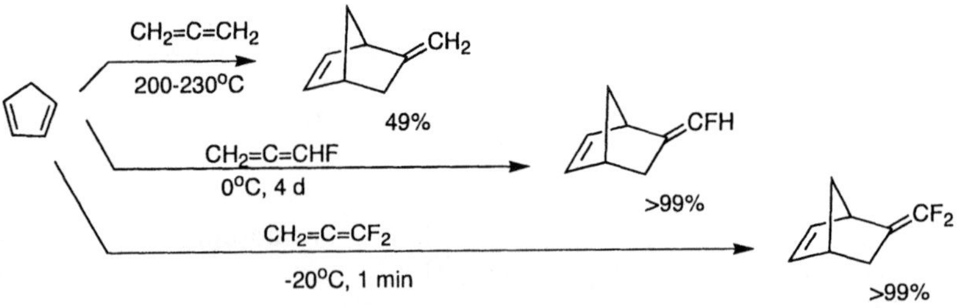

CH_2=C=CH_2, 200-230°C, CH_2, 49%; CH_2=C=CHF, 0°C, 4 d, CFH, >99%; CH_2=C=CF_2, -20°C, 1 min, CF_2, >99%

Scheme 1. Comparative fluoroallene reactivities toward cyclopentadiene

DFA was found to be highly reactive with 1,3-dipoles, and it underwent a totally regiospecific cycloaddition to its C2-C3 double bond. In contrast, its cycloadditions with substituted diazomethanes were not regiospecific but led to two adducts **3** and **4** (eq 6).

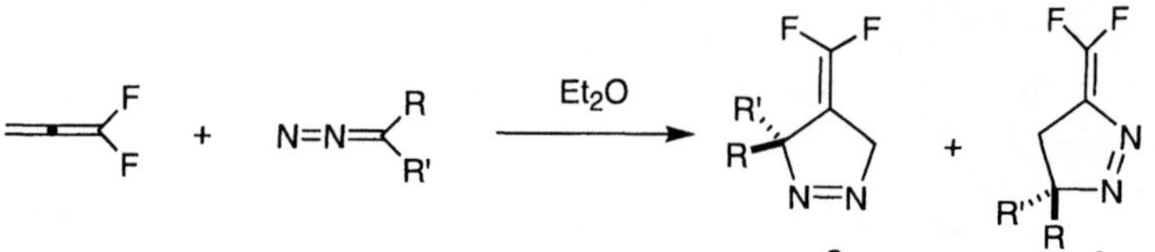

(eq 6)

95-99%

An intrinsic feature of MFA cycloadditions is the issue of π-facial diastereoselectivity. Face selectivity (syn/anti) in additions to trigonal carbon systems is a fundamental question in stereogenesis. Here, fluorine can be a very useful probe. Greater syn/anti selectivity was observed in Diels-Alder and 1,3-dipolar cycloadditions (eq 7). However, no definitive answer can be given to the cause of the regiochemical and stereochemical results obtained in cycloadditions of unsubstituted DFA and MFA. A change of reaction course including a possible radical mechanism (*7*) has been observed with unsymmetrical dienes are used (see also eq 15).

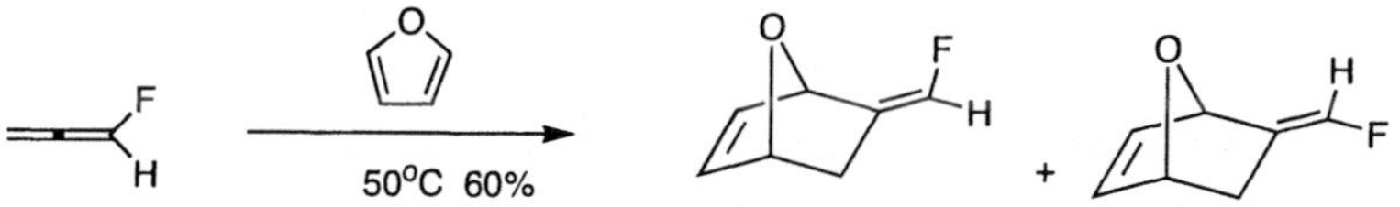

(eq 7)

70:30

An interesting use of *gem*-difluoroallene as ligand was reported recently.(*8*)

Functionalized Difluoroallenes

Our plan to construct a functionalized difluoroallene needed a readily available *gem*-difluoropropargyl starting material capable of undergoing isomerization to an allene. Compound **5** was deemed the candidate of choice because it could be prepared in multigram quantities in a single, high yielding step from a monosubstituted acetylene and CF_2Br_2, an industrial fluorinated feedstock.(*9*) Indium(*10*) in predominantly aqueous media brings about the conversion of **5** to a room-temperature stable species in solution (^{19}F NMR δ –88 ppm) which darkened immediately after solvent removal. ^{19}F NMR analysis of the residue showed complete disappearance of the signal at δ –88 ppm. This result led us to speculate that this signal probably corresponded to a difluoroallenyl indium intermediate, to which we ascribed structure **6** (eq 8).(*11*)

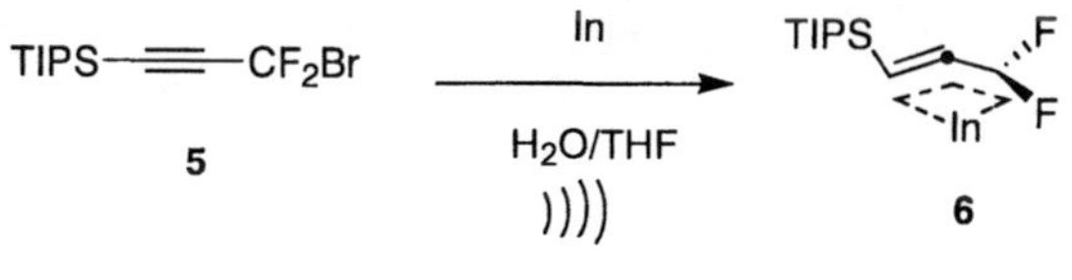

(eq 8)

The existence of an allenylindium species is not without precedence in the literature. Chan,(*12*) Marshall,(*13*) and Loh(*14*) had speculated that a transient species possessing allenic character was an intermediate in their respective synthesis of allenyl- and homopropargyl-alcohols. What is unprecedented is the room-temperature stability of **6** in solution. The effect of other metals on the formation of **6** was investigated but only indium produced a stable metal complex. Other experimental conditions (temperature, solvent, rate of addition, etc.) were studied with the goal of minimizing a competing side reaction (dimerization). Sonication, low reaction temperature (5°C), and dilute concentration of the starting material (0.15 M) are required to maximize production of **6**. Although the structure of this complex remains unknown, it is certain that the nature of the alkyl substituent on the silyl group does not influence its formation.(*15*) When the silyl group is substituted with an alkyl group (eq 9), the expected allene was not found, rather the corresponding reduced product **7** was obtained.(*16*)

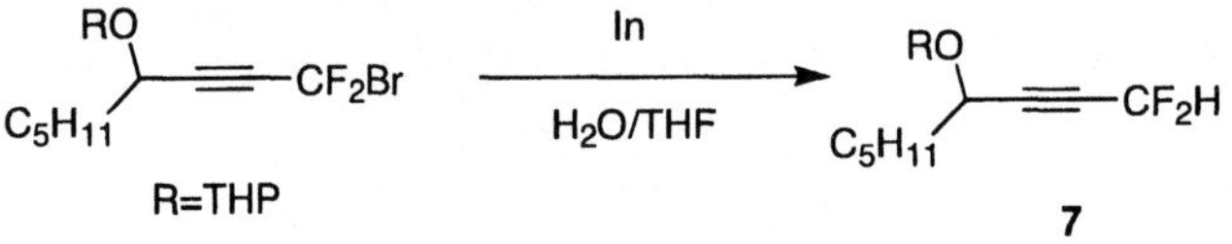

RO, C_5H_{11}, CF_2Br ; R=THP —In, H_2O/THF→ RO, C_5H_{11}, CF_2H **7** (eq 9)

Difluoroallenyl indium **6** (Scheme 2) reacted nucleophilically with aldehydes to yield the expected homopropargylic alcohols; but in the presence of aqueous formaldehyde, it produced allenyl alcohol **6a**. Conversion of the hydroxyl **8a** into its corresponding halide **8b** was achieved cleanly. Allene **8b** isomerized to the thermodynamically more stable conjugated diene isomer under Pd-mediated cross-coupling conditions--including Suzuki and Sonogashira--yielding *gem*-difluoro conjugated diene **9**.(*17*)

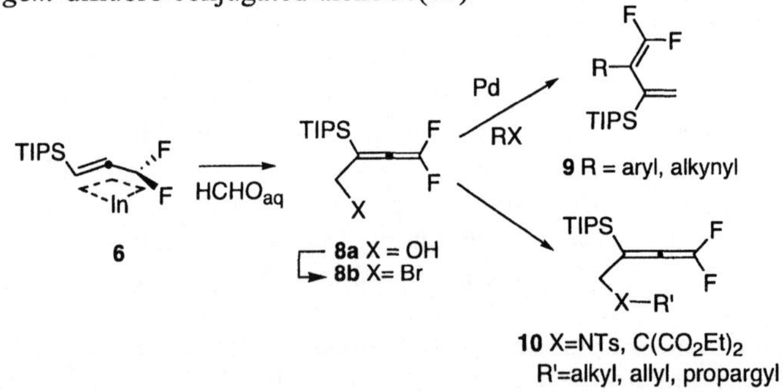

6 —$HCHO_{aq}$→ **8a** X = OH; **8b** X= Br —Pd, RX→ **9** R = aryl, alkynyl; → **10** X=NTs, $C(CO_2Et)_2$, R'=alkyl, allyl, propargyl

*Scheme 2. Reactions derived form difluoroallenyl alcohol **2a***

Similarly, displacement of bromide in **8b** with a phosphorus nucleophile produced the corresponding 1,3-dienylphosphonate **9** (R=P(O)$(OEt)_2$,) in what could be formally construed as a S_N2' reaction. The homologation of the side chain in **8b** was achieved(*18*) using soft nucleophiles and Trost's Pd-catalyzed alkylation methodology (**10** in Scheme 2). By exercising strict control on the reaction conditions the formation of the isomeric dienes was minimized. The oxygen in **8a** could be substituted with nitrogen using propargylic amines, under Mitsunobu's conditions. In these cases, the corresponding diene isomer **9** was not found. Prior to this work, microwave spectroscopy and theoretical calculations on the geometries of the parent 1,1-difluoroallene showed that the C=C and C-F bond lengths would decrease with increasing fluorine substitution, and FCF bond angles should resemble sp^3 hybridization. X-ray data,(*19*) obtained from crystals of **10**, provided the first crystallographic evidence to these statements.

Cyclization methodologies, especially intramolecular, are of paramount importance in the synthesis of organic compounds. Among the ring closure rules proposed by Baldwin 25 years ago, only very few exceptions of nucleophile-driven, disfavored *5-endo-trigonal* cyclization are known, among them the cyclization of allenyl alcohol **8a** to the hitherto unknown 2,2-difluoro-2,5-dihydrofuran **11**. This reaction needed mild conditions (eq 10).(*11*) The cyclization worked equally well using other primary aliphatic amines, but failed when a less basic amine such as *p*-iodoaniline was utilized.

8a —RNH$_2$, hexane, r.t., 20 h→ [TIPS, δ⁻, F, δ⁺, F, O δ⁻, δ⁺ H–NH$_2$R → TIPS, ⊖, F, F, O] —$R\overset{+}{N}H_3$ → RNH_2→ TIPS, H, F, F, O **11** (89%) (eq 10)

Although, in theory, an intramolecular cycloaddition of **10** could be used to prepare bicyclic fluorinated compounds, there are but a handful of reports detailing intramolecular metal-mediated fluoroallene cycloadditions, none of which involved fluoroallene-alkyne [2+2) cycloadditions. The Mo-mediated intramolecular cyclization of **10** produced a novel fused fluorinated cyclobutene **12**, in 90% yield (eq 11). No traces of the [3.2.0] exocyclic difluoromethylene regioisomer could be detected. The fact that this Mo-mediated [2+2]-intramolecular allene-alkyne cycloaddition is an intriguing departure from the expected Pauson-Khand (P-K) reaction prompted the authors to probe the reaction further.(*18*) In all cases, *gem*-difluorocyclobutenes were obtained in very good or excellent yield. TBAF desilylation of **10** produced diyne **13**.

HF_2C, X, R **13** X=NTs (86%) R=H ←TBAF, THF, 6hrs— TIPS, F, F, X, R **10** —$Mo(CO)_6$, DMSO/tol, 100°C→ TIPS, F, F, X, R **12** X=NTs, R=H (90%) (eq 11)

The cycloaddition of **10**, initially performed with 1.5 eq of molybdenum was repeated using catalytic amounts of molybdenum (10 mol%) with equal success. Mechanistically, this reaction represents a formal [2 + 2] cycloaddition, with the molybdenum metallocycle undergoing a reductive elimination, rather than a CO insertion.

The synthetic potential of functionalized difluoroallenes such as allenyl alcohol **8a** has been showcased by these facile synthesis of functionalized conjugated dienes, a mild 'Baldwin-disfavored' intramolecular 5-*endo-trig* cyclization, and a $Mo(CO)_6$ catalyzed [2+2] difluoro allene-alkyne intramolecular cyclization.

Functionalized Monofluoroallenes

The preparation of difluoroallenyl indium **6**, in predominantly aqueous media (Scheme 2), encouraged us to investigate whether monofluorinated allenyl indium could be similarly prepared from bromofluoropropyne **14**.(*20*) Although the corresponding indium species **15** could not be isolated, proof of its fleeting existence was the formation of TIPS-CH=C=CHF, accompanied by its propargylic isomer (1:1 ratio), and unreacted starting material. By eliminating water as co-solvent in the reaction, and using excess of aqueous formaldehyde, the desired **16** was obtained, accompanied by its propargyl isomer **17** in a 7:1 ratio, and dimer $(TIPS\text{-}C{\equiv}C\text{-}CHF)_2$ (Scheme 3)(*21*) TBAF treatment afforded dihydrofuran **18**, in 62% yield, by means of a 5-*endo-trig* cyclization path. The same product could be isolated in better yield using sodium hydride.

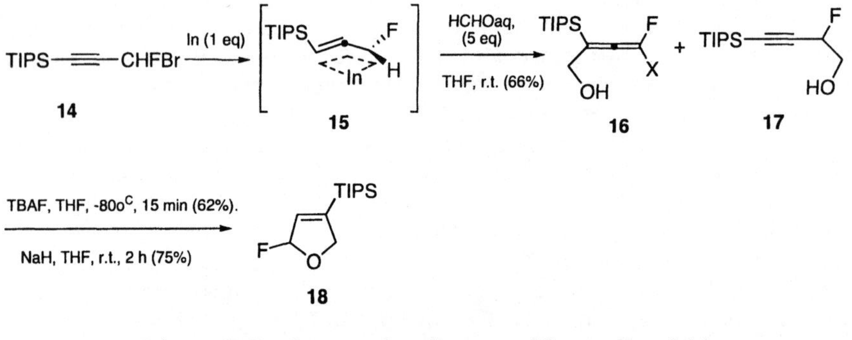

TIPS—≡—CHFBr (**14**) —In (1 eq)→ [**15**] —HCHOaq, (5 eq), THF, r.t. (66%)→ **16** + **17**

—TBAF, THF, -80o^C, 15 min (62%). / NaH, THF, r.t., 2 h (75%)→ **18**

*Scheme 3. Synthesis and cyclization of fluoroallenol **16***

Certain functional group interconversions could be carried out on **16** without compromising the integrity of the fluoroallenyl moiety. Oxidation, using Dess-Martin conditions, replacement of O by N under Mitsunobu conditions, resulted in a facile synthesis of **19** (Scheme 4). Conversion of **16** to its corresponding mesylate, followed by a S_N2' displacement with Br^- produced diene **20**. Coupling **20** with aryl boronic acids under Suzuki conditions, afforded substituted-1,3-butadienes **21**, mostly in good or very good yields, except when a strong electron-withdrawing substituent was employed on the aryl substituent. The stereochemistry of the fluorine-containing double bond in **21** has not yet been fully elucidated. Their spectral data showed that, in each case, the chemical shifts of C-1 and H-1, as well as $^1J_{HF}$ and $^1J_{CF}$ coupling constants, were in good agreement with the reported values for *E/Z* 1-fluoro-1,3-butadiene.(*22*)

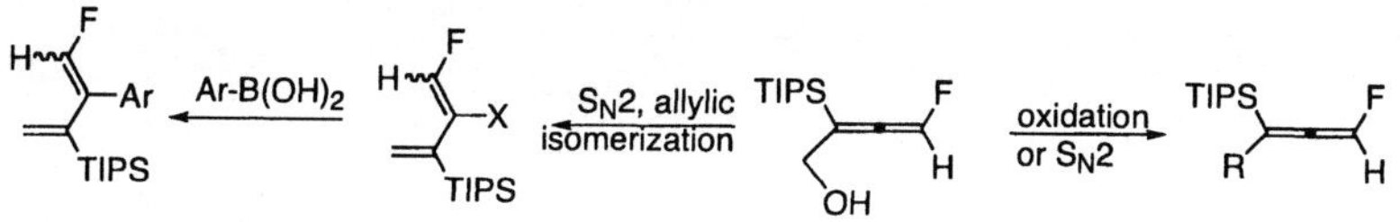

Ar-B(OH)$_2$; S_N2, allylic isomerization ; oxidation or S_N2

21 Ar=Ph, *p*-FC$_6$H$_4$; **20** X=Br, I ; **16** ; **19** R=CHO, CH$_2$NTs(BOC), CH$_2$NHTs, CH$_2$Br, CH$_2$OMs

*Scheme 4. Chemical modifications derived from monofluoroallenol **16**.*

In summary, fluoroallenol **16**, prepared from **14**, cyclized easily to **18** but preserved its fluoroallene integrity under oxidation and S_N2 displacement to yield fluoroallenyl derivatives with desirable synthetic handles—suitable for chain extension or further functionalization—such as aldehyde, amine, mesylate, halide **19**. Conversion of **16** to its corresponding mesylate, followed by a S_N2' displacement with Br^- produced diene **20**, which furnished 2-aryl-substituted 1,3-dienes **21** under Suzuki coupling conditions.

Monofluoroallenyl Phosphonates

α-Fluoroalkylphosphonates, in which the bridging C-O-P phosphate bond has been replaced with either a C-CFH-P or a C-CF$_2$-P bond are important resources for the study of biological phosphate mimics and other therapeutic applications.(*23*) A fluorinated allenylic scaffold, such as **23**, is uniquely equipped for the preparation of α-fluorinated phosphonates. Deprotection of the γ-silyl group in **22** with TBAF yielded fluoroallenyl phosphonate **23a**. A plausible mechanism is shown in Scheme 5 (*24*). The alkyl counterpart of **22** (R=C_5H_{11}), in the presence of benzaldehyde, produced allene **23c** (eq 12). Without benzaldehyde, the same experimental conditions led to the slow formation of α-fluoroallenylphosphonate **23b** in moderate yield. A moderate improvement to the synthesis of **22** (R=TIPS) was reported recently.(*25*)

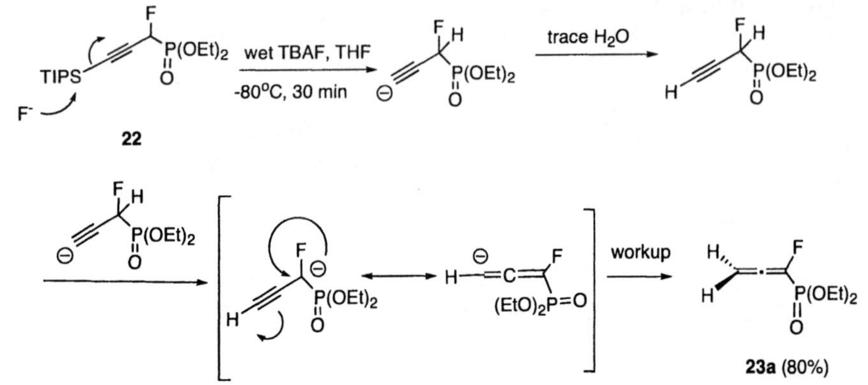

*Scheme 5. Plausible mechanism for the formation of **23a** via TBAF deprotection of **22**.*

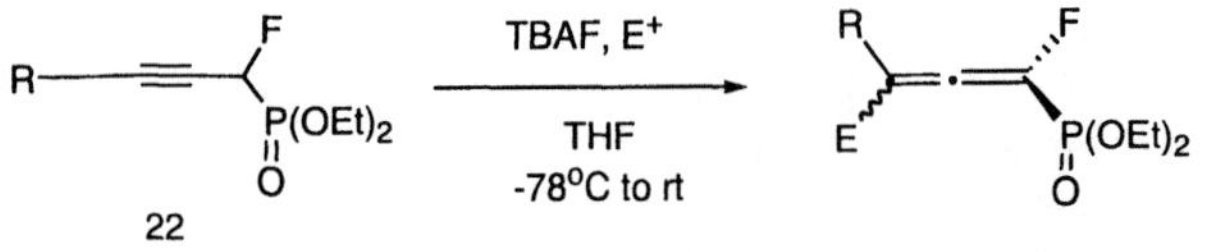

(eq 12)

	E	R	%
23b	H	C_5H_{11}	40
23c	CH(OH)Ph	C_5H_{11}	72

The electrophilic addition of iodine to **23a** furnished exclusively (*E*)-diiodo fluorovinylphosphonate **24** in excellent yield (eq 13) (*24*). This stereospecificity was surprising given the lack of stereocontrol reported by other workers during the halogenation of allenylphosphonates.(*26*) The synthesis of **24** does provide a stereoselective point of entry into a wide repertoire of α-fluorovinylphosphonates because it contains a vinyl iodide which might be capable of manipulation through transition metal catalyzed coupling reactions. In addition, nucleophilic displacement of the allylic iodide in **24** provided access to (*E*)-γ-substituted-α-fluoroalkenylphosphonates **25**.

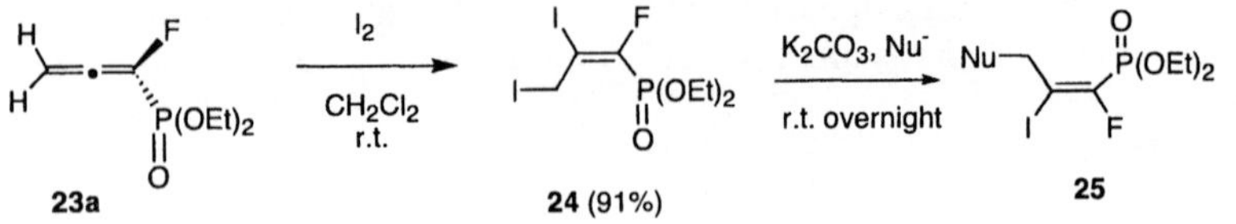

(eq 13)

The hydroamination of α-fluoroallenylphosphonate **23a** served as a conduit to the preparation of fluoroanalogs of enamines and β-enaminophosphonate, a key component in the synthesis of pyrimidone ring systems. With a secondary amine such as Et_2NH, where enamine-imine tautomerization is not possible, only the *(Z)*-isomer was produced.(*24*)

Heating **23a** alone in a sealed tube furnished dimer **26** exclusively in 73% yield (eq 14).(*27*) The Diels-Alder reaction of **23a** with cyclopentadiene at room temperature yields **27** exclusively. The formation of this *Z* isomer was attributed to the efficient orbital overlap that occurs when the phoshonyl group faces the diene in the transition state leading to **27**. The regiochemistry shown in the Diels-Alder reaction of **23c** with conjugated dienes is noteworthy because the corresponding allenecarboxylate (without fluorine) undergoes cycloaddition with cyclopentadiene at the proximal olefin rather than at the distant one.

(EtO)₂(O)P F **27** ← F P(O)(OEt)₂ **23a** → F P(O)(OEt)₂ (EtO)₂(O)P F **26** (eq 14)

Heating a solution of tri(*t*-butyl)silyloxyvinyl cyclohexene with **23a** afforded a 4:1 mixture of α-fluoromethylidenephosphonate **28a** and α-fluoromethylenephosphonate **28b**. Interestingly, prolonged heating reverses the ratio of both isomers but does not alter the isolated yield. Compound **28b** appears to be the thermodynamic product of this reaction, and its formation from **28a** could be explained by invoking a homolytic bond cleavage to give a stabilized bis(allyl) diradical **I**, which can then rearrange to **II** and recombine at the fluorophosphonate terminus to give **28b** (eq 15).

P(O)(OEt)₂ F TBSO **28a** Δ → [P(O)(OEt)₂ F TBSO] **I** ↓ TBSO P(O)(OEt)2 F **II** TBSO P(O)(OEt)₂ F **28b** (eq 15)

Other mechanistic explanations such as the intermediacy of an anion/cation pair, or the possibility of a retro-Diels-Alder reaction during prolonged heating cannot be discarded. A retro-Diels-Alder would reform allenephosphonate **23a**, which could then react as a dienophile using its proximal olefin, producing **28b**.

Conclusion

The study of difluoroallenes and monofluoroallenes has been restricted by the difficulty in preparing functionalized derivatives. The γ-functionalization of mono- and di-fluoroallenes has been achieved using propargyl-containing starting materials. The few representative examples reported so far have yielded important synthetic intermediates. This underscores the importance of the allene template as a useful fluorinated synthon. Future work in this area will concentrate on understanding the role of fluorine in the chemistry of allenes, and diversifying the syntheses of allenes using various activating groups on the γ-carbon af the fluorinated allene.

Acknowledgements

The generous financial support of the National Science Foundation (CHE-0213502) and the Petroleum Research Fund (36602-AC1) is gratefully acknowledged.

Literature Cited

1. Selected examples: Hoffmann-Roder, A.; N. Krause, *Angew. Chem. Int. Ed.* **2004**, *43*, 1196-1216. Nagao, Y.; S. Sano, *J. Synt. Org. Chem. Japan* **2003** *61*, 1088-1098. Wei, L.L.; H. Xiong; R.P. Hsung, *Acc. Chem. Res*, **2003**, *36*, 773-782. Sydnes, L. K., *Chem. Rev.* 2003. 103, p. 1133-1150. Trofimov, B. A., *Curr. Org. Chem.*, **2002**. *6*, 1121-1162. Bates, R. W.; V. Satcharoen, *Chem. Soc. Rev.* **2002**, *31*, 12-21. Marshall, J. A., *Organometallics In Synthesis. A Manual*, M. Schlosser, Ed. 2002, J. Wiley & Sons, NY. Marshall, J. A., *Chem. Rev.* **2000**, *100*, 3163-3185. Marshall, J. A., *Chemtracts* **1997**, *10*, 481-496. Ma, S.; A. Zhang, *J. Org. Chem.* **2002**, *67*, 2287-2294. Ma, S. M., *Eur. J. Org. Chem.* **2004**, 1175-1183. Ma, S. M., *Acc. Chem. Res.* **2003**, *36*, 701-712.
2. Dolbier, Jr., W. R.; Burkholder, C.; Piedrahita, C. *J. Fluorine Chem.* **1982**, *20*, 637-647.
3. Shi, G.; Xu, Y. *J. Fluorine Chem.* **1989**, *44*, 161-166.
4. Castelhano, A. L.; Krantz, A. *J. Am. Chem. Soc.* **1987**, *109*, 3491-3493.
5. Dolbier, Jr. W. R. *Acc. Chem. Res.* **1991**. 24, p. 63-69; and references therein.
6. Dixon, D.; Smart, B. *J. Phys. Chem.* **1989**, *93*, 7772-7780.
7. Dolbier, Jr., W. R. *Advances in Detailed Reaction Mechanisms,* **1991**, *1*, 127-179.
8. Lentz, D.; Willemsen, S. *Angew. Chem. Int. Ed.* **2001**, *40*, 2087-2091.
9. Wang, Z.; Hammond, G.B. *J. Chem. Soc. Chem. Commun.* **1999**, 2545-2546.

10. Recent reviews: Podlech, J.; Maier, T. C. *Synthesis* **2003**, 633-655. Pae, A. N.; Cho, Y.S. *Curr. Org. Chem.* **2002**, *6*, 715-737.
11. Wang, Z.; Hammond, G. B. *J. Org. Chem.* **2000**, *65*, 6547-6552.
12. Chan, T. H.; Yang, Y. *J. Am. Chem. Soc.* **1999**, *121*, 3228-3229. Miao, W. S.; Lu, W. S.; Chan, T. H. *J. Am. Chem. Soc.*, **2003**, *125*, 2412-2413.
13. Marshall, J. A. *Chemtracts* **1997**, *10*, 481-496.
14. Lin, M. J.; Loh, T. P. *J. Am. Chem. Soc.* **2003**, *125*, 13042-13043.
15. Arimitsu, S.; Xu, B.; Kishbaugh, T. L. S.; Griffin, L.; Hammond, G. B. *J. Fluorine Chem.* **2004**, *125*, 641-645.
16. Shen, Q. L. M.Sc. Thesis, University of Massachusetts Dartmouth, Dartmouth, MA, 2002.
17. Shen, Q. L.; Hammond, G. B. *Org. Lett.* **2001**, *3*, 2213-2215.
18. Shen, Q.L.; Hammond, G. B. *J. Am. Chem. Soc.* **2002**, *124*, 6534-6535.
19. Shen, Q. L.; Chen, C.-H.; Hammond, G. B. *J. Fluorine Chem.* **2002**, *117*, 131-135.
20. Lan, Y. F.; Hammond, G.B. *J. Org. Chem.* **2000**, *65*, 4217-4221.
21. Lan, Y.; Hammond, G. B. *Org. Lett.* **2002**, *4*, 2437-2439.
22. Dolbier, Jr., W. R.; Gray, T. A.; Keaffaber, J. J.; Celewicz, L.; Koroniak, H. *J. Am. Chem. Soc.* **1990**, *112*, 363-367.
23. For background information see O'Hagan, D.; Rzepa, H. S. *J. C. S. Chem. Commun.* **1997**, 645-652.
24. Zapata, A. J.; Gu, Y.; Hammond, G. B. *J. Org. Chem.* **2000**, *65*, 227-234.
25. Wang, Z.; Gu, Y.; Zapata, A. J.; Hammond, G. B. *J. Fluorine Chem.* **2001**, *107*, 127-132.
26. Xu, Z. Q.; Zemlicka, J. *Tetrahedron* **1997**, *53*, 5389-5396.
27. Gu, Y.; Hama, T.; Hammond, G. B. *J. Chem. Soc. Chem. Commun.* **2000**, 395-396.

Fluorine-Containing Aromatic and Heteroaromatic Synthons

Chapter 11

Regioexhaustive Functionalizations: To Miss an Isomer Means to Miss a Chance

Manfred Schlosser

Institute for Chemical Sciences and Engineering, Ecole Polytechnique Fédérale, CH–1015 Lausanne, Switzerland

To exploit the chemical potential inherent in a simple, aromatic or heterocyclic, precursor compound optimally one should be able to introduce any kind of substituents, in particular functional groups, into any vacant position. To achieve the latter objective, a specific kit ("toolbox") of advanced organometallic methods and reaction sequences has been elaborated and tested. The concept is illustrated by typical applications in the benzene, naphthalene and pyridine areas.

Only a few commercial organofluorine compounds are endowed with structural complexity whereas most others exhibit deceptively monotonous substituent patterns. To make a maximum out of the opportunities offered by such bulk chemicals has motivated the present investigation. The procedures developed proved most valuable in order to achieve the structural proliferation of fluorinated key compounds but were also applied, with equal success, to substrates carrying other heteroelements such as chlorine, bromine, oxygen or nitrogen (*1*).

There are good reasons for insisting on *regioexhaustiveness* if the unrestricted chemical modification of a readily available starting material rather than the target-oriented synthesis is the aim. First of all, it is dictated by economical considerations to get a maximum of derivatives out of the same precursor compound. Moreover, one needs the complete selection of isomers to extract valid structure/activity relationships and to find out whether the individual contributions of substituents are additive or depend on their environment.

When performing regiochemically exhaustive substitutions, our research is entirely tributary to the *organometallic approach* (*2*). This, of course, has to do with its versatility. Most importantly, we can access all kinds of derivatives by merely choosing the appropriate electrophile to be combined with the organometallic intermediate (*3*). Although it is most common to quench such reactions with dry ice and to isolate subsequently the corresponding carboxylic acid, there are dozens if not hundreds of other, similarly attractive reagents that open the entry to virtually all types of functionalities. Thus the problem of deciding what derivative to make exactly can be deferred to the very end of the reaction sequence. This unique *product flexibility* is a most precious hallmark of organometallic chemistry.

The Toolbox Methods

The "*ortho*-directed metalation" of substituted benzenes (a somewhat vague translation of Wittig's "gezielter *ortho*-Metallierung") has become a commonplace. To live up to our quest for regioselectivity we have to find ways to selectively introduce substituents, including functional groups, also in *meta* and *para* positions. The practical realization of this goal relies on the " 2 x 3 toolbox methods". As already outlined elsewhere (*1*), they comprise three subcategories of permutational hydrogen/metal interconversions ("metalations") and three subcategories of permutational halogen/metal interconversions ("exchanges"). These methods have been applied to the regioexhaustive substitution of fluorinated or otherwise halogenated benzenes, biphenyls, naphthalenes, *O*-protected phenols, *N*-protected anilines and a series of heterocyclic substrates such as indoles, quinolines, pyridines, pyrazoles and pyrroles (*1*, *4*). In the following, each of the existing options will be illustrated by typical examples.

Isomerization by Transmetalation

The first subcategory in the metalation sector takes advantage of a transmetalation mechanism to convert a kinetically favored, more basic species into a less basic isomer by equilibration (the "second-chance approach"). When treated with lithium diisopropylamide (LIDA) in tetrahydrofuran (THF) at –75 °C, 4-chloro-3-fluoropyridine undergoes deprotonation mainly at the coordinatively activated 2-position but to a minor extent also at the more acidic 5-position as evidenced by the carboxylic acids **1a** and **1b** obtained after the reaction with dry ice. However, in the course of 20 h, the 4-chloro-3-fluoro-2-pyridyllithium progressively attacks unconsumed substrate at the 5-position until it has been completely converted into 4-chloro-5-fluoro-3-pyridyllithium (*5*). In diethyl ether (DEE) the hydrogen/metal permutation process requires the neighboring group assistance of the lone pair of the nitrogen atom at which the lithium can dock. Therefore, with lithium 2,2,6,6-tetramethylpiperidide (LITMP) as the base, metalation occurs quantitatively at the 2-position in this solvent (*5*).

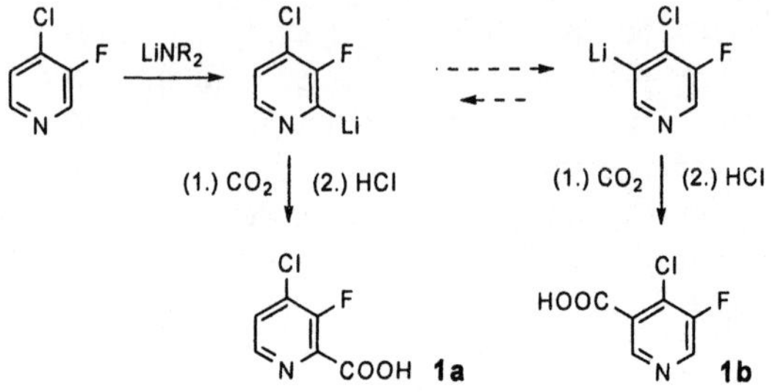

	1a	1b
LIDA / THF 20 h -75°C	0%	71%
LITMP / DEE 2 h -75°C	91%	0%

Reagent-Controlled Optional Site Selectivity of Metalations

Optional site selectivity in metalation reactions is hardly attained by lucky gambling with the reaction conditions but is better based on mechanistically guided substrate/reagent matching (*6*). Ordinary alkyllithiums prefer to abstract a proton from a site in the vicinity of an electron donor substituent which can coordinate the metal by virtue of a lone pair and thus lower the energy of the transition state. In contrast, if butyllithium is coordinatively saturated by powerful donor ligands such as potassium *tert*-butoxide or *N,N,N',N",N"*-pentamethyldiethylenetriamine (PMDTA), it just seeks to travel along the steepest acidity gradient, in other words to generate the least basic of all possible intermediates. This would imply the proton abstraction from a fluorine-adjacent rather than oxygen- or nitrogen-adjacent position if there is a choice. For this reason 1-fluoro-4-(methoxymethoxy)benzene (like its *ortho* isomer too), was exclusively deprotonated at the 3- or the 2-position affording respectively the acids **2a** and **2b**, depending on whether butyllithium was employed in the absence or presence of additives such as potassium *tert*-butoxide or PMDTA (*7, 8*).

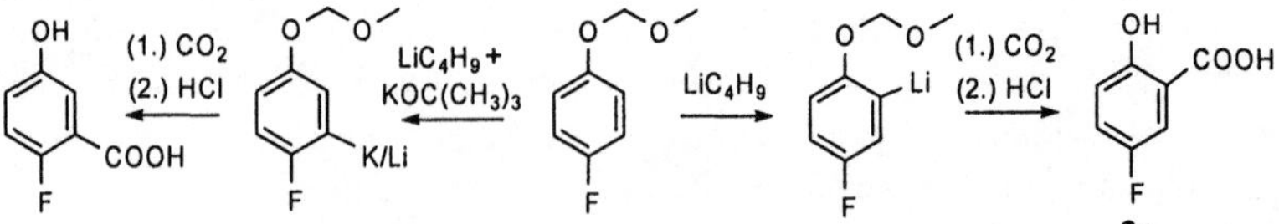

The weaker the base, the more it tends to generate the "least expensive", thermodynamically most stable intermediate (*9*). Thus, proton abstraction from the *O*-

methoxymethyl protected 2-chloro-3,4-difluorophenol occurs exclusively at the 6- or 5-position and gives subsequently rise to the acids **3a** or **3b** depending on whether butyllithium in diethyl ether or lithium 2,2,6,6-tetramethylpiperidide (LITMP) in tetrahydrofuran is employed.

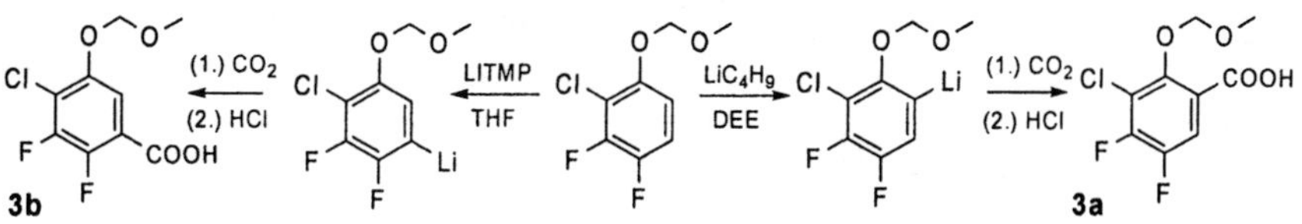

Deploying Site Protective Groups

The concept of deflecting the metalation from the top reactive site to a neighboring position by using a suitable perturbation has been described in a recent review article (*4*). "If optional site selectivity and basicity-driven halogen shuffling fail to outwit the thermodynamically most acidic position of the substrate, there is still one last chance left which appears at the same time to be the most obvious one. One merely has to bow to the inevitable, deprotonate the most reactive site and then to block it with an atom or a group electronegative and slim enough to activate its immediate vicinity for further deprotonation. The desired electrophilic substitution once accomplished, all what remains to be done is to remove the temporary substituent (*4*)." This description of a would-be helper reads like a warrant of apprehension for chlorine.

Actually chlorine perfectly fulfills all the given requirements. It has a considerable activating effect as shown by the comparison between the very facile metalation of 3-chlorobenzotrifluoride with that of benzotrifluoride which reacts quite sluggishly unless superbases are used (*11*). When the mixture is quenched by reaction with dry ice and neutralized, the carboxylic acid **4a** is isolated in high yield. The dechlorination to the 2-(trifluoromethyl)benzoic acid can be readily accomplished by catalytic hydrogenation or treatment with zinc powder in alkaline medium (*5*). Remarkably, the regioselectivity of the metalation is completely altered when butyllithium is replaced by the bulkier *sec*-butyllithium. The deprotonation takes now place exclusively at the CF_3-remote 4-position to afford the acid **4b** and, after subsequent reduction, 4-(trifluoromethyl)benzoic acid. The 3-isomer can be obtained starting from either 2- or 4-chlorobenzotrifluoride and passing through the chlorinated acids **4c** and **4d**. The three (trifluoromethyl)benzoic acids could of course also be readily prepared from the equally commercial, though more expensive 2-, 3- and 4-bromobenzotrifluorides by submitting them to a consecutive halogen/metal permutation and carboxylation.

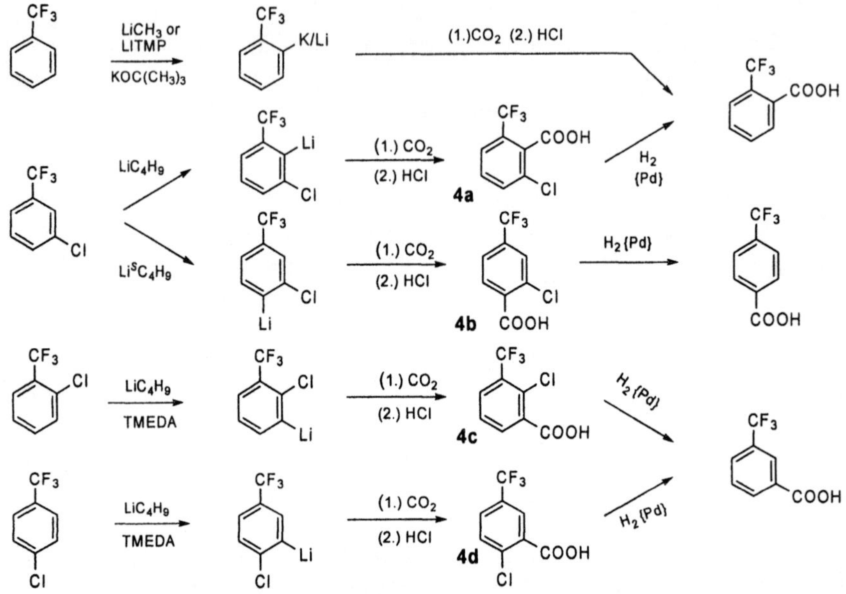

Another, if rarely exploited, option is to replace the "fill-in" chlorine substituent by a nucleophile rather than by hydrogen. 1,2,3-Trifluoro-4-nitrobenzene (**5**), a key intermediate on the route to the antibacterial Ofloxacin (*12*) is technically produced from 1,3-dichloro-2-fluoro-4-nitrobenzene which in turn is made from 1,3-dichloro-2-fluorobenzene (*13*). This material is generally prepared in a multi-step sequence. It can be more readily obtained in a one-pot procedure from the inexpensive fluorobenzene by repetitive metalation with *sec*-butyllithium and chlorination with 1,1,2-trichloro-1,2,2-trifluoroethane (*14*).

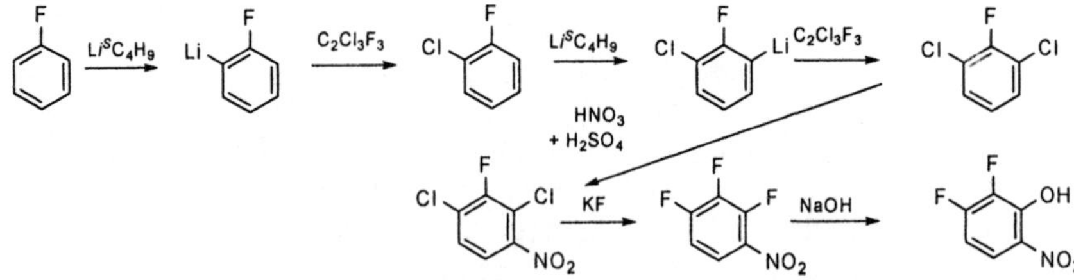

Unlike chlorine, organic bromine is not inert toward organometallic reagents but rather entertains a halogen/metal permutation with them. 2,3,6-Trifluorophenol can be very easily converted into the acid **6a** by selective *para* bromination (using molecular bromine), acetalization and reaction with butyllithium followed by carboxylation and neutralization. On the other hand, the acid **6b** is isolated if butyllithium is replaced by LIDA, thus promoting deprotonation, and the bromine protective group is reductively removed after the carboxylation (*14*).

LiC$_4$H$_9$ (1.) CO_2 (2.) HCl HCl **6a**
LIDA (1.) CO_2 (2.) HCl (1.) H_2 (Pd) (2.) HCl **6b**

Deprotonation-Triggered Heavy Halogen Hopping

2-Bromobenzotrifluoride represents the most impressive model case in the arene series. A proton is abstracted from the bromine-adjacent 3-position with LITMP in tetrahydrofuran at -100°C, leading to the acid **7a** upon carboxylation. However, when the mixture is brought to -75°C before quenching, bromine and lithium swap places by halogen/metal permutation involving traces of accidentally formed 2,3-dibromobenzotrifluoride as catalytic turntables and ending up, after carboxylation with the acid **7b** (*11*).

LITMP/THF -100°C (1.) CO_2 (2.) HCl **7a**
-75°C
(1.) CO_2 (2.) HCl **7b**

The nature of the halogen is critical for such basicity gradient driven isomerizations. Chlorine proves to be immobile in this respect at any temperature below -25°C whereas bromine and iodine migrate even at -125°C. Thus, when consecutively treated with the "Faigl mix" (*15*), the mixture of LITMP, PMDTA and potassium *tert*-butoxide, at -125°C and dry ice, 2-chloro-1,3-difluorobenzene is converted into the acid **8a** whereas 2-bromo-1,3-difluorobenzene and 1,3-difluoro-2-iodobenzene give the acids **8b** and **8c** (*16*).

LITMP + PMDTA
+ $KOC(CH_3)_3$
THF -125°C
(1.) CO_2 (2.) HCl
X = Cl
X = Br, I
(1.) CO_2 (2.) HCl
COOH

8a

8b: X = Br
8c: X = I

Discrimination Between Bromine and Iodine

When the readily accessible 1,5-dibromo-2,3,4-trifluorobenzene and 1-bromo-2,3,4-trifluoro-5-iodobenzene are subjected to a heavy halogen dislocation under the action of LIDA, the isomers 1,2-dibromo-3,4,5-trifluorobenzene and 2-bromo-3,4,5-trifluoro-1-iodobenzene result in high yield. The dibromo compound reacts with butyllithium in tetrahydrofuran at -100°C to generate cleanly an organolithium species which forms the acid **9a** upon carboxylation. However, under the same conditions, the iodobromo precursor affords the acid **9b** as the heavier halogen is displaced exclusively (*17*).

(1.) CO_2 (2.) HCl
LiC_4H_9 THF -100°C
X = I
LiC_4H_9 THF -100°C
X = Br
(1.) CO_2 (2.) HCl

9b **9a**

Discrimination Between Seeming Equal Halogens

For a long while the only example known was procured by 2,5-dibromopyridine as the substrate. With butyllithium in toluene the halogen/metal permutation occurs at the 2-position where the lone pair of the nitrogen atom can coordinate the metal and thus stabilize the transition state (*18*). Such a neighboring group participation is unnecessary in the more polar solvent tetrahydrofuran and, as a consequence, the exchange process is now oriented to the 5-position in order to produce the less basic intermediate (*19*).

Quite similarly, 2,3,5-tribromopyridine, made in a two-step protocol from 2-aminopyridine (*20*, *21*), can be converted in the two isomeric acids **10a** and **10b.** In toluene the coordination-seeking butyllithium attacks again the 2-position whereas isopropylmagnesium chloride in tetrahydrofuran promotes the permutational exchange at the most acidified 3-position (*22*).

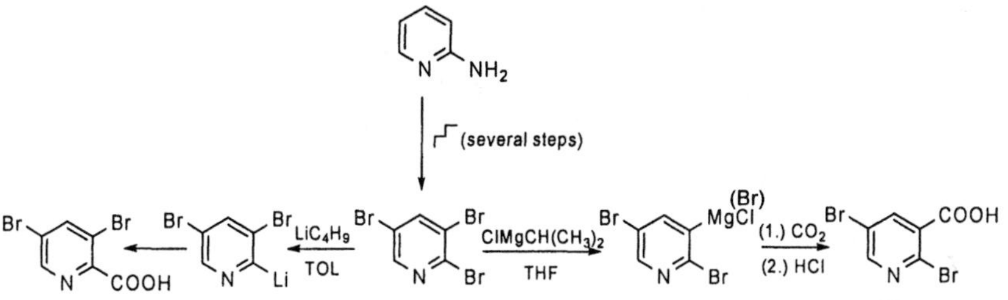

Which Tool to Select When ?

One of the six "educated methods" featured in the foregoing part is particular dear to the heart of the present author. The *reagent-modulated optional site selectivity* is, when it works, an extremely effective and straight-forward solution to the problem. In fortunate cases it enables the substrate to metamorphose to three different isomeric identities by merely varying the metalating base (*1*). Moreover, to be successful in this area happens rarely by mere coincidence, but is rather guided by deep mechanistic insight and hence provides a feeling of triumph. However, optional site selectivity depends on electronic differences between competing substituents which may be too subtle or too unbalanced to be exploitable. In other words, it is not an all-round option.

The faithful workhorses among the toolbox methods are deprotonation-triggered *halogen hopping* and the metalation-rerouting by *protective groups.* Before someone embarks on any of these possibilities, it would be helpful to dispose of some criteria of evaluation that allow one to estimate the performance in advance. Therefore we shall focus in the following two subchapters on some fundamentals and their boundary conditions.

Bromine or Iodine for Halogen Dislocations?

The key step of deprotonation-triggered heavy halogen migration being a halogen/metal permutation, iodine should outperform bromine by far. Instead, there are cases where an *ortho*-lithiated bromoarene proves perfectly stable whereas the iodoarene analog isomerizes instantanously. The 6-fluoroindole family offers an instructive example. Treatment of the unprotected starting material with two equivalents of the superbasic "LIC-KOR" mixture (butyllithium in the presence of potassium *tert*-butoxide; *2*, *23*, *24*) and subsequent carboxylation provides the pure 6-fluoroindole-7-carboxylic acid (**11a**). Conversely, the *N*-triisopropylsilyl congener undergoes metalation with PMDTA-activated *sec*-butyllithium at the 5-position. Both the 5-bromo and the 5-iodo derivative, obtained by treatment of the organolithium intermediate with the elemental halogen, are deprotonated by LIDA at the 4-position. However, only the iodinated species isomerizes to afford, after carboxylation and reduction, the acid **11c** whereas the acid **11b** can be accessed on the bromine route (*25*).

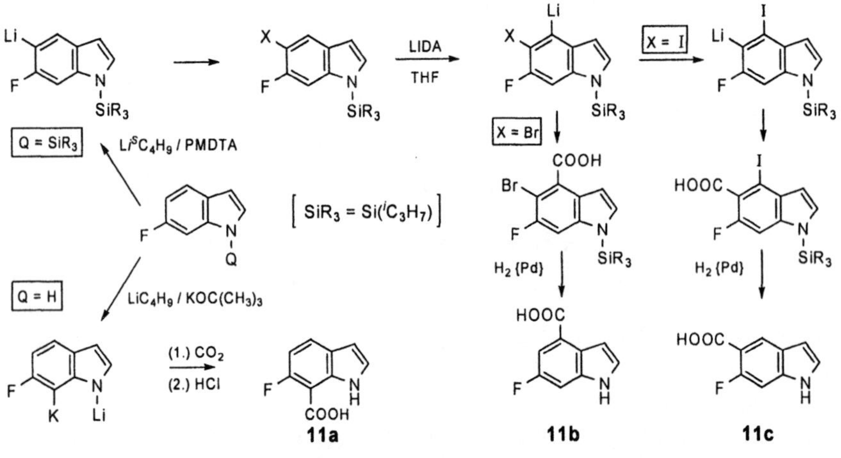

Iodine as a migrating substituent nevertheless suffers from some shortcomings which bromine exhibits only to a lesser extent. First of all, it sometimes becomes the prey of a reductive side reaction which simply replaces it by hydrogen (*26*). Alternatively or simultaneously, the isomerization may stop abruptly leaving some 15 - 25% of the starting material unconsumed. Fortunately, such an accident can easily be repaired by a *kinetic purification*. Addition of a little more than the stoichiometrically required amount (20 - 30%) of butyllithium to the mixture containing the main product (*e.g.* **12**) along with the undesired precursor **13** selectively eliminates the latter by preferential halogen/metal permutation. After hydrolysis, the deiodinated compound can readily be separated from the iodine-containing main product (*27*).

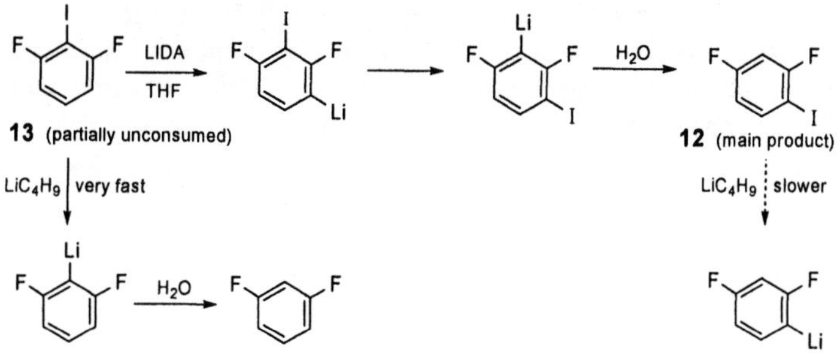

13 (partially unconsumed)

12 (main product)

The Protective Group Contest: Chlorine *vs.* Trialkylsilyl

Chlorine is the first choice as a metalation-rerouting substituent for several good reasons. It reliably blocks the initially most acidic position and activates the neighboring one if vacant. It can be used repetitively as illustrated by the metalation of 1-fluoronaphthalene where it deflects the attack of *sec*-butyllithium from the 2- to the 3- and, after accumulation, to the 4-position as mirrored by the isolation of the acids **14a**, **14b** and **14c** after reductive dechlorination (*2*). 1,2,3-Trichloro-4-fluoronaphtalene does no longer reacts with organometallics by metalation but rather by chlorine/lithium permutation. If one wishes to move the battlefield to the annulated ring, one may introduce at the fluorine-opposite position an easily removable group such as amino (*28*) or aminosulfonyl (*29*), both known to promote metalation at the *peri* position.

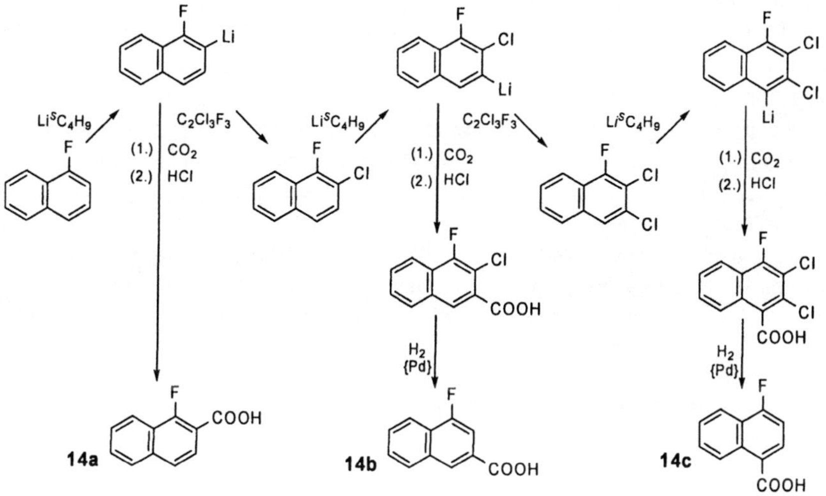

14a **14b** **14c**

Chlorine can be used as a metalation-rerouting group also in the heterocyclic area as the following example demonstrates convincingly. The CH-acidity of 1-substituted imidazoles is highest at the 2- and lowest at the 4-position (*30*). Protection of the firstly deprotonated 2-position by a chlorine atom enables one to functionalize subsequently the 5-position selectively in order to obtain the product **15** after appropriate electrophilic substitution and removal of the protective groups (*31*).

C_2Cl_6 — LiC_4H_9 — (1.) $HCON(CH_3)_2$ (2.) H_2O — H_2 {Pd} — LiC_4H_9

[OR = $OCH_2C_6H_5$]

15

Next to chlorine, trialkylsilyl is the favorite metalation-preventing protecttive entity. Although often satisfactory results with tri*methyl*silyl groups have been achieved (*32*, *33*), it deems advisable to use tri*ethyl*silyl groups instead in order to avoid any risk of $SiCH_3$ metalation (*34* - *36*).

Compared to chloro substituents, trialkylsilyl groups are more flexible as far as their removal is concerned. They can not only be replaced by hydrogen ("protodesilylation" using tetrabutylammonium fluoride hydrate as the reagent), but also by bromine or iodine ("halodesilylation" using molecular bromine or iodine chloride) (*37*).

There is another major difference between the two standard protective groups. Whereas chlorine just silences the position it occupies and activates the neighborhood toward metalation, trialkylsilyl in addition shields sterically its immediate vicinity. As shown in the last scheme one can take advantage of this contrasting behaviour by adapting the protective group to the desired regioselectivity.

3-Chloro-2-fluoropyridine, made from 2-fluoropyridine by metalation with LIDA and chlorination with 1,1,2-trichloro-1,2,2-trifluoroethane, is cleanly deprotonated at the 4-position when treated in turn with LIDA. The resulting organolithium species makes a bifurcation. If another chlorine atom is introduced, it will again activate the 5-position and thus pave the way to the acid **16a**. Contrarily, the metalation occurs only at the 6-position if the nitrogen-remote part of the heterocycle is shielded by a trimethylsilyl entity, as evidenced by the formation of the acid **16b** (*5*).

LIDA THF -75°C; (1.) CO_2 (2.) HCl; H_2(Pd)

16a

$C_2Cl_3F_3$

LIDA THF -75°C

[$SiR_3 = Si(CH_3)_3$; LIDMAE = $LiOCH_2CH_2N(CH_3)_2$]

$ClSiR_3$

LiC_4H_9 LIDMAE; (1.) CO_2 (2.) HCl; TBAF H_2(Pd)

16b

Conclusions

Hydrogen/metal permutation (metalation) reactions do not necessarily follow the motto "the stronger acid and the stronger base give the weaker acid and the weaker base". In fact, the kinetic and thermodynamic preferences for a given species depend on numerous factors such as the site inherent acidity, inductive effects, polarization, anomeric interactions, metal coordination, reagent structure, steric hindrance and electron-electron repulsion including buttressing effects (*38 - 40*). By manipulating these variable parameters skillfully one can generate various organometallic intermediates at will. The *Toolbox Methods* bring new momentum in *diversity-oriented* (as opposed to target-oriented) *synthesis* and, at the same time, secure *logistic economy*. The underlying new concepts are universally applicable and will hence prove valuable in many areas. As reflected by the results reported above, the toolbox approach has established itself already as an efficacious means to "breed" fluorinated building blocks from readily available bulk materials.

Acknowledgment

Our research work was supported by the *Schweizerische Nationalfonds zur Förderung der wissenschaftlicher Forschung*, Bern (grants 20-63'584-00 and 20-100'336-02). The author is indebted to all his collaborators who were involved in the given topics, in particular to the staff members Carla Bobbio, Fabrice Cottet, Elena Marzi and Frédéric Leroux.

References

1. Schlosser, M.; review to appear in *Angew.Chem.*
2. Schlosser, M. (ed.); *Organometallics in Synthesis: A Manual*, Wiley, Chichester, **2002**.
3. Schlosser, M.; Gorecka, J.; Castagnetti, E.; *Eur. J. Org. Chem.* **2003**, 452 - 462.
4. Schlosser, M.; *Eur. J. Org. Chem.* **2001**, 3975 - 3984.
5. Bobbio, C.; Schlosser, M.; unpublished results, **2003-2004**.
6. Mongin, F.; Maggi, R.; Schlosser, M.; *Chimia* **1996**, *50*, 650 - 652.
7. Katsoulos, G.; Takagishi, S.; Schlosser, M.; *Synlett* **1991**, 731 - 732.
8. Marzi, E.; Mongin, F.; Spitaleri, A.; Schlosser, M.; *Eur.J.Org.Chem.* **2001**, 2911 - 2915.
9. Takagishi, S.; Schlosser, M.; *Synlett* **1991**, 119 - 121.
10. Marzi, E.; Gorecka, J.; Schlosser, M.; *Synthesis* **2004**, 1609 – 1618.
11. Mongin, F.; Desponds, O.; Schlosser, M.;*Tetrahedron Lett.* **1996**, *37*, 2767 - 2770.
12. Hayakawa, I.; Hiramitsu, T.; Tanaka, Y.; *Chem. Pharm. Bull.* **1984**, *32*, 4907 - 4913.
13. Klaasens, K.H.; Schoot, C.J.; *Recl. Trav. Chim. Pays-Bas* **1956**, *75*, 186 - 189.
14. Marzi, E.; Schlosser, M.; unpublished results, **2004**.
15. Faigl, F.; Marzi, E.; Schlosser, M.; *Chem. Eur. J.* **2000**, *6*, 771 - 777.
16. Heiss, C.; Schlosser, M.; unpublished results, **2003**.
17. Heiss, C.; Schlosser, M.; *Eur. J. Org. Chem.* **2003**, 447 - 451.
18. Wang, X.; Rabbat, P.; O'Shea, P.; Tillyer, R.; Grabowski, E.J.J.; Reider, P.J.; *Tetrahedron Lett.* **2000**, *41*, 4335 - 4388.
19. Parham, W.E.; Piccirilli, R.M.; *J. Org. Chem.* **1977**, *42*, 257 - 260.
20. Craig, L.C.; *J. Am. Chem. Soc.* **1934**, *56*, 231 - 232.
21. Ife, R.; Catchpole, K.W.; Durant, G.J.; Ganellin, C.R.; Harvey, C.A.; Meeson, M.L.; Owen, D.A.A.; Parsons, M.E.; Slingsby, B.P.; Theobald, C.J.; *Eur. J. Med. Chem.* **1989**, *24*, 249 - 257; *Chem. Abstr.* **1989**, *112*, 138988r.
22. Cottet, F.; Schlosser, M.; unpublished results, **2002 - 2003**.
23. Schlosser, M.; *J. Organomet. Chem.* **1967**, *8*, 9 - 16.
24. Schlosser, M.; *Pure Appl. Chem.* **1988**, *60*, 1627 - 1634.
25. Schlosser, M.; Ginanneschi, A.; Leroux, F.; manuscript in preparation.
26. Mongin, F.; Marzi, E.; Schlosser, M.; *Eur. J. Org. Chem.* **2001**, 2771 - 2777.
27. Rausis, T.; Schlosser, M.; *Eur. J. Org. Chem.* **2002**, 3351 - 3358.
28. Eaborn, C.; Golborn, P.; Taylor, R.; *J. Organomet. Chem.* **1967**, *10*, 171 - 174.

29. Lambardino, J.G.; *J. Org. Chem.* **1971**, *36*, 1843 - 1845.
30. Iddon, B.; *Heterocycles* **1985**, *23*, 417 - 443; *Chem. Abstr.* **1985**, *102*, 166631f.
31. Langer Eriksen, B.; Vedsö, P.; Begtrup, M.; *J. Org. Chem.* **2001**, *66*, 8344 - 8348.
32. Mills, R.J.; Snieckus, V.; *J. Org. Chem.* **1989**, *54*, 4372 - 4386.
33. Marull, M.; Schlosser, M.; *Eur. J. Org. Chem.* **2004**, 1008 - 1013.
34. Mordini, A.; Schlosser, M.; *Chimia* **1986**, *40*, 309 - 310.
35. Macdonald, J.E.; Poindexter, G.S.; *Tetrahedron. Lett.* **1987**, *28*, 1851 - 1852.
36. Gorecka, J.; Leroux, F.; Schlosser, M.; *Eur. J. Org. Chem.* **2004**, 64 - 68.
37. Schlosser, M.; Rausis, T.; *Eur. J. Org. Chem.* **2004**, 1018 - 1024.
38. Schlosser, M.; *Angew. Chem.* **1998**, *110*, 1538 - 1556; *Angew. Chem. Int. Ed. Engl.* **1998**, *37*, 1496 - 1513.
39. Castagnetti, E.; Schlosser, M.; *Chem. Eur. J.* **2002**, *8*, 799 - 804.
40. Leroux, F.; Jeschke, P.; Schlosser, M.; review to appear in *Chem. Rev.*

Chapter 12

Perfluoroketene Dithioacetals: Versatile Building Blocks toward Trifluoromethyl Heterocycles

Charles Portella[1] and Jean-Philippe Bouillon[1,2]

[1]UMR 6519, Réactions Sélectives et Applications, CNRS-Université de Reims Champagne-Ardenne, B.P. 1039, 51687 Reims Cedex 2, France
[2]Current address: Sciences et Méthodes Séparatives, EA 2659, Université de Rouen, IRCOF, F–76821 Mont-Saint-Aignan Cedex, France

Perfluoroketene dithioacetals with three or four carbon atoms are easily prepared from the corresponding perfluoroaldehydes or esters. They are useful synthons, due to their ability to behave as simple masked carboxylic acid derivatives and to exhibit umpolung reactivity in combination with the vinyl fluorine substitution. With enolates as nucleophiles, multifunctional building blocks such as α-trifluoromethyl-γ-keto acid derivatives or α-trifluoromethyl-succinic acid derivatives are prepared, opening the access to a wide range of trifluoromethyl heterocycles. Two complementary series of heterocycles are synthesized according to the starting synthon, which may be regarded as an equivalent of a bis- or tris-electrophilic system.

Fifteen years ago, we discovered a reaction that was very interesting from a scientific and academic point of view, but was of poor preparative value. This reaction consisted of a selective photoreduction of an α-C-F bond of alkyl perfluorocarboxylic esters (*1*). The reductive species, excited HMPA, had to be used as solvent because of its low molecular absorption. This was a major drawback for the development of the reaction on a large scale owing to the cost and toxicity of HMPA. Undoubtedly this was also the major reason that this reaction remained ignored.

$$R_F\text{-}CF_2\text{-}CO_2R \xrightarrow[\text{HMPA}]{h\nu} R_F\text{-}CHF\text{-}CO_2R$$

Nethertheless, this transformation allowed us to explore the use of 2-hydro perfluorocarboxylic esters as intermediates for the elaboration of different types of organofluorine compounds. Indeed, a hydrogen atom on a perfluoroalkyl chain actually behaves as a functional group, and 2-hydro perfluoroesters proved to be valuable precursors towards polyfluorinated 3-oxoesters (*2*), enaminoesters (*3*) and heterocycles (*4,5*). Thus, it became relevant to consider a more practical method to prepare intermediates **1**. We thought that 2-hydro perfluoroesters could be considered as the result of the alcoholysis of a perfluoroketene. Perfluoroketenes being difficult to handle, we chose perfluoroketene dithioacetals as more convenient synthetic equivalents.

$$\underset{\mathbf{1}}{R_F\text{-}CHF\text{-}COY} \Longrightarrow \underset{\mathbf{2}}{(R_F)(F)C{=}C(SR)_2}$$

At the time we undertook this study, only the first representative of the series, $CF_2{=}C(SR)_2$, was known in the literature (*6*). As this paper is devoted to the synthesis of trifluoromethyl heterocycles, we will not further address these compounds. Peculiar fluorinated ketene dithioacetals of the type $(CF_3)_2C{=}C(SR)_2$ (*7*) and $R_FC(R){=}C(SR')_2$ (*8*) are also excluded, despite our significant contribution to the study of the latter family. This review focusses on the applications of synthons **2a,b**. After discussing their preparation, a second generation of more elaborated synthons derived from **1** will be described as intermediates in the synthesis of trifluoromethylated heterocycles.

Preparation of Perfluoroketene Dithioacetals

The main route towards synthons **2** consists of the thioacetalization of perfluoroaldehydes followed by HF elimination (Scheme 1). Both hydrated or hemiketal form of the aldehyde may be used. Initially described for higher homologues (R_F = $H(CF_2)_4$) using concentrated sulfuric acid and potassium hydroxide for the two steps, respectively (*9*), we found that titanium tetrachloride was the reagent of choice for performing the thioacetalization. Indeed, its reaction with the water or alcohol content releases hydrogen chloride which itself is a good thioacetalization catalyst. The second step is carried out under mild conditions using phase transfer catalysis (*10*). Well adapted to the first terms of the series owing to commercial availability of the perfluoroaldehydes,this methodology remains general and convenient since the aldehydes may be prepared by LAH reduction of perfluorocarboxylic esters (*11*).

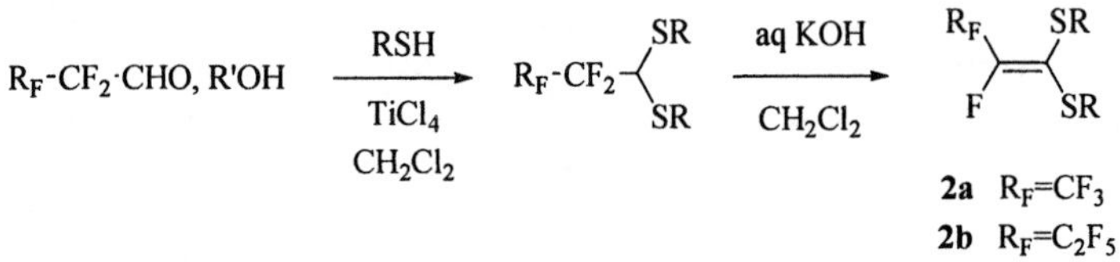

R_F-CF_2-CHO, R'OH → (RSH; $TiCl_4$, CH_2Cl_2) → R_F-CF_2-CH(SR)(SR) → (aq KOH; CH_2Cl_2) → R_F(F)C=C(SR)(SR)

2a R_F=CF_3
2b R_F=C_2F_5

Scheme 1

A thiophilic organometallic addition-elimination on perfluorodithio-carboxylic esters is the key reaction sequence in a different approach (Scheme 2) (*12*). Lithium and magnesium reagents work. In the reaction with a magnesium reagent, the distillation must be carried out after careful elimination of magnesium salts to avoid further reaction. The regioselectivity of this nucleophilic attack is obviously favored by the electron withdrawing effect of fluorine and the concomitant β-fluoride elimination. The prior preparation of the perfluorodithioesters is needed in this approach (*13*).

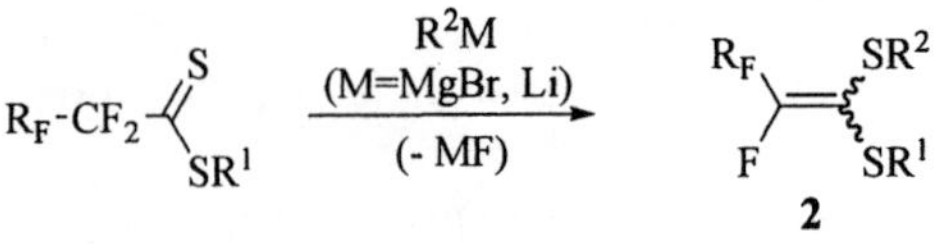

R_F-CF_2-C(=S)SR^1 → (R^2M (M=MgBr, Li); (- MF)) → R_F(F)C=C(SR^2)(SR^1) **2**

Scheme 2

Second Generation Synthons Derived from 2

α-Hydroperfluorocarboxylic Derivatives

Perfluoroketene dithioacetals **2** are simply a masked form of a carboxylic acid derivative. Indeed, acid hydrolysis leads to the corresponding 2-hydro thiol ester, which can be transesterified into an ester (Scheme 3) (*10*). Direct hydrolysis into an α-hydroperfluoroacid under stronger conditions has been exemplified for higher homologues (9). α-Hydroperfluoroacid derivatives have also been prepared by reactions of base with perfluoroalkenes (*14*).

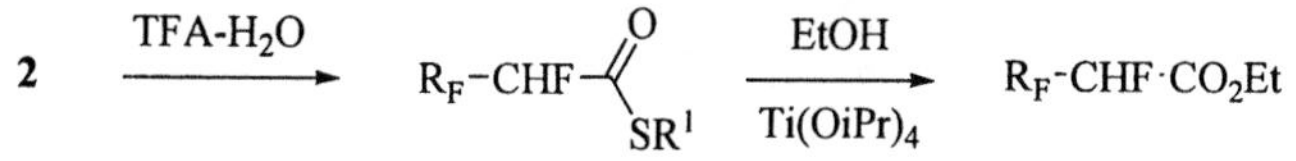

$$\mathbf{2} \xrightarrow{\text{TFA-}H_2O} R_F\text{-CHF-C(=O)SR}^1 \xrightarrow[\text{Ti(OiPr)}_4]{\text{EtOH}} R_F\text{-CHF}\cdot CO_2Et$$

Scheme 3

γ-Keto α-Perfluoroalkyl Acid Derivatives

The versatility of synthons **2** is mainly due to their ability to exhibit both a normal and an umpolung reactivity. The latter is of particular interest here due to the presence of a vinyl fluorine atom. Indeed, prior to any hydrolysis (normal reactivity), it is easy to convert **2** into a perfluoroalkyl derivative by a nucleophilic addition-elimination process. This fluoride substitution, which can be carried out with simple heteroatom or carbon centered nucleophiles (*15*), is very useful with functionalized ones such as enolates. In this way access to a second generation of bifunctionalized synthons is possible, opening the field of trifluoromethyl heterocycles.

Potassium ketone enolates were effective nucleophiles in a THF solution, giving a variety of γ-keto derivatives **3** (Scheme 4). Compounds **3** may be hydrolyzed into the corresponding γ-keto α-perfluoroalkyl thiol esters **4** or further derivatized before hydrolysis (*16*)(*17*)(*18*)(*19*).

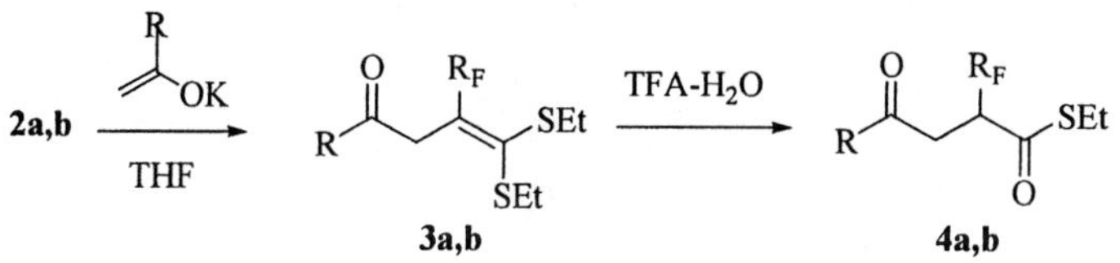

$$\mathbf{2a,b} \xrightarrow[\text{THF}]{R\text{-C(OK)=CH}_2} R\text{-CO-CH}_2\text{-C(}R_F\text{)=C(SEt)}_2\ (\mathbf{3a,b}) \xrightarrow{\text{TFA-}H_2O} R\text{-CO-CH}_2\text{-CH(}R_F\text{)-CO-SEt}\ (\mathbf{4a,b})$$

Scheme 4

α-Perfluoroalkyl Succinic Acid Derivatives

Surprisingly, the reaction of **2** with an ester enolate failed. However, two different routes were found giving access to α-perfluoroalkyl 1,4-dicarboxylic acid derivatives or synthetic equivalents. The first one consists of a chain reaction using ethyl 2-trimethylsilyl acetate as the acetate enolate source. The reaction is initiated by tetramethyl ammonium fluoride (TMAF). A minor amount (5%) is enough for a high yielding reaction where the driving chain transfer step consists of the formation of the strong Si-F bond (Scheme 5) (*20*). The ketene dithioacetal moiety of the ester **5** is hydrolyzed under standard acidic conditions to give the diester **6**. The ester function of **5** may also be selectively hydrolyzed and the resulting acid **7** transformed into the mono esterified succinic derivative **8** (Scheme 5).

Curiously, whereas lithium ethyl acetate enolate failed to react with **2**, a direct access to acid 7 and 3-substituted 2-trifluoromethyl succinic acid derivatives equivalents **9** was found using lithium enediolates as nucleophiles (Scheme 6) (*21*). Although such reactions generally allowed simple work-up procedures using pH controlled extractions (*22*), specific conditions had to be found to overcome some solubility problems inherent to these fluorinated species (*21*).

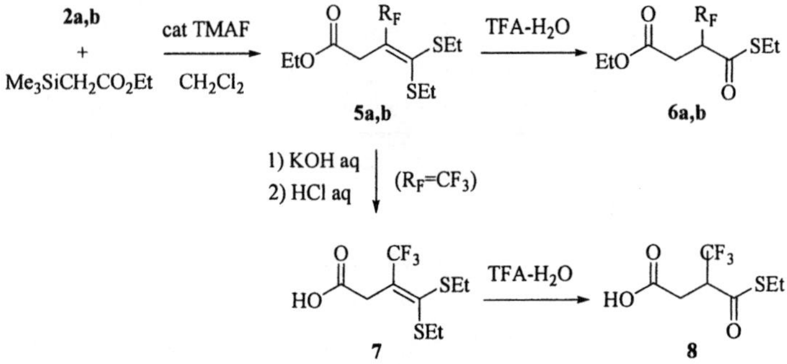

2a,b + $Me_3SiCH_2CO_2Et$ —(cat TMAF, CH_2Cl_2)→ **5a,b** —(TFA-H_2O)→ **6a,b**

5a,b —(1) KOH aq, 2) HCl aq; (R_F=CF_3))→ **7** —(TFA-H_2O)→ **8**

Scheme 5

Interestingly, compounds **5-9** are all 2-trifluoromethyl or 2-pentafluoroethyl succinic acid derivatives bearing two differentiated carboxylic functions, a useful feature for further applications.

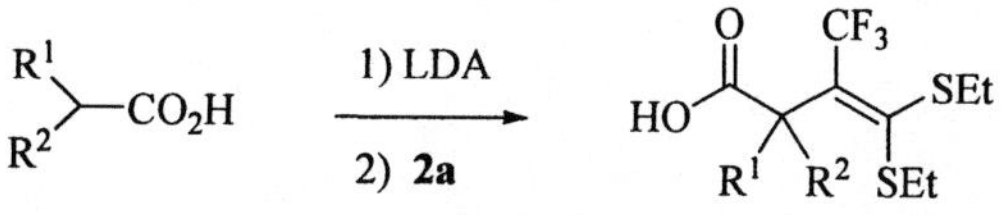

Scheme 6

Ethyl β-Bromoperfluorocrotonate

While investigating the thiophilic alkylation of perfluorodithiocarboxylic esters as a route to perfluoroketene dithioacetals (*12*), it was observed that a subsequent transformation of **2b** occured while heating in the presence of magnesium salts. This reaction was studied and optimized to eventually provide the new perhalo unsaturated dithioester **10** (Scheme 7) (*23*)(*24*), which proved to be an interesting building block for the synthesis of trifluoromethyl thiaheterocycles (*vide infra*). Compound **10** is the result of a bromide attack on the ethyl group giving a magnesium perfluorodithioester enolate which easily loses the β-fluorine. A subsequent addition-elimination sequence results in the substitution of the vinyl fluorine for bromine (*23*).

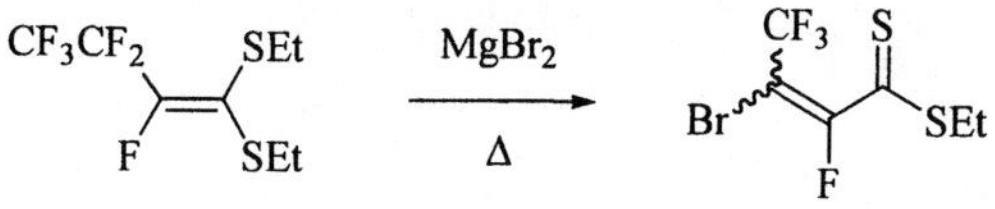

Scheme 7

Applications to the Synthesis of Trifluoromethylated Heterocycles

The various building blocks described above are divided into two categories bearing either a CF_3 or a C_2F_5 group. Each of these categories leads to specific applications, which are presented accordingly.

The Bis(ethylsulfanyl) Tetrafluoropropene Line

2-Trifluoromethyl γ-Lactones

The two synthons **3a** and **4a** are *a priori* suitable bifunctionalized candidates for heterocycle synthesis. As depicted in Scheme 8, carbonyl methylation or reduction of **3a** (R=Me) followed by acid hydrolysis leads to the lactones **12** in high yields (Scheme 8). The intermediate **3a** (R=Ph) derived from the reaction of **2a** with the acetophenone enolate did not lead to the corresponding lactone under these conditions, because of the different reactivity of the benzylic alcohol **11** (R=Ph) (*16*). Reversed acid hydrolysis-reduction sequence is preferable in this case, but leads to poor yields of the corresponding lactone and of the over-reduced lactol (*16*). Another approach to 2-trifluoromethyl γ-lactones was proposed later (*25*).

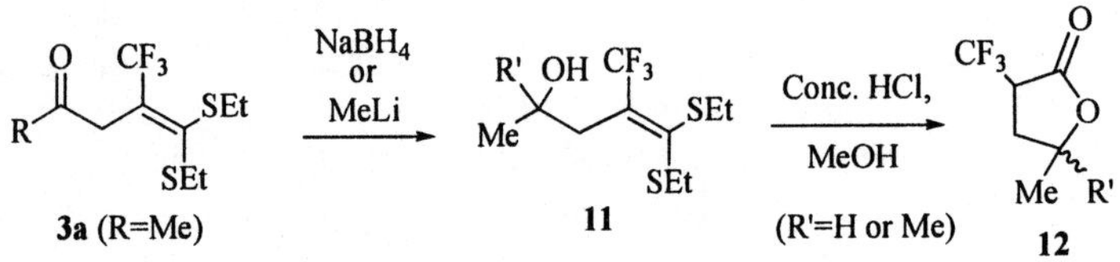

3a (R=Me) — $NaBH_4$ or MeLi → **11** — Conc. HCl, MeOH (R'=H or Me) → **12**

Scheme 8

α-Trifluoromethyl γ-Lactams

2-Trifluoromethyl γ-lactams are easily prepared by reductive amination of 2-trifluoromethyl γ-keto thiol ester **4a** ($R=CH_3$) (Scheme 9) (*17*). Excellent yields of *N*-aryl or *N*-alkyl lactams **13** (R'=Ph, Bn) are achieved using the BH_3-Pyridine complex as reducing agent. The N-unsubstituted compound **14** can also be prepared with the ammonium acetate-$NaBH_3CN$ system. N-substituted lactams **13** are obtained as a mixture of diastereomers in almost quantitative yields, whereas the γ-lactam **14** is formed as a single diastereomer (undetermined configuration) and is accompanied by a minor amount of the hemi thioaminal **15** (Scheme 9). This seems to be due to a cyclocondensation of the intermediate imine prior to diastereoselective reduction or addition of ethanethiol. The isolation of **15** as a unique product and as a single diastereomer (undetermined configuration) when **4a** (R=Me) reacted with ammonium acetate without any reductor is a good argument for such a pathway (*17*). γ-Lactams are interesting bioactive coupounds (*26*). There are very few reports of the 2-trifluoromethyl analogues (*27*). The strategy starting from **2a** provides a general and effective access to these compounds.

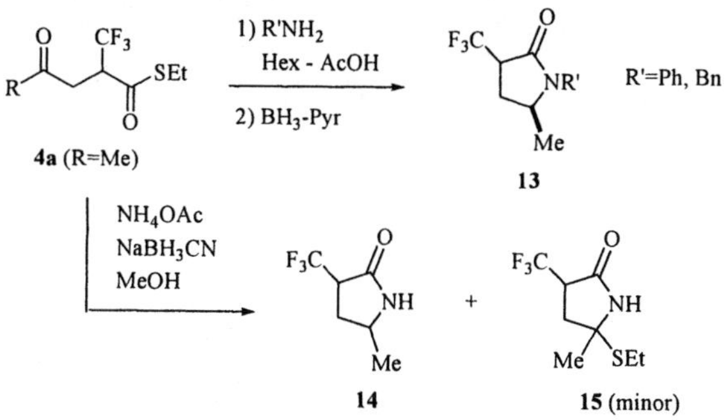

4a (R=Me)

1) $R'NH_2$ Hex - AcOH 2) BH_3-Pyr

13 R'=Ph, Bn

NH_4OAc $NaBH_3CN$ MeOH

14 + **15** (minor)

Scheme 9

4-Trifluoromethyl Pyridazine Derivatives

Heterocyclic compounds of the pyridazine family constitute an important class of biologically active derivatives (*28*). One of the main preparative methods starts from γ-keto acid derivatives (*29*). It was natural that 2-trifluoromethyl analogues were intermediates of choice to access 4-trifluoromethyl pyridazine derivatives. The unique representative found in the literature (a fluorinated analogue of minaprine) (*30*) came from a completely different building block: trifluoroethylidenacetophenone. The perfluoroketene dithiocetal **2a**, via the γ-keto thiolesters **4a**, proved to be an effective route towards a variety of 4-trifluoromethyl pyridazine derivatives, including the 4,5-dihydropyridazinone series **16**, the pyridazinones **17**, and the 3-chloropyridazines **18** as precursors of 3-aminopyridazines **19** (Scheme 10) (*19*).

The versatility of perfluoroketene dithioacetals is particularly well illustrated with these pyridazine derivatives. Indeed, a wide structural diversity may be envisaged, depending on the starting ketone enolate and the final nucleophile (amine in these examples).

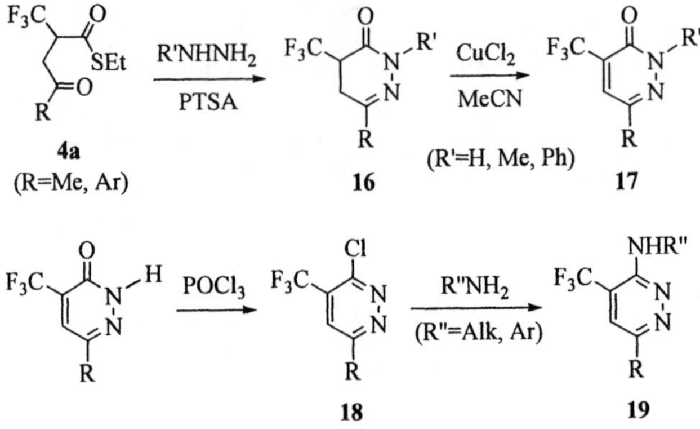

Scheme 10

2-Trifluoromethylsuccinimides

Whereas the succinic diester **6a** was supposed to be converted into pyridazinediones by reaction with hydrazines, the ethyloxycarbonyl group proved to be highly resistant to condensation, and only displacement of the ethylsulfanyl group was observed. In contrast to **6a**, the acid **8a** behaves as a bis(electrophilic) species, but fails to react with hydrazines to give pyridazine diones. Instead, reaction of **8a** with amines (hydrazines included) gives 2-trifluoromethylsuccinimides **21** via the acid **20** (Scheme 11). N-Alkyl, N-aryl

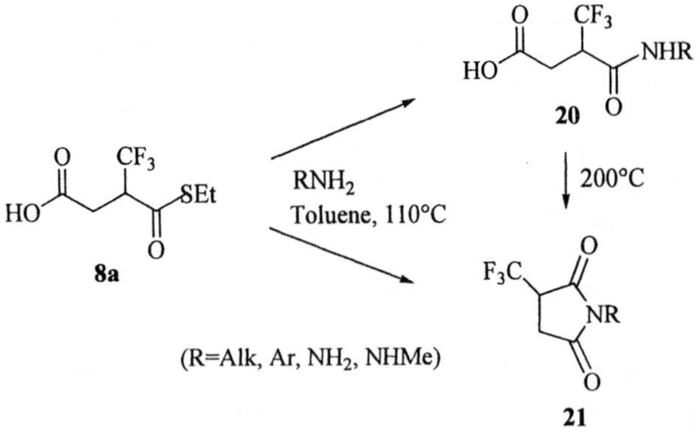

Scheme 11

and *N*-aminosuccinimides may be prepared (*20*). It is noteworthy that phenyl hydrazine gave the same product as aniline, corresponding to the formal lost of ammonia. A similar reaction occured with *N,N'*-dimethylhydrazine, giving *N*-methylamino succinimide (*20*).

The Bis(ethylsulfanyl) Hexafluorobutene Line

2-Trifluoromethyl Furans and Pyrroles

The β-elimination of a fluoride anion is much easier from a pentafluoroethyl group than from a trifluoromethyl group. Hence the behavior of the pentafluoroethyl building blocks **4b** with basic reagents such as amines is different, the first step being generally the elimination of HF. As a consequence, the reactions of **4b** with amines followed by subsequent cyclocondensation give trifluoromethyl heterocycles different than from **4a**, according to the reaction pathways depicted in Scheme 12 (*18*).

The result of HF elimination is the new highly functionalized intermediate **22**, which in turn can react with amines according to a basic or a nucleophilic pathway (Scheme 12). The amines strong enough to abstract the highly

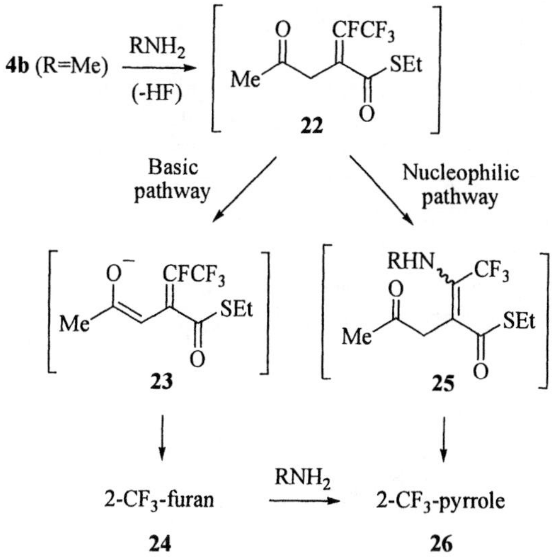

4b (R=Me) $\xrightarrow[\text{(-HF)}]{RNH_2}$ [Me–CO–CH$_2$–C(=CFCF$_3$)–CO–SEt] **22**

Basic pathway → [Me–C(O$^-$)=CH–C(=CFCF$_3$)–CO–SEt] **23** → 2-CF_3-furan **24**

Nucleophilic pathway → [Me–CO–CH$_2$–C(=C(NHR)CF$_3$)–CO–SEt] **25** → 2-CF_3-pyrrole **26**

2-CF_3-furan **24** $\xrightarrow{RNH_2}$ 2-CF_3-pyrrole **26**

Scheme 12

activated α-proton of **22** convert it into the enolate **23**. An intramolecular Michael type addition followed by a fluoride elimination provide 2-trifluoromethyl furan derivatives **24**. On the other hand, amines can act as nucleophiles to directly displace the vinyl fluorine. This leads to the β-enaminoester type intermediate **25**, the cyclocondensation of which gives 2-trifluoromethyl pyrrole derivatives **26**. The chemoselectivity of the reaction of **4b** with amines correlates with both their basicity and steric hindrance. Taking into account that furane can react with primary amines to be converted in pyrrole, under suitable conditions, good yields of either pyrrole or furan derivatives are obtained, as examplified in Scheme 13 (*18*).

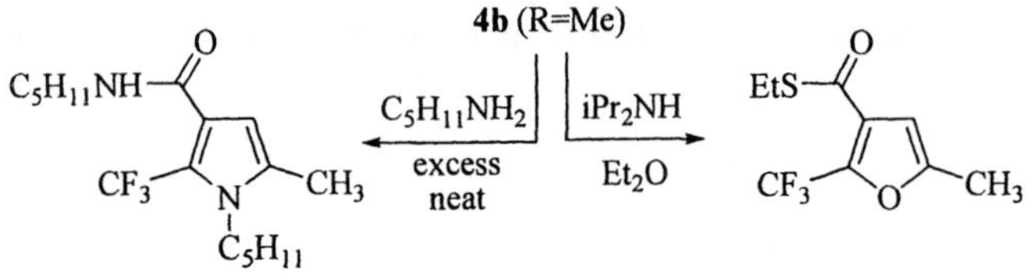

Scheme 13

Trifluoromethyl Pyrazoles and Pyrimidines

The reactivity of the succinic type synthon **6b** is currently under investigation. However, we can disclose here that, as observed for **6a**, the ethoxycarbonyl moiety exhibits the same resistance to condensation reactions. Under basic conditions, **6b** easily loses HF to convert into the corresponding α,β-unsaturated intermediate, which has a high tendency to tautomerize into a higher conjugated form. According to preliminary experiments with hydrazines and amidines, there are three electrophilic centers competing for giving two isomeric heterocycles (Scheme 14) (*31*).

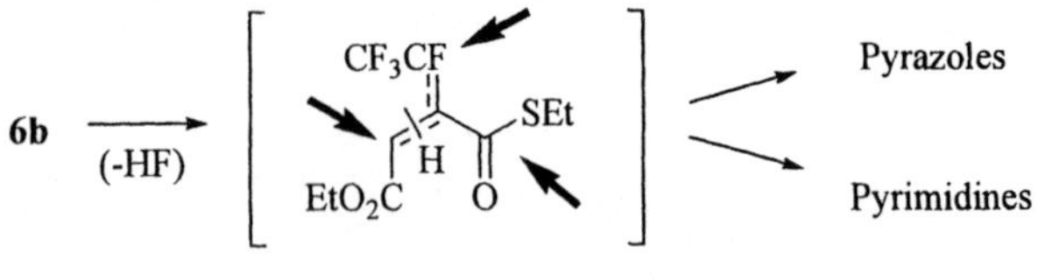

Scheme 14

5-Membered CF_3-Heterocycles from 2-Hydroperfluorobutanoates

The 2-hydroperfluorobutanoates derived from **2b** are synthetic equivalents of perfluorocrotonic acid derivatives and of 2-hydro-3-oxo-perfluoroesters, which are interesting precursors for pyrazole **27** (*4*) and oxazo- or imidazolidines **28** (*5*) type heterocycles (Scheme 15). The electron-withdrawing effect of the trifluoromethyl group favors an intramolecular Michael attack. It also stabilizes the hydrated form of pyrazolidinones (*4*).

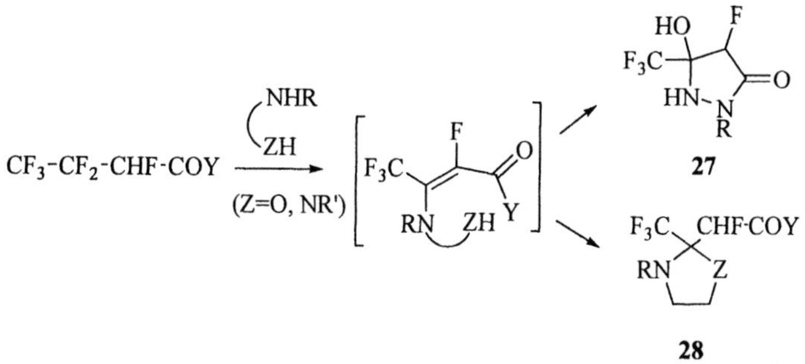

Scheme 15

5-Trifluoromethyl 1,2-Dithiole-3-Thione

During the optimization of the preparation of perhalodithiocrotonic ester **10** it was observed that some decomposition occurred with over heating, giving minor amounts of the dithiolethione **29**. This compound is indeed the result of a thermal reaction between molecular sulfur and the dithiocrotonic ester. The conditions were optimized to achieve a high yielding preparation of the 4-fluoro-5-trifluoromethyl-1,2-dithiole-3-thione **29** in a one-pot process from the corresponding perfluoroketene dithioacetal **2b** (Scheme 16) (*24*). Compounds **29** behave as effective dienophile via the C=S bond, as reported earlier for perfluorodithioesters (*13a*).

1,3-Dipolar Cycloaddition of Dithiocarboxylic Derivatives with Dimethylacetylendicarboxylate

As in non-fluorinated series, 1,2-dithiole-3-thione **29** reacts as a 1,3-dipole with dimethylacetylene dicarboxylate (DMAD) to give the adduct **30** which in turn reacts as an heterodiene with DMAD to give the 1:2 adduct **31**

(Scheme 16). In contrast to the non-fluorinated series, where this transformation is known to be a thermal process, the second [4+2] cycloaddition is a chain process initiated by a photoinduced single electron transfer to oxygen (*32*). The 1:2 adduct reacts with water by simple filtration over silica gel, leading to the trithianaphtalene type compound **32**. The detailed study of the reaction conditions gave access to an effective direct synthesis of either **31** or **32** from the dithiolethione **29** (*32*).

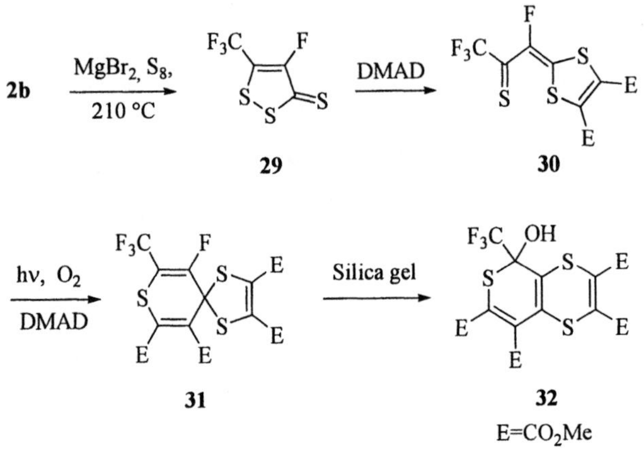

2b $MgBr_2$, S_8, 210 °C
F_3C F S S S
29
DMAD
F F_3C S S S E E
30

hν, O_2 DMAD
F_3C F S S S E E E E
31
Silica gel
F_3C OH S S S E E E E
32
E=CO_2Me

Scheme 16

CF_3 S SEt Br F
10
E—≡—E
(E = CO_2Me)
F S E F_3C S E S E F_3C F S E
33

110 °C PhMe
E F S F_3C E S S F_3C E F S E
34
Δ 190 °C
F F_3C S E F_3C S E F
35

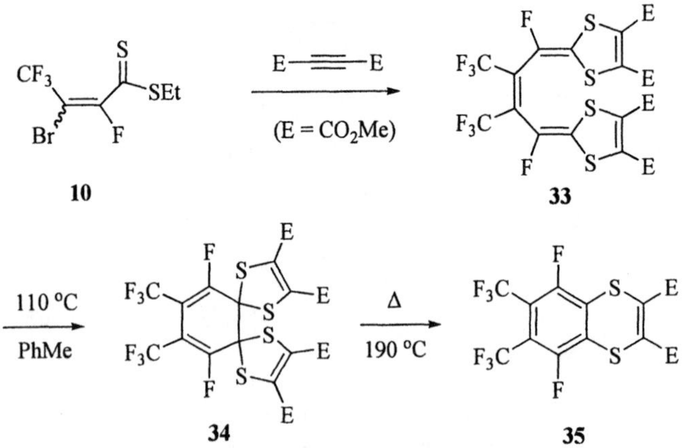

Scheme 17

The perhalodithiocrotonate **10** reacts similarly with DMAD in a 1,3-dipolar cycloaddition, but owing to the perhalo character of the substrate, an unexpected pathway is followed. The interesting vinylogue of tetrathiafulvalene (TTF) **33** is the result of a multistep process involving the condensation of a primary adduct with the substrate **10** and a subsequent cycloaddition with DMAD (*33*). This TTF derivative is quantitatively converted into the bis(spiro) compound **34** which under stronger heating may be transformed into the benzodithiine **35**. The latter may be directly prepared from the crotonic derivative **10** (Scheme 17).

Conclusion

Perfluoroketene dithioacetals **2a,b** are versatile fluorinated synthons for the synthesis of a wide range of trifluoromethyl heterocycles including oxygen, nitrogen and sulfur, via more elaborated intermediates **4-10.** Each of the two basic synthons open their own line of applications. Compounds **2a** and **2b** behave as equivalents of a polyelectrophile bearing two or three contiguous positive charges adjacent to a trifluoromethyl group. The ethylsulfanyl groups allow the "umpolung" reactivity, but the sulfur atoms play a more essential role since they can also be involved in the applications to thia-heterocycles.

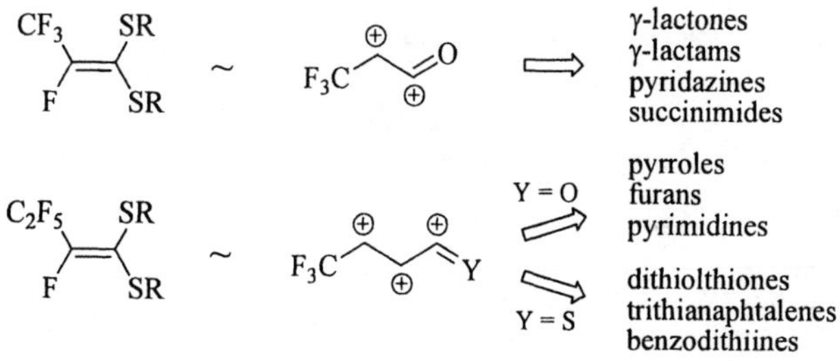

CF_3 SR / F SR ~ F_3C (⊕, ⊕, O) ⟹ γ-lactones, γ-lactams, pyridazines, succinimides

C_2F_5 SR / F SR ~ F_3C (⊕, ⊕, ⊕, Y) Y = O ⟹ pyrroles, furans, pyrimidines; Y = S ⟹ dithiolthiones, trithianaphtalenes, benzodithiines

The chemistry of perfluoroketene dithioacetals and derived synthons undoubtedly remains an open field for other extensions confirming their high versatility.

Acknowledgments. We warmly thank Dr Murielle Muzard, Dr Jean-François Huot, Dr Cédric Brulé and Béatrice Hénin, for their contribution to this work. We also thank Professor Yuriy Shermolovich and Dr Vadim Timoshenko for a fruitful collaboration in the thia-heterocycle syntheses. Dr Karen Plé has kindly offered his expertise for English correction. This work has been supported by CNRS, the Ministry of Education and Research, and by CEREP company.

References

1. Portella, C.; Iznaden, M. *Tetrahedron* **1989**, *45*, 6467-6478.
2. Iznaden, M.; Portella, C. *J. Fluorine Chem.* **1989**, *43,* 105-118.
3. Portella, C.; Iznaden, M. *J. Fluorine Chem.* **1991**, *51*, 1-20.
4. Portella, C.; Iznaden, M. *Synthesis* **1991**, 1013-1014.
5. Dondy, B.; Doussot, P.; Iznaden, M.; Muzard, M.; Portella, C. *Tetrahedron Lett.* **1994**, *35*, 4357-4360.
6. (a) Tanaka, K.; Nakai, T.; Ishikawa, N. *Chem. Lett.* **1979**, 175-178. (b) Gimbert, Y.; Moradpour, A. *Tetrahedron Lett.* **1991**, *32*, 4897-4900. (c) Gimbert, Y.; Moradpour, A.; Dive, G.; Dehareng, D.; Lahlil, K. *J. Org. Chem.* **1993**, *58*, 4685-4690. (d) Purrington, S. T.; Samaha, N. F. *J. Fluorine Chem.* **1989**, *43*, 229-234. (e) Purrington, S. T.; Sheu, K.-W. *Tetrahedron Lett.* **1992**, *33*, 3289-3292. (f) Purrington, S. T.; Sheu, K.-W. *J. Fluorine Chem.* **1993**, *65*, 165-167. (g) Piettre, S.; De Cock, C.; Merenyi, R.; Viehe, H. G. *Tetrahedron*, **1987**, *43*, 4309-4319.
7. Sterlin, S. R.; Izmailov, V. M.; Isaev, V. L.; Shal, A. A.; Sterlin, R. N.; Dyatkin, B. L.; Knunyants, I. L. *Zh. Vses. Khim. Obshch.* **1973**, *18*, 710-712; *Chem. Abstr.* **1974**, *80*, 81973.
8. (a) Solberg, J.; Benneche, T.; Undheim, K. *Acta Chim. Scand.* **1989**, 69-73. (b) Muzard, M.; Portella, C. *Synthesis* **1992**, 965-968. (b) Bergeron, S.; Brigaud, T.; Foulard, G.; Plantier-Royon, R.; Portella, C. *Tetrahedron Lett.* **1994**, *35*, 1985-1989. (c) Foulard, G.; Brigaud, T.; Portella, C. *J. Org. Chem.* **1997**, *62*, 9107-9113. (d) Foulard, G.; Brigaud, T.; Portella, C. *J. Fluorine Chem.* **1998**, *91*, 179-183.
9. Markovskii, L. N.; Slyusarenko, E. I.; Timoshenko, V. M.; Kaminskaya, E. I.; Kirilenko, A. G.; Shermolovich, Y. G. *Zh. Org. Khim.* **1992**, *28*, 14-22; *Chem. Abstr.* **1992**, *117*, 150502.
10. Muzard, M.; Portella, C. *J. Org. Chem.* **1993**, *58*, 29-31.
11. Pierce, O. R.; Kane, T. G. *J. Am. chem. Soc.* **1954**, *76*, 300-301.
12. Portella, C.; Shermolovich, Y. G. *Tetrahedron Lett.* **1997**, *38,* 4063-4064.
13. (a) Portella, C.; Shermolovich, Y. G.; Tschenn, O. *Bull. Soc. Chim. Fr.* **1997**, *134*, 697-702. (b) Babadzahanova, L. A.; Kirij, N. V.; Yagupolskii, Y. L. *J. Fluorine Chem.* **2004**, *125*, 1095-1098.
14. Selected significant references: (a) Nguyen, T.; Wakselman, C. *J. Fluorine Chem.* **1995**, *74*, 273-277. (b) Hu, C.; Tu, M. *Chinese Chem. Lett.* **1992**, *3*, 87-90; *Chem. Abstr.* **1992**, *117*, 89814. (c) Ishihara, T.; Kuroboshi, M.; Yamaguchi, K. *Chem. Lett.* **1990**, 211-214. (d) Nguyen, T.; Rubinstein, M.; Wakselman, C. *J. Fluorine Chem.* **1978**, *11*, 573-589. (e) Wakselman, C.; Nguyen, T. *J. Org. Chem.* **1977**, *42*, 565-566. (f) Rendall, J. L.; Pearlson, W. H. *US* **1958**, 2862024 19581125; *Chem. Abstr.* **1959**, *53*, 50780. (g)

Rendal, J. L.; Pearlson, W. H. *US* **1957**, 2795601 19570611; *Chem. Abstr.* **1957**, *51*, 90875. (h) Rendall, J. L.; Pearlson, W. H. *GB* **1955**, 737164 19550921; *Chem. Abstr.* **1956**, *50*, 74163.

15. Muzard, M. *PhD Dissertation*, Reims, 1992.
16. Huot, J.-F.; Muzard, M.; Portella, C. *Synlett* **1995**, 247-248.
17. Hénin, B.; Huot, J.-F.; Portella, C. *J. Fluorine Chem* **2001**, *107*, 281-283.
18. Bouillon, J.-P.; Hénin, B.; Huot, J.-F., Portella, C. *Eur. J. Org. Chem.* **2002**, 1556-1561.
19. Brulé, C.; Bouillon, J.-P.; Nicolaï, E.; Portella, C. *Synthesis* **2003**, 436-442.
20. Brulé, C.; Bouillon, J.-P.; Portella, C. *Tetrahedron*, **2004**, in press.
21. Sotoca, E.; Bouillon, J.-P.; Gil, S.; Parra, M.; Portella, C. *Tetrahedron Lett.* **2004**, in press.
22. (a) Brun, E. M.; Gil, S.; Mestres, R.; Parra, M. *Synthesis* **2000**, 1160-1165. (b) Brun, E. M.; Gil, S.; Mestres, R.; Parra, M. *Tetrahedron Lett.* **1998**, *54*, 15305-15320.
23. Bouillon, J. -P.; Shermolovich, Y. G.; Portella, C. *Tetrahedron Lett.* **2001**, *42*, 2133-2135.
24. Timoshenko, V. M.; Bouillon, J. -P.; Shermolovich, Y. G.; Portella, C. *Tetrahedron Lett.* **2002**, *43*, 5809-5812.
25. Tellier, F.; Audouin, M.; Sauvêtre, R. *J. Fluorine Chem.* **2002**, *113*, 167-175.
26. Baldwin, J. E.; Lynch, G.; Pitlik, J. *J. Antibiot.* **1991**, *44*, 1-50.
27. (a) Suzuki, M.; Okada, T.; Taguchi, T.; Hanzana, Y.; Iitoka, Y. *J. Fluorine Chem.* **1992**, 57, 239-243. (b) Paulini, K.; Reissig, H. –U. *J. Prakt. Chem./Chem.-Ztg* **1995**, *337*, 55-61.
28. (a) Steiner, G.; Gries, J.; Lenke, D. *J. Med. Chem.* **1981**, *24*, 59-72. (b) Wermuth, C. G. *Farmaco* **1993**, 48, 253-259.
29. Albright, J. D.; Moran, D. B.; Wright, W. B.; Collins, J. B.; Beer, B.; Lippa, A. S.; Greenblatt, E. N. *J. Med. Chem.* **1981**, *24*, 592-601.
30. Contreras, J. M.; Rival, Y. M.; Chayer, S.; Bourguignon, J. J.; Wermuth, C. G. *J. Med. Chem.* **1999**, *42*, 730-741.
31. Brulé, C.; Bouillon, J.-P.; Portella, C. *unpublished results.*
32. Timoshenko, V. M.; Bouillon, J.-P.; Chernega, A. N.; Shermolovich, Y. G.; Portella, C. *Eur. J. Org. Chem.* **2003**, 2471-2474.
33. Timoshenko, V. M.; Bouillon, J.-P.; Chernega, A. N.; Shermolovich, Y. G.; Portella, C. *Chem. Eur. J.* **2003**, *9*, 4324-4329.

Chapter 13

Highly Functionalized Heterocycles and Macrocycles from Pentafluoropyridine

Graham Sandford

Department of Chemistry, University of Durham, South Road, Durham, DH1 3LE, United Kingdom

Pentafluoropyridine is a synthetically versatile, pluripotent 'building block' which reacts sequentially with a variety of nucleophiles to provide a range of poly substituted pyridine and macrocyclic derivatives, with potential applications to the life science and supramolecular fields.

Highly fluorinated heteroaromatic systems are very susceptible towards nucleophilic attack and chemistry involving substitution of fluorine by a variety of nucleophiles continues to emerge (*1,2*). In particular, the chemistry of pentafluoropyridine **1** has been widely studied (*3*) and many reactions involving the regioselective replacement of the fluorine atom located at the 4-position by a range of nucleophiles has been established (*2,3*). Reactions proceed by the familiar two-step nucleophilic aromatic substitution pathway involving Meisenheimer complexes as intermediates (*4*) (Scheme 1). Mono-substituted pyridines **2** are, of course, highly activated systems towards nucleophilic attack but the utility of such derivatives as substrates for the synthesis of more synthetically elaborate multi-substituted pyridine systems has not been developed to any great extent, despite the almost limitless range of new polyfunctional systems that could be accessed if several or all fluorine atoms present in **2** are replaced regioselectively. Reports describing the formation of a pentasubstituted system **3** from the reaction between pentafluoropyridine and sodium thiophenolate (*5*) gives an indication (Scheme 1) of the reactivity of the parent heterocycle and all intermediate fluoropyridines towards nucleophilic attack.

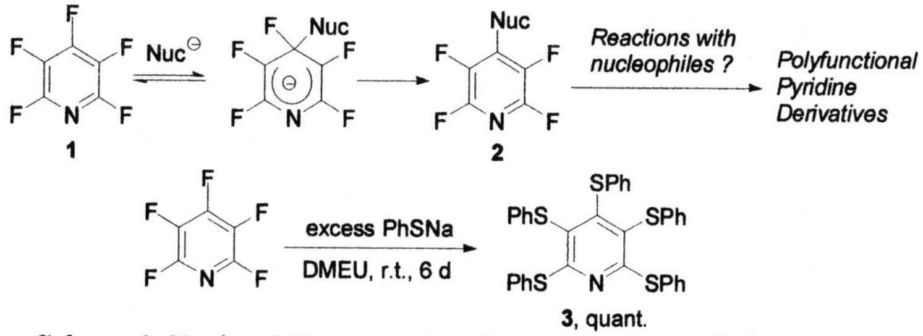

Scheme 1. Nucleophilic aromatic substitution of pentafluoropyridine

Here, we discuss the use of pentafluoropyridine as a building-block for the synthesis of a variety of polyfunctional pyridines and macrocycles that may be obtained for applications in the life science industries and supramolecular chemistry by a sequence of nucleophilic aromatic substitution processes.

A variety of highly halogenated heterocyclic derivatives are commercially available and, indeed, some are prepared on the industrial scale for use in the fibre reactive dye industry and as intermediates, e.g. 3,5-dichloro-2,4,6-trifluoropyridine is a herbicide precursor manufactured by the Dow company (*6*). No special handling procedures are required for syntheses involving perfluoroheterocyclic derivatives and, consequently, the chemistry of perfluorinated heterocycles could readily be used by the general organic chemistry community and scaled up by industry.

Highly Substituted Pyridine Derivatives

In the ongoing search for novel biologically active "lead" compounds, the life-science industries have extensive discovery programmes focused upon the synthesis of a wide range of structurally diverse, multi-functional systems that, ideally, can be accessed by parallel synthesis. It is now widely accepted that molecules that are most likely to possess drug-like physiochemical properties that may be administered orally, fall within the empirical "rules of 5" described by Lipinski (*7*). Consequently, there is a continuing great need for the development of methodology that allows the ready synthesis of libraries of small molecules that may be screened against new and existing biological targets for the purpose of lead generation.

The synthesis of libraries of Lipinski-type molecules is based upon the rapidly developing '*privileged structures*' strategy (*8*). Privileged structures are typically rigid polycyclic systems possessing a range of functionalities in a well defined structural system that can, potentially, interact with a variety of unrelated receptor sites and libraries of molecules that bear such privileged structures as sub-units have been targeted for drug discovery. Examples of privileged structures (*9*) that have been used for library generation (*10*) include benzamidines, indoles, *spiro*-piperidines, benzyl piperidines, benzopyrans (*8*) and various carbohydrates (*11*). An ideal privileged structure for lead generation must be easily synthetically manipulated to provide great structural diversity (*8*).

Heteroaromatic systems have, of course, a vast chemistry (*12*) and are excellent candidates for privileged structures given the criteria stated above. A very significant number of pharmaceuticals and other life-science products are heteroaromatic derivatives and the existence of many natural products and drugs that contain heteroaromatic cores bearing several pendant substituents exemplify these facts. Furthermore, the low molecular weight of heteroaromatic units provide suitable platforms for the synthesis of molecules that fall within the Lipinski rules. Heteroaromatic derivatives may form the basis of polyfunctional core scaffolds but, in order to be valuable synthetic scaffolds, such systems must be easily synthesized in high yield and selectivity. However, the problems associated with the use of established synthetic procedures, such as low reactivity of heteroaromatics towards electrophiles and nucleophiles and low selectivity, are well known and, of course, restrict the variety of heteroaromatic derivatives that can be rapidly synthesized using this strategy.

Methodology for the synthesis of ranges of structurally diverse heteroaromatic derivatives is, therefore, under continuing development and application of, for example, sequential electrophilic substitution and palladium catalyzed coupling reactions to the synthesis of many heterocyclic analogues (rapid analogue synthesis, RAS) has been reviewed recently (*13,14*).

In order to synthesize large numbers of structurally diverse systems, the principles of the newly emerging field of Diversity Oriented Synthesis (*15-17*) may be considered. DOS is aimed at the synthesis of collections of many molecules having structural complexity and diversity rather than the synthesis of specific target molecules. Successful DOS takes a starting material 'building-block' that may undergo several complexity generating reactions in which the product of the first reaction acts as a substrate of the second (the so-called

'*Libraries from libraries*' approach). Further diversity may be introduced into a system by varying stereochemistry and the presence of 'branch points' whereby a common substrate may be transformed into several products bearing different polyfunctional molecular skeletons. For both RAS and DOS strategies, however, the requirement for short, high yielding, regioselective and flexible routes to multiply functionalised heteroaromatic derivatives must be emphasized. A sequence of substitution processes involving the functionalisation of a heteroaromatic "core scaffold" is a strategy frequently employed and this idea is illustrated below (Scheme 2) in which pyridine is the parent of heterocyclic systems bearing up to five different substituents $R_1 - R_5$. Our complimentary approach utilises pentafluoropyridine **1** as the starting material (Scheme 2).

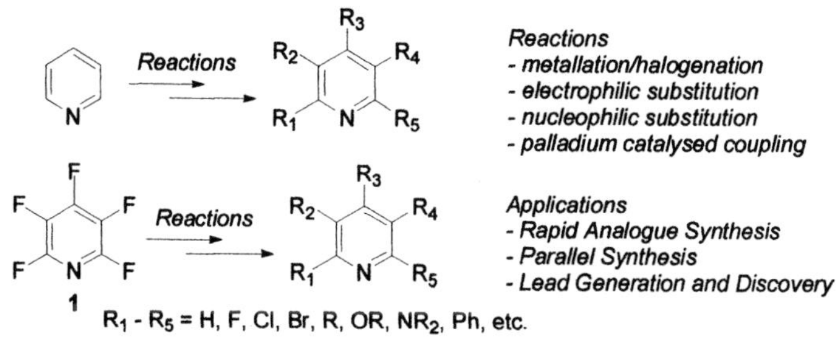

Reactions
- *metallation/halogenation*
- *electrophilic substitution*
- *nucleophilic substitution*
- *palladium catalysed coupling*

Applications
- *Rapid Analogue Synthesis*
- *Parallel Synthesis*
- *Lead Generation and Discovery*

R_1 - R_5 = H, F, Cl, Br, R, OR, NR_2, Ph, etc.

Scheme 2. Approaches to the synthesis of polysubstituted pyridine derivatives

Consequently, we envisaged that a range of polyfunctional pyridine derivatives could be derived from pentafluoropyridine by a series of nucleophilic aromatic substitution processes. Furthermore, it is well established (1,2) that, in general, the order of activation towards nucleophilic attack follows the sequence 4-fluorine > 2-fluorine > 3-fluorine and so, for a succession of five nucleophilic substitution steps, where Nuc_1 is the first nucleophile, Nuc_2 is the second, etc., the order of substitution is potentially selective as outlined below (Scheme 3). However, a few exceptions to these general rules have been reported (*18*) and this general scheme for regioselectivity may be effected by the growing number of non-fluorine substituents present on the pyridine ring.

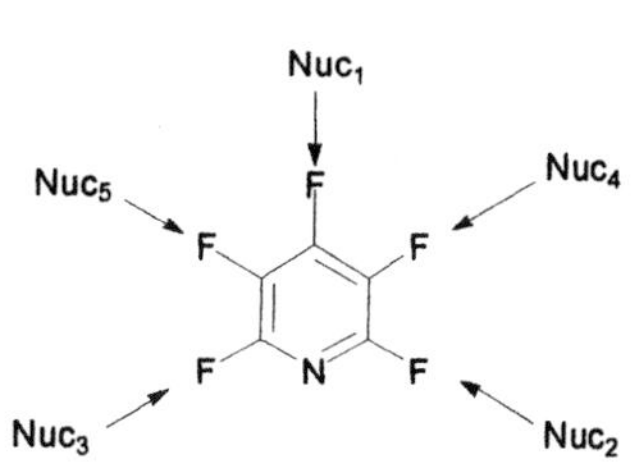

Scheme 3. Potential orientation of polysubstitution in pentafluoropyridine

At Durham, the viability of this idea was explored by first preparing perfluoro-4-isopropylpyridine **4** by nucleophilic substitution of the 4-fluorine atom by perfluoroisopropyl anion (i.e. $Nuc_1 = (CF_3)_2CF^-$, Scheme 3), generated *in situ* from hexafluoropropene and trimethylamine (*19*). A range of oxygen, nitrogen and carbon centred nucleophiles reacted with **4** (Scheme 4) to give di- and tri-substituted products (Nuc_2 and Nuc_3, Scheme 3) depending on the reaction stoichiometry and following the regiochemistry predicted earlier (*20*).

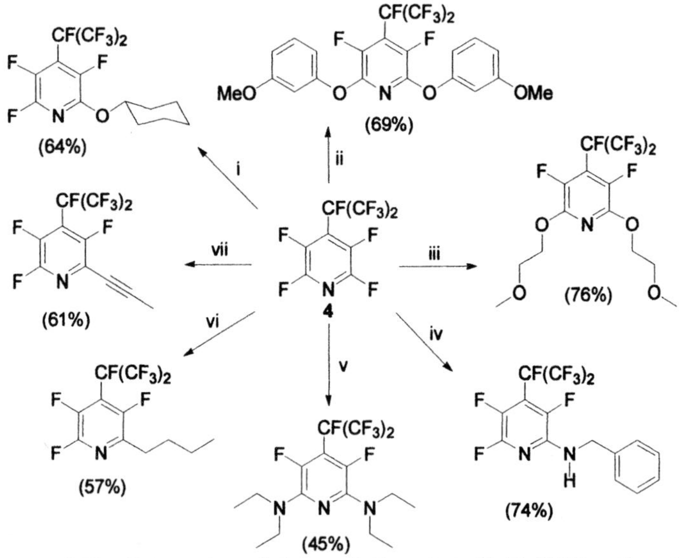

Reagents and Conditions: i, cyclohexanol (1 equiv.), NaH, THF, reflux, 24 h ; ii, $MeO\text{-}C_6H_4O^-Na^+$ (4 equiv.), THF, reflux, 24 h; iii, $CH_3OCH_2CH_2O^-Na^+$, THF, reflux, 24 h; iv, $PhCH_2NHMe$ (1 equiv.), THF, reflux, 30 min; v, Et_2NH, THF, reflux, 24 h; vi, BuLi (1 equiv.), Et_2O, -78°C, 45 min ; vii, $CH_3C{\equiv}C\text{-}MgBr$ (1 equiv.), THF, reflux, 24 h.

*Scheme 4. Reactions of perfluoro-4-isopropylpyridine **4***

As we can see, **4** is a useful building block but, since reactions are limited to nucleophilic substitution processes, we sought to develop chemistry of other related core scaffolds that possessed more versatile functionality. Therefore, we synthesized 2,4,6-tribromo-3,5-difluoro-pyridine **5** and 2,6-dibromo-perfluoro-4-isopropyl pyridine **6** in high yield by heating **1** and **4** with a mixture of hydrogen bromide and aluminium tribromide at 140°C respectively (*21*).

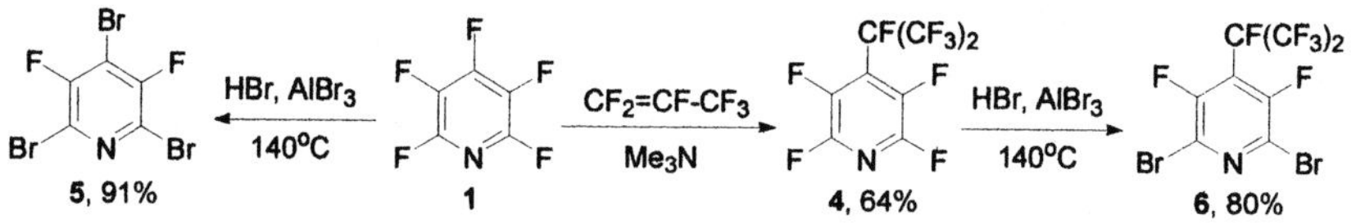

Scheme 5. Synthesis of key bromo-fluoropyridine building blocks.

We envisaged that not only would nucleophilic substitution of C-F and C-Br bonds be possible, but also the C-Br bonds could potentially be used for metallation, followed by trapping of the carbanion generated by an appropriate electrophile, and palladium catalysed coupling processes. Indeed, model poly-bromo-fluoro-heterocyclic derivative **5** reacts with a variety of nucleophiles and it was established (21) that 'hard' nucleophiles (e.g. oxygen centred nucleophiles) selectively replace fluorine whereas 'soft' nucleophiles (e.g., sulfur, nitrogen, etc.) selectively replace bromine (Scheme 6). Furthermore, debromo-lithiation of **5** occurs selectively at the 4-position and the resulting lithiated pyridine species was trapped by a variety of electrophiles (*22*) while palladium catalysed Sonogashira-type coupling reactions allowed the synthesis (*23*) of various alkynyl derivatives (Scheme 6).

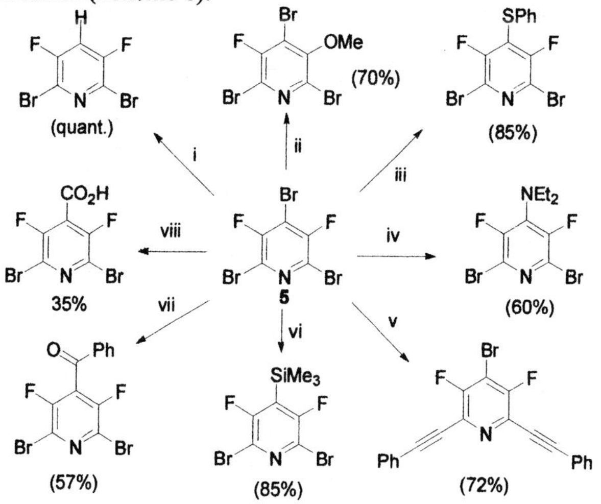

Reagents and Conditions: i,(a) n-BuLi, Et_2O, -78°C; (b) EtOH, -78°C – r.t.; ii, MeONa (1 equiv.) MeOH, r.t.; iii, PhSNa, MeCN, reflux, 24 h; iv, Et_2NH, 50°C, 3 d; v, Ph-C≡C-H, CuI, $Pd(Ph_3P)_2Cl_2$, Et_3N, r.t.; vi, (a) n-BuLi, Et_2O, -78°C; (b) Me_3SiCl, -78°C – r.t.; vii, ,(a)n-BuLi, Et_2O, -78°C; (b) PhCOCl, -78°C – r.t.; viii, (a) n-BuLi, Et_2O, -78°C; (b) CO_2, -78°C – r.t.

Scheme 6. Reactions of 2,4,6-tribromo-3,5-difluoropyridine ***5***

Of note is the palladium catalysed Sonagashira coupling reaction (reaction v, Scheme 6) which, upon addition of an excess of acetylene derivative provides 2,6-disubstituted pyridine systems (*21*). When 2-chlorophenylacetylene is used as the substrate, two different polymorphic, crystalline forms of product **7** may be engineered depending upon the solvent that is used as the crystal growth medium (Scheme 7).

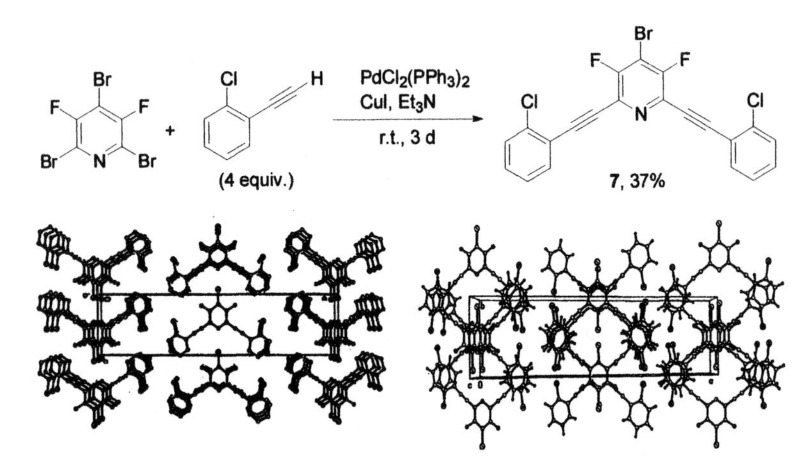

Scheme 7. Synthesis of 7 and crystal lattices of two polymorphs of 7.

Carbanions may also be prepared *in situ* from the perfluoroisopropyl pyridine derivative **6** and subsequent addition of the electrophilic heterocycle **4** furnishes (*24*) polyfunctional bipyridine system **8**. Bipyridine **8** reacts selectively with sodium methoxide to give product **9** arising from substitution of fluorine located *ortho* to ring nitrogen, rather than bromine (Scheme 8).

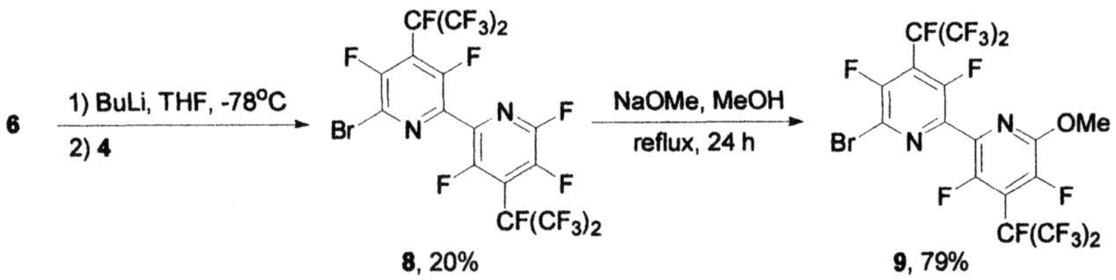

6 —1) BuLi, THF, -78°C; 2) **4**→ **8**, 20% —NaOMe, MeOH, reflux, 24 h→ **9**, 79%

*Scheme 8. Synthesis and reactions of bipyridine **8**.*

With the reactivity profile of **5** in mind we were able to synthesize (*23*) representative examples of pyridine derivatives bearing five different substituents from key intermediate **6** (Scheme 9). The structures of the pentasubstituted products obtained were governed by the order of nucleophilic substitution. For example, reaction of **6** with methoxide followed by piperidine gave a 1:1 mixture of pentasubstituted products **9** and **10** rendering this particular route inappropriate for efficient analogue synthesis whereas **9** was obtained as the major product (structure of **9** confirmed by X-ray crystallography) by simply changing the order of nucleophilic substitution (piperidine followed by methoxide). Palladium catalysed Sonagashira coupling of **6** with phenylacetylene gives intermediate **11** which furnishes pentasubstituted pyridine derivative **12** upon reaction with methoxide.

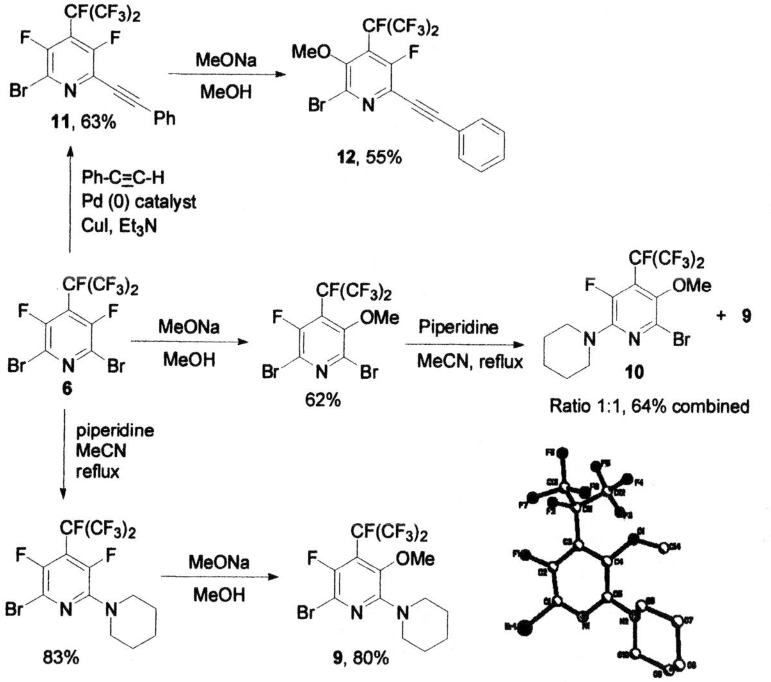

CF(CF3)2 F F Br N Ph **11**, 63% — MeONa / MeOH → CF(CF3)2 MeO F Br N **12**, 55%

Ph-C≡C-H / Pd (0) catalyst / CuI, Et3N

CF(CF3)2 F F Br N Br **6** — MeONa / MeOH → CF(CF3)2 F OMe Br N Br 62% — Piperidine / MeCN, reflux → CF(CF3)2 F OMe N N Br **10** + **9**

Ratio 1:1, 64% combined

piperidine / MeCN / reflux

CF(CF3)2 F F Br N N 83% — MeONa / MeOH → CF(CF3)2 F OMe Br N N **9**, 80%

Scheme 9. Synthesis of pyridine derivatives that bear five different functionalities

In summary, the sequential replacement of the fluorine atoms of pentafluoropyridine may be utilised for the synthesis of many multifunctional pyridine derivatives by reaction with the great number of nucleophilic species that are available. The application of the parallel synthesis techniques widely practised in the pharmaceutical industry would enable rapid access to a considerable number of pyridine analogues.

Macrocycles

Polyfunctional building blocks are, of course, used for many applications such as the synthesis of polymers, dendrimers and biopolymers as well as for various applications in supramolecular chemistry. The synthesis and properties of macrocyclic derivatives has become a major research field (*25-27*) since the pioneering work of Pedersen (*28*) and Lehn (*29*), whose syntheses of crown ethers and cryptands respectively, and their observations of intermolecular complexation phenomena, has rapidly developed into the more general field of supramolecular and biomimetic chemistry. Furthermore, macrocyclic compounds are now used as sensors, imaging agents for MRI treatment, catalysts and for ion analysis (*26*), providing further stimuli for the development of this field. In particular, the incorporation of heteroaromatic sub-units into a macrocyclic ring has been the focus of much attention due to the often unusual chemical, physical and biological properties of such systems.

By a consideration of the reactivity of pentafluoropyridine **1** towards nucleophiles (Scheme 3), a versatile two step macrocyclic ring forming process may be envisaged, in which the ring is constructed by processes involving reactions with a variety of a difunctional nucleophiles (Scheme 10). This strategy offers, in principle, a general route to many structurally diverse macrocycles with specific function that is dependent upon, for example, structural variation and functionality of exocyclic ring substituents.

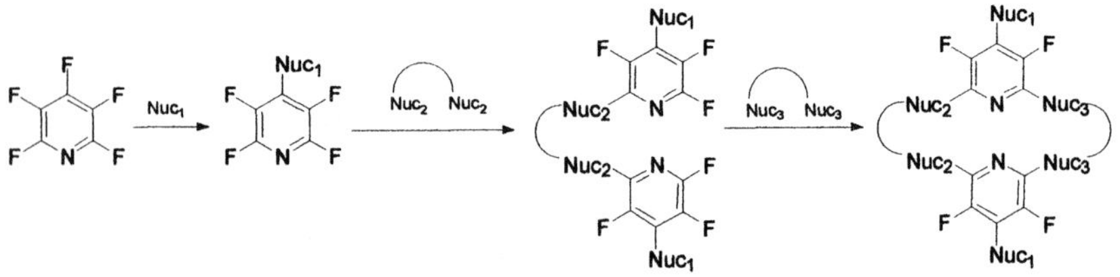

Scheme 10. General strategy for the synthesis of macrocyclic derivatives from pentafluoropyridine

Syntheses of macrocycles that involve a nucleophilic aromatic substitution process as the ring forming step have not been widely adopted (*30*). However, macrocycles may be prepared upon reaction of 2,6-dihalopyridine building blocks with various polyethylene glycol derivatives (*31,32*), although yields could be low. Recently, in related processes, trichloro-*s*-triazine provided access to a range of macrocyclic and cage derivatives (*33-35*).

By adapting the chemistry described above (Scheme 4), perfluoro-4-isopropylpyridine **4** was found to be an excellent building block for macrocycle synthesis (*36,37*). Nucleophilic substitution reaction of **4** with di-oxyanions, prepared *in situ* from bis-trimethylsilyl derivatives of diols and catalytic quantities of fluoride ion, firstly gives a bridged system **13**, in which two pyridine sub-units are connected by a polyether chain. Ring closure to macrocycle **14**, which was characterized by X-ray crystallography, is readily achieved by addition of a further equivalent of dinucleophile (Scheme 11).

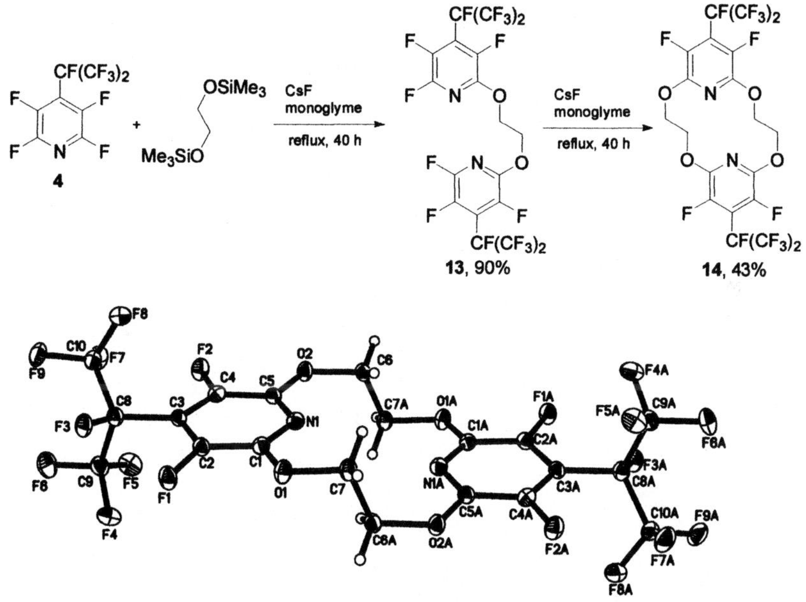

*Scheme 11. Synthesis and X-ray crystallographic structure of macrocycle **14**.*

This stepwise methodology allows the synthesis of many structurally diverse macrocycles (Scheme 12) from pentafluoropyridine, depending upon the choice of nucleophile and di-nucleophile (*37*). By analogy to the synthesis outlined in Scheme 11, 20-membered ring system **15** may be synthesized in two steps from **4** and a longer chain polyether linking unit. Variation of substituents that are present on the pyridine ring section of the macrocycle is very simple. For example, **16** is prepared from **1**, an appropriate polyfluoroalcohol (Nuc_1 = $C_6F_{13}CH_2CH_2O^-$, Scheme 10) and N,N'-dimethyethylene-diamine (Nuc_2, Nuc_3 = $MeNHCH_2CH_2NHMe$, Scheme 10) in three stages while **17** is synthesized from **1**, sodium methoxide and ethylene glycol by a similar procedure (*38*). Variation of the structural unit linking the two pyridine moieties is also readily achieved. For example, **18**, in which the two pyridine rings are connected by different linking groups, is synthesized from **4**, di-ethylene glycol (Nuc_2-Nuc_2 =

$HOCH_2CH_2OCH_2CH_2OH$, Scheme 10) and N,N'-dimethylethylene-diamine (Nuc_3-Nuc_3 = $MeNHCH_2CH_2NHMe$, Scheme 10).

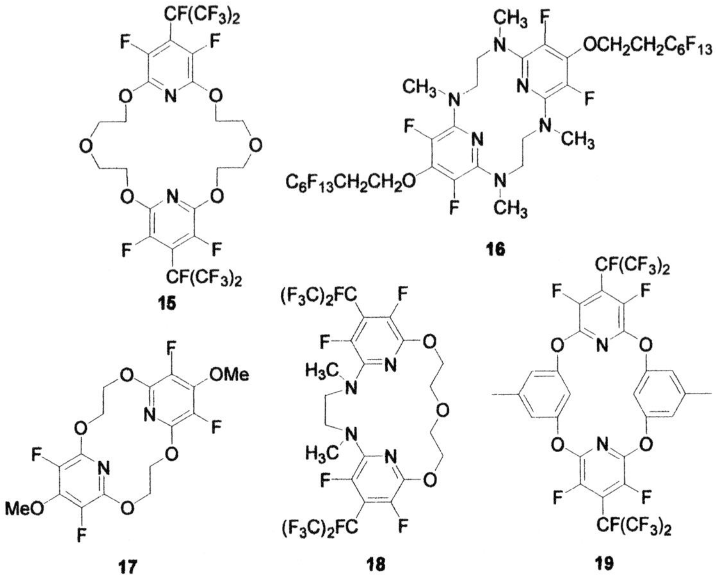

Scheme 12. Structurally diverse macrocycles derived from pentafluoropyridine

Analysis by X-ray crystallography reveals that macrocycles **15** and **16** have some unusual structural features (Scheme 13). For **15**, we would expect the two very bulky perfluoroisopropyl groups to occupy positions within the crystal lattice that are distant from each other. However, we observe that this is not the case and the macrocyclic ring buckles so that these two perfluorinated groups may occupy positions that are as close to each other as possible (Scheme 13). Inspection of the crystal lattice reveals that the individual molecules adopt head-to-head positions which results in the presence of fluorine- and hydrocarbon-rich 'domains' within the crystal, reflecting the solubility characteristics of fluorocarbon systems that has been used so effectively in fluorous phase chemistry (*39,40*). The segregation of perfluorinated 'domains' within the crystal lattice is also seen in the structure of macrocycle **16** in which the long perfluoroalkyl substituent groups overlap to form fluorine-rich environments.

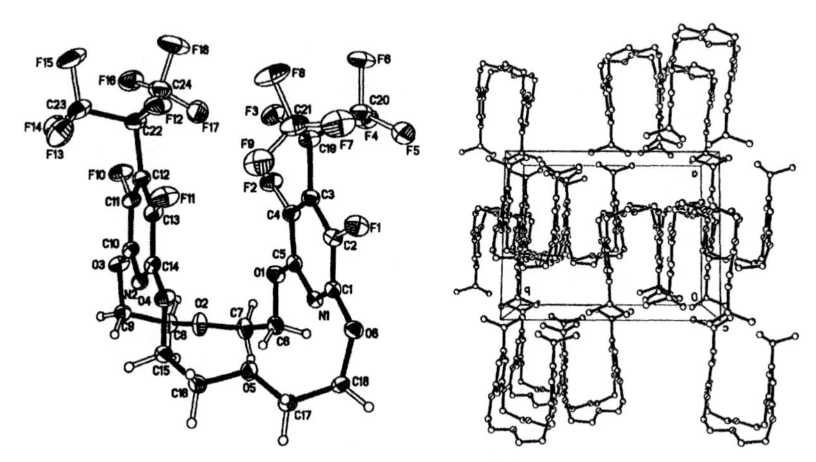

Scheme 13. Single crystal X-ray molecular structures of macrocycle ***15*** *(left and crystal lattice showing head-to-head arrangement (right; fluorine atoms have been removed for clarity).*

Mixing a solution of macrocycle **19** with a mixture of aqueous sodium chloride, bromide and iodide led to the observation of macrocycle/anion complexes by negative ion electrospray mass spectrometry. Whilst the number of macrocyclic derivatives that bind cations is now quite extensive, the number of macrocycles capable of anion recognition is considerably smaller. Systems that form complexes with anionic guests, such as polyammonium, guanidinium and pyrrole based systems (*41*) usually contain sites that are capable of directed hydrogen bonding, a situation that is not present in **19**. However, recent theoretical studies have indicated that interactions between the centre of electron deficient heterocyclic rings, such as trifluoro-s-triazine, and anions are possible and this may provide a clue to the type of complexation occurring in these cases.

In summary, pentafluoropyridine may be used for the construction of a variety of structurally diverse macrocycles by a sequence of nucleophilic aromatic substitution processes. The availability of perfluoroaromatic systems and a great many suitable difunctional nucleophiles makes the number of supramolecular systems that are available using this approach virtually limitless. The structural and complexation phenomena exhibited by just the relatively few systems that have been reported so far, indicate the fascinating opportunities that are provided by these multi-functional systems.

References

(1) Chambers, R. D.; Sargent, C. R. *Adv. Heterocycl. Chem.* **1981**, *28*, 1-71.
(2) Brooke, G. M. *J. Fluorine Chem.* **1997**, *86*, 1-76.
(3) Chambers, R. D.; Hutchinson, J.; Musgrave, W. K. R. *J. Chem. Soc.* **1964**, 3573-3576.
(4) Chambers, R. D. *Fluorine in Organic Chemistry*; John Wiley and Sons: New York, 1973.
(5) Gilmore, C. J.; MacNicol, D. D.; Murphy, A.; Russell, M. A. *Tetrahedron Lett.* **1984**, *25*, 4303-4306.
(6) Banks, R. E.; Smart, B. E.; Tatlow, J. C. *Organofluorine Chemistry. Principles and Commercial Applications*; Plenum: New York, 1994.
(7) Lipinski, C. A.; Lombardo, F.; Dominy, B. W.; Feeney, P. J. *Adv. Drug Delivery Rev.* **1997**, *23*, 3-25.
(8) Nicolaou, K. C.; Pfefferkorn, J. A.; Roecker, A. J.; Cao, G. Q.; Barluenga, S.; Mitchell, H. J. *J. Am. Chem. Soc.* **2000**, *122*, 9939-9953.
(9) Mason, J. S.; Morize, I.; Menard, P. R.; Cheney, D. L.; Hulme, C.; Labaudiniere, R. F. *J. Med. Chem.* **1999**, *42*, 3251-3264.
(10) Dolle, R. E.; Nelson, K. H. *J. Comb. Chem.* **1999**, *1*, 235-282.
(11) Hirschmann, R.; Ducry, L.; Smith, A. B. *J. Org. Chem.* **2000**, *65*, 8307-8316.
(12) Katritzky, A. R.; Rees, C. W. *Comprehensive Heterocyclic Chemistry*; Pergamon Press: Oxford, 1984;Vols. 1 - 8.
(13) Collins, I. *J. Chem. Soc., Perkin Trans. 1* **2000**, 2845-2861.
(14) Collins, I. *J. Chem. Soc., Perkin Trans. 1* **2002**, 1921-1940.
(15) Schreiber, S. L. *Science* **2000**, *287*, 1964-1969.
(16) Burke, M. D.; Lalic, G. *Chem. Biol.* **2002**, *9*, 535-541.
(17) Schreiber, S. L. *Chem. Eng. News* **2003**, *March 3*, 51-61.
(18) Banks, R. E.; Jondi, W.; Tipping, A. E. *J. Chem. Soc., Chem. Commun.* **1989**, 1268-1269.
(19) Chambers, R. D.; Gray, W. K.; Korn, S. R. *Tetrahedron* **1995**, *51*, 13167-13176.
(20) Chambers, R. D.; Hassan, M. A.; Hoskin, P. R.; Kenwright, A.; Richmond, P.; Sandford, G. *J. Fluorine Chem.* **2001**, *111*, 135-146.
(21) Chambers, R. D.; Hall, C. W.; Hutchinson, J.; Millar, R. W. *J. Chem. Soc., Perkin Trans. 1* **1998**, 1705-1713.
(22) Benmansour, H.; Chambers, R. D.; Sandford, G.; McGowan, G.; Dahaoui, S.; Yufit, D. S.; Howard, J. A. K. *J. Fluorine Chem.* **2001**, *112*, 349-355.
(23) Chambers, R. D.; Hoskin, P. R.; Sandford, G.; Yufit, D. S.; Howard, J. A. K. *J. Chem. Soc., Perkin Trans. 1* **2001**, 2788-2795.
(24) Chambers, R. D.; Hoskin, P. R.; Sandford, G. *Arkivoc* **2002**, *6*, 279-283.

(25) Dietrich, B.; Viout, P.; Lehn, J. M. *Macrocyclic Chemistry*; VCH: Weinheim, 1993.
(26) Lehn, J. M.; Atwood, J. L.; Davies, J. E. D.; MacNicol, D. D.; Votgle, F., Eds. *Comprehensive Supramolecular Chemistry*; OUP: Oxford, 1996; Vol. 1-11.
(27) Schneider, H. J.; Yatsimirsky, A. K. *Principles and Methods in Supramolecular Chemistry*; John Wiley and Sons: New York, 2000.
(28) Pedersen, C. J. *J. Inclusion Phenom.* **1988**, *6*, 337-350.
(29) Lehn, J. M. *J. Inclusion Phenom.* **1988**, *6*, 351-396.
(30) Sandford, G. *Chem. Eur. J.* **2003**, *9*, 1464-1469.
(31) Newkome, G. R.; McClure, G. L.; Simpson, J. B.; Danesh-Khoshboo, F. *J. Am. Chem. Soc.* **1975**, *97*, 3232-3234.
(32) Singh, H.; Kumar, S.; Jain, A.; Singh, P. *J. Chem. Soc., Perkin Trans. 1* **1990**, 965-968.
(33) Anelli, P. L.; Lunazzi, L.; Montanari, F.; Quici, S. *J. Org. Chem.* **1984**, *49*, 4197-4203.
(34) Lowik, D. W. P. M.; Lowe, C. R. *Tetrahedron Lett.* **2000**, *41*, 1837-1840.
(35) Lowik, D. W. P. M.; Lowe, C. R. *Eur. J. Org. Chem.* **2001**, 2825-2839.
(36) Chambers, R. D.; Hoskin, P. R.; Khalil, A.; Richmond, P.; Sandford, G.; Yufit, D. S.; Howard, J. A. K. *J. Fluorine Chem.* **2002**, *116*, 19-22.
(37) Chambers, R. D.; Hoskin, P. R.; Kenwright, A. R.; Khalil, A.; Richmond, P.; Sandford, G.; Yufit, D. S.; Howard, J. A. K. *Org. Biomol. Chem.* **2003**, *1*, 2137-2147.
(38) Chambers, R. D.; Khalil, A.; Richmond, P.; Sandford, G.; Yufit, D. S.; Howard, J. A. K. *J. Fluorine Chem.* **2004**, *125*, 715-720.
(39) Horvath, I. T. *Acc. Chem. Res.* **1998**, *31*, 641-650.
(40) Hope, E. G.; Stuart, A. M. *Adv. Inorg. Fluorides* **2000**, 403-436.
(41) Beer, P. D.; Gale, P. A. *Angew. Chem. Intl. Ed. Engl.* **2001**, *40*, 486-516.

Chapter 14

Synthetic Methods for the Preparation of Ring-Fluorinated Heterocycles via Intramolecular Vinylic Substitution of *gem*-Difluoroalkenes

Junji Ichikawa

Department of Chemistry, Graduate School of Science, The University of Tokyo, Hongo, Bunkyo-ku, Tokyo 113–0033, Japan

Ring-fluorinated heterocyclic compounds attract widespread attention as important components of agrochemicals, pharmaceuticals, and materials. To provide a general access to these compounds, a new methodology for the construction of ring-fluorinated heterocycles has been developed based on an intramolecular ring closure of 2,2-difluorovinylic compounds via an S_NV (addition–elimination) process. In this chapter, the synthesis of five- and six-membered ring-fluorinated heterocycles including one or two oxygen, sulfur, or nitrogen atoms will be covered.

Introduction

Selectively fluorinated heterocyclic compounds are widely used as important components in the pharmaceutical, agrochemical, and dyestuffs industries and also in material science (*1*). They have attracted much attention in terms of biological activities and other properties, which are often dramatically enhanced or altered by the introduction of fluorine onto the original molecules. Although introduction of fluorine into heterocyclic compounds can be achieved by employing classical Balz–Schiemann (fluorodediazotization) and Halex (halogen exchange) fluorinations, electrochemical fluorination, and fluorinating reagents (*2*), there still remain problems in the regioselectivity of fluorination, the requirement of multistep procedures, and the difficulty in handling the fluorinating reagents. Especially for ring-fluorinated heterocyclic compounds, so far only a limited number of methods have been reported for their synthesis in spite of their substantial potential as components (*3*) and intermediates (*4*). The introduction of fluorine atoms onto heterocyclic ring carbons is known to be more difficult than the introduction of fluorine onto the carbons of fused benzene rings. Consequently, a general method for the synthesis of selectively ring-fluorinated heterocycles is a highly desirable goal.

Synthetic Strategy for Ring-Fluorinated Heterocycles

gem-Difluoroalkenes possess remarkable reactivity toward nucleophilic substitution of their vinylic fluorines via addition–elimination processes (SNV process) (*5*). The orientation of attack is strictly governed so that the fluorines are placed at the position β to the electron-rich carbon in the transition state in order to avoid electron-pair repulsion. This reactivity is due to (i) the electrophilic activation of the C–C double bond by the two fluorine atoms, (ii) the stabilization of the intermediary carbanion by the β-anion stabilizing effect of fluorine, and (iii) the leaving group ability of the fluoride ion (Scheme 1).

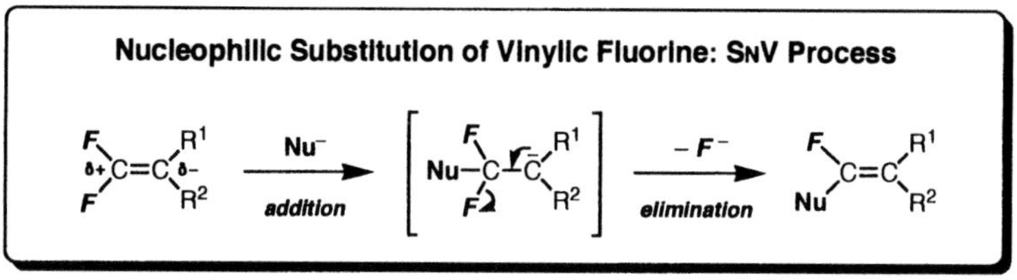

Nucleophilic Substitution of Vinylic Fluorine: SNV Process

Scheme 1

These unique properties of *gem*-difluoroalkenes can be utilized to give access to ring-fluorinated heterocycles simply by conducting this substitution in an intramolecular fashion (Scheme 2). Depending on the nucleophilic Y–Z moiety, *O*, *S*, *N*-containing heterocyclic systems with a five, six, or larger-membered ring can be made. In β,β-difluorostyrene derivatives as an example, nucleophiles with a Y–Z single bond (sp^3 nucleophiles) afford dihydroheteroarenes, and this ring formation could be followed by aromatization via elimination or oxidation, leading to heteroaromatic systems. On the other hand, nucleophiles with a Y–Z double bond (sp^2 nucleophiles) would give rise to the direct construction of heteroaromatic rings. Thus, this strategy should provide a potent methodology for the synthesis of diverse ring-fluorinated heterocycles.

In this chapter we describe our investigations on the "intramolecular substitution" concept for the synthesis of ring-fluorinated heterocycles starting from *gem*-difluoroalkenes. Other methods involving the vinylic fluorine substitution process provide ring-fluorinated quinolines (*6*), quinoxaline derivatives (*7*), pyrans (*8*), pyrazoles (*9*), and furan derivatives (*10*).

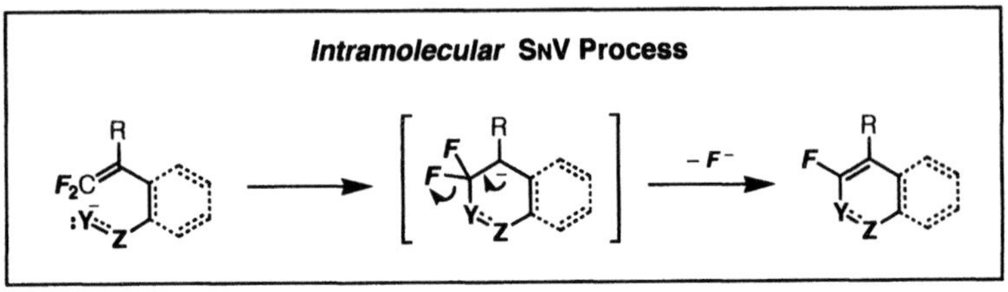

Intramolecular **SNV Process**

Scheme 2

Preparation of *gem*-Difluoroalkenes Bearing a Nucleophile

ortho-Functionalized *gem*-difluorostyrenes are the substrates for the construction of benzene-ring-fused heterocyclic systems. They are readily obtained from 2,2,2-trifluoroethyl *p*-toluenesulfonate by using the two methods that we have previously established as depicted in Scheme 3 (*11,12*). Each method consists of the following two processes, which can be conducted in a one-pot operation: (i) the *in situ* generation of *gem*-difluorovinylboranes or -zirconocene and (ii) their palladium-catalyzed cross-coupling reaction with aryl iodides bearing an *ortho*-functional group such as NH_2, OMe, CHO, CH_2OH, CH_2OMs, or CN. Moreover, further transformations provide other substrates with an appropriate *ortho*-functionality.

The substrates for heterocycles without fused benzene rings can also be prepared by these two sequences. A similar cross-coupling reaction with vinylic halides affords *gem*-difluoro-1,3-dienes, which undergo regioselective hydroboration to afford homoallylic alcohols (*12*). Their Mitsunobu reaction or the substitution of their tosylates affords *gem*-difluoroalkenes bearing a functional group at the homoallylic position.

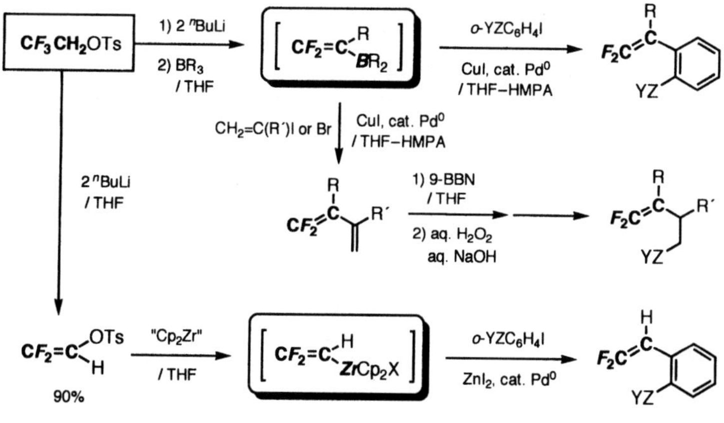

Scheme 3

Syntheses of 6-Membered Ring-Fluorinated Heterocycles

Cyclization of β,β-Difluorostyrenes Bearing an sp^3 Heteroatom Nucleophile

3-Fluorinated Isochromenes, Isothiochromenes, and Isoquinoline Derivatives

gem-Difluorostyrenes **1** and **3** were designed as substrates for the intramolecular substitution of sp^3 oxygen and sulfur nucleophiles, so that 6-membered rings would be formed. Treatment of **1** or **3** with a base such as NaH or NaOMe generated the corresponding alkoxides or thiolates, which in turn readily underwent the expected cyclization (Scheme 4) (*13*). Thus, 3-fluoro-isochromenes **2** and -thioisochromenes **4** were obtained in high to excellent yield (*14*).

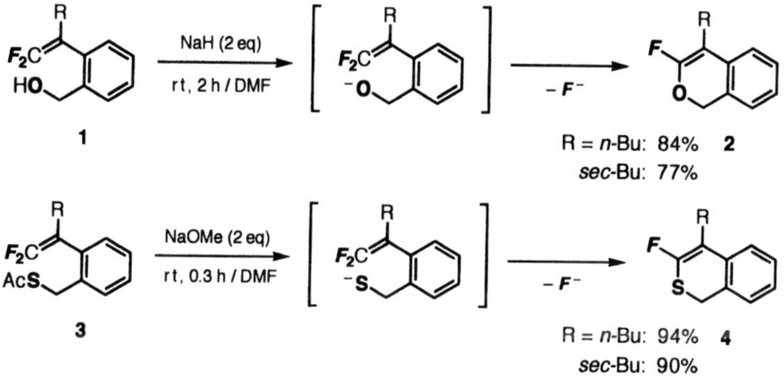

Scheme 4

Similarly, the reaction of the substrates bearing an sp^3 nitrogen nucleophile was successfully carried out. On treatment of styrene **5** with NaH, intramolecular cyclization readily proceeded to afford 3-fluoro-1,2-dihydroisoquinolines **6**. Employing 2 equiv of base successively promoted elimination of a sulfinic acid after the cyclization, leading to 3-fluoroisoquinolines **7** in excellent yield (Scheme 5) (*15,16*).

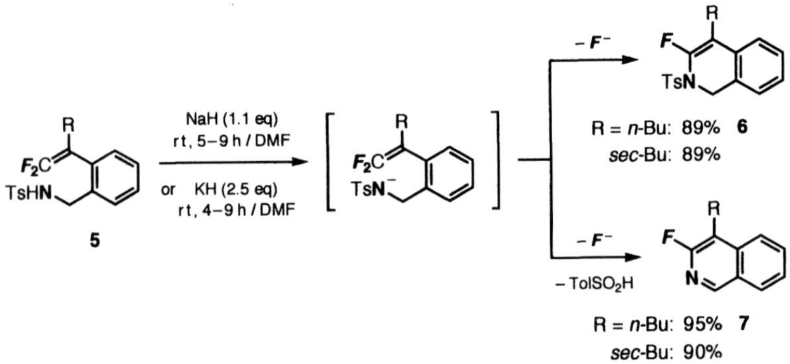

Scheme 5

Cyclization of β,β-Difluorostyrenes Bearing an sp^3 Carbon Nucleophile

3-Fluorinated Quinolines

For the construction of a quinoline framework, carbon nucleophiles should be employed instead of the nitrogen nucleophiles in the above isoquinoline synthesis. Thus, difluorostyrenes **8** were designed to bear a cyanomethylamino group as a precursor for a *formal* sp^3 carbon nucleophile. In the reaction of **8**, aromatization of the cyclized products is effected by elimination of (i) HCN or (ii) HX (X: a leaving group on the nitrogen) (Scheme 6).

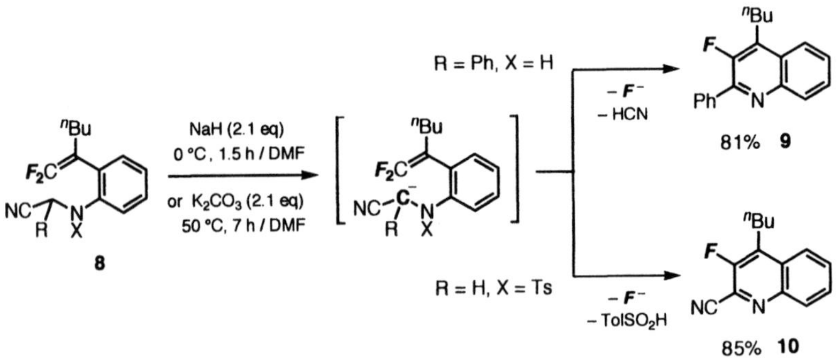

Scheme 6

When styrenes **8** (X = H) were treated with NaH (R ≠ H) or lithium 2,2,6,6-tetramethylpiperidide (LiTMP) (R = H), the intramolecular substitution of the *in situ* generated carbanions successfully proceeded, followed by elimination of HCN to give the desired fluoroquinolines **9** in high yield. In contrast, treatment of **8** (X = Ts) with 2 equiv of K_2CO_3 promoted cyclization, and successive elimination of sulfinic acid afforded 2-cyano-3-fluoroquinoline **10**. Its cyano group can be easily converted into carboxy, aminomethyl, and amino groups by the usual procedures (*17,18*).

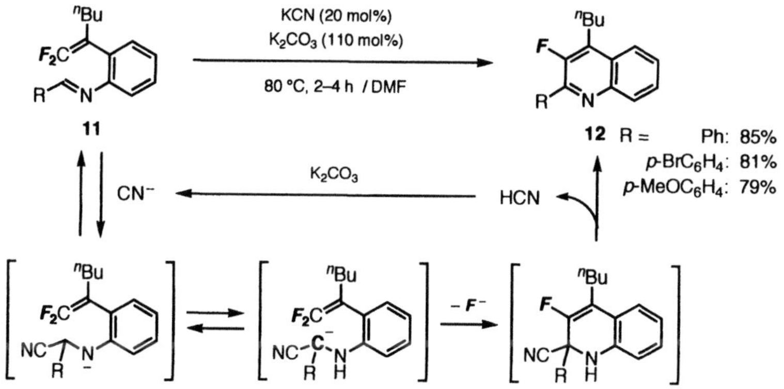

KCN (20 mol%)
K_2CO_3 (110 mol%)
80 °C, 2–4 h / DMF

12 R = Ph: 85%; *p*-BrC_6H_4: 81%; *p*-$MeOC_6H_4$: 79%

Scheme 7

The reaction of **8** requires (i) a strong base (NaH or LiTMP) for the generation of the carbanions and (ii) a stoichiometric amount of KCN for the preparation of **8** from the corresponding *o*-aminostyrenes. These drawbacks are overcome by a benzoin-type condensation of imines as shown in Scheme 7. Addition of CN^- to imines **11** followed by proton transfer provided the desired carbon nucleophiles. The generated carbanions were smoothly trapped by the intramolecular difluoroalkene moiety, and loss of HCN gave 3-fluoroquinolines **12**. Moreover, this process proceeds under KCN catalysis in the presence of K_2CO_3, which represents a quite rare example of benzoin-type condensation of imines, C–C bond formation at the imino carbon with electrophiles (*19*).

Cyclization of β,β-Difluorostyrenes Bearing an sp^2 Nitrogen Nucleophile

3-Fluorinated Isoquinoline Derivatives and Cinnolines

In the reactions mentioned so far, aromatization was effected by elimination after cyclization. For the direct construction of aromatic rings, a similar replacement of the vinylic fluorine was attempted with sp^2 nitrogen nucleophiles: oxime and imine nitrogens (-N=CH-). Since suitable starting materials, *o*-difluorovinyl-substituted benzaldehydes **13** were not stable enough, the combined process of difluorostyrene synthesis and their cyclization was examined. When the crude products of the coupling reaction were treated with NH_4OAc or NH_2OH, the expected dehydration and successive cyclization were readily induced (Scheme 8). 3-Fluorinated isoquinolines **14** or their *N*-oxides **15** were obtained in good yield based on *o*-iodobenzaldehyde (*16,20*).

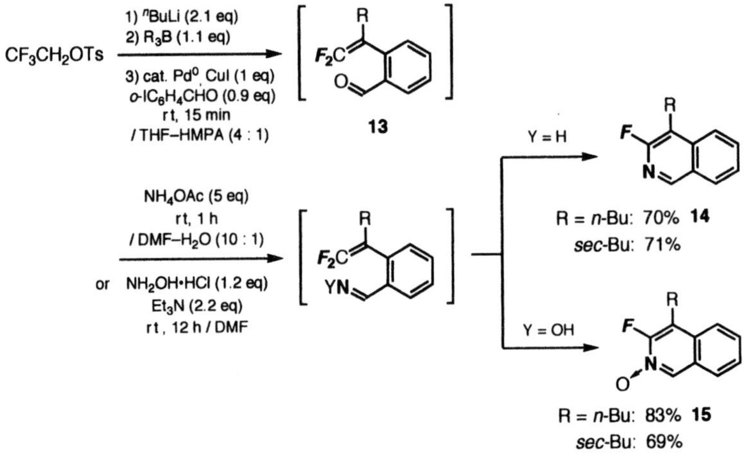

Scheme 8

As a further example of sp^2 nitrogen as a nucleophile, diimides (-**N**=N-) have been employed. Aminostyrene **16** was treated with isoamyl nitrite (*i*-AmONO) for diazotization, and then reduced with benzenethiol. The expected intramolecular substitution of the diimide nitrogen smoothly proceeded to give 3-fluorocinnolines **17** (Scheme 9) (*20,21*).

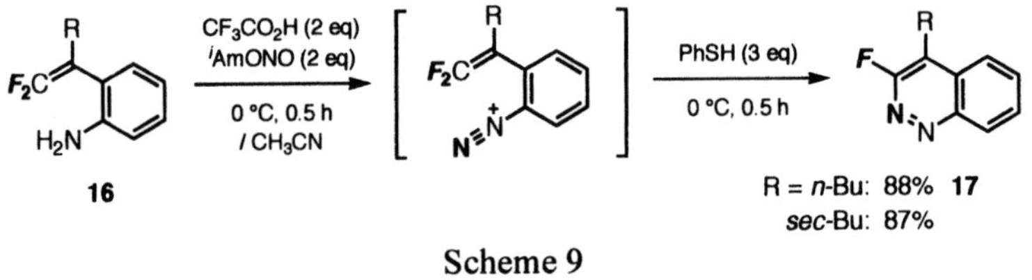

Scheme 9

Another method for the generation of intramolecular sp^2 nitrogen nucleophiles is the addition of external nucleophiles to C–N multiple bonds such as that of the cyano group. Thus *in situ* generated nucleophiles are expected to promote similar cyclizations to afford ring-fluorinated isoquinolines, which would allow introduction of a substituent on the C-1 ring carbon (Scheme 10). Since there are two reaction sites to be attacked by nucleophiles in **18**, the external nucleophiles must exclusively react with the cyano group without affecting the difluoromethylene moiety.

It was found that the addition of organolithiums in Et_2O or toluene regioselectively occurred to the cyano group of styrene **18**. The generated iminometals promoted cyclization to afford isoquinolines **19** in high yield. This sequence provides a facile access to 1,4-disubstituted 3-fluoroisoquinolines (*16,22*).

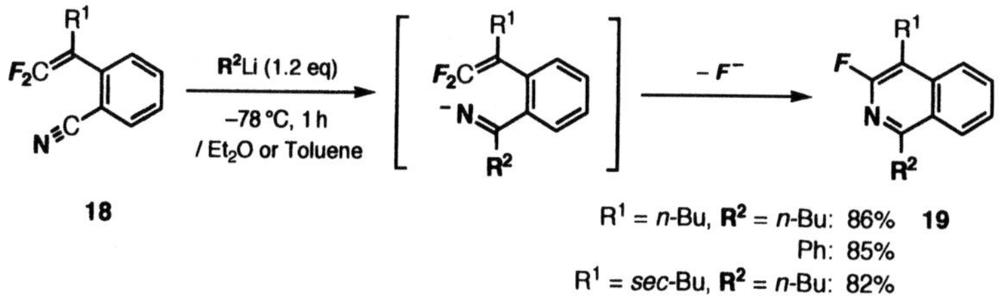

R^2Li (1.2 eq), −78 °C, 1 h / Et_2O or Toluene; − F^-

18

R^1 = *n*-Bu, R^2 = *n*-Bu: 86% **19**
Ph: 85%
R^1 = *sec*-Bu, R^2 = *n*-Bu: 82%

Scheme 10

Cyclization of β,β-Difluorostyrenes Bearing an sp^2 Carbon Nucleophile

3-Fluorinated Quinolines

In the construction of the isoquinoline framework, the cyano group has proved to be a powerful precursor of sp^2 nitrogen nucleophiles. By means of a similar tactics, quinoline synthesis is effected by changing the order of carbon and nitrogen in the *ortho* substituent of the substrates. The reaction of *o*-isocyano-substituted styrenes **20** with external nucleophiles proceeds in a similar manner via the corresponding sp^2 carbon nucleophiles to provide 3-fluoroquinolines in an appropriate solvent system (Scheme 11). When isocyanide **20** was treated with Grignard reagent in toluene, addition selectively occurred to the isocyano carbon. Then, HMPA was added to raise the reactivity of the intermediary carbanion, which led to 2,4-disubstituted 3-fluoroquinolines **21** (*18,22*). This reaction allows the introduction of a substituent at the 2 position of the quinoline ring with nucleophiles.

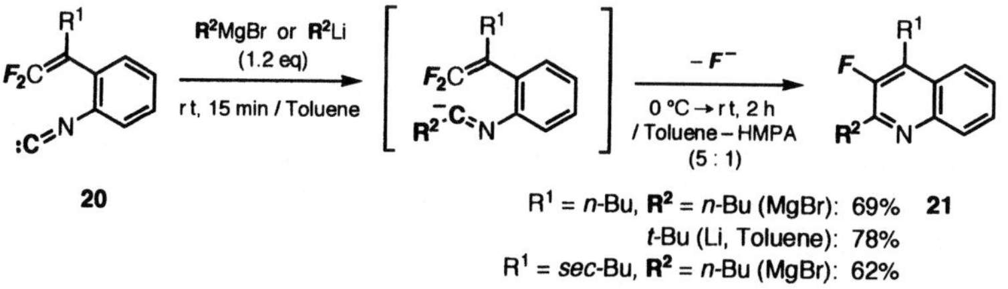

R^2MgBr or R^2Li (1.2 eq), rt, 15 min / Toluene; − F^-, 0 °C → rt, 2 h / Toluene – HMPA (5 : 1)

20

R^1 = *n*-Bu, R^2 = *n*-Bu (MgBr): 69% **21**
t-Bu (Li, Toluene): 78%
R^1 = *sec*-Bu, R^2 = *n*-Bu (MgBr): 62%

Scheme 11

The introduction of a C-2 substituent on the quinoline ring has been achieved not only with nucleophiles but also electrophiles (Scheme 12). One-electron reduction of the isocyano group of **20** with tributylstannyllithium generates the radical anion, which in turn undergoes cyclization. The cyclized intermediates, 2-quinolyl radicals are reduced further to 2-quinolyl anions, which are trapped by electrophiles to allow introduction of hydroxymethyl, acyl, and iodo groups. The introduction of aryl groups is also effected by palladium-catalyzed cross-coupling reaction via the transmetalation to zinc (*23*).

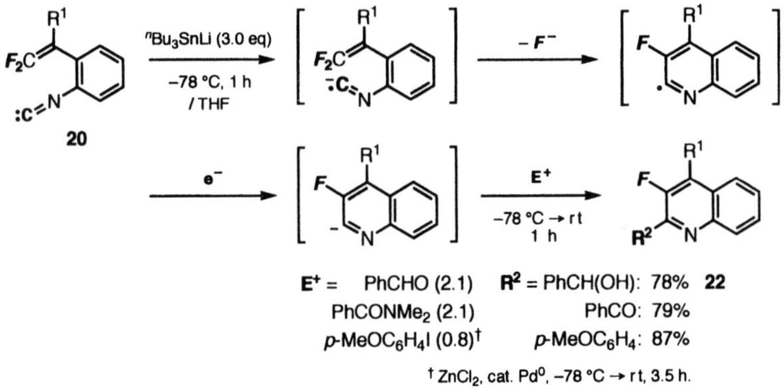

E^+ =	R^2 =	
PhCHO (2.1)	PhCH(OH): 78%	**22**
$PhCONMe_2$ (2.1)	PhCO: 79%	
p-$MeOC_6H_4I$ (0.8)†	*p*-$MeOC_6H_4$: 87%	

† $ZnCl_2$, cat. Pd^0, −78 °C → rt, 3.5 h.

Scheme 12

By the two complementary methods (Schemes 11, 12), a variety of substituents can be introduced at the 2-position of 3-fluoroquinolines with either nucleophiles or electrophiles.

Syntheses of 5-Membered Ring-Fluorinated Heterocycles

As shown above, the syntheses of 6-membered ring-fluorinated heterocycles have successfully been accomplished on the basis of the "intramolecular substitution" concept starting from *gem*-difluoroalkenes. According to the Baldwin's rules (*24*), this type of cyclization is classified as a 6-*endo-trig* ring closure, which is favored following these rules. On the other hand, for the construction of 5-membered rings by this methodology, the disfavored 5-*endo-trig* ring closure has to be effected.

The unique properties of 1,1-difluoro-1-alkenes, however, were expected to make a nucleophilic approach to 5-*endo-trig* cyclization feasible. Specifically, we thought that (i) the highly polarized difluorovinylidene double bond (^{13}C NMR: ca. 150 ppm and 90 ppm for $CF_2{=}C$) would allow initial ring formation by electrostatic attraction between the CF_2 carbon and the internal nucleophile and (ii) the successive elimination of fluoride ion could suppress the reverse ring opening, thus functioning as a "lock" (Scheme 13).

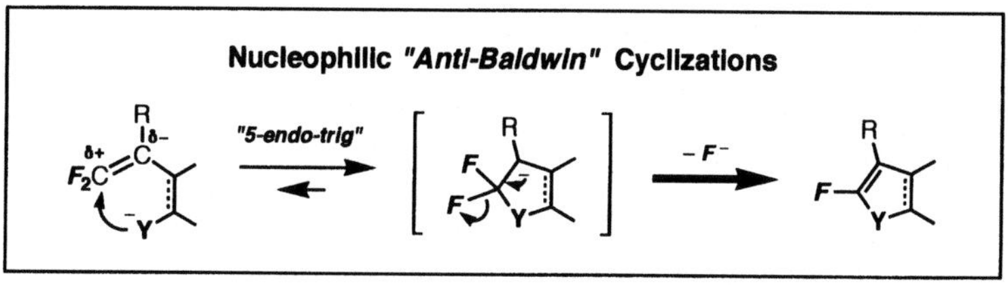

Scheme 13

Cyclization of β,β-Difluorostyrenes Bearing an sp^3 Heteroatom Nucleophile

2-Fluorinated Benzofurans, Benzothiophenes, Indoles

The reaction of an oxygen nucleophile was first attempted. When hydroxystyrene **23** was treated with NaH, the normally "disfavored" 5-*endo-trig* cyclization was successfully achieved to afford 2-fluorobenzo[*b*]furan **24** in high yield. The corresponding thiolate also undergoes the cyclization to provide 2-fluorobenzo[*b*]thiophene **26** (Scheme 14) (*12,25–27*).

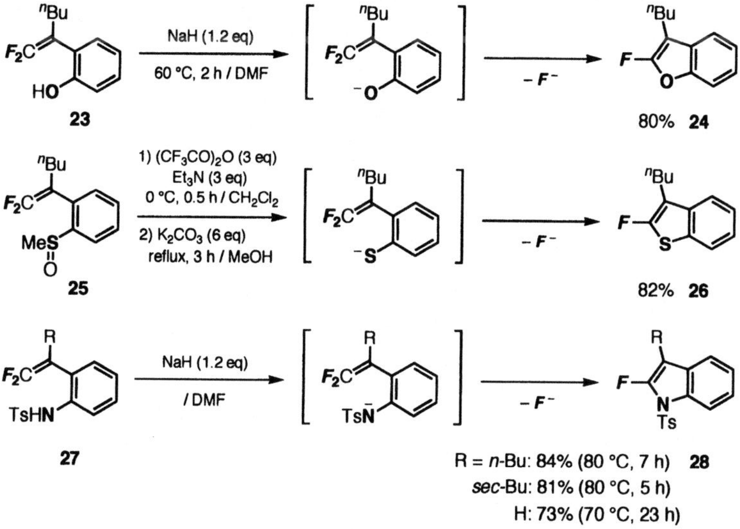

Scheme 14

Moreover, this type of cyclization can be applied to intramolecular nitrogen nucleophiles. On treatment of amidostyrenes **27** with NaH, the 5-*endo-trig* cyclizations again successfully occurred to afford 2-fluoroindoles **28** in high yield (Scheme 14) (*12,25,28*).

Cyclization of *gem*-Difluorobutenes Bearing an sp^3 Heteroatom Nucleophile

2-Fluorinated Dihydrofurans, Dihydrothiophenes, Pyrrolines

The scope of this nucleophilic *5-endo-trig* cyclization has been broaden by the reaction of difluoroalkenes **29**, **31**, **33**, where the nucleophilic functional groups are linked by two sp^3 carbons to the olefin moieties (*12,29*). The 5-*endo-trig* closures are successfully promoted by the intramolecular oxygen, sulfur, and nitrogen nucleophiles, leading to the desired ring-fluorinated dihydrofuran **30** (*30*), dihydrothiophene **32** (*31*), and pyrroline **34** (*32,33*) (Scheme 15). These results rule out the possibility of a 6π-electrocyclization mechanism.

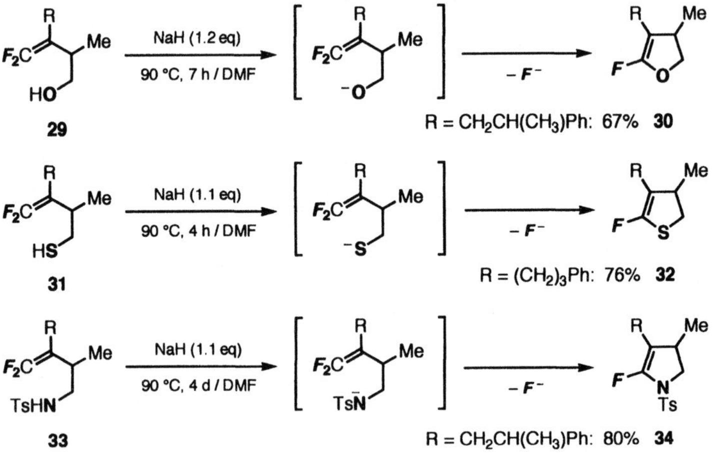

R Me F_2C HO **29** NaH (1.2 eq) 90 °C, 7 h / DMF [R F_2C Me $^-$O] – F^- R Me F O

R = $CH_2CH(CH_3)Ph$: 67% **30**

R Me F_2C HS **31** NaH (1.1 eq) 90 °C, 4 h / DMF [R F_2C Me $^-$S] – F^- R Me F S

R = $(CH_2)_3Ph$: 76% **32**

R Me F_2C TsHN **33** NaH (1.1 eq) 90 °C, 4 d / DMF [R F_2C Me $Ts\bar{N}$] – F^- R Me F N Ts

R = $CH_2CH(CH_3)Ph$: 80% **34**

Scheme 15

The favored nature of the 5-*endo-trig* cyclization in *gem*-difluoroalkene substrates was demonstrated by conducting a competitive reaction as shown in Scheme 16. The reaction proceeded not in a 5-*exo-trig* fashion but in a normally disfavored 5-*endo-trig* fashion, affording 2-fluoropyrroline **36** exclusively (*12*).

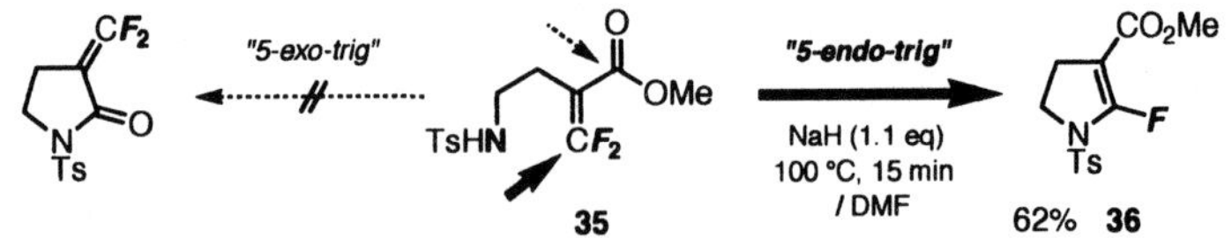

CF_2 O N Ts "5-exo-trig" O OMe TsHN CF_2 **35** "5-endo-trig" NaH (1.1 eq) 100 °C, 15 min / DMF CO_2Me F N Ts 62% **36**

Scheme 16

Summary

The reactivities of *gem*-difluoroalkenes allow the cyclization via an intramolecular substitution of vinylic fluorines. This methodology is quite simple and efficiently provides a variety of ring-fluorinated heterocyclic compounds such as isochromenes, isothiochromenes, quinolines, isoquinolines, cinnolines, indoles, pyrrolines, furans, and thiophenes depending on intramolecular nucleophiles (*34*). Concerning most of these fluorinated heterocycles, their properties have not been sufficiently explored due to the lack of general methods for their syntheses. Now that they are available with ease of preparation, a wide range of applications can be expected.

Acknowledgements

I thank my coworkers, Dr. Y. Wada, Mr. T. Mori, Mr. H. Miyazaki, Dr. M. Fujiwara, Mr. H. Kuroki, Ms. H. Moriyama, Mr. J. Mihara, Mr. K. Sakoda, Mr. T. Katsume, and Ms. T. Nohiro for their contributions to this project.

References

1. (a) *Organofluorine Chemistry, Principles and Commercial Applications*; Banks, R. E.; Smart, B. E.; Tatlow, J. C., Eds.; Plenum: New York, 1994. (b) *Biomedicinal Aspects of Fluorine Chemistry*; Filler, R.; Kobayashi, Y., Eds.; Kodansha and Elsevier Biomedical: Tokyo, 1982. (c) Welch, J. T.; Eswarakrishnan, S. *Fluorine in Bioorganic Chemistry*; John Wily & Sons: New York, 1991. (c) Silvester, M. J. *Adv. Heterocyclic Chem.* **1994**, *59*, 1–38. (d) Silvester, M. J. *Aldrichimica Acta* **1991**, *24*, 31–38.
2. *Chemistry of Organic Fluorine Compounds II*; Hudlicky, M.; Pavlath, A. E., Eds.; American Chemical Society: Washington, DC, 1995.
3. For example, see: Kato, T.; Saeki, K.; Kawazoe, Y.; Hakura, A. *Mutat. Res.* **1999**, *439*, 149–157 and references cited therein.
4. For example, see: (a) Mongin, F.; Mojovic, L.; Guillamet, B.; Trécourt, F.; Quéguiner, G. *J. Org. Chem.* **2002**, *67*, 8991–8994. (b) Arzel, E.; Rocca, P.; Marsais, F.; Godard, A.; Queguiner, G. *Tetrahedron* **1999**, *55*, 12149–12156.
5. (a) Smart, B. E. In *Organofluorine Chemistry, Principles and Commercial Applications*; Banks, R. E.; Smart, B. E.; Tatlow, J. C., Eds.; Plenum: New York, 1994, Chapter 3. (b) Lee, V. J. In *Comprehensive Organic Synthesis*; Trost, B. M., Ed.; Pergamon: Oxford, 1991, Vol. 4, Chapter 1.2.
6. (a) Strekowski, L.; Kiselyov, A. S.; Hojjat, M. *J. Org. Chem.* **1994**, *59*, 5886–5890 and references cited therein. (b) Kiselyov, A. S.; Strekowski, L. *Tetrahedron Lett.* **1994**, *35*, 7597–7600. (c) Burdon, J.; Coe, P. L.; Haslock, I. B.; Powell, R. L. *J. Fluorine Chem.* **1997**, *85*, 151–153.
7. Kubota, T; Yamamoto, K.; Tanaka, T. *Chem. Lett.* **1983**, 167–168.
8. Feiring, A. E. *J. Org. Chem.* **1980**, *45*, 1962–1964.

9. (a) Ichikawa, J.; Kobayashi, M.; Noda, Y.; Yokota, N.; Amano, K.; Minami, T. *J. Org. Chem.* **1996**, *61*, 2763–2769. (b) Volle, J.-N.; Schlosser, M. *Eur. J. Org. Chem.* **2000**, 823–828.
10. (a) Burger, K.; Helmreich, B. *J. Chem. Soc., Chem. Commun.* **1992**, 348–349. (b) Coe, P. L.; Burdon, J.; Haslock, I. B. *J. Fluorine Chem.* **2000**, *102*, 43–50. See also for addition reaction: (c) Yamazaki, T.; Hiraoka, S.; Sakamoto, J.; Kitazume, T. *J. Phys. Chem. A*, **1999**, *103*, 6820–6824. (d) Wang, Z.-G.; Hammond, G. B. *J. Org. Chem.* **2000**, *65*, 6547–6552.
11. Ichikawa, J. *J. Fluorine Chem.* **2000**, *105*, 257–263.
12. Ichikawa, J.; Wada, Y.; Fujiwara, M.; Sakoda, K. *Synthesis* **2002**, 1917–1936.
13. Wada, Y.; Ichikawa, J.; Katsume, T.; Nohiro, T.; Okauchi, T.; Minami, T. *Bull. Chem. Soc. Jpn.* **2001**, *74*, 971–977.
14. For the synthesis of fluorochromenes, see: (a) Hanamoto, T.; Shindo, K.; Matsuoka, M.; Kiguchi, Y.; Kondo, M. *J. Chem. Soc., Perkin Trans. 1* **2000**, 103–107. (b) Camps, F.; Coll, J.; Messeguer, A.; Pericás, M. A. *J. Heterocycl. Chem.* **1980**, *17*, 1377–1379.
15. Ichikawa, J.; Sakoda, K.; Moriyama, H.; Wada, Y. Manuscript in preparation. Starting from 2′-(3,3-difluoroallyl)anilides, 2-fluorinated quinolines were obtained.
16. For the synthesis of fluoroisoquinolines, see: Bellas, M.; Suschitzky, H. *J. Chem. Soc.* **1964**, 4561–4564.
17. Wada, Y.; Mori, T.; Ichikawa, J. *Chem. Lett.* **2003**, 1000–1001.
18. For the synthesis of fluoroquinolines, see: (a) Chambers, R. D.; Parsons, M.; Sandford, G.; Skinner, C. J.; Atherton, M. J.; Moilliet, J. S. *J. Chem. Soc., Perkin Trans. 1* **1999**, 803–810 and references cited therein. (b) Shi, G.-q.; Takagishi, S.; Schlosser, M. *Tetrahedron* **1994**, *50*, 1129–1134. (c) Ref. 6.
19. Mori, T.; Ichikawa, J. *Chem. Lett.* **2004**, 590–591.
20. Ichikawa, J.; Wada, Y.; Kuroki, H.; Mihara, J.; Mori, T. Manuscript in preparation.
21. For the synthesis of 3- or 4-fluorocinnolines, see: Chambers, R. D.; MacBride, J. A. H.; Musgrave, W. K. R. *J. Chem. Soc., Chem. Commun.* **1970**, 739–740.
22. Ichikawa, J.; Wada, Y.; Miyazaki, H.; Mori, T.; Kuroki, H. *Org. Lett.* **2003**, *5*, 1455–1458.
23. Ichikawa, J.; Mori, T.; Miyazaki, H.; Wada, Y. *Synlett*, **2004**, 1219–1222.
24. (a) Baldwin, J. E. *J. Chem. Soc., Chem. Commun.* **1976**, 734–736. (b) Baldwin, J. E.; Cutting, J.; Dupont, W.; Kruse, L.; Silberman, L.; Thomas, R. C. *J. Chem. Soc., Chem. Commun.* **1976**, 736–738. (c) Baldwin, J. E.; Thomas, R. C.; Kruse, L. I.; Silberman, L. *J. Org. Chem.* **1977**, *42*, 3846–3852.

25. Ichikawa, J.; Wada, Y.; Okauchi, T.; Minami, T. *Chem. Commun.* **1997**, 1537–1538.
26. For the synthesis of 2- or 3-fluorobenzo[*b*]furans, see: (a) Martín-Santamaría, S.; Carroll, M. A.; Carroll, C. M.; Carter, C. D.; Pike, V. W.; Rzepa, H. S.; Widdowson, D. A. *Chem. Commun.* **2000**, 649–650. (b) Barton, D. H. R.; Hesse, R. H.; Jackman, G. P.; Pechet, M. M. *J. Chem. Soc., Perkin Trans. 1*, **1977**, 2604–2608. (c) Bailey, J.; Plevey, R. G.; Tatlow, J. C. *Tetrahedron Lett.* **1975**, 869–870.
27. For the synthesis of 2- or 3-fluorobenzo[*b*]thiophenes, see: (a) Shirley, I. M. *J. Fluorine Chem.* **1994**, *66*, 51–57. (b) Nussbaumer, P.; Petranyi, G.; Stütz, A. *J. Med. Chem.* **1991**, *34*, 65–73. (c) Bensoam, J.; Mathey, F. *Tetrahedron Lett.* **1977**, 2797–2800.
28. For the synthesis of 2- or 3-fluoroindoles, see: Torres, J. C.; Garden, S. J.; Pinto, A. C.; da Silva, F. S. Q.; Boechat, N. *Tetrahedron* **1999**, *55*, 1881–1892 and references cited therein.
29. Ichikawa, J.; Fujiwara, M.; Wada, Y.; Okauchi, T.; Minami, T. *Chem. Commun.* **2000**, 1887–1888.
30. For the synthesis of fluorofurans, see: Forrest, A. K.; O'Hanlon, P. J. *Tetrahedron Lett.* **1995**, *36*, 2117–2118 and references cited therein.
31. For the synthesis of fluorothiophenes, see: (a) Chambers, R. J.; Marfat, A. *Synth. Commun.* **2000**, *30*, 3629–3632. (b) Kobarfard, F; Kauffman, J. M. *J. Heterocyclic Chem.* **1999**, *36*, 1247–1251. (c) Andrés, D. F.; Laurent, E. G.; Marquet, B. S. *Tetrahedron Lett.* **1997**, *38*, 1049–1052 and references cited therein. (d) Burger, K.; Helmreich, B. *Heterocycles* **1994**, *39*, 819–832 and references cited therein.
32. For the synthesis of fluoropyrrolines, see: Novikov, M. S.; Khlebnikov, A. F.; Shevchenko, M. V. *J. Fluorine Chem.* **2003**, *123*, 177–181.
33. For the synthesis of fluoropyrroles, see: (a) Wang, Y.; Zhu, S. *Org. Lett.* **2003**, *5*, 745–748 and references cited therein. (b) Tajima, T.; Ishii, H.; Fuchigami, T. *Tetrahedron Lett.* **2001**, *42*, 4857–4860. (c) Woller, E. K.; Smirnov, V. V.; DiMagno, S. G. *J. Org. Chem.* **1998**, *63*, 5706–5707. (d) Barnes, K.; Hu, Y.; Hunt, D. *Synth. Commun.* **1994**, *24*, 1749–1755.
34. For the synthesis of ring-fluorinated carbocycles, see: (a) Ichikawa, J.; Sakoda, K.; Wada, Y. *Chem. Lett.* **2002**, 282–283. (b) Ichikawa, J.; Miyazaki, H.; Sakoda, K.; Wada, Y. *J. Fluorine Chem.* **2004**, *125*, 585–593.

Chapter 15

Selective Anodic Fluorination of Heterocyclic Compounds

Toshio Fuchigami and Toshiki Tajima

Department of Electronic Chemistry, Tokyo Institute of Technology, Nagatsuta, Midori-ku, Yokohama 226–8502, Japan

Recent developments in regio- and diastereoselective anodic fluorination of heterocyclic compounds including α-amino acid derivatives are described. Marked effects of supporting fluoride salts and solvents on the fluorination are discussed. Synthetic applications of the anodically fluorinated heterocycles are also illustrated.

Introduction

Many heterocyclic compounds have unique biological activities. On the other hand, introduction of fluorine atom(s) into organic substances sometimes markedly enhances or dramatically changes their biological activities. Hence, partially fluorinated heterocycles are the current focus of much biological interest. However, limited examples of selective anodic fluorination reactions of heterocycles have been reported prior to early 1990s (*1-5*). These reactions are limited to only nitrogen- and oxygen-containing heterocycles, and the yields are generally quite low. With these facts in mind, the authors have developed conditions for highly selective anodic fluorination reactions of various heterocyclic compounds (*6-8*). In this chapter, marked effects of supporting fluoride salts and solvents are discussed and successful examples of regio- and diastereoselective fluorination are illustrated.

Effect of Supporting Fluoride Salts

Supprting fluoride salts are one of the main factors affecting the selective anodic fluorination. Et_3N-3HF has been mainly used as a supporting salt and fluoride ion source for partial anodic fluorination. However, this salt is not always effective because of its use often causes anode passivation, which suppresses electrolytic current. Generally, the oxidation potentials of fluoride salts increase with an increase of the content of HF in the salts. Thus, Et_3N-nHF(n=4, 5) and Et_4NF-nHF(n=3, 4) are suitable for the anodic fluorination of heterocycles having high oxidation potentials.

Highly regioselective anodic monofluorination of 2-aryl-4-thiazolidinones can be performed by using pulse electrolysis in Et_3N-3HF/MeCN (Scheme 1) (*9*). However, this electrolytic system is not suitable for anodic monofluorination of 2-substituted 1,3-dithiolan-4-ones and 1,3-oxathiolan-4-ones owing to severe anode passivation. In contrast, Et_4NF-4HF provides monofluorinated products selectively (Scheme 1) (*10, 11*). In these cases, benzylic fluorination dose not take place at all, although anodic benzylic substitution easily takes place in general. The high regioselectivity can be explained in terms of facilitation of deprotonation of fluorosulfonium ion intermediate of the substrates by the electron-withdrawing carbonyl group (i.e., kinetic acidity control), as shown in Scheme 1.

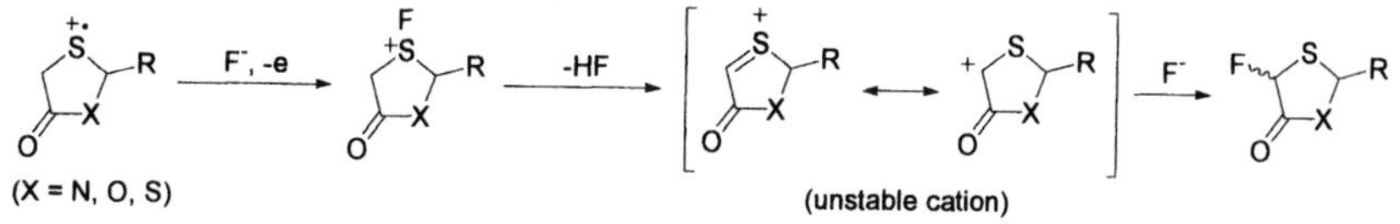

F^-, -e   -HF F^-

(X = N, O, S)   (unstable cation)

X = N, R = Ph (84%; *trans/cis* : 57/43) (in Et_3N-3HF/MeCN)
X = O, R = Et (86%; *trans/cis* : 53/47) (in Et_4NF-4HF/MeCN)
X = S, R = *n*-Pr (78%; *trans/cis* : 52/48) (in Et_4NF-4HF/MeCN)

Scheme 1

Fluorinated product selectivity is also affected greatly by supporting fluoride salts. As shown in Scheme 2, anodic fluorination of flavone by using Et_3N-3HF affords mainly 3-monofluoroflavone, while the use of Et_4NF-4HF results in formation of 2,3-difluoroflavone (*12*). Since Et_3N-3HF contains a considerable amount of free Et_3N, the difluorinated product once formed is converted to the monofluoroflavone by the dehydrofluorination with free Et_3N.

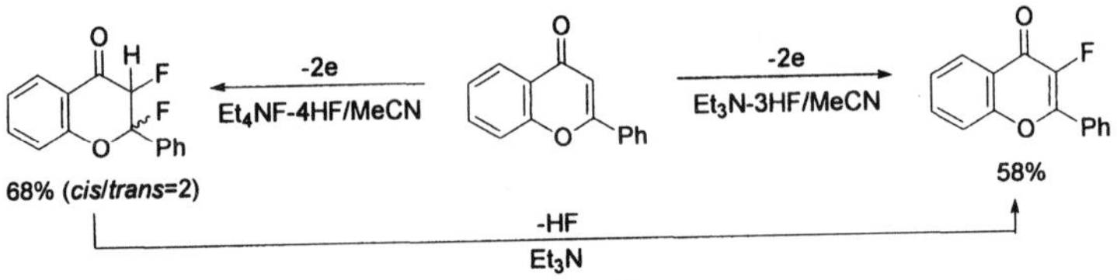

Scheme 2

Solvent effect on Anodic Fluorination

Acetonitrile (MeCN) has been commonly employed as an electrolytic solvent for anodic partial fluorination of organic compounds since MeCN is rather difficult to be oxidized. However, the use of MeCN very often causes anode passivation and/or acetamidation as a side reaction.

As shown in Scheme 3, solvent effects on electrolytic partial fluorination of α-(2-benzoxazolylthio)-, α-(2-benzothiazolylthio)-, and α-(2-pyrimidylthio)acetates were investigated systematically using acetonitrile (MeCN) and 1,2-dimethoxyethane (DME) containing Et_4NF-nHF (n=3, 4) (*13-15*). MeCN was not suitable for the fluorination of these heterocyclic sulfides since formation of nonconductive polymers on the anode took place to result in low yields or no formation of fluorinated products. In contrast, the use of DME did not cause such polymer film formation and provided much better or excellent fluorinated product yields. The superiority of DME can be explained in terms of its ability to solvate the cationic part of the fluoride salt in which a reactive fluoride ion is left for easy attack on the anodically generated cationic intermediate. It is noted that addition of DME into MeCN electrolytic solution markedly increased the product yield and current efficiency since the polymer film formation on the anode and overoxidation of the fluorinated products were effectively suppressed.

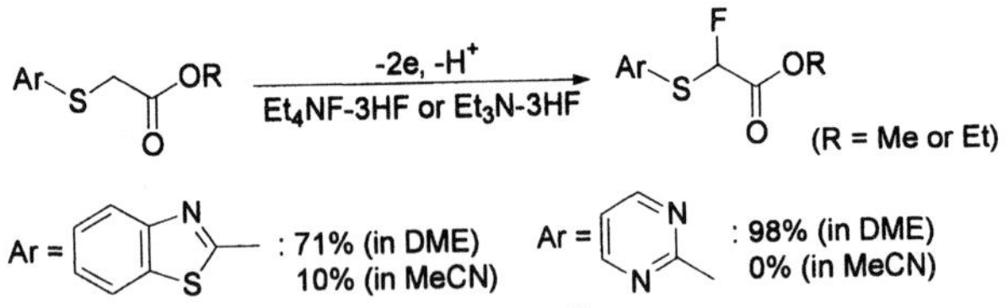

Scheme 3

A similar solvent effect was observed in the anodic fluorination of other heterocyclic compounds such as 4-quinolinyl and 2-pyrimidyl sulfides (*16, 17*). Thus, the active methylenethio group attached to various heterocycles are selectively fluorinated to give the corresponding α-fluorinated products (*18, 19*).

Heterocyclic propargyl sulfides are also regioselectively fluorinated in DME to provide the α-fluorinated sulfides (*20*).

The fluorinated product selectivity is also markedly changed by electrolytic solvents. As shown in Scheme 4, fluorodesulfurization proceeds selectively in CH_2Cl_2 while α-fluorination without desulfurization takes place preferentially in DME (*21, 22*). This interesting phenomenon can be explained as follows. The fluorination proceeds *via* a radical cation intermediate **A** as shown in Scheme 4. CH_2Cl_2 has a poor ability to solvate carbocations, therefore, **A** seems to be unstable in CH_2Cl_2. Consequently it is reasonable to assume that desulfurization mainly takes place prior to α-fluorination of **A**. On the other hand, DME is known to strongly coordinate cations. Therefore, DME should stabilize the intermediate **A** and also enhances the fluoride ion nucleophilicity. Then, the deprotonation of **A** with fluoride ions takes place prior to desulfurization followed by further oxidation to generate cation **B** and this cation reacts with a fluoride ion to provide the α-fluorinated product.

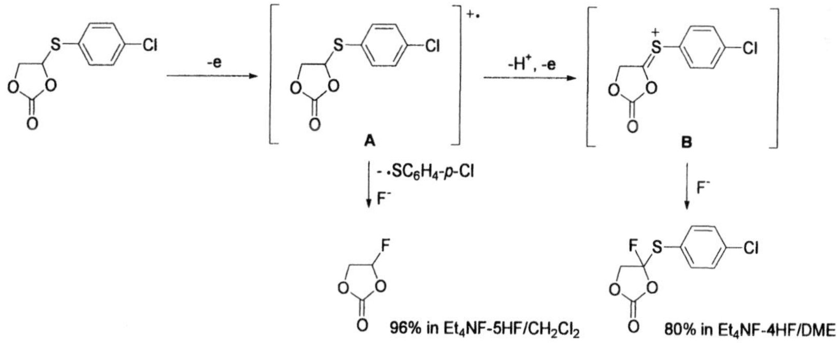

Scheme 4

Anodic Fluorination of Five-Membered Heteroaromatic Compounds

As mentioned above, anodic fluorination of heterocyclic sulfides usually takes place at the α-position to the sulfur atom of the side chain. In contrast, anodic fluorination of five-membered heteroaromatic sulfides was found to occur at the heteroaromatic rings. Anodic fluorination of 2-thiazolyl methyl sulfide, 2-thiazolyl propargyl sulfide, and 2-thiazolyl acetonyl sulfide provides the corresponding 5-fluorothiazole and 2,5,5-trifluorothiazoline derivatives (Scheme 5) (*23*). The latter products are readily hydrolyzed to give isolable 5,5-difluoro-2-hydroxythiazoline derivatives.

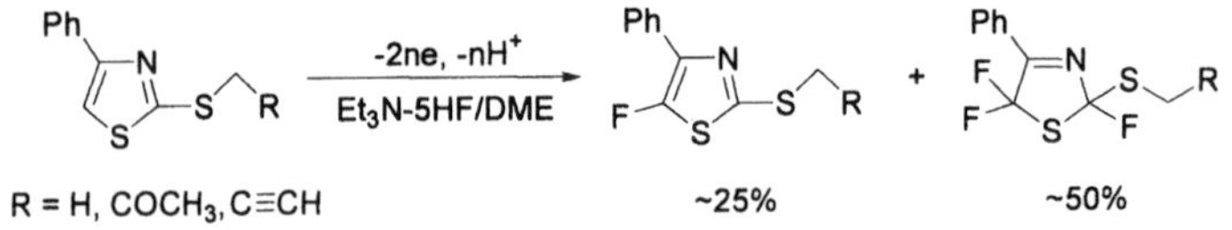

Ph
N
S
S R
-2ne, -nH+
Et3N-5HF/DME
R = H, COCH3, C≡CH
~25%
~50%

Scheme 5

On the other hand, anodic fluorination of 2-thiazolyl cyanomethyl sulfide affords 5-fluorothiazole and α-fluorinated thiazole derivatives (Scheme 6). Thus, the product selectivity is greatly changed by the electron-withdrawing ability of substituents at the side-chain of the thiazole ring. This is the first report of successful anodic fluorination of a thiazole ring.

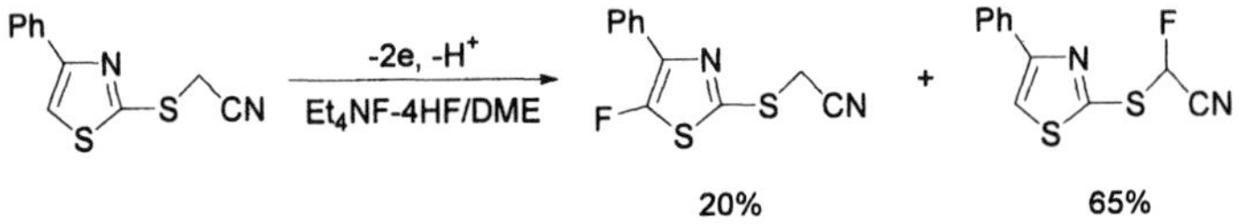

-2e, -H+
Et4NF-4HF/DME
20%
65%

Scheme 6

In contrast, anodic fluorination of oxazolyl sulfides provides 4,5-difluoro-2-oxazoline derivatives selectively and the corresponding 5-fluorooxazoles are not formed at all (Scheme 7) (*24*). No formation of 5-fluorooxazoles seems to be attributable to the less aromatic stability of an oxazolyl ring compared with that of a thiazolyl ring.

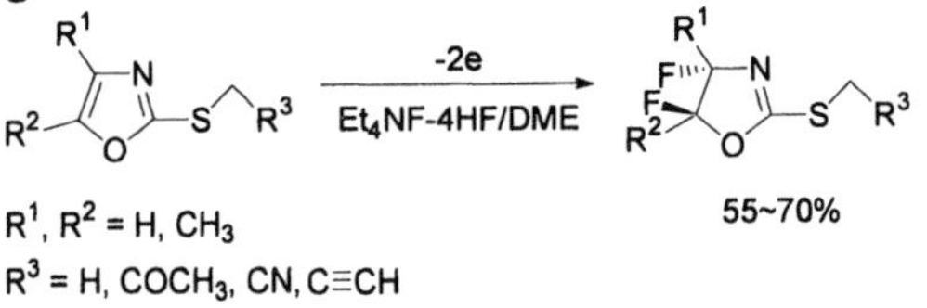

-2e
Et4NF-4HF/DME
R^1, R^2 = H, CH_3
R^3 = H, $COCH_3$, CN, C≡CH
55~70%

Scheme 7

On the other hand, anodic fluorination of α-(2-thiazolyl)acetonitrile provides mainly non-aromatized fluorinated products along with a small amount of the corresponding 5-fluoro- and α-fluorothiazole derivatives (Scheme 8) (*25*). Although the main products are tautomers of 5-fluorinated thiazole derivatives, their aromatization is difficult. They were found to be thermodynamically more stable than the fluorinated aromatic compounds based on their calculations using AM1.

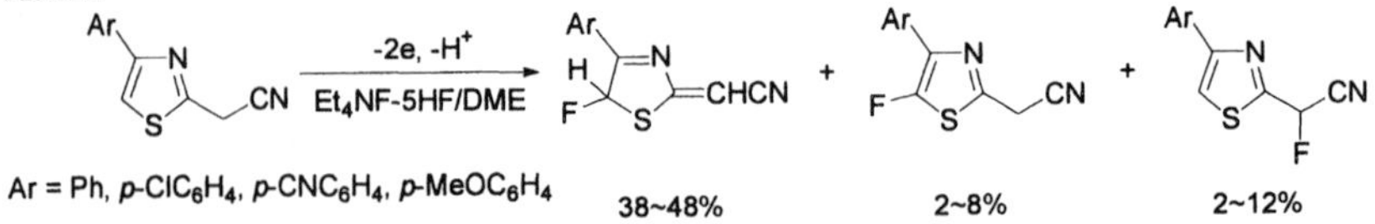

-2e, -H+
Et4NF-5HF/DME
Ar = Ph, p-ClC_6H_4, p-CNC_6H_4, p-$MeOC_6H_4$
38~48%
2~8%
2~12%

Scheme 8

Anodic Ring Fluorination of Pyrrole Derivative

Fluorinated pyrroles and thiophenes seem to be useful for not only precursors to biological compounds but also functional materials like conducting polymers. However, their direct fluorination is not straightforward. For example, the chemical direct fluorination results in extremely low yields (less than 6%) along with an unsatisfactory level of selectivity (*26, 27*). Anodic fluorination of pyrrole is also unsuccessful because its polymerization takes place exclusively. Previously, it was found that electron-withdrawing groups markedly enhanced anodic α-fluorination of various sulfides (*28-30*).

Therefore, we investigated anodic fluorination of 2-cyano-1-methylpyrrole **1** using various fluoride salts. As shown in Table 1, anodic fluorination of **1** took place efficiently to provide four fluorinated products **2**, **3**, **4**, and **5** depending on the supporting fluoride salts (*31, 32*). It is noted that 2-cyano-5-fluoro-1-methylpyrrole **2** was formed in a considerable amount by using Et_3N-2HF regardless of the solvents. When Et_3N-3HF was used in MeCN, trifluorinated product **3** was selectively and exclusively formed. On the other hand, the use of Et_3N-5HF and Et_4NF-4HF in MeCN provided difluorinated product **4** preferentially. In all cases, the *N*-methyl group was not fluorinated at all.

Table 1. Anodic Fluorination of 2-Cyano-1-methylpyrrole

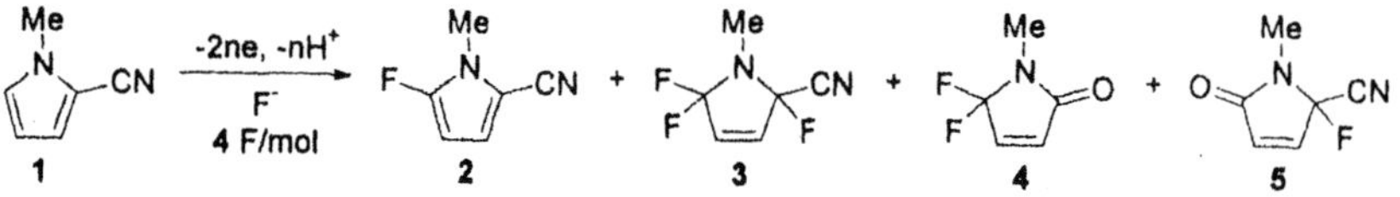

Entry	Solvent	Supporting Electrolyte (1 M)	Yield (%)[a]			
			2	**3**	**4**	**5**
1	MeCN	Et_3N-2HF	20	32	trace	trace
2	DME	Et_3N-2HF	20	12	trace	0
3	CH_2Cl_2	Et_3N-2HF	19	51	trace	trace
4	MeCN	Et_3N-3HF	5	65	3	2
5	MeCN	Et_3N-5HF	0	5	54	12
6	MeCN	Et_4NF-4HF	0	6	28	21

[a] Determined by ^{19}F NMR Spectroscopy.

The oxygen source of **4** seemed to be H_2O which is contained in the supporting fluoride salts. Indeed, the trifluoro product **3** was readily hydrolyzed to provide **4** efficiently, therefore, **3** was found to be a precursor to **4** (Scheme 9).

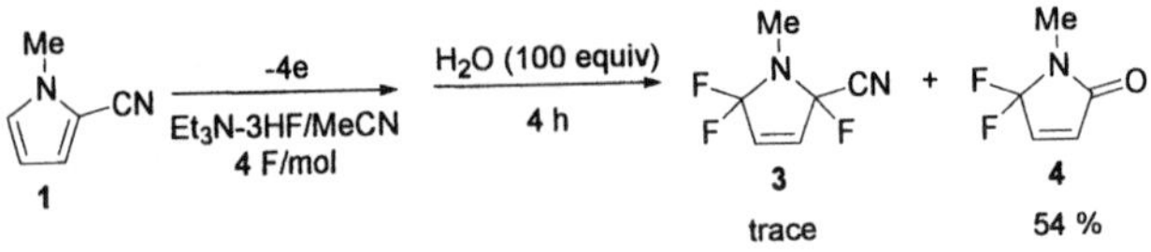

Scheme 9

The oxidation potentials of **1** and **2** were measured by cyclic voltammetry in 0.1 M $Bu_4N\text{-}ClO_4$/anhydrous acetonitrile. The first peak oxidation potential of **2** (E_p^{ox}=1.72 V *vs.* SCE) was found to be slightly less positive compared with that of **1** (E_p^{ox}=1.77 V *vs.* SCE). It was also comfirmed that anodic fluorination of **2** in $Et_3N\text{-}3HF$/MeCN provided **3** selectively in 66% yield.

From these results, the reaction seems to proceed as shown in Scheme 10. Since **2** is more easily oxidized than **1**, the trifluorinated product **3** is preferentially formed by the further electrochemical oxidation of **2** once formed during the electrolysis. However, since $Et_3N{\bullet}2HF$ is easily oxidized, the further oxidation of **2** seems to be suppressed by simultaneous oxidation of $Et_3N{\bullet}2HF$. In support of this hypothesis, the use of $Et_3N{\bullet}5HF$ and $Et_4NF{\bullet}4HF$, which are stable for oxidation, did not provide **2**. The trifluorinated product **3** is unstable; however, it is easily converted to the difluorinated product **4** efficiently by the hydrolysis of **3**. In sharp contrast to the case of **1**, the anodic fluorination of 1-methylpyrrole gave only a polymerized product and no fluorinated product was formed. Therefore, electron-withdrawing cyano group promotes the reactions of radical cations **A** with fluoride ions. In other words, the electron-withdrawing cyano group promotes anodic fluorination of **1**.

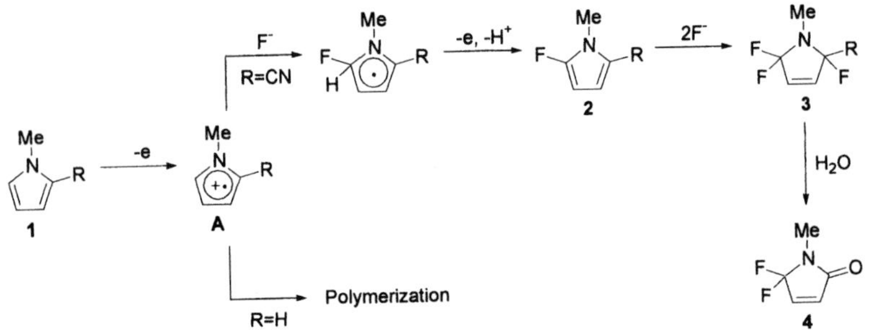

Scheme 10

The difluorinated product **4** has both a biologically interesting *gem*-difluoromethylene unit and an activated olefine in the heterocyclic ring. Therefore, we attempted the Diels-Alder reaction of **4** as the dienophile with various dienes.

As shown in Table 2, the Diels-Alder reaction of **4** with open-chain and cyclic dienes provided the cycloaddition products in excellent yields (*33*). Thus, **4** was found to be a highly useful building block containing a *gem*-difluoromethylene unit.

Table 2. Diels-Alder Reaction of 4 with Open-Chain Dienes

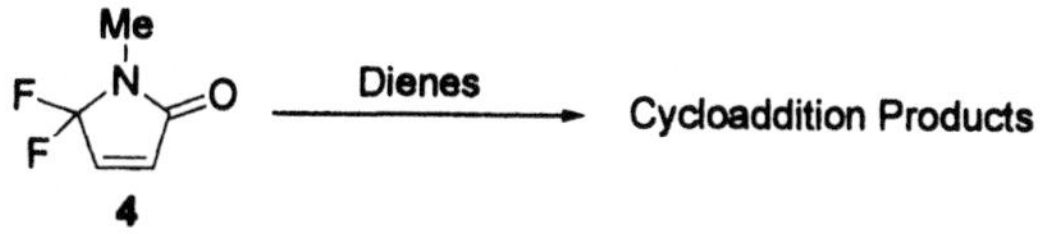

Diene	Solvent	Temparature (°C)	Time (d)	Cycloaddition products	Yield (%)[a]
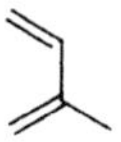 (5 equiv)	toluene	reflux	2.2	 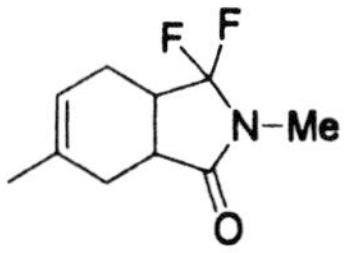	93 (88)[b] [3:2][c]
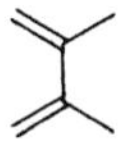 (5 equiv)	toluene	reflux	2		quant. (96)
 (100 equiv)	–	60	3	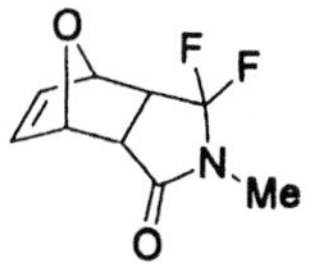	quant. (*endo*/*exo*=20/13)
 (5 equiv)	toluene	reflux	0.5	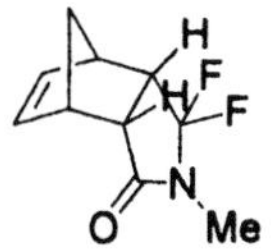	quant. (92) (*endo* only)
 (5 equiv)	toluene	reflux	1.5	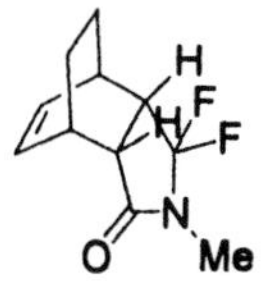	quant. (96) (*endo* only)

[a] Determined by ^{19}F NMR Spectroscopy.

[b] The figures in parentheses are isolated yields.

[c] Regioisomeric Ratio.

Diastereoselective Anodic Fluorination of Heterocycles

Asymmetric fluorination of organic compounds is of much importance, especially for medicinal and agrochemical applications. However, asymmetric anodic fluorination is very difficult in general, due to the small size of a fluoride ion. Therefore, only a few studies have been reported on asymmetric anodic fluorination so far (*34, 35*).

In fact, even intramolecular 1,3-asymmetric induction as shown in Scheme 1 is low. However, anodic fluorination of *N*-protected thiazolines derived from L-cysteine in an undivided cell proceeds with moderate to high diastereoselectivity as shown in Scheme 11(*36*).

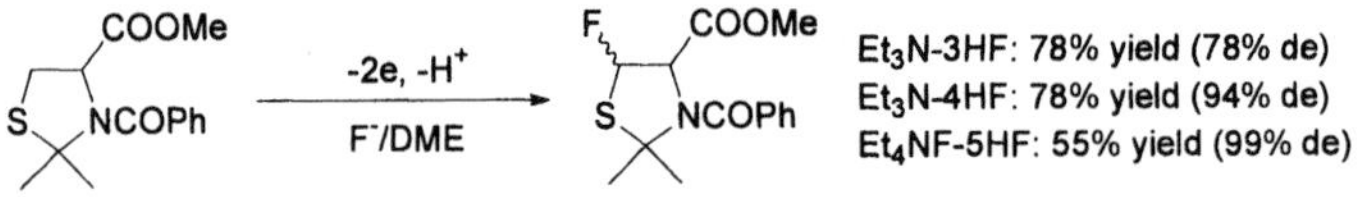

COOMe, S, NCOPh → ($-2e, -H^+$; F^-/DME) → F, COOMe, S, NCOPh

Et_3N-3HF: 78% yield (78% de)
Et_3N-4HF: 78% yield (94% de)
Et_4NF-5HF: 55% yield (99% de)

Scheme 11

Interestingly, the diastereoselectivity increases with an increase of HF content in the supporting fluoride salts and almost 100% de is obtained by using Et_4NF-5HF. Notably, the diastereoselectivity is also greatly affected by the bulkiness of the substituent on the nitrogen atom, and *N*-benzoylthiazoline provides much higher diastereoselectivity compared with *N*-formyl derivative as shown in Scheme 12.

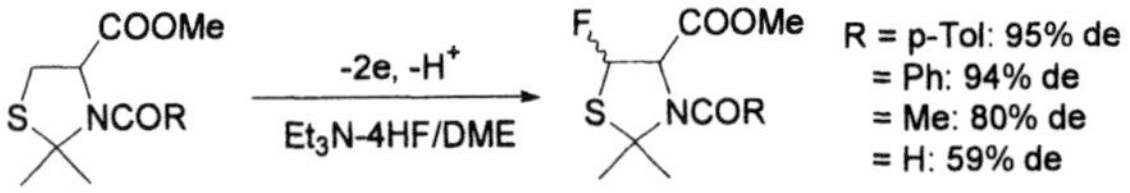

COOMe, S, NCOR → ($-2e, -H^+$; Et_3N-4HF/DME) → F, COOMe, S, NCOR

R = p-Tol: 95% de
= Ph: 94% de
= Me: 80% de
= H: 59% de

Scheme 12

Diastereoselective anodic fluorination of 1,3-oxazolidine derived from L-threonine is also successful as shown in Scheme 13 (*37*).

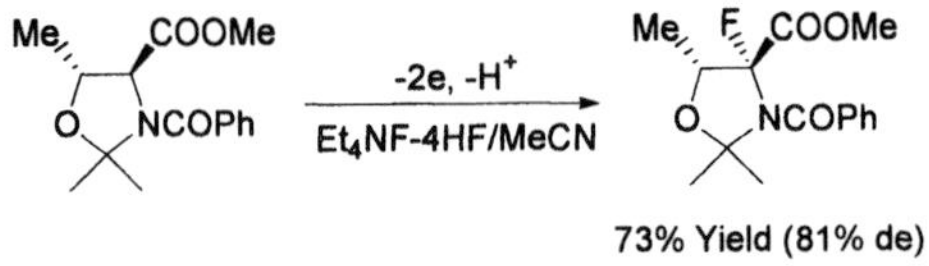

Me, COOMe, O, NCOPh → ($-2e, -H^+$; Et_4NF-4HF/MeCN) → Me, F, COOMe, O, NCOPh

73% Yield (81% de)

Scheme 13

Diastereoselective anodic fluorination of sulfides having various oxygen-containing heterocyclic substituents at the β-position was comparatively studied.

Among the oxygen-containing heterocyclic substituents, the 2-spirocyclohexyl-1,3-dioxolan-4-yl group gave the best diastereoselectivity (80% de) as shown in Scheme 14 (*38*). The diastereoselectivity was also affected by supporting fluoride salts and solvents. Chemical fluorination of **6** using selectfluor resulted in much lower diastereoselectivity (32% de) and extremely poor yield (5%).

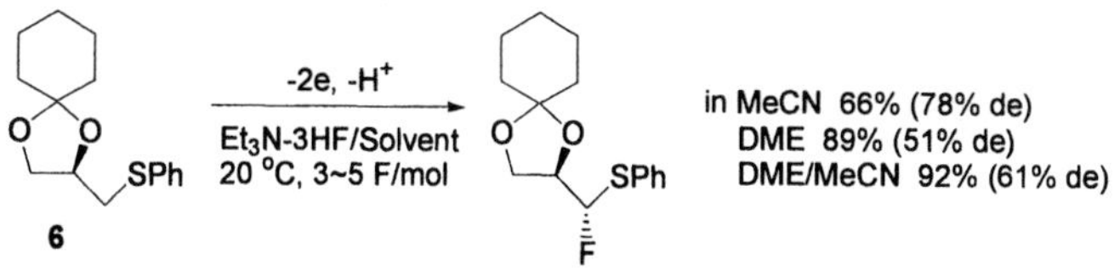

-2e, $-H^+$; Et_3N-3HF/Solvent, 20 °C, 3~5 F/mol

6

in MeCN 66% (78% de)
DME 89% (51% de)
DME/MeCN 92% (61% de)

Scheme 14

The major diastereoisomer of the fluorinated products of **6** was readily converted into the corresponding fluorinated diol in good yield by acidic hydrolysis as shown in Scheme 15.

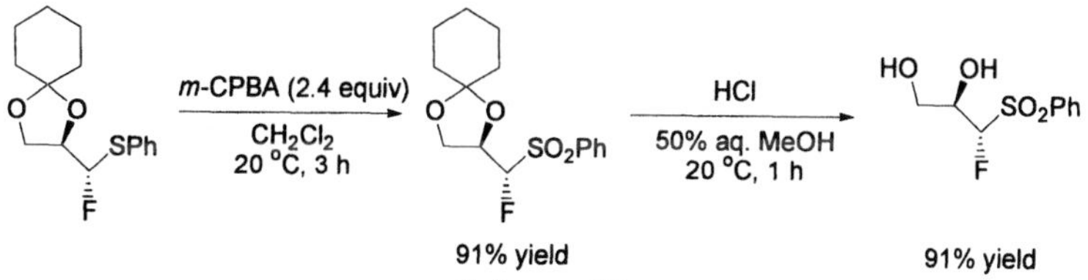

m-CPBA (2.4 equiv), CH_2Cl_2, 20 °C, 3 h — 91% yield

HCl, 50% aq. MeOH, 20 °C, 1 h — 91% yield

Scheme 15

Anodic fluorination of 4-(*p*-chlorophenylthio)methyl-1,3-dioxolan-2-one was successfully carried out in a mixture of MeCN and propylene carbonate (PC) (1:1) containing Et_3N-3HF using an undivided cell to provide the corresponding α-monofluorinated product in quantitative yield and with almost 100% current efficiency as shown in Scheme 16 (*39*). The diastereoisomeric mixture of the fluorinated products was readily converted into the corresponding fluorinated allyl alcohol and oxirane in good to moderate yields by treatment with an alkaline solution (Scheme 17).

-2e, $-H^+$; Et_3N-3HF, PC-MeCN (1:1)

quant. (16% de)

Scheme 16

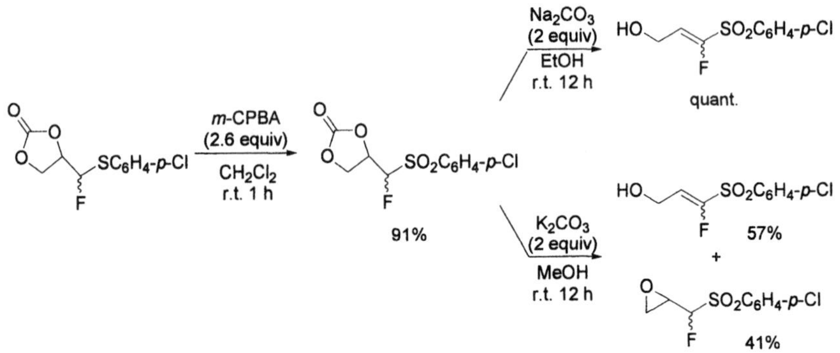

Scheme 17

Conclusion

In this chapter, we have described our recent results on the regio- and diastereoselective anodic fluorination of various heterocycles as well as the synthetic applications of the anodically fluorinated heterocycles. It is our hope that the information in this chapter will be helpful in the preparation of related other useful heterocyclic compounds.

References

1. Gambaretto, G. P.; Napoli, M.; Franccaro, C.; Conte, L. *J. Fluorine Chem.* **1982**, *19*, 427-436.
2. Ballinger, J. R.; Teare, F. W. *Electrochim. Acta* **1985**, *30*, 1075-1077.
3. Makino, K.; Yoshioka, H. *J. Fluorine Chem.* **1988**, *39*, 435-440.
4. Meurs, J. H. H.; Eilenberg, W. *Tetrahedron* **1991**, *47*, 705-714.
5. Sono, M.; Toyoda, N.; Shimizu, Y.; Tori, M. *Tetrahedron Lett.* **1994**, *35*, 9237-9238.
6. Fuchigami, T.; Higashiya, S.; Hou, Y.; Dawood, K. M. *Rev. Heteroatom Chem.* **1999**, *19*, 67-78.
7. Fuchigami, T. In *Advances in Electron-Transfer Chemistry*; Mariano, P. S., Ed.; JAI Press: Conneticut, 1999; vol. 6, p 41-130.
8. Fuchigami, T. In *Organic Electrochemistry,* 4th ed.; Lund, H.; Hammerich, O., Eds.; Marcel Dekker: New York, 2001; Chapter 25.
9. Fuchigami, T.; Narizuka, S.; Konno, A. *J. Org. Chem.* **1992**, *57*, 3755-3757.
10. Higashiya, S.; Narizuka, S.; Konno, A.; Maeda, T.; Momota, K. Fuchigami, T. *J. Org. Chem.* **1999**, *64*, 133-137.

11. Fuchigami, T.; Narizuka, S.; Konno, A.; Momota, K. *Electrochim. Acta* **1998**, *43*, 1985-1989.
12. Hou, Y.; Higashiya, S.; Fuchigami, T. *J. Org. Chem.* **1999**, *64*, 3346-3349.
13. Hou, Y.; Higashiya, S.; Fuchigami, T. *J. Org. Chem.* **1997**, *62*, 9173-9176.
14. Hou, Y.; Fuchigami, T. *J. Electrochem. Soc.* **2000,** *147*, 4567-4572.
15. Dawood, K. M.; Higashiya, S.; Hou, Y.; Fuchigami, T. *J. Fluorine Chem.* **1999**, *93*, 159-164.
16. Dawood, K. M.; Fuchigami, T. *J. Org. Chem.* **1999**, *64*, 138-143.
17. Shaaban, M. R.; Ishii, H.; Fuchigami, T. *J. Org. Chem.* **2000**, *65*, 8685-8689.
18. Shaaban, M. R.; Ishii, H.; Fuchigami, T. *J. Org. Chem.* **2001**, *66*, 5633-5636.
19. Dawood, K. M.; Higashiya, S.; Hou, Y.; Fuchigami, T. *J. Org. Chem.* **1999**, *64*, 7935-7939.
20. Riyadh, S. M.; Ishii, H.; Fuchigami, T. *Tetrahedron* **2001**, *57*, 8817-8821.
21. Ishii, H.; Yamada, N.; Fuchigami, T. *Chem. Commun.* **2000,** 1617-1618.
22. Ishii, H.; Yamada, N.; Fuchigami, T. *Tetrahedron* **2001**, *57*, 9067-9072.
23. Riyadh, S. M.; Fuchigami, T. *J. Org. Chem.* **2002**, *67*, 9379-9383.
24. Riyadh, S. M.; Ishii, H.; Fuchigami, T. *Tetrahedron* **2002**, *58*, 9273-9278.
25. Riyadh, S. M.; Fuchigami, T. *Heterocycles* **2003,** *60*, 15-22.
26. Gozzo, F. C.; Ifa, D. R.; Eberlin, M. N. *J. Org. Chem.* **2000**, *65*, 3920-3925.
27. Cerichelli, G.; Crestoni, M. E.; Fornarini, S. *Gazz. Chim. Ital.* **1990**, *120*, 749-755.
28. Konno, A.; Nakagawa, K.; Fuchigami, T. *J. Chem. Soc., Chem. Commun.* **1991,** 1027-1029.
29. Fuchigami, T.; Konno, A.; Nakagawa, K.; Shimojo, M. *J. Org. Chem.* **1994,** *59*, 5937-5941.
30. Fuchigami, T.; Shimojo, M.; Konno, A. *J. Org. Chem.* **1995**, *60*, 3459-3464.
31. Tajima, T.; Ishii, H.; Fuchigami, T. *Tetrahedron Lett.* **2001**, *42*, 4857-4860.
32. Tajima, T.; Ishii, H.; Fuchigami, T. *Electrochem. Commun.* **2001**, *3*, 467-471.
33. Tajima, T.; Fuchigami, T. *Synthesis* **2002**, *17*, 2597-2600.
34. Kabore, L.; Chebli, S.; Faure, R.; Laurent, E.; Marquet, B. *Tetrahedron Lett.* **1990**, *31*, 3137-3140.
35. Narizuka, S.; Koshiyama, H.; Konno, A.; Fuchigami, T. *J. Fluorine Chem.* **1995**, *73*, 121-127.
36. Baba, D.; Ishii, H.; Higashiya, S.; Fujisawa, K.; Fuchigami, T. *J. Org. Chem.* **2001**, *66*, 7020-7024.
37. Baba, D.; Fuchigami, T. *Tetrahedron Lett.* **2002**, *43*, 4805-4808.
38. Suzuki, K.; Fuchigami, T. *J. Org. Chem.* **2004**, *69*, 1276-1282.
39. Suzuki, K.; Fuchigami, T. *Electrochem. Commun.* **2004**, *6*, 183-187.

Chapter 16

Polyfluorinated Binaphthol Ligands in Asymmetric Catalysis

Yu Chen and Andrei K. Yudin*

Davenport Research Laboratories, Department of Chemistry, University of Toronto, 80 St. George Street, Toronto, Ontario M5S 3H6, Canada

Introduction

Modern asymmetric synthesis relies on new and improved catalytic transformations. Understanding the balance of steric and electronic factors is a prerequisite to fine-tuning a catalyst to achieve optimal selectivity in a particular reaction. Among the chiral ligands developed so far, BINOL **1** and related molecules with axial chirality have found wide utility in asymmetric catalysis (*1*).

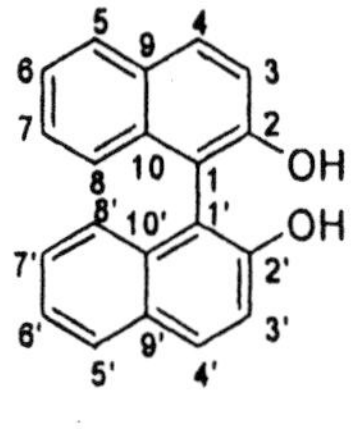

Figure 1. BINOL

BINOL was first synthesized in 1926 (*2*), however, its potential as a ligand for metal-mediated catalysis was left unrecognized until 1979 when Noyori demonstrated its utility in the reduction of aromatic ketones and aldehydes (*3*). Since Noyori's discovery, many modifications of the BINOL skeleton aimed at changing its steric and electronic properties have been reported (*4*). For example, partially hydrogenated BINOL was used in enantioselective alkylation of aldehydes (*4a*); conjugate addition of diethylzinc to cyclic enones (*4b*); and ring opening of epoxides (*4c*). Selective bromination at the 6 and 6' positions of the binaphthyl ring was shown to increase the enantioselectivity of the corresponding titanium catalysts in glyoxylate-ene reactions (*4d*). Bulky triarylsilyl groups at the 3 and 3'positions of the binaphthyl ring led to increased levels of enantiofacial discrimination of aldehydes in asymmetric Diels-Alder reactions (*4e*). 3,3'-Dinitrooctahydrobinaphthol was successfully applied in titanium-catalyzed asymmetric oxidation of methyl *p*-tolyl sulfide (*4f*).

Our explorations in the field of binaphthyl catalysis led us to synthesize partially fluorinated species. This interest was driven by the known stability of C-F bond towards oxidation (*5*); propensity of fluoroaromatics to engage in stabilizing stacking interactions with electron-rich aromatic rings (*6*), and well documented nucleophilic aromatic substitution chemistry of fluoroaromatic compounds (*7*). We envisaged that substitution of hydrogens by fluorines at the selected positions of the binaphthol ring should induce considerable change in the electronic character of the aromatic system. In this review we will discuss our recent efforts in the development of chiral polyfluorinated binaphthols as well as their application in asymmetric catalysis.

Preparation of F_8BINOL and F_4BINOL

The racemic form of F_8BINOL **2** was prepared according to Scheme 1. Tetrafluorobenzyne **3**, generated by treating chloropentafluorobenzene with *n*-butyllithium at -15°C, was reacted with 3-methoxythiophene (*8*) in a Diels-Alder reaction. Upon in situ extrusion of sulfur, 2-methoxy-5,6,7,8-tetrafluoronaphthalene **4** was obtained in 52% yield. Demethylation with BBr_3 gave 5,6,7,8-tetrafluoronaphthol **5**, which failed to undergo the $FeCl_3$-catalyzed oxidative coupling, commonly used for the preparation of BINOL from 2-naphthol. This lack of reactivity is believed to result from a relatively high oxidation potential of **5** (1.84V vs Ag/AgCl). The reductive Ullmann coupling was therefore chosen through the intermediacy of the 1-brominated precursor **6** to give the desired bis(methoxy) product **7** in 85% yield. Demethylation of **7** produced **2** in 90% yield, which was resolved through fractional crystallization of the bis[(-)-menthoxycarbonyl] derivatives **8** to afford the enantiomerically pure F_8BINOL **2** (Scheme 2) (*9*).Our initial attemps to prepare F_4BINOL **9** through the coupling methods incorporating standard Suzuki and Stille protocols failed. Fortunately, the oxidative cross-coupling using Cu(OH)Cl·TMEDA as the catalyst was successful (for selected examples on Ullmann and oxidative coupling see (*2,10,11*), however, only moderate yield (40%) was obtained. The optimal reaction condition involved mixing 1 eq of 2-naphthol, 0.7 eq of 5,6,7,8-tetrafluoronaphthol **5**, and 10 mol% Cu(OH)Cl·TMEDA at 140°C for 24 hours. The main by-products of this reaction were BINOL and F_8BINOL, which were isolated in 50% yield and less than 10% yield, respectively (Scheme 3).

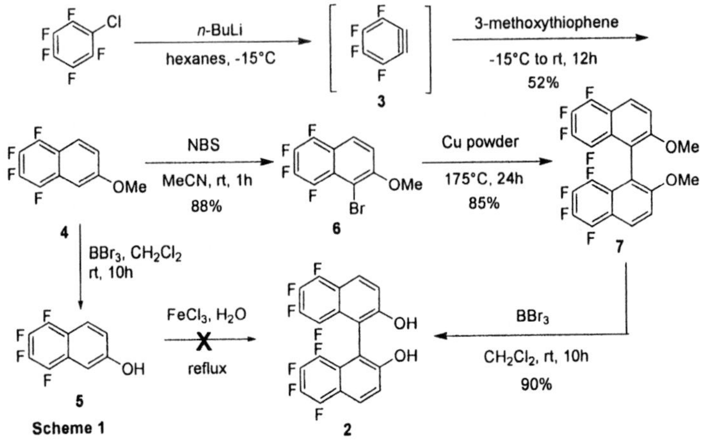

n-BuLi

hexanes, -15°C

3

3-methoxythiophene

-15°C to rt, 12h

52%

4

NBS

MeCN, rt, 1h

88%

6

Cu powder

175°C, 24h

85%

7

BBr_3, CH_2Cl_2

rt, 10h

5

$FeCl_3$, H_2O

reflux

2

BBr_3

CH_2Cl_2, rt, 10h

90%

Scheme 1

2 +

DMAP (catalytic)

pyridine, CH_2Cl_2, rt.

8

R=

fractional recrystallization in MeOH

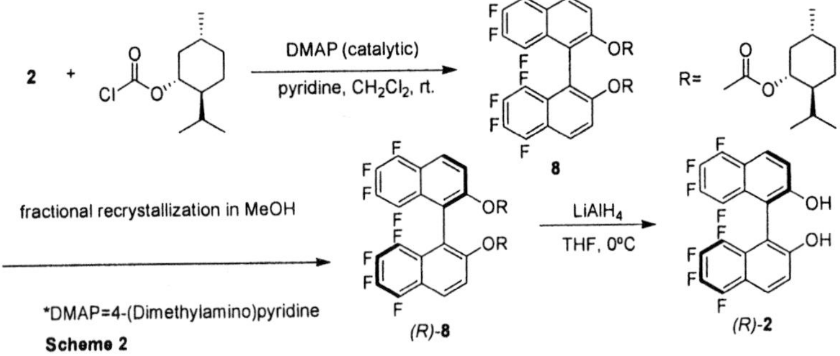

*DMAP=4-(Dimethylamino)pyridine

Scheme 2

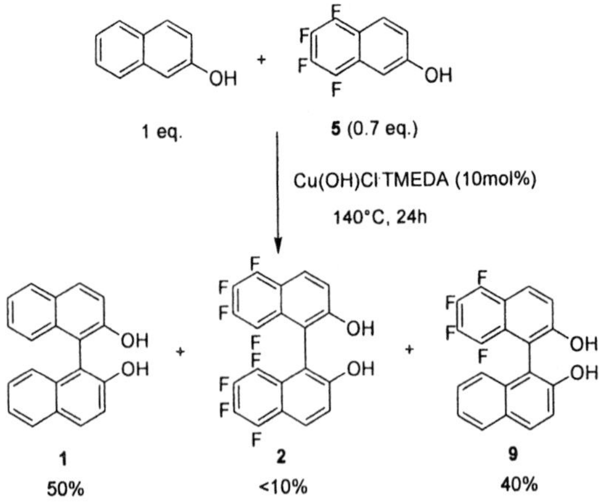

(R)-**8**

$LiAlH_4$

THF, 0°C

(R)-**2**

1 eq.

5 (0.7 eq.)

Cu(OH)Cl·TMEDA (10mol%)

140°C, 24h

1	**2**	**9**
50%	<10%	40%

Scheme 3 *TMEDA = tetramethyl ethylenediamine

A reductive pathway similar to our approach to F_8BINOL proved to be even less successful. The Ullmann coupling between **6** and 1-bromo-2-methoxynaphthol afforded only 5% yield of the desired product, and the major product observed was 2,2'-bismethoxy-F_8BINOL obtained in 85% yield. Since reductive couplings typically proceed in favor of electron-poor substrates, aldehyde **10** was utilized in place of 1-bromo-2-methoxynaphthol (Scheme 4). The desired product **11** was obtained in 48% yield, which was then oxidized through a Baeyer-Villiger reaction, followed by hydrolysis to give alcohol **12** in 64% yield (two steps). Demethylation with BBr_3 afforded racemic F_4BINOL **9** in 92% yield. The resolution of homochiral F_4BINOL was accomplished through a similar procedure to that used with F_8BINOL (*12*).

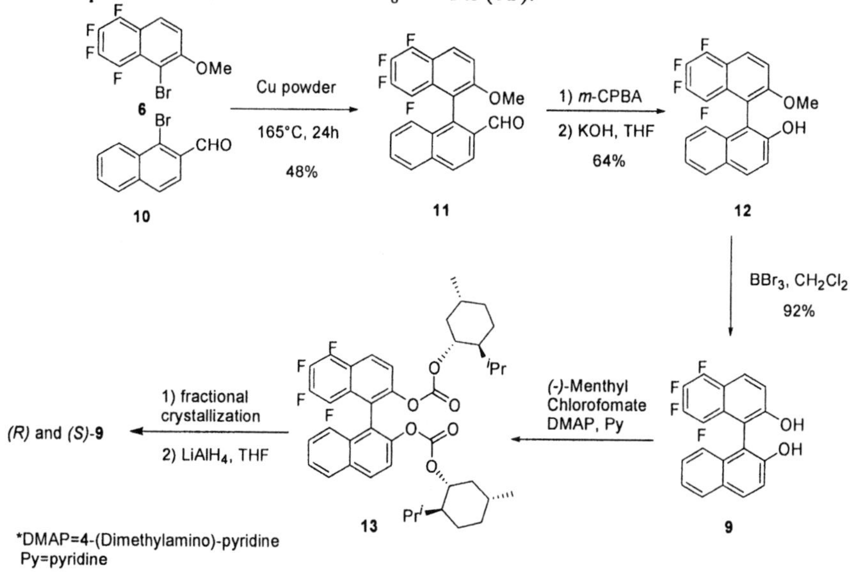

*DMAP=4-(Dimethylamino)-pyridine
Py=pyridine

Scheme 4

Regioselective Nucleophilic Aromatic Substitution of Fluorine on Polyfluorinated Binaphthols

The electron-deficient nature of the polyfluorinated aromatic rings was found to raise the oxidative stability of compound **2** compared to **1** as well as to increase the acidity of the ring-bound hydroxyl groups. Furthermore, the pKa' of the hydroxyl group in **1** decreases by 1 unit upon fluorine substitution at the aromatic rings (**1**, pKa'=10.28; **2**, pKa'=9.29). The average pKa' of F_4BINOL **9** is 9.8, just between that of F_8BINOL and BINOL (*13*). Since fluorine atom is only 0.27Å larger than hydrogen atom, one could anticipate only a small increase in the barrier to axial torsion upon fluorination. Such a combination of steric and electronic alterations in fact was found to lead to dramatic increase in configurational stability of homochiral **2** with electronic effects playing a decisive role. Contrary to **1**, which readily undergoes racemization in both acidic and basic media (*14*), a dramatic increase in configurational stability of (-)-**2** was observed under both acidic and basic conditions. When (-)-**2** was subjected to reflux in a 1:1 mixture of THF and 13% aq HCl, no racemization was detected after 24h. In comparison, (-)-**1** was racemized under these conditions decreasing enantiomeric excess of (-)-**1** from 99% to 13%. On the other hand, the enantiomeric excess of (-)-**1** decreased from 99% to 0% after 12h boiling in 5% aqueous NaOH, however, no racemization was observed when (-)-**2** was refluxed in aqueous NaOH after 24h.

The nucleophilic displacement of aromatic fluorine is a well-known reaction with a wide scope and utility (7). A variety of nucleophiles are known to participate in this chemistry. We thought that this chemistry might provide a versatile and mild route to functionalized fluorinated ligands with axial chirality. Thus, various alkoxides have been used as nucleophiles, and the corresponding products **15** were obtained in moderate to good yields with no racemization observed in the course of any of these substitution processes (*12,15*) (Scheme 5, eq 1 and 2). It should be noted that the 2,2'-hydroxyl groups on either F_8BINOL or F_4BINOL require protection since the reaction of sodium alkoxide with unprotected polyfluorinated binaphthols can produce a complicated mixture of polymethoxylated products. Alkyllithium species have also been used as nucleophiles to modify the 7-position of the fluorinated ring system. In this case, the unprotected polyfluorinated binaphthols can be used directly. For instance, the reaction between unprotected F_4BINOL and ^{t}BuLi occurs at -78°C with the desired product **20** obtained in 40% yield. (Scheme 5, eq 3)

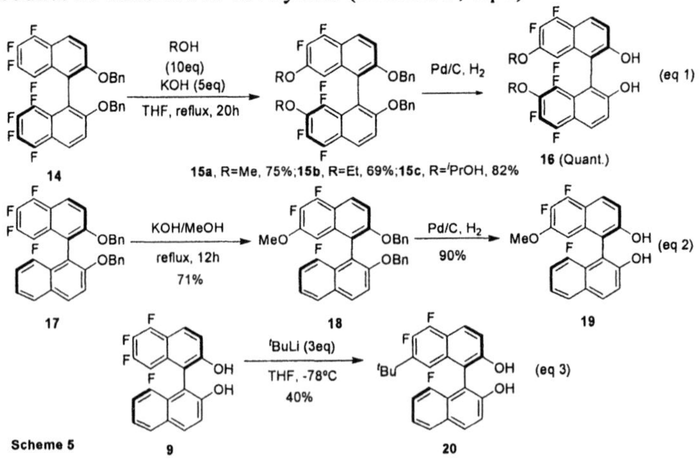

14 → ROH (10eq), KOH (5eq), THF, reflux, 20h → 15 → Pd/C, H_2 → 16 (Quant.) (eq 1)

15a, R=Me, 75%; **15b**, R=Et, 69%; **15c**, R=^{i}PrOH, 82%

17 → KOH/MeOH, reflux, 12h, 71% → 18 → Pd/C, H_2, 90% → 19 (eq 2)

9 → ^{t}BuLi (3eq), THF, -78°C, 40% → 20 (eq 3)

Scheme 5

Application of Polyfluorinated Binaphthols in Asymmetric catalysis

Oxidation of Sulfides to Sulfoxides

Enantiomerically pure sulfoxides are important compounds which have provided increasing application in asymmetric synthesis (*16*). To illustrate the utility of the fluorinated binaphthol-derived catalysts in asymmetric catalysis, we first chose sulfide oxidation as a model reaction (*17,12,15*). The results of this study are summarized in Scheme 6. Compared with *(R)*-BINOL, both *(R)*-F_8BINOL and *(R)*-F_4BINOL proved to be more efficient, and the % ee of the sulfoxide product **21** was improved (Scheme 6, entry 1, 2 and 3). Further introduction of steric bulk to the 7 and/or 7'-position(s) of the binaphthol rings resulted in decreased enantioselectivity as well as chemical yield (Scheme 6, entry 4-6). Slightly increased enantioselectivity was observed when lowering the reaction temperature, however, chemical yield was decreased (Scheme 6, entry 7). Similar to the previous results reported by Kagan and co-workers (*18*), water was essential to ensure rapid turnover in the titanium/diethyl tartrate oxidation system and two equivalents of water with respect to sulfide were found necessary for the turnover to take place in our case. It should also be noted that the most intriguing difference between the fluorinated binaphthols and BINOL systems in this sulfoxidation reaction is that a reversal of asymmetric induction upon fluorine substitution was observed (Scheme 6, entry 1 to entries 2-7) (*19*).

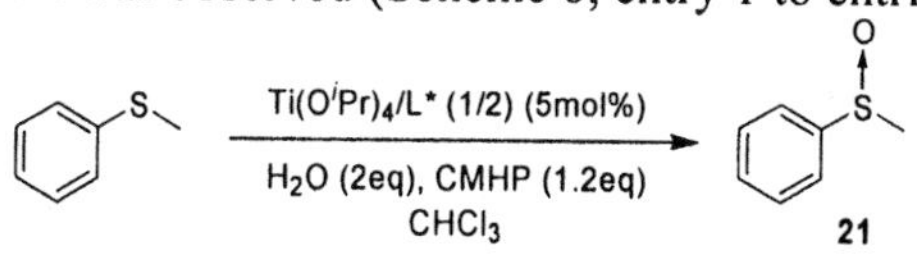

$Ti(O^iPr)_4$/L* (1/2) (5mol%)

H_2O (2eq), CMHP (1.2eq)

$CHCl_3$

21

Entry	Ligand	Time (h)	Temperature (°C)	Yield (%)	ee (%)
1	*(R)*-BINOL	42	0	69	3 *(R)*
2	*(R)*-F_8BINOL	4.5	0	77	75 *(S)*
3	*(R)*-F_4BINOL	2	0	78	80 *(S)*
4	*(R)*-**19**	3	0	49	28 *(S)*
5	*(R)*-**20**	2	0	32	27 *(S)*
6	*(R)*-**16a**	5	0	47	51 *(S)*
7	*(R)*-F_4BINOL	18	-20	46	84 *(S)*

Scheme 6

Diethylzinc Addition to Aldehydes

Enantioselective addition of dialkylzinc to aldehydes is one of the most reliable methods to prepare chiral *sec*-alcohols (*20*). It is also a standard reaction to test the reactivity and enantioselectivity of newly designed chiral ligands. Thus, the efficiency of the fluorinated binaphthol ligands were tested in the asymmetric addition of diethylzinc to naphthaldehyde (*21*). The catalysts were formed by mixing $Ti(^iPrO)_4$ with *(R)*-BINOL **1**, *(R)*-F_8BINOL **2**, or 7,7'-disubstituted *(R)*-F_8-BINOL **16** under the conditions where formation of monomeric catalyst precursors of 1:1 composition is favored (7:1 $Ti(^iPrO)_4$/ligand ratio). The results showed that all catalysts give similar enantioselectivities (Scheme 7), which indicates that both steric (substitution at the 7,7'-positions) and electronic effects are relatively insignificant in this case.

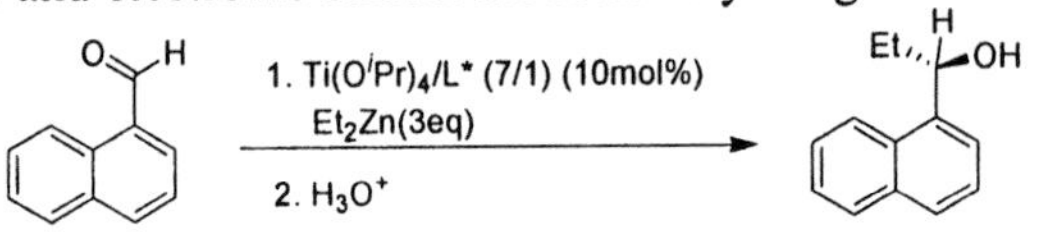

1. $Ti(O^iPr)_4$/L* (7/1) (10mol%) Et_2Zn(3eq)
2. H_3O^+

22

Entry	Ligand	ee (%)
1	*(R)*-BINOL	89 (*R*)
2	*(R)*-F_8BINOL	89 (*R*)
3	*(R)*-**16a**	88 (*R*)
4	*(R)*-**16b**	91 (*R*)
5	*(R)*-**16c**	90 (*R*)

Scheme 7

Glyoxylate-Ene Reaction

The ene reaction, first recognized in 1943 by Alder (*22*), is now among the simplest and most reliable ways to form C-C bonds by converting readily available olefins into more functionalized products with activation of an allylic C-H bond and allylic transposition of the C=C bond (*23*). Both asymmetric amplification and enantiomer-selective activation of racemic catalysts were recently described for the glyoxylate-ene process (*24, 25, 26*). A remarkable nonlinear effect (*27*) observed was attributed to the increase in activity of the homochiral BINOL/Ti adducts **23** and **24** compared to the more stable *meso*-adduct **25** (*23*).

$$M + rac\text{-}L \longrightarrow \underset{\mathbf{23}}{ML_RL_R} + \underset{\mathbf{24}}{ML_SL_S} + \underset{\mathbf{25}}{ML_RL_S} \quad \text{(eq 4)}$$

M=Ti(IV); L=BINOL

We reasoned that if one of the enantiomers of BINOL is replaced by its fluorinated analogue, a "pseudo-*meso*" aggregate might be formed for similar geometrical reasons as in the BINOL case. Homochiral F_8BINOL was thus investigated in titanium-catalyzed asymmetric glyoxylate-ene process using literature conditions (*28*). The results showed that the same sense of asymmetric induction was observed for both Ti/*(S)*-F_8BINOL and Ti/*(S)*-BINOL catalysts, whereas, initial rate studies indicated that the catalyst derived from F_8BINOL was approximately 4 times slower than the one derived from BINOL. The "pseudo-*meso*" aggregate *(R)*-F_8BINOL/*(S)*-BINOL/Ti(O^{i}Pr)$_4$ demonstrated excellent enantioselectivity. In the reaction between ethyl glyoxylate and α-methyl styrene, 99% enantiomeric excess was observed, while only 92% ee was produced in the presence of *(S)*-F_8BINOL/Ti(O^{i}Pr)$_4$. The most intriguing observation was a significant yield improvement in the case of a "pseudo-*meso*" compared to those for either F_8BINOL or BINOL alone. No conversion was observed in the reactions between aliphatic olefins and ethyl glyoxylate when either BINOL/Ti(O^{i}Pr)$_4$ or F_8BINOL/Ti(O^{i}Pr)$_4$ was used alone as a catalyst.

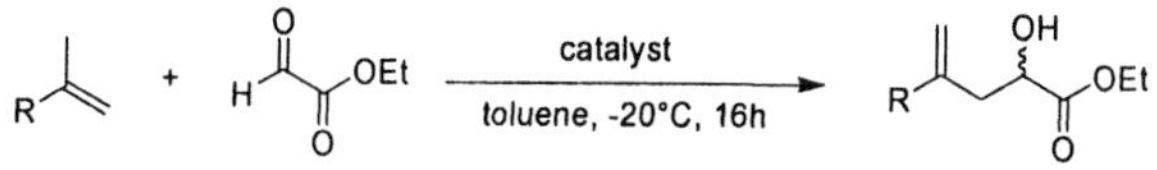

catalyst; toluene, -20°C, 16h → **26**

Entry	R	Catalyst	Yield (%)	ee (%)
1	phenyl	*(S)*-**2** (10mol%)/Ti(O^{i}Pr)$_4$ (5mol%)	53	92 *(S)*
2	phenyl	*(R)*-**2** (5mol%)/*(S)*-**1** (5mol%)/Ti(O^{i}Pr)$_4$ (5mol%)	95	99 *(S)*
3	methyl	*(R)*-**2** (5mol%)/*(S)*-**1** (5mol%)/Ti(O^{i}Pr)$_4$ (5mol%)	57	99 *(S)*
4	ethyl	*(R)*-**2** (5mol%)/*(S)*-**1** (5mol%)/Ti(O^{i}Pr)$_4$ (5mol%)	71	99 *(S)*
5	cyclopentyl	*(R)*-**2** (5mol%)/*(S)*-**1** (5mol%)/Ti(O^{i}Pr)$_4$ (5mol%)	59	99 *(S)*

Scheme 8

Friedel-Crafts Reaction

Friedel-Crafts reaction is a powerful C-C bond-forming processes in organic synthesis (*29*). In spite of its great synthetic value, only limited attention has been paid to its catalytic variants. The breakthrough in this area was recently achieved by Mikami (*30*) and Jørgensen (*31*), respectively. Since then, research on catalytic variants of asymmetric Friedel-Crafts reaction has attracted extensive interest (*32, 33*). In the area of titanium catalysis, Mikami and Ding showed that the activity of the BINOL/Ti(^{i}PrO)$_4$ catalysts were strongly influenced by the electron-withdrawing substituents on the binaphthol ring. Upon switching the catalyst from *(R)*-BINOL/Ti(^{i}PrO)$_4$ to the more Lewis acidic *(R)*-6,6'-Br$_2$-BINOL **27**/Ti(^{i}PrO)$_4$, the %ee was increased from 69% to 80% (Scheme 9, entry 1 and 2). Due to the highly electron-defficient nature of fluorinated binaphthols, both tetrafluorinated binaphthol (F$_4$BINOL) and octafluorinated binaphthol (F$_8$BINOL) were then investigated in the reaction between aromatic amines and ethyl glyoxylate (Scheme 9) (*34*). When *(R)*-F$_4$BINOL was employed in this reaction, a similar result to *(R)*-6,6'-Br$_2$-BINOL was observed (Scheme 9, entry 2 and 3). However, it was found that further increase in the Lewis acidity of the catalyst system by replacing *(R)*-F$_4$-BINOL with *(R)*-F$_8$BINOL caused a decrease in both chemical yield and enantioselectivity of the desired product **28**(Scheme 9, entry 4). On the other hand, the product **29** was obtained in 38% yield, which is believed to be formed through dehydration of the initially formed **28** followed by another Friedel-Crafts reaction with *N,N*-dimethylaniline. A "mixed" catalyst system *(R)*-BINOL/*(R)*-F$_8$BINOL/Ti(^{i}PrO)$_4$ was also tested under the same reaction conditions. The product **28** was obtained in 40% yield with 32% ee, while product **29** was obtained in 51% yield (Scheme 9, entry 5).

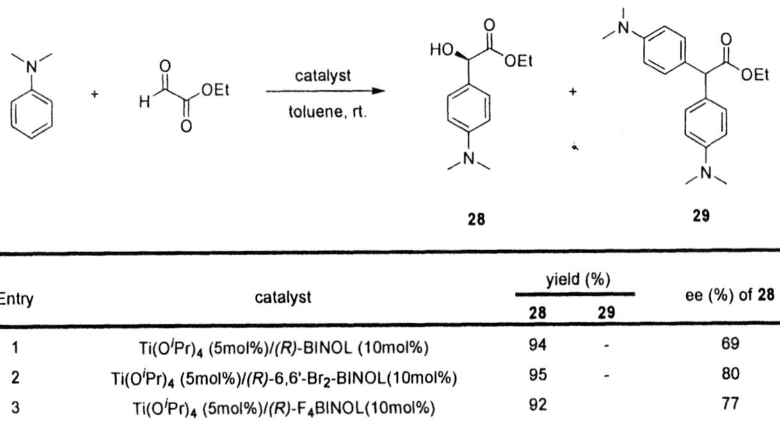

N,N-dimethylaniline + H–C(O)–C(O)OEt (ethyl glyoxylate) —catalyst, toluene, rt.→ **28** + **29**

Entry	catalyst	yield (%) 28	yield (%) 29	ee (%) of 28
1	Ti(O^{i}Pr)$_4$ (5mol%)/*(R)*-BINOL (10mol%)	94	-	69
2	Ti(O^{i}Pr)$_4$ (5mol%)/*(R)*-6,6'-Br$_2$-BINOL(10mol%)	95	-	80
3	Ti(O^{i}Pr)$_4$ (5mol%)/*(R)*-F$_4$BINOL(10mol%)	92		77
4	Ti(O^{i}Pr)$_4$ (5mol%)/*(R)*-F$_8$BINOL(10mol%)	54	38	53
5	Ti(O^{i}Pr)$_4$ (5mol%)/*(R)*-F$_8$BINOL(5mol%)/*(R)*-BINOL(5mol%)	40	51	32

Scheme 9

Epoxidation

The study of asymmetric epoxidation of olefins commenced in 1965 (*35*). In the 1980s, Sharpless and coworkers developed an efficient system that systematically produced either enantiomer of an epoxide from an allylic alcohol using substoichiometric amounts of titanium isopropoxide and diethyl tartrate.(*36*) Since then, much progress has been made towards asymmetric epoxidation of other types of olefins (*37,38*). Due to the fact that chiral α,β-epoxy ketones are useful intermediates in the synthesis of a wide range of natural products and pharmaceutically relevant molecules (*39*), great efforts have been recently devoted to developing efficient methods for enantioselective epoxidation of α,β-enones (*40*). In 1997 Shibasaki and co-workers reported a general catalytic asymmetric epoxidation of enones using alkali metal-free lanthanide-BINOL complexes (*41*). A few years later, both triphenylarsine oxide (*37*) and thiphenylphosphine oxide (*42*) were reported to be effective additives in the La-BINOL catalyzed asymmetric epoxidation. In the presence of 10 to 15mol% of the additive, the catalyst complex La-(*R*)-BINOL-Ph_3PO or La-(*R*)-BINOL-Ph_3AsO exhibited both higher reactivity and enantioselectivity than La-(*R*)-BINOL. Qian and co-workers demonstrated an efficient asymmetric epoxidation reaction of enones catalyzed by $Ln(^iPrO)_3$-6,6'-R_2-BINOL complexes (Ln=Yb, Gd; R=Br, Ph) (*39*). Their results showed that the steric bulk of substituents at 3,3'-positions on binaphthol rings is detrimental to the enantioselectivity of the reaction. Although the enantioselectivity was also decreased by introducing electron-donating groups at 6,6'-positions of the binaphthol rings, it was found that excellent enantioselectivity could be achieved by introducing electron-withdrawing substituents at these positions. Since fluorine is more electronegative than bromine, we reasoned that the fluorinated binaphthols might be better suited for the Lanthanide-BINOL catalyzed asymmetric epoxidation. The results showed that both *(R)*-F_8BINOL and *(R)*-F_4BINOL are superior to *(R)*-6,6'-Br_2-BINOL in terms of enantioselectivity in the epoxidation reaction of *trans*-chalcone **30a**. Whlie the C_2-unsymmetric F_4BINOL dramatically improved the ee to 93% (Scheme 10, entry 3), the best ee value was still generated by the most electron-deficient ligand F_8BINOL (Scheme 10, entry 2). In this same reaction, Gd-*(R)*-F_8BINOL system proved to be more efficient than Yb-*(R)*-F_8BINOL in both reactivity and enantioselectivity (Scheme 10, entry 2 and 6). The desired product **31a** was obtained in 86% yield with 96% ee under the optimal condition (Scheme 10, entry 2).

Due to our observation of the superiority of F_8BINOL as an efficient ligand in the asymmetric epoxidation of *trans*-chalcone, we further explored the generality

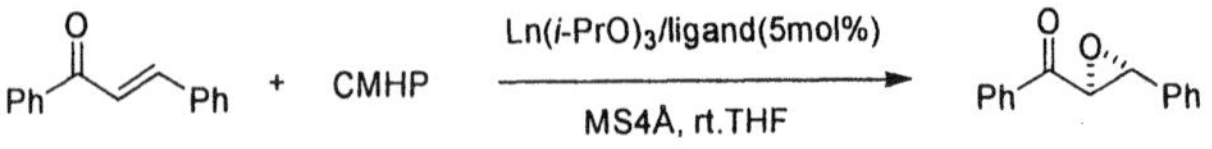

$Ln(i\text{-}PrO)_3$/ligand(5mol%), MS4Å, rt.THF: Ph-CO-CH=CH-Ph (**30a**) + CMHP → epoxide **31a**

Entry	$Ln(i\text{-}PrO)_3$	ligand	time	yield (%)	ee (%)
1	$Gd(i\text{-}PrO)_3$	(*R*)-BINOL	8h	89	83
2	$Gd(i\text{-}PrO)_3$	**(*R*)-F_8BINOL**	**6h**	**86**	**96**
3	$Gd(i\text{-}PrO)_3$	(*R*)-F_4BINOL	8h	84	93
4	$Gd(i\text{-}PrO)_3$	(*R*)-3,3'-Br_2BINOL	48h	36	27
5	$Gd(i\text{-}PrO)_3$	(*R*)-6,6'-Br_2BINOL	8h	95	78
6	$Yb(i\text{-}PrO)_3$	(*R*)-F_8BINOL	6h	60	70

*CMHP=Cumene hydroperoxide

Scheme 10

of this catalyst system by reacting a range of α,β-unsaturated ketones **30b-g** under the optimal reaction conditions. Good to excellent ee's and yields were obtained on a broad scope of substrates bearing either electro-withdrawing or electron-donating functionalities (Scheme 11); however, no reaction occurred in the case of *trans*-4-phenyl-3-buten-2-one **30h** and *trans*-4-methoxy-chalcone **30i** (Scheme 12, entry 1 and 4).

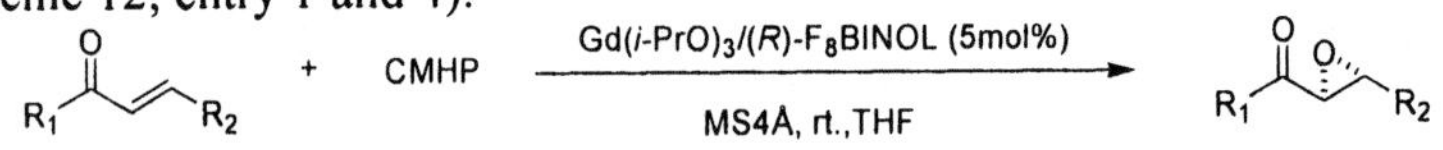

$Gd(i\text{-}PrO)_3$/(*R*)-F_8BINOL (5mol%), MS4Å, rt.,THF: R_1-CO-CH=CH-R_2 (**30a-g**) + CMHP → epoxide **31a-g**

Entry		R_1	R_2	time	Yield(%)	ee(%)
1	**31a**	Ph	Ph	6h	86	96
2	**31b**	Ph	2-Cl-C_6H_4	16h	91	90
3	**31c**	4'-Cl-C_6H_4	4-Cl-C_6H_4	3h	93	97
4	**31d**	Ph	4-NO_2-C_6H_4	10h	88	99
5	**31e**	4'-MeO-C_6H_4	Ph	12h	82	97
6	**31f**	4'-Cl-C_6H_4	Ph	7h	91	96
7	**31g**	Ph	4-Cl-C_6H_4	7h	90	93

(CMHP: cumene hydroperoxide)

Scheme 11

Further study indicated that excellent reactivity and enantioselectivity could be achieved by employing Gd-*(R)*-BINOL-Ph_3PO (1:1:3) system as the catalyst (Scheme 12, entry 2 and 5). On the other hand, the analogue Gd-*(R)*-F_8BINOL-Ph_3PO (1:1:3) was less efficient than Gd-*(R)*-BINOL-Ph_3PO (1:1:3) in terms of both reactivity and enantioselectivity (Scheme 12, entry 3 and 6).

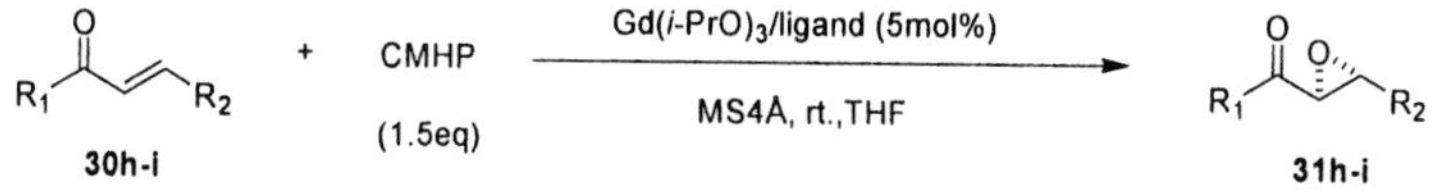

30h-i + CMHP (1.5eq) → [Gd(*i*-PrO)$_3$/ligand (5mol%), MS4Å, rt.,THF] → 31h-i

Entry		R_1	R_2	ligand	Ph_3PO	time	Yield(%)	ee(%)
1	30h	CH_3	Ph	*(R)*-F_8BINOL	-	48h	0	-
2	**30h**	**CH_3**	**Ph**	***(R)*-BINOL**	**15mol%**	**6h**	**92**	**98**
3	30h	CH_3	Ph	*(R)*-F_8BINOL	15mol%	20h	65	76
4	30i	Ph	4-MeO-C_6H_4	*(R)*-F_8BINOL	-	48h	0	-
5	**30i**	**Ph**	**4-MeO-C_6H_4**	***(R)*-BINOL**	**15mol%**	**16h**	**83**	**90**
6	30i	Ph	4-MeO-C_6H_4	*(R)*-F_8BINOL	15mol%	2days	62	78

*CMHP=Cumene hydroperoxide

Scheme 12

The activity of Gd-*(R)*-BINOL-Ph_3PO (1:1:3) and Gd-*(R)*-F_8BINOL were then compared in the epoxidation of substrate **30j**. The latter proved to be more efficient in terms of both reactivity and enantioselectivity (Scheme 13) (*43*).

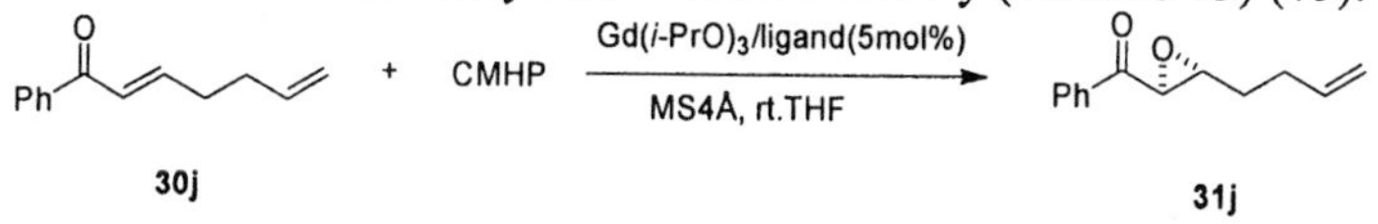

30j + CMHP → [Gd(*i*-PrO)$_3$/ligand(5mol%), MS4Å, rt.THF] → 31j

Entry	ligand	Ph_3PO	time	yield(%)	ee(%)
1	*(R)*-BINOL	-	48h	63	30
2	*(R)*-BINOL	15mol%	48h	40	16
3	***(R)*-F_8BINOL**	-	**16h**	**85**	**78**

*CMHP=Cumene hydroperoxide

Scheme 13

Conclusions

This review discussed the synthesis of highly electron-deficient polyfluorinated binaphthols as well as their applications as chiral ligands in metal mediated asymmetric catalysis. By employing polyfluorinated binaphthols instead of other commercially available BINOL family derivatives, improved chemical reactivity and enantioselectivity were observed in several useful catalytic asymmetric transformation processes. The polyfluorinated binaphthols are believed to be a class of useful binaphthol ligands and a complement to the BINOL family, especially in reactions involving highly acidic and/or oxidative conditions.

References:

1. For reviews see: (a) Noyori, R. *Chem. Soc. Rev.* **1989**, *18*, 187-208. (b) Noyori, R. *Asymmetric Catalysis in Organic Synthesis*; Wiley: New York, 1994. (c) Pu, L. *Chem. Rev.* **1998**, *98*, 2405-2494. (d) Chen, Y; Yekta, S.; Yudin, A.K. *Chem. Rev.* **2003**, *103*, 3155-3211. (e) Kočovský, P.; Vyskočil, Š.; Smrčina, M. *Chem. Rev.* **2003**, *103*, 3213-3245.
2. Pummerer, R.; Prell, E.; Rieche, A. *Chem. Ber.* **1926**, *59*, 2159-2161.
3. Noyori, R.; Tomino, I.; Tanimoto, Y. *J. Am. Chem. Soc.* **1979**, *101*, 3129-3131.
4. For selected examples see: (a) Chan, A. S. C.; Zhang, F. Y.; Yip, C. W. *J. Am. Chem. Soc.* **1997**, *119*, 4080-4081. (b) Zhang, F. Y.; Chan, A. S. C. *Tetrahedron: Asymmetry* **1998**, *9*, 1179-1182. (c) Iida, T.; Yamamoto, N.; Matsunaga, S.; Woo, H. G.; Shibasaki, M. *Angew. Chem., Int. Ed.* **1998**, *37*, 2223-2226. (d) Terada, M.; Motoyama, Y.; Mikami, K. *Tetrahedron Lett.* **1994**, *35*, 6693-6696. (e) Maruoka, K.; Itoh, T.; Shirasaka, T.; Yamamoto, H. *J. Am. Chem. Soc.* **1988**, *110*, 310-312. (f) Reetz, M. T.; Merk, C.; Naberfeld, G.; Rudolph, J.; Griebenow, N.; Goddard, R. *Tetrahedron Lett.* **1997**, *38*, 5273-5276.
5. Aizenberg, M.; Milstein, D. *Science* **1994**, *265*, 359-361.
6. (a) West, A. P.; Mecozzi, S.; Dougherty, D. A. *J. Phys. Org. Chem.* **1997**, *10*, 347-350. (b) Williams, J. H. *Acc. Chem. Res.* **1993**, *26*, 593-598.
7. Welch, J. T.; Peters, D.; Miethchen, R.; Il'chenko, A. Y.; Rudiger, S.; Podlech, J. in: Baasner, B.; Hagemann, H.; Tatlow, J. C. (Eds.), *Organo-Fluorine Compounds* 4th ed., vol. E10b/Part2, Georg Thieme Verlag, Stuttgart, 2000, pp. 293-459.
8. Gronowitz, S. *Ark. Kemi* **1958**, *12*, 239-246.
9. Yudin, A. K.; Martyn, L. J. P.; Pandiaraju, S.; Zheng, J.; Lough, A. *Org. Lett.* **2000**, *2*, 41-44.
10. (a) Newman, M. S.; Cella, J. A. *J Org. Chem.* **1974**, *39*, 2084-2087. (b) Miyano, S.; Tobita, M.; Nawa, M.; Sato, S.; Hashimoto, H. *J. Chem. Soc. Chem. Commun.* **1980**, *24*, 1233-1234. (c) Miyano, S.; Tobita, M.; Hashimoto, H. *Bull. Chem. Soc. Jpn.* **1981**, *54*, 3522-3526. (d) Hong, R.; Hoen, R.; Zhang, J.; Lin, G.-Q. *Synlett* **2001**, *10*, 1527-1530. (e) Lin, G.-Q.; Hong, R. *J. Org. Chem.* **2001**, *66*, 2877-2880.
11. (a) Noji, M.; Nakajima, M.; Koga, K. *Tetrahedron Lett.* **1994**, *35*, 7983-7984. (b) Nakajima, M.; Miyoshi, I.; Kanayama, K.; Hashimota, S.; Noji, M.; Koga, K. *J. Org. Chem.* **1999**, *64*, 2264-2271. (c) Xin, Z.; Da, C.; Dong, S.; Liu, D.; Wei, J.; Wang, R. *Tetrahedron: Asymmetry* **2002**, *13*, 1937-1940.
12. Yekta, S.; Krasnova, L. B.; Mariampillai, B.; Picard, C. J.; Chen, G.; Pandiaraju, S.; Yudin, A. K. *J. Fluorine Chem.* **2004**, *125*, 517-525.

13. Chen, G. M.Sc. thesis, University of Toronto, Toronto, ON, 2001.

14. Kyba, E. P.; Gokel, G. W.; de Jong, F.; Koga, K.; Sousa, L. R.; Siegel, M. G.; Kaplan, L.; Sogah, G. D. Y.; Cram, D. J. *J. Org. Chem.* **1977**, *42*, 4173-4184.

15. Chen, Y.; Yekta, S.; Martyn, L. J. P.; Zheng, J.; Yudin, A. K. *Org. Lett.* **2000**, *2*, 3433-3436.

16. (a) Di Furia, F.; Modena, G.; Seraglia, R. *Synthesis* **1984**, *4*, 325-326. (b) Kagan, H. B. In : Ojima, I. (Ed), *Catalytic Asymmetric Synthesis*, VCH, New York, 2000, pp 327-356. (c) Komatsu, N.; Hashizume, M.; Sugita, T.; Uemura, S. *J. Org. Chem.* **1993**, *58*, 4529-4533. (d) Carreno, M. C. *Chem. Rev.* **1995**, *95*, 1717-1760. (e) Katsuki, T.; Martin, V. S. *Org. React.* **1996**, *48*, 1-299. (f) Baker, R. W.; Thomas, G. K.; Rea, S. O.; Sargent, M. V.; *Aust. J. Chem.* **1997**, *50*, 1151-1157. (g) Lattanzi, A.; Bonadies, F.; Senatore, A.; Soriente, A.; Scettri, A. *Tetrahedron: Asymmetry* **1997**, *8*, 2473-2478. (h) Yamanoi, Y.; Yamamoto, T. *J. Org. Chem.* **1997**, *62*, 8560-8564. (i) Bolm, C.; Dabard, O. A. G. *Synlett* **1999**, *3*, 360-362.

17. Martyn, L. J. P.; Pandiaraju, S.; Yudin, A. K. *J. Organomet. Chem.* **2000**, *603*, 98-104.

18 (a) Pitchen,. P.; Kagan, H. B. *Tetrahedron Lett.* **1984**, *25*, 1049-1052. (b) Pitchen, P.; Dunach, E.; Deshmukh, M. N.; Kagan, H. B. *J. Am. Chem. Soc.* **1984**,. *106*, 8188-8193.

19. For examples where reversal of asymmetric induction was observed upon introduction of different substituents into the catalyst backbone see: (a) 4f. (b) Superchi, S.; Donnoli, M. I.; Rosini, C. *Tetrahedron Lett.* **1998**, *39*, 8541-8544.

20. For reviews see: (a) Noyori, R.; Kitamura, M. *Angew. Chem., Int. Ed.* **1991**, *30*, 34-48. (b) Soai, K.; Niwa, S. *Chem Rev.* **1992**, *92*, 833-856.

21. Reference 15 and some unpublished results by Chen, Y. and Yudin, A. K.

22. Alder, K.; Pascher, F.; Schmitz, A. *Ber. Dtsch. Chem. Ges.* **1943**, *76*, 27-53.

23. Reviews on ene reaction: (a) Mikami, K.; Nakai, T. In *Catalytic Asymmetric Synthesis*; Ojima, I. (Ed), VCH: New York, 2000, pp. 543-568. (b) Dias, L. C. *Curr. Org. Chem.* **2000**, *4*, 305-342. (c) Mikami, K.; Shimizu, M. *Chem. Rev.* **1992**, *92*, 1021-1050. (d) Snider, B. B. In *Comprehensive Organic Synthesis*; Trost, B. M.; Fleming, I. (Ed); Pergamon: London, **1991**; Vol. 2, pp. 527 and Vol. 5, pp. 1. (e) Mikami, K.; Terada, M.; Shimizu, M.; Nakai, T. *J. Synth. Org. Chem. Jpn.* **1990**, *48*, 292-303. (f) Hoffmann, H. M. R. *Angew. Chem., Int. Ed.* **1969**, *8*, 556-577.

24. Mikami, K.; Terada, M. *Tetrahedron* **1992**, *48*, 5671-5680.

25. (a) Mikami, K.; Matsukawa, S. *Nature* **1997**, *385*, 613-615. (b) Mikami, K.; Terada, M.; Korenaga, T.; Matsumoto, Y.; Matsukawa, S. *Acc. Chem. Res.* **2000**, *33*, 391-401.

26. Chavarot, M.; Byrne, J. J.; Chavant, P. Y.; Pardillos-Guindet, J.; Vallée, Y. *Tetrahedron: Asymmetry* **1998**, *9*, 3889-3894.

27. For reviews on nonlinear effect see: (a) Girard, C.; Kagan, H. B. *Angew. Chem., Int. Ed.* **1998**, *37*, 2922-2959. (b) Heller, D.; Drexler, H.-J.; Fischer, C.; Buschmann, H.; Baumann, W.; Heller, B. *Angew. Chem., Int. Ed.* **2000**, *39*, 495-499. (c) Blackmond, D. G. *Acc. Chem. Res.* **2000**, *33*, 402-411. (d) Todd, M. H. *Chem. Soc. Rev.* **2002**, *31*, 211-222.

28. Pandiaraju. S.; Chen. G.; Lough, A.; Yudin, A. K. *J. Am. Chem. Soc.* **2001**, *123*, 3850-3851.

29. (a) Olah, G. A. in *Friedel-Crafts and Related Reactions*, Wiley, New York, 1963. (b) Olah, G. A. in *Friedel-Crafts Chemistry*, Wiley, New York, 1973.
30. (a) Ishii, A.; Kojima, J.; Mikami, K. *Org. Lett.* **1999**, *1*, 2013-2016. (b) Ishii, A.; Soloshonok, V. A.; Mikami, K. *J. Org. Chem.* **2000**, *65*, 1597-1599.
31. Gathergood, N.; Zhuang, W.; Jørgensen, K. A. *J. Am. Chem. Soc.* **2000**, *122*, 12517-12522.
32. For a review see: Bandini, M.; Melloni, A.; Umani-Ronchi, A. *Angew. Chem., Int. Ed.* **2004**, *43*, 550-556.
33. For selected examples on asymmetric Friedel-Crafts reaction see: (a) Yuan, Y.; Wang, X.; Li, X.; Ding, K. *J. Org. Chem.* **2004**, *69*, 146-149. (b) Evans, D.; Scheidt, K. A.; Fandrick, K. R.; Lam, H. W.; Wu, J. *J. Am. Chem. Soc.* **2003**, *125*, 10780-10781. (c) Bandini, M.; Fagioli, M.; Melchiorre, P.; Melloni, A.; Umani-Ronchi, A. *Tetrahedron Lett.* **2003**, *44*, 5843-5846. (d) Zhou, J.; Tang, Y. *J. Am. Chem. Soc.* **2002**, *124*, 9030-9031. (e) Paras, N. A.; MacMillan, W. C. *J. Am. Chem. Soc.* **2001**, *123*, 4370-4371.
34. Chen, Y.; Yen, P. H.; Yudin, A. K. unpublished results.
35 (a) Henbest, H. B. *Chem. Soc., Spec. Publ.* **1965**, *19*, 83-92. (b) Ewins, R. C.; Henbest, H. B.; McKervey, M. A. *J. Chem. Soc., Commun.* **1967**, 1085-1086.
36. (a) Katsuki, K.; Sharpless, K. B. *J. Am. Chem. Soc.* **1980**, *102*, 5974-5976. (b) Hanson, R. M.; Sharpless, K. B. *J. Org. Chem.* **1986**, *51*, 1922-1925.
37. For reviews see: (a) Johnson, R. A.; Sharpless, K. B., in *Catalytic Asymmetric Synthesis*, 2nd ed.; Ojima, I., Ed. Wiley: New York, **2000**, 231-280. (b) Porter, M. J.; Skidmore, J. *Chem. Commun.* **2000**, 1215-1225. (c) Aggarwal, V. K. in *Comprehensive Asymmetric Catalysis*; Jacobsen, E. N.; Pfaltz, A.; Yamamoto, H., Eds.; Springer: New York, **1999**, 679-693.
38. For selected examples on asymmetric epoxidation of olefins see: (a) Zhang, W.; Loebach, J. L.; Wilson, S. R.; Jacobsen, E. N. *J. Am. Chem. Soc.* **1990**, *112*, 2801-2803. (b) Irie, R.; Noda, K.; Ito, Y.; Matsumoto, N.; Katsuki, T. *Tetrahedron Lett.* **1990**, *31*, 7345-7348. (c) Yamada, T.; Imagawa, K.; Nagata, T.; Mukaiyama, T. *Chem. Lett.* **1992**, 2231-2234. (d) Wang, Z.-X.; Miller, S. M.; Anderson, O. P.; Shi, Y. *J. Org. Chem.* **1999**, *64*, 6443-6458 and references therein. (e) Nemoto, T.; Ohshima, T.; Yamaguchi, K.; Shibasaki, M. *J. Am. Chem. Soc.* **2001**, *123*, 2725-2732 and references therein.
39. Roberts, S.; Skidmore, J. *Chem. Br.* **2000**, 31-33.
40. Chen, R.; Qian, C.; de Vries, J. G. *Tetrahedron* **2001**, *57*, 9837-9842 and references therein.
41. Bougauchi, M.; Wtanabe, S.; Arai, T.; Sasai, H.; Shibasaki, M. *J. Am. Chem. Soc.* **1997**, *119*, 2329-2330.
42. Daikai, K.; Hayano, T.; Kino, R.; Furuno, H.; Kagawa, T.; Inanaga, J. *Chirality* **2003**, *15*, 83-88.
43. Chen Y.; Yudin, A. K. unpublished results.

Chapter 17

Selective Fluorination of Biologically Important Imidazoles and Indoles

Kenneth L. Kirk

Laboratory of Bioorganic Chemistry, National Institute of Diabetes, and Digestive and Kidney Diseases, National Institutes of Health, DHHS, Bethesda, MD 20892

Heterocyclic rings play an important role in the design of synthetic biochemical and pharmacological probes and medicinal agents. Owing to the special properties of fluorine, fluorinated heterocyclic compounds have proven to be particularly important in this approach. This review will cover our early focus on the synthesis of ring-fluorinated analogues of imidazoles, such as histidine and histamine, and of indoles such as 5-hydroxytryptamine (serotonin) as well as more recent work on side-chain fluorinated imidazoles and indoles.

Introduction--The Discovery of a New Fluorination Procedure

The importance of the imidazole ring in biological structure and function would be evident only from importance of the amino acid histidine. Indeed, the histidine residue often plays critical structural and functional roles in enzymes and other peptides and proteins. Histidine also serves as the biological precursor of histamine, an important amine that functions as a neurotransmitter and a mediator of the immune response. In addition to the presence of imidazole as a structural element of histidine and histamine, certain imidazoles are important intermediates in the biosynthesis of purines. It was the successful quest for the first example of a ring-fluorinated imidazole that propelled us into the exciting field of fluorine chemistry. In the 1960's the author's early mentor, the late Louis A. Cohen, recognized: 1) the importance of fluorine substitution in the design of analogues of biologically important molecules, 2) the importance of the imidazole ring in biological structure and function, and 3) the notable absence of any examples of a ring-fluorinated imidazoles reported in the literature. Repeated unsuccessful attempts to prepare this elusive species using the usual fluorination procedures (e.g. thermal and metal-catalyzed decomposition of diazonium fluoroborates, nucleophilic displacement of an activated leaving group with fluoride) provided a possible explanation for the absence of fluoroimidazoles in the literature. All attempts led to decomposition products or gave no reaction. Indeed, persistent efforts over several years produced no example of a ring-fluorinated imidazole.

The failure of the Schiemann and related reactions was particularly discouraging, since a stable diazonium fluoroborate **1** could be prepared readily from methyl 4-amino-5-imidazole carboxylate **2** (Scheme 1). The failure of this diazonium salt to extrude nitrogen below compound-destruction temperature apparently reflects the strength of this particular diazo-heteroaromatic bond. In contrast, photochemical experiments that we initiated in 1970 with the related neutral species **3** (prepared by treatment of the diazonium salt with sodium carbonate) in organic solvents revealed that ultraviolet-light catalyzed loss of nitrogen was quite facile. Radical substitution products were isolated (Scheme 1). In addition, photolysis of a thin film of the diazonium fluoroborate gave indication that nitrogen loss was occurring with the formation of trace amounts of a new, less polar compound (K. L. Kirk, unpublished results). In order to carry out this reaction in solution, the same compound was irradiated in 50% aqueous fluoroboric acid. We were extremely pleased to discover that, following neutralization and extraction, methyl 4-fluoroimdazlole-5-carboxylate (**4**) could be isolated cleanly as a white crystalline solid in very acceptable yields (ca 35 - 40%). We assumed that the other product(s) of the reaction resulted from competing attack of water on the presumed carbonium ion intermediate and were water soluble (scheme 1). Thus, in 1971 we published the first examples of ring-fluorinated imidazoles, including 4-fluoro-L-hisitidine (Fig. 1, **5b**) (*1,2*).

This photochemical variant of the Schiemann reaction proved to be quite general and provided access to a series of important 2- and 4-fluorinated imidazole derivatives. An important advantage of this procedure is that the intermediate diazonium salt need not be isolated. A 2- or 4-amino-substituted imidazole can be dissolved in 50% fluoroboric acid, diazotized, and irradiated *in situ*. Based on the development of this reaction, we launched an extensive program to synthesize and evaluate the biology of a wide range of fluoroimidazole derivatives. This work has been the subject of previous reviews (*3,4,5*). The success of this program led us to synthesize and evaluate ring-fluorinated analogues of other important classes of compounds, including indoles and catechols (catecholamines and amino acids). We have also reviewed this work (*5,6*). In this report, an overview of earlier work with ring-fluorinated imidazoles and indoles will be given. This will be followed by a review of more recent work, wherein we have expanded our efforts to include side-chain fluorinated analogues of important bioimidazoles and bioindoles.

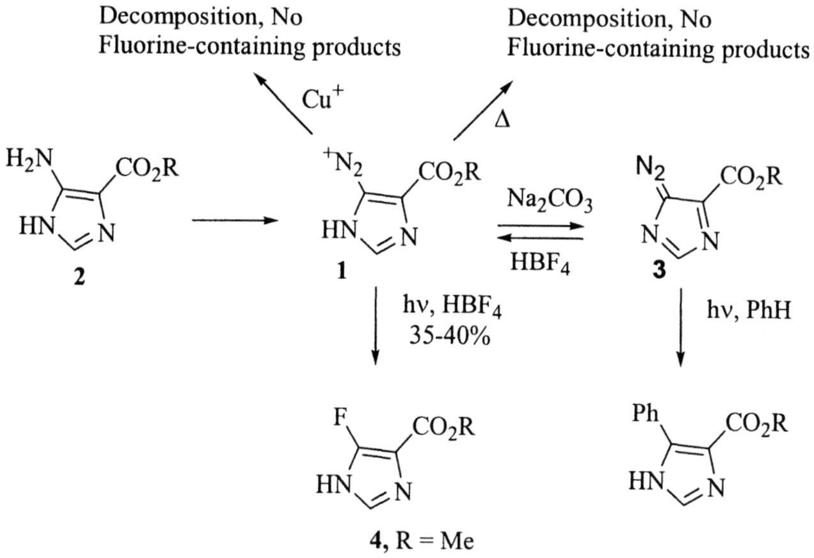

Decomposition, No Fluorine-containing products
Cu^+
Decomposition, No Fluorine-containing products
Δ
H_2N CO_2R HN N **2**
$^+N_2$ CO_2R HN N **1**
Na_2CO_3 / HBF_4
N_2 CO_2R N N **3**
hν, HBF_4 35-40%
hν, PhH
F CO_2R HN N **4**, R = Me
Ph CO_2R HN N

Scheme 1

Ring-Fluorinated Imidazoles

Reduction of 2-arylazoimidazoles provided the precursor 2-aminoimidazoles (Scheme 2) (*7,8*). 4-Aminoimidazoles cannot be isolated unless stabilized by an electron withdrawing substituent on the ring. More complex systems either were elaborated from the carboxylate prepared as above, or the amine was generated in situ, either by reduction of a 4-nitroimidazole with Zn in fluoroboric acid (*9*), or by fluoroboric acid catalyzed deprotection of a Boc-protected amine (Scheme 3) (*1,2,10*).

Figure 1 gives examples of ring-fluorinated imidazoles we prepared by these strategies. These include the parent 2-fluoro- and 4-fluoroimidazoles (**7a,b**) (*1,2,8*) , 4,5-diflouoroimidazole (**7c**) (*10*), 2- and 4-fluorohistidines (**5a,b**) and histamines (**6a,b**) (*1,2, 8,9*), 4-fluoro-5-imidazole carboxamide riboside (7) (*11*), and ring-fluorinated urocanic acids (**8a,b**) (*12,13*). The difluoroester **9** is a rare

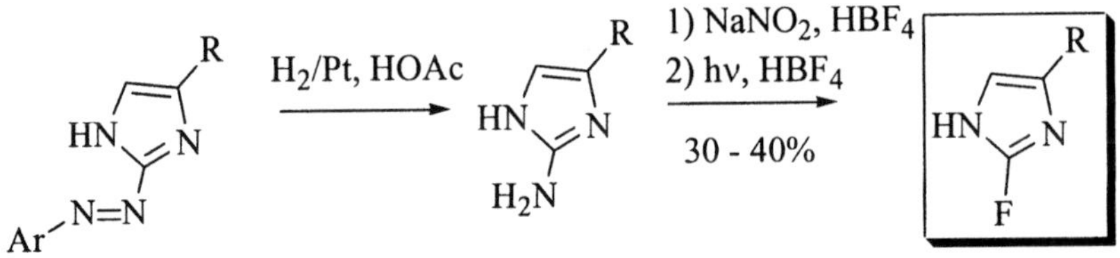

Scheme 2

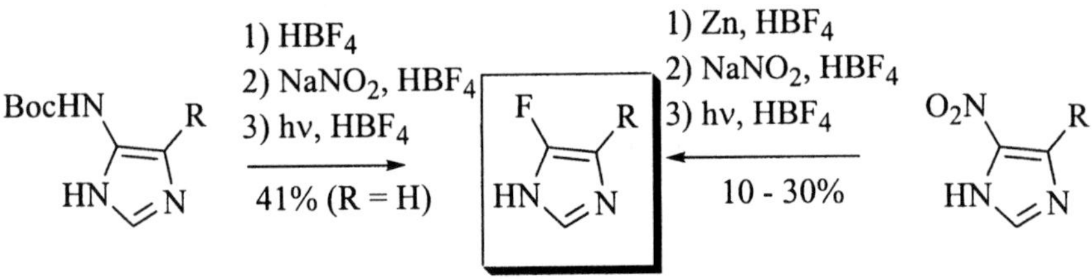

Scheme 3

example of a 2,4-difluorinated imidazole derivative (*14*). The profound effect that fluorine has on the physical properties of this system can be seen readily by examining pK_a's listed in Figure 1. Of the many biological effects of ring-fluorination, the broad spectrum of activities shown by 2-fluorohisitidine is particularly striking. Included are potent antibacterial and antiviral activities, possibly associated with the fact that the analogue is incorporated into bacterial and viral protein (*15*). It is also incorporated into mammalian protein (*15*), inhibits leukocytopoiesis in the mouse (*16*), is incorporated into murine protein in vivo (*17*), has antimalarial activity (*18*), and inhibits enzyme induction in vivo (*4*). Also apparent is the importance of regiochemistry, since 4-fluorohisitidine is inactive towards the same targets. We recently have initiated new studies in an attempt to understand the fundamental mechanism(s) of the potent biological activities of 2-fluorohistidine.

Ring-Fluorinated Indoleamines

The Abramovitch adaptation of the Fischer indole synthesis was used to prepare 6-fluoroserotonin (**10a**) and 4,6-difluoroserotonin (**10b**), as well as the corresponding melatonins (**11a,b**) (*19*) (Figure 2). In the Fischer cyclization, ring closure is regiospecific, and none of the 4-fluorinated indole could be detected. We subsequently used an alternative procedure to install fluorine in the indole 4-position, based on regioselective electrophilic fluorination using *N*-fluorobenzenesulfinimide, in order to synthesize **10c** and **11c** (*20*).

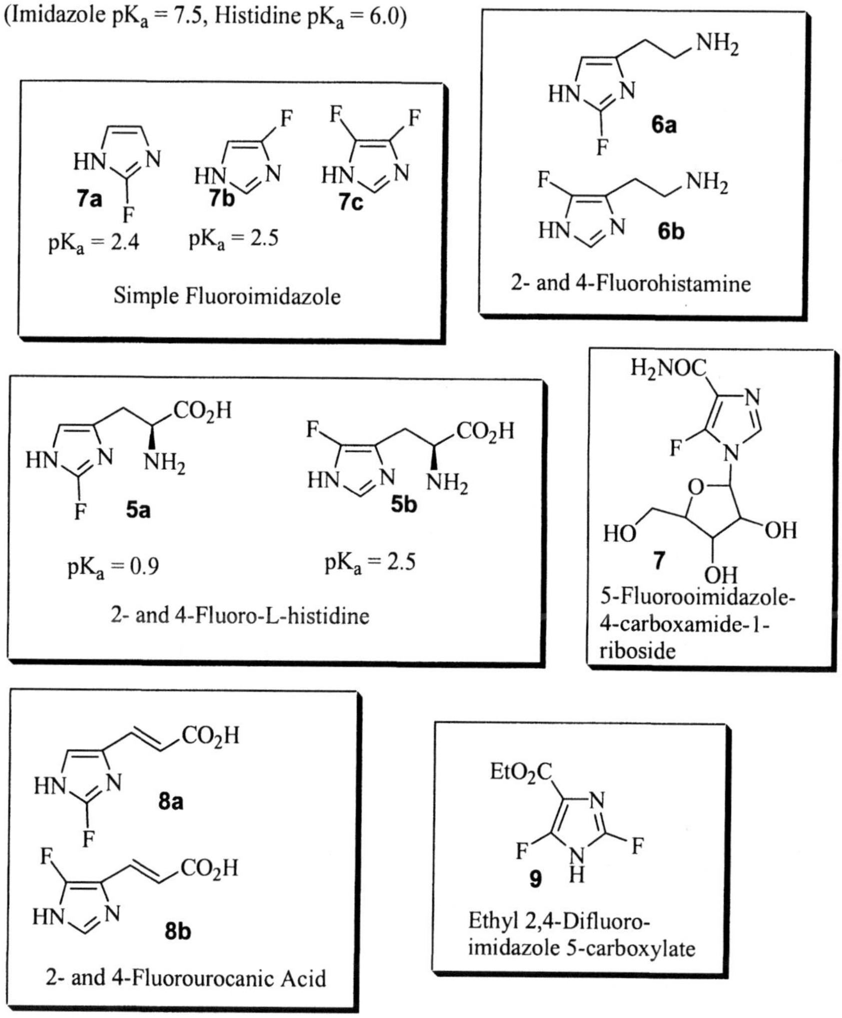

(Imidazole $pK_a = 7.5$, Histidine $pK_a = 6.0$)

7a $pK_a = 2.4$ 7b $pK_a = 2.5$ 7c

Simple Fluoroimidazole

6a 6b

2- and 4-Fluorohistamine

5a $pK_a = 0.9$ 5b $pK_a = 2.5$

2- and 4-Fluoro-L-histidine

7

5-Fluorooimidazole-4-carboxamide-1-riboside

8a 8b

2- and 4-Fluorourocanic Acid

9

Ethyl 2,4-Difluoro-imidazole 5-carboxylate

Figure 1. Examples of ring-fluorinated imidazoles

Side-Chain Fluorinated Imidazoles

In our earlier work, we also prepared ring trifluoromethylated imidazoles, including 2- and 4-trifluoromethylhistidine (**12a,b**) (Figure 3) (*21*,22). Based on the very interesting biological properties we had discovered with fluoro- and trifluoromethyl imidazoles, we next considered imidazoles substituted with a fluoromethyl- or difluoromethyl substituent. A potential advantage to be found in

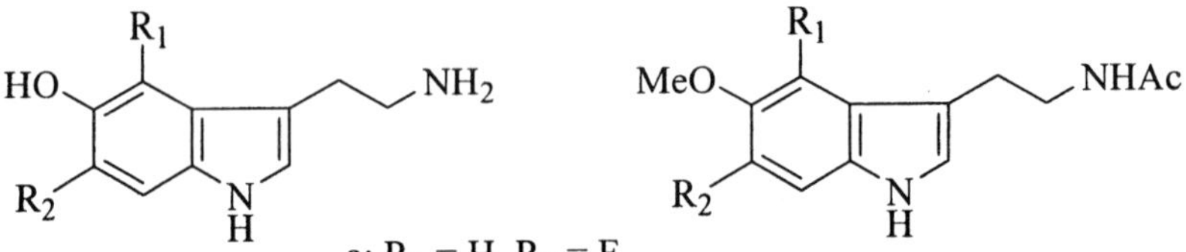

10a-c

a: $R_1 = H, R_2 = F$
b: $R_1 = R_2 = F$
c: $R_1 = F, R_2 = H$

11a-c

Figure 2. Ring-fluorinated indole derivatives.

this series would be the available site on the carbon bearing the fluorine for chain extension (as, for example, in **13**). Thus, this could represent fluorine substituted on the side-chain of histidine, of particular interest since this is the site of the reaction catalyzed by histidine ammonia lyase in the first step of metabolic degradation of histidine.

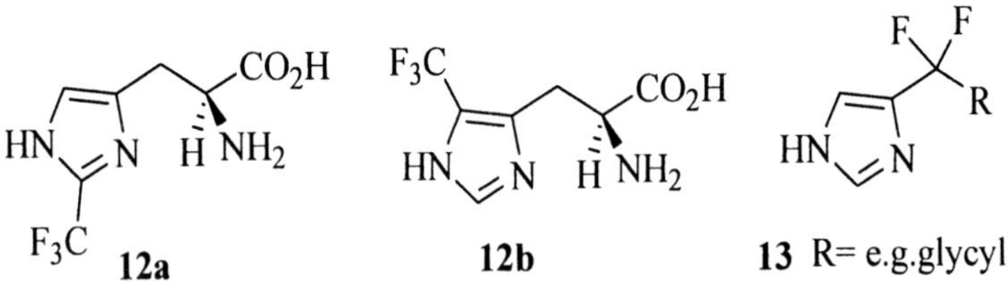

12a 12b 13 R= e.g.glycyl

Figure 3. Trifluoromethylhistidines and a difluoromethylimidazole **13**.

It must be pointed out that imidazole derivatives with fluorine on the side chain have been prepared previously. In contrast to our targets, these usually are the α-fluoromethyl or α–difluoromethyl-histidine and histamine derivatives, developed as inhibitors of PLP-dependent enzymes (*23*). β-Fluorohistamine was prepared by Kollonitich in 1970 by sulfur tetrafluoride deoxyfluorination of the corresponding β-hydroxy-histamine (*24*).

Deoxyfluorination—Synthesis of 2- And 4-Fluoromethylimidazole and 2- and 4-Difluoromethylimidazole

Deoxyfluorination of 1-trityl-2- or 4-imidazole carboxaldehyde produced the corresponding difluoromethylimidazole (**14a,b**) following de-tritylation

(Scheme 4) (*25*). The corresponding fluoromethylimidazoles (**15a,b**) were similarly prepared from hydroxymethylimidazole precursors (*25*).

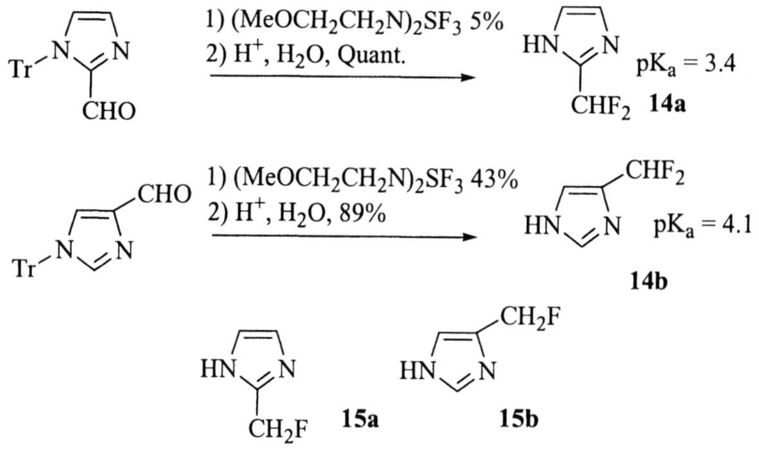

Scheme 4

In an approach to more complex molecules, 1-trityl-4-imidazole carboxaldehyde was converted to the trimethylsilylcyanohydrin by known procedures. Following the strategy developed by McCarthy et al., (*26*) this was treated with DAST to produce α–fluoronitrile, albeit in low yield. All attempts to reduce this to the amine met with failure (Scheme 5). This approach was set aside and new strategies were considered.

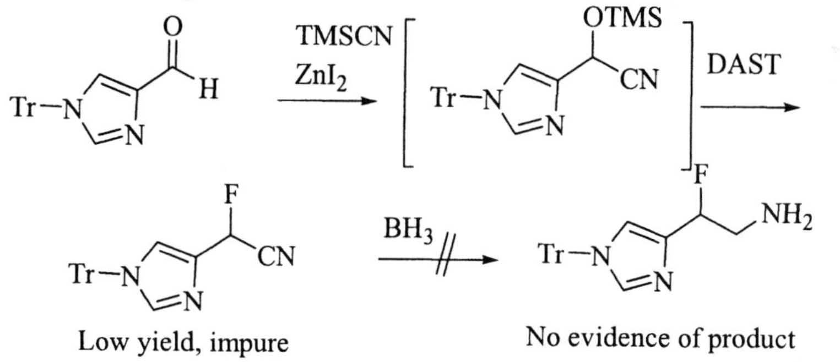

Scheme 5

Addition of "FBr" to Vinyl Imidazoles--α-Fluorohistamines

The addition of an FBr equivalent to a double bond has been developed as an effective route to fluorinated aliphatic compounds. The addition can be stereospecific and follows Markovnikov regiochemistry to give a vicinal fluoro, bromo adduct. Bromine can be substituted to provide additional functionality, or, alternatively, it can be eliminated to produce a vinyl fluoride. Adding value to this procedure is the fact that the "FBr" addition can be repeated to produce geminal difluoro substitution. Haufe and co-workers (*27*) and Schlosser and co-workers (*28*) have developed this extensively.

This approach was used to prepare β-fluorohistamine (**16**) and β,β-difluorohistamine (**17**) from 1-trityl-4-vinylimidazole, as outlined in scheme 6 (*29*).

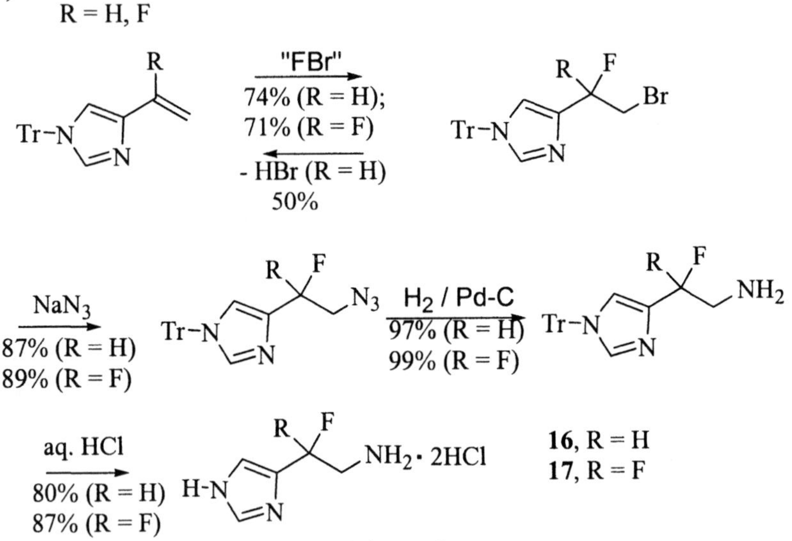

R = H, F

"FBr"
74% (R = H);
71% (R = F)
- HBr (R = H)
50%

NaN_3
87% (R = H)
89% (R = F)

H_2 / Pd-C
97% (R = H)
99% (R = F)

aq. HCl
80% (R = H)
87% (R = F)

$NH_2 \cdot 2HCl$

16, R = H
17, R = F

Scheme 6

Using similar strategies β-fluoro- and β,β-difluorohistidinol (**18, 19**) were prepared from 3-(1-trityl-1*H*-imidazol-4-yl)-prop-2-en-1-ol (*30*) In this case the stereochemistry of addition of "FBr" and elimination of HBr becomes important Scheme 7).

Attempts to prepare the corresponding fluorinated histidines using this approach were thwarted by our inability to oxidize the primary hydroxyl to the carboxylic acid. We are exploring other approaches to the amino acids.

Side-Chain Fluorinated Urocanic Acids.

Trans-urocanic acid formed by enzymatic elimination of ammonia is an important metabolite of histidine. Photochemical isomerization of *trans*-urocanic acid produces *cis*-urocanic, a compound that *in vivo* has been associated with

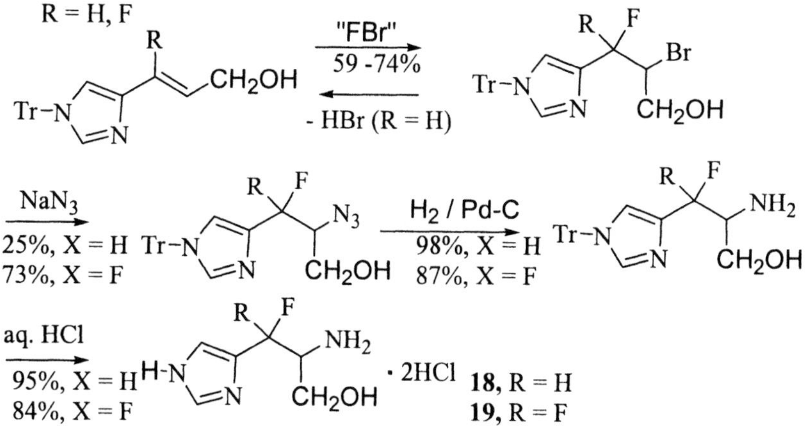

R = H, F

"FBr" 59 -74%

- HBr (R = H)

Tr–N, CH2OH, R, F, Br

NaN_3: 25%, X = H; 73%, X = F

N_3, CH_2OH

H_2 / Pd-C: 98%, X = H; 87%, X = F

NH_2, CH_2OH

aq. HCl: 95%, X = H; 84%, X = F

H-N, NH_2, CH_2OH · 2HCl

18, R = H
19, R = F

Scheme 7

photo-immunosupression. *Trans*-urocanic acid is further degraded by urocanase to give fomaminoglutamic acid, a source of biological C-1 units (*31*). 2-Fluorourocanic acid is a potent inhibitor of urocanase (*32*). Since the side chain of urocanic acid is modified by the enzyme, we were particularly intrigued by the potential consequences to the enzymatic reaction of fluorine situated on the double bond.

In addition to 2- and 4-fluorourocanic acids described above (Figure 1), we have prepared α– and β–fluorourocanic acid (**20** and **21**, respectively). The former were prepared by olefination of 1-tritylimidazole-4-carboxyaldehyde with triethyl 2-fluoro-2-phosphonoacetate (Scheme 8) (*33*).

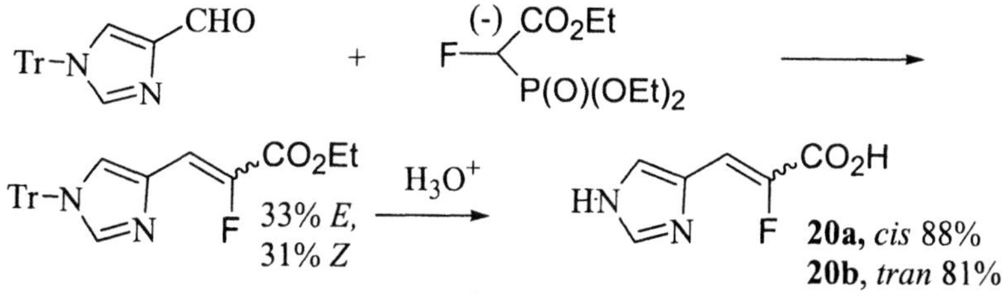

Tr–N, CHO + F–(-)–CO_2Et, $P(O)(OEt)_2$

Tr–N, CO_2Et, F: 33% *E*, 31% *Z*

H_3O^+

HN, CO_2H, F

20a, *cis* 88%
20b, *tran* 81%

Scheme 8

Cis- and *trans*-β-fluorourocanic acids (**21a**, **21b**) were prepared by "FBr" addition to 3-(1-trityl-1*H*-imidazol-4-yl)-prop-2-en-1-ol followed by elimination and oxidation (Scheme 9) through the same intermediate used to make fluorohistidinols (*34*). Diminished stereoselectivity in the addition of "FBr" to the *E*-olefin led to problems in producing the *Z*-isomer. There was a compensatory loss of stereoselectivity in the second oxidation step that partially alleviated this problem. Nonetheless, we devised another approach to β-fluorourocanic acids using "FBr" additions to alkynes (Scheme 10) (*35*).

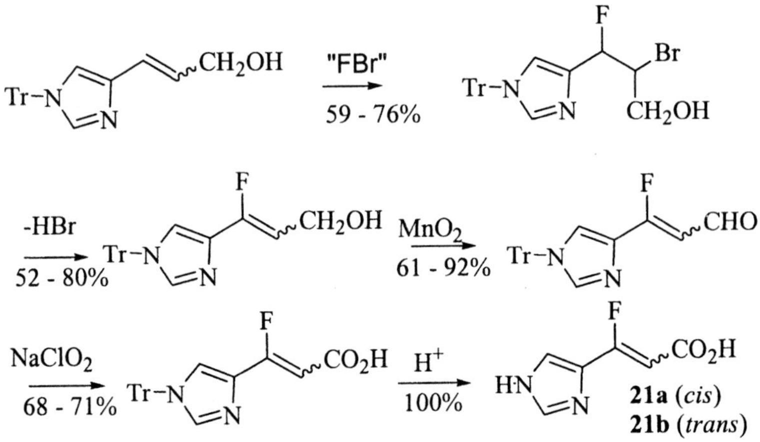

(Scheme 9)

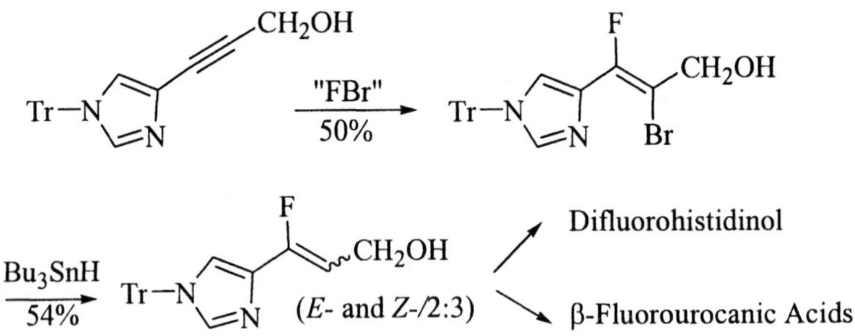

(Scheme 10)

Side Chain Fluorinated Indoles

Attempts to use "FBr" addition to vinyl indoles as a route to side-chain fluorinated indoles were unsuccessful because of our inability to displace bromine (or iodine) from the original FBr or FI addition product with a nitrogen nucleophile. Therefore we returned to our original plan to use deoxyfluorination to introduce fluorine. 1-Boc-3-azidoacetyl indoles were prepared from readily available chloroacetyl indoles. Treatment with deoxofluor proceeded cleanly to give the 1,1-difluoro-2-azidoethyl-substituted indole. Following reduction of the azido group to the amine, deprotection gave the target β,β-difluoro tryptamine derivatives (**22a-c**). Acylation of the 5-methoxy derivative following azide reduction provided a route to β,β-difluoromelatonin (**23**) (*36*) (Scheme 11).

R = H, OBn, OMe

Deoxofluor

R = H, OBn, OMe

R = H, OH, OMe

H_2 / Pd-C

(R = OMe)
1. Acylation
2. Deprotection

HCl (g), EtOAc

22a, R = H
22b, R = OH
22c, R = OMe

23

(Scheme 11)

Biological Activity

Certain aspects of the biological activities of ring fluorinated imidazoles were discussed above, and have been reviewed. To date we have limited data on the biological activity of side-chain fluorinated imidazoles and indoles. In one example, initial attempts to use β,β-difluorotryptamine (**22a**) in an enzymatic assay were thwarted by the instability of this analogue under conditions of the assay. We have not yet examined interactions with 5-HT receptors. β,β–Difluorohistamine showed no activity at a histamine receptor. Both α- and β-difluorourocanic acids (**20b, 21b**) (*trans*) were weak non-competitive inhibitors of urocanase. Neither β-fluoro- nor β,β-difluorohistidinol had any effect on normal or tumor cells either in the presence or absence of cis-platin.

There remain several areas of investigation. In particular, we will examine the effects of side-chain fluorination on the interactions of side-chain fluorinated amines with monoamine oxidases and on PLP-dependent enzymes such as decarboxylases. The latter study will be initiated upon completion of the syntheses of the corresponding amino acids.

Summary and Outlook

Syntheses of side-chain fluorinated analogues of imidazoles and indoles add to the inventory of fluorinated derivatives of these important biomolecules. There remains much to do regarding the effects of fluorination on biological activity. The most immediate goal for future synthetic work is completion of the syntheses of side-chain fluorinated analogues of histidine and tryptophan. Routes under investigation include fluorodeoxygenation of intermediate hydroxy or carbonyl precursors. Enantioselective approaches to these precursors are being pursued. Results of this synthetic work will be published in the near future.

Acknowledgements

The work presented in this review is the result of the dedicated work of students, postdoctoral fellows and collaborators over many years. The chemistry presented also relies on the previous discoveries and publications of fluorine chemists throughout the world. The author gratefully acknowledges his reliance on these contributions.

References

1. Kirk, K. L.; Cohen, L. A. *J. Am. Chem. Soc.* **1971**, *93*, 3060-3061.
2. Kirk, K. L.; Cohen, L. A. *J. Am. Chem. Soc.* **1973**, *95*, 4619-4624.
3. Kirk, K. L.; Cohen, L. A. *Biochemistry of the Carbon-Fluorine Bond,* Filler, R., Ed., ACS Symposium Series, American Chemical Society, **1976**, pp 23-36.

4. Klein, D. C.; Kirk, K. L. *Biochemistry of the Carbon-Fluorine Bond,* Filler, R., Ed., ACS Symposium Series, American Chemical Society, **1976**, pp 37-56.
5. Kirk, K. L.; Nie, J.-Y. *Biomedical Frontiers of Fluorine Chemistry*, Ojima, I.; McCarthy, J. R.; Welch, J. T., Eds., ACS Symposium Series, American Chemical Society, **1996**, pp 312-327.
6. Kirk, K. L.; Cantacuzene, D.; Creveling, C. R. *Biomedicinal Aspects of Fluorine Chemistry,* Filler, R.; Kobayashi, Y. Eds., Kodansha, Tokyo, **1982**, pp 75-91.
7. Nagai, W.; Kirk, K. L.; Cohen, L. A. *J. Org. Chem.*, **1973**, *38*, 1971-1974.
8. Kirk, K. L.; Nagai, W.; Cohen, L. A. *J. Am. Chem. Soc.* **1973**, *95*, 8389-8392.
9. Kirk, K. L.; Cohen, L. A. *J. Org. Chem.* **1973**, *38*, 3647-3648.
10. Dolensky, B.; Takeuchi, Y.; Cohen, L. A.; Kirk, K. L. *J. Fluorine Chemistry* **2001**, *107*, 147-148.
11. Reepmeyer, J. C.; Kirk, K. L.; Cohen, L. A. *Tetrahedron Lett.* **1975**, *47*, 4107-4110.
12. Klee, C. B.; La John, L. E.; Kirk, K. L.; Cohen, L. A. *Biochem. Biophys. Res. Commun.* **1977**, *75*, 674-681.
13. Fan, J.-F.; Dolensky, B.; Kim, I.-H.; Kirk, K. L. *J. Fluorine Chem.* **2002**, *115*, 137-142.
14. Takahashi, K.; Kirk, K. L.; Cohen, L. A. *J. Org. Chem.* **1984**, *49*, 1951-1954.
15. De Clercq, E.; Billiau, A.; Edy, V. G.; Kirk, K. L.; Cohen, L. A. *Biochem. Biophys. Res. Commun.* **1978**, *82*, 840-846.
16. Creveling, C. R.; Kirk, K. L.; Highman, B. *Res. Commun. Chem. Path. Pharmacol.* **1977**, *16*, 507-522.
17. Creveling, C. R.; Padgett, W. L.; McNeal, E. T.; Cohen, L. A.; Kirk, K. L. *Life Sciences*, **1992**, *51*, 1197-1204.
18. Howard, R. J.; Andrutis, A. T.; Leech, J. H.; Ellis, W. Y.; Cohen, L. A.; Kirk, K. L. *Biochem. Pharmacol.* **1986**, *35*, 1589-1596.
19. Kirk, K. L. *J. Heterocyclic Chem.* **1976**, *13*, 1253-1256.
20. Hayakawa, Y.; Singh, M.; Shibata, N.; Kirk, K. L. *J. Fluorine Chemistry,* **1999**, *97*, 161-164.
21. Kimoto, H.; Kirk, K. L.; Cohen, L. A. *J. Org. Chem.* **1978**, *43*, 3403-3405.
22. Kimoto, H.; Fujii, S.; Cohen, L. A. *J. Org. Chem.* **1984**, *49*, 1060-1064.
23. Kollonitsch, J. *Biomedicinal Aspects of Fluorine Chemistry,* Filler, R.; Kobayashi, Y., Eds., Kodansha, Tokyo, **1982**, pp 93-122.
24. Kollonitsch, J.; Marburg, S.; Perkins, L. M. *J. Org. Chem.* **1979**, *44*, 771-777.
25. Dolensky, B.; Kirk, K. L. *Coll. Czech Chem. Commun*, **2002,** *67*, 1335-1344.

26. LeTourneau, M. E.; McCarthy, J. R. *Tetrahedron Lett.* **1984**, *25,* 5227-5230.
27. Alvernhe, G.; Laurent, A.; Haufe, G. *Synthesis*, **1987**, 562-564.
28. Matsubara, S.; Matsuda, H.; Hamatani, T.; Schlosser, M. *Tetrahedron,* **1988**, *44*, 2855-2863.
29. Dolensky, B.; Kirk, K. L. *J. Org. Chem.* **2001**, *66*, 4687-4691.
30. Dolensky, B.; Narayanan, J.; Kirk, K. L. *J. Fluorine Chemistry,* **2003,** *123,* 95-99.
31. Rétey, J. *Arch. Biochem. Biophys.* **1994**, *314,* 1-16.
32. Klee, C. B.; La John, L. E.; Kirk, K. L.; Cohen, L. A. *Biochem. Biophys. Res. Commun.* **1977**, *75*, 674-681.
33. Percy, E.; Singh, M.; Takahashi, T.; Takeuchi, Y.; Kirk, K. L. *J. Fluorine Chem.* **1998**, *91*, 5-7.
34. Dolensky, B.; Kirk, K. L. *J. Org. Chem.* **2002**, *67*, 3468-3473.
35. Dolensky, B.; Kirk, K. L. *J. Fluorine Chem.* **2003,** *124*, 105-110.
36. Deng, W.-P.; Nam, G.; Fan, J.; Kirk, K. L. *J. Org. Chem.* **2003**, *68*, 2798-2802.

Trifluoromethyl-Containing Synthons

Chapter 18

Functional Group Transformations at α-Carbon to Trifluoromethyl Group

Toshimasa Katagiri and Kenji Uneyama

Department of Applied Chemistry, Faculty of Engineering, Okayama University, Tsushimanaka 3–1–1, Okayama 700–8530, Japan

This account summarizes the building methods of chiral α-carbon to trifluoromethyl group and further substitutions on that carbon. The preparations of the chiral trifluoromethylated compounds are relatively well studied. Meanwhile, the stereospecific substitutions on the α-carbon to trifluoromethyl group should be developed to increase the versatility of synthetic strategies.

1. Introduction

The importance of optically active fluoro-organic compounds in the fields of medicinal chemistry, agrochemicals, and material science is well known [1]. The poor availability of optically active fluoro-organic starting materials, which stems from a lack of natural fluoro-organics, has made the preparations of optically active fluoro-organics only *via* synthetic methodologies [2]. The unique reactivity of such fluoro-organics makes the construction of the desired structure challenging. Therefore, a smart strategy and an effective device have been needed to achieve this goal. Thus, a stereocontrolled preparation of fluoro-organics has been of interest not only to biochemists and medicinal chemists, but also to synthetic organic chemists.

Among the second-row elements, nucleus of the fluorine atom has biggest positive charge next to neon atom. Thus, the electrons surrounding the fluorine atom are attracted near by the nucleus, due to lack of the shielding effects by inner sphere electrons. This strongest electron withdrawing effect of the fluorine atom characterizes properties and reactivities of the fluoro-organics. It is well known that the orbital energy level of fluorine 2*p* is ca. 5 eV lower than that of hydrogen 1*s* [*3*]. Thus, both the bonding and the antibonding orbitals generated by the interaction of parent organic SOMO and fluorine 2*p* (right hand side) are always lower than those generated with hydrogen 1*s* (left hand side), as illustrated in Scheme 1.

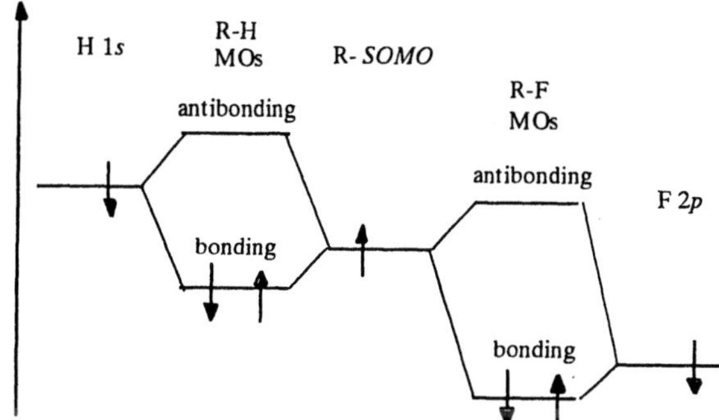

Scheme 1. Molecular orbital energy diagram for fluorinated compounds

This MO energy level diagram explains some properties of fluoro-organics also. One may find the energy level of bonding (occupied) orbital resembles to that of fluorine 2*p*. Thus, the wave function of the bonding orbital resembles to that of fluorine 2*p*. That is, the electrons in this orbital are attracted around the fluorine atom. Moreover, the lower level of the bonding orbital suggests smaller fluctuation of electrons, due to strong interaction by nuclei of the molecule. In other words, the fluoro-organic has smaller electric polarizabilities and refractive index; thus, causes a smaller van der Waals interactions [*4*].

The σ_m of the fluorine atom is 0.34 and σ_p is 0.06, thus σ_I (inductive effect) is 0.50 while σ_R (resonance effect) is -0.44 [*5*]. That is, the fluorine atom itself is an electron withdrawing group from inductive point of view, while electron donating group from resonance. Meanwhile, trifluoromethyl group is a pure electron withdrawing group. The σ_m of the trifluoromethyl group is 0.43 and σ_p is 0.54; thus the σ_I of the group is 0.41 and σ_R is 0.14 [*5*].

Trifluoromethyl group withdraws electrons anyway, which makes the surroundings of the group electron deficient and makes the trifluoromethyl group itself negatively charged. Of course, the extent of the negative charge on the trifluoromethyl group is depending on the residual moieties; some MO calculation estimated it to be 0.1 eV [*6*]. One may consider that the trifluoro-methyl group is a concentrated negative charge, thus hinders access of negatively charged nucleophiles [*7*]. The effect of the trifluoromethyl group to the reactivity is such simple, but strong.

2. Construction of trifluoromethylated chiral carbon center

The most popular structures of optically active fluoro-organics, 3-substituted-1,1,1-trifluoro-2-propanol, **1** [*8*] and 2-amino-3-substituted-1,1,1-trifluoropropane, **2** [*9*], are shown in Scheme 2. The merit for the utilization of **1** and **2** is not only limited in their easy preparations but also their possible optical purification *via* recrystallizations.

Scheme 2

2.1 Trifluoromethylated carbinols

Higher than 96% ee (2% impurity) is required for optically pure materials, and >99.6% ee (0.2% impurity) is required for pharmaceuticals. Thus, the optically active fluoro-organics with moderate enantiomeric excess cannot be used practically as they are. They need some methods for optical purifications. It is very fortunate that the β-fluorinated alcohols have higher melting points than non-fluorinated alcohols; melting point of 2,2,2-trifluoroethanol (-44 °C) is 86 °C higher than that of ethanol (-130 °C) and that of 1,1,1,3,3,3-hexafluoro-2-propanol (-4 °C) is 85 °C higher than that of 2-propanol (-90 °C) [*10*]. This would be due to higher acidities of fluorinated alcohols. Optical purification of 3-piperidyl-1,1,1-trifluoro-2-propanol, an amino alcohol with intramolecular hydrogen bond acceptor, was reported [*11*]. The hydrogen bondings of 3,3,3-trifluorolactates recognize their chirality even in the liquid state [*12*]. The structure of the hydrogen bonding systems in the crystal had been studied by X-ray crystallographic analyses [*13*]. Thus, some fluoro-alcohols with moderate optical purities may be purified in almost optically pure form *via* recrystallizations [*14-17*].

2.1.1 Biochemical resolution of racemates (Scheme 3).

Lipase acylation [*18-20*] as well as deacylation [*20-27*] resolves optically active α-trifluoromethylated carbinols **1**. Deacylation of the ester seems much effective than acylation [*20*]. Even optically active *tert*-alcohols have been prepared by the process [*23b,26*].

Scheme 3

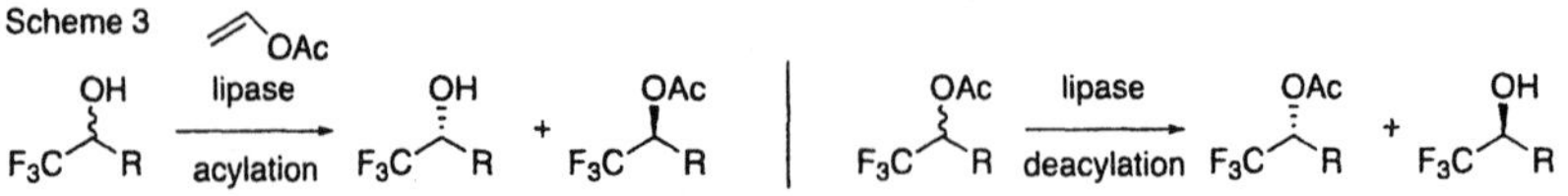

2.1.2 Stereocontrolled reductions (Scheme 4).

Stereocontrolled reduction of trifluoromethylated ketones would be the most popular preparative method for 3-substituted-1,1,1-trifluoro-2-propanols [*28*].

Scheme 4

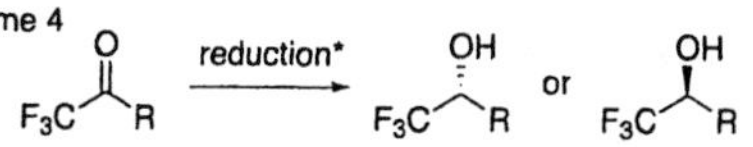

Noteworthy is that enantioselective reduction of trifluoroacetophenones seems more difficult than that of trifluoroacetones. This would be due to similarity between the electrostatic structure of negatively charged trifluoromethyl group and that of negatively charged π-system of aromatic rings (Scheme 5) [*6*].

Scheme 5

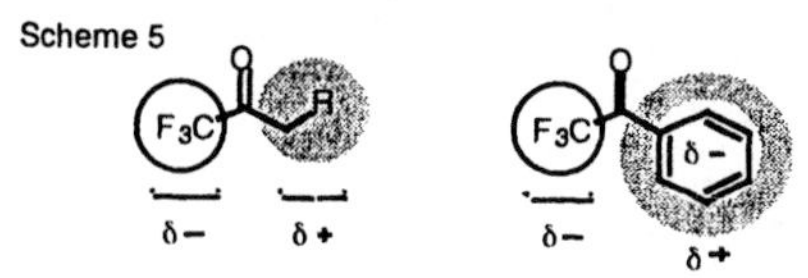

Enzymatic [*29*] and microbial [*30-33*] reductions of the ketones have been studied very well. Due to high structure specificity of the enzymes, applicability of the process for the aimed fluoro-organics seems to be tough to predict. Even the substituent far from the carbonyl reaction center affected to the enantiomeric excess [*31b*].

Many diastereoselective reductions of carbonyl compounds with chiral auxiliary [*34.35*], and enantioselective reduction with using chiral reducing agents [*36-42*] or catalytic hydrogenations [*43-46*] have been reported. They give almost optically pure alcohols in good yields.

Catalytic enantioselective alkylation of fluoral, trifluoroacetaldehyde gave a variety of trifluoromethylated carbinols in good enantioselectivities [*47*]. Alkylation with optically active carbon nucleophiles also generated a new stereogenic center effectively [*34,48,49*].

Strong electron withdrawing effect of trifluoromethyl group stabilizes the hemiacetal form. Thus, the nucleophilic alkylation of simple trifluoroacetate, such as CF_3-COOEt gives trifluoromethyl alkyl (or aryl) ketone predominantly. Further catalytic or stoichiometric enantioselective alkylation gave *tert*-alcohols in good yields [*50,51*]. Similarly, catalytic enantioselective Friedel-Crafts type alkylation of trifluoropyruvate which was prepared from hexafluoropropene oxide, gave a variety of α-hydroxy acids, the family of MTPA [*52*]. Diastereoselective alkylation of ketones by carbon nucleophiles with chiral auxiliaries gave trifluoromethylated *tert*-alcohols in good yields [*53-55*]. Chromatographic purification may remove the by-produced diastereomer, thus the method would be of practical.

Diastereoselective trifluoromethylation by trifluoromethyl carbanion species to optically active aldehydes gave trifluoromethylated *sec*-alcohols [*56,57*].

2.1.3 Stereocontrolled oxidations (Scheme 6).

Sharpless dihydroxylation of the trifluoromethylated olefins gave highly functionalized optically active diols [*58-60*] in good yields with good enantioselectivities. Products were purified by recrystallizations.

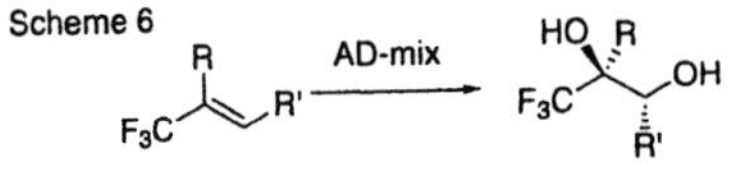

Scheme 6

F_3C–C(R)=CH–R' —AD-mix→ F_3C–C(HO)(R)–CH(OH)–R'

2.2. Trifluoromethylated amines

Due to their biological interests [*9*], there have been so many reports on the preparations of optically active 2-amino-3-substituted-1,1,1-trifluoropropanes, **2.** Recrystallization of the derived diastereomers gave optically pure products [*61,62*]. Enzymatic resolutions also worked [*63,64*].

2.2.1 Stereocontrolled reductions (Scheme 7).

Compared to that of the carbonyl compounds, there have been only a few trials on enantioselective reduction of trifluoromethylated imines [*65-67*]. There still remain a lot of interesting subjects to study. Among the stereocontrolled reductions, 1,3-proton shift method (Scheme 7b) seems to be have a synthetic merit for easy removing of auxiliary [*68*]. Prevention of the further racemization would be a remained problem which should be solved [*69*].

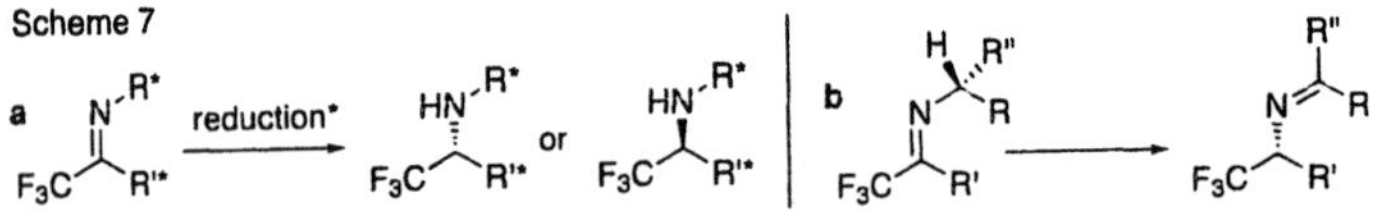

Scheme 7

a F_3C–C(R''*)=N–R* —reduction*→ F_3C–CH(R''*)–HN–R* or F_3C–CH(R''*)–HN–R*

b F_3C–C(R')=N–CH(R'')(R) → F_3C–CH(R')–N=C(R'')(R)

There have been some reports on diastereoselective reductions of imines bearing chiral auxiliaries on their nitrogen atoms [*70-72*].

Diastereoselective reduction of *N,O*-ketal also gave trifluoromethylated amine [*73*]. Although the starting *N,O*-ketal has a chirality on the trifluoromethylated carbon, the chirality of the product was independent from that of starting *N,O*-ketal (Scheme 8). The chirality of the product was depending on other chiralities on the carbon frameworks and rings.

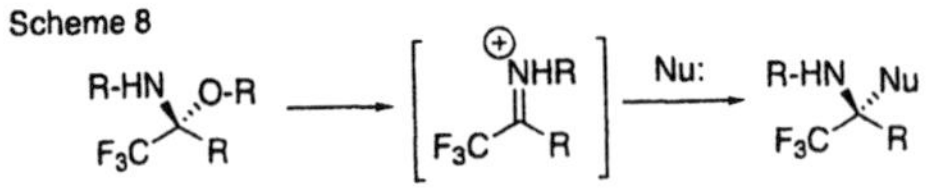

Scheme 8

R-HN, O-R, F_3C, R → [F_3C–C(R)=$\overset{\oplus}{N}$HR] —Nu:→ R-HN, Nu, F_3C, R

2.2.2 Stereocontrolled alkylations (Scheme 9).

Recently, highly enantioselective alkylations of trifluoromethylated imine was reported [*74,75*]. Full optimization of the chiral ligand enabled the highly enantioselective reactions. The alkylated imine was in a hetero 6-membered aromatic ring, thus fixation of *E-Z* structure of the imine seems to be a key for the high enantioselectivity.

Scheme 9

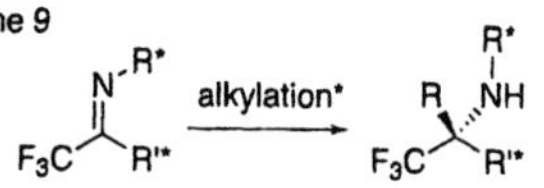

N-R* / F_3C–C(=N)–R'' —alkylation*→ F_3C–C(R)(NH–R*)–R''

Diastereoselective alkylation of imines from fluoral or trifluoromethylated ketones have been studied well. Mother chiral center was on a trifluoro-methylated carbon framework [76], on an imine nitrogen as a chiral auxiliary [77-80], or on a nucleophile [81-83]. Nucleophilic trifluoromethylation of imines with chiral auxiliary also gave optically active trifluoromethylated amines [84].

2.2.3 Diastereoselective aminations (Scheme 10).

Diastereoselective 1,4-amination of trifluoromethylated-acrylamide [85] or trifluoromethylated-nitroolefin [86] gave optically active β,β,β-trifluorinated amines with new chiral center. Diastereoselectivity has been remained moderate.

Scheme 10

F_3C–CH=CH–C(=O)–NR'_2 —R*-NH_2→ F_3C–CH(NH–R*)–CH_2–C(=O)–NR'_2 | F_3C–CH=CH–NO_2 —R*-NH_2→ F_3C–CH(NH–R*)–CH_2–NO_2

2.3. Preparation of other optically active trifluoromethylated compounds

2.3.1 Acetals and ketals.

The hemiacetal form of the fluoral has a trifluoromethylated stereogenic center. Enantioselective catalytic addition of oxygen nucleophile gave optically active hemiacetals [87] (Scheme 11). Further activation of the hydroxy group gave an optically active sulfonate which is a general precursor of trifluoromethylated *sec*-alcohols *via* stereoselective substitution of the sulfonate by carbon nucleophiles.

Scheme 11

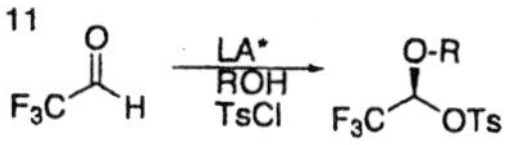

F_3C–CHO —LA*, ROH, TsCl→ F_3C–CH(O-R)–OTs

Some *N,O*-ketals [73,88-90] and *N,S*-ketals [91] have been prepared diastereoselectively. Their further substitutions on the stereogenic center could be regarded as elimination-addition reactions *via* imine species. Metal coordination to the oxygen moiety controlled stereochemistry of further addition [92].

2.3.2 Epoxides and aziridines.

Optically active (75% ee, *S*)-2,3-epoxy-1,1,1-trifluoropropane (TFPO) was industrially produced by a microbial oxidation of trifluoropropene (Scheme 12) [93]. The optically active epoxide has been prepared *via* many routes. Recent enantioselective hydrolysis of racemate gives optically pure epoxide very easily [94]. The nucleophilic ring opening reaction of this epoxide gives a variety of 3-substituted-1,1,1-trifluoro-2-propanols. Details on the preparations and the ring opening reactions of the TFPO were summarized in reviews [7a,b,95].

Scheme 12

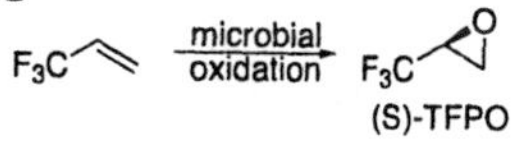

F_3C–CH=CH_2 —microbial oxidation→ F_3C-epoxide (S)-TFPO

Highly substituted trifluoromethylated epoxides were prepared by catalytic enantioselective epoxidation of the corresponding olefins [*96,97*] and diastereoselective carbene addition to the trifluoromethylated ketone [*98*].

Optically active aziridine was prepared *via* addition of carbenoid with chiral auxiliary to trifluoromethylated *N,O*-ketal (imine synthon) [*99*].

2.3.3 Compounds with trifluoromethylated chiral carbon without hetero atom.

Diastereoselective reduction of trifluoromethylated olefin gave trifluoromethylated chiral carbon without heteroatom [*100*]. Diastereoselective 1,4-addition of carbon nucleophiles to trifluoromethylated olefin with carbonyl or sulfoxide also gave trifluoromethylated chiral carbons without heteroatoms [*101,102*]. An intramolecular S_N2' reaction also gave optically pure lactone [*103,104*]. Preparations of optically pure quaternary carbons with trifluoromethyl group were attained by direct fluorinations of optically pure carboxylic acids by SF_4 [*105*].

2.3.4 A new resolution by halogen bonding.

Recently, a new optical resolution using halogen bonding between optically active fluorinated bromide and chiral amine was developed [*106*]. This interaction, halogen bonding, is a unique interaction of bromo- or iodo-fluoro-organics. Further utilization of the interaction for optical resolution is expected.

2.4. Conclusion of survey on preparation of optically active trifluoromethylated compounds.

The literature survey of preparative method for optically enriched (or pure) trifluoromethylated compounds showed us remarkable developments in these ten years. We now can obtain some optically pure trifluoromethylated compounds. Still we have limitations that many of available structures have heteroatom substitutions at chiral center, -OH or $-NR_2$.

Catalytic enantioselective preparation is ideal. However, we have only a limited number of such reactions with high enough enantioselectivities. Furthermore, we have to purify optically any target compounds for practical use. The method for optical purification of enantiomer is very limited. We have some chance to purify the crystalline fluoro-organics, but we have to use expensive chiral chromatographic methods for oily or wax compounds. Meanwhile, diastereomers can be separated *via* simple and rather cheap chromatographic methods. This possible purification of the diastereomers made diastereoselective reaction rather practical. However, the diastereoselective strategy needs protection by chiral auxiliary and deprotection processes.

Right now, we synthesize the aimed compounds from available non-chiral fluoro-organics, then give a stereogenic center almost at the final stage of the syntheses, respectively. In general organic synthesis, we start a total synthesis of the aimed compound from easily available chiral compound, so called "chiral pool." To adopt this strategy, we also have to develop reliable chiral fluoro-organics and stereospecific transformations at their chiral center.

3. Stereospecific transformation of functional groups on α-carbon to trifluoromethyl group

We have to discriminate stereospecific reactions from diastereoselective reactions, at first. Some stereoselective substitution of *N,O*-ketal proceeded *via* tentative recovery of imine moiety followed by nucleophilic addition. Thus, the chirality of the product was independent from that of the reaction center, at all. The stereoselectivity seems to controlled by chirality of neighboring moieties (Scheme 8,13) [*73,77,88-90,107,108*].

Scheme 13

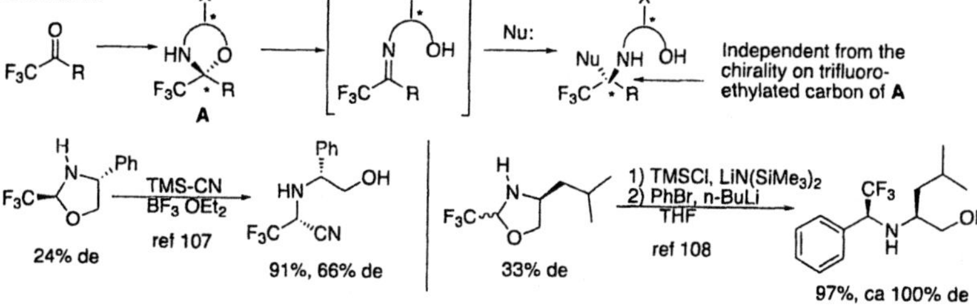

Independent from the chirality on trifluoroethylated carbon of **A**

TMS-CN, $BF_3\ OEt_2$, ref 107; 24% de → 91%, 66% de

1) TMSCl, $LiN(SiMe_3)_2$ 2) PhBr, n-BuLi, THF, ref 108; 33% de → 97%, ca 100% de

Meanwhile, chirality of the product in another nucleophilic substitution of *N,O*-ketal was depending on the chirality of the ketal carbon (Scheme 14) [*92,109*].

Scheme 14

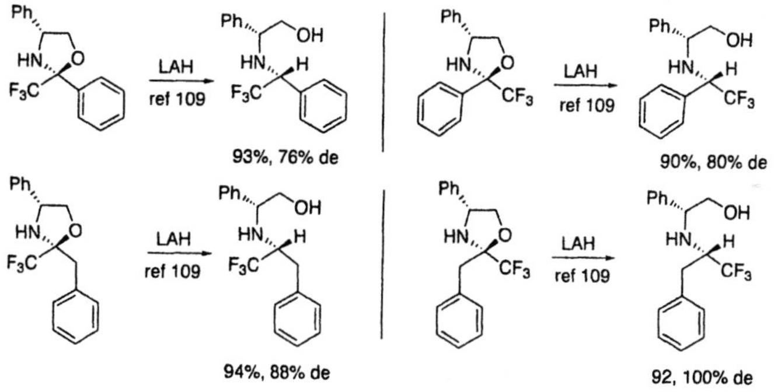

LAH, ref 109: 93%, 76% de

LAH, ref 109: 90%, 80% de

LAH, ref 109: 94%, 88% de

LAH, ref 109: 92, 100% de

A coordination of organometallic reagent to the oxygen was suggested for explanation of retained chirality of the product [*92*].

3.1. Nucleophilic substitution

Fortunately, a wide variety of preparations for the enantiomerically enriched α-trifluoromethylated carbinol unit, CF_3-CH(OH)-, have been developed (see above). These compounds **1** had been considered to be a potential optically active fluorinated starting materials with a wide applicability. However, the stereospecific nucleophilic substitution of the hydroxy group in the unit structure was always notoriously difficult [*110*]. Thus, the synthetic application of the compounds had been limited to their use without recombination of the carbon-oxygen bonds (Scheme 15).

Scheme 15

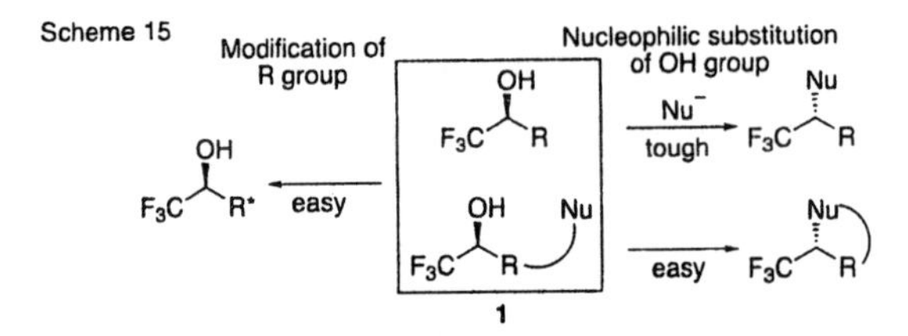

The prevention of the nucleophilic substitution of the hydroxyl group in the CF_3-CH(OH)- unit has been attributed to the strong electron withdrawing effect as well as the steric hindrance of the trifluoromethyl group. The former effect strengthens and shortens the C-O bond. The steric hindrance of the group is much larger than that of the methyl group and has sometimes been compared with the isopropyl and/or *tert*-butyl groups [*111,112*]. The origin of the steric hindrance is the electrostatic repulsion of the negatively charged trifluoromethyl group to the negatively charged nucleophiles; thus the group would be repulsed by a greater distance than that results from the van der Waals repulsion. We believed that the Coulombic repulsion effect would be suppressed in an intramolecular substitution, because a nucleophile would be readily put near the reaction center. In fact, we succeeded in preparing optically active aziridine [*113*], and cyclopropanes [*114*] *via* the intramolecular nucleophilic substitutions of the hydroxy group in the CF_3-CH(OH)- unit.

Until 1995, a few nucleophilic substitutions of the hydroxy group of the CF_3-CH(OH)- unit had been reported [*115-121*]. However, the S_N2-type reactions without participation of the neighboring heteroatom or π-conjugations had been a very few [*115,117,118*].

We had made a study on the halogenations of 3-isopropyloxy-1,1,1-trifluoro-2-propanol **3** to chlorinated product **4** with a series of reagents (Scheme 16). Our trials, as well as previous reports, found that only PPh_3/CCl_4 was effective [*118,122*]. After the cautious separation from starting fluoro-alcohol **3**, the product **4** was submitted to hydrolysis under a basic condition. A trace (ca. 3%) amount of fluoro-alcohol **3** was isolated and chiral GC analysis of the product was identified to be the starting fluoro-alcohol **3** with 75% ee (*S*). Thus, the halogenation with PPh_3/CCl_4 would be enantiospecific process with inversion of configuration.

Scheme 16

F_3C-CH(OH)-CH_2-O-^{i}Pr (**3** 75% ee) → reagents / conditions (refs 118,122) → F_3C-CH(X)-CH_2-O-^{i}Pr (**4**) → KOH / H_2O → F_3C-CH(OH)-CH_2-O-^{i}Pr (**3** 75% ee)

PPh_3/CCl_4 Δ: **4**, 40%
HCl/H_2SO_4 Δ: no reaction
HCl/AcOH Δ: no reaction
$SOCl_2$/Py Δ: no reaction
SO_2Cl_2/Py Δ: no reaction
PCl_3/CH_2Cl_2 Δ: no reaction
PCl_5/CH_2Cl_2 Δ: no reaction
$(PhO)_3P$, $BnBr/CH_2Cl_2$ Δ: no reaction
Ph_2PCl, Et_3N Δ: no reaction
Ph_3PCl_2/CH_2Cl_2 Δ: no reaction

Contrary to the reaction of hydroxyether **3**, reactions of hydroxysulfides **5** with PPh_3Cl_2/CH_3CN or PPh_3/CCl_4 produced two different halo sulfides, **6** and **7**, depending on the reagents and conditions (Scheme 17) [*123*]; the reaction with PPh_3Cl_2/CH_3CN mainly produced rearranged halo sulfide **7**, while the reaction with PPh_3/CCl_4 gave a directly substituted halo sulfide **6**. When the latter reagent was used in a more polar solvent, such as CH_3CN, the main product of the reaction became the rearranged 7 (Scheme 17).

Scheme 17

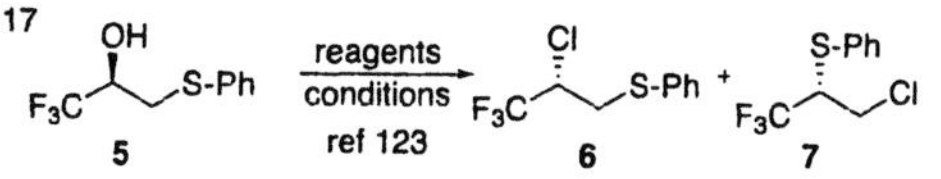

reagent, conditions	6 : 7	yield
PPh_3Cl_2/CH_3CN, Δ, 22 h	10 : 90	(92%)
PPh_3 / CCl_4, Δ, 2.5 h	60 : 40	(64%)
PPh_3, CCl_4 / CH_3CN, Δ, 24 h	23 : 77	(77%)

Scheme 18

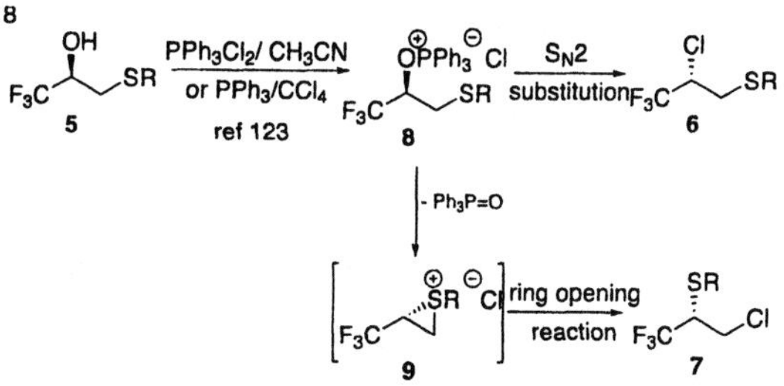

Both compounds would be produced *via* common intermediate **8** in Scheme 18. The halo sulfide **6** would be produced *via* direct S_N2 through the action of the counter chloride ion of intermediate **8**, while the halo sulfide 7 would be produced *via* intramolecular S_N2 through the action of the sulfur atom to give an episulfonium **9**, followed by a successive ring-opening reaction at the less-hindered methylene position by counter chloride ion.

The solvent effect would play an important role in determining the reaction course as shown in Scheme 18. When the solvent is less polar, the ion pair should be "contact." Thus, the counter chloride ion would be located close to the reaction center; it would be near enough to undertake the direct S_N2 to give **6**. In the case of more polar solvent, on the other hand, the ion pair would be somewhat "separated." Thus, the intramolecular S_N2 would be initially preferable at producing electronically unstable episulfonium **9**, which would then ring-opened by counter anion, to give 7. The total transformation of **4** to either **6** or 7 involves a stereospecific inversion at the trifluoromethylated carbon atom. Here, we can propose that the key item for nucleophilic substitution at α-carbon to trifluoromethyl group is "how to put the nucleophile near by the reaction center". On this basis, an intramolecular reaction would be the most simple and reliable way to put a nucleophile near by the reaction center.

3.1.1 Intramolecular strategy

Appel's condition with enantiomerically pure aminoalcohol **10** with PPh_3Cl_2/CH_3CN produced a corresponding *N*-benzyl-2-trifluoromethylaziridine **11** in good yield (Scheme 19) [*113*]. The stereochemistry of the aziridine was confirmed to be (*R*); the reaction also completed with perfect inversion of configuration. With a similar manner, *N*-tosyl-2-trifluoromethyl aziridine was produced [*124*]. These aziridine can be a general precursor for β,β,β-trifluoroamines. Ring opening reaction of the aziridine has been reported, elsewhere [*125*].

Scheme 19

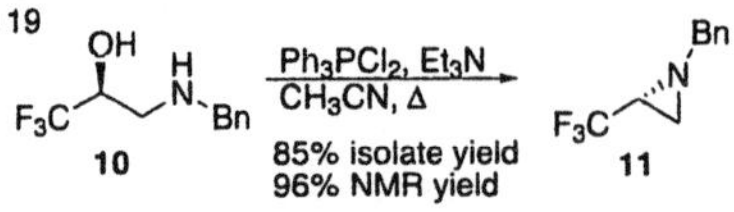

Scheme 20

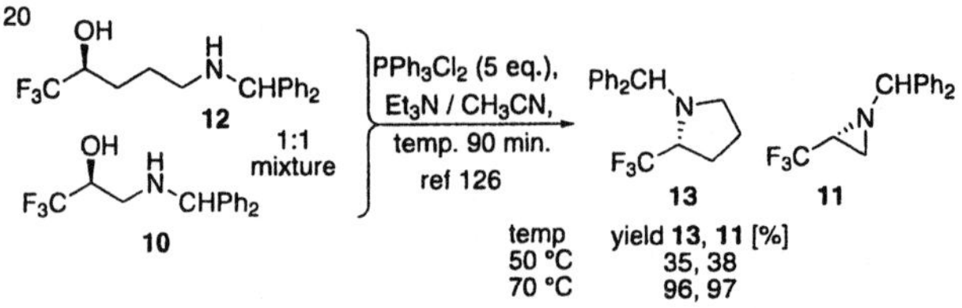

A similar ring closure reaction using PPh_3Cl_2/CH_3CN to produce pyrrolidine **13** was successful [*126*]. A relative rate study on the ring-closure reactions of **12** and **10** showed that the reaction rates of these two cyclizations were comparable (Scheme 20).

Intramolecular nucleophilic substitution by carbanion species had been regarded as being much more difficult than that with nitrogen nucleophiles, because of there being much lager repulsion between the trifluoromethyl group and the localized negative charge of the carbanion than between the trifluoromethyl group and the lone pair on nitrogen nucleophile. To prevent this large unwanted repulsion, the negative charge on the carbanion moiety should be delocalized. The reaction of cyanohydrin **14a-h** with aromatic moieties was achieved easily to produce trifluoromethylated cyclopropanes in moderate to good yields (Scheme 21, 40-85%) [*6, 114a*].

Scheme 21

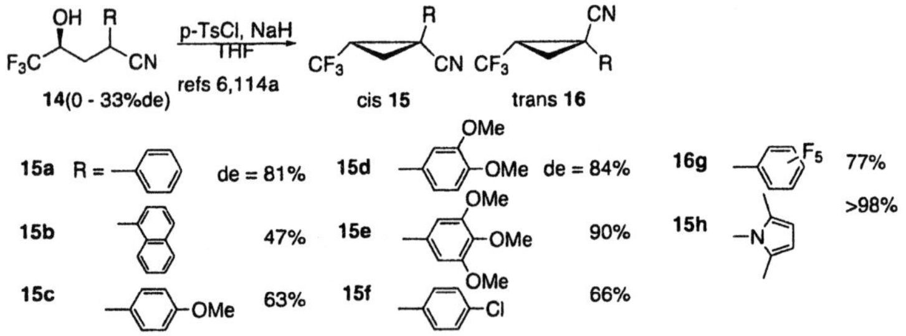

The carbanion center of intermediate anion species **17** and **18** would be sp^2-configurational thus prochiral, and the de of the starting material would not affect to the de of the product. The de was affected by electrostatic repulsion between trifluoromethyl group and cyano moiety or aromatic moieties (Scheme 22) [*6*]. Absolute configuration of the product was controlled by chiral carbon center attached by trifluoromethyl group. The product was converted to 2-trifluoromethylcyclopropane amino acids [*114b*].

Scheme 22

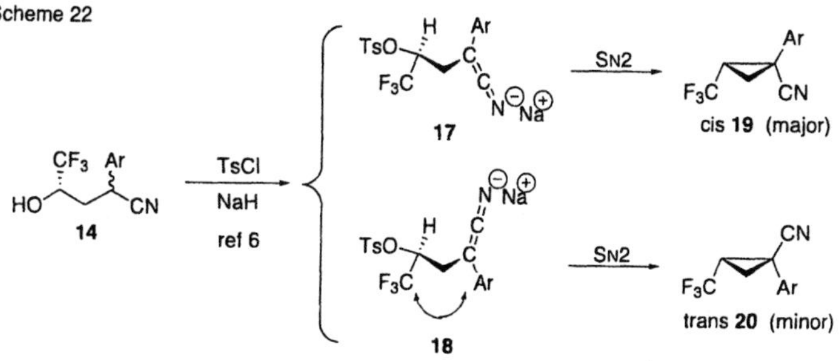

CF_3 Ar; HO ~ CN; **14** — TsCl, NaH, ref 6 → TsO, H, Ar, F_3C, **17** (SN2) → cis **19** (major); TsO, H, F_3C, Ar, **18** (SN2) → trans **20** (minor)

3.1.2 Intermolecular reactions

As was described in our previous reviews and in the previous section, the intermolecular S_N2 at α-carbon to trifluoromethyl group is notorious for its difficulty [7]. Thus, the reaction needs some device to attract nucleophiles near by the reaction center. At the same time, the reaction should be so called "pure S_N2", without participation of cationic S_N1 mechanism throughout the reaction, due to the strong electron withdrawing effect of trifluoromethyl group. Thus, the reaction would undergo with complete inversion of configuration. The reaction is tough to do but is stereochemically pure process.

To date, a nucleophilic substitution of hydroxyl group at the primary carbon of 2,2,2-trifluoroethanol has been reported. The reaction gave the amino acid derivatives in moderate yield (Scheme 23) [*117*]. A reaction of much hindered α-trifluoromethylated secondary alcohol with organometallic reagents has resulted in recovery of starting alcohol (Scheme 24) [*110,127*].

Scheme 23

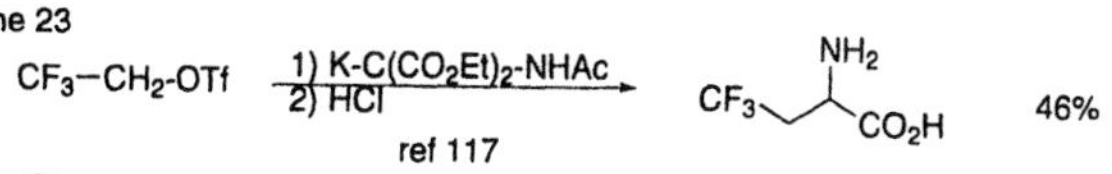

CF_3–CH_2-OTf — 1) K-C(CO_2Et)$_2$-NHAc, 2) HCl, ref 117 → CF_3CH$_2$CH(NH_2)CO_2H 46%

Scheme 24

R–CH(CF_3)–OTf — Metal$^+$ $^-CH_2$-R, ref 110,127 → No Reaction

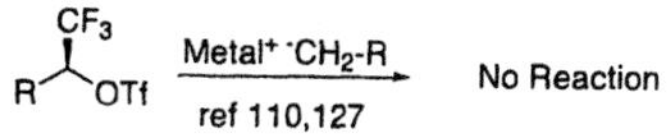

Some stereospecific intermolecular substitutions at α–carbon to trifluoromethyl group by heteroatom nucleophiles have been reported; the substitutions by halogens were illustrated in Scheme 16 [*118,122*]. Nucleophilic substitutions of a triflate with benzoic acid successfully underwent with complete inversion of the absolute configuration of the reaction center with CsF catalyst to produce the other diastereomer stereoselectively (Scheme 25, upper) [*128*]. Similar nucleophilic substitutions by oxygen, nitrogen, and sulfur nucleophiles utilizing KF or CsF catalysts gave the products with inversion of the absolute configurations in moderate yields (Scheme 25, lower) [*129*].

In both cases, the CsF seems to be a key agent for the intermolecular nucleophilic substitutions.

Ammonolysis of tosylate under a high pressure gave optically pure amine in stereospecific manner (Scheme 26) [*130*]. Some electronic participation of phenyl ring is expected in this reaction. The aid of aromatic moiety in Friedel-Crafts alkylation had been confirmed [*116a*].

Scheme 25

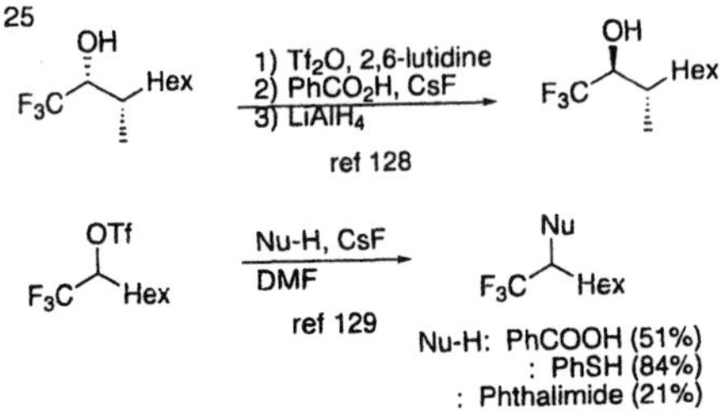

OH F$_3$C Hex → 1) Tf_2O, 2,6-lutidine 2) $PhCO_2H$, CsF 3) $LiAlH_4$ ref 128 → OH F$_3$C Hex

OTf F$_3$C Hex → Nu-H, CsF DMF ref 129 → Nu F$_3$C Hex

Nu-H: PhCOOH (51%)
: PhSH (84%)
: Phthalimide (21%)

Scheme 26

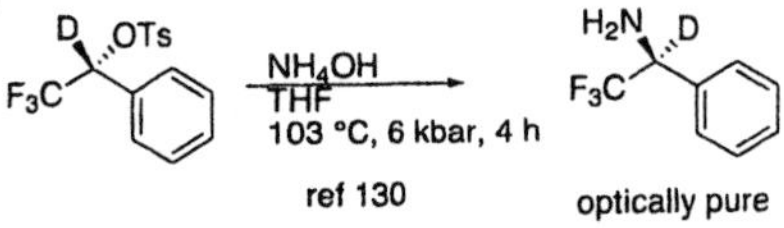

D OTs F$_3$C → NH_4OH THF 103 °C, 6 kbar, 4 h ref 130 → H_2N D F$_3$C

optically pure

Neighboring oxygen atom could assist formation of cationic species with contribution of oxonium structure. Thus, the substitution of one oxygen in acetal should be easier than that of alcohols. Actually, alkylations by carbanions have been succeeded (Scheme 27) [*87, 131*]. The use of aluminum ate complex as the nucleophile seems to be a key for this process. The strong interaction between organo-aluminium and trifluoromethyl group has been sometimes suggested [*132*]. Chirality did not transfer perfectly; some epimerization occurred.

Scheme 27

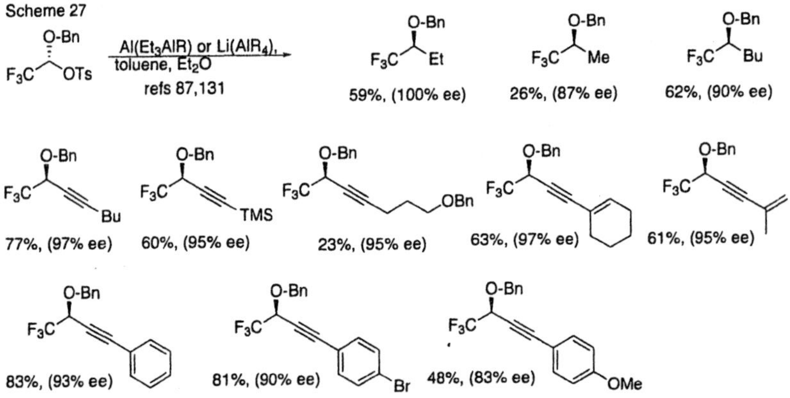

O-Bn F$_3$C OTs → $Al(Et_3AlR)$ or $Li(AlR_4)$, toluene, Et_2O refs 87,131

O-Bn F$_3$C Et 59%, (100% ee)
O-Bn F$_3$C Me 26%, (87% ee)
O-Bn F$_3$C Bu 62%, (90% ee)

O-Bn F$_3$C Bu 77%, (97% ee)
O-Bn F$_3$C TMS 60%, (95% ee)
O-Bn F$_3$C OBn 23%, (95% ee)
O-Bn F$_3$C 63%, (97% ee)
O-Bn F$_3$C 61%, (95% ee)

O-Bn F$_3$C 83%, (93% ee)
O-Bn F$_3$C Br 81%, (90% ee)
O-Bn F$_3$C OMe 48%, (83% ee)

A substitution of chlorine by fluorine under a strong oxidative condition also underwent with inversion of absolute configuration, but with some racemization (Scheme 28) [*133*].

Scheme 28

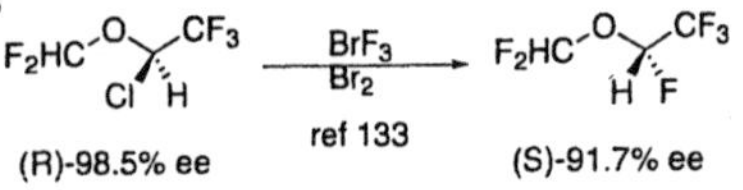

F_2HC O CF_3 Cl H → BrF_3 Br_2 ref 133 → F_2HC O CF_3 H F

(R)-98.5% ee (S)-91.7% ee

3.1.3 S_N2' strategy.

An S_N2' type reaction of γ-trifluoromethylated allylalcohols gave *tert*-carbon framework with complete chirality transfer (Scheme 29) [*134*]. The reaction could be an alternative method for stereospecific preparation of optically active trifluoromethylated carbon without heteroatom substituents.

Scheme 29

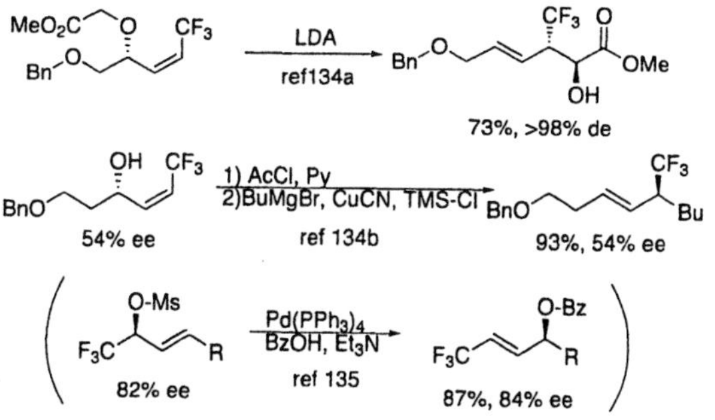

LDA, ref134a — 73%, >98% de

1) AcCl, Py; 2)BuMgBr, CuCN, TMS-Cl, ref 134b — 54% ee → 93%, 54% ee

$Pd(PPh_3)_4$, BzOH, Et_3N, ref 135 — 82% ee → 87%, 84% ee

The starting compound was produced from trifluoroacetylenes but would be prepared from optically pure α-trifluoromethylated allylalcohols *via* an S_N2' substitution [*135*].

3.1.4 Scope and remained problems on enantiospecific S_N2 strategy

Among the nucleophilic substitutions at α-carbon of trifluoromethyl group, intramolecular nucleophilic substitution seems to work, somewhat. However, sever limitation is that the products have been limited to the three-membered ring compounds. Production of five membered ring compounds became tough due to the higher degree of freedom of longer methylene chain.

Present scope of the stereospecific intermolecular substitution is still narrow. Trifluoromethylated carbinols have experienced heteroatom substitutions with the aid of CsF salt, but no carbanion substitutions, yet. Neighboring group participation by oxygen enabled substitutions by carbanions, but causes some epimerizations.

3.2 Substitution of protons by electrophiles: trifluoromethylated carbanion chemistry.

A carbanion with sp^3 carbon center can hold its chirality in the course of some additions to electrophiles. Moreover, trifluoromethyl group is a strong electron withdrawing group, thus it should stabilize the carbanion at α-carbon. Therefore, the α-trifluoromethylated carbanions from optical active trifluoromethylated compounds could be good synthetic units which posses both trifluoromethyl group and chirality. However, there have been a few reports on such sp^3 carbanion species [*136,137*]. This present situation is due to unstable nature of such carbanion, it spontaneously gives off a fluorine atom as a fluoride to produce difluoroolefins [*1b,2*]. To suppress this unwanted side reaction, many trifluoromethylated carbanions have been prepared as conjugate enol form with sp^2 carbanion center [*137,138*], where they have no chance for stereospecific reaction at that carbon. One of the possible access to the stereochemically stable carbanion is to utilize a ring fused carbanion, in particular, a carbanion on a small ring such as cyclopropanes, oxiranes, and aziridines. In actual, only the

Scheme 30

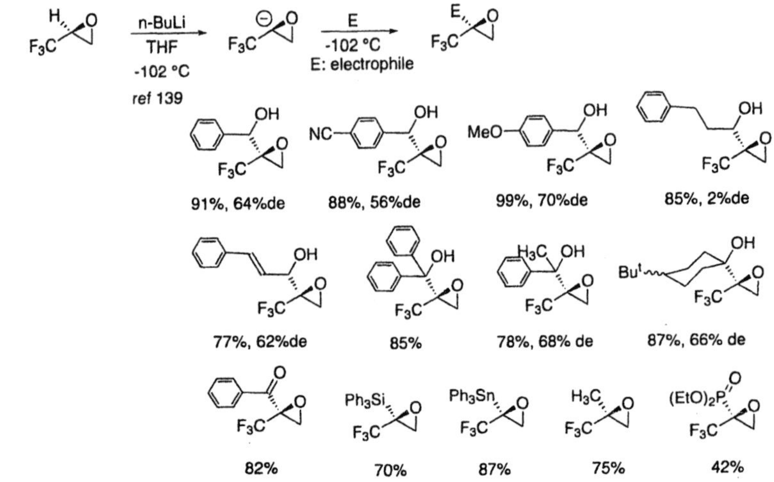

F_3C-oxirane (H) → n-BuLi, THF, -102 °C, ref 139 → carbanion → E, -102 °C (E: electrophile) → E-substituted F_3C-oxirane

91%, 64%de — 88%, 56%de — 99%, 70%de — 85%, 2%de

77%, 62%de — 85% — 78%, 68% de — 87%, 66% de

82% — 70% — 87% — 75% — 42%

oxiranyl anion [*139*] and the aziridinyl anion [*124, 139b*] retained their chirality on the α-carbon, among the reported α-trifluoromethylated carbanions.

The carbanion from optically active TFPO reacted with a variety of electrophiles to give optically active 2-substituted-2-trifluoromethyl oxiranes (Scheme 30) [*139*].

Anion center (HOMO) of sp^2 carbanion delocalizes in it's π-system (Figure 1 left). Thus, negative charges on the fluorine atoms in trifluoromethyl group are almost the same extent as that of neutral species. That of trifluoromethylated oxiranyl anion also delocalizes in its epoxy ring (Figure 1 right). One may found the HOMO lobes of the oxiranyl anion resemble to that of the sp^2

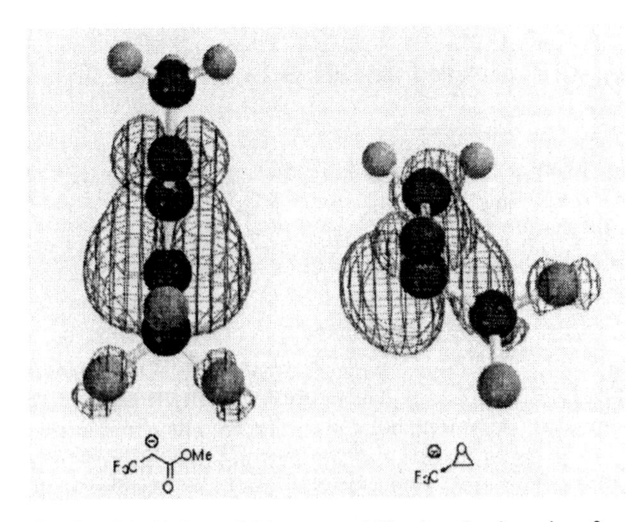

Figure 1: HOMO orbital lobes of (a) ester stabilized carbanion view from carbonyl oxygen, and (b) oxiranyl anion view from oxygen of epoxy ring.

carbanion. This similar distribution would make the both carbanion species stable enough to be usable for organic syntheses [*140*].

Similar to the oxiranyl anion, optically pure 2-trifluoromethylated aziridinyl anion reacted with a variety of electrophiles to give optically pure trifluoromethylated compounds (Scheme 31) [*124,139b*].

Scheme 31

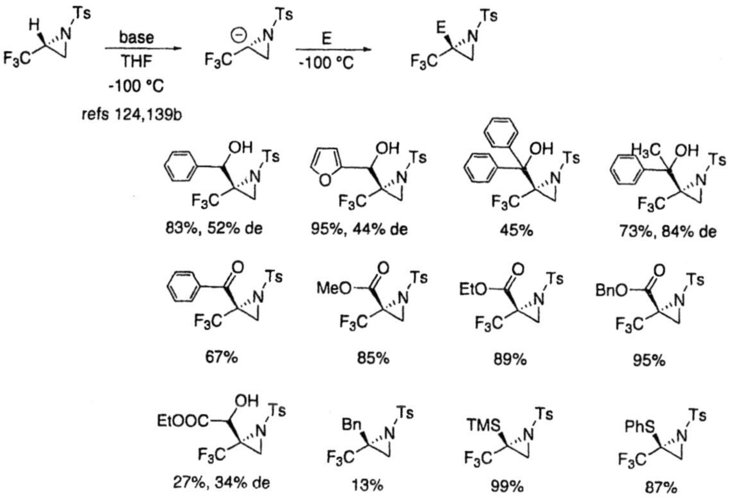

base, THF, -100 °C; E, -100 °C

refs 124,139b

83%, 52% de — 95%, 44% de — 45% — 73%, 84% de

67% — 85% — 89% — 95%

27%, 34% de — 13% — 99% — 87%

4. Conclusion.

Even the most common unit reaction in organic chemistry, stereospecific nucleophilic substitution and preparation of optically active carbanion, has not yet been done with α-trifluoromethylated compounds. We have to consider extreme effect of the fluorine and have to develop a new device to overcome the problems. The eventual desired developments and devices in the field of fluorine chemistry may have a wide ripple effect in the promotion of organic synthesis, as one of a strongest synthetic methodology.

Acknowledgments

We thank the Ministry of Education, Science, Sports and Culture of the Government of Japan (Grant-in-Aid for Scientific Research on Priority Areas, No. 706 Dynamic Control of Stereochemistry and No 12650854).

References

1. (a) Soloshonok, V. A. Ed, *Enantiocontrolled Synthesis of Fluoro-Organic Compounds,* Wiley, Chichester (1999). (b) Hiyama, T.; Kanie, K.; Kusumoto, T.; Morizawa, Y.; Shimizu, M., *Organofluorine Compounds,* Springer, Berlin (2000).

2. Kitazume, T.; Yamazaki, T. *Experimental Methods in Organic Fluorine Chemistry*, Kodansha and Godon & Breach, Tokyo (1998).
3. Atkins, P. Physical Chemistry, 5th ed., Freeman, New York (1994), pp488.
4. Katagiri, T.; Uneyama, K. *Bull. Chem. Soc. Jpn.* **2001**, *74*, 1409-1410.
5. (a) Taft, R. W. Jr., *J. Am. Chem. Soc.* **1957**, *79*, 1045-1049. (b) Hansch, C., Leo, A.; Taft, R. W. *Chem. Rev.* **1991**, *91*, 165-195.
6. Katagiri, T.; Yamaji, S.; Handa, M.; Irie, M.; Uneyama, K. *Chem. Commun.* **2001**, 2054-2055.
7. (a) Katagiri, T. Synthesis of Stereochemically Defined Trifluoromethyl-Containing Compounds through (*S*)-3,3,3-Trifluoropropene Oxide, in *Enantiocontrolled Synthesis of Fluoro-Organic Compounds: Stereochemical Challenges and Biomedicinal Targets.* Soloshonok, V. A. Ed., Wiley: Chichester (1999), p 161-178. (b). Katagiri, T.; Uneyama, K. *J. Fluorine. Chem.* **2000**, *105*, 285-293. (c) Katagiri, T.; Uneyama, K. *Chirality* **2003**, *15*, 4-9.
8. Iseki, K. *Tetrahedron* **1998**, *54*, 13887-13914.
9. Bravo, P.; Crucianelli, M.; Ono, T.; Zanda, M. *J. Fluorine Chem.* **1999**, *97*, 27-49.
10. *Aldrich Catalog Handbook of Fine Chemicals 1996-1997,* Aldrich, Milwaukee, (1996); *CRC Handbook of Chemistry and Physics, 74th ed.* CRC Press, Boca Raton, (1993);
11. (a) Katagiri, T.; Fujiwara, Y.; Takahashi, S.; Ozaki, N.; Uneyama, K. *Chem. Commun.* **2002**, 986-987. (b) Fujiwara, Y.; Katagiri, T.; Uneyama, K. *Tetrahedron Lett.* **2003**, *44*, 6161-6163.
12. Katagiri, T.; Yoda, C.; Furuhashi, K.; Ueki, K.; Kubota, T. *Chem. Lett.* **1996**, 115-116.
13. (a) Katagiri, T.; Uneyama, K. *Chem. Lett.* **2001**, 1330-1331. (b) Katagiri, T.; Duan, M.; Mukae, M.; Uneyama, K. *J. Fluorine Chem.* **2003**, *120*, 165-172.
14. Jacques, J; Collet, A.; Wilen, S. H. *Enantiomers, Racemates, and Resolutions,* John Wiley & Sons, New York (1981).
15. Ishizuka, T.; Miura, H.; Nohira, H. *Nihon Kagaku Kaishi* **1990**, 1171-1177.
16. Bussche-Hunnefeld, C. von dem; Cescato, C.; Seebach, D. *Chem. Ber.*, **1992**, *125*, 2795-2802
17. Ishii, A.; Kanai, M.; Higashihara, K.; Mikami, K. *Chirality*, **2002**, *14*, 709-712.
18. Gasper, J.; Guerrero, A. *Tetrahedron: Asymm.* **1995**, *6*, 231-238.
19. Hamada, H.; Shiromoto, M.; Funahashi, M.; Itoh, T.; Nakamura, K. *J. Org. Chem.* **1996**, *61*, 2332-2336
20. Komatsu, Y.; Sasaki, F.; Takei, S.; Kitazume, T. *J. Org. Chem.* **1998**, *66*, 8058-8061.
21. Lin, J. T.; Yamazaki, T.; Kitazume, T. *J. Org. Chem.* **1987**, *52*, 3211-3217.

22. Laumen, K.; Schneider, M. P. *J. Chem. Soc., Chem. Commun.* **1988**, 598-600.
23. (a) O'Hagen, D.; Zaidi, N. A.; Lamont, R. B. *Tetrahedron: Asymm.* **1993**, *4*, 1703-1708. (b) O'Hagen, D.; Zaidi, N. A. *Tetrahedron: Asymm.* **1994**, *5*, 1111-1118. (c) O'Hagen, D.; Zaidi, N. A. *J. Chem. Soc., Perkin Trans. 1* **1992**, 947-949.
24. Shimizu, M.; Sugiyama, K.; Fujisawa, T.; *Bull. Chem. Soc. Jpn.* **1996**, *69*, 2655-2659.
25. Itoh, T.; Shiromoto, M.; Inoue, H.; Hamada, H.; Nakamura, K. *Tetrahedron Lett.* **1996**, *37*, 5001-5002.
26. Konigsberger, K.; Prasad, K.; Repic, O. *Tetrahedron: Asymm.* **1999**, *10*, 679-687.
27. Zhang, Y.; Li, J.-F.; Yuan, C.-Y. *Tetrahedron* **2003**, *59*, 473-479.
28. Ramachandran, P. V. Asymmetric Reduction of Fluorine-Containing Carbonyl Compounds, in *Enantiocontrolled Synthesis of Fluoro-Organic Compounds: Stereochemical Challenges and Biomedicinal Targets.* Soloshonok, V. A. Ed., Wiley: Chichester (1999), p 179-228.
29. (a) Nakamura, K.; Matsuda, T.; Itoh, T.; Ohno, A. *Tetrahedron Lett.* **1996**, *37*, 5727-5730. (b) Nakamura, K.; Matsuda, T. *Tetrahedron* **1998**, *54*, 8393-8402. (c) Matsuda, T.; Hanada, T.; Nakajima, N.; Itoh, T.; Nakamura, K. *J. Org. Chem.* **2000**, *65*, 157-163.
30. Bucciarelli, M.; Forni, A.; Moretti, I.; Torre, G. *J. Chem. Soc., Chem. Commun.* **1978**, 456-457.
31. (a) Fujisawa, T.; Ichikawa, K.; Shimizu, M. *Tetrahedron: Asymm.* **1993**, *4*, 1237-1240. (b) Fujiwara, T.; Onogawa, Y.; Sato, A; Mitsuya, T.; Shimizu, M. *Tetrahedron*, **1998**, *54*, 4267-4276. (c) Fujiwara, T.; Tanaka, S.; Onogawa, Y.; Shimizu, M. *Tetrahedron Lett.* **1999**, *40*, 1953-1956.
32. (a) Arnone, A.; Biagini, G.; Cardillo, R.; Resnati, G. Begue, J. P.; Bonnet-Delpon, D.; Kornilov, A. *Tetrahedron Lett.* **1996**, *37*, 3903-3906. (b) Arnone, A.; Bernardi, R.; Blasco, F.; Cardillo, R.; Resnati, G.; Gerus, I. I.; Kukhur, V. P. *Tetrahedron*, **1998**, *54*, 2809-2818.
33. Nakamura, K.; Yamanaka, R. *Tetrahedron: Asymm.* **2002**, *13*, 2529-2533.
34. Tomoyasu, T.; Tomooka, K.; Nakai, T. *Synlett* **1998**, 1147-1149.
35. Arnone, A.; Bravo, P.; Panzeri, W.; Viani, F.; Zanda, M. *Eur. J. Org. Chem.* **1999**, 117-127.
36. Morrison, J. D.; Ridgway, R. W. *J. Org. Chem.* **1974**, *39*, 3107-3110.
37. Nasipuri, D.; Bhattacharya, P. K. *J. Chem. Soc., Perkin Trans. 1* **1977**, 576-578.
38. (a) Ramachandran, P. V.; Teodorovic, A. V.; Brown, H. C. *Tetrahedron* **1993**, *49*, 1725-1738. (b) Ramachandran, P. V.; Teodorovic, A. V.; Gong, G.; Brown, H. C. *Tetrahedron: Asymm.* **1994**, *5*, 1075-1086. (c)

Ramachandran, P. V.; Gong, B.; Brown, H. C. *J. Org. Chem.* **1995**, *60*, 41-46
39. Prakash, G. K. S.; Mandal, M.; Schweizer, S.; Petasis, N. A.; Olah, G. A. *Org. Lett.* **2000**, *2*, 3173-3176.
40. Gran, U.; Wennerstrom, O.; Westman, G. *Tetrahedron: Asymm.* **2000**, *11*, 3027-3040.
41. Corbett, M. S.; Liu, X.; Sanyal, A.; Snyder, J. K. *Tetrahedron Lett.* **2003**, *44*, 931-935.
42. Yong, K. H.; Chong, J. M. *Org. Lett.* **2002**, *4*, 4139-4142.
43. (a) Kuroki, Y.; Asada, D. Sakamaki, Y.; Iseki, K. *Tetrahedron Lett.* **2000**, *41*, 4603-4607. (b) Kuroki, Y.; Sakamaki. Y.; Iseki, K. *Org. Lett.* **2001**, *3*, 457-459
44. (a) Arx, M. von; Mallat, T.; Baiker, A. *Angew. Chem. Int. Ed. Engl.* **2001**, *40*, 2303-2305. (b) Arx, M. von; Bürgi, T.; Mallat, T.; Baiker, A. *Chem. Eur. J.* **2002**, *8*, 1430-1437.
45. Marcu, I. C.; Millet, J. M. M.; Herrmann, J. M. *Cat. Lett.* **2002**, *78*, 267-271.
46. Sterk, D.; Stephan, M. S.; Mohar, B. *Tetrahedron Lett.* **2004**, *45*, 535-537.
47. (a) Mikami, K.; Yajima, T. Synthesis of enantiomerically pure α-trifluoromethyl alcohols *via* chiral Lewis acid-catalyzed carbonyl-ene reactions and their application in the design of new liquid crystals, in *Enantiocontrolled Synthesis of Fluoro-Organic Compounds: Stereochemical Challenges and Biomedicinal Targets*. V. A. Soloshonok, Ed., Wiley: Chichester (1999), p 557-574. (b) Mikami, K.; Yajima, T.; Terada, M.; Kato, E.; Maruta, M. *Tetrahedron: Asymm.* **1994**, *5*, 1087-1090. (c) Mikami, K.; Yajima, T.; Takasaki, T.; Matsukawa, S.; Terada, M.; Utimaru, T.; Maruta, M. *Tetrahedron* **1996**, *52*, 85-98. (d) Mikami, K.; Yajima, T.; Terada, M.; Suzuki, Y.; Kobayashi, I. *J. Chem. Soc., Chem. Commun.* **1997**, 57-58. (e) Ishii, A.; Mikami, K. *J. Fluorine Chem.* **1999**, *97*, 51-55. (f) Mikami, K.; Yajima, T.; Siree, N.; Terada, M.; Suzuki, Y.; Takanishi, Y.; Takezoe, H. *Synlett* **1999**, *12*, 1895-1898. (g) Ishii, A; Soloshonok, V. A.; Mikami, K. *J. Org. Chem.* **2000**, *65*,1597-1599.
48. Soloshonok, V. A.; Kukhar, V. P.; Galushko, S. V.; Svistunova, N. Y.; Avilov, D. V.; Kuzmina, N. A.; Raevski, N. I.; Struchkov, Y. T.; Pysarevsky, A. P.; Belokon, Y. N. *J. Chem. Soc. Perkin Trans 1* **1993**, 3143-3155.
49. Bravo, P.; Frigerio, M.; Melloni, A.; Panzeri, A.; Pesenti, A.;Viani, F.; Zanda, M. *Eur. J. Org. Chem.* **2002**, 1895-1902.
50. Tan, L.; Chen, C.; Tillyer, R. D.; Grabowski, E. J. J.; Reider, P. J. *Angew. Chem. Int. Ed.* **1999**, *38*, 711-713.
51. Loh, T.-P.; Zhou, J.-R.; Li, X.-R. *Tetrahedron Lett.* **1999**, *40*, 9333-9336.
52. Zhuang, W.; Gathergood, N.; Hazell, R. G.; Jorgenson, K. A. *J. Org. Chem.* **2001**, *66*, 1009-1013.

53. Soloshonok, V. A.; Avilov, D. V.; Kukhur, V. P. *Tetrahedron: Asymm.* **1996**, *7*, 1547-1550.
54. Fernandez, R.; Martin-Zamora, E.; Pareja, C.; Vazquez, J.; Diez, E.; Monge, A.; Lassaletta, J. M. *Angew. Chem. Int. Ed.* **1998**, *37*, 3428-3430.
55. Sani, M.; Belotti, D.; Giavazzi, R.; Panzeri, W.; Volonterio, A.; Zanda, M. *Tetrahedron Lett.* **2004**, *45*, 1611-1615.
56. Hanzawa, Y.; Uda, J.; Kobayashi, Y.; Ishido, Y.; Taguchi, T.; Shiro, M. *Chem. Pharm. Bull.* **1991**, *39*, 2459-2461.
57. Lavaire, S.; Plantier-Royon, R.; Potella, C. *Tetrahedron: Asymm.* **1997**, *9*, 213-226.
58. Vanhessche, K. P. M.; Sharpless, K. B. *Chem. Eur. J.* **1997**, *3*, 517-522;
59. Aladro, F. J.; Guerra, F. M.; Moreno-Dorado, F. J.; Bustamante, J. M.; Jorge, Z. D.; Massanet, G. M. *Tetrahedron Lett.* **2000**, *41*, 3209-3213.
60. Jiang, Z. X.; Qin, Y. Y.; Qing, F. L. *J. Org. Chem.* **2003**, *68*, 7544-7547.
61. Laurent, P.; Hennig, L.; Burger, K.; Hiller, W.; Neumayeb, M. *Synthesis* **1998**, 905-909.
62. Magnus, N. A.; Confalone, P. N.; Storace, L.; Patel, M.; Wood, C. C.; Davis, W. P.; Parsons, R. L., Jr. *J. Org. Chem.* **2003**, *68*, 754-761.
63. Soloshonok, V. A.; Kirilenko, A. G.; Fokina, N. A.; Shishkina, I. P.; Galushko, S. V.; Kukhar, V. P. *Tetrahedron: Asymm.* **1994**, *5*, 1119-1126.
64. Davoii, P.; Forni, A.; Franciosi, C.; Moretti, I.; Prati, F. *Tetrahedron Asymm.* **1999**, *10*, 2361-2371.
65. (a) Sakai, T.; Yan, F.; Uneyama, K. *Synlett*, **1995**, 753-754. (b) Sakai, T.; Yan, F.; Kashino, S.; Uneyama, K. *Tetrahedron* **1996**, *52*, 233-244.
66. Demir, A. S.; Sesenoglu, O.; Gercek-Arkin, Z. *Tetrahedron: Asymm.* **2001**, *12*, 2309-2313.
67. Abe, H.; Amii, H.; Uneyama, K. *Org. Lett.* **2001**, *3*, 313-315.
68. (a) Soloshonok, V. A.; Ono, T. *J. Org. Chem.* **1997**, *62*, 3030-3031. (b) Soloshonok, V. A.; Ono, T.; Soloshonok, I. V. *J. Org. Chem.* **1997**, *62*, 7538-7539.
69. Soloshonok, V. A.; Kukhar, V. P. *Tetrahedron* **1997**, *53*, 8307-8314.
70. Bravo, P.; Cavicchio, G.; Cmcianelli, M.; Markovsky, A. L.; Volonterio, A.; Zanda, M. *Synlett*, **1996**, 887 -889.
71. Pirke, W. H.; Sikkenga, D. L.; Pavlin, M. S. *J. Org. Chem.* **1977**, *42*, 384-387.
72. Torok, B.; Prakash, G. K. S. *Adv. Synth. Catal.* **2003**, *345*, 165-168.
73. Jiang, J.; DeVita, R. J.; Doss, G. A.; Goulet, M. T.; Wyvratt, M. J. *J. Am. Chem. Soc.* **1999**, *121*, 593-594.
74. Kauffman, G. S.; Harris, G. D.; Dorow, R. L.; Stone, B. R. P.; Parsons, R. L. Jr.; Pesti, J. A.; Magnus, N. A.; Fortunak, J. M.; Confalone, P. M.; Nugent, W. A. *Org. Lett.* **2000**, *2*, 3119-3121.
75. Jiang, B.; Si, Y.-G. *Angew. Chem. Int. Ed.* **2004**, *43*, 216-218.

76. Bravo, P.; Capelli, P.; Meille, S. V.; Seresini, P.; Volonterio, A.; Zanda, M. *Tetrahedron: Asymm*. **1996**, *7*, 2321-2332.
77. Magnus, N. A.; Confalone, P. N.; Storace, L. *Tetrahedron Lett.* **2000**, *41*, 3015-3019.
78. (a) Bravo, P.; Crucianelli, M.; Vergani, B.; Zanda, M. *Tetrahedron Lett.* **1998**, *39*, 7771-7774. (b) Lazzaro, F.; Crucianelli, M.; De Angelis, F.; Frigerio, M.; Malpezzi, L.; Volonterio, A.; Zanda, M. *Tetrahedron: Asymm.* **2004**, *15*, 889-893.
79. (a) Enders, D.; Funabiki, K. *Org. Lett.* **2001**, *3*, 1575-1577. (b) Funabiki, K.; Nagamori, M.; Matsui, M.; Enders, D. *Synthesis*, **2002**, 2585-2588.
80. Gong, Y.; Kato, K. *Tetrahedron: Asymm*. **2001**, *12*, 2121-2127.
81. Soloshonok, V. A.; Avilov, D. V.; Kukhar', V. P.; Meervelt, L. V.; Mischenko, N. *Tetrahedron Lett.* **1997**, *38*, 4691-4674.
82. Ojima, I.; Slater, J. C. *Chirality* **1997**, *9*, 487-494.
83. (a) Bravo, P.; Crucianelli, M.; Farina, A.; Volonterio, A.; Zanda, M. *Eur. J. Org. Chem.* **1998**, 435-440. (b) Bravo, P.; Guidetti, M.; Viani, F.; Zanda, M.; Markovsky, A. L.; Sorochinsky, A. E.; Soloshonok, I. V.; Soloshonok, V. A. *Tetrahedron* **1998**, *54*, 12789-12806. (c) Bravo, P.; Fustero, S.; Guidetti, M.; Volonterio, A.; Zanda, M. *J. Org. Chem.* **1999**, *64*, 8731-8735.
84. Prakash, G. K. S.; Mandal, M.; Olah, G. A. *Angew. Chem. Int. Ed.* **2001**, *40*, 589-590.
85. (a) Volonterio, A.; Bravo, P.; Zanda, M. *Org. Lett.* **2000**, *2*, 1827-1830. (b) Volonterio, A.; Chiva, G.; Fustero, S.; Piera, J.; Rosello, M. S.; Sani, M.; Zanda, M. *Tetrahedron Lett.* **2003**, *44*, 7019-7022.
86. Molteni, M.; Volonterio, A.; Zanda, M. *Org. Lett.* **2003**, *5*, 3887-3890.
87. (a) Poras, H.; Matsutani, H.; Yaruva, J.; Kusumoto, T.; Hiyama, T. *Chem. Lett.* **1998**, 665-666. (b) Matsutani, H.; Poras, H.; Kusumoto, T.; Hiyama, T. *J. Chem. Soc. Chem. Commun.* **1998**, 1259-1260.
88. Ishii, A.; Miyamoto, F.; Higashiyama, K.; Mikami, K. *Tetrahedron Lett.* **1998**, *39*, 1199-1201.
89. Ishii, A.; Miyamoto, F.; Higashiyama, K.; Mikami, K. *Chem. Lett.* **1998**, 119-120.
90. Magnus, N. A.; Confalone, P. N.; Storace, L. *Tetrahedron Lett.* **2000**, *41*, 3015-3019.
91. Bravo, P.; Crucianelli, M.; Fronza, G.; Zanda, M. *Synlett*; **1996**, 249-250.
92. Ishii, A.; Miyamoto, F.; Higashiyama, K.; Mikami, K. *Tetrahedron Lett.* **1998**, *39*, 1199-1201.
93. Furuhashi, K. *Chirality in Industry,* A. N. Collins, G. N. Sheldrake, J. Crosby (eds.), John Wiley & Sons (1992).
94. Schaus, S. E.; Brandes, B. D.; Larrow, J. F.; Tokunaga, M.; Hansen, K. B.;Gould, A. E.;Furrow, M. E.; Jacobsen, E. N. *J. Am. Chem. Soc.* **2002**, *124*, 1307-1315.

95. Katagiri, T. 1,1,1-Trifluoro-2,3-epoxypropane in *Encyclopedia of Reagents in Organic Synthesis,* ed by Paquette, L. A. Wiley, in press (RN00506)
96. Begue, J. P.; Bonnet-Delpon, D. *ACS Symposium Series* **1996**, *639*, 59-72.
97. Brandes, B. D.; Jacobsen, E. N. *J. Org. Chem.* **1994**, *59*, 4378-4380.
98. (a) Bravo, P.; Bruche, L.; Panzeri, W.; Viani, F.; Vichi, L. *J. Fluorine Chem.* **1998**, *91*, 27-36. (b) Arnone, A.; Bravo, P.; Frigerio, M.; Viani, F.; Soloshonok, V. A. *Tetrahedron,* **1998**, *54*, 11841-11860. (c) Bravo, P.; Corradi, E.; Frigerio, M.;Meille, S. V.; Panzeri, W.; Pesenti, C.; Viani, F. *Tetrahedron Lett.* **1999**, *40*, 6317-6320.
99. Akiyama, T.; Ogi, S.; Fuchibe, K. *Tetrahedron Lett.* **2003**, *44*, 4011-4013.
100. Yamazaki, T.; HIraoka, S.; Kitazume, T. *Tetrahedron: Asymm.* **1997**, *8*, 1157-1160.
101. (a) Yamazaki, T.; Shinohara, N.; Kitazume, T.; Sato, S. *J. Fluorine Chem.* **1999**, *97*, 91-96. (b) Kitazume, T.; Tamura, K.; Jiang, Z.; Miyake, N.; Kawasaki, I. *J. Fluorine Chem.* **2002**, *115*, 49-53.
102. (a) Soloshonok, V. A.;Avilov, D. V.; Kukhar',V. P.; Meervelt, L. V.; Mischenko, N. *Tetrahedron Lett.* **1997**, *38*, 4903-4904. (b) Soloshonok, V. A.; Cai, C.; Hruby, V. J. *Tetrahedron,* **1999**, *55*, 12031-12044.
103. Ogura, K.; Ogu, K; Ayabe, T.; Sonehara, J.; Akazome, M. *Tetrahedron Lett.* **1997**, *38*, 5173-5176.
104. Bravo, P.; Arnone, A.; Bandiera, P.; Bruche, L.; Ohashi, Y.; Ono, T.; Sekine, A.; Zanda, M. *Eur. J. Org. Chem.* **1999**, 111-115.
105. (a) Dmowski, W.; Piasecka-Maciejewska, K. *J. Fluorine Chem.* **1999**, *97*, 97-100. (b) Dmowski, W.; Piasecka-Maciejewska, K. *J. Fluorine Chem.* **2000**, *105*, 77-82. (c) Dmowski, W. *J. Fluorine Chem.* **2001**, *109*, 33-37.
106. Farina, A.; Meille, S. V.; Messina, M. T.; Metrangolo, P.; Resnati, G.; Vecchio, G. *Angew. Chem. Int. Ed.* **1999**, *38*, 2433-2436.
107. Lebouvier, N.; Laroche, C.; Huguenot, F.; Brigaud, T. *Tetrahedron Lett.* **2002**, *43*, 2827-2830.
108. Gosselin, F.; Roy, A.; O'Shea, P. D.; Chen, C.; Volante, R. P. *Org. Lett.* **2004**, *6*, 641-644.
109. Ishii, A.; Miyamoto, F.; Higashiyama, K.; Mikami, K. *Chem. Lett.* **1998**, 119-120.
110. Shinohara, N.; Yamazaki, T.; Kitazume, T. *Rev. Heteroatom Chem.* **1996**, *14*, 165-182.
111. Nagai, T.; Nishioka, G.; Koyama, M.; Ando, A.; Miki, T.; Kumadaki, I. *J. Fluorine Chem.* **1992**, *57*, 229-237.
112. Schlosser, M.; Michel, D. *Tetrahedron* **1996**, *52*, 99-108.
113. Katagiri, T.; Ihara, H.; Takahashi, M.; Kashino, S.; Furuhashi, K; Uneyama, K. *Tetrahedron: Asymm.* **1997**, *8*, 2933-2937.

114. (a) Katagiri, T.; Irie, M.; Uneyama, K. *Tetrahedron: Asymm.* **1999**, *10*, 2583-2589. (b) Katagiri, T.; Irie, M.; Uneyama, K. *Org. Lett.* **2000**, *2*, 2423-2425.
115. Klaubunde, K. J.; Burton, D. J. *J. Am. Chem. Soc.* **1972**, *94*, 5985-5990.
116. (a) Bonnet-Delpon, D.; Cambillau, C.; Charpentier-Morize, M.; Jacquot, R.; Mesureur, D.; Ourevitch, M. *J. Org. Chem.* **1988**, *53*, 754-758. (b) Bonnet-Delpon, D.; Charpentier-Morize, M.; Jacquot, R. *J. Org. Chem.* **1988**, *53*, 759-762.
117. Tsushima, T.; Kawada, K.; Ishihara, S.; Uchida, N.; Shiratori, O.; Higaki, J.; Hirata, M. *Tetrahedron* **1988**, *44*, 5375-5387.
118. (a) Hanack, M.; Ullmann, J. *J. Org. Chem.* **1989**, *54*, 1432-1435. (b) Ullmann, J.; Hanack, M. *Synthesis* **1989**, 685-687.
119. Uneyama, K.; Momota, M. *Tetrahedron Lett.* **1989**, *30*, 2265-2266.
120. Katayama, M.; Kimoto, H.; Gautam, R. K.; Nishida, M.; Fujii, S. *Report Gov. Indust. Res. Inst. Nagoya* **1992**, *41*, 185-195.
121. Kubota, T.; Iijima, M.; Tanaka, T. *Tetrahedron Lett.* **1992**, *33*, 1351-1354.
122. Katagiri, T. *Jpn Kokai Tokkyo Koho* **1996**, *JP08*-27044 (*CA* **1996**, *124*, 342627).
123. Katagiri, T.; Hirai, Y.; Kusunoki, N.; Fujiwara, Y.; Takahashi, M.; Uneyama, K. *Abstr. 27th Symp. Heteroatom. Chem. Jpn.* **2000**, 1-3.
124. Yamauchi, Y.; Kawate, T.; Itahashi, H.; Katagiri, T.; Uneyama, K. *Tetrahedron Lett.* **2003**, *44*, 6319-6322.
125. Katagiri, T.; Takahashi, M.; Fujiwara, Y.; Ihara, H.; Uneyama, K. *J. Org. Chem.* **1999**, *64*, 7323-7329.
126. Katagiri, T.; Yamashita, H.; Uneyama, K. *Abstr. 28th Symp. Heteroatom Chem. Jpn.* **2001**, 298-301.
127. Yamazaki, T.; Iwatsubo, H.; Kitazume, T. *private communication,*
128. (a) Mikami, K.; Yajima, T.; Terada, M.; Kawauchi, S.; Suzuki, Y.; Kobayashi, I. *Chem. Lett.* **1996**, 861-862. (b) Mikami, K.; Yajima, T.; Terada, M.; Suzuki, Y.; Kobayashi, I. *J. Chem. Soc., Chem. Commun.* **1997**, 57-58.
129. (a) Hagiwara, T.; Tanaka, K.; Fuchikami, T. *Tetrahedron Lett.* **1996**, *37*, 8187-8190. (b) Hagiwara, T.; Ishizuka, M.; Fuchikami, T. *Nippon Kagaku Kaishi* **1998**, 750-756.
130. Pirkle, W. H.; Hauske, J. R.; Eckert, C. A.; Scott, B. A. *J. Org. Chem.* **1977**, *42*, 3101-3103.
131. (a) Matsutani, H.; Poras, H.; Kusumoto, T.; Hiyama, T. *Synlett* **1998**, 1353-1354. (b) Matsutani, H.; Kusumoto, T.; Hiyama, T. *Chem. Lett.* **1999**, 529-530.
132. (a) Ooi, T.; Uraguti, D.; Kagoshima, N.; Maruoka, K. *Tetrahedron Lett.* **1997**, *38*, 5679-5682. (b) Ooi, T.; Furuya, M,; Maruoka, K. *Chem. Lett.* **1998**, 817-818.

133. Rozov, L. A.; Huang, C. G.; Halpern, D. F.; Vernice, G. G.; Ramig, K. *Tetrahedron: Asymm.* **1997**, *8*, 3023-3025.

134. (a) Konno, T.; Umetani, H.; Kitazume, T. *J. Org. Chem.* **1997**, *62*, 137-150. (b) Yamazaki, T.; Umetani, H.; Kitazume, T. *Tetrahedron Lett.* **1997**, *38*, 6705-6708.

135. Konno, T.; Nagata, K.; Ishihara, T.; Yamanaka, H. *J. Org. Chem.* **2002**, 67, 1786-1775.

136. (a) Fujita, M.; Hiyama, T. *Bull. Chem. Soc. Jpn.* **1987**, *60*, 4377-4384. (b) Shimizu, M.; Fujimoto, T.; Liu, X.; Minezaki, H.; Hata, T.; Hiyama, T. *Tetrahedron* **2003**, *59*, 9811-9823.

137. (a) Jiang, B.; Xu, Y. *Tetrahedron Lett.* **1992**, *33*, 511-514. (b) Jiang, B.; Xu, Y. *J. Org. Chem.* **1991**, *56*, 7336-7740. (c) Shi, G. -Q.; Huang, X. -H.; Hong, F. *J. Org. Chem.* **1996**, *61*, 3200-3204. (d) Watanabe, H.; Yan, F. -Y.; Sakai, T.; Uneyama, K. *J. Org. Chem.* **1994**, *59*, 758-761. (e) Watanabe, H.; Yamashita, F.; Uneyama, K. *Tetrahedron Lett.* **1993**, *34*, 1941-1944. (f) Uneyama, K.; Noritake, C.; Sadamune, K. *J. Org. Chem.* **1996**, *61*, 6055-6057. (g) Heinze, P. L.; Burton, D. J. *J. Org. Chem.* **1988**, *53*, 2714-2720. (h) Morken, P. A.; Lu, H.; Nakamura, A.; Burton, D. J. *Tetrahedron Lett.* **1991**, *32*, 4271-4274. (j) Spawn, T. D.; Burton, D. J. *Bull. Soc. Chim. Fr.* **1986**, 876-880. (k) Miller, W.; Snider, R. H.; Hummel, R. J. *J. Am. Chem. Soc.* **1969**, *91*, 6532-6534. (l) Banks, R. E.; Haszeldine, R. N.; Taylor, D. R.; Webb, G. *Tetrahedron Lett.* **1970**, 5215-5216. (m) Drakesmith, F. G.; Stwert, O. J.; Tarrant, P. *J. Org. Chem.* **1968**,*33*, 280-285. (n) Fuchikami, T.; Ojima, I. *Tetrahedron Lett.* **1982**, *23*, 4099-4100. (o) Fuchikami, T.; Yamanouchi, A.; Ojima, I. *Synthesis* **1984**, 766-768.

138. (a) Ishikawa, N.; Yokozawa, T. *Bull. Chem. Soc. Jpn.* **1983**, *56*, 724-726. (b) Seebach, D.; Beck, A. K.; Renaud, P. *Angew. Chem., Int. Ed. Engl.* **1986**, *25*, 98-99. (c) Fuchigami, T.; Nakagawa, Y. *J. Org. Chem.* **1987**, *52*, 5276-5277. (d) Fuchigami, T.; Nakagawa, Y.; Nonaka, T. *J. Org. Chem.* **1987**, *52*, 5489-5491. (e) Uneyama, K.; Momota, M. *Bull. Chem. Soc. Jpn.* **1989**, *62*, 3378-3379. (f) Qian, C. -P.; Nakai, T. *Tetrahedron Lett.* **1990**, *31*, 7043-7046. (g) Morken, P. A.; Lu, H. Y.; Nakamura, A.; Burton, D. J. *Tetrahedron Lett.* **1991**, *34*, 4271-4274. (h) Beck, A. K.; Seebach, D. *Chem. Ber.* **1991**, *124*, 2897-2911. (i) Komatsu, Y.; Sakamoto, T.; Kitazume, T. *J. Org. Chem.* **1999**, *64,* 8369-8374.

139. (a) Yamauchi, Y.; Katagiri, T.; Uneyama, K. *Org. Lett.* **2002**, *4*, 173-176. (b) Yamauchi, Y.; Kawate, T.; Katagiri, T.; Uneyama, K. *Tetrahedron* **2003**, *59*, 9839-9847.

140. (a) Satoh, T. *Chem. Rev.* **1996**, *96*, 3303-3325. (b) McCoull, W., Davis, F. A. *Synthesis* **2000**, 1347-1365. (c) Hodgeson, D. M.; Gras, E. *Syntheses* **2002**, 1625-1642.

Chapter 19

Use of Trifluoroacetaldehyde Ethyl Hemiacetal in a Simple and Practical Synthesis of β-Hydroxy-β-trifluoromethylated Ketones

Kazumasa Funabiki

Department of Materials Science and Technology, Faculty of Engineering, Gifu University, 1–1, Yanagido, Gifu 501–1193, Japan

The reaction of trifluoroacetaldehyde ethyl hemiacetal or hydrate with an equimolar amount of enamines or imines, derived from various methyl ketones with aliphatic, aromatic, and heteroaromatic substituents, *via* the in-situ generation of trifluoroacetaldehyde, affording high yields of the corresponding β-hydroxy-β-trifluoromethyl ketones is described. Extended studies using chiral auxiliaries or catalyst are also discussed.

Introduction

α-Trifluoromethylated alcohols are some of the most valuable compounds in organofluorine synthesis, because they can serve as the useful core system of liquid crystals (*1*) or antidepressants (*2*) imparted by a trifluoromethyl group. Most syntheses of these compounds utilize the α-trifluoromethylated building blocks such as trifluoroacetaldehyde (CF_3CHO), its derivatives (*3*), and α-trifluoromethyl ketones as the starting substrates. Among them, CF_3CHO is especially the most attractive compound for the construction of α-trifluoromethylated alcohol units. Therefore, it widely employed in Aldol (*4*), Mukaiyama (*5*), Ene (*6*), Friedel-Crafts (*7*), Morita-Baylis-Hillman reactions (*8*), etc. However, just before employing CF_3CHO, it should be generated from its hemiacetal or hydrate using an excess amount of conc. sulfuric acid at a high reaction temperature (*9*). Moreover, careful treatment of the aldehyde is required due to its troublesome properties such as being a gas at room temperature, highly reactive leading to self-polymerization, and extremely hygroscopic (*10*) (Scheme 1).

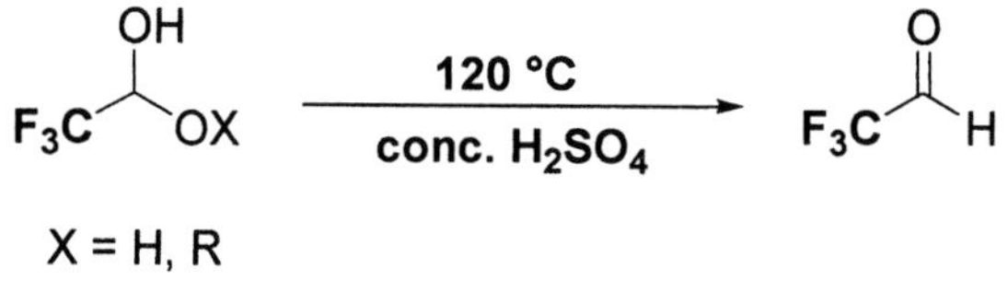

$CF_3CH(OH)OX$ —(120 °C, conc. H_2SO_4)→ CF_3CHO

X = H, R

Scheme 1. Previous method for the generation of trifluoroacetaldehyde.

Therefore, it is quite important to find and develop much more convenient and environmentally-benign methods for the effective in-situ generation and tandem (asymmetric) carbon-carbon bond formation reaction, although there are some successful examples using hydrazones (*3a*), active methylene compounds (*3d*), and so on (*3c, h*). (Scheme 2).

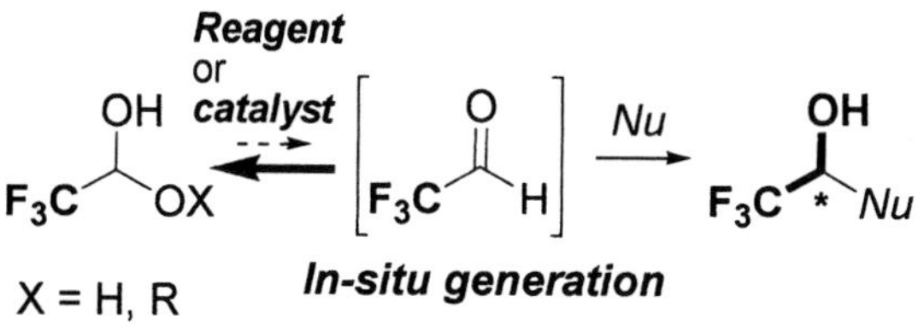

$CF_3CH(OH)OX$ ⇄ (*Reagent* or *catalyst*) [CF_3CHO] —*Nu*→ $CF_3C^*H(OH)Nu$

X = H, R

In-situ generation

Scheme 2.

This article reviews our novel protocol for the effective generation of CF_3CHO and the simultaneous stereoselective carbon-carbon bond forming reaction leading to β-trifluoromethylated aldol units.

Reaction of Trifluoroacetaldehyde Ethyl Hemiacetal with Enamines or Imines

Reaction with Various Enamines

The reaction of trifluoroacetaldehyde ethyl hemiacetal **1a** with an equimolar amount of an enamine, derived from acetophenone with morpholine, at room temperature in hexane for 1 h, followed by hydrolysis, produced the β-hydroxy-β-trifluoromethyl ketone **3a** in 88% yield (*11*) (Table 1, entry 1).

Table 1. Reaction of Trifluoroacetaldehyde Ethyl Hemiacetal or Hydrate with Enamines Derived from Various Methyl Ketones[a]

1) 1 h
2) H^+
hexane

1a : X = Et
1b : X = H

2

3

Entry	1	Enamine	R^1	Temp.	Product	Yield (%)[b]
1	**1a**	**2a**	Ph	rt	**3a**	88
2	**1a**	**2b**	4-MeC_6H_4	rt	**3b**	87
3	**1a**	**2c**	4-$MeOC_6H_4$	rt	**3c**	72
4	**1a**	**2d**	4-ClC_6H_4	rt	**3d**	86
5[c]	**1a**	**2e**	4-$NO_2C_6H_4$	rt	**3e**	52
6	**1a**	**2f**	2-MeC_6H_4	rt	**3f**	87
7	**1a**	**2g**	2-thienyl	rt	**3g**	75
8[d]	**1a**	**2h**	*i*-Pr	rt	**3h**	25
9[c]	**1b**	**2a**	Ph	reflux	**3a**	84

[a] The reaction was carried out with trifluoroacetaldehyde ethyl hemiacetal **1a** or hydrate **1b** (1 mmol) with enamine **2** (1 mmol) in hexane (4 ml) at ambient temperature. [b] Yields of isolated pure products. [c] Toluene in place of hexane was used as a solvent. [d] A mixture of **2h** and 4-(3-methylbut-1-en-2-yl)morpholine (29:71) was used.

Other various aromatic enamines as well as the thienyl one easily participated in the reaction to give the corresponding β-hydroxy-β-trifluoromethyl ketone **3** in good to excellent yields (entries 2-7). However, the reaction of 4-nitropheyl substituted enamine **2e** with **1a** gave the only a 13% yield of the product **3e**, probably due to the low solubility of **2e**. Using toluene as the solvent improved the yield of **3e** (entry 5). The reaction of the hemiacetal **1a** with the mixture of aliphatic enamines, such as **2i** and 4-(3-methybut-1-en-2-yl)morpholine (29 : 71) gave **3h** in 25% yield, and there is no detectable amount of the product, which reacts with the 4-(3-methybut-2-en-2-yl)morpholine, in the crude reaction mixture (entry 8). Trifluoroacetaldehyde hydrate (75 wt%) also reacted with the enamine **2a** to produce **3a** in 84% yield (entry 9).

Reaction with Various Imines

The results of the reaction of trifluoroacetaldehyde ethyl hemiacetal **1a** with imines **4** derived from various methyl ketones are summarized in Table 2 (*12*).

Table 2. Reaction of Trifluoroacetaldehyde Ethyl Hemiacetal or Hydrate with Imines Derived from Various Methyl Ketones[a]

F_3C–CH(OH)–OX (**1a** : X = Et; **1b** : X = H) + CH_3–C(=Nc-Hex)–R^1 (**4**) → 1) 1 h, 2) H^+, hexane → F_3C–CH(OH)–CH_2–C(=O)–R^1 (**3**)

Entry	**1**	Imine	R^1	Product	Yield (%)[b]
1	**1a**	**4a**	Ph	**3a**	91
2	**1a**	**4b**	4-MeC_6H_4	**3b**	88
3	**1a**	**4c**	4-$MeOC_6H_4$	**3c**	94
4	**1a**	**4d**	4-ClC_6H_4	**3d**	89
5	**1a**	**4e**	4-$NO_2C_6H_4$	**3e**	79
6	**1a**	**4f**	4-$EtOCOC_6H_4$	**3i**	83
7	**1a**	**4g**	2-thienyl	**3g**	78
8	**1a**	**4h**	*i*-Pr	**3h**	65
9	**1a**	**4i**	*c*-Hex	**3j**	93
10	**1a**	**4j**	*t*-Bu	**3k**	61
11	**1b**	**4a**	Ph	**3a**	90

[a] The reaction was carried out with trifluoroacetaldehyde ethyl hemiacetal **1a** or hydrate **1b** (1 mmol) with imine **4** (1 mmol) in hexane (4 ml) at reflux temperature. [b] Yields of isolated pure products.

Similar to the reaction of enamines **2**, the reaction between the hemiacetal **1a** and various imines **4a-g** with aromatic substituents including the thienyl group proceeded smoothly at reflux temperature to provide the corresponding β-hydroxy-β-trifluoromethyl ketone **3a-e,g,i** in good to excellent yields (entries 1-7). Of much significance is that the reaction of hemiacetal **1a** with aliphatic imines **4h-j** carrying *i*-propyl, cyclohexyl, and *t*-butyl group, produced the products **3h-k** in high yields (entries 8-10). For the imines with an *i*-propyl or *c*-hexyl group as R^1, although there are two kinds of tautomeric enamines in equilibrium, the reaction of the imines **4h,i** with CF_3CHO proceeded with complete regioselectivity at the sterically less hindered α-position of the kinetic enamines to furnish the corresponding β-hydroxy-β-trifluoromethyl aliphatic ketones **3h,j** in good yields (Figure 1). No products **3h',j'** derived from the more highly substituted enamines were observed in the reaction mixture based on the ^{19}F NMR measurement.

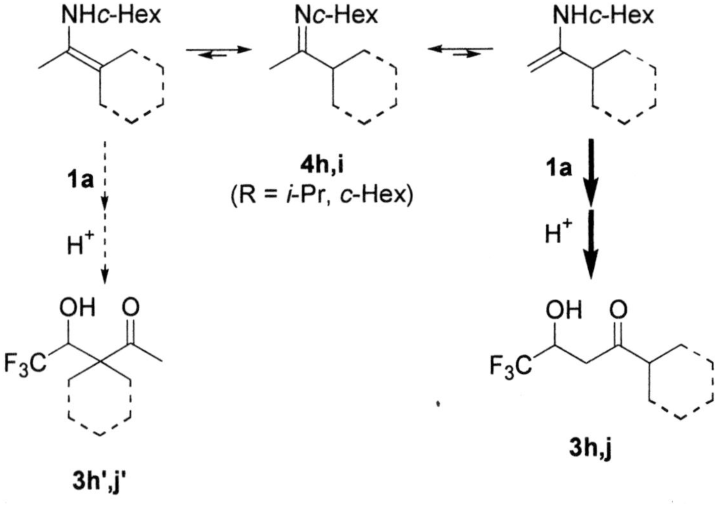

Figure 1.

Reaction of Various Polyfluoroalkylaldehyde Derivatives

This methodology can be applied to the difluoroacetaldehyde ethyl hemiacetal **1c** or pentafluoropropionaldehyde hydrate **1d**. Thus, the treatment of **1c** or **1d** with the enamine **3a** or imine **4a** gave the corresponding β-hydroxy-β-difluoromethyl or pentafluoroethyl ketones **5** or **6** in good yields (Table 3, entries 1-4).

Table 3. Reaction of Polyfluoroalkylaldehyde Hemiacetal or Hydrate with Enamines or Imines Derived from Various Methyl Ketones[a]

1c: Rf = CHF_2, X = Et
1d: Rf = CF_3CF_2, X = H

hexane, 1 h; H^+

5: Rf = CHF_2
6: Rf = CF_3CF_2

Entry	1	2 or 4	R^1	Temp.	Product	Yield (%)[b]
1	**1c**	**2a**	Ph	rt	**5a**	65
2	**1c**	**4a**	Ph	reflux	**5a**	67
3	**1d**	**2a**	Ph	rt	**6a**	78
4	**1d**	**4a**	Ph	reflux	**6a**	70

[a] The reaction was carried out with polyfluoroalkylaldehyde hemiacetal or hydrate **1** (1 mmol) with enamine **2** or imine **4** (1 mmol) in hexane (4 ml).

Reaction of Trifluoroacetaldehyde Ethyl Hemiacetal with Chiral Imines or Enamines

Reaction with Various Chiral Imines

Next, we carried out the reaction between the trifluoroacetaldehyde ethyl hemiacetal **1a** and imines **7** carrying chiral auxiliaries (*13*). The results of the asymmetric reactions using chiral imines **7** under various conditions are summarized in Table 4. When the hemiacetal **1a** was allowed to react with an equimolar amount of the chiral imine **7a**, derived from acetophenone and (*R*)-1-phenylethylamine, in hexane at room temperature for 7 h, the aldol product **3a** was obtained in 62% yield with a good enantiomer ratio (*S*:*R* = 80.1:19.9) (entry 1). The absolute configuration of the major isomer of **3a** was assigned as *S* by comparison of the reported value of the optical rotation (*14*).

Polarity of the solvent affects both the yields and enantiomer ratios of **3a**. Less polar solvents, such as hexane and toluene (entries 1-2), are more suitable for the reaction than polar ones, such as THF, dichloromethane, and acetonitrile (entries 3-5). Among the examined chiral auxiliaries, the (*R*)-1-(1-naphthyl)ethyl group was the most effective for the reaction to give **3a** in 66% yield with the best enantioselectivity (entry 8). The reaction at 0 °C gave the product with a higher selectivity (entry 9). Lowering the reaction temperature to -15 °C did not have a significant effect on increasing the enantioselectivity (entry 10).

The results of the reaction between the polyfluoroalkylaldehyde hemiacetal or hydrate **1** and various chiral imines **2** under the optimized conditions are summarized in Table 5. Aromatic, aliphatic, and heteroaromatic-substituted imines except for the 2-methylphenyl one participated nicely in the reaction with the trifluoroacetaldehyde ethyl hemiacetal **1a** or hydrate **1b** to afford the corresponding β-hydroxy-β-trifluoromethyl ketones **3** in good yields with good enantioselectivities (entries 1-6, 8-11). However, the reaction of the imine **2i** containing the 2-methylphenyl group did not smoothly proceed with the only a 14% yield of the product **3f** along with a low *ee* (entry 7). The reason for the decrease in both the yield and *ee* is unclear at present.

Additionally, the present protocol is nicely applicable to the difluoroacetaldehyde ethyl hemiacetal **1c** or pentafluoropropionaldehyde hydrate **1d**, providing the good yields of the products **5a** and **6a** (entries 12 and 13). Compared with the trifluroacetladehyde ethyl hemiacetal, the use of difluoroacetaldehyde ethyl hemiacetal **1c** resulted in a reduction of the *ee* (entry 12).

Furthermore, in the case of the trifluoromethylated or pentafluoroethylated products **3** or **6** with the aromatic group, the *ee* values of these products could be improved by a simple recrystallization method (*15*). However, this method was not effective for difluoromethylated ketone **4** due to its lower melting point compared to those of the trifluoromethylated ones.

Table 4. Reaction of Trifluoroacetaldehyde Ethyl Hemiacetal with Chiral Imines Derived from Acetophenone under Various Conditions[a]

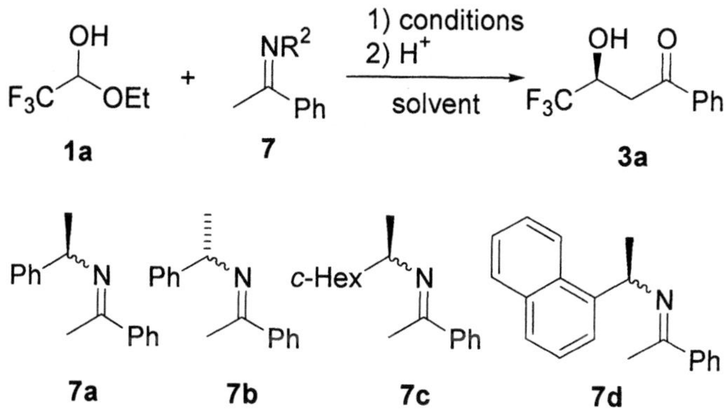

Entry[a]	Imine	Solvent	Conditions	Yield (%)[b]	Isomer Ratio (*S* : *R*)[c]
1	**7a**	hexane	rt, 7 h	62	80.1 : 19.9
2	**7a**	PhMe	rt, 7 h	64	77.6 : 22.4
3	**7a**	THF	rt, 7 h	8	73.3 : 26.5
4	**7a**	CH_2Cl_2	rt, 7 h	61	71.3 : 28.7
5	**7a**	MeCN	rt, 7 h	56	71.9 : 28.1
6	**7b**	hexane	rt, 7 h	73	20.6 : 79.4
7	**7c**	hexane	rt, 7 h	92	62.5 : 37.5
8	**7d**	hexane	rt, 7 h	66	85.5 : 14.5
9	**7d**	hexane	0 °C, 7 d	57	90.5 : 9.5
10	**7d**	hexane	-15 °C, 7 d	48	89.3 : 10.7

[a] All the reaction was carried out with trifluoroacetaldehyde ethyl hemiacetal **1a** (0.5 mmol) and chiral imine 7 (0.5 mmol) in hexane (2 ml). [b] Yields of isolated pure products. [c] Determined by HPLC analysis with CHIRALCEL OD (hexane:*i*-PrOH=95/5).

Table 5. Reaction of Polyfluoroalkylaldehyde Ethyl Hemiacetal or Hydrate with Various Chiral Imines

1) 0 °C, 7 d
2) H^+
hexane

1a: Rf = CF_3, X = Et
1b: Rf = CF_3, X = H
1c: Rf = CHF_2, X = Et
1d: Rf = CF_3CF_2, X = H

7

3: Rf = CF_3
5: Rf = CHF_2
6: Rf = CF_3CF_2

7d: R^1 = Ph
7e: R^1 = 4-MeC_6H_4
7f: R^1 = 4-ClC_6H_4
7g: R^1 = 4-$MeOC_6H_4$
7h: R^1 = 2-thienyl
7i: R^1 = 2-MeC_6H_4
7j: R^1 = 3-MeC_6H_4
7k: R^1 = *c*-Hex
7l: R^1 = *i*-Pr
7m: R^1 = *t*-Bu

Entry[a]	1	Imine	Product	Yield (%)[b]	*Er* (*S* : *R*)[c]	*Ee*[c]	*Ee*[c,d]
1	**1a**	**7d**	**3a**	57	90.5 : 9.5	81.0	92.8
2	**1b**	**7d**	**3a**	57	89.1 : 10.9	78.2	-
3	**1a**	**7e**	**3b**	68	89.4 : 10.6	78.8	>99.9
4	**1a**	**7f**	**3c**	64	87.6 : 12.4	75.2	>99.9
5[e]	**1a**	**7g**	**3d**	51	86.0 : 14.0	72.0	93.8
6	**1a**	**7h**	**3e**	37	90.3 : 9.7	80.6	>99.9
7	**1a**	**7i**	**3f**	14	56.9 : 43.1	13.8	-
8	**1a**	**7j**	**3g**	70	90.6 : 9.4	81.2	-
9	**1a**	**7k**	**3h**	73	93 : 7[f]	86[f]	-
10	**1a**	**7l**	**3i**	59	88 : 12[f]	76[f]	-
11	**1a**	**7m**	**3j**	24	92 : 8[f]	84[f]	-
12	**1c**	**7d**	**5a**	53	75.5 : 24.5	51.0	-
13	**1d**	**7d**	**6a**	51	89.7 : 10.3	79.4	95.6

[a] All the reaction was carried out with **1** (0.5 mmol) and chiral imine 7 (0.5 mmol) in hexane (2 ml) at 0 °C for 7 d. [b] Yields of isolated pure products. [c] Determined by HPLC analysis with CHIRALCEL OD (hexane:*i*-PrOH=95/5). [d] After recrystallization. [e] Toluene was used as a solvent. [f] Determined by ^{19}F NMR after the formation of Mosher's ester.

Reaction with Chiral Enamine

The treatment of trifluoroacetaldehyde ethyl hemiacetal **1a** with the chiral enamine **8**, prepared from acetophenone with (*S*)-2-(methoxymethyl)pyrrolidine (SMP) (*16*), in hexane under various reaction temperatures, unfortunately, gave **3a** in 60-77% yields with very low selectivities, as shown in Table 6.

Table 6. Reaction of Trifluoroacetaldehyde Ethyl Hemiacetal with Chiral Enamine Derived from Acetophenone under Various Conditions[a]

Entry[a]	Conditions	Yield (%)[b]	Isomer Ratio (*S* : *R*)[c]
1	reflux, 1 h	69	42.8 : 57.2
2	rt, 1 h	77	50.1 : 49.9
3	0 °C, 24 h	60	57.1 : 42.9

[a] All the reaction was carried out with trifluoroacetaldehyde ethyl hemiacetal **1a** (0.5 mmol) and chiral enamine **8** (0.5 mmol) in hexane (2 ml). [b] Yields of isolated pure products. [c] Determined by HPLC analysis with CHIRALCEL OD (hexane:*i*-PrOH=95/5).

Proline-catalyzed Direct Aldol Reaction with Ketones

Finally, a first novel catalytic in-situ generation of CF_3CHO from its hemiacetal as well as its successive direct asymmetric aldol reaction with some ketones was examined (*17*). The reaction of trifluoroacetaldehyde ethyl hemiacetal **1a** in the presence of a catalytic amount (30 mol%) of (*L*)-proline with acetone afforded β-hydroxy-β-trifluoromethyl ketone **3l** with good enantioselectively. As summarized in Table 7, a survey of the reaction media revealed that various solvents could be used for this reaction. However, the employment of DMSO resulted in a significant loss of enantioselectivity (entry 1). The reaction in THF or hexane was sluggish and produced low yields of **3l** (entries 2 and 7). The best enatioselectivity of **3l** was obtained when benzene was used as the solvent but with a decreased yield (entry 6). Importantly, acetone can be used as the solvent as well as a ketone donor with excellent yield and good enantioselectivity (entry 3).

Both trifluoroacetaldehyde hydrate **1b** and pentafluoropropionaldehyde hydrate **1d** participated well in the (*L*)-proline-catalyzed direct aldol reaction to afford **3l** or **6b** in good yields with good enantioselectivities (entries 8 and 9).

Table 7. Direct Aldol Reaction of Trifluoroacetaldehyde Ethyl Hemiacetal with Acetone under Various Conditions[a]

cat. (*L*)-proline, solvent, rt. 48 h

1a: Rf = CF_3, X = Et
1b: Rf = CF_3, X = H
1d: Rf = CF_3CF_2, X = H

3l : Rf = CF_3
6b : Rf = CF_3CF_2

Entry[a]	1	Solvent	Yield (%)[b]	Isomer Ratio (*S* : *R*)[c]	*Ee*[c]
1	**1a**	DMSO	96	51.4 : 48.6	2.8
2	**1a**	THF	19	68.7 : 31.3	37.4
3	**1a**	acetone[d]	97	67.6 : 32.4	35.2
4	**1a**	MeCN	64	70.7 : 29.2	41.5
5	**1a**	CH_2Cl_2	45	73.9 : 26.1	47.8
6	**1a**	benzene	32	75.8 : 24.2	51.6
7	**1a**	hexane	19	73.0 : 27.0	46.0
8	**1b**	acetone	64	68.9 : 31.1	37.8
9	**1c**	acetone	69	72.0 : 28.0	44.0

[a] All the reaction was carried out with **1** (1 mmol) and 30 mol% of (*L*)-proline in the mixed solvent of acetone (2 ml) and solvent (8 ml). [b] Determined by ^{19}F NMR using benzotrifluoride as an internal standard. [c] Determined by HPLC analysis with CHIRALCEL OD-H (hexane:*i*-PrOH=95/5) after *p*-chlorobenzoylation. [d] Acetone, dehydrated (Kanto Chemical Co., Ltd.) was used.

Surprisingly, in lieu of acetone, cyclopentanone or cyclohexanone reacted smoothly with the trifluoroacetaldehyde ethyl hemiacetal **1a** in the presence of 30 mol% of (*L*)-proline to give 57 to then quantitatively yield product **9** or **10** with high *anti*-selectivities as well as excellent enantioselectivities (*18*) (Scheme 3).

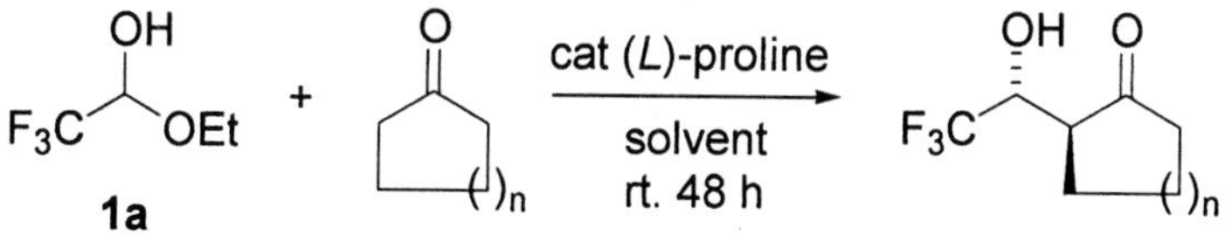

n = 1; **9** (quant[a]; *syn*/*anti* = 8/92, 97%*ee*)
n = 2; **10** (57%[a]; *syn*/*anti* = 4/96, >98%*ee*)
[a] Measured by ^{19}F NMR.

Scheme 3.

In summary, we have determined that the novel stoichiometric in-situ generation of trifluoroacetaldehyde as well as its simultaneous asymmetric carbon-carbon bond formation reaction with enamines or imines, produced the corresponding β-hydroxy-β-trifluoromethyl ketones in good yields with high enantioselectivities. The major advantages of these processes are the use of only a stoichiometric amount of enamines or imines, good yields as well as high enantioselectivities, and no step required for the generation of trifluoroacetaldehyde. For the reaction of chiral imines, the recovery of the chiral auxiliary is quite easy.

Furthermore, our new protocol can be nicely extended to the direct aldol reaction *via* the catalytic in-situ generation of trifluoroacetaldehyde from its hemiacetal using a catalytic amount of (*L*)-proline under extremely mild conditions.

Acknowledgment

This work was financially supported by a Grant-in-Aid for Encouragement of Young Scientists (B) (Grant No.14750665) from the Ministry of Education, Culture, Sports, Science, the Gifu University, Nagoya Industrial Science Research Institute, the OGAWA Science and Technology Foundation, and the Central Glass Co., Ltd. We also thank the Central Glass Co., Ltd., for the gift of trifluoroacetaldehyde ethyl hemiacetal and hydrate. Finally, the author is grateful to Professor Dr. Masaki Matsui for his continued support and useful discussions.

References

1. Mikami, K. *Asymmetric Fluoroorganic Chemistry: Synthesis, Application, and Future Directions*, Ramachandran, P. V., Ed.; American Chemical Society, Washington, DC, 1999, p 255-269.

2. Wouters, J.; Moureau, F.; Evrard, G.; Koenig, J.-J.; Jegham, S.; George, P.; Durant, F. *Bioorg. Med. Chem.* **1999**, *7*, 1683-1693.
3. For hemiacetal or hydrate, see: (a) Fernández, R.; Martín-Zamora, E.; Pareja, C.; Alcarazo, M.; Martín, J.; Lassaletta, J. M. *Synlett* **2001**, 1158-1160. (b) Gong, Y.; Kato, K.; Kimoto, H. *J. Heterocyclic Chem.* **2001**, *38*, 25-28. (c) Shirai, K.; Onomura, O.; Maki, T.; Matsumura, Y. *Tetrahedron Lett.* **2000**, *41*, 5873-5876. (d) Kuwano, R.; Miyazaki, H.; Ito, Y. *J. Organomet. Chem.* **2000**, *603*, 18-29. (e) Gong, Y.; Kato, K.; Kimoto, H. *Bull. Chem. Soc. Jpn.* **2000**, *73*, 249-250. (f) Loh, T. -P.; Li, X. –R. *Tetrahedron* **1999**, *55*, 5611-5622. (g) Sakumo, K.; Kuki, N.; Kuno, T.; Takagi, T.; Koyama, M.; Ando, A.; Kumadaki, I. *J. Fluorine Chem.* **1999**, *93*, 165-170. (h) Gong, Y. F.; Kato, K.; Kimoto, H. *Synlett* **1999**, 1403-1404. (i) Loh, T. –P.; Xu, K. C.; Ho, D. S. C.; Sim, K. Y. *Synlett* **1998**, 369-370. (j) Omote, M.; Ando, A.; Takagi, T.; Koyama, M.; Kumadaki, I. *Tetrahedron* **1996**, *52*, 13961-13970. (k) Ishihara, T.; Hayashi, H.; Yamanaka, H. *Tetrahedron Lett.* **1993**, *34*, 5777-5780. (l) Kubota, T.; Iijima, M.; Tanaka, T. *Tetrahedron Lett.* **1992**, *33*, 1351-1354. (m) Guy, A.; Lobgeois, M. *J. Fluorine Chem.* **1986**, *32*, 361-366. For trifluoroacetaldehyde other derivatives, see: (n) Matsutani, H.; Poras, H.; Kusumoto, T.; Hiyama, T. *Chem. Commun.* **1998**, 1259-1260. (o) Matsutani, H.; Poras, H.; Kusumoto, T.; Hiyama, T. *Synlett* **1998**, 1353-1354. (p) Matsutani, H.; Kusumoto, T.; Hiyama, T. *Chem. Lett.* **1999**, 529-530. (q) Xu, Y.; Dolbier, Jr., W. R. *Tetrahedron Lett.* **1998**, *39*, 9151-9154.
4. For the reaction with boron enolates, see: (a) Iseki, K.; Kobayashi, Y. *Chem. Pharm. Bull.* **1996**, *44*, 2003-2008. (b) Iseki, K.; Oishi, S.; Kobayashi, Y. *Tetrahedron* **1996**, *52*, 71-84. (c) Makino, Y.; Iseki, K. Oishi, S.; Hirano, T.; Kobayashi, Y. *Tetrahedron Lett.* **1995**, *36*, 6527-6530. For with a lithium enolate, see: (d) Qian, C.-P.; Liu, Y. –Z.; Tomooka, K.; Nakai, T. *Org. Synth.* **1999**, *76*, 151-158. (e) Qian, C. P.; Nakai, T.; Dixon, D. A.; Smart, B. E. *J. Am. Chem. Soc.* **1990**, *112*, 4602-4604. (f) Patel, D. V.; Rielley-Gauvin, K.; Ryono, D. E; Free, C. A.; Smith, S. A.; Petrillo, E. W. *J. Med. Chem.* **1993**, *36*, 2431-2441. (g) Patel, D. V.; Rielley-Gauvin, K.; Ryono, D. E. *Tetrahedron Lett.* **1988**, *29*, 4665-4668. (h) Seebach, D.; Juaristi, E.; Miller, D. D.; Schickli, C.; Weber, T. *Helv. Chem. Acta* **1987**, *70*, 237-261. (i) Yamazaki, T.; Takita, K.; Ishikawa, N. *Nippon Kagaku Kaishi* **1985**, 2131-2139. (j) Tius, M. A.; Savariar, S. *Tetrahedron Lett.* **1985**, *26*, 3635-3638. For with a zinc enolate, see: (k) Watanabe, S.; Sakai, Y.; Kitazume, T.; Yamazaki, T. *J. Fluorine Chem.* **1994**, *68*, 59-61. For with a nickel enolate, see: (l) Soloshonok, V. A.; Kukhar, V. P.; Galushko, S. V.; Svistunova, N. Y.; Avilov, D. V.; Kuz'mina, N. A,; Raevski, N. I.; Struchkov, Y. T.; Pysarevsky, A. P.; Belokon, Y. N. *J. Chem. Soc., Perkin Trans. 1* **1993**, 3143-3155.

5. (a) Mikami, K.; Yajima, T.; Takasaki, T.; Matsukawa, S.; Terada, M.; Uchimaru, T.; Maruta, M. *Tetrahedron* **1996**, *52*, 85-98. (b) Ishii, A.; Kojima, J.; Mikami, K. *Org. Lett.* **1999**, *1*, 2013-2016. (c) Ishii, K.; Mikami, K. *J. Fluorine Chem.* **1999**, *97*, 51-55.
6. (a) Hayashi, E.; Takahashi, Y.; Itoh, H.; Yoneda, N. *Bull. Chem. Soc. Jpn.* **1994**, *67*, 3040-3043. (b) Ogawa, K.; Nagai, T.; Nonomura, M.; Takagi, T.; Koyama, M.; Ando, A.; Miki, T.; Kumadaki, I. *Chem. Pharm. Bull.* **1991**, *39*, 1707-1712. (c) Pautrat, R.; Marteau, J.; Cheritat, R. *Bull. Soc. Chim. Fr.* **1968**, 1182-1186. (d) Mikami, K. Yajima, T.; Siree, N.; Terada, M.; Suzuki, Y.; Kobayashi, I. *Synlett* **1996**, 837-838.
7. (a) Shermolovich, Y. G.; Yemets, S. V. *J. Fluorine Chem.* **2000**, *101*, 111-111. (b) Ishii, A.; Soloshonok, A. V.; Mikami, K. *J. Org. Chem.* **2000**, *65*, 1597-1599.
8. Reddy, M. V. R.; Rudd, M. T.; Ramachandran, P. V. *J. Org. Chem.* **2002**, *67*, 5382-5385.
9. (a) Braid, M.; Iserson, H.; Lawlore, F. E. *J. Am. Chem. Soc.* **1954**, *76*, 4027. (b) Henne, A. L.; Pelley, R. L.; Alm, R. M. *J. Am. Chem. Soc.* **1950**, *72*, 3370-3371. (c) Shechter, H.; Conrad, F. *J. Am. Chem. Soc.* **1950**, *72*, 3371-3373.
10. Ishii, A.; Terada, Y. *J. Syn. Org. Chem. Jpn.* **1999**, *57*, 898-899.
11. For a preliminary communication, see: (a) Funabiki, K.; Nojiri, M.; Matsui, M.; Shibata, K. *Chem. Commum.* **1998**, 2051-2052. For a full paper, (b) Funabiki, K.; Matsunaga, K.; Nojiri, Hashimoto, W.; Yamamoto, H.; Shibata, K.; M.; Matsui, M. *J. Org. Chem.* **2003**, 68, 2853-2860.
12. For a preliminary communication, see: (a) Funabiki, K.; Matsunaga, K.; Matsui, M.; Shibata, K. *Synlett* **1999**, 1477-1479. For a full paper, ref. 11b.
13. Shibata, K.; Matsui, M.; Funabiki, K. *Jpn. Kokai Tokkyo Koho*, JP 2001226308 (2001). Funabiki, K.; Hashimoto, W.; Matsui, M. *Chem. Commun.* in press.
14. Lin, J.-T.; Yamazaki, T.; Kitazume, T. *J. Org. Chem.* **1987**, *52*, 3211-3217.
15. Ishii, A.; Kanai, M.; Higashiyama, K.; Mikami, K. *Chirality* **2002**, *14*, 709-712.
16. Enders, D.; Klatt, M. *Synthesis* **1996**, 1403-1418.
17. Funabiki, K.; Yamamoto, H.; Nagamori, M.; Matsui, M.; *Jpn. Kokai Tokkyo Koho*, submitted (application No. 2003－403268) (2003).
18. Very recently, Saito and Yamamoto have published a similar result of cyclopentanone using other organocatalyst, see: Torii, H.; Nakadai, M.; Ishihara, K.; Saito, S.; Yamamoto, H. *Angew. Chem., Int. Ed.* **2004**, *43*, 1983-1986.

Chapter 20

Fluorinated Synthon: Asymmetric Catalytic Reactions

Koichi Mikami, Yoshimitsu Itoh, and Masahiro Yamanaka

Department of Applied Chemistry, Tokyo Institute of Technology, Meguro-ku, Tokyo 152–8552, Japan

1. Introduction

Organofluorine compounds continue to attract much attention, having important applications as physiologically active agents, liquid crystals, and in other areas (*1*). Frequently, mono-fluoro- or trifluoromethyl-containing compounds having specific absolute configuration are synthesized to produce new analogues that exhibit particularly high physiological activity and remarkable physical properties (*2*).

Methods for the synthesis of fluorine-containing compounds can be broadly classified into two types: carbon-fluorine bond forming reactions (fluorination with fluorinating reagents)(*1 a,b,c, 3*) and carbon-carbon bond forming reactions employing fluorinated synthons, such as perfluoro-alkanes and -alkenes or fluorine-containing carbonyl compounds.

In this article, we summarize our research on catalytic asymmetric syntheses using fluorinated synthons such as fluoral and trifluoropyruvate. Formally, the reaction involves electrophilic replacement of a C-H bond of an olefin, silyl enolether or aromatic compound with a C-C bond appended with a trifluoromethyl group. (Figure 1).

2. Catalytic enantioselective C-C bond formation of fluorinated carbonyl compounds

Catalytic enantioselective carbon-carbon bond forming reactions with prochiral fluorinated carbonyl compounds represent very efficient and economical approaches to chiral organo-fluorine compounds. Therefore, there is much current interest in developing procedures for asymmetric catalytic carbon-carbon bond forming reactions involving fluorinated synthons.

2-1. Carbonyl-ene reaction

C-H Bond activation and C-C bond formation are the clues to synthetic exploitation in organic synthesis. In principle, the ene reaction converts readily available alkenes into more functionalized products with activation of an allylic C-H single bond and transposition of the C=C double bond (4). This intermolecular [1,5]-hydrogen shift is one of the simplest atom-economical and green processes available for C-C bond formation (5). In particular, the class of the ene reactions involving carbonyl compound as the enophile is referred to as the 'carbonyl-ene reaction' (6). When carbonyl compounds are used as enophiles, alcohols are exclusively formed in a stereoselective manner. In the carbonyl-ene reaction with fluoral, binaphthol (BINOL)-derived titanium (BINOLate-Ti) catalyst (obtained from BINOL and $TiCl_2(O^iPr)_2$) in the presence of MS4A gave the ene product **1** and the Friedel-Crafts product **2** (Table 1) (7). The ratio of **1** and **2** was higher for fluoral than for chloral.

Table 1. BINOLate-Ti catalyzed carbonyl-ene *vs.* Friedel-Crafts reactions of fluoral and chloral.

BINOL (10 mol%) + $TiCl_2(O^iPr)_2$ (10 mol%), MS4A; alkene + H(C=O)CX_3 (2.0 eq.), CH_2Cl_2, 0 °C, 30 min → **1** + **2**

n	*X*	*% yield*	ratio (% ee) **1**	: **2**
0	F	78	62 (>95% *ee*)	38 (>95% *ee*)
1		93	76 (>95% *ee*)	24 (>95% *ee*)
0	Cl	57	55 (26% *ee*)	45 (75% *ee*)
1		49	52 (34% *ee*)	48 (66% *ee*)

This difference can be explained based on the balance of the LUMO energy level of the aldehyde and the charge distribution on the carbonyl carbon. We calculated the charge and LUMO energy in the complex between the aldehyde and acid H^+ used as the chemical model of the Lewis acid at the RHF/6-31G** level. These calculation show that the LUMO energy is lower for fluoral than for chloral, and thus the positive charge on the carbonyl carbon is higher with chloral (Figure 2). The frontier orbital interaction between the ene HOMO and enophile LUMO is the principal interaction in the ene reaction. Consequently, fluoral, having the lower LUMO energy, could show a greater tendency for concerted ene reaction. In contrast, chloral, having larger partial positive charge at the carbonyl carbon, more readily undergoes stepwise cationic reactions, in this case the Friedel-Crafts (F-C) reaction (Scheme 1) (8).

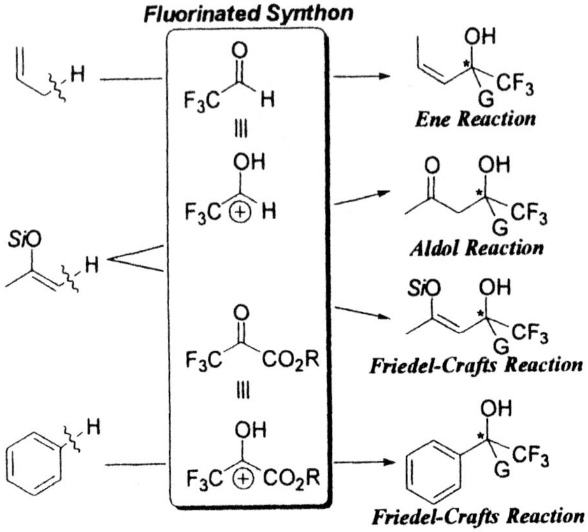

Figure 1. Catalytic asymmetric reaction using fluorinated synthons

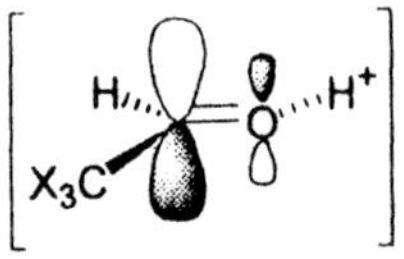

	LUMO (eV)	Charge of carbonyl carbon
X = F	-5.40	+0.61
X = Cl	-4.88	+0.64

Figure 2. LUMO energy levels and charge distributions of carbonyl carbon of fluoral and chloral (RHF/6-31G**)

Scheme 1

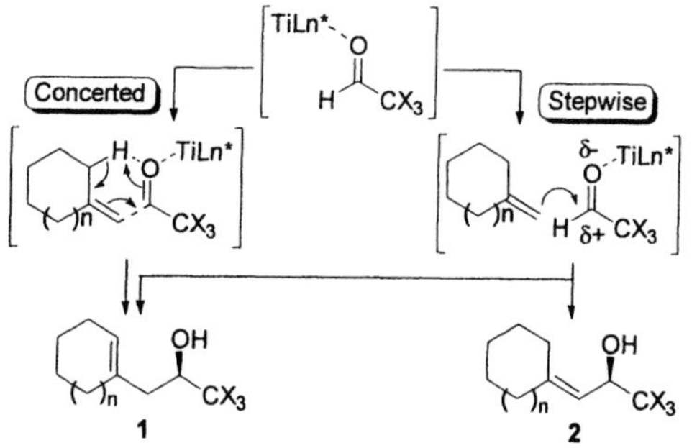

By introducing an electron donating methyl group on the ene components to increase the HOMO level, the ene reaction was facilitated and the F-C reaction was retarded (Scheme 2) (9). In regard to the diastereoselectivity of the ene reaction, the *syn* product is obtained with nearly perfect selectivity. This selectivity can be explained by considering the 6-membered ring transition state (10) as indicated in Scheme 2. The *syn* isomer is preferentially produced due to destabilization by 1,3-diaxial interactions in the transition state that produces the *anti* isomer.

Scheme 2

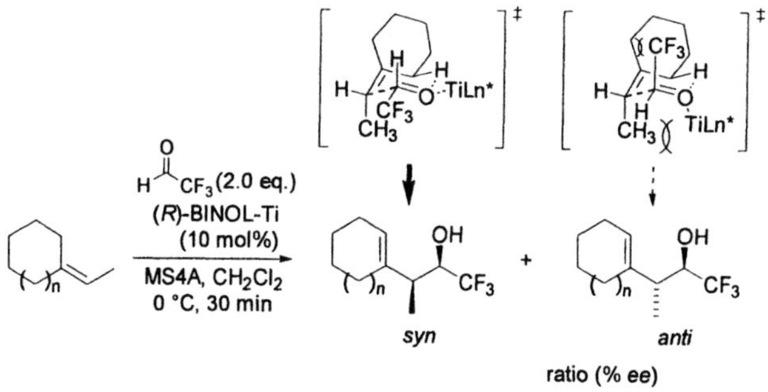

	% yield	ratio (% ee) *syn* : *anti*
n=1	94	98 (96% ee) : 2
2	76	94 (95% ee) : 6

Hemiacetals of such aldehydes can also be employed in these carbon-carbon bond-forming reactions. In the reaction using difluoroacetaldehyde ethyl hemiacetal, MS5A rather than MS4A can be used in order to preferentially trap ethanol generated from the difluoro acetal. This leads to a higher ene selectivity along with high enantioselectivity (> 95% *ee*) (Table 2) (11).

Table 2. BINOLate-Ti catalyzed carbonyl-ene reaction of hemiacetal.

EtO(OH)CHF$_2$ (1.2 eq.), (*R*)-BINOL-Ti (10 mol%), 0°C, CH_2Cl_2, 2h

additive	% yield	ratio(% ee) *syn* : *anti*
-	23	90 (>95% *ee*) : 10
MS4A	30	93 (>95% *ee*) : 7 (>95% *ee*)
MS5A	47	91 (>95% *ee*) : 9

We recognized that a carbonyl-ene reaction with CF_3-ketones would be a synthetically important process, providing a short route to chiral tertiary α-CF_3-carbinols with homo-allylic functionality. However, there had been essentially no successful examples of asymmetric catalysis of ketone-ene reactions (12), because of low ene reactivity of ketones compared to aldehydes. Recently we reported the first successful example of asymmetric catalysis of the trifluoromethylpyruvate-ene reaction using the "naked" dicationic SEGPHOS-Pd(II) complex in CH_2Cl_2, derived from SEGPHOS [(4,4'-bi-1,3-benzodioxole)-5,5'-diylbis(diarylphosphine)] (13), $PdCl_2(CH_3CN)_2$, and $AgSbF_6$, to construct the corresponding quaternary carbon center (14). The presence of the cationic SEGPHOS-Pd(II) complex (**3**) leads to a high chemical yield, (*E*)-olefin selectivity, *anti*-diastereoselectivity, along with high enantioselectivity in this much less reactive carbonyl-ene reactions with ketones, even with less reactive mono- and 1,2-disubstituted olefins (Scheme 3) (15).

2-2. Aldol reaction

The Mukaiyama-aldol reaction of silyl enol ethers is one of the most important carbon-carbon bond forming reactions in organic synthesis (16). With fluoral, this aldol reaction can readily proceed even in the absence of a catalyst, presumably due to the high electrophilicity of fluoral and an intermolecular interaction between Si and F (Scheme 4) (11).

However, it is possible to suppress the uncatalyzed reaction process and achieve a catalytic asymmetric reaction simply by adding ketene silyl acetal (KSA) and fluoral simultaneously to a solvent containing the BINOLate-Ti catalyst, (Scheme 5). Using this simple procedure, a high level of enantioselectivity was achieved (up to 96% *ee* of the (*R*)-enantiomer).

Scheme 3

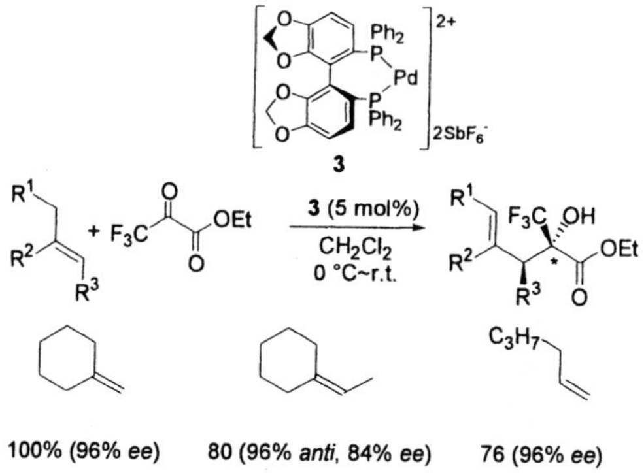
2+
Ph2
P
Pd
P
Ph2
2SbF6-
3
R1
R2
R3
F3C
OEt
3 (5 mol%)
CH2Cl2
0 °C~r.t.
F3C
OH
OEt
C3H7
100% (96% ee)
80 (96% anti, 84% ee)
76 (96% ee)

Scheme 4

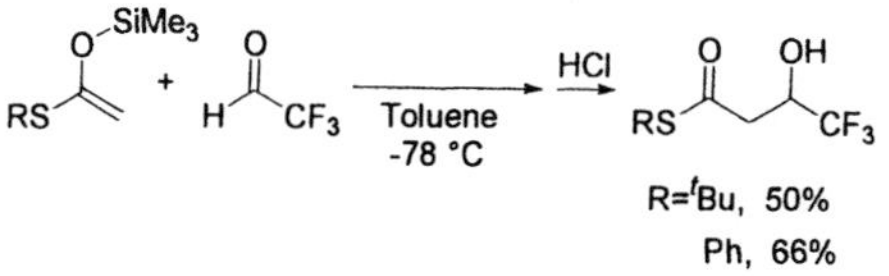
SiMe3
RS
H
CF3
Toluene
-78 °C
HCl
OH
RS
CF3
R=tBu, 50%
Ph, 66%

Scheme 5

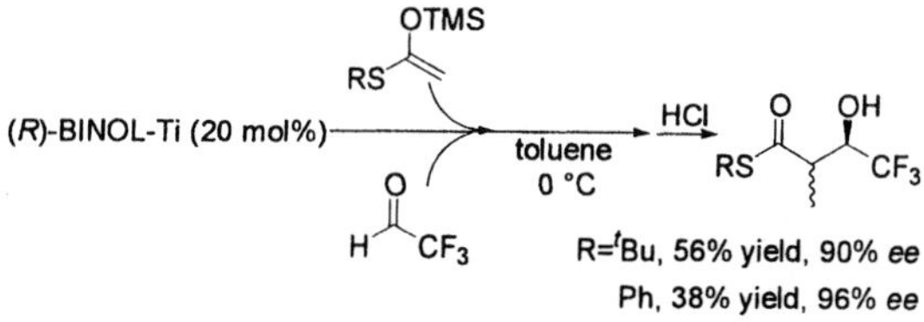
OTMS
RS
(R)-BINOL-Ti (20 mol%)
toluene
0 °C
HCl
OH
RS
CF3
H
CF3
R=tBu, 56% yield, 90% ee
Ph, 38% yield, 96% ee

In addition, with a ketene silyl thioacetal having a methyl group at the α-position, the reaction proceeded with high enantioselectivity, despite having low diastereoselectivity (Scheme 6).

Scheme 6

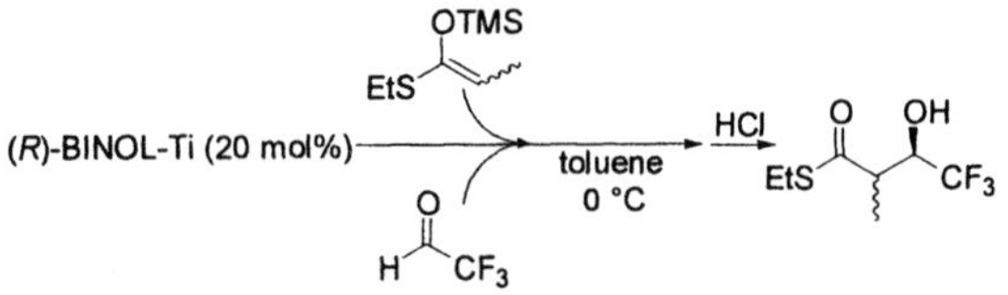

(*R*)-BINOL-Ti (20 mol%), EtS(OTMS)C=CHCH$_3$, CF$_3$CHO, toluene, 0 °C; HCl → EtSC(O)CH(CH$_3$)CH(OH)CF$_3$

KSA	% yield	*syn* : *anti*
95%*Z*	64	48 (55% *ee*) : 52 (64% *ee*)
98%*E*	48	44 (89% *ee*) : 56 (83% *ee*)

2-3. Friedel-Crafts reaction

Friedel-Crafts (F-C) reactions also constitute one of the most useful carbon-carbon bond-formation processes in organic synthesis (*17*). However, there have been only very few reports of asymmetric catalytic F-C reactions. Recently, we investigated the asymmetric catalytic F-C reactions of the prochiral fluorine-containing carbonyl compounds leading to the chiral tertiary α-CF_3-carbinols. We reported the asymmetric catalytic F-C reaction of silyl enol ethers under the Mukaiyama-aldol reaction conditions. The reaction of *tert*-butyldimethylsilyl or triisopropylsilyl enol ether with fluoral was carried out using the (*R*)-BINOLate-Ti complex to afford the F-C product **4** with high % *ee* rather than the usual aldol product **5** (Table 3) (*18*).

The ratio of the F-C product depends very much on the bulkiness of the silyl group. Indeed, triisopropylsilyl enol ether overwhelmingly afforded **4** in contrast to trimethylsilyl enol ether, which gave **5** instead of **4**. The Mukaiyama aldol reaction, namely Lewis acid-promoted carbonyl addition reaction of a silyl enol ether to aldehydes or ketones, is known to proceed via the desilylated β-metaloxy carbonyl chelate intermediates. However, the silyl enol ether F-C product can also be generated under Lewis acid-catalyzed Mukaiyama aldol conditions. This results from the ability of the bulky silyl group to inhibit the nucleophilic substitution reaction on the silyl group (path a in Figure 3). This allows the deprotonation reaction pathway to proceed to produce a silylenol ether (path b in Figure 3). Furthermore, the strong electron-withdrawing CF_3 group could lower the nucleophilicity of the titanium alkoxide in the zwitterionic intermediate to further retard the reaction path a.

Adding value to this FC reaction is the fact that subsequent diastereoselective reactions of the silyl enol ether F-C product with electrophiles can yield highly functionalized fluorinated aldols that are of material and pharmaceutical interests.

Table 3. Asymmetric Friedel-Crafts reactions of silyl enol ethers with fluoral

(*R*)-BINOL-Ti, CH_2Cl_2, 0 °C

SiR_3	R^1	R^2	cat. (mol %)	% yield **4** (E:Z)	% yield **5**	% ee[a]
TMS	4'-Me-Ph	Me	20	0	27[b]	0
TBDMS	Ph	H	5	67 (1:5)	14	98
TIPS	Ph	H	1	90 (1:5)	4	96

[a]The enantiomeric excess of (*Z*)-**11**. [b]The usual aldol product was obtained as the TMS ether. The diastereomeric ratio=1:4.

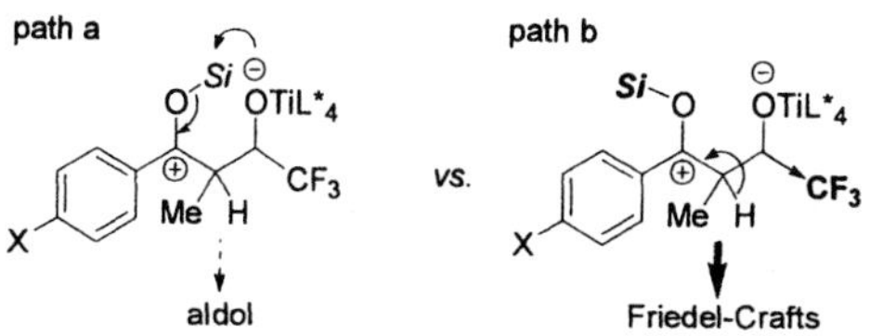

Figure 3. Comparison of aldol pathway (path a) and Friedel-Crafts pathway (path b) in the zwitterionic intermediate.

In addition, the oxidation by *m*-CPBA or desilylation by TBAF of **4** leads selectively to mono-protected *syn*-diol **6** and *anti*-aldol **7** with high diastereoselectivity, respectively (Scheme 7).

Scheme 7

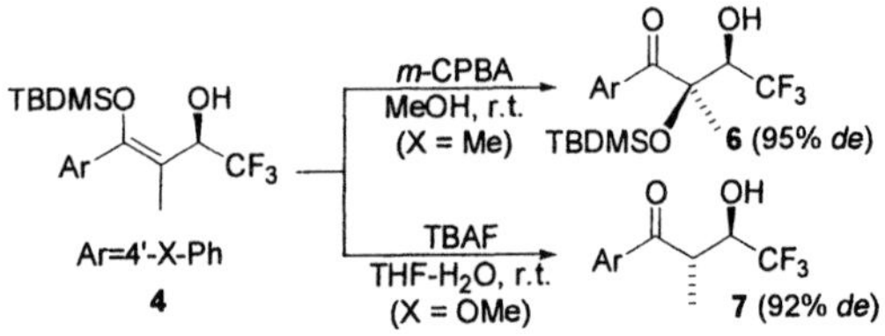

TBDMSO OH Ar CF3 Ar=4'-X-Ph **4** — *m*-CPBA, MeOH, r.t. (X = Me) → **6** (95% *de*); TBAF, THF-H_2O, r.t. (X = OMe) → **7** (92% *de*)

We also studied the asymmetric F-C reaction of a vinyl ether with fluoral catalyzed by (*R*)-BINOLate-Ti complex derived from BINOL and $TiCl_2(O^iPr)_2$. In a similar manner to the F-C reaction of silyl enol ether, a reactive vinyl ether was obtained as the F-C product. This can sequentially react with *m*-CPBA to afford highly functionalized organofluorine compounds. The reaction with the enol ether gave the F-C product almost exclusively and with high enantioselectivity (up to 85% *ee*), with essentially no aldol product being obtained (Table 4). When employing a vinyl ether possessing a *β*-methyl substituent, (*E*)-**8** was predominantly obtained irrespective of the geometry of the substrate. The subsequent diastereoselective oxidation of the F-C products by *m*-CPBA provided diastereoselectively *syn*-α,β-dihydroxy ketones **9** in high chemical yields (Scheme **8**) (*19*)

Table 4. Asymmetric Friedel-Crafts reactions of vinyl ether with fluoral

MeO R R1 + H(C=O)CF3 — (*R*)-BINOL-Ti, CH_2Cl_2, 0 °C → MeO OH R CF3 R1 **8**

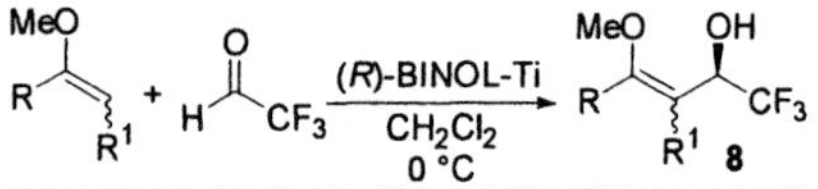

R	*R^1*	*cat.*	*8 (E:Z)*	*% ee*
Ph	Me	10mol %	54 % (1:2)	72
4'-Me-Ph	H	20mol %	64 % (5:1)	85

Scheme 8

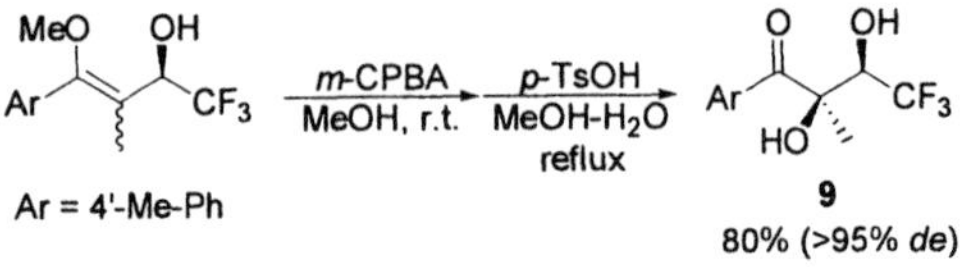

MeO OH Ar CF3, Ar = 4'-Me-Ph — *m*-CPBA, MeOH, r.t. → *p*-TsOH, MeOH-H_2O, reflux → O OH Ar HO CF3 **9** 80% (>95% *de*)

We also investigated the asymmetric F-C reation of fluoral with aromatic substrates. Aromatic compounds have low nucleophilicity relative to silyl enol ethers or vinyl ethers, and hence the asymmetric F-C reaction needs Lewis acidic metal catalyst such as BINOLate-Ti. Both the yield and the enantioselectivity of the F-C product are increased when a catalyst with high Lewis acidity is used and an electron-withdrawing group (Br) is introduced onto BINOL. The asymmetric F-C reaction of phenyl ethers selectively gave *p*-**10** rather than *o*-**10**. The regioselectivity of the F-C product was increased by using *n*-butyl phenyl ether (*p*-**10**: *o*-**10** = 8:1) (*20*). When (*R*)-6,6'-Br_2-BINOL is added as an additive (), the yield and enantioselectivity are further increased (89% yield, 90% *ee*), and an asymmetric activation effect is observed (Table 5) (*21,22*).

Table 5. Asymmetric Friedel-Crafts reactions of aromatic compounds with fluoral

RO-C6H5 + H-C(=O)-CF3 —(*R*)-BINOL-Ti, CH_2Cl_2, 0 °C→ *p*-10 (RO-C6H4-CH(OH)-CF3) + *o*-10

R	*cat. (mol %)*	*additive (mol %)*	*% yield*	*p-10:o-10*	*% ee*[a]
Me	30	-	82	4 : 1	73
Me	5	-	94	4 : 1	84
n-Bu	15	-	85	8 : 1	83
Ph	10	-	90	3 : 1	54
Me	10	(*R*)-6,6'-Br_2-BINOL (10)	89	4 : 1	90
n-Bu	10	(*R*)-6,6'-Br_2-BINOL (10)	90	8 : 1	90

[a]The enantiomeric excess of *p*-10.

Recently, we developed an asymmetric F-C reaction of ethyl trifluoropyruvate with aromatic compounds catalyzed by Pd(II) BINAP or SEGPHOS complexes. This reaction can proceed at lower reaction temperature (–30 °C) than the reaction with fluoral and gives products with high enantioselectivity (*23*). The F-C product obtained using the Pd(II) catalyst gave higher chemical yield and enantioselectivity than that using a Cu(II) catalyst (*24*). In marked contrast to the carbonyl-ene reaction, the BINAP ligand provides higher enantioselectivity than does the SEGPHOS ligand (Scheme 9).

References

1 a) Mikami, K.; Itoh, Y.; Yamanaka, M. *Chem. Rev.* **2004**, *104*, 1-16. b) Mikami, K.; Itoh, Y.; Yamanaka, M. *Fine Chemical* **2003**, *32(1)*, 35-50. c) Mikami, K.; Itoh, Y.; Yamanaka, M. *Fine Chemical* **2003**, *32(2)*, 11-20. d) Hiyama, T.; Kanie, K.; Kusumoto, T.; Morizawa, Y.; Shimizu, M. *Organofluorine Compounds*; Springer-Verlag: Berlin Heidelberg, 2000. e) Soloshonok, V. A. Ed. *Enantiocontrolled Synthesis of Fluoro-*

Scheme 9

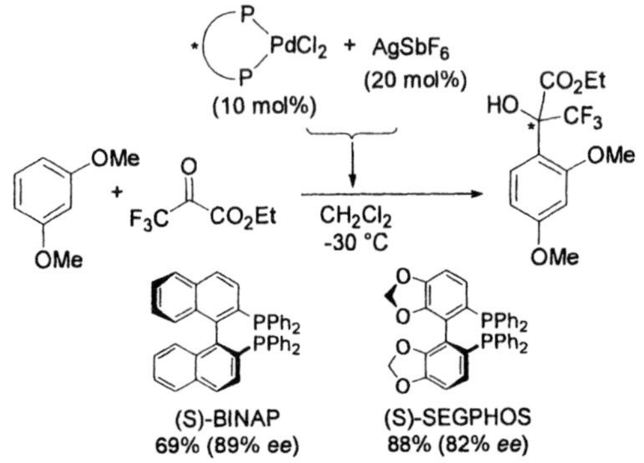

(S)-BINAP
69% (89% *ee*)

(S)-SEGPHOS
88% (82% *ee*)

OrganicCompounds; Wiley: Chichester, 1999. f) Chambers, R. D. Ed. *Organofluorine Chemistry*, Springer, Berlin 1997. g) Iseki, K. *Tetrahedron* **1998**, *54*, 13887-13914. h) Ojima, I.; McCarthy, J. R.; Welch, J. T. Eds. *Biomedical Frontiers of Fluorine Chemistry*; American Chemical Society: Washington DC, 1996. i) Smart, B. E., Ed. *Chem Rev.* **1996**, *96* , No. 5 (Thematic issue of fluorine chemistry). j) Banks, R. E.; Smart, B. E.; Tatlow, J. C. Eds. *Organofluorine Chemistry: Principles and Commercial Applications*; Plenum Press: New York. 1994. k) Kitazume, T.; Ishihara, T.; Taguchi, T. *Chemistry of Fluorine*; Koudansha: Tokyo, 1993. l) Ishikawa, N. Ed. *Synthesis and Reactivity of Fluorocompounds*; CMC: Tokyo, Vol. 3, 1987. m) Ishikawa, N.; Kobayashi, Y. *Fluorine Compounds*; Koudansha: Tokyo, 1979. n) Hudlicky, M. *Chemistry of Organic Fluorine Compounds, 2nd edn*; Ellis Horwood: Chichester, 1976.

2 a) Smart, B. E. *J. Fluorine Chem.* **2001**, *109*, 3-11. b) Schlosser, M. *Angew. Chem. Int. Ed.* **1998**, *37*, 1496-1513.

3 For asymmetric fluorination, see: a) Togni, A. *Chem. Rev.* in press. b) Gouverneur, V.; Greedy, B. *Chem. Eur. J.* **2002**, *8*, 767-771. c) Muñiz, K. *Angew. Chem. Int. Ed.* **2001**, *40*, 1653-1656. d) Resnati, G. *Tetrahedron* **1993**, *49*, 9385-9445. e) Bravo, P.; Resnati, G. *Tetrahedron: Asymmetry* **1990**, *1*, 661-692.

4 Comprehensive reviews on ene reactions: a) Mikami, K.; Shimizu, M.; *Chem. Rev.* **1992**, *92*, 1021-1050. b) Snider, B. B. In *Comprehensive Organic Synthesis*; Trost, B. M.; Fleming, I. Eds., Pergamon: London, 1991; Vol 2, p 527-561 and Vol 5, p 1-27.

5 a) Trost, B. M. *Science* **1991**, *254*, 1471-1477. b) Trost, B. M. *Angew. Chem. Int. Ed. Engl.* **1995**, *34*, 259-281.

6 Review on 'carbonyl-ene reactions': Mikami, K.; Terada, M.; Shimizu, M.; Nakai, T. *J. Synth. Org. Chem. Jpn.* **1990**, *48*, 292-303.

7 Milkami, K.; Yajima, T.; Terada, M.; Uchimaru, T. *Tetrahedron Lett.* **1993**, *34*, 7591-7594.

8 Akihiro, I.; Mikami, K. *J. Synth. Org. Chem. Jpn.* **2000**, *58*, 324-333.

9 Mikami, K.; Yajima, T.; Terada, M.; Kato, E.; Maruta, M. *Tetrahedron: Asymmetry* **1994**, *5*, 1087-1090.

10 a) Mikami, K.; Loh, T. –P.; Nakai, T. *Tetrahedron Lett.* **1988**, *29*, 6305-6308. b) Yamanaka, M.; Mikami, K. *Helv. Chim. Acta* **2002**, *85*, 4264-4271.
11 Mikami, K.; Yajima, T.; Takasaki, T.; Matsukawa, S.; Terada, M.; Uchimaru, T.; Maruta, M. *Tetrahedron* **1996**, *52*, 85-98.
12 Evans reported the excellent example of asymmetric catalysis of glyoxylate-ene reaction with a variety of olefins by chiral bis-oxazoline Cu^{2+} complexes: Evans, D. A.; Tregay, S. W.; Burgey, C. S.; Paras, N. A.; Vojkovsky, T. *J. Am. Chem. Soc.* **2000**, *122*, 7936-7943.
13 SEGPHOS = (4,4'-bi-1,3-benzodioxole)-5,5'-diylbis(diphenylphosphine). a) Saito, T.; Yokozawa, T.; Ishizaki, T.; Moroi, T.; Sayo, N.; Miura, T.; Kumobayashi, H. *Adv. Synth. Catal.* **2001**, *343*, 264-267. EP 850945A **1998**, US 5872273 **1999**.
14 Reviews: a) Martin, S. F. *Tetrahedron* **1980**, *36*, 419-460. b) Fuji, K. *Chem. Rev.* **1993**, *93*, 2037-2066. c) Corey, E. J.; Guzman-Perez, A. *Angew. Chem. Int. Ed.* **1998**, *37*, 388-401.
15 Aikawa, K.; Kainuma, S.; Hatano, M.; Mikami, K. *Tetrahedron Lett.* **2004**, *1*, 183-185.
16 Reviews: a) Carreira, E. M. In *Comprehensive Asymmetric Catalysis*; Jacobsen, E. N.; Pfaltz, A.; Yamamoto, H., Eds.; Springer: Berlin Heidelberg, 1999; Vol. 3, p 997-1065. b) Mekelburger, H. B.; Wilcox, C. S. In *Conprehensive Organic Synthesis*; B. M. Trost, B. M.; Fleming, I., Eds,; Pergamon press: Oxford, 1991; Vol 2, p 99-131. c) Heathcock, C. H. In *Conprehensive Organic Synthesis*; B. M. Trost, B. M.; Fleming, I., Eds,; Pergamon: Oxford, 1991; Vol 2, p 133-179 & 181-238. d) Kim, B. M.; Williams, S. F.; Masamune, S. In *Conprehensive Organic Synthesis*; B. M. Trost, B. M.; Fleming, I., Eds,; Pergamon: Oxford, 1991; Vol 2, p 239-275. e) Rathke, M. W.; Weipert, P. In *Conprehensive Organic Synthesis*; B. M. Trost, B. M.; Fleming, I., Eds,; Pergamon: Oxford, 1991; Vol 2, p 277-299. f) Paterson, I. In *Comprehensive Organic Synthesis*; B. M. Trost, B. M.; Fleming, I., Eds,; Pergamon: Oxford, 1991; Vol 2, p 301-319. g) Mukaiyama, T.; *Org. React.* **1982**, *28*, 203-331.
17 Reviews: a) Smith, M. B. *Organic Synthesis*; McGraw-Hill: New York, 1994; p 1313-1349. b) Heaney, H. In *Comprehensive Organic Synthesis*; Trost, B. M., Fleming, I., Eds.; Pergamon Press: Oxford, 1991; Vol. 2, p 733-752. c) Roberts, R. M.; Khalaf, A. A. In *Friedel-Crafts Alkylation Chemistry. A Century of Discovery*; Dekker: New York, 1984. d) Olah, G. A. *Friedel-Crafts Chemistry*; Wiley-Interscience: New York, 1973.
18 Ishii, A.; Kojima, J.; Mikami, K. *Org. Lett.* **1999**, *1*, 2013-2016.
19 Ishii, A.; Mikami, K. *J. Fluorine. Chem.* **1999**, *97*, 51-55.
20 Ishii, A.; Soloshonok, V. A.; Mikami, K. *J. Org. Chem.* **2000**, *65*, 1597-1599.
21 Review for asymmetric activation see: Mikami, K.; Terada, M.; Korenaga, T.; Matsumoto, Y.; Ueki, M.; Angelaud, R. *Angew. Chem. Int. Ed.* **2000**, *39*, 3532-3556.
22 Mikami, K.; Yamanaka, M. *Chem. Rev.* **2003**, *103*, 3369-3400.
23 Aikawa, K.; Mikami, K. in preparation
24 Zhuang, W.; Gathergood, N.; Hazell, R. G.; Jørgensen, K. A. *J. Org. Chem.* **2001**, *66*, 1009-1013.

Chapter 21

Industrial Synthetic Application of Fluoral and 3,3-Dichloro-1,1,1-trifluoroacetone as Fluorinated Synthons

Akihiro Ishii, Masatomi Kanai, and Yutaka Katsuhara

Chemical Research Center, Central Glass Company, Ltd., Kawagoe, Saitama 350–1151, Japan

Practical synthetic methods of enantiopure 4,4,4-trifluoro-3-hydroxybutyric acid derivatives, trifluorolactic acid derivatives and enantiopure 1-methyl-2,2,2-trifluoroethylamine using fluoral and 3,3-dichloro-1,1,1-trifluoroacetone are described.

Introduction

Fluoral and 3,3-dichloro-1,1,1-trifluoroacetone (DCTFA) are regarded as fluorinated synthons consisting of two or three carbons with a trifluoromethyl group. Fluoral and DCTFA are industrially produced through a modified Swarts reaction from chloral and pentachloroacetone (PCA), respectively (Scheme 1) (*1,2*). The latter DCTFA is further chlorinated to produce 3,3,3-trichloro-1,1,1-trifluoroacetone (TCTFA) (*3*). On the other hand, DCTFA is reductively dechlorinated to produce 1,1,1-trifluoroacetone (TFA) (*4*). However, industrial synthetic application of these fluorinated synthons, particularly DCTFA and TCTFA, has not been well investigated (*5,6*). In this review, practical synthetic methods of industrially important fluorinated intermediates using these fluorinated synthons are described. In particular, our attention is focused to practical asymmetric syntheses of 4,4,4-trifluoro-3-hydroxybutyric acid derivatives (*7,8,9*) and 1-methyl-2,2,2-trifluoroethylamine (*10,11,12*), and alternative synthetic method of trifluorolactic acid derivatives without using alkali cyanide (*3,13*).

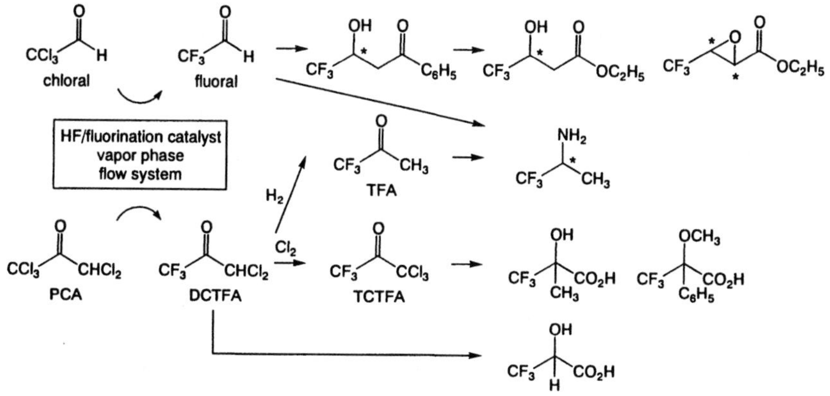

Scheme 1. *Industrial synthetic application of fluoral, DCTFA, TCTFA and TFA as fluorinated synthons.*

Practical synthetic method of enantiopure 4,4,4-trifluoro-3-hydroxybutyric acid derivatives using fluoral

Optically active 4,4,4-trifluoro-3-hydroxybutyric acid derivatives are versatile intermediates in pharmaceutical chemistry and optoelectronic material science (*14*). However, asymmetric synthesis of these derivatives on a large scale has been quite limited (*15*). On the other hand, enantio-enriched 4,4,4-trifluoro-3-hydroxybutyrophenone (**1**) can be readily prepared by asymmetric Friedel-Crafts reaction of vinyl ether with fluoral gas catalyzed by a chiral binaphthol-derived titanium catalyst (Scheme 2) (*16*). Recently, Funabiki et al. also have developed a practical asymmetric synthesis of this compound (**1**), in which easily-handling fluoral ethyl hemiacetal is advantageously used instead of fluoral gas (*8*). If subsequent Baeyer-Villiger reaction of enantiopure **1** would proceed, a practical

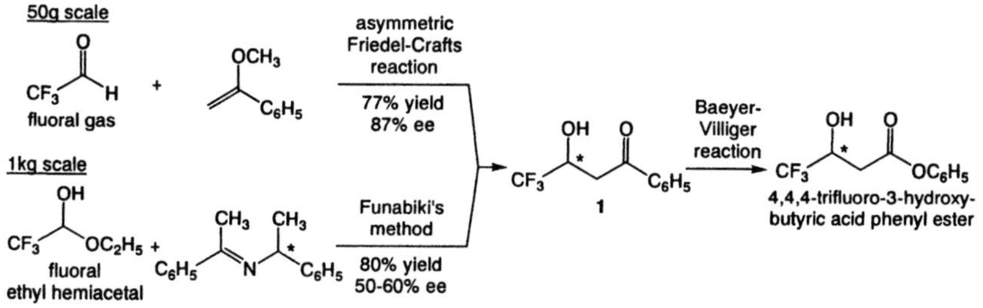

Scheme 2. *Practical asymmetric synthesis of 4,4,4-trifluoro-3-hydroxybutyric acid derivatives.*

asymmetric synthesis of 4,4,4-trifluoro-3-hydroxybutyric acid derivatives could be achieved. Enantio-enriched **1** had a unique property in crystallization, wherein heterochiral crystal (*R-S*, racemate) precipitated preferentially to homochiral crystal (*R-R* or *S-S*) (Figure 1). Therefore, enantiomeric excess (ee) of mother liquor dramatically increased up to 97% ee through a recrystallization. Enantiopure **1** was obtained by following homochiral crystallization.

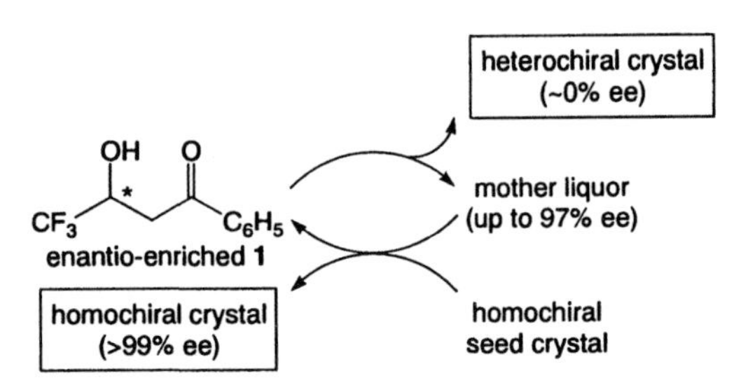

Figure 1. Optical purification of enantio-enriched 4,4,4-trifluoro-3-hydroxybutyrophenone (1) by combination of heterochiral and homochiral crystallization.

In X-ray analysis, *R-S* heterochiral crystal had double hydrogen bonds between hydroxy groups and carbonyl groups in a head to tail fashion. On the other hand, *R-R* homochiral crystal had a single hydrogen bond in a head to head fashion (Figure 2). Therefore, heterochiral crystal was considered to be thermodynamically more stable than homochiral crystal. In fact, much higher melting point of heterochiral crystal (81ºC) was observed than that of homochiral crystal (49ºC).

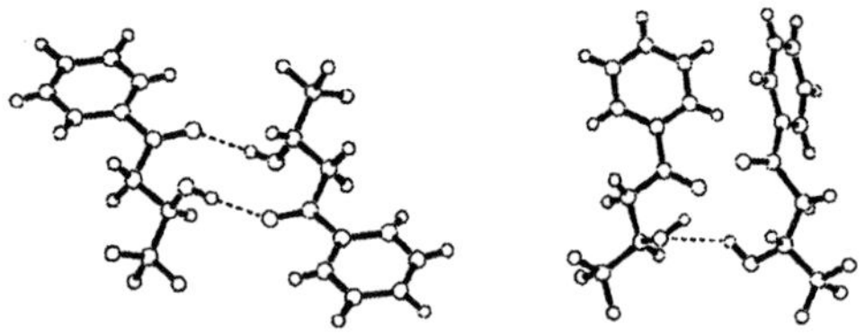

Figure 2. ORTEP drawings of R-S heterochiral crystal and R-R homochiral crystal.

Baeyer-Villiger reaction of enantiopure 4,4,4-trifluoro-3-hydroxybutyrophenone (**1**) proceeded smoothly using inexpensive permonophosphoric acid (H_3PO_5) (*17*) to produce novel phenyl ester (**2**) in a high yield (Scheme 3).

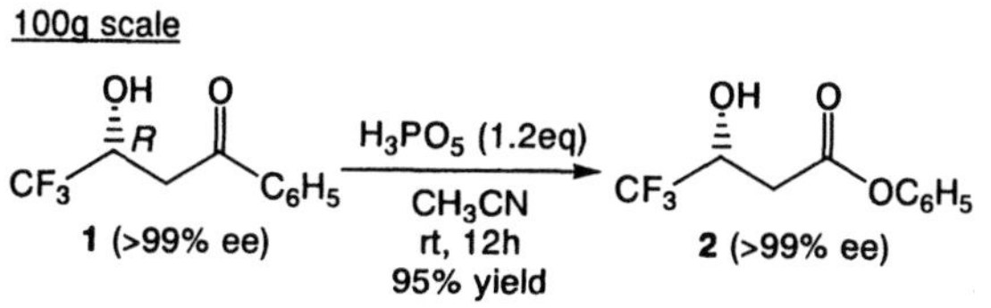
100g scale
OH
O
R
CF3
C6H5
1 (>99% ee)
H3PO5 (1.2eq)
CH3CN
rt, 12h
95% yield
OH
O
CF3
OC6H5
2 (>99% ee)

Scheme 3. Baeyer-Villiger reaction of enantiopure 4,4,4-trifluoro-3-hydroxybutyrophenone (1).

Obtained phenyl ester (**2**) was expected to play as a reactive ester. Transesterification of **2** to ethyl ester (**3**), transamination to benzyl amide (**4**) and hydride reduction using sodium borohydride to primary alcohol (**5**) occurred smoothly (Scheme 4).

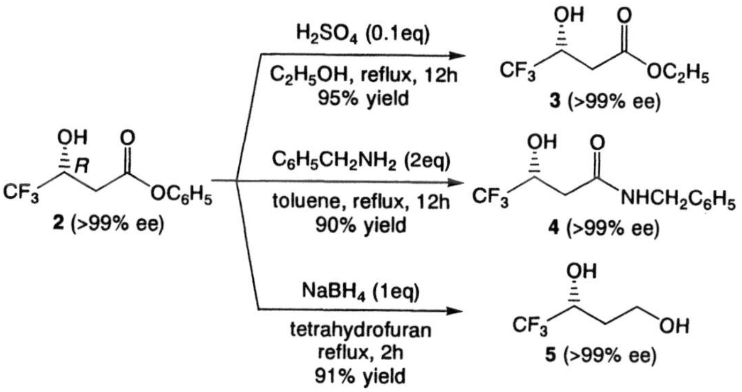
OH
O
R
CF3
OC6H5
2 (>99% ee)
H2SO4 (0.1eq)
C2H5OH, reflux, 12h
95% yield
OH
O
CF3
OC2H5
3 (>99% ee)
C6H5CH2NH2 (2eq)
toluene, reflux, 12h
90% yield
OH
O
CF3
NHCH2C6H5
4 (>99% ee)
NaBH4 (1eq)
tetrahydrofuran
reflux, 2h
91% yield
OH
CF3
OH
5 (>99% ee)

Scheme 4. Reactivity of enantiopure 4,4,4-trifluoro-3-hydroxybutyric acid phenyl ester (2) as a reactive ester.

Optically active 4,4,4-trifluoro-2,3-epoxybutyric acid derivatives also are important intermediates for further complex chiral organofluorine compounds (*18*). Our synthetic strategy is formation of epoxide ring at initial stage, followed by a similar Baeyer-Villiger reaction. Chlorination of enantiopure 4,4,4-trifluoro-3-hydroxybutyrophenone (**1**) was carried out using chlorine gas to produce *syn*-isomer (**6**) in a highly diastereoselective manner (Scheme 5). Ring closure reaction in the presence of triethylamine gave *trans*-epoxide (**7**) in a high yield. In this ring closure reaction, epimerization at α-position might occur first, and then *trans*-**7** would be produced through *anti*-**6**. Subsequent Baeyer-Villiger reaction of *trans*-**7** proceeded without ring opening of epoxide to produce epoxy phenyl ester (**8**). Further transesterification of **8** gave expected epoxy ethyl ester (**9**) totally without any racemization.

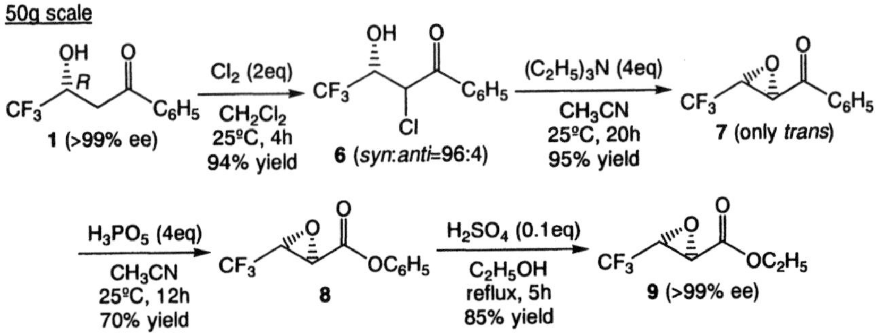

50g scale

$CF_3CH(OH)CH_2COC_6H_5$ 1 (>99% ee) — Cl_2 (2eq), CH_2Cl_2, 25ºC, 4h, 94% yield → 6 (*syn*:*anti*=96:4) — $(C_2H_5)_3N$ (4eq), CH_3CN, 25ºC, 20h, 95% yield → 7 (only *trans*)

— H_3PO_5 (4eq), CH_3CN, 25ºC, 12h, 70% yield → 8 (OC_6H_5) — H_2SO_4 (0.1eq), C_2H_5OH, reflux, 5h, 85% yield → 9 (>99% ee) (OC_2H_5)

Scheme 5. Practical asymmetric synthesis of 4,4,4-trifluoro-2,3-epoxybutyric acid ethyl ester (9).

Thus, practical synthetic method of enantiopure 4,4,4-trifluoro-3-hydroxybutyric acid derivatives using fluoral has been developed through a unique property of enantio-enriched butyrophenone (**1**) in crystallization, followed by Baeyer-Villiger reaction of enantiopure **1** and epoxy phenyl ketone (**7**) using H_3PO_5.

Practical synthetic method of trifluorolactic acid derivatives using DCTFA and TCTFA

Transformation of non-fluorinated 1,1-dichloroacetone to lactic acid has been already reported (*19*). If a similar reaction would proceed in fluorinated DCTFA, trifluorolactic acid (**10**) could be directly produced in a practical manner. Expected reaction proceeded smoothly using hydrate form of DCTFA under basic conditions (Scheme 6).

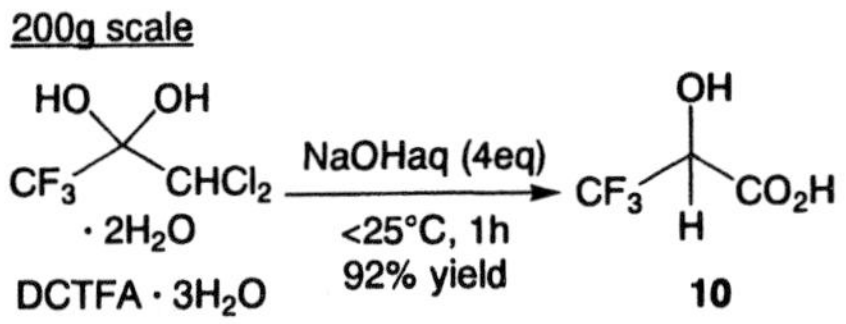

200g scale

$CF_3C(OH)_2CHCl_2 \cdot 2H_2O$ (DCTFA · $3H_2O$) — NaOHaq (4eq), <25°C, 1h, 92% yield → $CF_3CH(OH)CO_2H$ **10**

Scheme 6. Practical synthetic method of trifluorolactic acid (10) using DCTFA.

In this reaction, a formal intramolecular redox reaction occurred, namely a reduction of *gem*-diol and an oxidation of dichloromethyl group. From this consideration, a tentative 1,2-hydride transfer mechanism of *in situ* formed epoxide might be proposed (Figure 3).

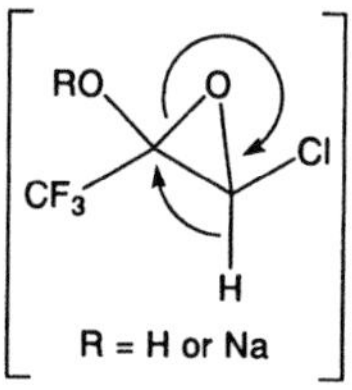

RO O Cl CF$_3$ H

R = H or Na

Figure 3. 1,2-Hydride transfer mechanism.

DCTFA was readily chlorinated by chlorine gas in the presence of a catalytic amount of quinoline to produce TCTFA in a quantitative yield. Addition reaction between TCTFA and carbon nucleophile has never been reported until now. If an adduct product would be obtained, subsequent transformation of trichloromethyl group could provide α-substituted trifluorolactic acid derivatives. Among them, (*S*)-α-methyltrifluorolactic acid (**13**) is a very important intermediate in development of therapeutic agent for urinary incontinence (*20*). Grignard reaction between TCTFA and methylmagnesium chloride gave expected tertiary alcohol (**11**) in a usual manner (Scheme 7). Methanolysis of **11** under basic conditions produced methoxy methyl ester (**12**) in a good yield. Subsequent demethylation of **12** using hydrobromic acid gave racemic α-methyltrifluorolactic acid (**13**) in a quantitative yield. Enantiopure (*S*)-**13** was obtained through a fractional recrystallization of 1:1 diastereomeric mixture of methoxy amide (**14**) (*21*), derived from **12** and (*R*)-1-phenylethylamine.

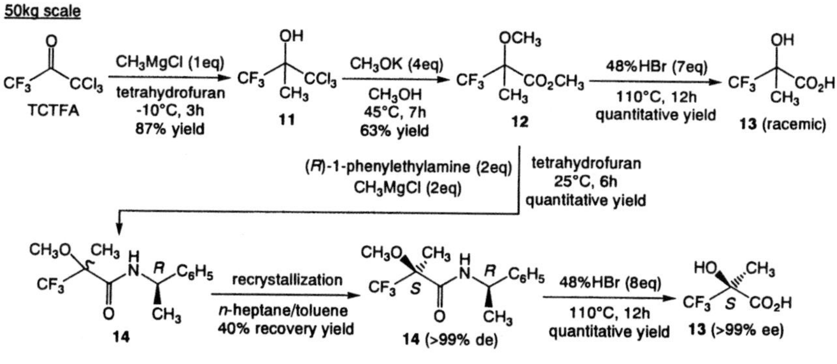

50kg scale

TCTFA → CH$_3$MgCl (1eq), tetrahydrofuran, -10°C, 3h, 87% yield → **11** → CH$_3$OK (4eq), CH$_3$OH, 45°C, 7h, 63% yield → **12** → 48%HBr (7eq), 110°C, 12h, quantitative yield → **13** (racemic)

12 → (*R*)-1-phenylethylamine (2eq), CH$_3$MgCl (2eq), tetrahydrofuran, 25°C, 6h, quantitative yield → **14**

14 → recrystallization, *n*-heptane/toluene, 40% recovery yield → **14** (>99% de) → 48%HBr (8eq), 110°C, 12h, quantitative yield → **13** (>99% ee)

Scheme 7. Practical synthetic method of α-methyltrifluorolactic acid (racemic-13, (S)-13) using TCTFA.

This synthetic strategy was applied to a practical synthetic method of Mosher's acid (racemic-**16**). However, only poor result was obtained in Grignard reaction of TCTFA with phenylmagnesium halide. After detailed examination, Friedel-

Crafts reaction between TCTFA and benzene gave expected adduct product (**15**) in an acceptable yield (Scheme 8). Critical reaction conditions were use of aluminium chloride as a Lewis acid and control of reaction temperature, wherein -30ºC must be maintained for a relatively long reaction time. A similar methanolysis of **15**, followed by hydrolysis of *in situ* formed methyl ester produced Mosher's acid (racemic-**16**) in a high yield.

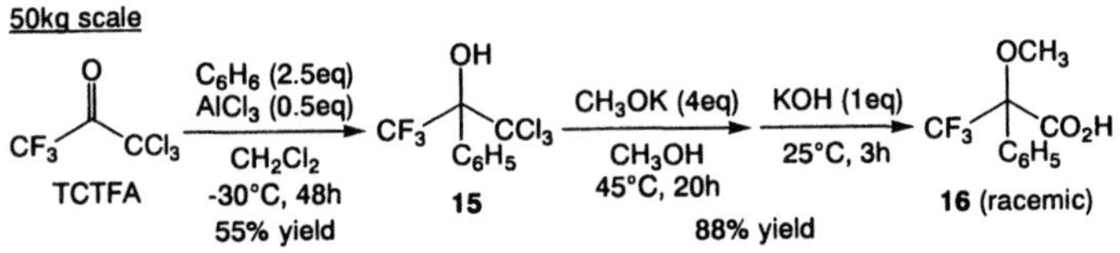

50kg scale
$CF_3C(O)CCl_3$ (TCTFA) → [C_6H_6 (2.5eq), $AlCl_3$ (0.5eq), CH_2Cl_2, -30°C, 48h, 55% yield] → $CF_3C(OH)(C_6H_5)CCl_3$ (**15**) → [CH_3OK (4eq), CH_3OH, 45°C, 20h; KOH (1eq), 25°C, 3h; 88% yield] → $CF_3C(OCH_3)(C_6H_5)CO_2H$ (**16** (racemic))

Scheme 8. Practical synthetic method of Mosher's acid (racemic-16) using TCTFA.

In both methanolyses, ring opening reaction of *in situ* formed epoxide occurred regioselectively on carbon attached to a trifluoromethyl group (Figure 4).

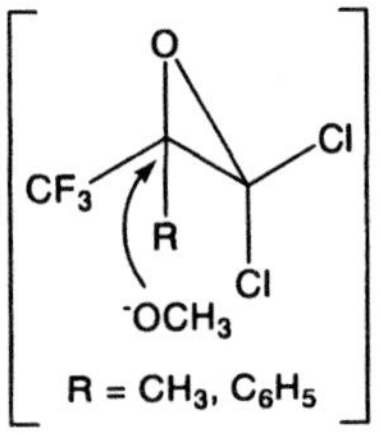

O, CF_3, R, Cl, Cl, $^-OCH_3$
R = CH_3, C_6H_5

Figure 4. Regioselective ring opening.

Thus, practical synthetic method of trifluorolactic acid derivatives using DCTFA and TCTFA has been developed. This alternative synthetic method does not demand use of troublesome alkali cyanide.

Practical synthetic method of enantiopure 1-methyl-2,2,2-trifluoroethylamine using TFA and fluoral

DCTFA could be reductively dechlorinated in the presence of water to produce TFA in a hydrate form. Synthetic application of TFA has been relatively well investigated, therefore, one example of current topics in our company is described, which is a practical synthetic method of enantiopure 1-methyl-2,2,2-trifluoroethylamine (**19**) (*22*). Our synthetic strategy is 1,3-induced asymmetric reduction of chiral imine or 1,3-oxazolidine derived from TFA. Chiral imine

(**17**) was prepared with *trans* form through a dehydration between TFA and commercially available (*S*)-1-phenylethylamine (Scheme 9). Diastereoselectivity of asymmetric hydrogenation of **17** was critically influenced by reaction temperature to improve up to 75:25 ratio (*S,S*:*R,S*) at 0ºC. Obtained secondary amine (**18**) was purified to >99.5% diastereomeric excess (de) by a fractional recrystallization of hydrobromic acid salt. Hydrogenolytic cleavage of chiral auxiliary group in (*S,S*)-**18** was carried out only by changing reaction temperature to 60ºC. Enantiopure hydrochloric acid salt of (*S*)-**19** was obtained without any racemization in a total yield of 37%.

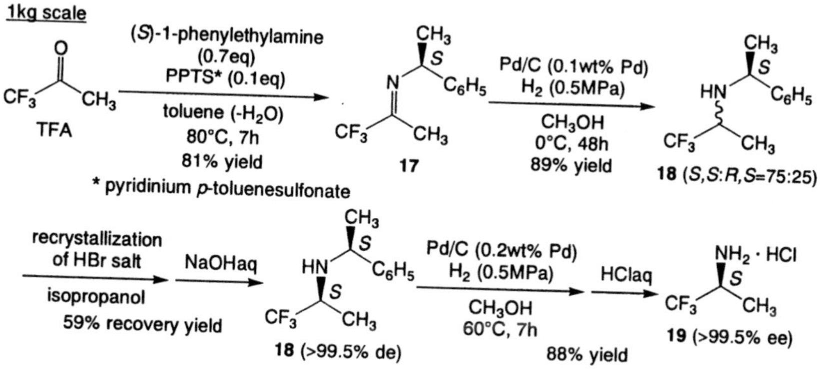

Scheme 9. Practical asymmetric synthesis of (S)-1-methyl-2,2,2-trifluoroethylamine hydrochloric acid salt (19) using TFA (I).

Chiral 1,3-oxazolidine (**20**) was prepared from TFA and (*S*)-phenylglycinol in a highly diastereoselective manner (*R,S*:*S,S*=97:3) (Scheme 10). Asymmetric hydride reduction of **20** proceeded through a retention mechanism to produce preferentially (*S,S*)-diastereomer (**21**) (*S,S*:*R,S*=75:25). (*S,S*)-Diastereomer more easily precipitated than (*R,S*)-diastereomer in a recrystallization to improve to >99.5% de. In a similar hydrogenolysis of (*S,S*)-**21**, same enantiopure hydrochloric acid salt of (*S*)-**19** was obtained in a total yield of 36%. General asymmetric synthesis of 1-substituted-2,2,2-trifluoroethylamine using fluoral ethyl hemiacetal also has been reported from our company. 1,3-Induced asymmetric methylation of chiral amino hemiacetal (**22**), *in situ* prepared from fluoral ethyl hemiacetal and (*S*)-phenylglycinol, gave same secondary amine (**21**) in a highly diastereoselective manner (*S,S*:*R,S*=97:3) through a metallo-imine mechanism (Scheme 11).

Thus, practical asymmetric synthesis of 1-methyl-2,2,2-trifluoroethylamine using TFA and fluoral has been developed through 1,3-induced asymmetric reduction and methylation, respectively. From an industrial point of view, use of inexpensive 1-phenylethylamine was preferred to that of phenylglycinol as a chiral auxiliary group (Scheme 9 > Scheme 10,11).

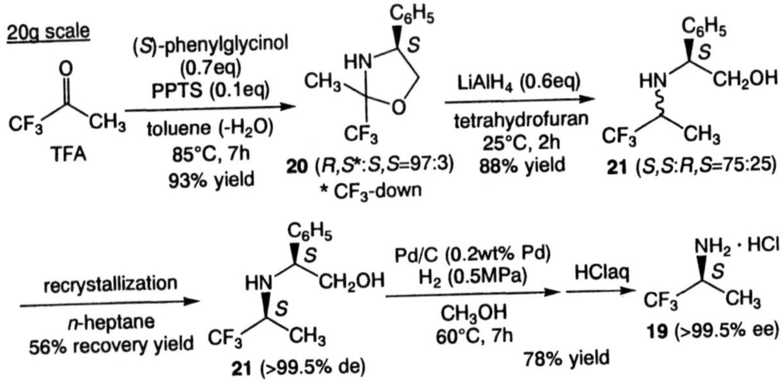

Scheme 10. Practical asymmetric synthesis of (S)-1-methyl-2,2,2-trifluoroethylamine hydrochloric acid salt (19) using TFA (II).

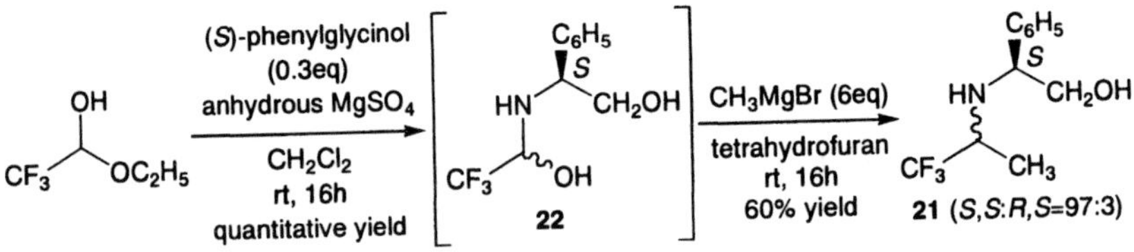

Scheme 11. General asymmetric synthesis of 1-substituted-2,2,2-trifluoroethylamine using fluoral.

Conclusion

Our research activity has been concentrated on industrial synthetic application of fluoral and DCTFA as fluorinated synthons. Above-mentioned key fluorinated intermediates can be now purchased on a kg scale from our company. Our contribution in this field will accelerate development of further complex organofluorine drugs.

Acknowledgment

We thank Prof. Koichi Mikami (Tokyo Institute of Technology), Prof. Kazumasa Funabiki (Gifu University), Prof. Kimio Higashiyama (Hoshi University) and Tamejiro Hiyama (Kyoto University) for fruitful discussions.

Literature Cited

1. Antonini, A.; Putters, R.; Wetroff, G. US Patent 3,787,489, 1974.
2. Kanai, M.; Sakaya, T.; Watanabe, M.; Goto, Y.; Nadano, R. US Patent 5,905,174, 1999.
3. Ishii, A.; Kanai, M.; Yasumoto, M.; Inomiya, K.; Kuriyama, Y.; Katsuhara, Y. *J. Fluorine Chem.* **2004**, *125*, 567.
4. Tsukamoto, M.; Yoshikawa, F.; Fujimoto, M.; Takada, N.; Sugimori, Y.; Negishi, J. JP Patent 2003-342221.
5. Oda, Y.; Yanakawa, M. JP Patent 2000-063306.
6. Panetta, C. A.; Casanova, T. G. *J. Org. Chem.* **1970**, *35*, 4275.
7. Ishii, A.; Kanai, M.; Higashiyama, K.; Mikami, K. *Chirality* **2002**, *14*, 709.
8. Funabiki, K.; Ishii, A.; Kanai, M.; Hayami, T.; Shibata, K.; Matsui, M.; Kuriyama, Y.; Yasumoto, M. US Patent 6,639,100, 2003.
9. Ishii, A.; Kanai, M.; Kuriyama, Y.; Yasumoto, M.; Inomiya, K.; Ootsuka, T.; Ueda, K. JP Patent Application 2003-283118.
10. Ishii, A.; Kuriyama, Y.; Yasumoto, M.; Kanai, M.; Inomiya, K.; Ootsuka, T.; Ueda, K. JP Patent Application 2003-166525.
11. Ishii, A.; Miyamoto, F.; Higashiyama, K.; Mikami, K. *Chem. Lett.* **1998**, 119.
12. Ishii, A.; Higashiyama, K.; Mikami, K. *Synlett* **1997**, 1381.
13. Ishii, A.; Yasumoto, M.; Kanai, M.; Kuriyama, Y.; Inomiya, K.; Ootsuka, T.; Sutoh, K.; Ueda, K. JP Patent Application 2004-001313.
14. Sting, A. R.; Seebach, D. *Tetrahedron* **1996**, *52*, 279.
15. Jeulin, S.; Duprat de Paule, S.; Ratovelomanana-Vidal, V.; Genet, J.-P.; Champion, N.; Dellis, P. *Angew. Chem. Int. Ed.* **2004**, *43*, 320.
16. Ishii, A.; Mikami, K. *J. Fluorine Chem.* **1999**, *97*, 51.
17. Ogata, Y.; Tomizawa, K.; Ikeda, T. *J. Org. Chem.* **1978**, *43*, 2417.
18. v. d. Bussche-Hunnefeld, C.; Seebach, D. *Chem. Ber.* **1992**, *125*, 1273.
19. Tanaka, T.; Kuroda, T.; Kishimoto, H.; Kamimori, S. *Yukagaku* **1979**, *28*, 501.
20. Kumazawa, T.; Yamagata, T.; Suzuki, K.; Aono, S.; Atsuki, K.; Karasawa, A.; Seishi, T.; Takai, H.; Yoshida, M. EP Patent 979,821, 2000.
21. Yasuhara, F.; Takeda, M.; Ochiai, Y.; Miyano, S.; Yamaguchi, S. *Chem. Lett.* **1992**, 251.
22. Pfrengle, W.; Pees K.-J.; Albert G.; Carter P.; Rehnig A.; Cotter H. v. T. US Patent 6,204,269, 2001.

Chapter 22

Synthesis and Utility of Fluorinated Acylsilanes

John T. Welch, Woo Jin Chung, Seiichiro Higashiya, Silvana C. Ngo, and Dong Sung Lim

Department of Chemistry, University at Albany, State University of New York, 1400 Washington Avenue, Albany, NY 12222

A variety of fluorinated acylsilanes were synthesized. Difluoroacetyltrialkylsilanes (**3**) were prepared through retro-Brook rearrangement from reaction of trifluoroethanol and chlorotrialkylsilanes in the presence of LDA. Sequential Mg-promoted defluorination in the presence of chloro-trimethylsilane, followed by acidic hydrolysis resulted in the formation of monofluoroacetyltrialkylsilanes (**4**). Electrophilic fluorination of 1,1-difluoro-2-trialkylsilyl-2-trialkylsilyloxy ethenes (**2**) with Selectfluor® gave trifluoro-acetyltrialkyl silanes (**5**). The synthetic utility of fluorinated acylsilanes in olefination and allylation reactions under a variety of reaction conditions is also described.

Since A. G. Brook's first observation at late 50's that a carbonyl group located α to silicon atom displays unusual reactivity (1), acylsilanes have attracted interest. In response to the potential preparative utility of these materials various synthetic methods for the formation of acylsilanes have been developed (2).

Brook originally discovered and investigated the intramolecular 1,2-anionic migration of a silyl group from carbon to oxygen in the late 50's and early 60's (3). The migratory aptitude of a silyl group is a general phenomenon, comprising a family of [1,n]-carbon to oxygen silyl migration events commonly referred to as Brook rearrangements (Figure 1) (4). The reverse process dubbed a retro-Brook rearrangement, an intramolecular migration of a silyl group from oxygen to carbon, was also reported and extensively studied (Figure 1) (5).

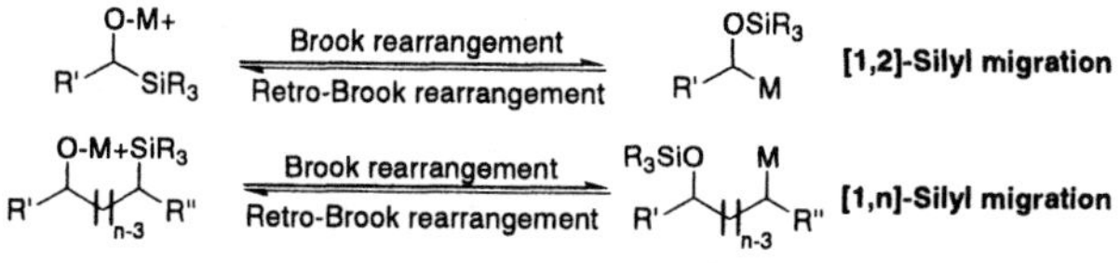

Figure 1. Brook and retro-Brook rearrangement

Thus, Brook and retro-Brook rearrangements of acylsilanes have found increasing use in organic synthesis for the formation of useful building blocks. A variety of applications and utilities as synthetic intermediates in the construction of complex natural products have been described (6).

The remarkable impact that fluorination has on agricultural, medicinal and materials chemistry has lead to a rapid growth in the development of methods for the regio- and stereospecific introduction of fluorine into organic molecules. Considerable effort has been devoted to both the preparation of fluorinated building blocks, which can be transformed to the desired target molecules and to the development of new fluorinating reagents for direct fluorination at an appropriate point in a synthesis (7).

Combining both fluorine and the unique reactivity of silicon, fluorinated acylsilanes, in particular, could easily be expected to be versatile fluorinated building blocks with diverse synthetic utility. However reports on either the synthetic methods necessary for the formation of fluorinated acylsilanes or the applications of these compounds have been very limited.

Discussion

Synthesis of fluorinated acylsilanes

Trifluoroacetyltriphenyl- and trifluoroacetyldimethylphenylsilane were synthesized by the reaction of the appropriate organosilyllithium reagent and trifluoroacetic anhydride in the presence of cuprous iodide. However, this method is limited to the formation of aryl substituted trifluoroacylsilanes due to the instability of silyllithium reagents (Figure 2a) (8a-b). Recently, a new two step synthetic method employing an electrochemical transformation for the construction of trifluoroacetyltrimethylsilane from ethyl trifluoroacetate was reported. Unfortunately the requirement for special electrochemical apparatus is a significant drawback to general use of this method (Figure 2b) (8c). Clearly, a more effective and general synthetic method for the formation of fluorinated acylsilanes from easily available starting materials under convenient reaction conditions is still highly desired.

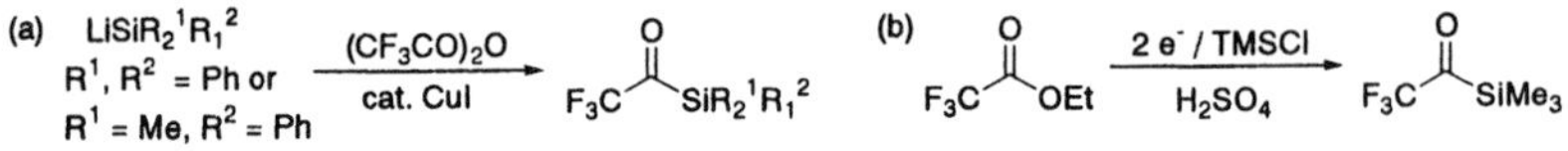

Figure 2. Synthesis of trifluoroacylsilanes

Synthesis of difluoroacetyltrialkylsilanes (3)

A new strategy was designed based on the reactivity of a difluorovinyl anion generated from protected trifluoroethanol by treatment with strong base (9). Silylation of the protected difluorovinyl anion [**I**] with a variety of chlorotrialkylsilanes would generate 1,1-difluoro-2-trialkylsilylenol ethers (**2**), which could be further transformed by deprotection to difluoroacetyltrialkylsilanes (**3**). In addition, reductive defluorination (10) of (**3**) or electrophilic fluorination (11) of (**2**) would lead to formation of either monofluoroacetyltrialkylsilanes (**4**) or trifluoroacetyltrialkylsilanes (**5**) (Figure 3).

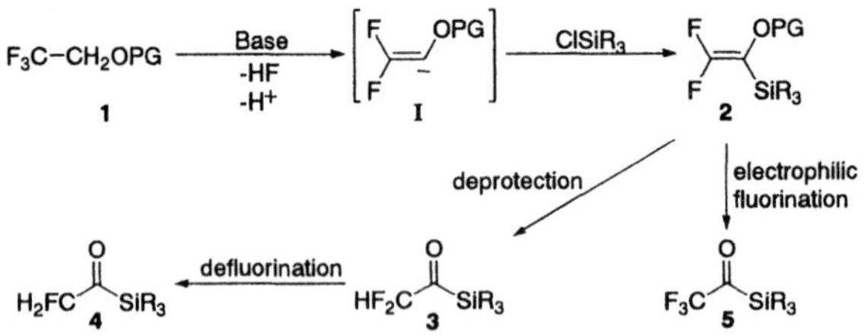

Figure 3. Newly designed synthetic method for the formation of fluorinated acylsilanes

In a model reaction, the triethylsilyl ether of trifluoroethanol (**1a**) (see Table I) was prepared and treated with lithium diisopropylamide (LDA).

Unexpectedly, the penultimate target difluoroacetyltriethylsilane (**3a**) was obtained directly instead of the anticipated 1,1-difluoro-2-trimethylsilyl-2-triethylsilyloxyethene (**2a**). It was postulated that difluoroacetyltriethylsilane (**3a**) was formed by retro-Brook rearrangement (Figure 4).

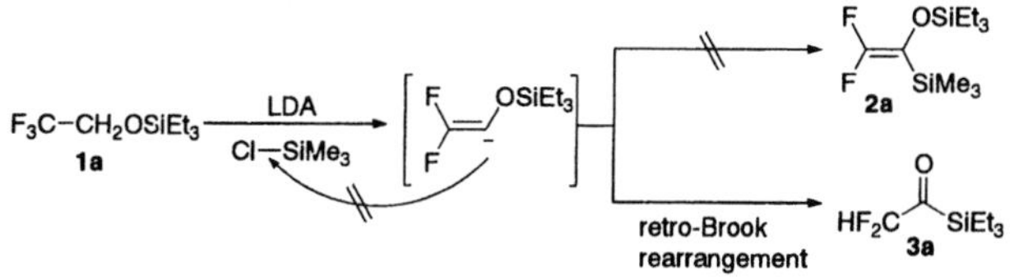

Figure 4. Unexpected silyl group migration

During the course of exploring the generality of the formation of difluoroacetyltrialkylsilanes (**3**), it was found that trialkylsilyl trifluoroethyl ethers (**1**), prepared *in situ*, could be used without purification prior to subsequent transformations. It was necessary for the addition of LDA to a mixture of trifluoroethanol and appropriate chlorotrialkylsilanes in tetrahydrofuran (THF) to be carried out carefully so that the reaction temperature was not allowed to exceed than –30 °C. The difluorovinyl anion [**I**] was apparently unstable at temperatures greater than –30 °C. The formation of difluoroacetyltrialkylsilanes (**3**) can be rationalized as occurring via two consecutive reactions; initial defluorination followed by a second deprotonation of trialkylsilyl trifluoroethyl ethers (**1**) to generate the difluorovinyl anion [**I**], which upon retro-Brook rearrangement generated intermediate [**II**], and secondly hydrolysis [**II**] to form the difluoroacetyltrialkylsilanes (**3**) (Figure 5).

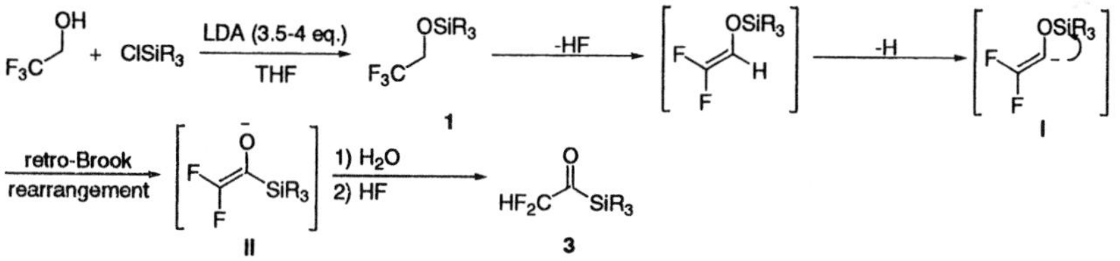

Figure 5. Plausible mechanism for the formation of (3)

As summarized in Table I, a variety of difluoroacetyltrialkylsilanes (**3**) were synthesized in moderate yield. Difluoroacetyltriethylsilane (**3a**) and difluoroacetyl-*tert*-butyldiphenylsilane (**3c**) were obtained in 60% and 74%, respectively. On the other hand, difluoroacetyltrialkylsilanes with more hindered silyl groups, such as triisopropylsilyl (**3b**) and *tert*-butyldimethylsilyl (**3d**), required the addition of hexamethylphosphoramide (HMPA). The much less hindered trimethylsilyl derivative (**3e**) and relatively labile triphenylsilyl derivative (**3f**) were formed in only low yields in the crude mixture and were not isolable by the same method employed in the preparation of (**3a-d**).

Table I. Synthesis of difluoroacetyltrialkylsilanes (3)

3	*SiR₃*	*Yield (%)*	*Condition*
3a	$SiEt_3$	60	0 °C, 3 hrs
3b	Si^iPr_3	63	0 °C, 36 hours, and then room temp. for 24 hours with HMPA
3c	Si^tBuPh_2	74	Room temp. 24 hours
3d	Si^tBuMe_2	13	-20 °C for overnight with HMPA
3e	$SiMe_3$	0	-15 °C for overnight
3f	$SiPh_3$	0	-15 °C for overnight
3f	$SiPh_3$	18	Acidic work-up / controlled hydrolysis

Synthesis of 1,1-difluoro-2-trialkylsilyl-2-trialkylsilyloxyethenes (2)

During the study of the formation of difluoroacetyltrialkylsilanes (**3**), it was possible to trap enolate intermediate **[II]** on introduction of another equivalent chlorotrialkylsilane to form the difluoroenoxysilanes (**2**) (Figure 6). In all cases (Table II), the 1,1-difluoro-2-trialkylsilyl-2-trialkylsilyloxyethenes (**2**) were obtained in good yields, even with the 2-*tert*-butyldimethylsilyl (**2d**). The 2-trimethylsilyl (**2e-f**), and 2-triphenylsilyl (**2g-h**) derivatives that had not previously given satisfactory results did form the corresponding acylsilanes (Figure 5, Table I).

Interestingly, the initial aliquot of chlorotrialkylsilane leads to *C*-silylation through retro-Brook rearrangement from intermediate **[I]** and an additional equivalent of a second silylating reagent results in *O*-silylation. This process generally occurs without interchange of the silyl groups to predictably form the desired products (**2**). These observations clearly establish that the intermediate exists as an enolate **[II]**, generated from retro-Brook rearrangement of silyl group from **[I]**, rather than as difluorovinyl anion **[I]**.

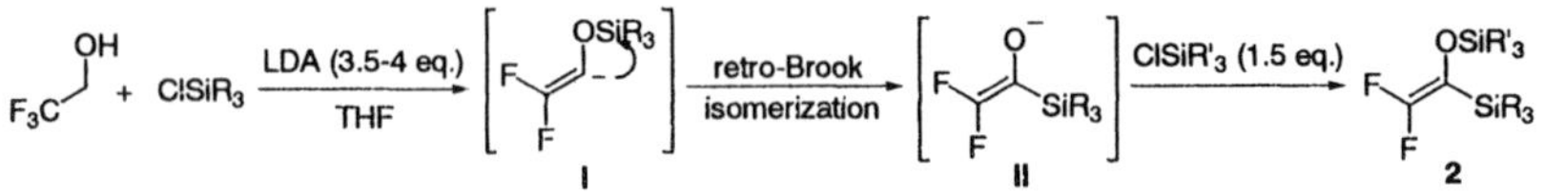

Figure 6. Formation of 1,1-difluoro-2-trialkylsilyl-2-trialkylsilyloxyethenes (2)

Table II. Synthesis of 1,1-difluoro-2-trialkylsilyl-2-trialkylsilyloxyethenes (2)

Entry	*2*	*SiR_3*	*SiR'_3*	*Yield (%)*
1	**2a**	$SiEt_3$	$SiMe_3$	80
2	**2b**	Si^iPr_3	$SiMe_3$	69
3	**2c**	Si^tBuPh_2	$SiMe_3$	71
4	**2d**	Si^tBuMe_2	$SiMe_3$	65
5	**2e**	$SiMe_3$	$SiMe_3$	62
6	**2f**	$SiMe_3$	$SiPh_3$	68
7	**2g**	$SiPh_3$	$SiMe_3$	67
8	**2h**	$SiPh_3$	$SiPh_3$	43

Synthesis of monofluoroacetyltrialkylsilanes (4) and corresponding monofluoroenoxysilanes (6)

The reductive defluorination by magnesium metal in the presence chlorotrimethylsilane, a method reported by Uneyama (10), was employed for the formation of monofluoroacetyltrialkylsilanes (**4**). The addition of HMPA prevented passivation of the magnesium surface. The monofluoroenoxysilanes (**6**) and the corresponding hydrolysis products, monofluoroacetyltrialkylsilanes (**4**), were prepared in good yields (Figure 7, Table III).

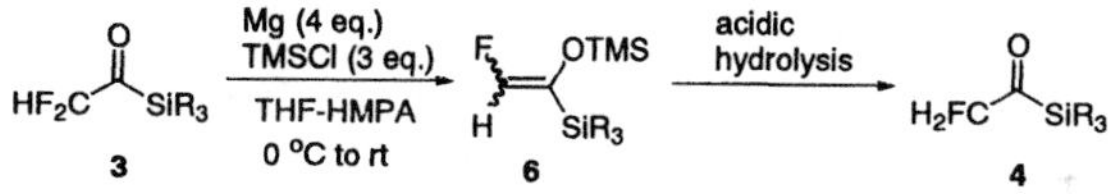

*Figure 7. Synthesis of monofluoroenoxysilanes (**6**) and monofluoroacetyltriakylsilanes (**4**)*

Table III. Synthesis of (4) and (6)

3	*SiR_3*	*6*	*Yield (%)*	*4*	*Yield (%)*
3a	$SiEt_3$	**6a**	74	**4a**	75
3b	Si^iPr_3	**6b**	76	**4b**	82
3c	Si^tBuPh_2	**6c**		**4c**	69
3d	Si^tBuMe_2	**6d**	69	**4d**	85
3f	$SiPh_3$	**6f**		**4f**	68

Electrophilic halogenation of (2)

Enoxysilanes are well known precursors to α-functionalized ketones (12). Of the α-functionalized ketones, α-haloketones are important synthetic building blocks for the introduction of the ketone moiety into molecules by Reformatsky reactions (13). Electrophilic halogenation reactions of difluoroenoxysilanes bearing either the triisopropylsilyl (**2b**) or the triphenylsilyl group (**2g**) on vinyl carbon were utilized for the synthesis of α,α-difluoro-α-halo-acetyltrialkylsilanes (**7**) (Figure 8).

Reaction of (**2b**) and (**2g**) with electrophilic halogenating reagents (2 equivalents), such as *N*-chlorosuccinimide (NCS), *N*-bromosuccinimide (NCS), and iodine (I_2), in THF at 0 °C for 30 minutes gave the corresponding α,α-difluoro-α-halo-acetyltrialkylsilanes (**7**), in good to moderate yields. Chlorination with NCS did require longer reaction times and higher reaction temperatures (room temperature) compared to bromination and iodination. Using 10 equivalents of NCS to affect chlorination of **2b** gave a satisfactory yield in shorter reaction times (2 days). However, when the excess of chlorinating reagent was decreased to 5 equivalents, not only did reaction times (4 days) increase, byproduct formation also increased lowering the yield of **7**.

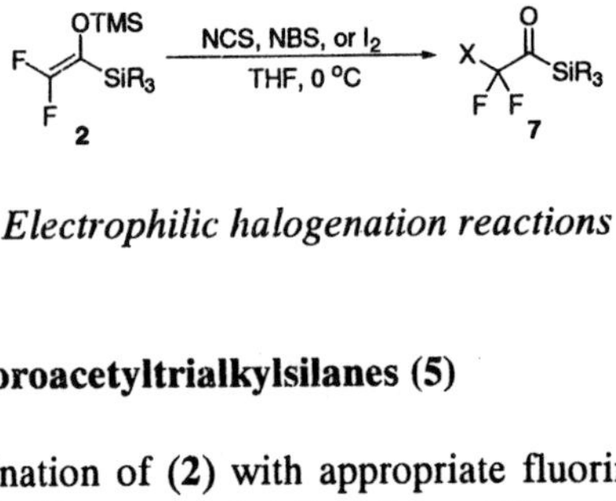

Figure 8. Electrophilic halogenation reactions of (2)

Synthesis of trifluoroacetyltrialkylsilanes (5)

Electrophilic fluorination of (**2**) with appropriate fluorinating reagents can lead to the formation of trifluoroacetyltrialkylsilanes (**5**), synthetic equivalents of trifluoroacetaldehyde. Highly volatile and reactive trifluoroacetaldehyde is commonly generated *in situ* from precursors such as hemiacetals, hemiaminals, or aminals under acidic conditions (14). With synthetic methods for the formation of trifluoroacetyltrialkylsilanes very limited (8), a general method for the facile formation of trifluoroacetyltrialkylsilanes was highly desirable.

Reaction of (**2**) with 1.5 equivalents of Selectfluor® in a mixture of dichloromethane and acetonitrile (1:4) at room temperature for 1 day resulted in the formation of trifluoroacetyltriakylsilanes (**5**) (Figure 9). In general, 1,1-difluoro-2-trialkylsilyl-2-trimethylsilyloxyethenes (**2**) bearing bulky silyl groups gave higher yields of **5** as consequence of reduced volatility. On work-up and evaporation of solvents, (**5**) was spectroscopically (^{1}H and ^{19}F NMR) pure and can be used in subsequent reactions without additional purification. Formation of trifluoroacetyltrimethylsilane (**5e**) was also observed by ^{19}F NMR with a resonance at δ -80.6 ppm corresponding CF_3 group. Unfortunately isolation was unsuccessful, presumably as a consequence of the volatile nature of (**5e**) (Table IV) (8c).

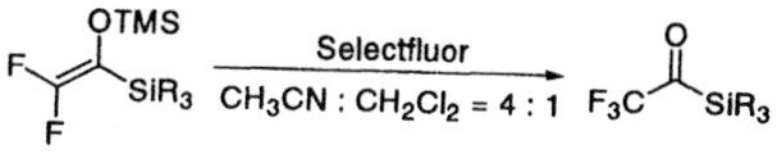

$$\text{2} \xrightarrow[\text{CH}_3\text{CN : CH}_2\text{Cl}_2 = 4:1]{\text{Selectfluor}} F_3C\text{-CO-SiR}_3 \ (\text{5})$$

Figure 9. Electrophilic fluorination of (2) with Selectfluor®

Table IV. Synthesis of trifluoroacetyltrialkylsilanes (5)

Entry	*5*	*Temperature*	*SiR_3*	*Yield (%)*
1	**5a**	-78 °C	Si^tBuPh_2	0
2	**5a**	0 °C	Si^tBuPh_2	0
3	**5a**	Room Temp.	Si^tBuPh_2	87
4	**5b**	Room Temp.	$SiPh_3$	73
5	**5c**	Room Temp.	Si^iPr_3	48
6	**5d**	Room Temp.	$SiEt_3$	35
7	**5e**	Room Temp.	$SiMe_3$	0

Source: Reproduced with permission from *J. Fluorine Chem.* **2004**, **125**, 543. Copyright 2004.

Synthetic utility of fluorinated acylsilanes

Olefination reactions of fluorinated acylsilanes

The reactions of acylsilanes with phosphorus and sulfur ylides often involve silyl group migration to form enoxysilanes (15). This tendency is even more pronounced in the transformations of fluorinated acylsilanes, giving fluorinated enoxysilanes as the major product (Figure 10) (16). However, no information on the influence of the nature of trialkylsilyl group on the reaction pathway has previously been reported.

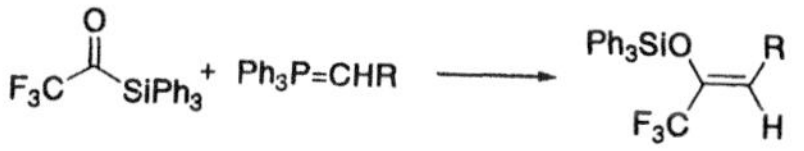

$$F_3C\text{-CO-SiPh}_3 + Ph_3P\text{=CHR} \longrightarrow (Ph_3SiO)(F_3C)C\text{=}CHR$$

Figure 10. Wittig reaction of fluorinated acylsilanes

Wittig reaction of difluoroacetyltrialkylsilanes (3)

Wittig reaction of difluoroacetyltriethylsilane (**3a**) with triphenylphosphonium methylide produced two fluorinated compounds: the expected alkene (**8a**), and as consequence of Brook rearrangement, the fluorinated enoxysilane (**9a**) (Figure 11, Table V). However, a similar reaction with difluoroacetyl-*tert*-butyldiphenylsilane (**3c**) yielded only fluorinated enoxysilane (**9b**). Furthermore, when (**3a**) was allowed to react with a resonance-stabilized triphenylphosphonium benzylide, the only product isolated was the normal Wittig product (**8c**).

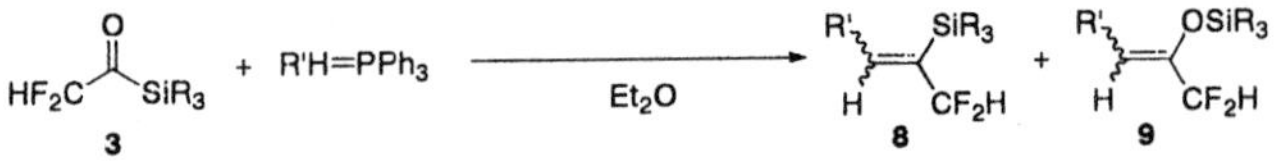

$$HF_2C{-}C(=O){-}SiR_3\ (\mathbf{3}) + R'H{=}PPh_3 \xrightarrow{Et_2O} R'(H)C{=}C(SiR_3)(CF_2H)\ (\mathbf{8}) + R'(H)C{=}C(OSiR_3)(CF_2H)\ (\mathbf{9})$$

Figure 11. Wittig reaction of difluoroacetyltrialkylsilanes (2)

Table V. Wittig reaction of (3)

Entry	*3*	*SiR_3*	*Time (h)*	*R'*	*Yield (%)[a]*	*Yield (%)*
1	**3a**	$SiEt_3$	1	H	**8a** (21)	**9a** (25)
2	**3c**	Si^tBuPh_2	2	H	**8b** (0)	**9b** (63)
3	**3a**	$SiEt_3$	24	Ph	**8c** (75)	**9c** (0)

Source: Reproduced with permission from *J. Fluorine Chem.* **2002**, **117**, 207. Copyright 2002.

It is clear that both the reactivity of ylides and the nature of substituents on silicon affected on the outcome of the reaction.

Horner-Emmons reaction of difluoroacetyltrialkylsilanes (3)

Reaction of (**3**) with stabilized ylides (17), Horner-Emmons type reagents, was also investigated (Figure 12) and results are summarized in Table VI.

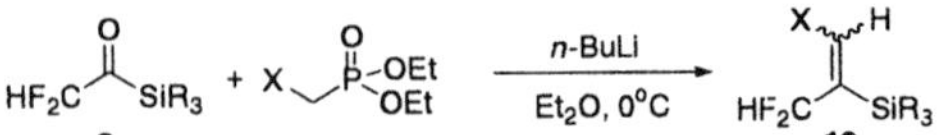

$$HF_2C{-}C(=O){-}SiR_3\ (\mathbf{3}) + X{-}CH_2{-}P(=O)(OEt)_2 \xrightarrow[Et_2O,\ 0^oC]{n\text{-}BuLi} X(H)C{=}C(HF_2C)(SiR_3)\ (\mathbf{10})$$

Figure 12. Horner-Emmons reaction of (3)

Table VI. Horner-Emmons reaction of (3)

Entry	*3*	*X*	*Time (h)*	*Yield (%)*	E/Z
1	**3a**	C(O)OEt	1	**10a** (61)	99/1
2	**3b**	C(O)OEt	4	**10b** (52)	99/1
3	**3c**	C(O)OEt	1	**10c** (59)	99/1
4	**3a**	CN	1	**10d** (76)	78/22
5	**3b**	CN	1	**10e** (59)	99/1
6	**3c**	CN	1	**10f** (42)	99/1

Reaction of (**3**) with the phosphonate anion, generated by deprotonation of either triethyl phosphonoacetate or diethyl (cyanomethyl)phosphonate, in diethyl ether at 0 °C gave only the normal Horner-Emmons product (**10**) with no isomerized product detected. These results are in agreement with those obtained with triphenylphosphonium benzylide, another stabilized ylide. The reaction of difluoroacetyltriisopropylsilane (**3b**) with triethyl phosphonoacetate (Table VI, entry 2) required longer reaction times than compounds (**3a**) and (**3c**), principally as a consequence of steric obstruction of the carbonyl carbon by the alkyl silane. With only one exception, reaction of (**3a**) with diethyl (cyanomethyl)phosphonate (Table VI, entry 4), the reaction was highly stereoselective as indication by the observation of only one vinyl proton around δ 6 ppm in ^{1}H NMR spectrum and a single fluorine resonance at δ –110 ppm in ^{19}F NMR. Extensive studies to determine an absolute stereochemistry of the product (**10**) by both NMR spectroscopy and single crystal X-ray diffraction studies revealed the absolute stereochemistry as (*E*).

The stereoselective formation of the (*E*)-isomer can be rationalized by the steric influences observed in the transition state model shown in Figure 13. The oxyanion and phosphonate have to occupy a synperiplanar relationship for *syn*-elimination. As steric congestion between silyl group and X (CN and C(O)OEt) in transition state [**IV**] is greater than between the silyl group and hydrogen in [**III**] path *a* is favored leading to the selective formation of the (*E*)-isomer.

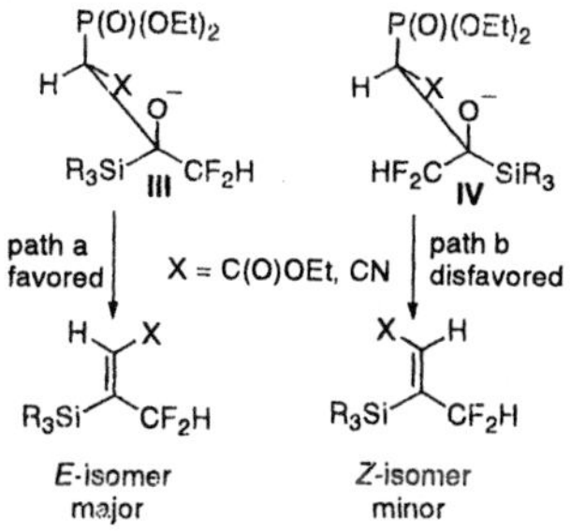

Figure 13. Transition state model for rational of stereoselectivity

Reaction of mono- (4) and difluoroacetyltrialkylsilanes (3) with sulfur ylides

In reactions of mono- (**4**) and difluroacetyltrialkylsilanes (**3**) with sulfur ylides (Figure 14), the least bulky silyl derivative, difluoroacetyltriethylsilane (**3a**), reactions with dimethylsulfonium methylide under various reaction conditions produced complex fluorinated mixtures. On the other hand, reactions with dimethylsulfoxonium methylide, in THF at 0 °C, cleanly formed a single fluorinated product. Dimethylsulfoxonium methylide reacted with all fluorinated acylsilanes (**3** and **4**) examined to give exclusively the enoxysilane products (**9** and **11**) in good yields (Table VII). The rearrangement was insensitive to either the steric demand of silyl group or the extent of fluorination of α-carbon. The low isolated yield of compound (**11a**) may be a consequence of the volatility of the compound. Unfortunately, monofluoroacetyltriethylsilane (**4a**) did not undergo reaction in a manner that lead to facile isolation of the product.

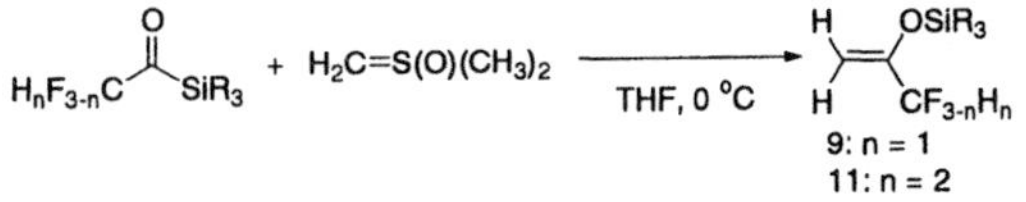

$H_nF_{3-n}C$–C(=O)–SiR_3 + $H_2C{=}S(O)(CH_3)_2$ →(THF, 0 °C) $H_2C{=}C(OSiR_3)(CF_{3-n}H_n)$

9: n = 1
11: n = 2

Figure 14. Reaction of fluorinated acylsilanes (3) and (4) with sulfur ylides

Reaction pathways for the formation of alkenes and enoxysilanes are depicted in Figure 15. Addition of the ylide to the fluorinated acylsilane would result in the formation of intermediate [V]. In cases where the ylides were stabilized by either resonance or the electron withdrawing effect of substituents(Y), such as phenyl (Table V, entry 3), carbonyl (Table VI, entries 1-3), or cyano groups (Table VI, entries 4-6), reaction path b is favored forming alkenes. On the other hand, reaction of ylides with substituents (Y) with no

Table VII. ***Reaction of fluorinated acylsilanes (3) and (4) with sulfur ylides***

Entry	*n*	*SiR₃*	*Yield (%)*
1	1	$SiEt_3$	**9a** (31)
2	1	Si^iPr_3	**9d** (70)
3	1	Si^tBuPh_2	**9b** (84)
4	2	Si^iPr_3	**11b** (69)
5	2	Si^tBuPh_2	**11c** (67)

Source: Reproduced with permission from *J. Fluorine Chem.* **2002**, **117**, 207. Copyright 2002.

substituents (Y) with no stabilizing influence, underwent Brook rearrangement to give the enoxysilanes (path a) (Table VII). Moreover, the effect of silicon substituents on the reaction pathways was clear. Reaction with a more electrophilic silyl group favored reaction path *a* and the enoxysilane major product (Table V, entry 2), whereas, reaction with less electrophilic silyl group preferred path *b* resulting in formation of the alkene (Table V, entry 1).

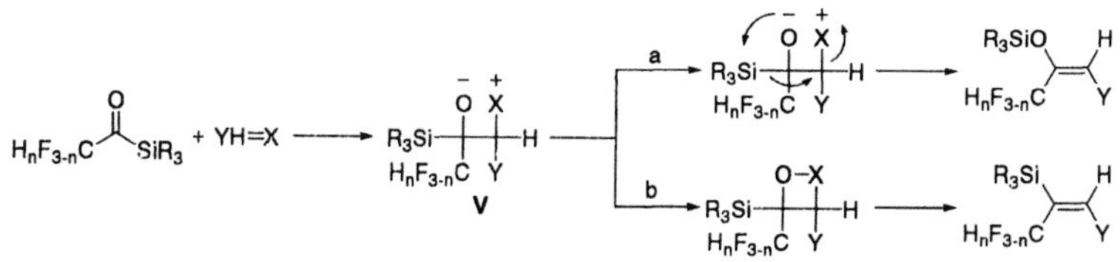

Scheme 15. Plausible reaction pathways for formation of alkenes and enoxysilanes

Allylation reaction of fluorinated acylsilanes

Metal mediated allylation reaction in aqueous media

The concurrent introduction of a hydroxy group and a carbon-carbon double bond, such as would occur during the allylation reaction of a carbonyl compound (18), might be especially useful in the preparation of multi-functional compounds. As such, homoallylic alcohols can be easily converted to β-hydroxy carbonyl compounds, such as aldehydes (18, 19) or ketones (20), and may undergo a facile one-carbon homologation to δ-lactones.[21]

The advantages of conducting metal-mediated allylation reactions of carbonyl compounds in aqueous media as opposed to organic solvents (22) include the avoidance of flammable or anhydrous solvents, reduced environmental and economic costs, and simplified processing in protection-deprotection schemes (23). A variety of allylation reagents containing Zn(24), Bi(25), Sn(26), Mg(27), Mn(28), Sb(29), Pd(30), Hg(31), and In(32) which react with carbonyl groups in aqueous media have been reported.

The reaction of organometallic reagents, for examples lithium and Grignard compounds, with fluorinated acylsilanes is an intriguing process. For instance, reaction of fluorinated acylsilanes with organometallic reagents such as these resulted in the formation of difluoroenoxysilanes. The formation of difluoroenoxysilanes may result from nucleophilic addition of the organometallic reagent, Brook rearrangement of silicon from carbon to oxygen followed by β-elimination of fluorine (Figure 16) (8a, 33).

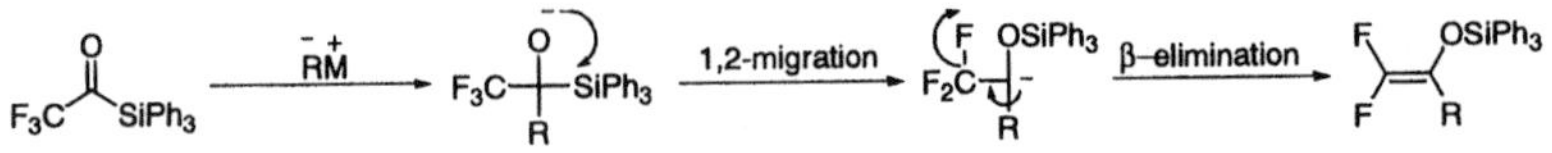

Figure 16. *Formation of difluoroenoxysilanes by addition of organometallic reagents to fluorinated acylsilanes*

To form fluorinated homoallylic alcohols with total suppression of Brook rearrangement and the consequent formation of fluoroenoxysilanes, the use of aqueous media was proposed.

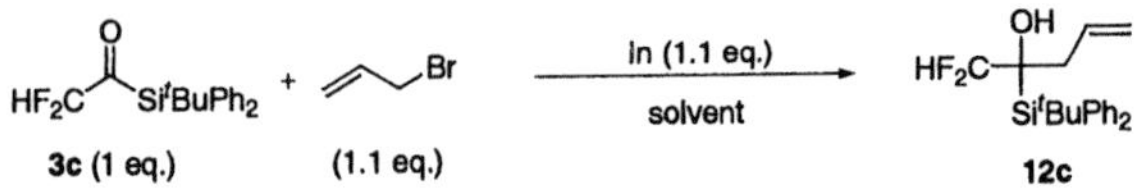

Figure 17. *Indium-mediated allylation reaction of (3c) in various concentration of water*

Indium is the metal of choice for an allylation reaction as it requires no activation and does not promote side reactions such as reduction and carbonyl coupling. Difluoroacetyl-*tert*-butyldiphenylsilane (**3c**) was treated with allyl bromide (1.1 eq.) and indium (1.1 eq.) in water at room temperature for two days to form homoallylic alcohol (**12c**) with recovery of unreacted (**3c**) possible. Addition of an organic solvent miscible with water influences the reaction apparently by improving the solubility of the reactants. To establish the most favorable reaction conditions, the indium-mediated allylation reaction of (**3c**) with allyl bromide was studied with various concentrations of water in THF (Figure 17)

Both water and THF are necessary for the allylation reaction; only trace quantities of homoallylic alcohol (**12c**) were formed when either pure water or anhydrous THF was employed. However, decreasing concentrations of THF relative to water led to both longer reaction times and lower conversion rates. This outcome may be rationalized by recognizing that solubility of (**3c**) decreases in increasingly aqueous solvents. The exclusive formation of the homoallylic alcohol with total suppression of the Brook rearrangement is particularly noteworthy.

The generality of the indium-mediated allylation reaction was explored with various difluoroacetyltrialkylsilanes (**3**) in a water and THF (1:1) mixture. The desired homoallylic alcohols (**12**) were synthesized in good yields with total suppression of Brook rearrangement in all cases tested. Silicon substituents had no effect on the product formation. However, in reaction with a substituted allyl bromide such as 4-bromo-2-methyl-2-butene (R' and R'' = CH_3), the desired homoallylic alcohol was not formed, rather, the fluorinated acylsilane was recovered (Figure 18)

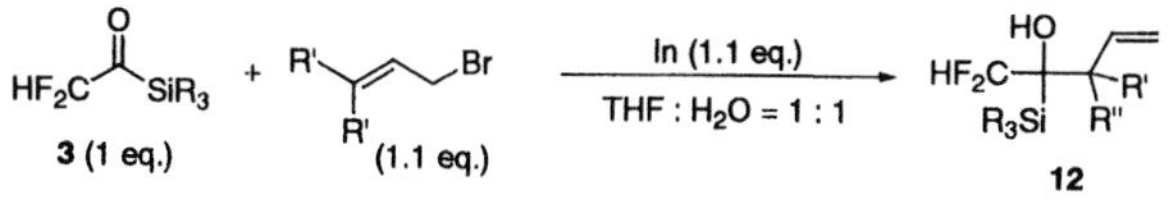

$$\underset{\mathbf{3}\ (1\ \text{eq.})}{HF_2C-C(=O)-SiR_3} + \underset{(1.1\ \text{eq.})}{R'R'C{=}CH-CH_2Br} \xrightarrow[\text{THF : } H_2O = 1:1]{\text{In (1.1 eq.)}} \underset{\mathbf{12}}{HF_2C-C(OH)(R_3Si)-C(R')(R'')-CH{=}CH_2}$$

Figure 18. Generality of indium-mediated reaction of (3)

To overcome the low reactivity found in the indium-mediated allylation reaction of substituted allyl bromides, a zinc-promoted process was investigated. The zinc-mediated allylation reaction gives better results in aqueous saturated ammonium chloride (NH_4Cl) solution than in pure water (24). Addition of NH_4Cl to the medium might not only increase the ionic strength of media, but also change the acidity (34a). NH_4Cl, while activating the carbonyl, also activates the metal surface to facilitate the formation of allylating reagent (34b).

The zinc-promoted allylation reaction of (**3c**) with allyl bromide was studied in various concentrations of saturated NH_4Cl in THF. The importance of saturated NH_4Cl is evident because the homoallylic alcohol (**12c**) was not formed with THF alone. The lower conversion of (**3c**), observed in reaction where the solvent was limited to saturated NH_4Cl, is presumably a consequence of the decreased solubility of (**3c**). Under the optimum reaction conditions, a 1:1 mixture of aqueous NH_4Cl and THF, the zinc-promoted allylation reaction appeared to be faster (30 minutes) than the corresponding indium-mediated process in aqueous THF (24 hours).

Zinc-promoted allylation reactions of difluoroacetyltrialkylsilanes (**3**) with various allyl bromides gave homoallylic alcohols (**12** and **13**) in good yields (Figure 19). The steric effect of substituents on silicon was again evident in reactions with 4-bromo-2-methyl-2-butene. Reaction with less sterically hindered acylsilane (**3a**, SiR_3 = $SiEt_3$) led to complete conversion into desired homoallylic alcohols (**12e** and **13e**) in less time than reaction with more hindered acylsilane (**3c**, SiR_3 = Si^tBuPh_2). The steric influence of allyl bromides was also seen in transformations where 4-bromo-2-methyl-2-butene was employed. In the case of (**3a**), reaction with 4-bromo-2-methyl-2-butene required longer reaction times (90 minutes) than those with non- or mono-substituted allyl bromides (30 minutes). Reaction of (**3a**) with 4-bromo-2-methyl-2-butene resulted in formation of a regio-isomeric mixture of the γ-adduct (**12e**) and the α-adduct (**13e**). However, γ-regioselectivity was observed with crotyl bromide.

In this work, the different reactivities observed for indium and zinc could be derived from the different ionization potentials of these two metals. Several explanations have been proposed in efforts to establish the nature of the reactive

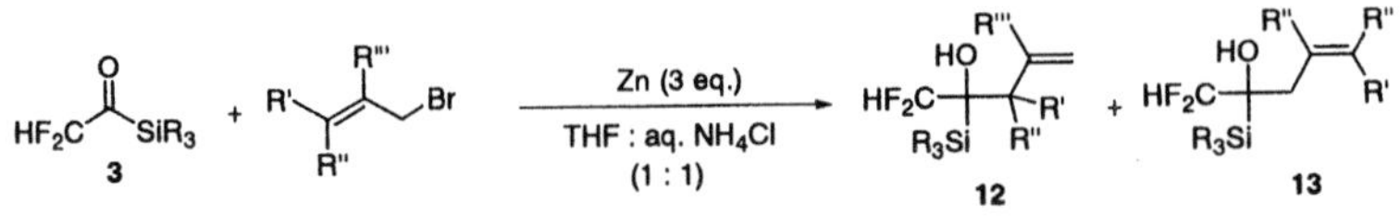

Figure 19. Zinc-promoted allylation reaction of (3)

species in metal-mediated allylation reaction. Li and Chan (22b) proposed a single electron transfer (SET) mechanism for the zinc-promoted allylation reaction. On the other hand, Chan and Yang (36) postulate allylindium(I) is preferentially formed in aqueous media as a consequence of the low ionization potential of indium.

Recovery of unreacted acylsilane (**3c**) from both indium- and zinc-mediated reaction in anhydrous THF might exclude the possible electron transfer from the metal surface to acylsilane. However, observation of gas evolution in an exothermic process in the zinc-promoted reaction, but not in indium-mediated transformation, suggests a possible electron process involving the metal surface in the zinc-promoted process. Such a route is also supported by the improved yields under acidic conditions or those conditions where ultrasound facilitated the process by activating the metal surface (24b, 37). Possible involvement of an indium surface - allyl radical process cannot be excluded. Unfortunately, from the information available at this time, no definitive conclusion on the origin of the different reactivities of indium and zinc is possible.

Lewis acid-promoted allylation reaction

The Lewis acid-promoted allylation reaction of allyltrimethylsilanes with carbonyl compounds, the Sakurai reaction, affords an excellent opportunity to compare a different reaction pathway with the aforedescribed metal-mediated reactions (18c, 38).

The $TiCl_4$-promoted reaction of difluoroacetyl-*tert*-butyldiphenylsilane (**3c**) and monofluoroacetyl-*tert*-butyldiphenylsilane (**4c**) with various allyltrimethylsilanes in dichloromethane was studied (Figure 20).

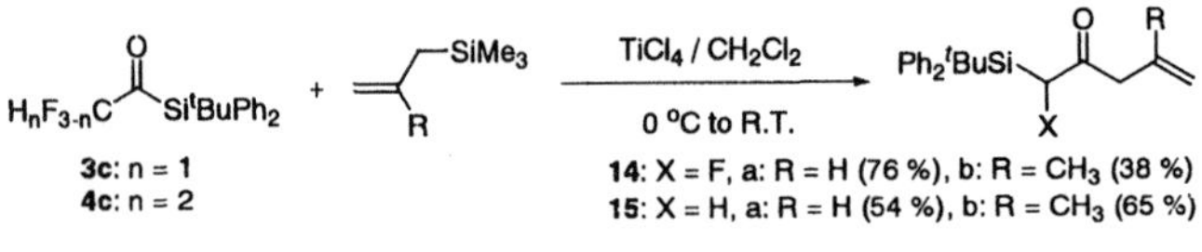

Figure 20. Sakurai-type allylation of mono- and difluorinated acylsilane (3c) and (4c)

Neither fluorinated enoxysilanes nor homoallylic alcohols were obtained. Instead, α-*tert*-butyldiphenylsilyl ketones (**14**) and (**15**) were synthesized in moderate yields.

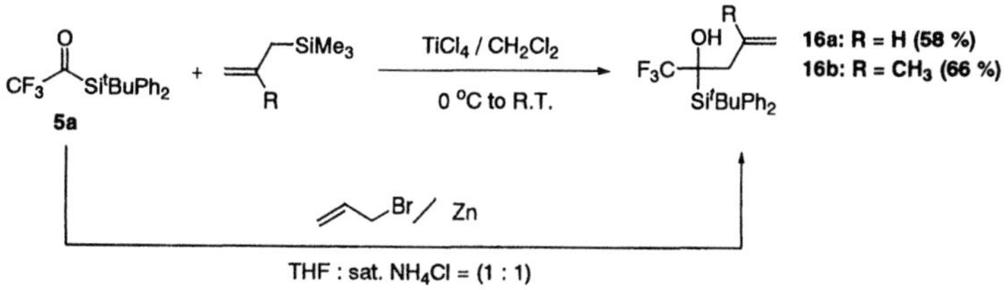

Figure 21. Allylation reaction of trifluoroacetyl-tert-butyldiphenylsilane (5a)

The reaction of trifluoroacetyl-*tert*-butyldiphenylsilane (**5a**) is also very intriguing (Figure 21). Reaction under the conditions employed with (**3c**) and (**4c**) resulted in a recovery of (**5a**) even at longer reaction times (1 day). However increased quantities of $TiCl_4$ (1.5 eq.) and allyltrimethylsilane (3 eq.) led to the total consumption of (**5a**). Under these conditions, homoallylic alcohols (**16**) were synthesized in moderate yields. The formation of homoallylic alcohols (**16**) was also confirmed by the identitied nature of product that resulting from the zinc-promoted allylation reaction of (**5a**) in aqueous media. In contrast, reaction of (**3c**) with an increased quantity of both $TiCl_4$ (1.5 eq.) and allyltrimethylsilane (3 eq.) resulted in the formation of α-*tert*-butyldiphenylsilyl ketone (**14a**) without the formation of a homoallylic alcohol, suggesting the extent of fluorination of the acylsilanes affects the reaction pathway.

The plausible mechanism for the formation of the α-*tert*-butyldiphenylsilyl ketones (**14** and **15**) and the homoallylic alcohol (**16**) is outlined in Figure 22. Addition of an allyl group to the carbonyl carbon could result in the formation of intermediate [**VI**]. An initial silyl group migration followed by defluorination would yield [**VII**]. A subsequent, second migration could generate α-*tert*-butyldiphenylsilyl ketones (**14** and **15**). The increasing extent of fluorination influences the reaction pathway diverting the formation of α-*tert*-butyldiphenylsilyl ketones (**14** and **15**) to homoallylic alcohol (**16**). Calculations of electron density employing a semiempirical method, PM3, shows that electron density on the oxygen anion of intermediate [**VI**] is decreased with increasing fluorination. This decrease retards silyl migration in the case of the trifluorinated intermediate leading to homoallylic alcohol (**16**) formation.

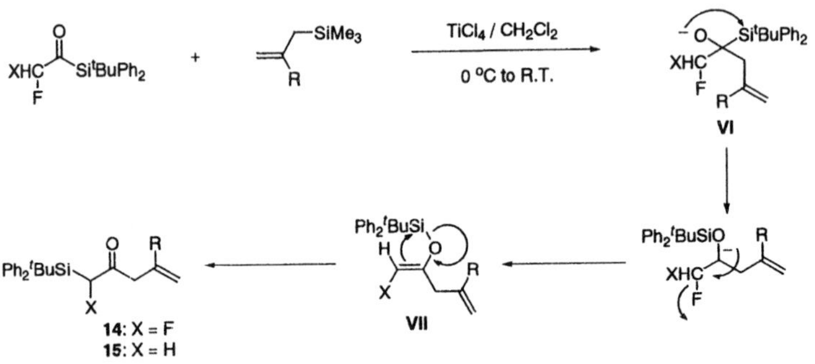

Figure 22. *Plausible mechanism for the formation of α-tert-butyldiphenylsilyl ketones (14 and 15)*

Conclusion

A variety of new fluorinated acetyltrialkylsilanes and the corresponding fluorinated enoxysilanes were synthesized in good to moderate yields. Reaction of trifluoroethanol with lithium diisopropylamide in the presence of chlorotrialkylsilanes gave difluoroacetyltrialkylsilanes (**3**). A plausible reaction pathway for the formation of difluoroacetyltrialkylsilanes was tested by the variation of silylating agents employed in a second silylation. The formation of a variety of 1,1-difluoro-2-trialkylsilyl-2-trialkylsilyloxyethens (**2**), supports the explanation of the suggested reaction mechanism. The reductive defluorination of difluoroacetyltrialkylsilanes (**3**) using magnesium in the presence of chlorotrimethylsilane formed 1-fluoro-2-trialkylsilyl-2-trimethylsilyloxyethenes (**6**), respectively. Moreover, careful acidic hydrolysis of 1-fluoro-2-trialkylsilyl-2-trimethylsilyloxyethenes (**6**) yielded the corresponding monofluoroacetyl trialkylsilanes (**4**). Electrophilic halogenation reactions of 1,1-difluoro-2-trialkylsilyl-2-trimethylsilyloxyethenes (**2**) with *N*-chlorosuccinimide (NCS), *N*-bromosuccinimide (NCS), and iodine (I_2) gave α-difluorohaloacetyl trialkyl silanes (**7**) in good to moderate yields. In addition, electrophilic fluorination of 1,1-difluoro-2-trialkylsilyl-2-trimethylsilyloxyethens (**2**) with Selectfluor® in a mixture of acetonitrile and dichloromethane resulted in the formation of trifluoroacetyltrialkylsilanes (**5**).

Reaction of fluorinated acetyltrialkylsilanes with phosphorus and sulfur ylides was also investigated. Reaction of difluoroacetyltrialkylsilanes (**3**) with Wittig reagents gave either expected alkenes, the normal Wittig products (**8**), or alkeneoxysilanes, the Brook rearrangement products (**9**), depending on the reaction conditions. However, reaction with stabilized ylides resulted in the exclusive formation of the expected alkenes (**10**) without Brook rearrangement.

The alkenes were formed with high *E*-stereoselectivity where the stereochemistry was confirmed by both NMR spectroscopy and single crystal X-ray diffraction studies. In contrast, reaction of fluorinated acetyltrialkylsilanes with sulfur ylides such as dimethylsulfoxonium methylide yielded exclusively alkeneoxysilanes (**9** and **11**) via Brook rearrangement of silyl groups.

The allylation reactions of fluorinated acylsilanes (**3**, **4**, and **5**) were investigated under a variety of reaction conditions. The metal-mediated allylation reaction of (**3**) in aqueous media resulted in the formation of homoallylic alcohols in good yield. Metal-mediated reactions lead to total suppression of Brook rearrangement. The importance of the concentration of water in THF was evident in the indium-mediated reaction. However, the presence of saturated NH_4Cl is more important than the concentration of THF in the zinc-promoted reaction. Overall the zinc-promoted allylation reaction is more effective than the indium-mediated reaction. Mono- and disubstituted allyl bromides gave the desired homoallylic alcohols (**12** and **13**) in good yields under zinc-promoted conditions whereas no formation of homoallylic alcohols was observed in the indium-mediated reaction. While no definitive conclusion can be drawn the different reactivities of zinc and indium might be a consequence of the different ionization potentials of these two metals. Not only the substituents on the allyl bromides, but also those on silicon affect the regio- and diastereochemistry.

The $TiCl_4$ mediated allylation reactions of fluorinated acylsilanes (**3**, **4**, and **5**) were also examined. Reaction of mono- and difluoroacetyl-*tert*-butyldiphenylsilanes (**4c** and **3c**) with various allyltrimethylsilanes resulted in the formation of α-silylated ketones (**14** and **15**) in moderate yields by means of sequential Brook- and retro-Brook rearrangements. On the other hand, reaction of trifluoroacetyl-*tert*-butyldiphenylsilanes (**5a**) under the same condition resulted in a recovery of **5a** as well as formation of homoallylic alcohols (**16**). Increasing the quantities of both the Lewis acid and the allyltrimethylsilanes relative to the minimum required stoichiometry resulted in formation of the homoallylic alcohols (**16**) in moderate yields with complete consumption of (**5a**). The influence of fluorine on the reaction pathways leading to the formation of either α-silylated ketones (**14** and **15**) or homoallylic alcohols (**16**) was deduced from computations using a semiempirical method, the PM3 model. Variation of the electron density on the oxy anion of intermediate [**VI**] modulated by fluorination leads to alternative reaction pathways.

References

1. Brook, A. G.; Mauris, R. J. *J. Am. Chem. Soc.* **1957**, *79*, 971-3.
2. (a) Ricci, A.; Degl'Innocenti, A. *Synthesis* **1989**, 647-63. (b) Page, P. C. B.; Klair, S. S.; Resenthal, S. *Chem. Soc. Rev.* **1990**, *19*, 147-95. (c)

Bonini, B. F.; Comes-Franchini, M.; Fochi, M.; Mazzanti, G.; Ricci, A. *J. Organometal. Chem.* **1998**, *567*, 181-89. (d) Capperucci, A.; Degl'Innocenti, A. *Recent Res. Dev. Org. Chem.* **1999**, *3*, 385-406.

3. (a) Brook, A. G. *J. Am. Chem. Soc.* **1958**, *80*, 1886-9. (b) Brook, A. G.; Warner, C. M.; McGriskin, M. A. *J. Am. Chem. Soc.* **1959**, *81*, 981-3. (c) Brook, A. G.; Schwartz, N. V. *J. Am. Chem. Soc.* **1960**, *82*, 2435-9. (d) Brook, A. G.; Iachia, B. *J. Am. Chem. Soc.* **1961**, *83*, 827-31. (e) Brook, A. G. *J. Org. Chem.* **1960**, *25*, 1072. (f) Brook, A. G.; Schwartz, N. V. *J. Org. Chem.* **1962**, *27*, 2311-15.
4. (a) Brook, A. G.; Bassindale, A. R. In *Rearrangement in Ground and Excited States*; De Mayo, P., Ed.; Academic Press: New York, **1980**; Vol. 2. 149-227. (b) Colvin, E. W. In *Silicon in Organic Synthesis*; Butterworths: London, **1981**. (c) Brook, A. G. *Acc. Chem. Res.* **1974**, *7*, 77-84. (d) Moser, W. H. *Tetrahedron* **2001**, *57*, 2065-84.
5. (a) Speier, J. L. *J. Am. Chem. Soc.* **1952**, *74*, 1003-10. (b) West, R.; Lowe, R.; Stewart, H. F.; Wright, A. *J. Am. Chem. Soc.* **1971**, *93*, 282-3.
6. (a) Jankowski, P.; Raubo, P.; Whicha, J. *Synlett* **1994**, 985. (b) Reich, H. J.; Rusek, J. J.; Olson, R. R. *J. Am. Chem. Soc.* **1979**, *101*, 2225-27. (c) Linghu, X.; Nicewicz, D. A.; Johnson, J. S. *Org. Lett.* **2002**, *4*, 2957-60. (d) Linghu, X.; Johnson, J. S. *Angew. Chem. Int. Ed.* **2003**, *42*, 2534-36. (e) Taketa, K.; Sawada, Y.; Sumi, K. *Org. Lett.* **2002**, *4*, 1031-33. (e) Hu, T.; Corey, E. J. *Org. Lett.* **2002**, *4*, 2441-43. (f) Zhang, J.; Corey, E. J. *Org. Lett.* **2001**, *3*. 3215-16.
7. (a) *Enantiocontrolled Synthesis of Fluoro-Organic Compounds*; Soloshonok, V. A., Ed.; Wiley: Chichester, **1999**. (b) *Organofluorine Chemistry*; Chambers, R. D., Ed.; Springer: Berlin, **1997**. (c) Smart, B. E., Ed. *Chem. Rev.* **1996**, 96, No. 5 (Thematic issue on fluorine chemistry). (d) Percy, J. M. *Top. Curr. Chem.* **1997**, *193*, 131-95. (e) *Selective Fluorination in Organic and Bioorganic Chemistry*; Welch, J. T., Ed.; ACS Symposium Series 456; American Chemical Society: Washington DC, **1991**. (f) Welch, J. T.; Eswarakrishnan, S. *Fluorine in Bioorganic Chemistry*; Wiley: New York, **1991**. (g) *Biomedical Frontiers of Fluorine Chemistry*; Ojima, I.; McCarthy, J. M.; Welch, J. T., Eds.; ACS Symposium Series 639; American Chemical Society: Washington DC, **1996**. (h) *Fluorine-containing Amino Acids*; Kukhar, V. P.; Soloshonok, V. A., Eds.; John Wiley & Sons: Chichester, **1995**. (i) Tozer, M. J.; Herpin, T. F. *Tetrahedron* **1996**, *52*, 8619-83. (j) Prakash, G. K. S.; Yudin, A. K. *Chem. Rev.* **1997**, *97*, 757-86. (k) Singh, R. P.; Shreeve, J. M. *Tetrahedron* **2000**, *56*, 7613-32. (l) Prakash, G. K. S., Mandal, M. *J. Fluorine Chem.* **2001**, *112*, 123-31.

8. (a) Jin, F.; Jiang, B.; Xu, Y. *Tetrahedron Lett.* **1992**, *33*, 1221-24. (b) Bonini, B. F.; Comes-Franchini, M.; Fochi, M.; Mazzanti, G.; Nanni, C.; Ricci, A. *Tetrahedron Lett.* **1998**, *39*, 6737-40. (c) Bordeau, M.; Clavel, P.; Barba, A.; Berlande, M.; Biran, C.; Roques, N. *Tetrahedron Lett.* **2003**, *44*, 3741-44.
9. (a) Patel, S. T.; Percy, J. M.; Wilkes, R. D. *Tetrahedron* **1995**, *51*, 9201-16. (b) Howarth, J. A.; Martin Owton, W.; Percy, J. M.; Rock, M. H. *Tetrahedron* **1995**, *51*, 10289-302. (c) Deboos, G. A.; Fullbrook, J. J.; Percy, J. M. *Org. Lett.* **2001**, *3*, 2859-61. (d) Ichikawa, J.; Wada, Y.; Fujiwara, M.; Sakoda, K. *Synthesis* **2002**, 1917-36.
10. (a) Amii, H.; Kobayashi, T.; Hatamoto, Y.; Uneyama, K. *J. Chem. Soc., Chem. Commun.* **1999**, 1323. (b) Uneyama, K.; Amii, H. *J. Fluorine Chem.* **2002**, *114*, 127-31.
11. Prakash, G. K. S.; Hu, J.; Alauddin, M. M.; Conti, P. S.; Olah, G. A. *J. Fluorine Chem.* **2003**, *121*, 239-43.
12. Sato, T.; Abe, T.; Kuwajima, I. *Tetrahedron Lett.* **1978**, *19*, 259-62.
13. Chattopadhyay, A.; Salaskar, A. *Synthesis* **2000**, 561-64.
14. (a) Kubota, T.; Iijima, M.; Tanaka, T. *Tetrahedron Lett.* **1992**, *33*, 1351-54. (b) Mispelaere, C.; Roques, N. *Tetrahedron Lett.* **1999**, *40*, 6411-14. (c) Xu, Y.; Dolbier, Jr., W. R. *Tetrahedron Lett.* **1998**, 39, 9151-54.
15. (a) Brook, A, G.; Fieldhouse, S. A. *J. Organometal. Chem.* **1967**, *10*, 235-46. (b) Soderquist, J. A.; Anderson, C. L. *Tetrahedron Lett.* **1988**, *29*, 2425-28. (c) Soderquist, J. A.; Anderson, C. L. *Tetrahedron Lett.* **1988**, *29*, 2777-78. (d) Larson, G. L.; Soderquist, J. A.; Caludio, M. R. *Synthetic Commun.* **1990**, *20*, 1095-104. (e) Nakajima, T.; Segi, M.; Sugimoto, F.; Hioki, R.; Yokota, S.; Miyashita, K. *Tetrahedron* **1993**, *49*, 8343-58.
16. Jin, F.; Xu, Y. *J. Fluorine Chem.* **1993**, *62*, 207-10.
17. (a) Trost, B. M.; Melvin, Jr., L. S. In *Sulfur Ylides*; Academic Press: New York, **1975**. (b) Block, E. In *Reactions of Organosulfur Compounds*; Academic Press: New York, **1978**.
18. (a) Hoffman, R. W. *Angew. Chem. Int. Ed.* **1982**, *21*, 555-66. (b) Yamamoto, Y. *Acc. Chem. Res.* **1987**, *20*, 243-49. (c) Yamamoto, Y.; Asao, N. *Chem. Rev.* **1993**, *93*, 2207-93.
19. Yamamoto, Y.; Maruyama, K. *Heterocycles* **1982**, *18*, 357-86.
20. Yatagai, H.; Yamamoto, Y.; Maruyama, K. *J. Am. Chem. Soc.* **1980**, *102*, 4548-50.
21. Wuts, P. G. M.; Obrzut, M. L.; Thompson, P. A. *Tetrahedron Lett.* **1984**, *25*, 4051-54.
22. (a) Li, C. J. *Chem. Rev.* **1993**, *93*, 2023-35. (b) Chan, T. H.; Li, C. J.; Lee, M. C.; Wei, Z. Y. *Can. J. Chem.* **1994**, *72*, 1181-92. (c) Li, C. J.

Tetrahedron **1996**, *52*, 5643-68. (d) Ribe, S; Wipf, P. *J. Chem. Soc., Chem. Commun.* **2001**, 299-307.

23. (a) Chan, T. H.; Li, C. J. *J. Chem. Soc., Chem. Commun.* **1992**, 747-8. (b) Kim. E.; Gordon, D. M.; Schmid, W.; Whitesides, G. M. *J. Org. Chem.* **1993**, *58*, 5500-7.
24. (a) Han, B. H.; Boudjouk, P. *J. Org. Chem.* **1982**, *47*, 751-2. (b) Petrier, C.; Luche, J. L. *J. Org. Chem.* **1985**, *50*, 910-12. (c) Chan, T. H.; Li, C. J. *Organometallics*, **1990**, *9*, 2649-50. (d) Bieber, L. W.; da Silva, M. F.; da, Costa, R. C.; Silva, L. O. S. *Tetrahedron Lett.* **1998**, *39*, 3655-8. (e) Lambardo, M.; Girotti, R.; Morganti, S.; Trombini, C. *J. Chem. Soc., Chem. Commun.* **2001**, 2310-11.
25. Wada, M.; Ohki, H.; Akiba, K. *J. Chem. Soc., Chem. Commun.* **1987**, 708-9.
26. Uneyama, K.; Kamaki, N.; Moriya, A.; Torii, S. *J. Org. Chem.* **1985**, *50*, 5396-99.
27. (a) Li, C. J.; Zhang, W. C. *J. Am. Chem. Soc.* **1998**, *120*, 9102-03. (b) Zhang, W. C.; Li, C. J. *J. Org. Chem.* **1999**, *64*, 3230-36. (c) Fukuma, T.; Lock, S.; Miyoshi, N.; Wada, M. *Chem. Lett.* **2002**, 376.
28. Li, C. J.; Meng, Y.; Yi, X. H. *J. Org. Chem.* **1998**, *63*, 7498-504.
29. (a) Ren, P. D.; Jin, Q. H.; Yao, Z. P. *Synth. Commun.* **1997**, *27*, 2761-7. (b) Li, L. H.; Chan, T. H. *Tetrahedron Lett.* **2000**, *41*, 5009-12.
30. Zhou, J. Y.; Jia, Y.; Sun, G. F.; Wu, S. H. *Synth. Commun.* **1997**, *27*, 1899-906.
31. Chan, T. H.; Yang, Y. *Tetrahedron Lett.* **1999**, *40*, 3863-66.
32. (a) Cintas, P. *Synlett* **1995**, 1087-96. (b) Li, C. J.; Chan, T. H. *Tetrahedron* **1999**, *55*, 11149-76. (c) Podlech, J.; Maier, T. C. *Synthesis* **2003**, 633-55. (d) Paquette, L. A. *Synthesis* **2003**, 765-74.
33. (a) Jin, F.; Xu, Y.; Huang, W. *J. Chem. Soc., Perkin Trans. 1* **1993**, 795-9. (b) Jin, F.; Xu, Y.; Huang, W. *J. Chem. Soc., Chem. Commun.* **1993**, 814-16.
34. (a) Paquette, L. A.; Mitzel, T. M. *J. Am. Chem. Soc.* **1996**, *118*, 1931-7. (b) Zhang, J.; Blazecka, P. G.; Berven, H.; Belmont, D. *Tetrahedron Lett.* **2003**, *44*, 5579-82.
35. (a) Wilson, S. R.; Guazzaroni, M. E. *J. Org. Chem.* **1989**, *54*, 3087-91. (b) Issac, M. B.; Chan, T. H. *Tetrahedron Lett.* **1995**, *36*, 8957-60.
36. Chan, T. H.; Yang, Y. *J. Am. Chem. Soc.* **1999**, *121*, 3228-9.
37. Li, C. J.; Chan, T. H. *Organometallics* **1991**, *10*, 2548-49.
38. (a) Fleming, I.; Dunogues, J.; Smithers, R. *Org. React.* **1989**, *37*, 57575. (b) Masse, C. E.; Panek, J. S. *Chem. Rev.* **1995**, *95*, 1293-316.

Chapter 23

The Role of Fluorine-Containing Methyl Groups toward Diastereofacial Selection

Takashi Yamazaki[1,2,3], Satoshi Takei[1], Tatsuro Ichige[1], Seiji Kawashita[1], Toshio Kubota[3], and Tomoya Kitazume[1]

[1]Graduate School of Bioscience and Biotechnology, Tokyo Institute of Technology, 4259 Nagatsuta, Midori-ku, Yokohama 226–8501, Japan
[2]Department of Materials Science, Ibaraki University, 4–12–1 Nakanarusawa, Hitachi 316–8511, Japan
[3]Current address: Department of Applied Chemistry, Tokyo University of Agriculture and Technology, 2–24–16 Nakamachi, Koganei 184–8588, Japan

Michael addition reactions of enolates derived from the selected ketone, ester, and amide to γ-$CH_{3-n}F_n$-α,β-unsaturated ketones (n=1~3) were proved to smoothly furnish the desired 1,4-adducts with high level of diastereofacial selectivity, which monotonously decreased by reduction in a number of fluorine involved. Although the Felkin-Anh model can accommodate the stereochemical outcome when *E*-acceptors were employed, the opposite stereoisomer was obtained from the corresponding trifluorinated *Z*-isomer. The hyper-conjugative stabilization of transition states by electron donation from the proximate allylic substituents (the Cieplak rule) successfully explains the observed π-facial preference of both *E*- and *Z*-acceptors.

Introduction

It is well documented (*2-4*) that introduction of a trifluoromethyl moiety into organic molecules gave rise to significant change in their physical properties which is due in part to its strong electron-withdrawing nature making the proximate bonds shorter and stronger. When this group directly attaches to π systems, electrophilicity is extraordinarily enhanced by decreasing the molecular orbital, especially LUMO energy levels (*5,6*). On the other hand, because of three fluorine atoms and each of which possessing three lone pairs, this moiety behaves like a mass of electrons which effectively prevents the access of appropriate nucleophiles.

Previously, we have reported (*7*) that *E*-4,4,4-trifluorocrotonate **1** readily underwent Michael addition reactions (*8-10*) with enolates derived from acylated oxazolidinones, and the intermediates trapped as ketene silyl acetals were found to follow highly stereoselective Ireland-Claisen rearrangements in the presence of a Pd catalyst, controlling three consecutive stereocenters in the product **2** (Scheme 1). The smooth conjugate addition at the first stage (*5,11,12*) would be elucidated in consequence of the significant stabilization of the Michael intermediates due to intramolecular F-Li chelation (*5,11*), and the stereochemical outcome of the Ireland-Claisen procedure (*13*) at the next step would be understood by the different steric requirement between the both π faces, allowing the rearrangement to occur from the *re*-face where the smaller CF_3 group occupies (compare **TS-*re*** and **-*si*** in Scheme 1). However, the results for the simplified derivative **3** was totally opposite to our expectation. Thus, in spite of closer steric bulkiness of *i*-Pr and CF_3 groups (*14-16*), good selectivity of *ca.* 9:1 was attained by the preferential rearrangement again occurring from the same *re*-face. This fact clearly demonstrated that the different electrostatic circumstance of the two π faces would be responsible for realization of this good

Scheme 1 Sequential Michael and Ireland-Claisen rearrangement

i) R-CH=C(OLi)Ox
ii) TMSCl
iii) $PdCl_2(PhCN)_2$

1

2 (single isomer)

≡ R'

Ox:

TS-*re* >> TS-*si*

3

4 74% yield, 88% *syn*

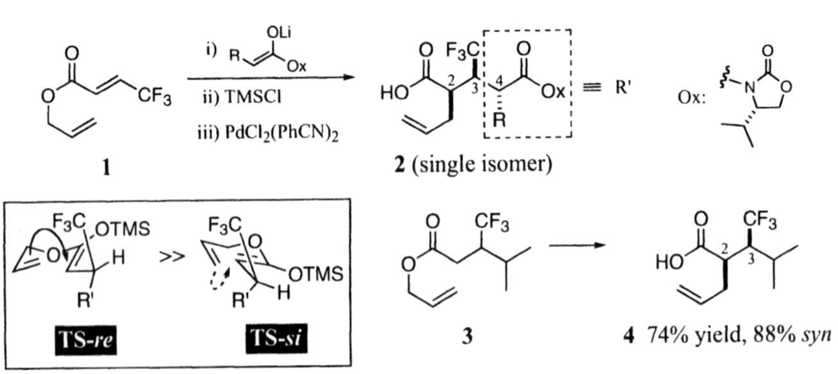

*Scheme 2 Preparation of the trifluorinated acceptors **E**- and **Z-6a***

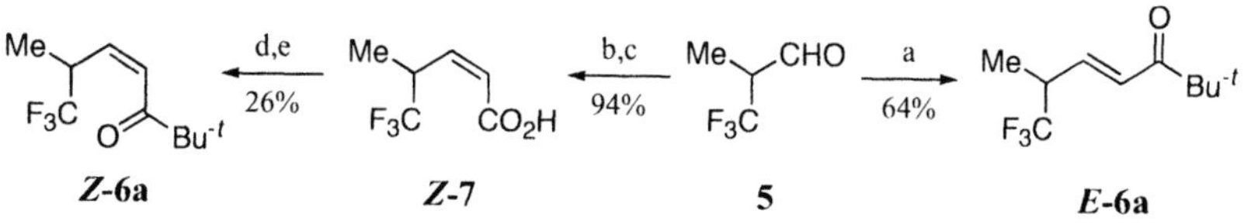

Me F3C O Bu-t **Z-6a** ←(d,e; 26%)— Me F3C CO2H **Z-7** ←(b,c; 94%)— Me CHO F3C **5** —(a; 64%)→ Me F3C O Bu-t **E-6a**

a) $(EtO)_2P(O)CH_2C(O)Bu^{-t}$, *n*-BuLi/$Et_2O$; b) $(PhO)_2P(O)CH_2CO_2Et$, NaH/THF; c) NaOH/THF-H_2O; d) AcCl, Et_3N/CH_2Cl_2; e) *tert*-BuMgCl/THF.

2,3-*syn* selectivity, not by the steric bulkiness of the two substituents occupied the both faces (*17*). Because this type of phenomenon has not been reported yet, we have decided to prepare the variously fluorinated substrates for the clarification of the role of $CH_{3-n}F_n$ groups for controlling the π-facial selectivity.

Preparation of the Starting Materials

The Michael acceptor with a CF_3 group, ***E*-6a**, was synthesized from the commercially available aldehyde **5** and the Horner-Wadsworth-Emmons (HWE) reagent derived from bromopinacolone (Scheme 2). On the other hand, the corresponding *Z*-isomer, ***Z*-6a**, was obtained by the literature method (*18*) using the HWE reagent with a phenoxy group on the phosphorus atom, enabling the selective formation of *Z*-α,β-unsaturated acid ***Z*-7** by the successive hydrolytic treatment. Its further conversion to the desired *Z*-acceptor was not clean and satisfactory yields have not been attained yet in spite of our extensive investigation of reaction conditions. Condensation of $(PhO)_2P(O)CH_2C(O)Bu^{-t}$ and **5** were also carried out as the directly accessible process to the target ***Z*-6a**, but the product was formed only in 30% yield in an unexpected *E* selective (*E*:*Z*≈8:2) manner.

At the next stage, the monofluorinated acceptor ***E*-6c** was prepared from 3-hydroxyisobutyrate **8** which was at first fluorinated (*19*) by the Ishikawa's reagent (*20*) and its "half" reduction by DIBAL, followed by the olefination with the bromopinacolone-based HWE reagent (*21-25*) (Scheme 3). Our initial plan was to construct ***E*-6c** *via* the intermediate ***E*-10** by taking into account the low boiling point and high volatility of the compound **9**, but it turned out that the final fluorination yielded the unexpected product ***E*-11**. This type of reaction was sometimes observed when allylic alcohols were treated with this type of reagents (*26*). The present homoallylic alcohol transformation allowed entry of fluorine to the most stable and least congested intermediary cation **Int-3** following to the S_N1 type mechanism.

The difluorinated acceptor was considered to be the most difficult to

Scheme 3 Preparation of the monofluorinated acceptor ***E*-6c**

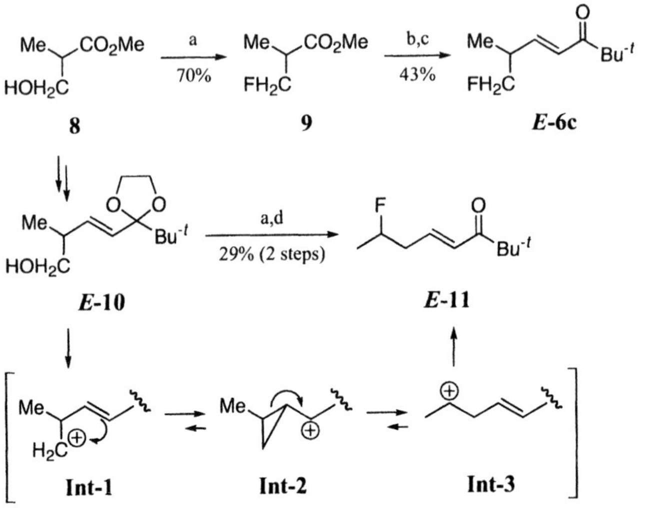

a) $Et_2NCF_2CHFCF_3/CH_2Cl_2$; b) DIBAL/$Et_2O$; c) $(EtO)_2P(O)CH_2C(O)Bu^{-t}$, *n*-BuLi/$Et_2O$; d) H^+.

construct among three acceptors required here from the standpoint of the availability of starting materials and/or limited preparation methods. Under such a circumstance, we have devised the synthetic sequences as shown in Scheme 4. Readily available trifluoromethacrylic acid **12** was first converted into the corresponding acid chloride whose esterification with 2-phenylethanol and further smooth hydrogenolysis yielded trifluorinated isobutyrate **13** in

Scheme 4 Preparation of the difluorinated acceptor ***E*-6b**

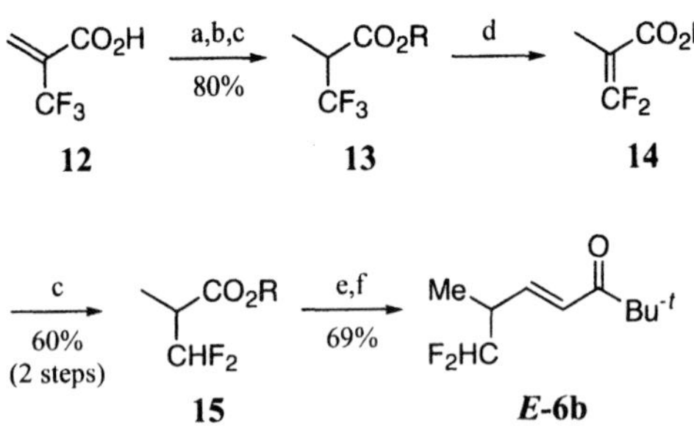

a) $C_6H_4(COCl)_2$; b) ROH, pyridine/CH_2Cl_2; c) 10% Pd/C/MeOH; d) LDA/THF; e) DIBAL/Et_2O; f) $(EtO)_2P(O)CH_2C(O)Bu^{-t}$, *n*-BuLi/$Et_2O$ (R: $PhCH_2CH_2$-)

excellent yield. During this study, we have found (*27*) that LDA effected the clean formal dehydrofluorination of **13** to furnish terminally difluorinated methacrylate **14**. The key to success of this transformation was to slowly add a small excess amount of LDA to keep the reaction temperature almost constant at-78 °C, which seemed to effectively retard the delivery of fluoride from the resultant enolate species. Otherwise the highly electrophilic olefinic carbon atom with two fluorine atoms (*28,29*) would accept the quite facile nucleophilic attack by this strong base. This inverse addition method eventually improved the yield of the product **14** to 86% from 25% when **13** was added to a solution of LDA. Hydrogenation of **14** required positive pressure of H_2 for conversion with an acceptable reaction rate (10 mol% of a Pd catalyst under 5 atm of H_2) which was in sharp contrast to the case of **12**, which was quite readily transformed into **13** in the presence of only a 1/20 amount (0.5 mol%) of the catalyst under atmospheric pressure of H_2. This method opened a new route to introduce a CHF_2 group at the 2 position of appropriate carbonyl compounds by way of trifluoromethylation of their ketene or enol silyl acetals (*30*) at the first step.

Table 1 Reaction of Enolates with *E*-6 or *Z*-6a

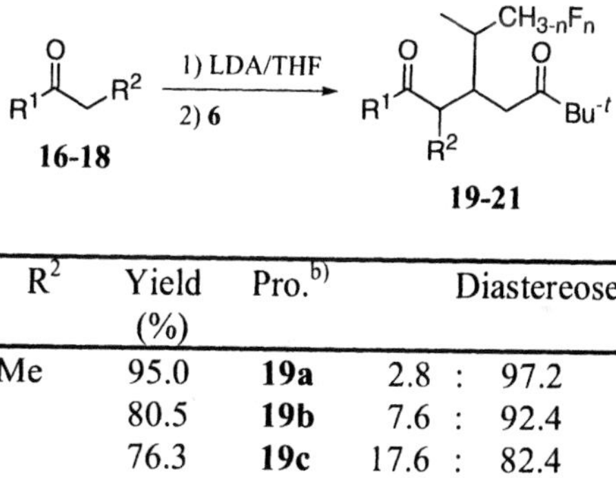

16-18 **19-21**

Accpt.[a]	R[1]	R[2]	Yield (%)	Pro.[b]	Diastereoselectivity[c]			
***E*-6a**	Ph	Me	95.0	**19a**	2.8 :	97.2		
***E*-6b**			80.5	**19b**	7.6 :	92.4		
***E*-6c**			76.3	**19c**	17.6 :	82.4		
***Z*-6a**			0					
***E*-6a**	EtO	MeS	quant	**20a**	7.7 :	92.3	[<1.0 : >99.0]	
***E*-6b**			86.0	**20b**	15.1 :	84.9	[16.7 : 83.3]	
***E*-6c**			76.3	**20c**	30.9 :	69.1	[26.3 : 73.7]	
***Z*-6a**			45.9	**20a**	<1.0 :	>99.0		
***E*-6a**	Me_2N	Me	87.5	**21a**	3.6 :	5.1 :	30.4 :	60.9
***E*-6b**			74.4	**21b**	5.0 :	13.8 :	30.9 :	50.3
***E*-6c**			78.8	**21c**	15.6 :	16.3 :	34.0 :	34.1
***Z*-6a**			87.6	**21a**	2.2 :	16.6 :	28.4 :	52.8

a) Acceptor. b) Product. c) In the bracket was shown the diastereomeric ratio after removal of the MeS group.

Michael Addition Reactions of Enolates

As described above, we have obtained all Michael acceptors ***E*-6** and ***Z*-6a** with $CH_{3-n}F_n$ groups (n=1–3) which were subjected to the reactions with enolates from propiophenone **16**, ethyl (methylthio)acetate **17**, and *N,N*-dimethyl-propionamide **18** (Table 1). As we expected, these donors experienced facile conjugate addition to yield stereoisomeric mixtures **19–21** in every instance. In principle, four diastereomers could be formed but this was the case only when the amide donor **18** was employed. The enolates from the ketone **16** and ester **17** selectively afforded two stereoisomers. Their chemical yields were generally increased in proportion to the number of fluorine involved in the acceptors **6**.

Stereochemistry of the adducts with a CF_3 group was determined as follows. First of all, the major diastereomer of **21a** from **18** and ***E*-6a** was successfully purified and separated by simple column chromatography and gave suitable crystal for X-ray crystallographic analysis which unambiguously manifested its relative stereochemistry as 2*R**,3*R**,5*S** (Figure 1). On the other hand, when the compound **20a** was treated with *n*-Bu_3SnH, a facile elimination of a MeS group occurred to furnish a single isomer **22a** (Scheme 5). This result indicated that two stereoisomers of **20a** stemmed from the epimer at 2 position. **22a** was further treated with an excess amount of LDA and the trapping of the resultant bisenolate by MeI realized the regioselective methylation at the 2 position. It was clarified from the ^1H as well as ^{19}F NMR analyses that **23a** consisted of a 26:74 mixture and its ester group was independently converted into phenyl ketone and *N,N*-dimethylamide to produce **24a** and **25a**, respectively, basically with retention of stereochemical integrity. The major and minor isomers of the latter amide **25a** were analytically proved to be the same as the ones obtained as the most and the second predominant stereoisomers of the amide Michael adduct

*Figure 1 3D Structure of the Most Predominant Diastereomer of **21a** by Crystallographic Analysis (some hydrogen atoms are omitted for clarity)*

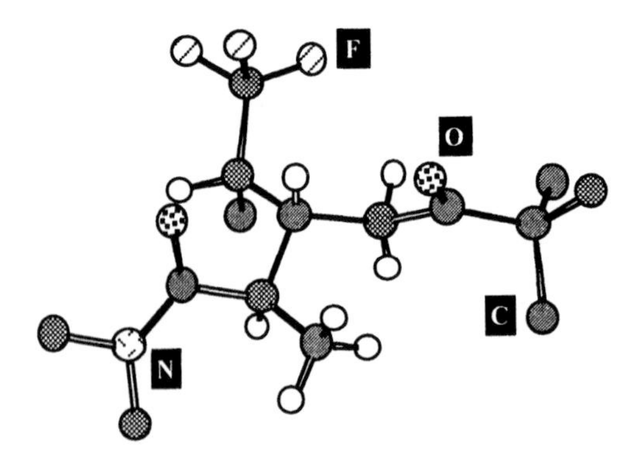

Scheme 5 Determination of relative stereochemistry

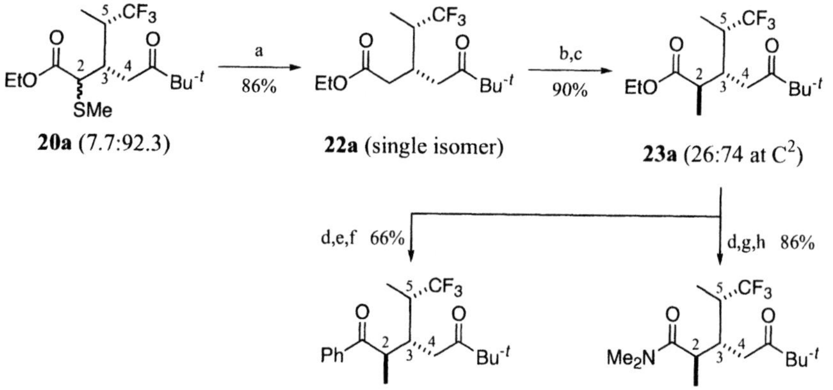

20a (7.7:92.3) → a, 86% → **22a** (single isomer) → b,c, 90% → **23a** (26:74 at C^2)

23a → d,e,f 66% → **24a** (27:73 at C^2)

23a → d,g,h 86% → **25a** (32:68 at C^2)

a) *n*-Bu_3SnH, cat. AIBN; b) LDA/THF; c) MeI, HMPA; d) LiOH; e) Et_2NSF_3; f) PhMgBr/THF; g) *t*-BuC(O)Cl, pyr.; h) 50% $HNMe_2$ aq.

21a. Thus, taking into account the afore-mentioned crystallographic data (Figure 1), the most abundant stereoisomer of **25a** should possess the identical relative stereochemistry to the one shown in Scheme 5, and the minor product should be the 2*S**,3*R**,5*S** form, epimeric at the 2 position. Stereochemical relationship between C^3 and C^5 in **20a** became apparent by tracing the scheme back in spite of uncertainty at the C^2 site. On the other hand, the identical stereoisomeric structure was confirmed between the major isomers of both **19a** derived from ketone enolate and **24a** again by NMR spectroscopy. However, their minor isomers were found to be different and the minor **19a** should be epimeric either at C^3 or C^5.

These structural determination allowed us to estimate the diastereofacial selectivity as 94.4% for **19a**, 100% for **20a**, and 91.3 and 81.2% for **21a** from ***E***- and ***Z*-6a**, respectively. Although our clarification of the relative configuration was only for CF_3-containing products, di- as well as monofluorinated acceptors ***E*-6b** and **-6c** would be deduced to exhibit the similar tendency for their stereoselectivity.

Explanation of the Stereochemical Outcome

As described above, we have succeeded in the diastereoselective enolate Michael addition reactions, and became interested in the factors responsible for such stereoselection. Usually, Felkin-Anh type transition state (TS) models (*31-*

Figure 2 Felkin-Anh Type TS Models

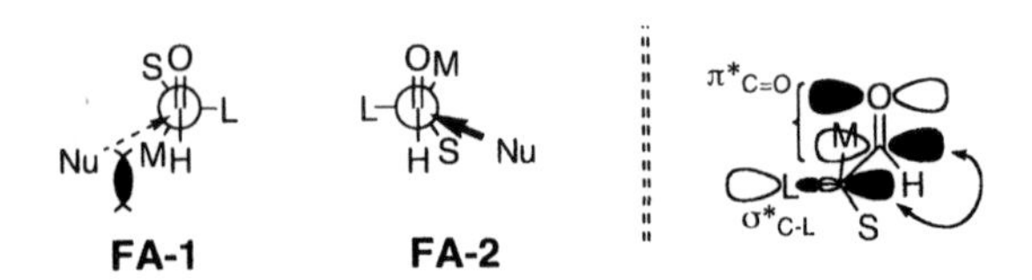

34) are employed for the explanation of these issues (*35*), which were originally applied to the carbonyl π-facial selectivity possessing α-asymmetric carbon atoms. The incoming nucleophiles (Nu) are considered to approach the reaction site to fulfill the O=C···Nu angle of 100-110 degree from the opposite side of the most sterically demanding and/or electronwithdrawing group L which occupies the perpendicular position to the C=O group (Figure 2). Such preference is ascribed to the $\sigma^*_{C\text{-}L}$-$\pi^*_{C=O}$ interaction effectively lowering the LUMO energy level and increasing electrophilicity. Although two TS models, **FA-1** and **2**, are possible, steric repulsion between the nucleophilic species and the carbonyl α-ligand allowed us to anticipate the latter more stable, leading to the occurrence of the reaction from the *re*-face preferentially.

Such TS models were further extended to the C=C system like Michael acceptors containing γ-chiral centers (Figure 3) (*36*). Replacement of oxygen for carbon gave rise to more or less steric congestion but, as long as R^1 being H, such small perturbation rendered the conformation of the γ-stereogenic center to be the same as the original case shown in Figure 2. Thus, the L group occupies the perpendicular position to the C=C bond, and Nu would attack the carbonyl β-carbon by way of **FA-4** due to smaller steric requirement rather than **FA-3**. However, this is not the case for compounds with *Z*-olefinic stereochemistry ($R^1 \neq H$) in which significant steric interaction would cause the least hindered substituent S to be located in the same plane as the C=C bond (dihedral angle S-C-C=C of ~0°) for minimizing allylic 1,3-strain (*37*). Then, a nucleophile would initiate a new bond formation from the face opposite to the bulkier ligand L (**AS-1**). As an exception, we should also consider **AS-2** type TS where Nu attacks with intermolecular interaction with the ligand which enables the access even from the sterically less favorable face.

Figure 3 Felkin-Anh and the Related TS Models

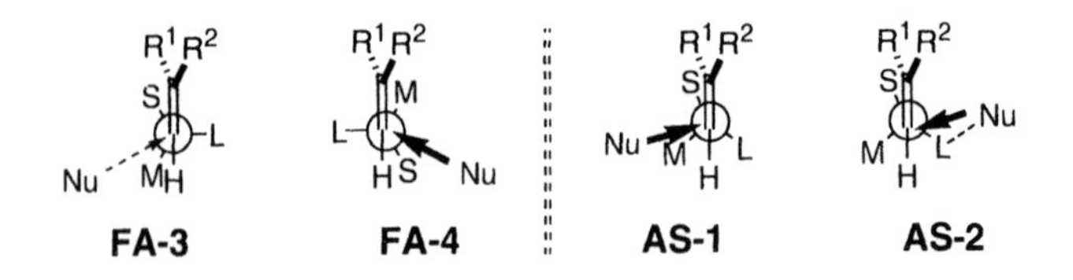

Scheme 6 Michael addition reactions under various conditions

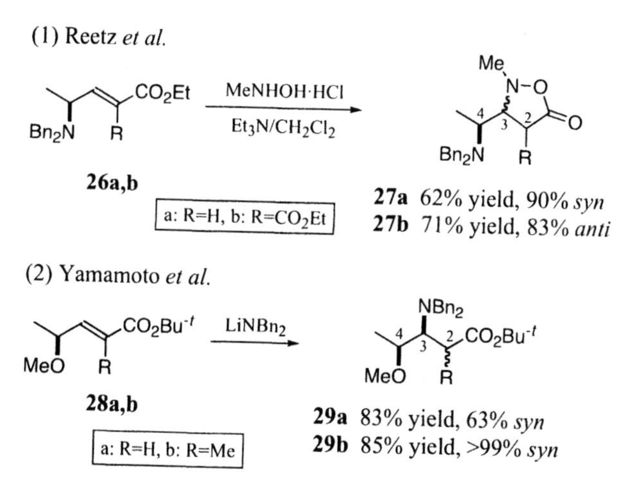

Previously, Reetz and his coworkers reported (*38*) the 1,2-oxazolidin-5-one formation *via* conjugate addition (Scheme 6). 3,4-*Syn* stereochemistry was constructed in a ratio of *syn*:*anti* being 90:10 from **26a**, while the diastereomeric *anti*-isomer became predominant for **26b** (R = CO_2Et). This sharp difference in the π-facial selection, highly dependent on the structure of the acceptors employed, would be explained by the TS model **FA-4** for the former but **AS-1** for the latter. On the other hand, Yamamoto *et al.* found (*39*) examples showing high level of *syn* selectivity even when substrates possess *cis* substituents like **28b**. The methoxy group played a pivotal role in this instance by forming chelation with the incoming nucleophiles which definitely determined the face where Michael addition really occurred (**AS-2**).

As shown in Table 1 and Scheme 5, both ***E***- and ***Z*-6a** predominantly furnished the Felkin-Anh type products and high affinity of fluorine with metals was considered to determine the latter stereoselection like the case of Yamamoto *et al.* Thus, these two Michael acceptors could be considered to follow **FA-4** and **AS-2**, respectively. However, on the basis of our recent results (*40*), this does not seem to be the case. Esters **30** with fluorine-containing auxiliaries were benzylated in a *re*-face selective fashion by construction of the rigid bicyclo-[3.3.1] type intermediates which allowed the phenyl and methyl moieties in the auxiliaries to effectively cover the opposite *si* face. As shown in Scheme 7, existence of fluorine seems to be responsible for fixing the conformation and the diastereoselectivity of the products **31** is almost constant although the chemical yields varied to some extent. It is interesting to note that the absence of fluorine is responsible to the apparent decrease of the de values, indicating the important role of this atom's inherent electronic characteristics. However, Table 1 exhibits

Scheme 7 Diastereoselective alkylation of ***30***

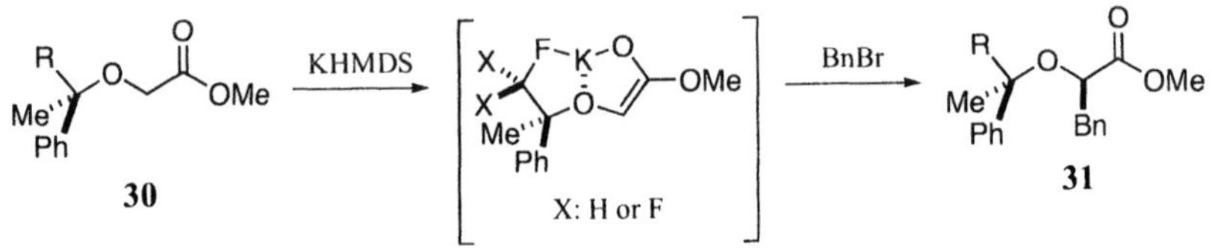

KHMDS → [X: H or F] → BnBr

a: R=CF_3, 79% yield, 80% de; **b**: R=CHF_2, 45% yield, 84% de;
c: R=CH_2F, 41% yield, 84% de; **d**: R=CH_3, 31% yield, 48% de;

clear dependence of diastereomeric selection on the number of fluorine, which made us suspicious for considering the chelation-controlled reaction and eventually we have excluded the possibility of this chelation mechanism.

Then, what can properly rationalize how our stereoselectivity was attained? Eventually, we have reached the conclusion that it is Cieplak rule (*41,42*) which can consistently explain the present stereochemical outcome. This rule is summarized as the hyperconjugative stabilization effect by electron donation from adjacent bonds to an incipient bond and thus reactions occur from the side opposite to the best electron donating substituent.

With this definition in mind, this rule is applied to our system. First of all, the H^4-C-C=C part of the stable conformation would keep planarity for minimizing allylic 1,3-strain especially in the case of R^2 not being H (**Int-4** in Figure 4). However, when a nucleophile approaches from the top side, the stereogenic center rotates clockwise so that the CF_3 group at the TS occupies the *anti* position to the incoming Nu and the incipient bond $\sigma^*_{\neq}$ orbital is able to accept electron donation from the neighboring $\sigma_{C\text{-}CF3}$ bond (**C-1**). On the other hand, **C-2** is the TS model where Nu attacks from the bottom side and, in this case, the CH_3 moiety is located on the other side of the nucleophile and the similar electrostatic interaction would occur between $\sigma_{C\text{-}CH3}$-$\sigma^*_{\neq}$ orbitals. Considering the strongly electronwithdrawing nature of a CF_3 group, it is quite apparent that electrostatic stabilization should be more pronounced for **C-2** rather than **C-1**, leading to predominant acceptance of nucleophilic reagents

Figure 4 Explanation of Stereoselectivity by the Cieplak Models

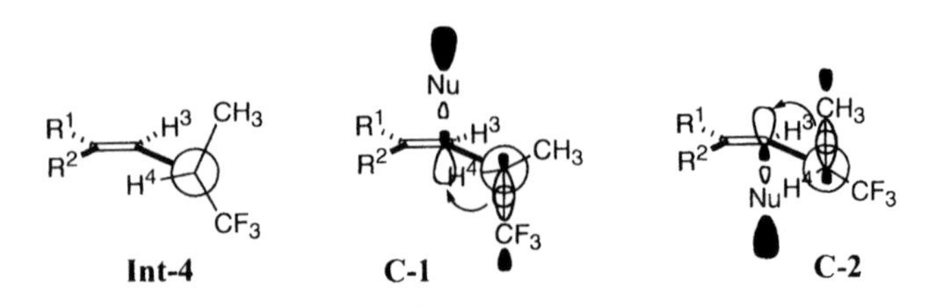

from the same olefinic face where a bulkier CF_3 group occupies. As a consequence, the TS model **C-2** leads to the production of the relative stereochemical relationship between C^3 and C^5 as shown in Scheme 5. Such experimental results could be alternatively understood by Felkin-Anh model in the case of ***E*-6a** (*ca* 120° counterclockwise rotation of the allylic chiral carbon in **C-2** gives rise to the identical conformation to **FA-4** as described in Figure 3), which appears to expect the access of Nu from the same π-face. However, as mentioned above, Felkin-Anh model cannot account for the correct stereoisomers when *Z*-acceptors are used. We believe this is also the case for ***E*-6A** because **Int-4** type conformation is most abundant too in the case of *E*-acceptors (*43*). This is the reason why we adapt the Cieplak rule because it is only the Cieplak rule at present which has led to the consistent explanation for the obtained diastereofacial selectivity irrespective of the stereochemistry of Michael acceptors, and there are such examples whose selectivity is explained on the basis of this theory (*44-48*).

As described above, we have successfully clarified the role of $CH_{3-n}F_n$ groups at the γ position of α,β-unsaturated ketones for diastereoselective enolate-Michael addition reactions. Their electron-withdrawing effect allowed the incoming nucleophiles to readily discriminate the two different π-faces. This phenomena were consistently explained by adapting the Cieplak rule for not only ***E*-6** but also the corresponding *Z*-isomer.

References and Notes

1. This manuscript is translated and reconstructed on the basis of the article already published. Yamazaki, T. *J. Synth. Org. Chem. Jpn.* **2004**, *62*, 911-918.
2. Smart, B. E. *J. Fluorine Chem.* **2001**, *109*, 3-11.
3. Hiyama, T. "*Organofluorine Compounds Chemistry and Application*" Springer, Berlin, 2000.
4. Kitazume, T.; Yamazaki, T. "*Experimental Methods in Organic Fluorine Chemistry*" Kodansha, Tokyo, 1998.
5. Shinohara, N.; Haga, J.; Yamazaki, T.; Kitazume, T.; Nakamura, S. *J. Org. Chem.* **1995**, *60*, 4363-4374.
6. Linderman, R. J.; Jamois, E. A. *J. Fluorine Chem.* **1991**, *53*, 79-91.
7. Yamazaki, T.; Shinohara, N.; Kitazume, T.; Sato, S. *J. Org. Chem.* **1995**, *60*, 8140-8141.
8. Perlmutter, P. "*Conjugate Addition Reactions in Organic Synthesis*" Pergamon, New York, 1992.
9. Oare, D. A.; Heathcock, C. H. *Top. Stereochem.* **1989**, *19*, 227-407.

10. Yamazaki, T. In "*Enantiocontrolled Synthesis of Fluoro-organic Compounds*" V. A. Soloshonok, Ed., John Wiley & Sons, New York, 1999, p. 263-286.
11. Yamazaki, T.; Haga, J.; Kitazume, T.; Nakamura, S. *Chem. Lett.* **1991**, 2171-2174.
12. Yamazaki, T.; Haga, J.; Kitazume, T. *Chem. Lett.* **1991**, 2175-2178.
13. Ziegler, F. E. *Chem. Rev.* **1988**, *88*, 1423-1452.
14. MacPhee, J. A.; Panaye, A.; Dubois, J.-E. *Tetrahedron* **1978**, *34*, 3553-3562.
15. Bott, G.; Field, L. D.; Sternhell, S. *J. Am. Chem. Soc.* **1980**, *102*, 5618-5626.
16. Weseloh, G.; Wolf, C.; König, W. A. *Chirality* **1996**, *8*, 441-445.
17. Recently, Fleming *et al.* reported a similar type of rearrangement by substrates containing trialkylsilyl groups instead of a CF_3 moiety. Betson, M. S.; Fleming, I. *Org. Biomol. Chem.* **2003**, *1*, 4005-4016.
18. Ando, K. *J. Org. Chem.* **1999**, *64*, 8406-8408.
19. O'Hagan, D. *J. Fluorine Chem.* **1989**, *43*, 371-377.
20. Takaoka, A.; Iwakiri, H.; Ishikawa, N. *Bull. Chem. Soc. Jpn.* **1979**, *52*, 3377-3380.
21. Takacs, J. M.; Helle, M. A.; Seely, F. L. *Tetrahedron Lett.* **1986**, *27*, 1257-1260.
22. Thenappan, A.; Burton, D. J. *J. Org. Chem.* **1990**, *55*, 4639-4642.
23. Lanier, M.; Haddach, M.; Pastor, R.; Reiss, J. G. *Tetrahedron Lett.* **1993**, *34*, 2469-2472.
24. Tsukamoto, T.; Kitazume, T. *J. Chem. Soc. Perkin Trans. 1* **1993**, 1177-1181.
25. Haas, A. M.; Hägele, G. *J. Fluorine Chem.* **1996**, *78*, 75-82.
26. Legoupy, S.; Cévisy, C.; Guillemin, J.-C.; Grée, R. *J. Fluorine Chem.* **1999**, *93*, 171-173.
27. Yamazaki, T.; Ichige, T.; Kitazume, T. *Collect. Czech. Chem. Commun.*, **2002**, *67*, 1479-1485.
28. Yamazaki, T.; Hiraoka, S.; Sakamoto, J.; Kitazume, T. *J. Phys. Chem. A* **1999**, *103*, 6820-6824.
29. Ichikawa, J.; Wada, Y.; Fujiwara, M.; Sakoda, K. *Synthesis*, **2002**, 1917-1936.
30. Miura, K.; Takeyama, Y.; Oshima, K.; Utimoto, K. *Bull. Chem. Soc. Jpn.* **1991**, *64*, 1542-1553.
31. Chérest, M.; Felkin, H.; Prudent, N. *Tetrahderon Lett.* **1968**, 2199-2204.
32. Chérest, M.; Felkin, H. *Tetrahderon Lett.* **1968**, 2205-2208.
33. Anh, N. T. *Top. Cur. Chem.* **1980**, *88*, 146-162.
34. Smith, R. J.; Trzoss, M.; Bühl, M.; Bienz, S. *Eur. J. Org. Chem.* **2002**, 2770-2775.

35. Mengel, A.; Reiser, O. *Chem. Rev.*, **1999**, *99*, 1191-1123.
36. Yamamoto, Y.; Chounan, Y.; Nishii, S.; Ibuka, T.; Kitahara, H. *J. Am. Chem. Soc.* **1992**, *114*, 7652-7660.
37. Hoffmann, R. W. *Chem. Rev.* **1989**, *89*, 1841-1860.
38. Reetz, M. T.; Röhrig, D.; Harms, K.; Frenking, G. *Tetrahedron Lett.* **1994**, *35*, 8765-8768.
39. Asao, N.; Shimada, T.; Sudo, T.; Tsukada, N.; Yazawa, K.; Gyoung, Y.-S.; Uyehara, T.; Yamamoto, Y. *J. Org. Chem.* **1997**, *62*, 6274-6282.
40. Yamazaki, T.; Ando, M.; Kitazume, T.; Kubota, T.; Omura, M. *Org. Lett.* **1999**, *1*, 905-908.
41. Cieplak, A. S. *Chem. Rev.* **1999**, *99*, 1265-1336.
42. Gung, B. W. *Tetrahedron* **1996**, *52*, 5263-5301.
43. **Int-4** type conformation is calculated to exist in 94% probability at the reaction temperature -78 °C by *ab initio* calculation (Gaussian 03W at the B3LYP/6-31+G* basis set).
44. Coxon, J. M.; McDonald, D. Q. *Tetrahedron* **1992**, *48*, 3353-3364.
45. Tsai, T.-L.; Chen, W.-C.; Yu, C.-H.; le Noble, W. J.; Chung, W.-S. *J. Org. Chem.* **1999**, *64*, 1099-1107.
46. Yadav, V. K.; Jeyaraj, D. A.; Parvez, M.; Yamdagni, R. *J. Org. Chem.* **1999**, *64*, 2928-29.
47. Ono, M.; Nishimura, K.; Nagaoka, Y.; Tomioka, K. *Tetrahedron Lett.* **1999**, *40*, 1509-1512.
48. Recently, Fleming *et al.* reported a similar type of Michael addition by substrates containing trialkylsilyl groups instead of a CF_3 moiety. Betson, M. S.; Fleming, I.; Ouzman, V. A. *Org. Biomol. Chem.* **2003**, *1*, 4017-4024.

Chapter 24

Reactions of Unsaturated Organometallic Reagents on Trifluoroacetaldimines

Danièle Bonnet-Delpon, Jean-Pierre Bégué, and Benoit Crousse

Centre National de la Recherche Scientifique, Université Paris-Sud, Biocis UPRES A 8076, Rue J.B. Clément, F–92296 Châtenay-Malabry, France

Allyl zinc reagents, vinyl magnesium bromide and lithium acetylides could easily react without any activation on trifluoroacetaldimines to afford homoallyl, allyl and propargyl amines in good yields. From chiral imines, these amines were obtained in excellent diastereoselectivity (>98%). After *N*-allylation, homoallyl amines could give access to piperidine derivatives by ring closing metathesis. The allyl amine allowed the synthesis of the trifluoromethyl epoxide **24**, fluorinated analog of a key precursor of aspartyl proteases inhibitors.

Introduction

Amino compounds constitute a large class of naturally occurring and biologically important molecules. One approach to these compounds is the addition of organometallic reagents to imines or imine derivatives. This well-known reaction allows the preparation of stereochemically defined amines, since the addition of organometallic reagents to the C=N bonds of imines or imine derivatives can be stereocontrolled. Despite the poor reactivity of imines towards nucleophilic reagents, a variety of methods have been developed which broaden the scope of this organometallic reaction. Classically the electrophilicity of the imino function bond can be increased by the addition of external promoters (proton or Lewis acid) to provide a more reactive iminium salt, or by the *N*-acylation or *N*-sulfonylation to give more activated imines or iminium salts (*1*).

N-oxidation is also a good way to increase the reactivity of imines through the formation of nitrones (*1*). Surprisingly, despite the electronwithdrawing effect of a fluoroalkyl group which can favor to addition of nucleophiles on C=N bond, the addition of an unsaturated organometallic reagent to fluorinated aldimines has not been explored so far.

Considering the unique consequences of fluorine substitution for hydrogen in organic molecules, the synthesis of selectively fluorinated and enantiomerically pure unsaturated amino compounds, as biological relevant targets, might be of particular interest (*2-4*).

Consequently our objective was to perform the addition of unsaturated organometallic reagents to trifluoromethyl aldimines, with no particular other type of activation. The success of such method would bring the possibility to prepare, by the easy introduction of a chiral *N*-substituent, non racemic amines, as already shown in the case of phenyl lithium addition on chiral fluorinated imine derivatives (*4f*). We develop in this chapter our studies on these reactions as an access to allyl, homoallyl and propargyl amines, and their use as synthons for the preparation of new biologically relevant fluorinated amino compounds, such as fluorinated peptidomimetic units and heterocyclic compounds (Scheme 1).

Scheme 1

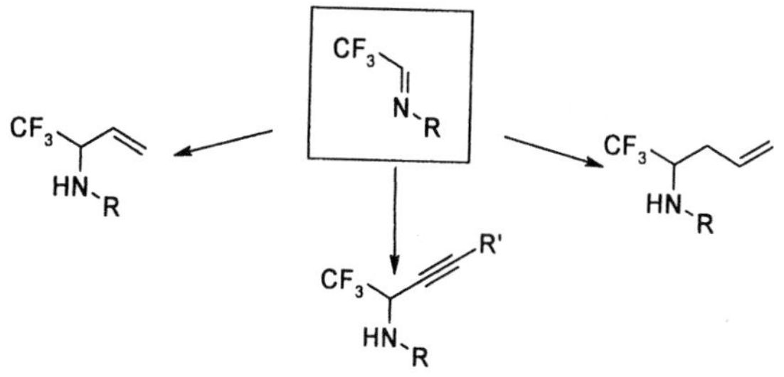

All starting trifluoromethyl aldimines were readily synthetized by the condensation between the hydrate or the hemiacetal of fluoral with the appropriate amine, and were always obtained as *anti* isomers.

Trifluoromethyl Homoallylic Amines: Allylation Reactions.

The preparation of homoallyl trifluoromethyl amines had not been intensively explored so far. Few approaches were described: - one synthesis from the *N*-(*p*-toluenesulfonyl)-trifluoromethyl aldimine by an ene reaction (*5*), - a Lewis acid-mediated addition of an allylsilane to aldimines derived from fluoral or related hemiaminals, including non racemic ones (*6, 4a*). Concerning the organometallic approach, to our knowledge few examples were reported and they concern imines activated by an electronwithdrawing *N*-substituent: the addition of the allylmagnesium on a trifluoromethyl sulfenimine generated by electrooxidation of $CF_3CH_2NR_2$ (*7*) and on the *N*-acyl-1-chloro-2,2,2-trifluoro ethylamine (*8*), and more recently the addition of allyllithium on α-trifluoromethyl hydrazones (*9*).

We explored the reaction between the *N*-benzyl trifluoroacetaldimine **1** and allyl bromides with different metals under Barbier conditions (*10*). This approach does not require the isolation of the organometallic reagents allowing thus a great diversity of structures for the allyl moiety. With Mg and In, the reaction was efficient only with the non substituted allyl bromide. With Zn, conditions were found (1.3 equiv. of allyl bromide, 1.2 equiv. of turnings of Zn activated by TMSCl, at room temperature in DMF) which allowed the complete reaction with allyl bromide and 3,3-dimethyl allyl bromide (table 1).

Table 1. Allylation Reaction under Barbier conditions

CF_3CH=N-Bn (**1**) + Allylic bromides —Metal→ Homoallylic amines **3a,b**

Allyl Br	*Metal*	*Time (h)*	*product*	*Yield (%)*
Br-CH$_2$CH=CH$_2$	Mg[a]	0.75	CF_3CH(NHBn)CH$_2$CH=CH$_2$ **3a**	75
	In[b]	1		96
	Zn	0.75		95
Br-CH$_2$CH=C(CH$_3$)$_2$	Mg[a]	1	CF_3CH(NHBn)C(CH$_3$)$_2$CH=CH$_2$ 3b	20
	In[b]	24		Traces
	Zn[b]	0.75		97

[a] in ether at r.t. [b] in DMF at r.t.

These latter conditions were then successfully applied to the reaction with various patterns of allyl bromides, starting from the *N*-benzyl aldimine **1** or the *N-para*-methoxyphenyl aldimine **2** (Table 2). Reactions can be conducted alternatively in DMF at room temperature or in THF at reflux. In all cases reactions were fast and clean, and yields in homoallylic amines **3, 4** were good to excellent (75-97%). Similarly imine **1** could react with propargyl bromide to afford the homopropargyl amine **3e** in 76% yield.

Table 2. Allylation Reaction with Zinc in DMF[a] and THF[b]

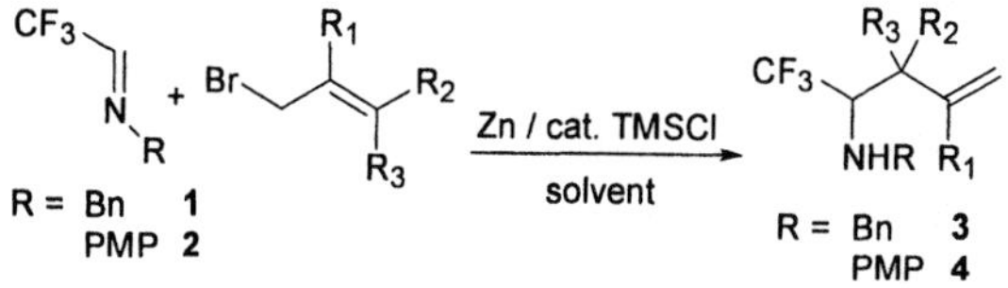

$CF_3CH{=}N{-}R$ (R = Bn **1**, PMP **2**) + $BrCH_2C(R_1){=}C(R_2)(R_3)$ —Zn / cat. TMSCl, solvent→ $CF_3CH(NHR)C(R_3)(R_2)C(R_1){=}CH_2$ (R = Bn **3**, PMP **4**)

Allyl Br	Time (h)	Solvent	3,4	Yield (%)
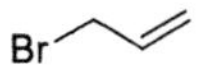 Br-allyl	0.75	DMF	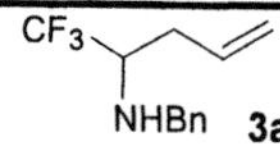 CF_3, NHBn **3a**	95
		THF		85
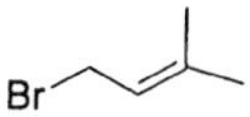 Br-prenyl	1	DMF	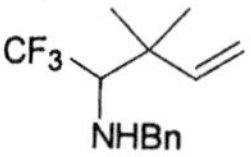 CF_3, NHBn **3b**	97
		THF		82
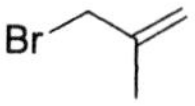 Br-methallyl	0.75	DMF	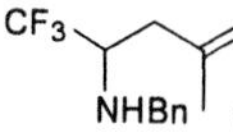 CF_3, NHBn **3c**	75
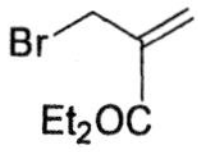 Br, Et_2OC	0.75	DMF	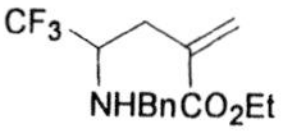 CF_3, NHBn CO_2Et **3d**	75
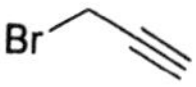 Br-propargyl	0.75	DMF	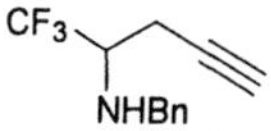 CF_3, NHBn **3e**	76
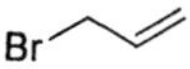 Br-allyl	0.75	DMF	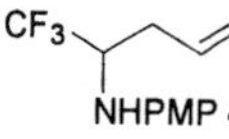 CF_3, NHPMP **4a**	79
	4	THF		92
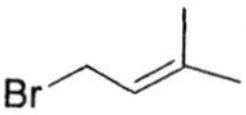 Br-prenyl	0.75	DMF	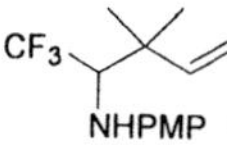 CF_3, NHPMP **4b**	82
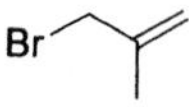 Br-methallyl	0.75	DMF	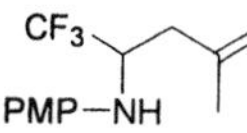 CF_3, PMP—NH **4c**	83

[a] At room temperature. [b] At reflux.

The success of this reaction which does not require an activation of the imine group by an *N*-acyl substituent, allowed us to envisage a chiral approach using a chiral *N*-substituent. With the (*R*)-*N*-phenethyl aldimine **5**, the homoallylic amine **6a** was obtained in good yield (87%), but with poor diastereoselectivity (60/40) (Scheme 2).

Scheme 2

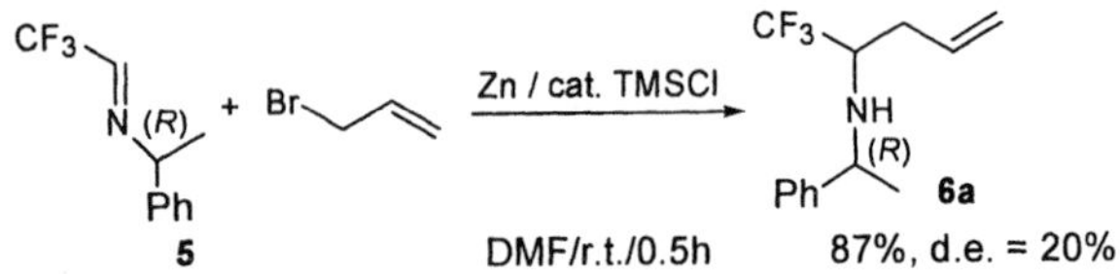

Zn / cat. TMSCl, DMF/r.t./0.5h: **5** → **6a**, 87%, d.e. = 20%

In order to improve diastereoselectivity, other chiral *N*-substituents were evaluated. Oxazolidines **7** (62/38) which are prepared from fluoral and phenyl glycinol (*4f*), were reported to provide a 70/30 mixture of diastereoisomers of chiral homoallyl amine in $BF_3.Et_2O$-promoted reaction with an allylsilane (*4a*). When placed under Barbier conditions in DMF at room temperature, with 1.5 equiv. of zinc, oxazolidines **7** provided amines in very poor yield. An increased amount of Zn (3-4 equiv.) was required to obtain the *N*-phenyl glycinol allyl amines **8** in 65%. Compared to reactions from **5**, the diastereoisomeric excess was improved (d.e.= 61%). The same reaction performed in THF (reflux) provided 45% of **8** with a very poor d.e. (9%. (Scheme 3).

Scheme 3

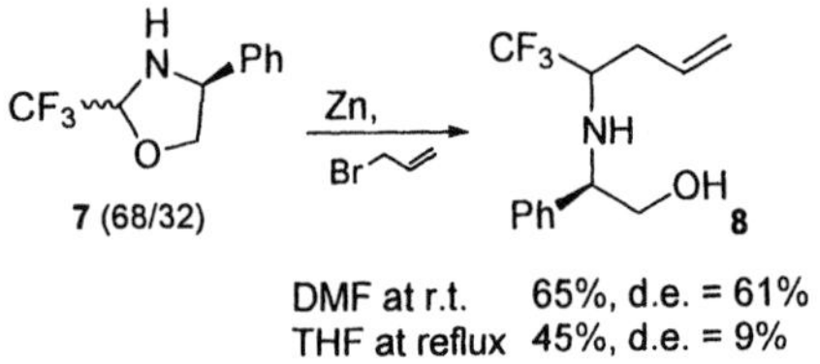

7 (68/32) → Zn, allyl bromide → **8**

DMF at r.t. 65%, d.e. = 61%
THF at reflux 45%, d.e. = 9%

The change in ratio of diastereoisomers in starting material and products suggests the intermediate formation of an iminium ion, as already postulated in reaction using Lewis acids (4a). This is probably due to the excess of zinc used in these reactions. These results indicate that chiral oxazolidines are not good substrates for reactions with allyl zinc reagents: they are less reactive than aldimines, and diastereoselectivity is not sufficiently high. In order to prevent the formation of oxazolidines, the (R)-phenylglycinol was converted into the corresponding methyl ether and the allylation reaction was performed with the CF_3-aldimine **9** in THF and DMF with allyl bromides (Table 3). The high reactivity was recovered : only 1.3 equiv. of Zn were required, and good yields in homoallylic amines **10** were obtained (65-85%). Furthermore diastereoselectivities were excellent. The absolute configuration could not be determined.

The stereoselectivity of the addition reaction may be accounted for by a chelation-controlled mechanism (*11*). From the aldimine **9**, the metal coordinates with N and O atoms of the auxiliary group to produce a rigid five-membered chelate (figure 1). The allyl group attacks the C=N bond from the less congested face (opposite to the phenyl group) to give the amine **10**. According to this transition state model, the addition of the allyl moiety occurs with creation of an R chiral center. This has been confirmed with our result obtained in reaction of vinylation (*vide infra*).

Table 3. Allylation Reaction with the Aldimine 9.

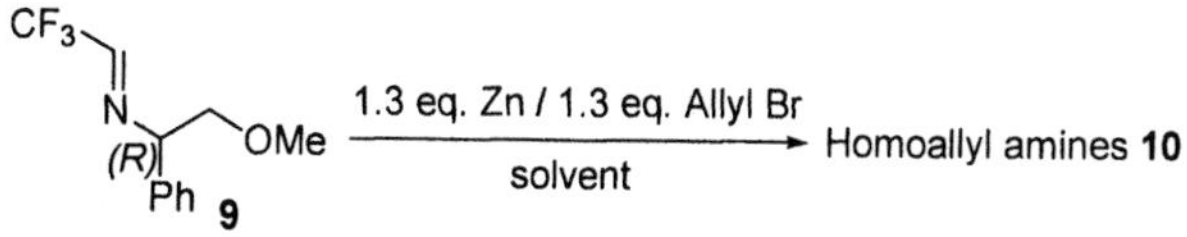

9 (CF3–CH=N–(R)-CH(Ph)CH2OMe) → 1.3 eq. Zn / 1.3 eq. Allyl Br, solvent → Homoallyl amines **10**

Allyl Br	*Solvent*	*Time (h)*	*10*	*yield (%)*	*d.e. (%)*[a]
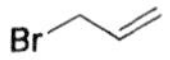 Br–CH2CH=CH2	THF	2	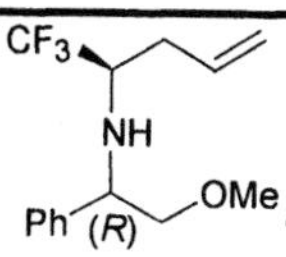 **10a**	85	96
	DMF	0.5		68	84
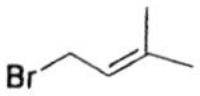 Br–CH2CH=C(CH3)2	THF	0.75	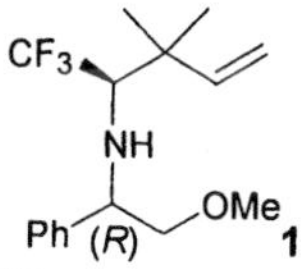 **10b**	65	98
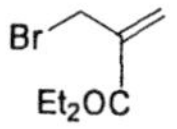 Br–CH2C(=CH2)CO2Et	DMF	2	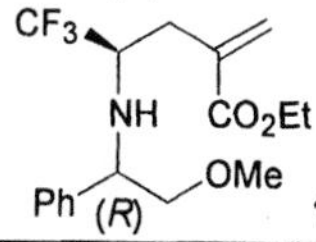 **10d**	72	85

[a] determined by ^{19}F NMR of the crude product.

Figure 1

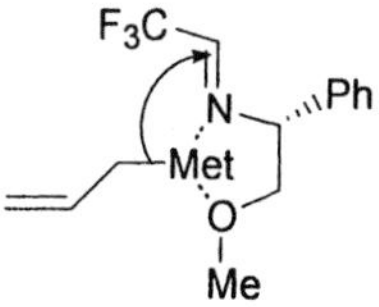

F3C, N, Ph, Met, O, Me

Among the various classes of nitrogen heterocyclic systems involved in biochemical reactions, pyrrolidines and piperidines received particular interest triggered by their isolation from various natural sources (*12*). These patterns are present in many therapeutic agents and considerable efforts have been focused to introduce novel substitutions in nitrogen containing elements. Among these, trifluoromethyl substituent, when strategically positioned can greatly modify the biological activity of the target molecules.

Recently, ω,ω-di-unsaturated trifluoromethyl amines have been prepared from protected allyl amines by an *N*-allylation after deprotection and *N*-acylation, and they were shown to easily undergo a ring closure metathesis (RCM) (*13*).

We searched for an easy methodology leading to di-unsaturated trifluoromethyl amines without protection-deprotection steps. The *N*-allylation was thus investigated directly with homoallyl amines **3, 4, 10** (*14*). The reaction was conducted under classical conditions with $NaHCO_3$ and allyl bromides, in acetonitrile in the presence of a catalytic amount of KI. However the reaction required a reflux, very long reaction times, and yields were highly dependent of the substrate. Since the Barbier allylation provides a zinc amide as primary product, the *N*-allylation could be envisaged directly *in situ*. DMF was chosen as solvent for this one pot Zn-catalyzed allylation / *N*-allylation sequence. Aldimines **1**, **2**, and **5** were placed at room temperature in the presence of Zn, and 5 equiv. of allyl bromide, and after 45 min the reaction medium was heated to reflux. Di-unsaturated amines **11a**, **12a**, and **15a** could be obtained in 70%, 61% and 54% respectively (Table 4). Moreover this one pot procedure could also allow the introduction of two different allyl substituents. Reactions were performed at room temperature in DMF with 1.4 equiv. of allyl bromide, using a slight excess of zinc (1.5 equiv.) to avoid a further *N*-allylation; then a second allyl bromide (4 equiv.) was added and the reaction was conducted at reflux.

This one pot procedure allowed realizing efficiently the *N*-allylation reaction, and thus to isolate di-unsaturated amines in reasonable to good yields.

Having in hand an efficient access to a varied range of di-unsaturated amines, RCM reaction was investigated with all *N*-allyl homoallyl amines **11-15** (table 5). They easily reacted at room temperature, in the presence of the Grubbs catalyst (2-10%) (*15*) to afford the expected CF_3-dehydropiperidine derivatives in excellent yields. However with the *N*- methallyl amines **12a** and **14a** reaction times were long and even required reflux (entries 4 and 6).

Trifluoromethyl Allyl Amines.

The preparation of allyl amines was previously described by reaction of an excess of vinylmagnesium bromide with the reactive iminium salt generated from an *N*-acyl-1-chloro-2,2,2-trifluoro ethylamine (*8*). A similar approach involving the addition of a vinyl trifluoroborate to a trifluoromethyl iminium species was recently reported (*6a*). An alternative approach is the trifluoromethylation of unsaturated *N*-sulfinimines or *N*-tosyl aldimines (*16*). Otherwise the unsaturation could also be later introduced from fluorinated β-amino sulfones by a methylenation-desulfonylation reaction (*17*).

Table 4. One-pot Synthesis of Unsaturated Amines.

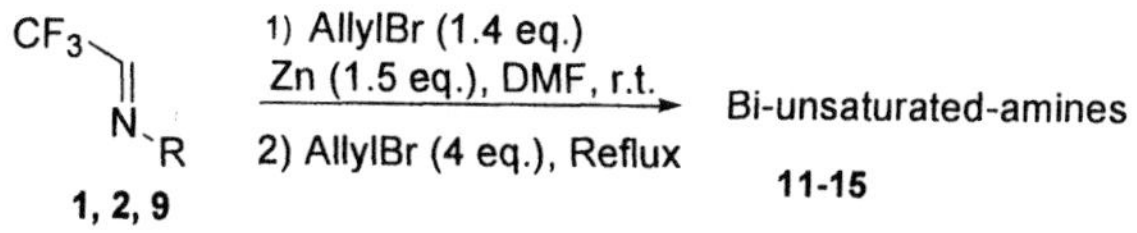

$CF_3CH{=}NR$ (1, 2, 9) → 1) AllylBr (1.4 eq.) Zn (1.5 eq.), DMF, r.t. 2) AllylBr (4 eq.), Reflux → Bi-unsaturated-amines 11-15

Entry	*imine*	*AllylBr*	*Time (h)*	*Product 11-15*	*Yield (%)*
1	**1**	1) Br-allyl 2) Br-allyl 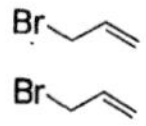	1 3	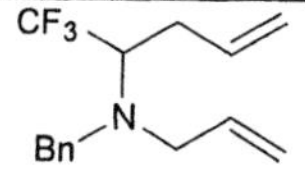 **11a** (Bn)	70
2	**1**	1) Br-methallyl 2) Br-allyl 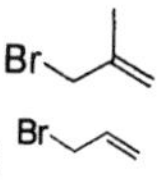	1 5	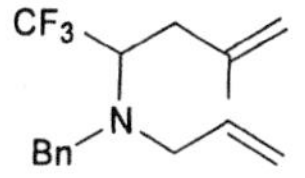 **11c** (Bn)	63
3	**1**	1) Br-allyl 2) Br-methallyl 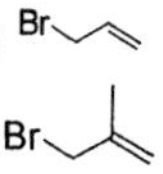	1 3	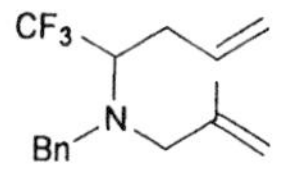 **12a** (Bn)	56[a]
4	**1**	1) Br-prenyl 2) Br-methallyl 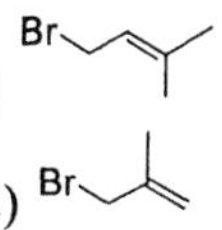	1 3	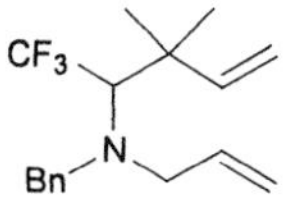 **11b** (Bn)	57[a]
5	**2**	1) Br-allyl 2) Br-allyl 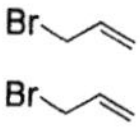	1 3	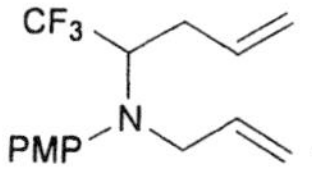 **13a** (PMP)	61[a]
6	**2**	1) Br-allyl 2) Br-methallyl 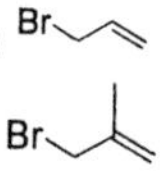	1 2	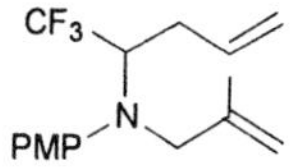 **14a** (PMP)	56[a]
7	**2**	1) Br-methallyl 2) Br-allyl 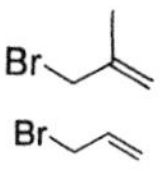	1 1	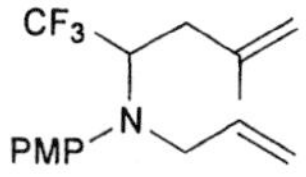 **13c** (PMP)	86[a]
8	**9**	1) Br-allyl 2) Br-allyl 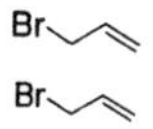	1 1	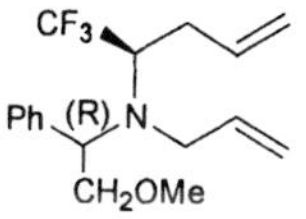 **15a** (Ph (R) CH₂OMe)	54[a,b]

[a] Presence of CuI was required. [b] 30% of starting material recovered.

Table 5. Ring Closing Metathesis[a]

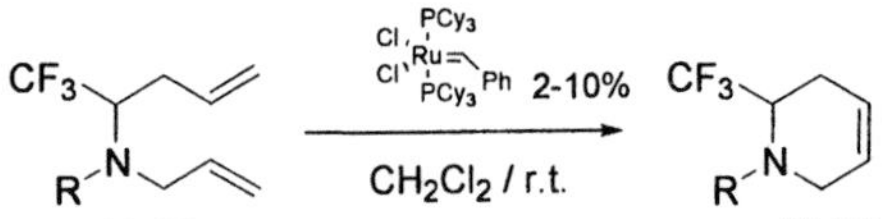

CF_3, R–N, **11-15** → $Cl_2Ru(PCy_3)_2(=CHPh)$ 2-10%, CH_2Cl_2 / r.t. → CF_3, R–N, **16-20**

Entry	*diene*	*Time (h)*	*product*	*Yield (%)*
1	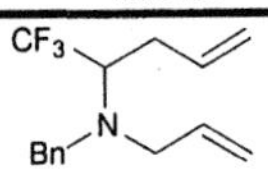 CF_3, Bn–N **11a**	2	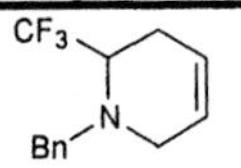 CF_3, Bn–N **16a**	95
2	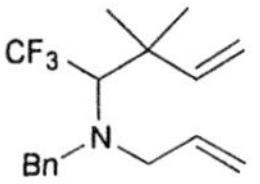 CF_3, Bn–N **11b**	2	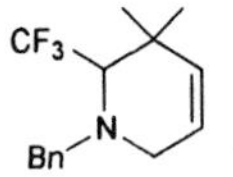 CF_3, Bn–N **16b**	96
3	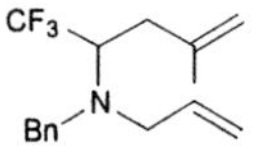 CF_3, Bn–N **11c**	2	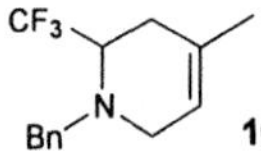 CF_3, Bn–N **16c**	93
4	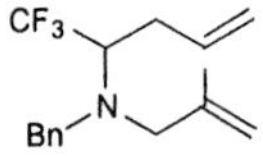 CF_3, Bn–N **12a**	48	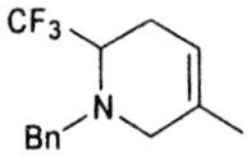 CF_3, Bn–N **17a**	89
5	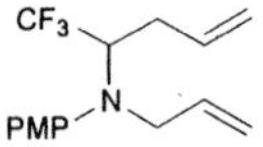 CF_3, PMP–N **13a**	2	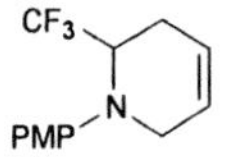 CF_3, PMP–N **18a**	90
6	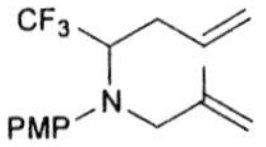 CF_3, PMP–N **14a**	48	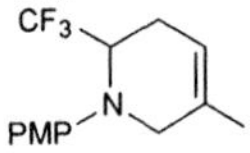 CF_3, PMP–N **19b**	41[a]
7	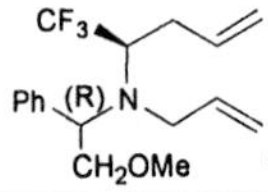 CF_3, Ph (R) N, CH_2OMe **15a**	3,5	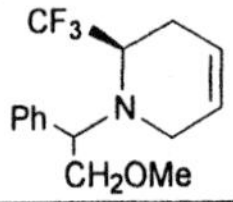 CF_3, Ph N, CH_2OMe **20a**	98

[a] at reflux

We first investigated the reaction of the commercially available vinyl magnesium bromide with the trifluoromethyl aldimines **1** (*N*-benzyl) and **2** (*N-para*-methoxyphenyl) in different solvents (toluene, THF, ether). Diethyl ether was found to be the best solvent for this reaction. Aldimine **1** and **2** could react with the vinyl Grignard, at 0°C, leading to the allyl amines **21** and **22** which were isolated in excellent yield (95%) (scheme 4). In this case again, the easiness of the method which does not require any activating *N*-substituent prompted us to investigate the reaction with an aldimine *N*-substituted with a chiral group. According to our results obtained in the allylation reaction, the vinylation was performed from the aldimine **9**. The addition of the vinyl Grignard proceeded smoothly at 0°C in 1h, and provided the allyl amine **23** in a high yield (86%) and an excellent diastereoselectivity (>98%).

Scheme 4

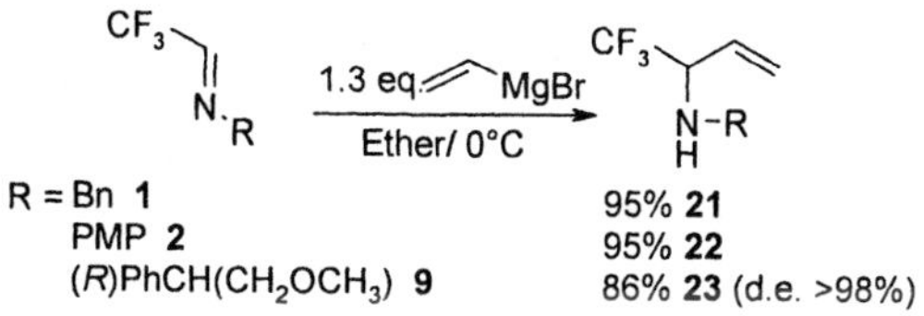

1.3 eq. vinyl-MgBr, Ether/ 0°C

R = Bn **1** — 95% **21**
PMP **2** — 95% **22**
(*R*)PhCH(CH_2OCH_3) **9** — 86% **23** (d.e. >98%)

These CF_3 allyl amines and in particular the non racemic one, are very interesting building blocks for the preparation of new polyfunctionalized trifluoromethyl amines, and new peptidomimetics. We were interested in the preparation of the amino epoxide **24**, a fluorinated analogue of the epoxide **25**. This latter is a very important key synthon for the synthesis of the central peptidomimetic diaminopropanol unit which is present in most HIV-1 protease inhibitors (ex. Saquinavir, figure 2) (*18*). This amino epoxide **25** can also be precursor of central units present in other aspartyl proteases such as cathepsin D (*19*) and plasmepsins (*20*).

Figure 2

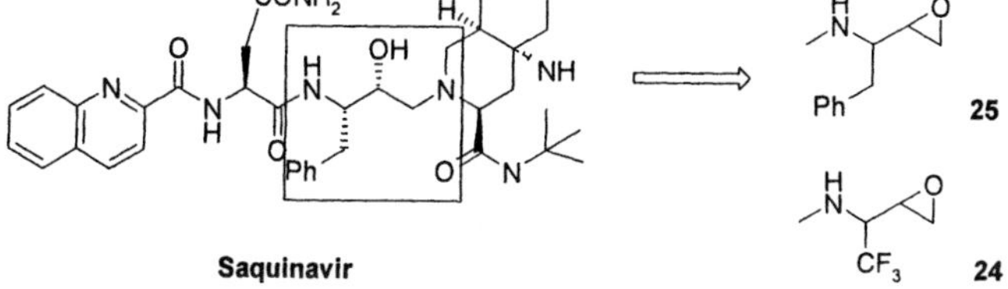

Saquinavir — **25** — **24**

Besides specific properties brought by fluorine substitution (*21-23*), largely exploited for the design of fluorine containing protease inhibitors (*24*), fluoroalkyl groups are able to mimic a large hydrophobic aliphatic or aromatic group (phenyl, benzyl or isopropyl) of the side chain of amino acids (*25*). This could be a powerful tool for optimizing the binding properties with proteases which exhibit preference for large hydrophobic P^1-residue.

Our efforts have thus been focused on the preparation of the trifluoromethyl epoxide **24**, as a precursor of the diaminopropanols **26** (scheme 5). While the preparation of the non fluorinated epoxide was largely described through different ways (*18, 26*), these latter could not be satisfactorily applied to the synthesis of the trifluoromethyl analogue. The easy availability of allyl amines could open a new access to the epoxide **24** and hence to diaminopropanols.

Scheme 5

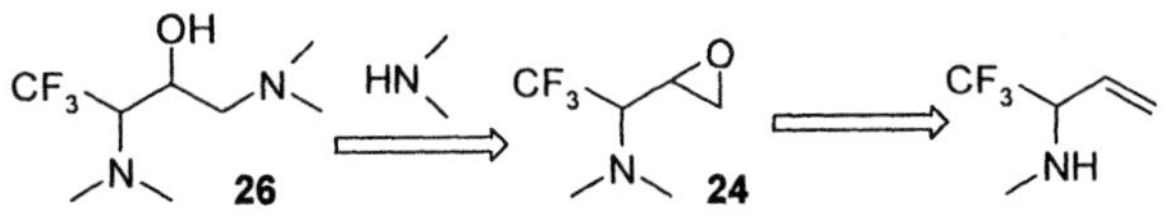

CF_3, OH, N, HN, N **26** ⟹ CF_3, O, N **24** ⟹ CF_3, NH

The epoxidation reaction of the double bond of allyl amines **21-23** with a peracid, required the previous protection of the amine (*27*). The classical protection reactions through urethanes (with Boc_2O or CbzCl) failed; starting amines were recovered unchanged. However, we succeeded to prepare *N*-acyl amines **27-29** which were obtained in good yield after treatment of **21-23** with acetic anhydride at reflux (Scheme 6).

Scheme 6

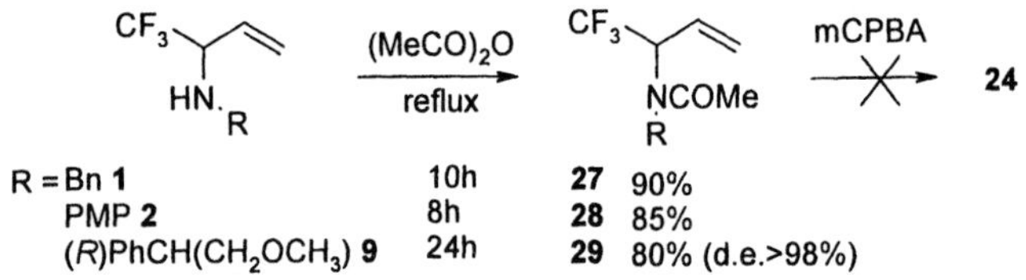

CF_3, HN, R —$(MeCO)_2O$, reflux→ CF_3, NCOMe, R —mCPBA ✗→ **24**

R = Bn **1** — 10h — **27** 90%
PMP **2** — 8h — **28** 85%
(*R*)PhCH(CH_2OCH_3) **9** — 24h — **29** 80% (d.e.>98%)

In the event, these protected trifluoromethyl amines were found to be completely unreactive toward epoxidation with mCPBA in dichloromethane, even at reflux. No trace of epoxides **24** could be detected. This poor reactivity of the double bond probably results from the presence of both electron withdrawing CF_3 and amide groups.

At this stage we explored an alternative way through the cyclisation of a bromhydrin. The treatment of the allyl amine **27** with Br_2 in CH_2Cl_2 followed by hydrolysis resulted in a clean reaction, providing remarkably only one bromhydrin **30** as a single regio- and stereo-isomer (86%). This unexpected regioselectivity strongly suggested the formation of an ionic intermediate resulting from an intramolecular ring opening of the bromonium ion by the *N*-acyl group (scheme 7), which was readily hydrolysed into the amino alcohol **30**.

This mechanism explains the regioselectivity of the reaction with a favored formation of a 5 member ring intermediate **B** (5-*exo-tet*). The stereoselectivity is the result of a non anticipated stereospecific formation of the initial bromonium ion, followed by a stereoselective bromonium ring opening, probably by an *anti* addition of the acyl group. Further experiments of chemical derivatives allowed us to assign the *syn* configuration to the amino alcohols (*vide infra*). Of course it is not possible to exclude that compound **B** could result from a concerted process and the bromonium ion **A** being not an discrete intermediate. The same reaction performed with the chiral vinyl amine **29** also provided the bromhydrin **32** in good yield and with an excellent regio- and stereo-selectivity (d.e.>98%) (scheme 8).

Scheme 7

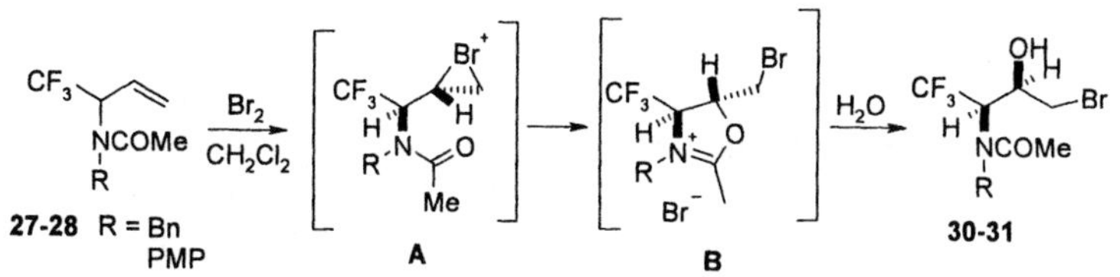

CF3
NCOMe
R
27-28 R = Bn
PMP
Br2
CH2Cl2
CF3
Br+
H
H
R
N
O
Me
A
CF3
H
Br
H
N
O
R
Br−
B
H2O
CF3
OH
H
Br
H
NCOMe
R
30-31

The cyclisation of bromhydrins **30-31** to the corresponding epoxides were successfully performed under basic conditions (*t*BuOK in THF at 0°C) (80%). From the chiral bromhydrin **32**, deacetylation occurred and epoxide **24** was obtained in 65% yield and with excellent diastereoselectivity (>98%) (scheme 8). Only one diastereoisomer was detected in the crude product.

Scheme 8

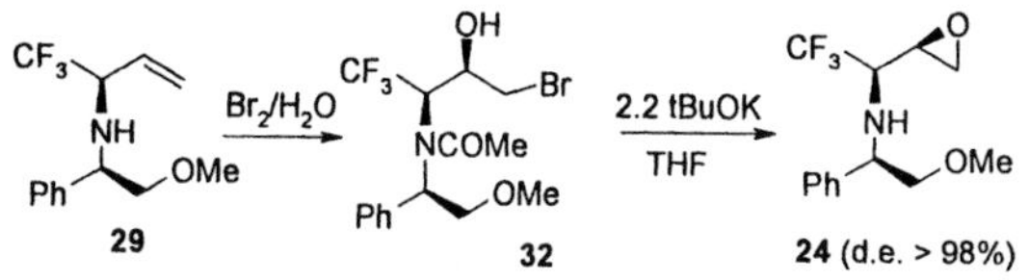

CF3
NH
Ph
OMe
29
Br2/H2O
CF3
OH
Br
NCOMe
Ph
OMe
32
2.2 tBuOK
THF
CF3
O
NH
Ph
OMe
24 (d.e. > 98%)

The relative and absolute configurations of chiral compounds **29**, **32** and **24** have been established after the complete conversion of the epoxide **24** into the azetidinol **33** (scheme 9) which crystallized and allowed X-Ray diffraction analysis. The structure of **33** (*28*) revealed the *syn R,R* configuration between carbons 2 and 3, and hence the *syn R,R* configuration of **24**. Consequently the vinylation of the aldimine **9** provided the allyl amine with creation of an *R* chiral center, confirming thus the hypothesis of a chelation controlled mechanism with an attack opposite to the phenyl group (figure 1).

The ring opening of the epoxide **24** with amines and amino acids is under investigation in order to obtain trifluoromethyl analogs of known aspartyl protease inhibitors.

Scheme 9

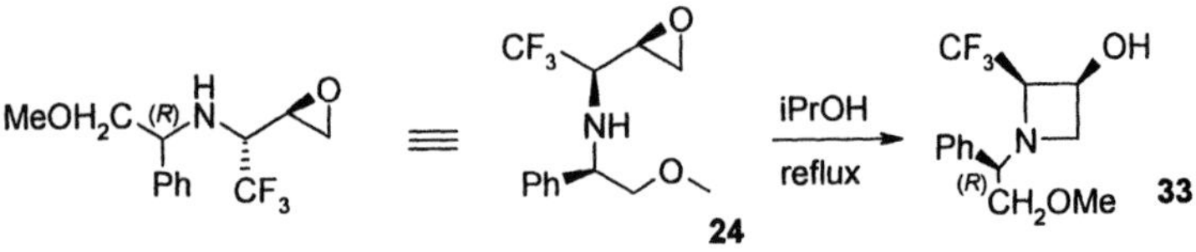

Trifluoromethylated Propargylic Amines.

Propargylic amines, like propargylic alcohols, can serve as important building blocks for organic synthesis (*29*). A classical access to propargylic amines is the addition of acetylides to imines. From trifluoromethyl aldimines, the addition of trimethylsilyl-acetylenes to trifluoromethyl iminium ions, generated from corresponding oxazolidines (*4a*) or aminal (*30*), has been described in the presence of Lewis acids. However to our knowledge the addition of acetylides to non activated trifluoromethylated *N*-aryl and *N*-alkyl aldimines has not been reported so far.

The reaction has been investigated with the *N*-benzylaldimine **1** and the lithium acetylide of trimethylsilylacetylene (1.2 equiv.), generated *in situ* at -78°C with *n*BuLi, in different solvents. The aldimine **1** was very reactive at -78°C in THF and in ether, but led to a mixture of products. In order to increase the selectivity the reaction was conducted in toluene at -78°C and proceeded slowly to give the propargyl amine **34a** (90%) (scheme 10).

Scheme 10

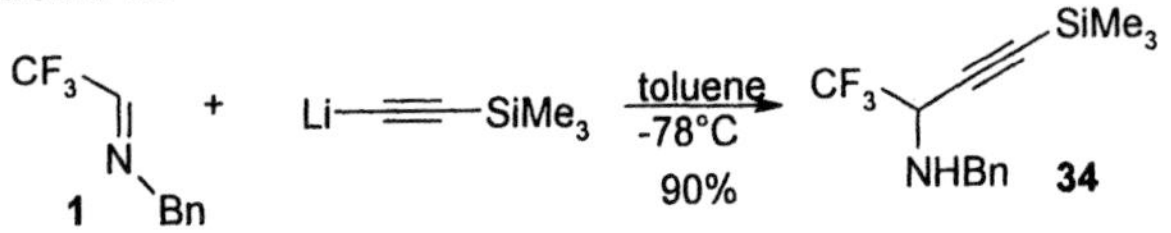

The reaction was extended to other acetylenic reagents starting from the *N*-benzyl and the *N-para*-methoxyphenyl aldimines **1**, **2** (Table 6). In all cases, reactions were clean, and yields in propargyl amines **34-35** were good to excellent. The methyl ether of phenylglycinol in **9** induced an excellent control of stereoselectivity: the addition of acetylides to the aldimine **9** provided the homo chiral amines **36a-c** (d.e.> 98%) which has been assumed to have the *R* configuration as for the allyl amine **23** (figure 4).

Table 6: Addition Reaction of Acetylides to Aldimines

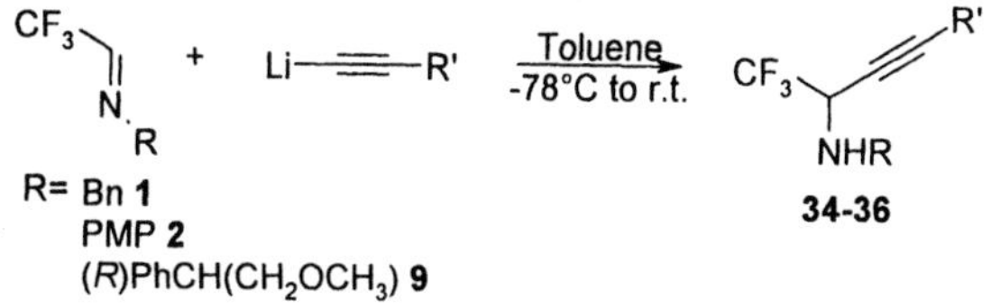

Toluene, -78°C to r.t.

R= Bn **1**
PMP **2**
(*R*)PhCH(CH_2OCH_3) **9**

34-36

Entry	*imine*	*R'*	*Products*	*Yields (%)*
1	**1**	$SiMe_3$	**34a**	61
2	**1**	Bu	**34b**	94
3	**1**	Ph	**34c**	98
4	**2**	$SiMe_3$	**35a**	84
5	**2**	Bu	**35b**	71
6	**2**	Ph	**35c**	78
7	**9**	$SiMe_3$	**36a**	60
8	**9**	Bu	**36b**	77
9	**9**	Ph	**36c**	95

Figure 4

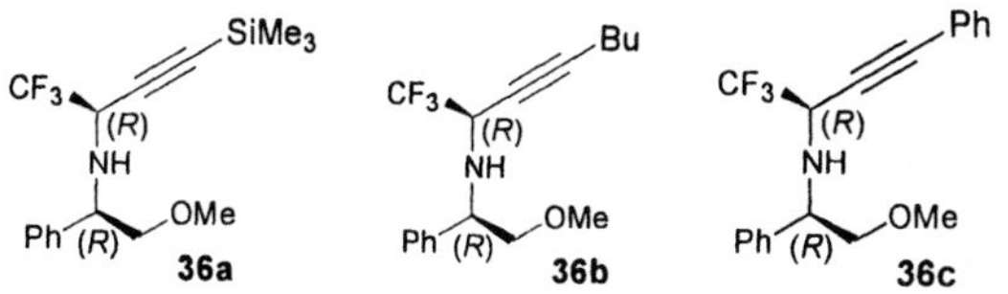

36a **36b** **36c**

Conclusion

A new general method for the preparation of various allyl, homoallyl and propargyl amines has been developed. It involves the easy addition of unsaturated organometallic reagents to simple trifluoromethyl aldimines without recourse to Lewis acids or to a *N*-activation by an electron withdrawing substituent. This method allows the use of various chiral *N*-substituent in the starting aldimine and methylether of phenylglycinol induced a high diastereoselectivity (d.e.>98%). Consequently, unsaturated amines could be produced in high yields and as pure enantiomers. Due to their possible multiple facets of reactivity they are themselves useful building-blocks, in particular for the preparation of fluorinated analogues of relevant bioactive compounds, such as piperidine derivatives and new peptidomimetics

Acknowledgments: We thank Julien Legros, Franck Meyer, Guillaume Magueur, N'guyen Thi Ngoc Tam, Matthieu Colliboeuf for their contribution and Michèle Ourévitch for NMR experiments. The authors thank Rhodia for post doctoral fellowship (F.M.), CNRS and MNERT for PhD grant (T.N.N. and G.M.), and Central Glass company for generous gift of fluoral.

References:

1. Bloch, R. *Chem. Rev.* **1998**, *98*, 1407-1438.
2. (a) Resnati, G.; Soloshonok, V. A., Eds.; Tetrahedron Symposia-in-print No 58; Fluoroorganic Chemistry: Synthetic Challenges and Biomedical Rewards. *Tetrahedron* **1996**, *52*, 1-330. (b) *Organofluorine Compounds in Medicinal Chemistry and Biochemical Applications*, Filler, R.; Kobayashi, Y.; Yagupolski, L. **1993**; Eds Elsevier, Amsterdam. (c) *Biomedical Frontiers of Fluorine Chemistry*, Ojima, I.; McCarthy, J.R.; Welch, J.T. Eds.A.C.S.:Washington, D.C., **1996**. (d) P.N. Edwards *Use of Fluorine in Chemotherapy. In Organofluorine Chemistry: Principles and commercial Applications*, Banks, R.E.; Tatlow, J.C., Plenum Press, **1994**, N.Y., 501.
3. For reviews on asymmetric synthesis of organofluorine compounds see: (a) Iseki, K. *Tetrahedron* **1998**, *54*, 13887-13914. (b) Bravo, P.; Resnati, G. *Tetrahedron: Asymmetry* **1990**, *1*, 661-692.
4. Asymmetric syntheses of perfluoroalkylamines include the following: (a) Lebouvier, N.; Laroche, C.; Huguenot, F.; Brigaud, T. *Tetrahedron Lett.* **2002**, *43*, 2827-2830. (b) Surya Prakash, G.K.; Mandal, M.; Olah, G.A. *Angew. Chem., Int. Ed.* **2001**, *40*, 589-590. (c) Surya Prakash, G.K.; Mandal, M.; Olah, G. A. *Org. Lett.* **2001**, *3*, 2847-2850. (d) Enders, D.; Funabiki, K. *Org. Lett* **2001**, *3*, 1575-1577. (e) Soloshonok, V. A.; Ono, T. *J. Org. Chem.* **1997**, *62*, 3030-3031. (f) Ishii, A.; Higashiyama, K.; Mikami, K. *Synlett* **1997**, *12*, 1381-1382. (g) Wang, Y.; Mosher, H.S. *Tetrahedron Lett.* **1991**, *32*, 987-990. (g) Pirkle, W.H.; Hauske, J. R. *J. Org. Chem.* **1977**, *42*, 2436-2439.
5. a) Kumadaki, I.; Jonoshita, S.; Harada, A.; Omote, M.; Ando, A. *J. Fluorine Chem.*, **1999**, 97, 61-63. b) Joshita, S.; Harada, A.; Omote, M.; Ando, A.; Kumadaki, I. *Chem. Pharm. Bull.*, **1999**, *47*, 656.
6. a) Billard, T.; Langlois, B.R. *J. Org. Chem.* **2002**, *67*, 997-1000. b) Langlois, B.R.; Billard, T. *Synthesis* **2003**, *2*, 185-194.
7. Fuchigami, T.; Ichikawa, S.; Konno, A. *Chem. Lett.* **1992**, 2405.
8. Weygand, F.; Steglich, W.; Pietta, P. *Chem. Ber.* **1967**, *100*, 3841-3849.
9. Funabiki, K.; Nagamori, M.; Matsui, M.; Enders, D. *Synthesis* **2002**, *17*, 2585-2588.

10. Legros, J.; Meyer, F.; Coliboeuf, M; Crousse, B.; Bonnet-Delpon, D.; Bégué, J.P. *J. Org. Chem.* **2003**, *68*, 6444-6446.
11. (a) Ukali; Y. ; Watai, T.; Sumi, T.; Fujisawa, T. *Chem. Lett.* **1991**, 1555-1558. (b) Fukuda, T.; Takehara, A.; Haniu, N.; Iwao M. *Tetrahedron: Asymmetry*, **2000**, *11*, 4083-4091. (c) Yanada, R.; Negoro, N.; Okaniwa, M.; Ibuka, T. *Tetrahedron* **1999**, 55, 13947-13956.
12. (a) O'Hagan, D. *Nat. Prod. Rep.* **1997,** *14,* 637-641. (b) O'Hagan, D. **2000**, *17*, 435-446. (c) Watson, P.S.; Jiang, B.; Scott, B. *Org. Lett.*, **2000**, *2*, 3679-3681.
13. Gille, S.; Ferry, A.; Billard, T.; Langlois, B.R. *J. Org. Chem.* **2003**, *68*, 8932-8935.
14. Patent France 02/04725, 16/04/2002. PCT R 02047G1, déposé le 15 Avril 2003.
15. For reviews on RCM: (a) Armstrong S. K. *J. Chem. Soc., Perkin Trans. 1* **1998**, 371. (b) Fürstner A. *Angew. Chem., Int. Ed. Engl.* **2000**, *39*, 3012-3043. (c) Trnka, T. M.; Grubbs, R. H. *Acc. Chem. Res.* **2001**, *34*, 18-29.
16. a) Prakash, G.K.S.; Mandal, M.; Olah, G.A. *Org. Lett.* **2001**, *3*, 2847-2850. b) Kumadaki, I.; Jonoshita, S.; Harada, A.; Omote, M.; Ando, A. *J. Fluorine Chem.* **1999**, *97*, 61-63. c) Nelson, D.W.; Owens, J.; Hiraldo, D. *J. Org. Chem.* **2001**, *66*, 2572-2582.
17. Fustero, S.; Garcia Soler, J.; Bartolomé, A.; Sanchez Rosello, M. *Org. Lett.* **2003**, *5*, 2707-2710.
18. (a) Leung, D.; Abbenante, G.; Fairlie, D.P. *J. Med. Chem.* **2000**, *43*, 305-341. (b) James, M.N.G. Ed. *Advances in Experimental Medicine and Biology*; 1998; Vol. 436. (c) Takahashi, K. Ed. *Advances in Experimental Medicine and Biology;* 1995; Vol. 362.
19. (a) Babine, R.E.; Bender, S.L. *Chem. Rev.* **1997**, *97*, 1359-1472. (b) Seife, C. *Science* **1997**, *277*, 1602-1603. (c) Craik, M.S. ; Debouck C. *Perspect. Drug Discovery Des.* **1995**, *2*, 1-125. (d) Wlodawer, A.; Erickson, J.W. *Annu. Rev. Biochem.* **1993**, *62*, 543-585.
20. (a) Haque, T.S.; Skillman, A.G.; Lee, C. E.; Habashita, H.; Gluzman, I.Y.; Ewing, T.J.A.; Goldberg, D.E.; Kuntz, I.D.; Ellman, J.A. *J. Med. Chem.* **1999**, *42*, 1428-1440. (b) Mineno, T.; Avery, M.A. *Chemstracts-Organic Chemistry* **2001**, *14*, 518. (c) Dahlgren, A.; Kvarnström, I.; Vrang, L.; Hamelink, E.; Hallberg, A.; Rosenquist, A.; Samuelsson, B. *Bioorg. & Medicinal Chem.* **2003**, *11*, 827-841. (d) Nöteberg, D.; Hamelink, E.; Hultén, J.; Wahlgren, M.; Vrang, L.; Samuelsson, B.; Hallberg, A. *J. Med. Chem.* **2003**, *46*, 734-746.
21. a) Park, B.K.; Kitteringham, N.R. *Drug Metabolism Rev.* **1994**, *26*, 605. b) Irurre, J.; Casas, J.; Messeguer, A. *Biorg. Med. Chem. Lett.* **1993**, *3*, 179-

182. c) Miller, J.A.; Coleman, M.C.; Matthews, R.S. *J. Org. Chem.* **1993**, *58*, 2637-2639.
22. Ojima, I.; Kato, K.; Jameison, F.A.; Conway, J.; Nakahashi, K.; Hagiwara, M.; Miyama, T. *Biorg. Med. Chem. Lett.* **1992**, *2*, 219-222.
23. Patel, D.V.; Rielly-Gauvin, K.; Ryono, D.E.; Free, C.A.; Smith, S.A.; Petrillo Jr, E.W. *J. Med. Chem.* **1993**, *36*, 2431-2447.
24. Sham, H.L. and Schirlin, D. in 2b.
25. a) Ojima, I.; Slater, J. C.; Veith, J.P.; Abouabdellah, A.; Bégué, J.P.; Bernacki, R.J. *Bioorg. & Medicinal Chem. Lett.* **1997**, *7*, 133-138. b) Myers, A. G.; Barbay, J.K.; Zhong, B.*J. Am. Chem. Soc.* **2001**, *123*, 7207-7219. c) Imperiali, B.; Abeles, R.H. *Biochemistry* **1986**, *25*, 3760-3767. d) Sani, M.; Bruché, L.; Chiva, G.; Fustero, S.; Piera, J.; Volontario, A.; Zanda, M. *Angew. Chem. Int. Ed.* **2003**, *42*, 2060-2063. e) Pesenti, C.; Arnone, A.; Bellosta, S.; Bravo, P.; Canavesi, M.; Corradi, E.; Frigerio, M. ; Meille, S.V.; Monetti, M.; Panzeri, W.; Viani, F.; Venturini, R.; Zanda, M. *Tetrahedron,* **2001**, *57*, 6511-6522.
26. (a) Parkes, K.E.B.; Bushnell, D.J.; Crackett, P.H.; Dunsdon, S.J.; Freeman, A.C.; Gunn, M.P.; Hopkins, R.A.; Lambert, R.W.; Martin, J.A.; Merret, J.H.; Redshaw, S.; Spurden, W.C.; Thomas, G.J. *J. Org. Chem.* **1994**, *59*, 3656-3664. (b) Brånalt, J.; Kvarnström, I.; Classon, B.; Samuelsson, B.; Nillroth, U.; Helena Danielson, U.; Karlén, A.; Hallberg, A. *Tetrahedron Lett.* **1997**, *38*, 3483. (c) Beaulieu, P.L.; Wernic, D. *J. Org. Chem.* **1996**, *61*, 3635. (d) Ghosh, A.K.; Hussain, K.A.; Fidanze, S. *J. Org. Chem.* **1997**, *62*, 6080-6082. (e) Badorrey, R.; Cativiela, C.; Díaz-de-Villegas, M.D.; Gálvez, J.A. *Tetrahedron* **2002**, *58*, 341.
27. Epoxidation reaction from non protected amine **5a** with MCPBA was attempted, but afforded mixture of products.
28. The authors have deposited atomic coordinates for this structure **9c** with the Cambridge Crystallographic. Copies of data (CCDC 230713) can be obtained from CCDC via www.ccdc.cam.ac.uk/conts/retrieving.html (or 12 Union Road, Cambridge CB2 1EZ, UK, e-mail: deposit@ccdc.cam.ac.uk).
29. (a) Porco, J.A., Jr; Schoenen, F.J.; Stout, T.J.; Clardy, J.; Schreiber, S.L. *J. Am. Chem. Soc.* **1990**, *112*, 7410-7411. (b) Corriu, R.J.P.; Bolin, G.; Moreau, J.J.E. *Tetrahedron Lett.* **1991**, *32*, 4121-4124. (c) Campi, E.M.; Jackson, W.R.; Nilsson, Y. *Tetrahedron Lett.* **1991**, *32*, 1093-1094. (d) Mandai, T.; Ryoden, K.; Kawada, M.; Tsuji, J. *Tetrahedron Lett.* **1991**, *32*, 7683-7686. (e) Matsuda, I.; Sakakibara, J.; Nagashima, H. *Tetrahedron Lett.* **1991**, *32*, 7431-7434. (f) Clive, D.L.J.; Cole, D.C.; Tao, Y. *J. Org. Chem.* **1994**, *59*, 1396-1406.
30. Xu, Y.; Dolbier, W.R. *J. Org. Chem.* **2000**, *65*, 2134-2137.

Difluoromethylene-Containing Synthons

Chapter 25

Optically Active *gem*-Difluorocyclopropanes: Synthesis and Application for Novel Molecular Materials

Toshiyuki Itoh

Department of Materials Science, Faculty of Engineering, Tottori University, 4–101 Koyama Minami, Tottori 680–8552, Japan

Chemo-enzymatic syntheses of a variety of novel *gem*-difluorocyclopropanes have been accomplished. Important aspects of the synthetic procedure include optical resolution based on lipase-catalyzed ester hydrolysis, difluorocarbene addition, and olefin metatheses. Several types of optically active multi-*gem*-diflurocyclopropane derivatives were prepared.

Introduction

The cyclopropyl group is a structural element present in a wide range of naturally occurring biologically active compounds found in both plants and microorganisms (*1*). Since substitution of two fluorine atoms on the cyclopropane ring is expected to alter both chemical reactivity and biological activity due to the strong electron-withdrawing nature of fluorine, *gem*-difluorocyclopropanes are expected to display unique biological and physical properties (*2,3*). To the best of our knowledge, the first synthesis of *gem*-difluorocyclopropane was reported by Sargeant in 1970 (*4*). Since then, a great deal of attention has been paid to the chemistry and unique properties of *gem*-difluorocyclopropanes (*5-12*). Numerous types of *gem*-difluorocyclopropanes have been synthesized such as biologically active compounds (*6-9*), liquid crystalline compounds (*11*), and even polymeric compounds (*12*). We have been interested in the unique properties of chiral bis- and multi-*gem*-difluorocyclopropanes. For example, we found the shape of 1,6-bis(hydroxymethyl)-2,2,5,5-tetrafluorobicyclopropane (**1**) (*9b*) to be slightly different from that of the simple bicyclopropane (**2**). MO calculation suggested a kinked form of two difluorocyclopropane groups for compound **1**, while no such twisted form was indicated for bicyclopropane **2**. The results of these calculation were confirmed by results of CD spectroscopic analysis of optically active **2** in that a large CD spectral change on the Cotton effect was observed (*9b*). The computational chemistry suggested a highly helical shape for petakis- *gem*-difluorocyclopropane **3**, as shown in Figure 1.

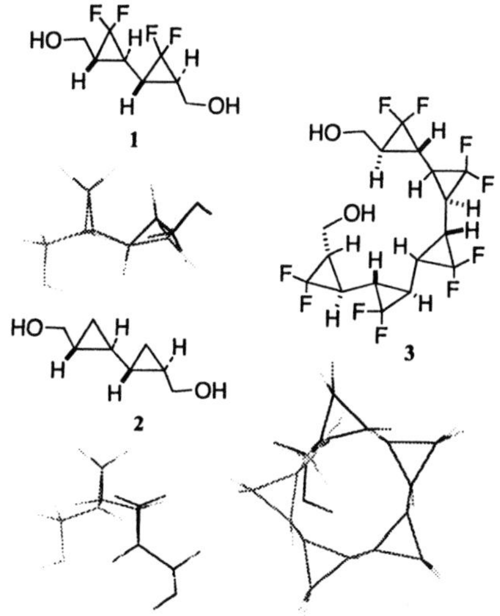

Figure 1. Result of MO(PM3) calculation of bis- and pentakis-gem-difluorocyclopropanes

Optically active multi-*gem*-difluorocyclopropanes are challenging targets for synthetic organic chemists. However, little attention has been paid to the chemistry of this material and when we undertook the present project, only two methods have been reported for the synthesis of optically active *gem*-difluoro- cycloropanes, both from the Taguchi group (*7,8*). The first synthesis of optically active *gem*-difluorocyclopropane was accomplished in 1994. Diastereoselective 1,4-addition of the enolate **4** with 4-bromodifluorobutanoate and subsequent oxidative cyclopropanation gave *gem*-difluorocyclopropane **5** with high diastereoselecivity (Eq. 1) (*7*). The second synthesis was based on the addition of difluorocarbene to the optically active olefin **6** and subsequent separation of the resulting two diastereo-isomers **7** using HPLC (Eq. 2)(*7*).

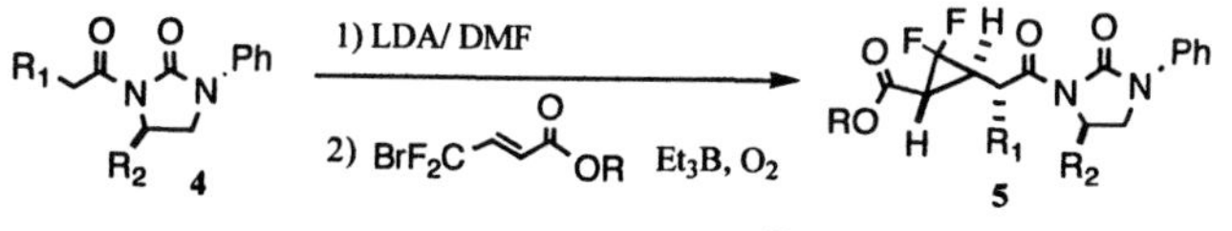

1) LDA/ DMF; 2) BrF_2C–CH=CH–C(O)OR, Et_3B, O_2 (1)

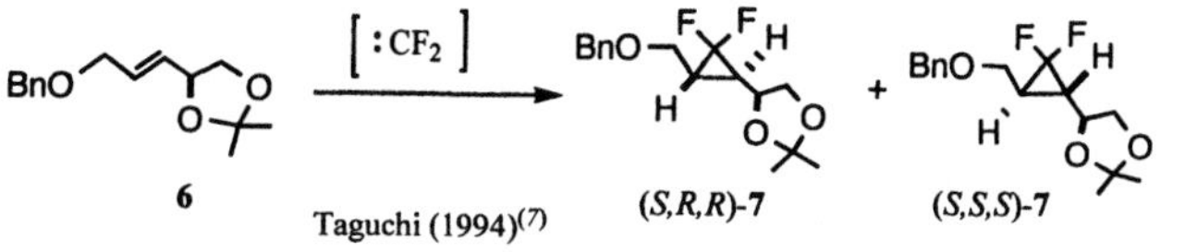

[:CF_2] Taguchi (1994)[7] (S,R,R)-7 + (S,S,S)-7 (2)

These two methods are excellent but are difficult to apply for large-scale preparation. For this reason, we developed a synthetic methodology for preparing optically active *gem*-difluorocyclopropanes using lipase technology

(*9,11b*). We review here the details of our synthesis project of optically active multi-*gem*-difluorocyclopropanes.

1. Preparation of optically active *gem*-difluorocyclopropanes

The lipase-catalyzed reaction protocol was used as the key process of our asymmetric synthesis for chiral *gem*-difluorocyclopropanes (*9,11b*). The value of an enzymatic reaction in organic synthesis is now well respected for its broad applicability and environmentally friendly nature. In particular, lipase-catalyzed enantioselective acetylation or deacetylation is known to be a useful means of preparing optically active alcohols (*13*).

(*cis*)-8 —Lipase QL, buffer (pH7.2), 35°C→ (*cis*)-9 >99%ee (3)

(*trans*)-(±)-8 —Lipase SL or Lipase QL, pH7.2, 35°C, 1 h, E* >11→ (*trans*)-(+)-9 (*S,S*) + (*trans*)-(-)-8 (*R,R*) >99%ee (4)

Initially we tested lipase-catalyzed enantioselective hydrolysis of diacetate (*cis*)-**8**. Twenty-eight commercially available lipases were screened for their activity and lipase QL (Meito) from *Alcaligenes* sp. provided the corresponding monoacetate (*cis*)-**9** in the highest enantiomeric excess (Eq 3) (*9a,9f*). The diacetate of (*trans*)-**8** is not a prochiral but a racemic form, so that optical resolution of (±)-(*trans*)-**8** was performed using the lipase-catalyzed reaction (Eq 4). The enzymatic reactions were tentatively evaluated by the "E* value" which was calculated by the results of enantiomeric excesses of the monoacetate (*trans*)-(*S,S*)-(+)-**9** and unreacted (*trans*)-(*R,R*)-(-)-**8** according to the same method of the "E value" (*14*), although it was the combined results of two enantioselective enzymatic reactions, the initial hydrolysis of (*trans*)-(±)-**8** and subsequent hydrolysis of the corresponding monoacetate **9**. In this reaction, the best result was recorded when *trans*)-(±)-**8** was reacted with lipase SL (*Pseudomonas cepacia* SL-25, Meito), and the diacetate (*trans*)-(*R,R*)-(-)-**8** remaining was obtained with >99% ee (Eq. 4) (*9a,9f*).

Chiral (*trans,trans*)-1,6-bishydroxylmethyl-2,2,5,5-tetrafluorobicyclo- propane (**1**) is an inviting target for synthetic organic chemists because computational chemistry suggests a kinked form of two *gem*-difluorocyclopropane groups for **1**. However, no example of the synthesis of even the racemic form of **1** had been reported when we undertook this project (*9b*). The dibenzylether of bisdifluorocyclopropane *(trans,trans)*-**1** was directly synthesized in 66% overall yield from (*E,E*)-1,8-dibenzyloxy-2,4-hexadiene using 5 eq. of difluorocarbene, produced by the thermolysis of sodium chlorodifluoroacetate. Subsequent hydrogenolysis of the benzyl group produced **1** (*9b,9f*). In the first step of difluorocyclopropanation, the benzyl protecting group was essential in order to achieve difluorocyclopropanation in acceptable yield (*9b*). The optical resolution of the diacetate, (*trans,trans*)-**10**, which is a mixture of the *dl* and *meso* forms, was very successfully accomplished by the lipase SL-catalyzed reaction. We found that the dibenzyl ether of *meso*-**1** was preferentially crystallized from hexane. The *dl*-rich dibenzylether (*dl*:*meso* = 7:3) was obtained by removing the *meso*-isomer by a single recrystallization from hexane,[9f] and this mixture was converted to the corresponding *dl*-rich-diacetate **10** and subjected to lipase-catalyzed hydrolysis. The lipase SL-catalyzed hydrolysis of the *dl*-rich-diacetate **10** (*dl*:*meso* = 7:3) gave the diol (*S,R,R,S*)-(+)-**1** in 30% yield with 99% ee, and the unreacted diacetate (*R,S,S,R*)-(-)-**10** in 24% yield with >99% ee. The *meso*-isomer was obtained as the monoacetate **11** in 30% yield. We thus succeeded in preparing both enantiomers of bis-*gem*-difluorocyclopropane **1** with sufficient optical purity (Eq. 5). This was the first example of the synthesis of a bis-*gem*-difluorocyclopropane in optically active form.

dl

meso

10

dl : *meso* (7:3)

Lipase SL

pH7.2

E* > 210

(*R,S,S,R*)-**11** Y= 24% >99%ee

(*S,R,R,S*)-**1** Y= 30% 99%ee

meso-**10** Y=30% (100%)

(5)

Recently two groups also reported efficient preparation of optically active *gem*-difluorocyclopropanes using lipase-catalyzed reaction as a key technology. Kirihara reported that enantioselective monohydrolysis of diacetate **12** gave the monoacetate **13** with excellent optical purity and this was converted to the novel cyclopropane amino acid **14** (Eq. 6)(*10*). Meijere and colleagues accomplished the synthesis of a unique spiro type *gem*-difluorcyclopropane **17** using lipase technology. The chiral spirobicyclopropane **16** was prepared by the optical resolution of lipase-catalyzed reaction and subsequent conversion by the addition of difluorocarbene to afford *gem*-difluorocyclopropane **17**. This was used as the core molecule for the liquid crystal compound **18** (Eq.7) (*11a*).

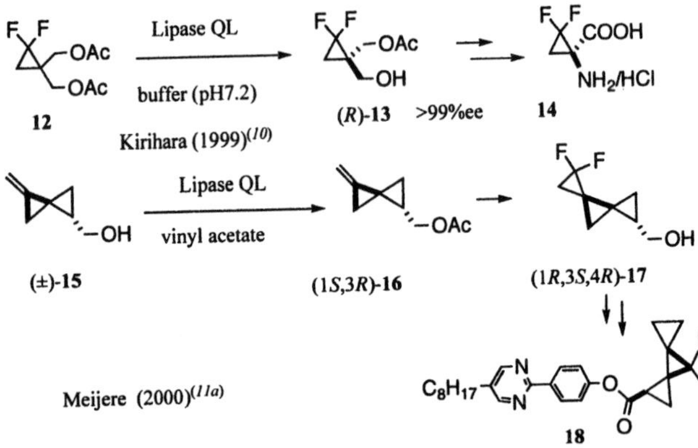

12 → Lipase QL, buffer (pH7.2) → (*R*)-13 >99%ee → 14 (6)

Kirihara (1999)[10]

(±)-15 → Lipase QL, vinyl acetate → (1*S*,3*R*)-16 → (1*R*,3*S*,4*R*)-17 → 18 (7)

Meijere (2000)[11a]

2. Synthesis of multi-*gem*-difluorocyclopropanes

Retrosynthetic analysis suggests two pathways to access the target multi-*gem*-difluorocyclopropanes. It is well documented that various functional groups can be derived from the tributylstannyl group with complete retention of configuration through metal exchange reactions (*15*). Therefore we postulated that tetrakis-*gem*-difluorocylopropane **19** may be accessible through cyclopropylstannane **20** *via* a palladium-catalyzed homo-coupling reaction (*16,17*). We initially investigated the homo-coupling strategy. Bis-*gem*-difluoro-cyclopropylstannane **20** was synthesized as follows: 3-tributylstannyl-2-propenol (**21**) was subjected to PDC oxidation which led to the corresponding aldehyde. Subsequent Honor-Wadworth-Emmons reaction (*18*) gave **22** which was converted to the diene **23** through DIBAL reduction and benzyl protection of the hydroxyl group. Diene **23** was treated with diflourocarbene to give the desired bisdiflurocyclopropylstannane **20** in 58% yield. Unfortunately, we were unable to find a catalyst to effect the homo-coupling reaction of **20** (Scheme 1).

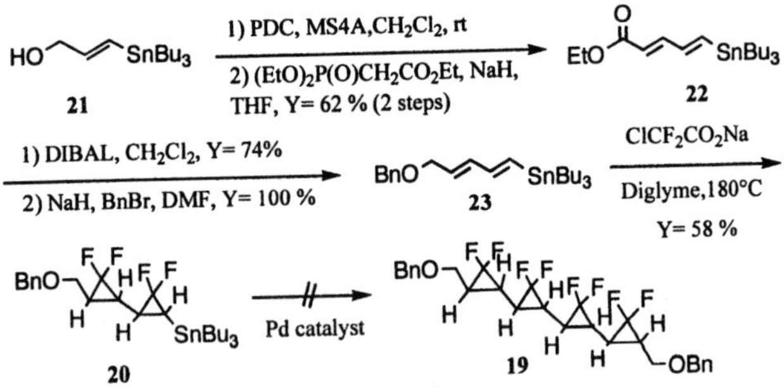

21 → 1) PDC, MS4A, CH_2Cl_2, rt; 2) $(EtO)_2P(O)CH_2CO_2Et$, NaH, THF, Y= 62 % (2 steps) → **22**

1) DIBAL, CH_2Cl_2, Y= 74%; 2) NaH, BnBr, DMF, Y= 100 % → **23** → $ClCF_2CO_2Na$, Diglyme, 180°C, Y= 58 %

20 → Pd catalyst (no reaction) → **19**

Scheme 1

Therefore we examined a second strategy for the synthesis of multi-*gem*-diflurocyclopropanes. This was based on a ruthenium complex catalyzed-olefin methathesis reaction protocol (Scheme 2). Olefin metathesis has been extensively studied and procedures leading to remarkably broad applicability have been developed (*19*). We first investigated the reaction of vinylcyclopropane **24** as a model compound using the Grubbs catalyst A (*19*). However, the coupling product was obtained in only 15 % yield in the presence of 15 mol % of the catalyst. Nonetheless, the newly formed olefin had perfect (E)-stereochemistry. We tested four solvent systems: dichloromethane, toluene, benzene and THF, and the desired product **25** was obtained only when the reaction was carried out in dichloromethane at room temperature. We then used a total of 1.0 equivalent of the catalyst by adding 20 mol% of the dichloromethane solution dropwise five times at 12 h intervals. This increased the chemical yield and we obtained the desired coupling product **25** in 35% yield. It is assumed that decomposition of a metallacyclobutane intermediate in this reaction releases ethylene gas and regenerates the ruthenium carbene complex to complete the catalytic cycle. In the present reaction, however, the retro decomposition of metallacyclobutene to vinylcyclopropane also seemed to occur due to the strong electron withdrawing nature of difluoromethylene moiety and this stopped the catalytic cycle. Though, another explanation of the poor results may be possible (*9c*). Second generation ruthenium-based olefin metathesis catalysts have been reported (*20*).

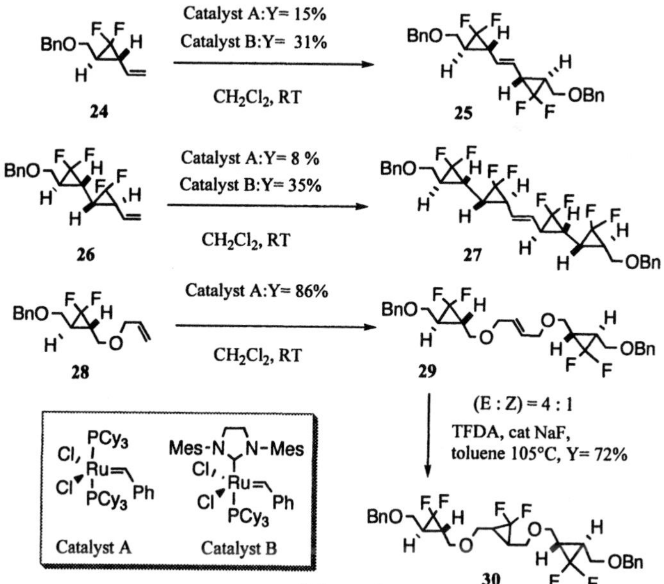

Scheme 2

We found that use of catalyst B indeed gave better results than A and the metathesis product could be obtained in 31 % yield even when 15 mol% of the catalyst was used. The most dramatic increase in chemical yield was recorded when bisdiflurocyclopropane **26** was subjected to the metathesis reaction. The desired tetrakisdifluorocylopropane derivative **27** was obtained in 35 % yield using catalyst B, while the chemical yield of the metathesis product was only 8% when a total of 1.0 equivalent of catalyst A was used (five additions of 20 mol% of the catalyst at 12 h intervals). In contrast, the olefin metathesis reaction of allylic ether **28** proceeded smoothly to give the coupling product **29** in excellent yield of 86 % using 15 mol% of the commercial catalyst A (*9c,9d*). Since compound **29** possess an olefinic unit between difluorocyclopropane moieties, we were able to add one more difluorocyclopropane group to the structure. Thus, trisdifluorocyclopropane (*trans,trans,trans*)-**30** was synthesized in 54% yield from **29** using excess amounts of difluorocarbene (50 eq.), produced by the thermolysis of sodium chlorodifluoroacetate (8 M in a diglyme solution) (*9c*). Recently, trimethylsilyl fluorosulfonyldifluoroacetate (TFDA) was developed as a good difluorocarbene source by the Dolbier group (*5i*). This reagent was applicable to our compound, and the desired tris-*gem*-difluorocyclopropane **30** was obtained in 72% yield using 10 eq. of the carbene source. Dolbier's procedure thus was found to be useful for preparation of polydifluorocyclopropanes possessing an odd number of difluorocyclopropane moieties (Scheme 2) (*9d*). Eight types of multi-*gem*-difluorinated cyclopropane derivatives have now been synthesized via this olefin metathesis reaction methodology (*9d*). As noted above, the products obtained by the olefin methathesis reaction protocol possess olefinic parts between the difluorocyclopropane moieties. This is an important advantage since this allows the addition of another difluorocyclopropane group or of a variety of other functional groups.

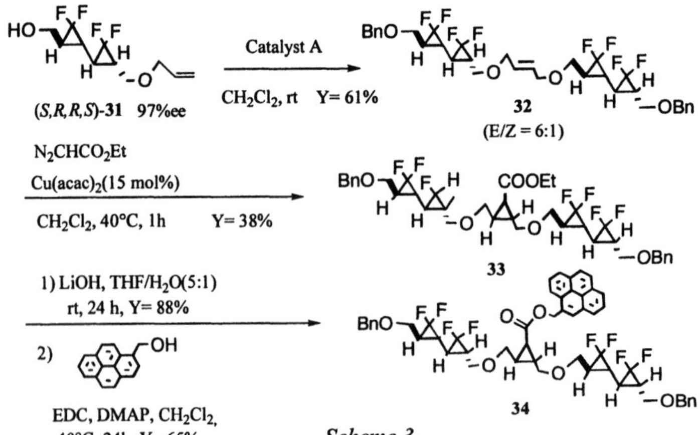

(*S,R,R,S*)-**31** 97%ee

Catalyst A

CH_2Cl_2, rt Y= 61%

32

(E/Z = 6:1)

N_2CHCO_2Et

$Cu(acac)_2$(15 mol%)

CH_2Cl_2, 40°C, 1h Y= 38%

33

1) LiOH, THF/H_2O(5:1)

rt, 24 h, Y= 88%

2)

EDC, DMAP, CH_2Cl_2,

40°C, 24h ,Y= 65%

34

Scheme 3

We are considering various applications of our multi-*gem*-difluoro-cyclopropanes. For example, tetrakis-*gem*-difluorocyclopropane **32** was easily prepared by the homo coupling reaction of (*S,R,R,S*)-**31** in optically active form starting from 97% ee of (*S,R,R,S*)-**1** in 47% overall yield (three steps) using the olefin methathesis protocol. This was treated with ethyl diazoacetate in the presence of 15 mol% of copper acetoacetate in CH_2Cl_2 at 40°C to give pentakis-cyclopropane **33** in 38% yield (*21*). Hydrolysis and subsequent esterification with pyrenylmethanol using EDC in the presence of DMAP afforded pyrenyl ester **34** in 65% yield (Scheme 3). Since compound **34** is actually a mixture of diastereomers, we are now attempting to purify each isomer. We are hoping that the compound will become a unique sensitive marking agent for DNA.

Conclusions

We have carried out chemo-enzymatic syntheses of optically active multi-*gem*-difluorocyclopropane using lipase-catalyzed ester hydrolysis, difluorocarbene addition, and olefin methatheses as key reactions. We first accomplished the synthesis of (*trans*)-**8** and (*trans,trans*)-**1** in optically active form. These compounds were used as the core molecules for synthesizing novel multi-*gem*-difluorocyclopropane derivatives, **30**, **21**, **34**. We are now able to prepare chiral *gem*-difluorocyclopropanes on a multi-gram scale. Since our building blocks have hydroxymethyl groups at each terminus, it should be emphasized that they are extremely versatile intermediates. We thus plan to synthesize various types of multi-*gem*-difluorocyclopropanes such as amino acids, carboxylic acids, polymeric compounds and hybrid types of difluorocyclopropanes using these versatile building blocks we have prepared.

Acknowledgments

This work was supported by a Grant-in-Aid for Scientific Research on Priority Areas (No. 283, "Innovative Synthetic Reactions") from The Ministry of Education, Science, Sports, Culture and Technology of Japan. The author is grateful to Professor Kenji Uneyama of Okayama University for the helpful discussions throughout this work. The author also gratefully acknowledges and his students, in particular, Dr. Koichi Mitsukura and Miss Nanane Ishida who helped to complete this study.

References

1. For reviews see: a) "Cyclopropanes and Related Rings", *Chem. Rev.* **2003**, *103*, 931-1648. (b) Tozer, M. J.; Herpin, T. F. *Tetrahedron,* **1996**, *52*, 8619-

8683. b) Wong, H. N. C.; Hon, M-Y.; Tse, C-H.; Yip, Y-C. *Chem. Rev.* **1989**, *89*, 165-198.

2. For reviews, see. (a) Welch, J. T. *Tetrahedron*, **1987**, *43*, 3123-3197. (b) Resnati, G. *Tetrahedron*,**1993**, *49*, 9385-9445. (c) "Enantiocontrolled Synthesis of Fluororganic Compounds: Stereochemical Challenges and Biomedicinal Targets", Ed. Soloshonok, V. A., Wiley, Chichester, UK (1999).
3. (a) Dolbier, Jr. W. R.; Battiste, M. A. *Chem. Rev.* **2003**, *103*, 1071-1098. (b) Fedory'nski, M. *Chem. Rev.* **2003**, *103*, 1099-1132. and references cited therein.
4. Sargeant, P. B. *J. Org. Chem.* **1970**, *35*, 678-682.
5. Unique physical and chemical properties have been reported for *gem*-difluorocyclopropanes. (a) Seyferth, D.; Hopper, S. P. *J. Org. Chem.* **1972**, *37*, 4070-4075. (b) Schlosser, M.; Chau, L-V.; Bojana, S. *Helv. Chim. Acta*, **1975**, *58*, 2575-2585. (c) Schlosser, M.; Spahi'c; B.; Chau, L-V. *Helv. Chim. Acta*, **1975**, *58*, 2586-2604. (d) Schlosser, M.; Bessard, Y. *Tetrahedron*, **1990**, *46*, 5222-5229. (e) Morikawa, T.; Uejima. M.; Kobayashi, Y. *Chem. Lett.* **1988**, 1407-1710. (f) Bessard, Y.; Müller, U.; Schlosser, M. *Tetrahedron*, **1990**, *46*, 5213-5221. (g) Schlosser, M.; Bessard, Y. *Tetrahedron*, **1990**, *46*, 5222-5229. (h) Tian, F.; Battiste, M. A.; Dolbier, Jr. W. R. *Org. Lett.* **1999**, *1*, 193-195. (i) Tian, F.; Kruger, V.; Bautista, O.; Duan, J-X.; Li, A-R.; Dolbier, Jr. W. R.; Chen, Q-Y. *Org. Lett.* **2000**, *2*, 563-564.
6. Biological activities were reported for difluorocyclopropane derivatives. (a) Boger, D. L.; Jenkins, T. J. *J. Am. Chem. Soc.* **1996**, *118*, 8860-8870. (b) Koizumi, N.; Takrgawa, S.; Mieda, M.; Shibata, K. *Chem. Pharm. Bull.* **1996**, *44*, 2162-2164. (c) Taguchi, T.; Kurishita, M.; Shibuya, A.; Aso, K. *Tetrahedron*, **1997**, *53*, 9497-9508. (d) Pechacek, J. T.; Bargar, T. M.; Sabol, M. R. *Bioorg. Med. Chem. Lett.* **1997**, *7*, 2665-2668. (e) Csuk, R.; Eversmann, L. *Tetrahedron*, **1998**, *54*, 6445-6459. (f) Csuk, R.; Thiede, G. *Tetrahedron,* **1999**, *55*, 739-750.
7. Taguchi, T.; Shibuya, A.; Sasaki, H.; Endo, J-i.; Morikawa, T. ; Shiro, M. *Tetrahedron: Asymmetry*, **1994**, *5*, 1423-1426.
8. (a) Shibuya, A.; Kurishita, M.; Ago, C.; Taguchi, T. *Tetrahedron*, **1996**, *52*, 271-278. (b) Taguchi, T.; Kurishita, M.; Shibuya, A.; Aso, K. *Tetrahedron*, **1997**, *53*, 9497-9508. (c) Shibuya, A.; Sato, A.; Taguchi, T. *Bioorganic & Medicinal Chem. Lett.* **1998**, *8*, 1979-1984. (d) Shibuya, A.; Okada, M.; Nakamura, Y.; Kibashi, M.; Horikawa, H.; Taguchi, T. *Tetrahedron*, **1999**, *55*, 10325-10340.
9. (a) Itoh, T.; Mitsukura, K.; Furutani, M. *Chem. Lett.* **1998**, 903-904. (b) Mitsukura, K.; Korekiyo, S.; Itoh, T. *Tetrahedron Lett.* **1999**, *40*, 5739-5742. (c) Itoh, T.; Mitsukura, K.; Ishida, N.; Uneyama, K. *Org. Lett.* **2000**, *2*, 1431-1434. (d) Itoh, T. *J. Synth. Org. Chem. Jpn.* **2000**, *58*, 316-326. (e)

Itoh, T.; Ishida, N.; Mitsukura, K.; Uneyama, K. *J. Fluorine Chem.* **2001**, *112*, 63-69. (f) Itoh, T.; Ishida, N.; Mitsukura, K.; Hayase, S.; Ohashi, K. *J. Fluorine Chem.* **2004**, 125, 775-783.

10. (a) Kirihara, M.; Takuwa, T.; Kawasaki, M.; Kakuda, H.; Hirokami, S-i.; Takahata, H. *Chem. Lett.* **1999**, 405-406.
11. Liquid crystals that have *gem*-difluorcyclopropane moiety. (a) Miyazawa, K.; Yufit, D. S.; Howard, J. A. K.; de Meijere, A. *Eur. J. Org. Chem.* **2000**, 4109-4117. (b) Itoh, T.; Ishida, N.; Ohashi, M.; Asep, R.; Nohira, H. *Chem. Lett.* **2003**, *32*, 494-495.
12. Polymer that possesses *gem*-diflurocyclopropane moiety. (a) Reisinger, J. J.; Hillmyer, M. A. *Prog. Polym. Sci.* **2002**, *27*, 971-1005. (b) Ren, Y.; Lodge, T. P.; Hillmyer, M. A. *J. Am. Chem. Soc.* **1998**, *120*, 6830-6831. (c) Ren, Y.; Lodge, T. P.; Hillmyer, M. A. *Macromolecules*, **2000**, *33*, 866-876.
13. For reviews, see: (a) Wong, C. H.; Whitesides, G. M. "Enzymes in Synthetic Organic Chemistry", Tetrahedron Organic Chemistry Series, Vol. 12, Eds, Baldwin J. E.; Magnus, P. D. Pergamon, (1994). (b) Theil, F. *Chem. Rev.* **1995**, *95*, 2203-2227. (c) Itoh, T.; Takagi, Y.; Tsukube, H. *J. Molecular Catalysis B: Enzymatic*, **1997**, *3*, 259-270. (d) Theil, F. *Tetrahedron*, **2000**, *56*, 2905-2919.
14. Chen, C. -S.; Fujimoto, Y.; Girdauskas, G.; Sih, C. J. *J. Am. Chem. Soc.* **1982**, *102*, 7294-7299.
15. For a review see: Nozaki, H. "Organotin Chemistry", Organometallics in Synthesis, A Manual, Ed. Schlosser, M. John Wiley & Sons, Chichester, 535-578 (1994)
16. McDermott, T. S.; Mortlock, A. A.; Heathcock, C. H. *J. Org. Chem.* **1996**, *61*, 700-709.
17. Itoh, T.; Emoto, S.; Kondo, M. *Tetrahedron*, **1998**, *54*, 5225-5232.
18. Wodsworth, Jr. W. S., *Organic Reaction*, **1977**, *25*, 73-253.
19. For reviews see: (a) Grubbs, R. H.; Chang, S. *Tetrahedron*, **1998**, *54*, 4413-4450. (c) Ulman, M.; Grubbs, R. H. *J. Org. Chem.* **1999**, *64*, 7207-7209.
20. Scholl, M.; Ding, S.; Lee, C. W.; Grubbs, R. H. *Org. Lett.* **1999**, *1*, 953-956.
21. For a review see. Doyle, M. P. *Chem. Rev.* **1986**, *86*, 919-939.

Chapter 26

General Methods for the Synthesis of Difluoromethylphosphonates

Thierry P. Lequeux

Laboratoire de Chimie Moléculaire et Thio-organique, UMR CNRS 6507, University of Caen, ENSICAEN, 6 Boulevard du Maréchal Juin, F14050 Caen, France

Due to the limitation of methylenephosphonates as biological models of phosphates, difluoromethylenephosphonates have been, and are still a center of interest as stable isopolar phosphate mimics. However, the major drawback to prepare potential enzyme inhibitors lies in the difficulties to introduce this function onto complex molecules. This chapter describes the most common methods reported in the literature to prepare such structures.

1. Fluorophosphonates as Isoacidic Analogs of Phosphates

It has been shown that the replacement of the bridging oxygen atom of a phosphate by a difluoromethylene adequately maintain the acidity of the phosphonic moiety (*1*). Moreover, the difluoromethylenephosphonate group appears to approximate most closely the geometry of the phosphate group (*2*). In the alkyl- or benzyl-phosphonates series, the introduction of a single fluorine atom is often enough to restore the iso-acidic character (*3*, *4*, *5*, *6*). It is generally assumed that the phosphate unit will be fully ionized on protein binding, and difluoromethylenephosphonates has emerged as excellent pyrophosphate and phosphate surrogates.

2. Synthesis of Difluoromethylenephosphonates

Up to date, numerous efforts have been, and are still focus on the development of general synthetic methodologies to prepare a variety of structures containing the difluoromethylenephosphonate function. The most common routes presented here are: (a) the nucleophilic or free radical introduction of the difluoromethylenephosphonate unit; (b) the elaboration of this functional group from *gem*-difluoroalkenes; (c) the introduction of fluorine atoms by fluorination of (keto-) phosphonates; (d) the elaboration of complex structures from building-blocks already fluorophosphorylated (figure 1).

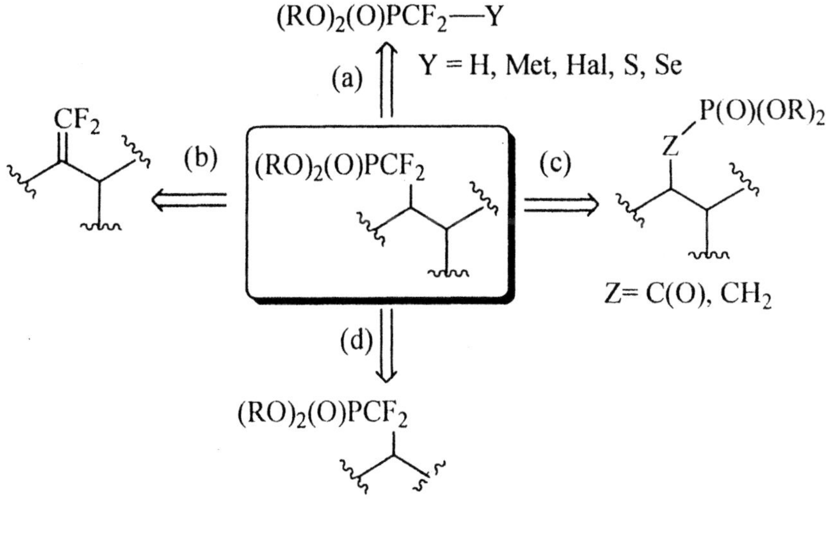

$(RO)_2(O)PCF_2$—Y

(a) Y = H, Met, Hal, S, Se

CF_2 (b) $(RO)_2(O)PCF_2$ (c) $P(O)(OR)_2$ Z

Z= C(O), CH_2

(d)

$(RO)_2(O)PCF_2$

Figure 1.

2.1 The Direct Introduction of the Difluoromethylenephosphonate Unit

Kondo's and Burton's groups have originally developed the direct introduction of the difluoromethylenephosphonate function from *O,O*-diethyl bromodifluoro- and difluoro-methylphosphonates (*7*). A variety of metalated derivatives were prepared (metal: Li, Zn, Cd, Mg) by their deprotonation with LDA (*8*), reduction with metals (*9, 10*), or by halogen-metal exchange reaction (*11*). Resulting organometallic reagents were used to introduce the function by nucleophilic addition or substitution, and by coupling or free radical reactions.

2.1.1 Nucleophilic Addition

The *O,O*-diethyl phosphonodifluoromethyllithium is thermally unstable but it can be trapped with various electrophiles (*12*). On the contrary the organozinc and -cadmium analogous reagents are stable at 20°C, and can be added onto strong electrophiles or reacted with halides in the presence of catalyst. Kondo and co-workers first reported that the lithium reagent reacts with aldehydes and ketones to give high yields of functionalized hydroxy-phosphonates (*8*). However, phosphorus transfer can occur during the addition. This can be limited by a careful control of the temperature (*13*), or by working in the presence of cerium(III) chloride (*14*). The organomagnesium reagent allows also the preparation of alcohols in good yields (*11*). The major disadvantage of these methods is the use of regulated HCFC or CFC as starting materials. In our group, a freon-free source of the lithium reagent has been developed as alternative method (scheme 1). By fluorination of a sulfide and nucleophilic displacement of sulfur atom, the lithiated carbanion can be formed and trapped with various electrophiles (*15*).

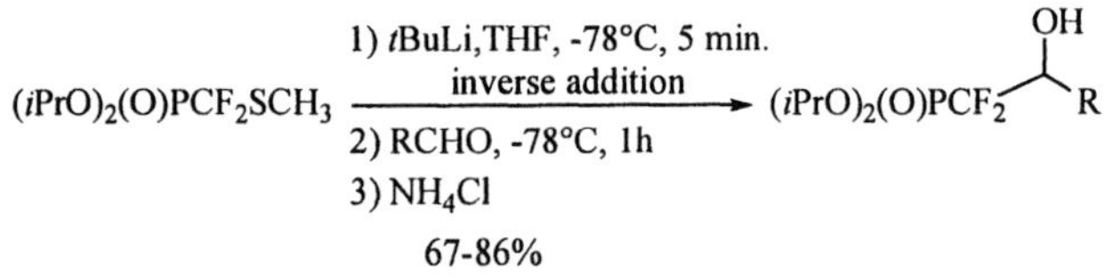

$$(iPrO)_2(O)PCF_2SCH_3 \xrightarrow[\text{2) RCHO, -78°C, 1h; 3) } NH_4Cl]{\text{1) } t\text{BuLi, THF, -78°C, 5 min. inverse addition}} (iPrO)_2(O)PCF_2CH(OH)R$$

67-86%

Scheme 1.

Resulting β–hydroxyphosphonates are generally used in the synthesis of "primary" difluoromethylenephosphonates by their deoxygenation (*16*), as described for the preparation of nucleotides (*17*), or peptide analogs (*18*).

2.1.2 Nucleophilic Substitution

The displacement of a leaving group by the lithium reagent has been reviewed (*19*). The alkylation of the carbanion with alkyl halides appears to be much more difficult than originally reported and moderate yields were observed (*3, 5, 16a, 20, 21, 22*). The displacement of a triflate is more efficient to prepare a variety of primary phosphonates (scheme 2). The reaction is spontaneous and modified glucofuranosides and pyranosides were obtained in good yields (*23, 24, 25*). This methodology was successfully applied to the synthesis of analogs of phosphoserine (*26*), phospholipids (*27*), and nucleotides (*28*).

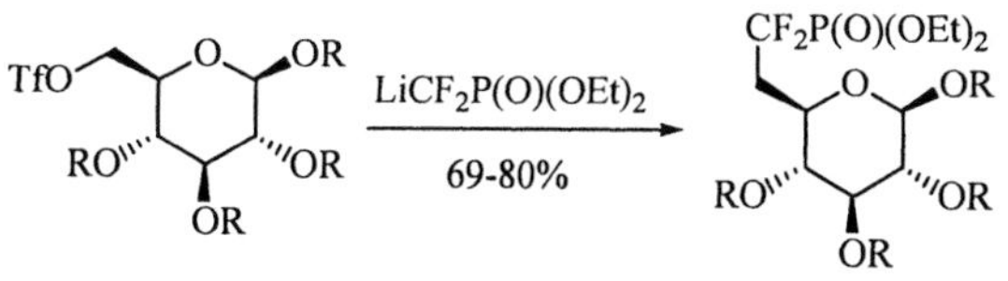

TfO, OR, RO, OR, OR → $LiCF_2P(O)(OEt)_2$, 69-80% → $CF_2P(O)(OEt)_2$, OR, RO, OR, OR

Scheme 2.

The synthesis of secondary difluoromethylenephosphonates by substitution of a leaving group was described recently (scheme 3). It has been shown that the displacement of a secondary allylic phosphate group proceeds through the formation of a π-allyl intermediate, and afforded a regioisomeric mixture of products in good yields (*29*).

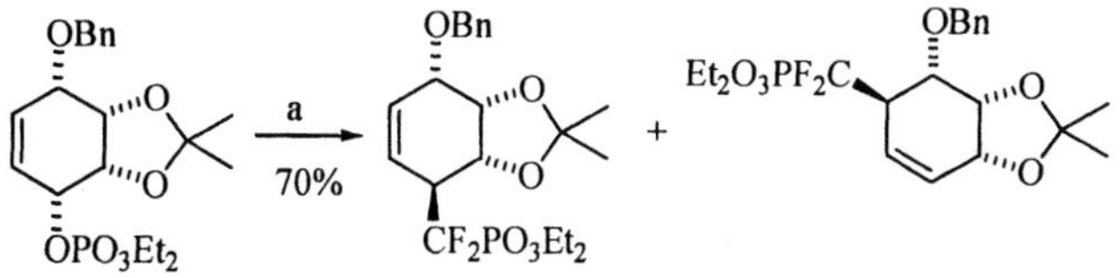

OBn, O, O, OPO_3Et_2 → a, 70% → OBn, O, O, $CF_2PO_3Et_2$ + $Et_2O_3PF_2C$, OBn, O, O

17 : 83

Scheme 3. (a) $(EtO)_2(O)PCF_2ZnBr$, CuBr

2.1.3 Nucleophilic Conjugate Addition

By contrast to the nucleophilic substitution, the Michael addition was less studied to prepare secondary difluorophosphonates. The conjugate addition of the lithium reagent, originally limited to aromatic nitroalkenes (*30*), has been extended by us, to aliphatic nitroalkenes or vinyl sulfones when addition is performed in the presence of cerium(III) chloride (scheme 4) (*31*, *32*, *33*).

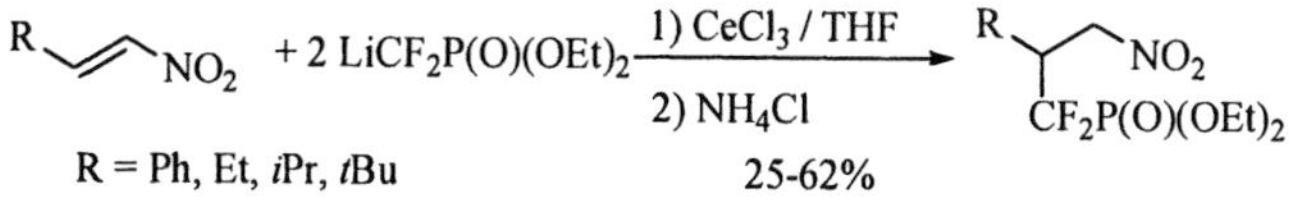

R, NO_2 + 2 $LiCF_2P(O)(OEt)_2$ → 1) $CeCl_3$ / THF, 2) NH_4Cl, 25-62% → R, NO_2, $CF_2P(O)(OEt)_2$

R = Ph, Et, *i*Pr, *t*Bu

Scheme 4.

The conjugate addition onto esters is more difficult to control and a competitive addition can occurred (*34*). However, when hindered esters were involved (scheme 5), functionalized phosphonates can be obtain in high yields (*35*).

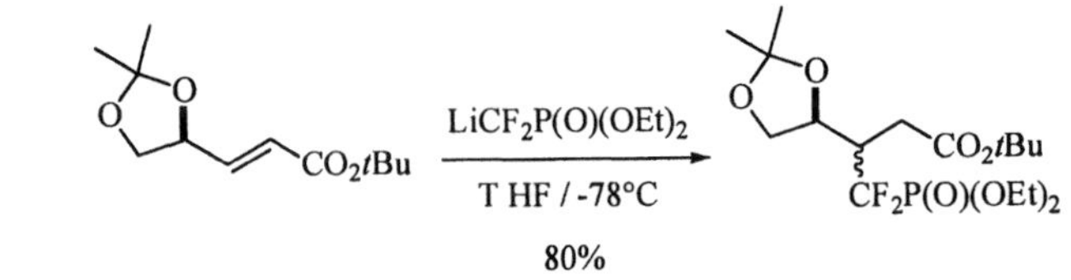

$$\text{(acetonide)}\!-\!CH{=}CH\!-\!CO_2tBu \xrightarrow[\text{T HF / -78°C}]{LiCF_2P(O)(OEt)_2} \text{(acetonide)}\!-\!CH(CF_2P(O)(OEt)_2)\!-\!CH_2\!-\!CO_2tBu$$

80%

Scheme 5.

2.1.4 Nucleophilic Substitution on Carbonyl Groups

Ishihara and co-workers reported an efficient preparation of 2-oxo-1,1-difluoroalkylphosphonates (β-ketophosphonates) from the organozinc reagent and acyl chlorides (*9*). This acylation can be also conducted in the presence of copper(I) bromide (*36, 37)*, and potential ATCase inhibitors were prepared by this way (scheme 6) (*38*).

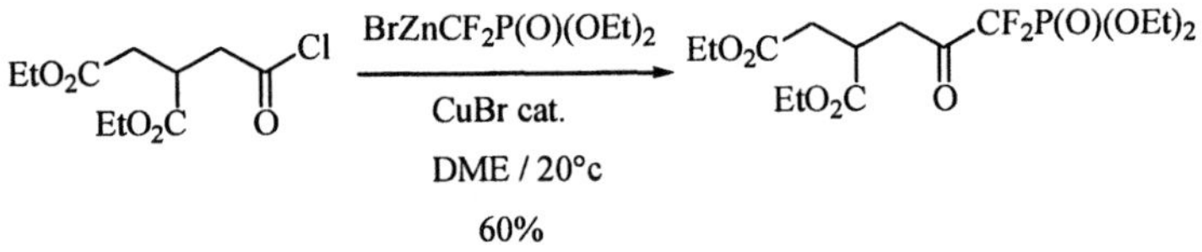

$$EtO_2C\!-\!CH_2\!-\!CH(CO_2Et)\!-\!CH_2\!-\!C(O)Cl \xrightarrow[\text{CuBr cat.}]{BrZnCF_2P(O)(OEt)_2} EtO_2C\!-\!CH_2\!-\!CH(CO_2Et)\!-\!CH_2\!-\!C(O)CF_2P(O)(OEt)_2$$

DME / 20°c

60%

Scheme 6.

Other direct approaches have been explored to prepare α,α-difluoro-β-ketophosphonates. In particular, the addition of the lithium reagent onto ester carbonyls is a straightforward method (*39, 40, 41, 42*). However, this addition is sometime difficult to reproduce. In our hand, good results were observed when reactions are conducted in the presence of cerium(III) chloride (*34, 43*). Cerium salts modulated the reactivity of the carbanion, and the addition to DMF, electrophile which failed to react with the lithium reagent (*12*), afforded the corresponding aldehyde hydrate when performed in the presence of cerium(III) chloride (scheme 7).

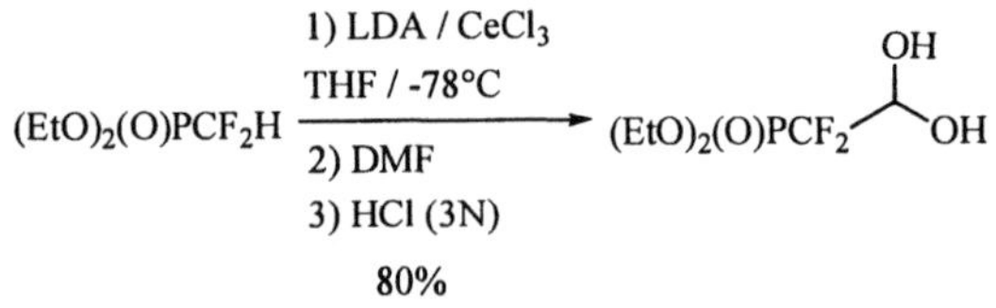

$$(EtO)_2(O)PCF_2H \xrightarrow[\text{2) DMF}]{\text{1) LDA / }CeCl_3\text{, THF / -78°C}} (EtO)_2(O)PCF_2CH(OH)_2$$

3) HCl (3N)

80%

Scheme 7.

2.1.5 *Free Radical Addition and Metal Promoted Coupling Reaction*

Primary phosphonates can be prepare in one step, by the addition of diethyl iododifluoromethylphosphonate onto alkenes in the presence of catalytic amount of palladium(0) (*44*), or copper (*45*) (scheme 8). Bimetallic redox system Co(III)/Zn can also promoted this addition onto both electron-deficient or -rich alkenes (*46*).

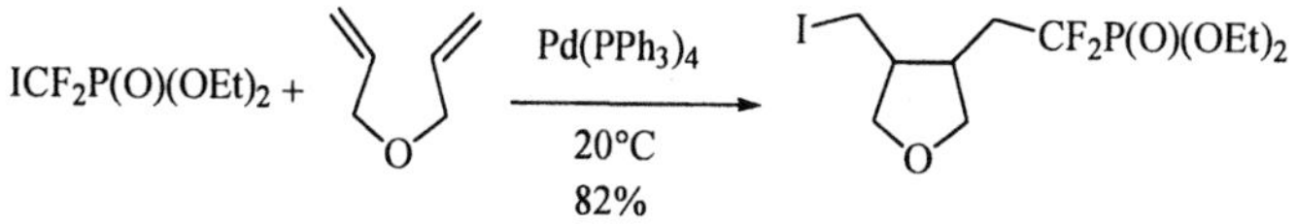

$ICF_2P(O)(OEt)_2$ + diallyl ether $\xrightarrow[\text{20°C, 82\%}]{Pd(PPh_3)_4}$ 3-(iodomethyl)-4-[$CH_2CF_2P(O)(OEt)_2$]tetrahydrofuran

Scheme 8.

Benzylic or allylic phosphonates can be also prepared by a coupling reaction from organometallic reagents and aryl or alkenyl halides. The cadmium reagent reacts with aryl iodides only in the presence of stoichiometric amount of copper(I) chloride (*47, 48*).The same reaction can be carried out under similar conditions from the organozinc reagent (*49, 50, 51, 52*). This method was extended to alkenyl halides, and α,α-difluoro allylic phosphonates were obtained in good yields (*53, 54, 55*) (scheme 9).

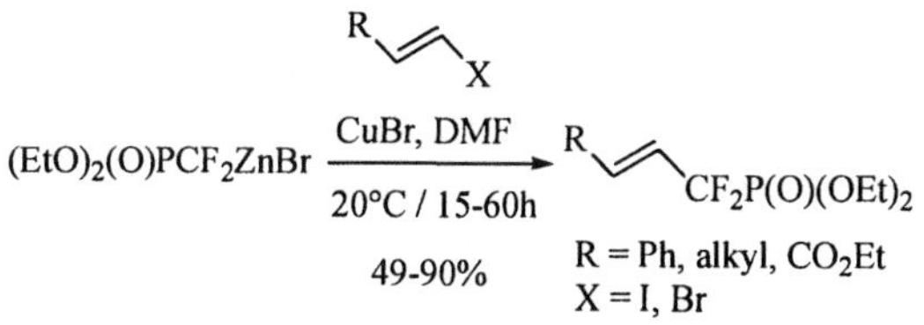

$(EtO)_2(O)PCF_2ZnBr$ $\xrightarrow[\text{20°C / 15-60h, 49-90\%}]{RCH{=}CHX,\ CuBr,\ DMF}$ $RCH{=}CHCF_2P(O)(OEt)_2$

R = Ph, alkyl, CO_2Et
X = I, Br

Scheme 9.

Alternative synthesis of difluoroallylphosphonates have been developed from terminal alkynes and free radical species (Scheme 10). The first method involves the sulfinatodehalogenation of the iododifluoromethylphosphonate (*56*), and the second method the thermal decomposition of the organocopper reagent to generate and trap the free radical intermediate (*55*).

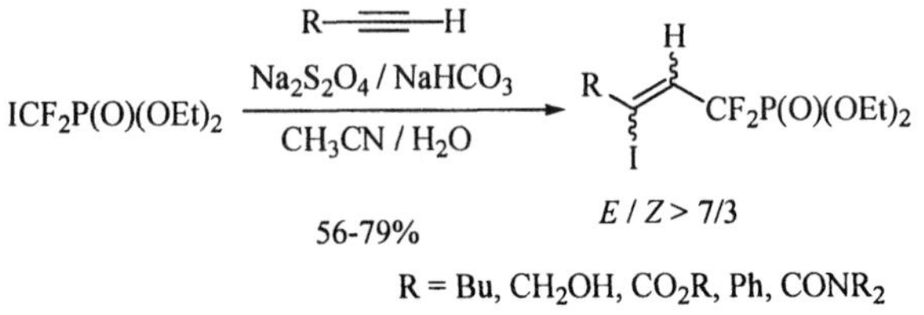

$$ICF_2P(O)(OEt)_2 \xrightarrow[CH_3CN\ /\ H_2O]{R{-}{\equiv}{-}H,\ Na_2S_2O_4\ /\ NaHCO_3} R(I)C{=}C(H)CF_2P(O)(OEt)_2$$

56-79%

$E\,/\,Z > 7/3$

R = Bu, CH_2OH, CO_2R, Ph, $CONR_2$

Scheme 10.

More recently, we showed that the addition of the phosphonodifluoromethyl radical onto alkenes, mediated with tributyltinhydride/initiator, allows the one-step synthesis of secondary phosphonates (57). Both secondary and primary difluoromethyl-phosphonates were prepared from alkylsulfanyl- or alkylselenyl-difluorophosphonates. Using this approach, free radical addition onto dihydrofuran, afforded secondary phosphonofuran derivative in one step and in 47% yield (scheme 11).

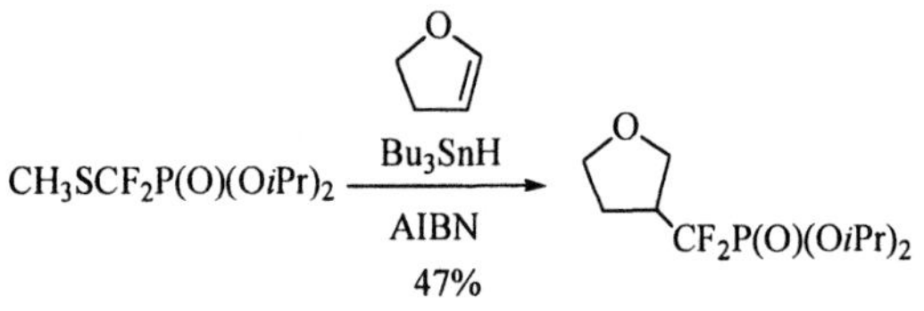

$$CH_3SCF_2P(O)(OiPr)_2 \xrightarrow[AIBN]{\text{dihydrofuran},\ Bu_3SnH} \text{3-tetrahydrofuranyl-}CF_2P(O)(OiPr)_2$$

47%

Scheme 11.

2.2 Construction of the Difluoromethylenephosphonate Function

The construction of this function can be realized by carbon-phosphorus or by carbon-fluorine bond formations. The first method involves the addition of a phosphonyl radical onto difluoroalkenes, and the second the electrophilic or the nucleophilic fluorination of phosphonates.

2.2.1 Carbon-Phosphorus Bond Formation

Few examples of carbon-phosphorus bond formation were reported. This method was applied to the synthesis of carbohydrates containing a difluoromethylenephosphonate. Products were obtained in a selective manner by the addition of a dialkoxyphosphonyl radical onto 2,2-disubstituted *gem*-difluoroalkenes (*58, 59*) (scheme 12).

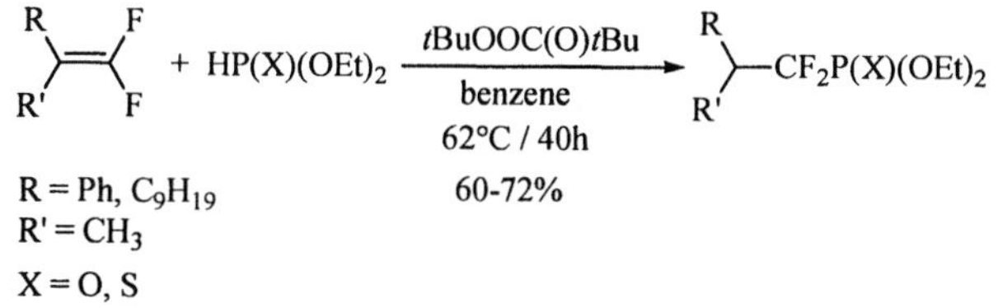

$$\mathrm{RR'C{=}CF_2} + \mathrm{HP(X)(OEt)_2} \xrightarrow[\text{benzene, } 62°\mathrm{C} / 40\mathrm{h}, \ 60\text{-}72\%]{t\mathrm{BuOOC(O)}t\mathrm{Bu}} \mathrm{RR'CH{-}CF_2P(X)(OEt)_2}$$

R = Ph, C_9H_{19}
R' = CH_3
X = O, S

Scheme 12.

From difluoroenol ethers the bimolecular homolytic substitution of diethoxyphosphonylphenyl selenide with tin radical furnished the best results (*60, 61*).

2.2.2 Carbon-Fluorine Bond Formation

The fluorination of α-ketophosphonates has been explored in the synthesis of phosphotyrosine mimics, in which the bridging oxygen was replaced by a difluoromethylene (*4*). The reaction needed harsh conditions and best results were observed by running the fluorination neat in the presence of excess of DAST (scheme 13). Protected amino-acids were tolerated allowing a one-step synthesis of modified phosphotyrosine (*62*).

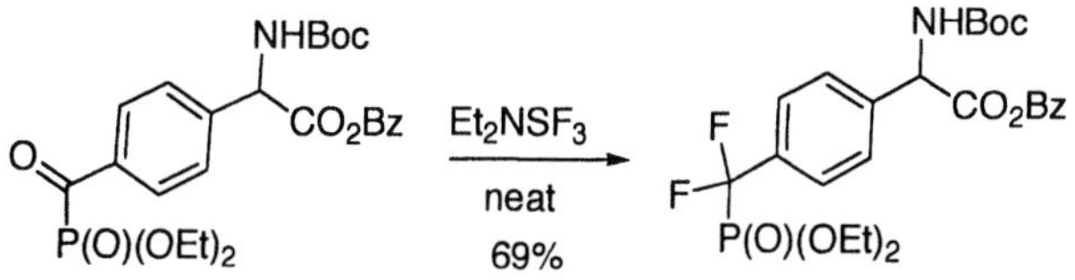

NHBoc, CO_2Bz, O, $P(O)(OEt)_2$ → (Et_2NSF_3, neat, 69%) → NHBoc, CO_2Bz, F, F, $P(O)(OEt)_2$

Scheme 13.

Another method is the fluorination of methylenephosphonates by using electrophilic fluorinating reagent (*63*). N-fluoro-benzenesulfonimide allows to obtain a variety of structures when the difluorination is conducted in two steps (*64*) (scheme 14). Nevertheless, from non-activated phosphonates, products were obtained in low yields (*65*). When F-TEDA was used, difluorination of activated phosphonates in one-pot sequence was preferred (*66,67*).

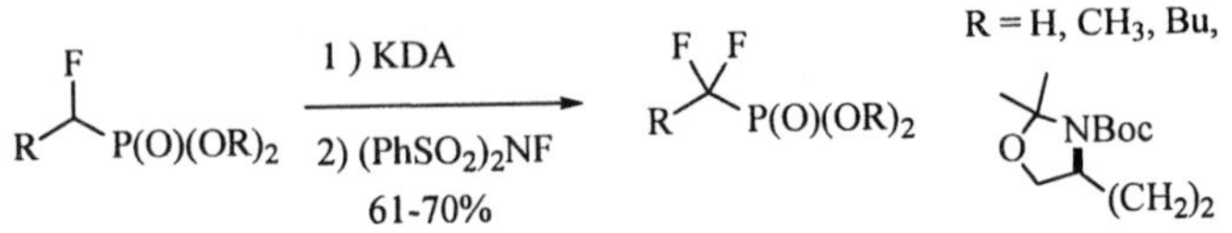

$$R\text{-}CHF\text{-}P(O)(OR)_2 \xrightarrow[\text{2) }(PhSO_2)_2NF\text{, 61-70\%}]{\text{1) KDA}} R\text{-}CF_2\text{-}P(O)(OR)_2$$

R = H, CH_3, Bu, (NBoc oxazolidine)$(CH_2)_2$

Scheme 14.

The synthesis of difluoromethylenephosphonate by using nucleophilic fluorinating reagents was less explored. The use of HF-base complex to introduce two fluorine atoms has been reported for the fluorination of sulfides through a halogen-exchange reaction or an electrolytic fluorination (*15, 68*). We showed that the chlore-fluor exchange can be performed with the non-corrosive 3HF-NEt_3 complex in the presence of zinc bromide. This method allows to prepare alkylsulfanyl α,α-difluoromethylphosphonate in good yields (scheme 15), as new sources of both phosphonodifluoromethyl free radical and carbanion (15, 57).

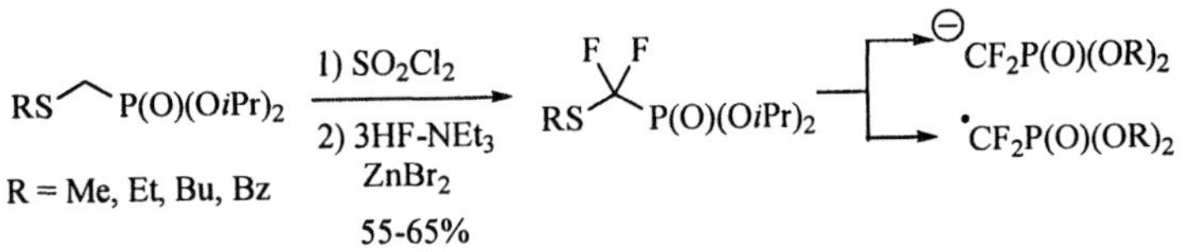

$$RS\text{-}CH_2\text{-}P(O)(OiPr)_2 \xrightarrow[\text{2) 3HF-}NEt_3\text{, }ZnBr_2\text{, 55-65\%}]{\text{1) }SO_2Cl_2} RS\text{-}CF_2\text{-}P(O)(OiPr)_2 \rightarrow {}^{\ominus}CF_2P(O)(OR)_2 \text{ / } {}^{\bullet}CF_2P(O)(OR)_2$$

R = Me, Et, Bu, Bz

Scheme 15.

2.3 The Building Block Approach

Due to the difficulties to run the selective introduction of the difluoromethylenephosphonate function onto complex molecules, the building-blocks approach can be used as alternative strategy to prepare fluorinated analogs of phosphates. Selected representative examples are described here.

Synthesis based on difluoromethylene-bisphosphonates and -phosphonoacetates

The most used in the series is the difluoromethylenebisphosphonate (*1, 12*), which was directly incorporated into nucleotides, and terpenes (*69*). On the other hand, phosphonoacetate derivatives were involved in multi-step synthesis (Scheme 16). For example, thymidine analogs were prepared through their derivation into functionalized chiral phosphonoalcohols (*70, 71*). Its oxidative cleavage, followed by a ring closure reaction of the intermediate aldehyde afforded the key lactones in good yield.

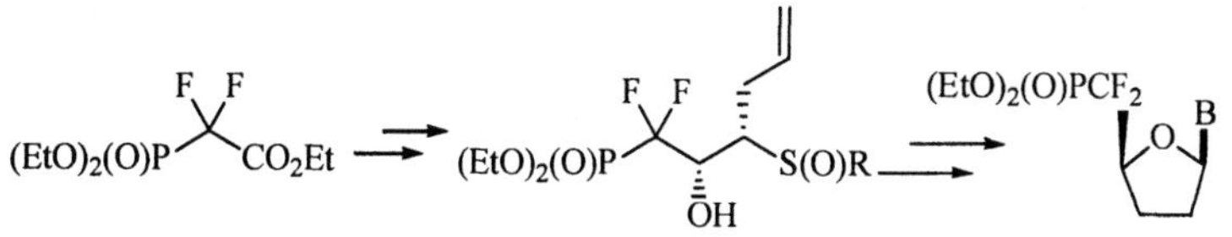

$(EtO)_2(O)P$ CF_2 CO_2Et → $(EtO)_2(O)P$ F F OH S(O)R → $(EtO)_2(O)PCF_2$ O B

Scheme 16.

Synthesis based on diethyl 2-oxo-1,1-difluoroethylphosphonate hydrate

The diethyl 2-oxo-1,1-difluoroethylphosphonate hydrate has been used as masked aldehyde (*34*). When involved in Horner-Wadworth-Emmons or Henry's reaction, a variety of activated difluoroallylic phosphonates were prepared (*43, 72*) (scheme 17).

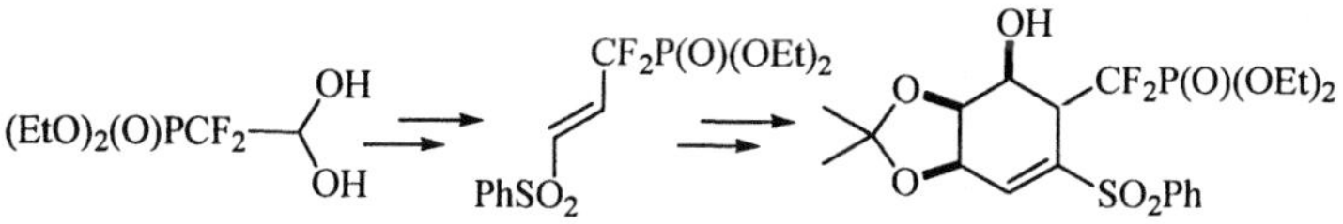

$(EtO)_2(O)PCF_2$ OH OH → $CF_2P(O)(OEt)_2$ $PhSO_2$ → OH $CF_2P(O)(OEt)_2$ O O SO_2Ph

Scheme 17.

These activated alkenes reacts with dienes to afford cyclic secondary difluoromethylenephosphonates as intermediate in the synthesis of DFMPA-functionalized cyclohexene derivatives of biological interest (*73, 74*). Cyclohexene derivatives can be also prepared by cycloaddition carried out from electron-rich phosphonodifluoromethyl-homodienes (*75*).

Synthesis based on the diethyl difluorophosphonomethyldithioacetate

Until recently, no multi-step syntheses were developed from fluorinated phosphonodithioesters. Derivation of the dithioacetate into thioamide was achieved in one-step from amines and amino-acids. By ring closure of intermediate amido-alcohols, we prepared functionalized thiazolines and thiazoles derivatives, which can be used as precursors of potent PNP inhibitors (*76*) (scheme 18).

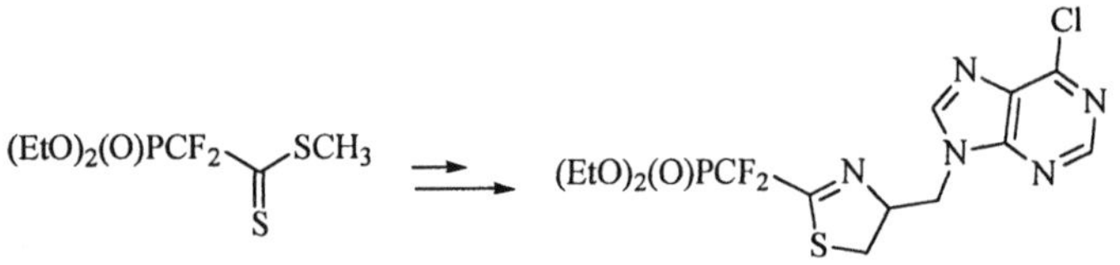

Scheme 18.

This activated dithioester can be also involved in hetero Diels-Alder reactions to prepare thiopyranosyl derivatives (*77*).

Synthesis based on allylic difluoromethylenenephosphonate derivatives

The diethyl 1,1-difluoro-3-butenylphosphonate, has been first derived into pivotal structures for the preparation glycerate analogs (*78*). Its use as free radical acceptor allows the preparation of thionucleotide analogs (*79*). The key step is the synthesis of a functionalized xanthate. The thiolactone was obtained by a ring closure reaction of the intermediate thiol (scheme 19).

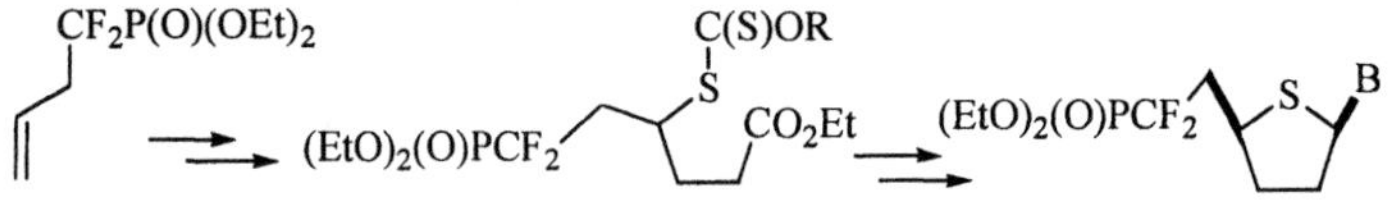

Scheme 19.

3. Conclusion

As summarized herein, numerous efforts have been made to prepare analogs of phosphates by different methods. However due to the presence of fluorine atoms the synthesis of complex structures is rather difficult, and the known methods described for the preparation of methylenephosphonates cannot be applied. New reagents and building blocks are still needful for the preparation of analogs of bio-active compounds, and, in particular, reagents easy to handle, and methods opening a rapid access to secondary difluoromethylphosphonates.

References

1. (a) Blackburn, G. M.; England, D. A.; Kolkmann, F. *J. Chem. Soc. Chem. Comm.* **1981**, 930-932. (b) Blackburn, G. M.; Kent, D. E.; Kolkmann, F. *J. Chem. Soc. Perkin Trans 1* **1984**, 1119-1125.
2. (a) Thatcher, G. R.; Campbell, A. S. *J. Org. Chem.* **1993**, *58*, 2272-2281. (b) Chambers, R. D.; O'Hagan, D.; Lamont, R. B.; Jain, S. C. *J. Chem. Soc. Chem. Commun.* **1990**, 1053-1054.
3. Kim, C. U. ; Luh, B. Y.; Misco, P. F.; Bronson, J. J.; Hitchcock, M. J. M.; Ghazoulli, I.; Matin, J. C. *J. Med. Chem.* **1990**, *33*, 1207-1213.
4. Smyth, M. S.; Ford, H.; Burke, T. R. *Tetrahedron Lett.* **1992**, *33*, 4137-4140.
5. Jakeman, D. L.; Ivory, A. J.; Williamson, M. P.; Blackburn, G. M. *J. Med. Chem.* **1998**, *41*, 4439-4452.
6. (a) Berkowitz, D.B.; Bose, M.; Pfannenstiel, T. J.; Doukov, T. *J. Org. Chem.* **2000**, *65*, 4498-4508. (b) Berkowitz, D. B.; Bose, M. *J. Fluorine Chem.* **2001**, *112*, 13-33. (c) O'Hagan, D.; Rzepa, H. S. *Chem. Commun.* **1997**, 645-652.
7. For a review *see* Burton, D. J.; Yang, Z. *Tetrahedron* **1992**, *48*, 189-275.
8. *Lithiated carbanion*: Obayashi, M.; Ito, E.; Kondo, K. *Tetrahedron Lett.* **1982**, *23*, 2323-2326.
9. *Organozinc reagent*: Burton, D. J.; Ishihara, T.; Maruta, M. *Chem. Lett.* **1982**, 755-758.
10. *Organocadmium reagent*: Burton, D. J.; Takei, R.; Seiji, S. *J. Fluorine Chem.* **1981**, *18*, 197-202.
11. *Organomagnesium reagent*: Wachbüsch, R.; Samadi, M.; Savignac, P. *J. Organomet. Chem.* **1997**, *529*, 267-278.
12. Blackburn, G. M.; Brown, D. ; Martin, S. J.; Parratt, M. J. *J. Chem. Soc. Perkin Trans 1* **1987**, 181-186.
13. Piettre, S. R.; Cabanas, L. *Tetrahedron Lett.* **1996**, *37*, 5881-5884.
14. Blades, K.; Cockerill G. S.; Eaterfield, H. J.; Lequeux, T. P.; Percy, J. M. *Chem. Commun.* **1996**, 1615-1616.
15. Henry-dit-Quesnel, A.; Toupet, L.; Pommelet, J. C.; Lequeux, T. *Org. Biomol. Chem.* **2003**, *1*, 2486-2491.
16. (a) Martin, S. F.; Dean, D. W.; Wagman A. S. *Tetrahedron Lett.* **1992**, *33*, 1839-1842. (b) Nieschalk, J., O'Hagan, D. *J. Chem. Soc. Chem. Commun.* **1995**, 719-720.
17. Levy, S. G., Wasson, D. B.; Carson, D. A.; Cottmam, H. B. *Synthesis* **1996**, 843-846.
18. Otaka, A.; Miyoshi, K.; Burke, T. R.; Roller, P. P.; Kubota, H.; Tamamura, H.; Fujii, N. *Tetrahedron Lett.* **1995**, *36*, 927-930.
19. For a review *see :* Burton, D. J.; Yang, Z.; Qiu, W. *Chem. Rev.* **1996**, 1641-1715.
20. Bigge, C. F.; Drummond, J. T.; Johnson, G. *Tetrahedron Lett.* **1989**, *30*, 7013-7016.

21. Halazy, S.; Ehrhard, A.; Danzin, C. *J. Am. Chem. Soc.* **1991**, *113*, 315-317.
22. Halazy, S.; Ehrhard, A.; Eggenspiller, A.; Bergès-Gross, V.; Danzin, C. *Tetrahedron* **1996**, *52*, 177-184.
23. Berkowitz, D. B.; Eggen, M.; Shen, Q.; Sloss, D. G. *J. Org. Chem.* **1993**, *58*, 6174-6176.
24. Berkowitz, D. B.; Bhuniya, D.; Peris, G. *Tetrahedron Lett.* **1999**, *40*, 1869-1872.
25. Berkowitz, D. B.; Sloss, D. G. *J. Org. Chem.* **1995**, *60*, 7047-7050.
26. Berkowitz, D. B.; Shen, Q.; Maeng, J. *Tetrahedron Lett.* **1994**, *35*, 6445-6448.
27. Vinot, T. K.; Griffith, O. H.; Keana, J. F. W. *Tetrahedron Lett.* **1994**, *35*, 7193-7196.
28. Matulic-Adamic J.; Haeberli, P.; Usman, N. *J. Org. Chem.* **1995**, *60*, 2563-2569.
29. Yokomatsu, T.; Kato, J.; Sakuma, C.; Shibuya, S. *Synlett* **2003**, 1407-1410.
30. Howson, W.; Hills, J. M.; Blackburn, G. M.; Broekman, M. *Bioorg. Med. Chem. Lett.* **1991**, *1*, 501-502.
31. Lequeux, T. P.; Percy, J. M. *Synlett* **1995**, 361-362.
32. Blades, K.; Lapôtre, D.; Percy J. M. *Tetrahedron Lett.* **1997**, *38*, 5895-5898.
33. Blades, K.; Percy, J. M. *Tetrahedron Lett.* **1998**, *39*, 9085-9088.
34. Blades, K.; Lequeux, T. P.; Percy, J. M. *Tetrahedron* **1997**, *53*, 10623-10632.
35. Murano, T.; Muroyama, S.; Yokomatsu, T.; Shibuya, S. *Synlett* **2002**, 1657-1660.
36. Burton, D. J.; Sprague, L. G.; Pietrzyk, D. J.; Edelmuth, S. H. *J. Org. Chem.* **1984**, *49*, 3438-3440.
37. Burton, D. J.; Sprague, L. G. *J. Org. Chem.* **1988**, *53*, 1523-1527.
38. Lindell, S. D.; Turner, R. M. *Tetrahedron Lett.* **1990**, *31*, 5381-5384.
39. Phillion, D. P.; Cleary, D. G. *J. Org. Chem.* **1992**, *57*, 2763-2764.
40. Berkowitz, D. B.; Eggen, M.; Shen, Q.; Shoemarker, R. K. *J. Org. Chem.* **1996**, *61*, 4666-4675.
41. Ladame, S.; Bardet, M.; Périé, J.; Willson, M. *Bioorg. Med. Chem.* **2001**, *9*, 773-783.
42. Bouvet, D.; O'Hagan, D. *Tetrahedron* **1999**, *55*, 10481-10486.
43. Lequeux, T. P.; Percy, J. M. *J. Chem. Soc. Chem. Commun.* **1995**, 2111-2112.
44. Yang, Z.; Burton, D. J. *Tetrahedron Lett.* **1991**, *32*, 1019-1022.
45. Yang, Z.; Burton, D. J. *J. Org. Chem.* **1992**, *57*, 4676-4683.
46. Hu, C.; Chen, J. *J. Chem. Soc. Perkin Trans 1* **1993**, 327-330.
47. Qiu, W.; Burton, D. J. *Tetrahedron Lett.* **1996**, *37*, 2745-2748.
48. Qabar, M. N.; Urban, J.; Kahn, M. *Tetrahedron* **1997**, *53*, 11171-11178.
49. Yokomatsu, T.; Murano, T.; Suemune, K.; Shibuya, S. *Tetrahedron* **1997**, *53*, 815-822.
50. Yokomatsu, T.; Minowa, T.; Murano, T.; Shibuya, S. *Tetrahedron* **1998**, *54*, 9341-9356.

51. Li, Z.; Yeo, S. L.; Palle, C. J.; Ganesan, A. *Bioorg. Med. Chem. Lett.* **1998**, *8*, 2443-2446.
52. Cockerill, G. S.; Easterfield, H. J.; Percy, J. M.; Pintat, S. *J. Chem. Soc. Perkin Trans 1* **2000**, 2597-2599.
53. (a) Yokomatsu, T.; Suemune, K.; Murano, T.; Shibuya, S. *J. Org. Chem.* **1996**, *61*, 7207-7211. (b) Yokomatsu, T.; Abe, H.; Yamagishi, T.; Suemune, K.; Shibuya, S. *J. Org. Chem.* **1999**, *64*, 8413-8418. (c) Yokomatsu, T.; Hayakawa, Y.; Suemune, K.; Kihara, T.; Soeda, S.; Shimeno, H.; Shibuya, S. *Bioorg. Med. Chem. Lett.* **1999**, *9*, 2833-2836. (d) Yokomatsu, T.; Yoshinobu, H.; Kihara, T.; Koyanagi, S.; Soeda, S.; Shimeno, H.; Shibuya, S. *Bioorg. Med. Chem. Lett.* **2000**, *8*, 2571-2579.
54. Otaka, A.; Mitsuyama, E.; Kinoshita, T.; Tamamura, H.; Fujii, N. *J. Org. Chem.* **2000**, *65*, 4888-4899.
55. (a) Zhang, X.; Burton, D. J. *Tetrahedron Lett.* **2000**, *41*, 7791-7794. (b) Zhang, X.; Burton, D. J. *J. Fluorine Chem.* **2002**, *116*, 15-18.
56. Li, A.; Chen, Q. *Synthesis* **1996**, 606-608.
57. Lequeux, T.; Lebouc, F.; Lopin, C.; Yang, H.; Gouhier, G. Piettre, S. *Org. Lett.* **2001**, *3*, 185-188.
58. Piettre, S. R. *Tetrahedron Lett.* **1996**, *37*, 2233-2236.
59. Piettre, S. R. *Tetrahderon Lett.* **1996**, *37*, 4707-4710.
60. Herpin, T. F.; Houlton, J. S., Motherwell, W. B.; Roberts, B. P.; Weibel, J. M. *J. Chem. Soc. Chem. Commun.* **1996**, 613-614.
61. Herpin, T. F.; Motherwell, W. B.; Roberts, B.; Roland, S.; Weibel, J. M. *Tetrahedron* **1997**, *53*, 15085-15100.
62. a) Wrobel, J.; Dietrich, A. *Tetrahedron Lett.* **1993**, *34*, 3543-3546. b) Burke, T. R.; Smyth, M. S.; Otaka, A.; Roller, P. P. *Tetrahedron Lett.* **1993**, *34*, 4125-4128. c) Ye, B.; Burke, T. R.; *Tetrahedron* **1996**, *52*, 9963-9970. d) Sols, D.; Hale, R.; Patel, D. V. *J. Org. Chem.* **1996**, *61*, 1537-1539. e) Yao, Z.; Ye, B.; Wu, X.; Wang, S.; Wu, L.; Zhang, Z.; Burke, T. R. *Bioorg. Med. Chem.* **1998**, *6*, 1799-1810.
63. For a review *see* Taylor, S. D.; Kotoris, C. C.; Hum, G. *Tetrahedron* **1999**, *55*, 12431-12477.
64. Differding, E.; Duthaler, R. O.; Krieger, A.; Rüegg, G. M.; Schmidt, C. *Synlett* **1991**, 395-396.
65. Chen, W.; Flavin, M. T.; Filler, R.; Xu, Z. *Tetrahedron Lett.* **1996**, *37*, 8975-8978.
66. Lal, G. S.; Pez, G. P.; Syvret, R. G. *Chem. Rev.* **1996**, *96*, 1737-1755.
67. Ladame, S. ; Willson, M.; Périé, J. *Eur. J. Org. Chem.* **2002**, 2640-2648.
68. Konno, A.; Fuchigami, T. *J. Org. Chem.* **1997**, *62*, 8579-8581.
69. For examples *see* a) Jo Davisson, V.; Woodside, A. B.; Neal, T. R.; Stremler, K. E.; Muehlbacher, M.; Poulter, C. D. *J. Org. Chem.* **1986**, *51*, 4769-4779. b) Bystrom, C. E.; Pettigrew, D. W.; Remington, S. J.; Branchaud, B. P. *Bioorg. Med. Chem. Lett.* **1997**, *7*, 2613-2616. c) Hamilton, C. J.; Roberts, S. M.; Shipitsin, A. *Chem. Commun.* **1998**, 1087-1088.

70. Arnone, A.; Bravo, P.; Frigero, M.; Viani, F.; Zappalà, C. *Synthesis* **1998**, 1511-1518.
71. Arnone, A.; Bravo, P.; Frigero, M.; Mele, A.; Vergani, B.; Viani, F. *Eur. J. Org. Chem.* **1999**, 2149-2157.
72. Blades, K.; Lequeux, T.; Percy, J. M. *Chem. Commun.* **1996**, 1457-1458.
73. Blades, K.; Butt, A. H.; Cockerill, G. S.; Easterfield, H. J.; Lequeux, T.; Percy, J. M. *J. Chem. Soc. Perkin Trans 1* **1999**, 3609-3614.
74. Butt, A. H.; Kariuki, B. M.; Percy, J. M.; Spencer, N. S. *Chem. Commun* **2002**, 682-683.
75. Yokomatsu, T.; Katayama, S.; Shibuya, S. *Chem. Commun.* **2001**, 1878-1879.
76. Pfund, E.; Lequeux, T.; Vazeux, M.; Masson, S. *Org. Lett.* **2002**, *4*, 843-846.
77. Pfund, E.; Lequeux, T.; Vazeux, M.; Masson, S. *Tetrahedron Lett.* **2002**, *43*, 2033-2036.
78. Chambers, R. D.; Jaouhari, R.; O'Hagan, D. *Tetrahedron* **1989**, *45*, 5101-5108.
79. Boivin, J.; Ramos, L.; Zard, S. Z. *Tetrahedron Lett.* **1998**, *39*, 6877-6880.

Chapter 27

Mg-Promoted Selective C–F Bond Cleavage: A General and Efficient Access to Difluoroenol Silyl Ethers and Their Derivatives

Hideki Amii and Kenji Uneyama

Department of Applied Chemistry, Faculty of Engineering, Okayama University, Tsushimanaka 3–1–1, Okayama 700–8530, Japan

In this chapter, we describe the progress of C-F bond cleavage reactions which afford an effective access to difluoroenol ethers and their derivatives, which are useful building blocks for a wide variety of difluoromethylene compounds.

Organofluorine compounds are receiving increasing attention in the medicinal, agricultural, and material sciences (*1*). Difluoromethylene compounds are important synthetic targets because of their unique biological activities (*2*). For instance, the introduction of a difluoromethylene unit into the peptides has led to the discovery of potent protease inhibitors mimicking the transition state for hydrolytic amide bond cleavage (*3*). Needless to say, the fluorinated building blocks play an important role in synthetic organofluorine chemistry.

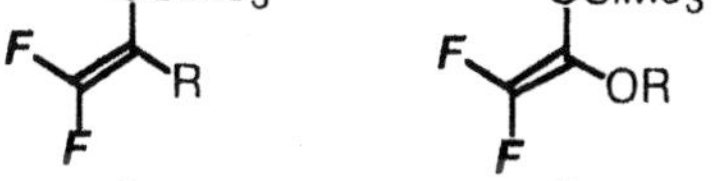

Figure 1.

Difluoroenol silyl ethers **1** and difluoroketene silyl acetals **2** are synthetic equivalents of an enolate of difluoromethyl ketones and difluoroacetates and are versatile building blocks for difluoro compounds (Fig. 1). Several synthetic applications of these compounds, including their enantioselective versions, have been reported (eqs 1-4) (*4-8*).

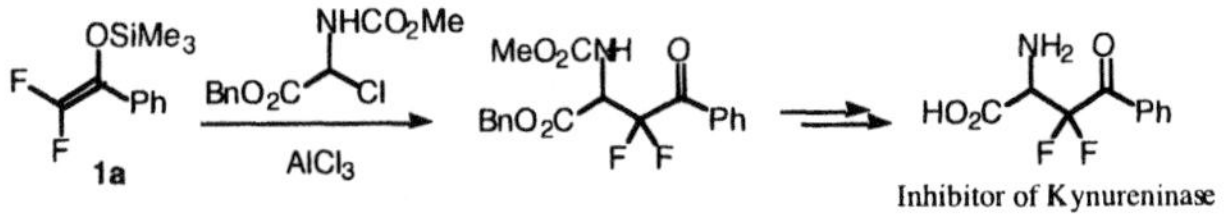

(1)

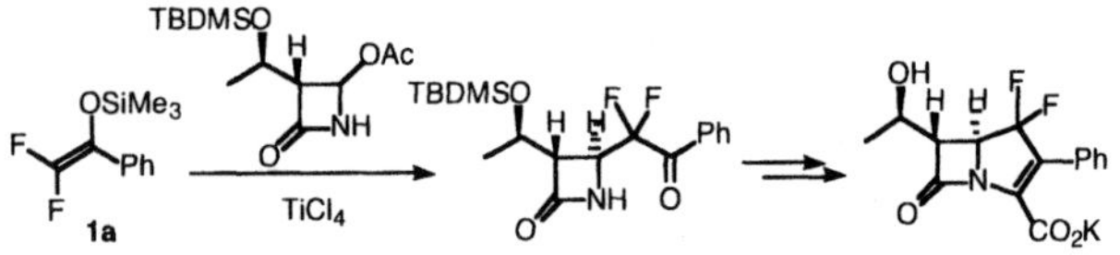

Inhibitor of Kynureninase

(2)

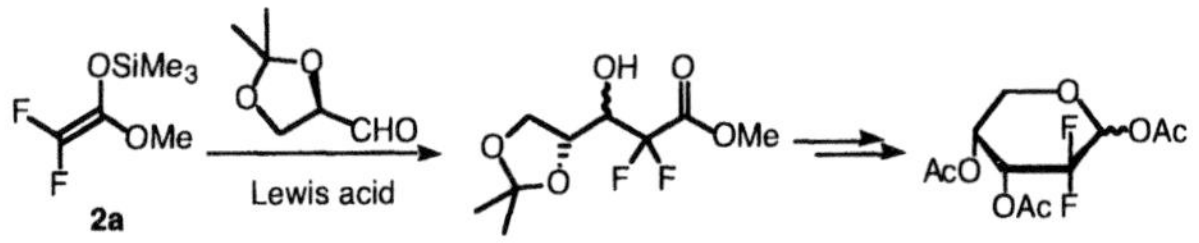

(3)

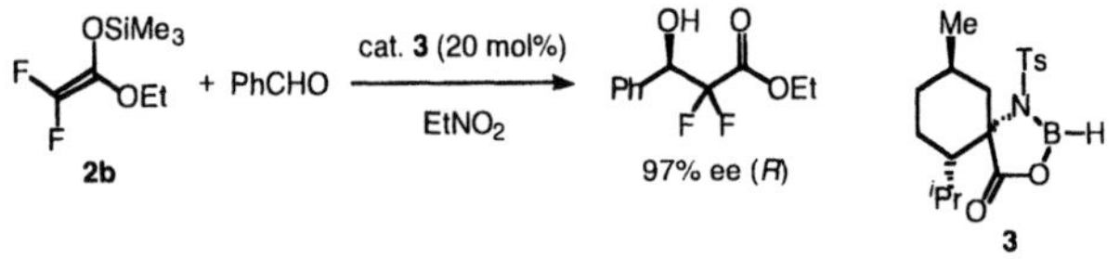

(4)

Preparative Methods of Difluorinated Enol Ethers and Their Derivatives. One of the well-established methods to prepare difluoroenol silyl ethers **1** and difluoroketene silyl acetals **2** is the dehalogenation from halodifluoromethyl groups (eqs 5-6) (*8a, 9*).

(A) Reformatsky-type reactions

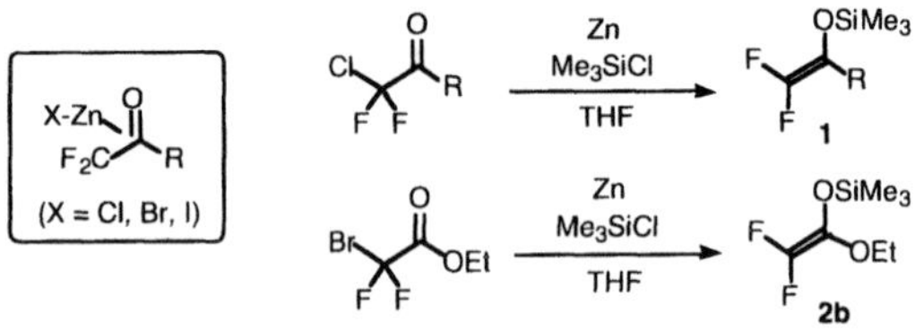

(5)

(6)

Due to the broad and easy availability of trifluoromethylated materials, selective defluorination of a trifluoromethyl group is one of the promising methods for difluoro compounds. The preparative routes to *gem*-difluorinated enol ethers and their homologs have been accomplished by (B) acid or base-catalyzed dehydrofluorination of fluorinated alcohol derivatives (eqs 7-11) (*10-15*), (C) reductive dehalogenation of vicinal halides (eq 12) (*16*), (D) Brook-type rearrangement of α-silyl-α-trifluoromethyl alkoxides (eq 13) (*17-19*) and retro-Brook rearrangement of α-siloxyvinyl lithium (eq 14) (*20*), and (E) S_N2'-type addition of nucleophiles to the heteroatoms of C-N and C-S double bonds of fluorinated iminoesters and dithioesters (eqs 15-16) (*21,22*).

(B) Acid or base-catalyzed dehydrofluorination

:B; H; F; F; OR'; R; F

F_3C–CH_2OTs $\xrightarrow{\text{LDA (2 eq)}}$ F_2C=C(OTs)Li $\xrightarrow{BR_3}$ F_2C=C(OTs)BR_2 (7)

F_3C–CH_2OMEM $\xrightarrow{\text{LDA (2 eq)}}$ F_2C=C(OMEM)Li $\xrightarrow{R^1R^2CO}$ F_2C=C(OMEM)C(R^1)(R^2)OH (8)

F_3C–CH(OH)–CF_3 $\xrightarrow{n\text{-BuLi (2 eq)}}$ F_2C=C(OLi)CF_3 $\xrightarrow{PhCOCl}$ F_2C=C(O-COPh)CF_3 (9)

$RfCF_2CH_2OTs$ $\xrightarrow[\text{THF}]{\text{BuLi}}$ Rf(F)C=C(OTs)H $\xrightarrow{\text{BuLi}}$ Rf(F)C=C(OTs)Li
(Rf = HCF_2, CF_3)
$\xrightarrow{ZnBr_2}$ Rf(F)C=C(OTs)ZnX $\xrightarrow[\text{THF}]{\text{Ar-I, cat. Pd(PPh}_3)_4}$ Rf(F)C=C(OTs)Ar (10)

F_3C–CH($OCOCF_3$)–CO_2Me $\xrightarrow{BF_3\cdot NEt_3}$ F_2C=C($OCOCF_3$)CO_2Me $\xrightarrow{MeOH}$ F_2C=C(OH)CO_2Me (11)

(C) Reductive dechlorofluorination

+2e; Cl; F; F; OR'; R; F

F_3C–C(Cl)(O-allyl)–CO_2Et $\xrightarrow[\text{DMF}]{\text{Zn, cat. CuI}}$ F_2C=C(O-allyl)CO_2Et (12)

(D) Brook rearrangement and retro-Brook rearrangement

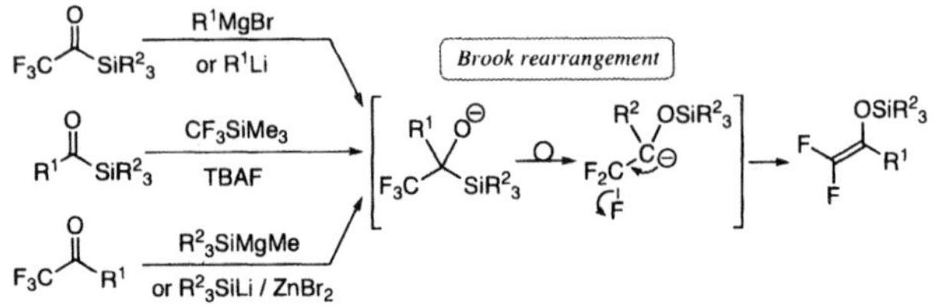

(13)

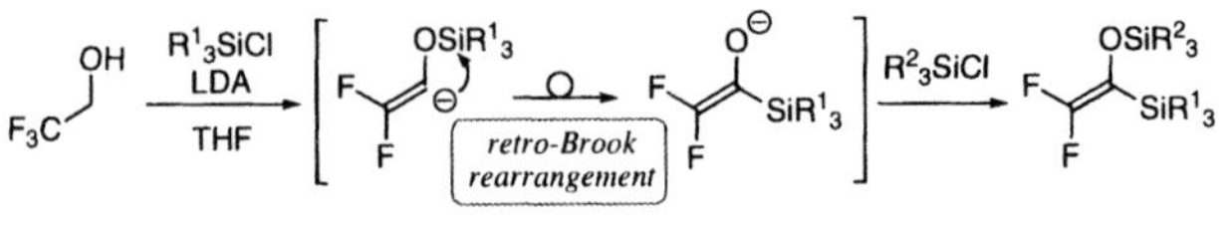

(14)

(E) S_N2'-type reactions

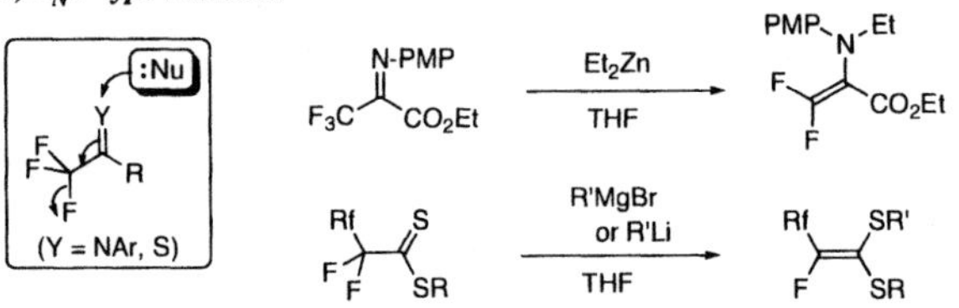

(15)

(16)

Mg-Promoted Selective Defluorination of Trifluoromethyl Ketones. Selective activation reactions of inert bonds, such as C-H, C-C, and C-F bonds, are a challenging topic of considerable scientific and technological interest (*23*). Recently, there have been several reports on the selective C-F bond cleavage reactions by the use of metal reagents (*24-29*). In general, the cleavage of a C-F bond is not easy due to the large bond energy (ca. 552 kJ mol^{-1}). However, the bond breaking does rather easily occur when a CF_3 group is attached to the π-electron system because electron acceptance into the π-system and subsequent extrusion of a fluoride ion may make large contributions to the driving force of the reaction (Scheme 1)

Scheme 1

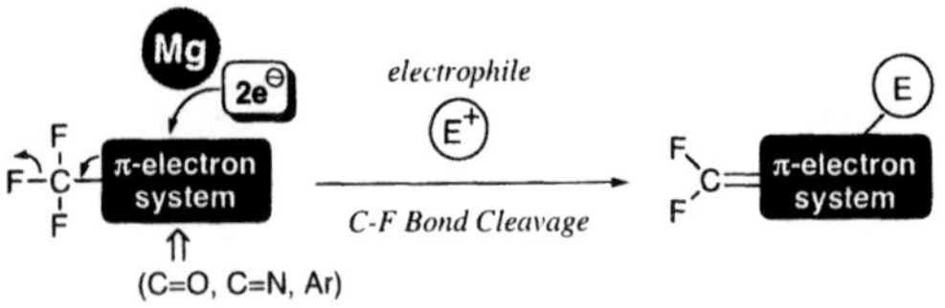

Electrochemical methods have hitherto been developed for reductive defluorination of a trifluoromethyl group (*30*), and they can be successfully applied to the preparation of difluoromethylene building blocks (*31-35*).

In 1999, we reported that metallic magnesium, which serves as a more convenient electron source, proves useful for the C-F bond breaking process of trifluoromethyl ketones **4** to provide a practical route to 2,2-difluoroenol silyl ethers **1** (Scheme 2) (*36,37*).

Scheme 2

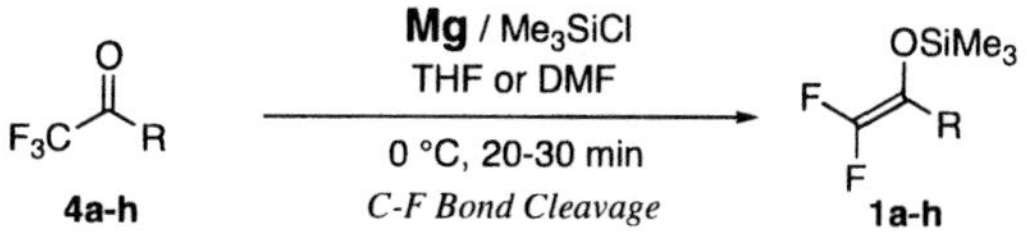

F_3C–CO–R (**4a-h**) → **Mg** / Me_3SiCl, THF or DMF, 0 °C, 20-30 min, *C-F Bond Cleavage* → F_2C=C($OSiMe_3$)R (**1a-h**)

R = Ph (a), 4-$MeOC_6H_4$ (b), 4-$CF_3C_6H_4$ (c), 4-ClC_6H_4 (d), 2-furyl (e), 2-thienyl (f), C_6H_{13} (g), Cy (h).

Yield: 91% (**1a**), 89% (**1b**), 87% (**1c**), 98% (**1d**), 97% (**1e**), 97% (**1f**), 56% (**1g**), 62% (**1h**).

Compared to previously available methods, the present methodology has several advantages: (i) the starting trifluoromethyl ketones **4** are readily available directly from trifluoroacetates; (ii) Mg as a reducing agent is inexpensive and easy to be handled (*38*); and (iii) selective formation of 2,2-difluoroenol silyl ethers **1** is achieved in a short reaction time (only 20-30 min) under mild reaction conditions. In light of operational simplicity and high efficiency, Mg(0)-promoted selective defluorination of **4** and the subsequent transformation of **1** gave a reliable route for preparing difluoro compounds.

As a synthetic application of difluoroenol silyl ethers **1**, Bonnet-Delpon and Bégué *et al.* reported fluoro artemisinins by means of Lewis acid-catalyzed reactions of dihydroartemisinin acetate with difluoroenoxysilanes **1**, which were prepared by our defluorination route (eq 17) (*39*).

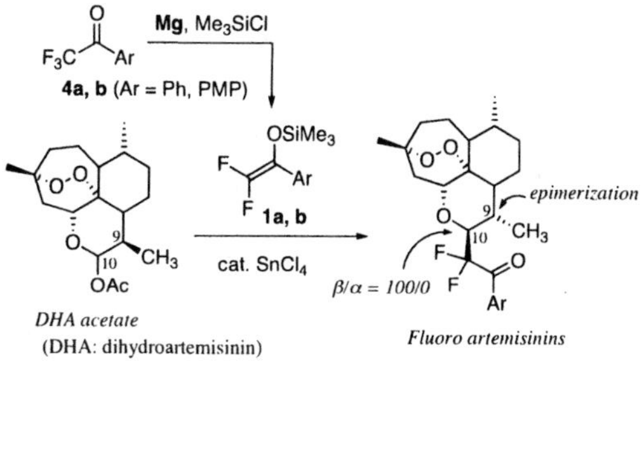

F_3C–CO–Ar, **4a, b** (Ar = Ph, PMP) → **Mg**, Me_3SiCl → F_2C=C($OSiMe_3$)Ar (**1a, b**)

DHA acetate (DHA: dihydroartemisinin) + **1a, b** → cat. $SnCl_4$ → *Fluoro artemisinins* (epimerization; *β/α = 100/0*) (17)

Electrophilic halogenation of 2,2-difluoroenol silanes **1** is highly useful for the preparation of α-halodifluoromethyl ketones (*40*). Prakash and Olah *et al.* presented the reaction of difluoro silyl enol ethers **1a** with halogens at low temperature which produced a high yield of α-halodifluoromethyl ketones and [^{18}F]-labeled trifluoromethyl ketone **5** (eq 18) (*41*).

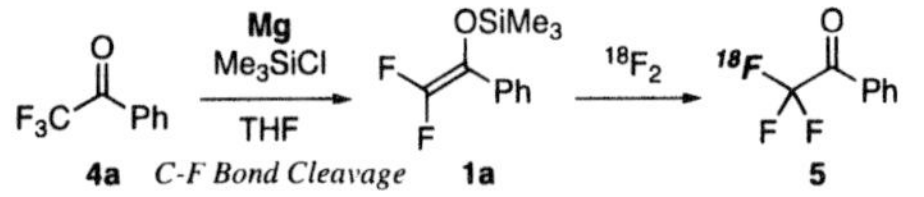

$$F_3C\text{-}CO\text{-}Ph\ (\mathbf{4a}) \xrightarrow[\text{THF}]{\text{Mg, } Me_3SiCl} F_2C{=}C(OSiMe_3)Ph\ (\mathbf{1a}) \xrightarrow{^{18}F_2} {}^{18}F\text{-}CF_2\text{-}CO\text{-}Ph\ (\mathbf{5}) \quad (18)$$

C-F Bond Cleavage

To demonstrate further synthetic utility of our C-F bond cleavage reactions, we developed 1-butoxy-4,4-difluoro-3-(trimethylsiloxy)-1,3-butadiene **6**, a fluorinated analog of Danishefsky's diene (*42*), acting as one of the most fascinating C_4 building blocks (Scheme 3) (*43*). The process shown in Scheme 3 provides access to difluoro Danishefsky's diene **6** in a simple and efficient fashion.

Scheme 3

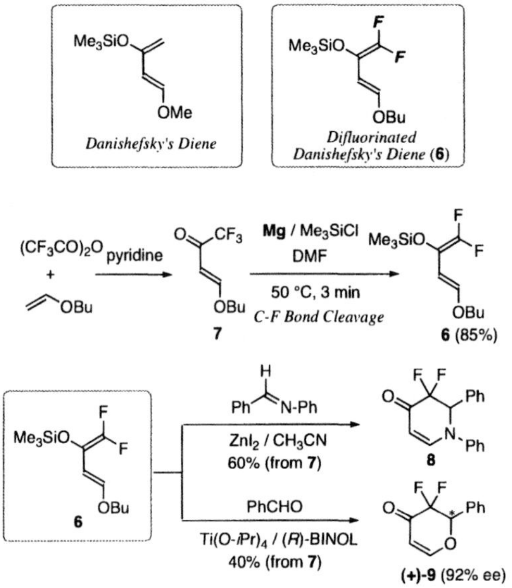

Me3SiO, OMe — *Danishefsky's Diene*; Me3SiO, F, F, OBu — *Difluorinated Danishefsky's Diene* (**6**)

(CF3CO)2O + vinyl OBu →(pyridine)→ **7** (CF3, O, OBu) →(Mg / Me3SiCl, DMF, 50 °C, 3 min, *C-F Bond Cleavage*)→ **6** (85%)

6 → (Ph-CH=N-Ph, ZnI2 / CH3CN, 60% (from **7**)) → **8**

6 → (PhCHO, Ti(O-*i*Pr)4 / (*R*)-BINOL, 40% (from **7**)) → **(+)-9** (92% ee)

The starting ketone **7** was easily prepared from the reaction of trifluoroacetic anhydride (TFAA) with vinyl ethers (*44*). Upon treatment with 8 equiv of Mg and Me_3SiCl in DMF, defluorinative silylation of trifluoromethyl ketone **7** proceeded to afford difluorobutadiene **6** in 85% yield. The subsequent hetero

Diels-Alder reactions of **6** gave fluorinated six-membered heterocycles. The reactions of diene **6** with aldehydes and aldimines in the presence of Lewis acid ($ZnBr_2$ or ZnI_2) gave the fluorinated dihydropyrones and difluoro dihydropyridones, respectively. Furthermore, the enantioselective hetero Diels-Alder reaction with benzaldehyde afforded corresponding dihydropyrone **(+)-9** in 92% ee in the presence of chiral titanium(IV)-BINOL system.

Mg-Promoted Selective Defluorination of Difluoromethyl Ketones. Similarly, difluoromethyl ketones underwent the selective defluorination reactions to provide the monofluorinated compounds. Prakash and Olah *et al.* reported a facile preparation of monofluoromethyl ketones. Metallic Mg-mediated reductive defluorination of difluoromethyl ketone **10** readily generated monofluorinated enol silyl ether **11**, which upon fluoride or acid assisted hydrolysis gave the respective ketone **12** in good yield (eq 19) (*45*).

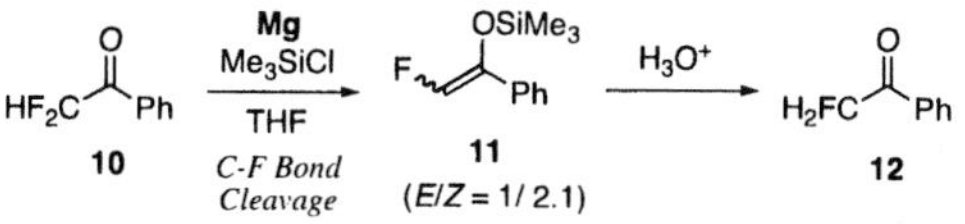

HF_2C–CO–Ph (**10**) → [Mg, Me_3SiCl, THF; *C-F Bond Cleavage*] → F–CH=C($OSiMe_3$)Ph (**11**, (*E*/*Z* = 1/2.1)) → [H_3O^+] → H_2FC–CO–Ph (**12**) (19)

α-Fluoro-α,β-unsaturated carbonyl compounds are promising precursors for fluoroalkene oligopeptide isosteres, due to the interesting character of fluoroolefin moieties. Replacement of the amide bond with a monofluorinated carbon-carbon double bond (CF=C) has been recognized to provide conformationally fixed peptide bond isosteres (Scheme 4) (*46,47*). Physical data such as bond length, dipole moment, and charge distribution of fluoroolefins also suggest the similarity between an amide moiety and a fluoroolefin (*48-51*).

Scheme 4

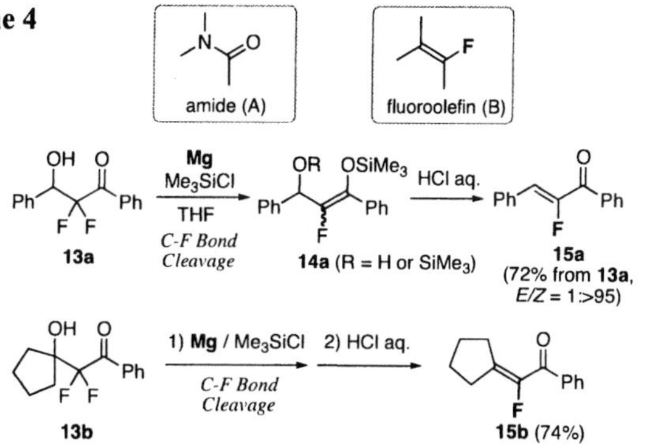

amide (A)   fluoroolefin (B)

13a → [Mg, Me_3SiCl, THF; *C-F Bond Cleavage*] → **14a** (R = H or $SiMe_3$) → [HCl aq.] → **15a** (72% from **13a**, *E*/*Z* = 1:>95)

13b → [1) **Mg** / Me_3SiCl 2) HCl aq.; *C-F Bond Cleavage*] → **15b** (74%)

C–F bond cleavage reactions were applied to the preparation of monofluoroenones (*52*). The action of Mg/Me_3SiCl in THF on difluoroaldol **13a** caused the defluorinative silylation to afford monofluoroenol silyl ether **14a**. The subsequent acid-catalyzed hydrolysis of the crude enol **14a** yielded α-fluoro-α,β-unsaturated ketone **15a**. β-Hydroxyketone **13b** from cyclopentanone underwent defluorination cleanly, affording exo-fluoromethylene cyclic compound **15b** in 74% yield.

Selective Defluorination of Trifluoroacetates, Trifluoromethyl Imines, and Aromatics. A Mg(0)/Me_3SiCl system was found to be also effective for the selective defluorination reactions of trifluoroacetates, trifluoromethyl imines, and aromatics. Aryl trifluoroacetates (**16**) were suitable substrates for Mg-promoted C-F bond cleavage reactions, to provide α-silyldifluoroacetates **17** in 55-66% yields (Scheme 5) (*53*). In these reactions, the initial products must be the *O*-silylated compounds (difluoroketene silyl acetals) which then rearranged to *C*-silylated difluoroacetates **17** under the reaction conditions. The resultant difluoroacetates **17** are versatile building block for highly functionalized difluoroacetates which can be prepared by TBAF or KF/CuI-catalyzed C-C bond formation at the α-position with a variety of electrophiles (*34,35*).

Scheme 5

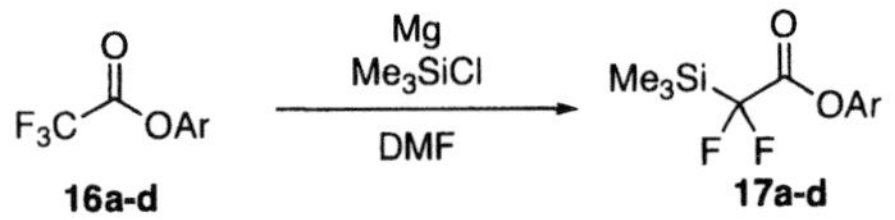

F_3C–C(=O)–OAr (**16a-d**) → [Mg, Me_3SiCl, DMF] → Me_3Si–CF_2–C(=O)–OAr (**17a-d**)

Ar = Ph (a), 4-$MeOC_6H_4$ (b), 4-MeC_6H_4 (c), 4-ClC_6H_4 (d).
Yield: 66% (**17a**), 64% (**17b**), 56% (**17c**), 55% (**17d**).

The present protocol was also applicable for the selective defluorination of trifluoromethyl imines **18**, to yield the corresponding *N*-silylated difluoroenamines **19** (Scheme 6) (*54*).

Scheme 6

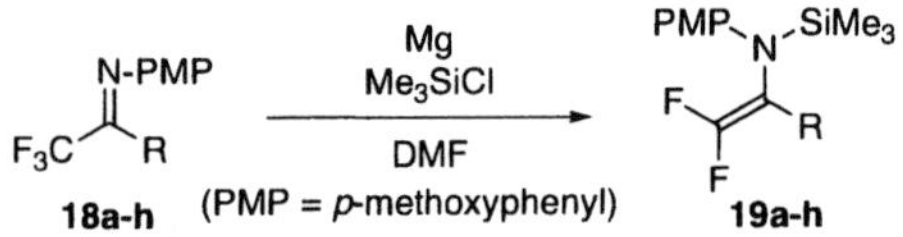

F_3C–C(=N-PMP)–R (**18a-h**) → [Mg, Me_3SiCl, DMF] → F_2C=C(R)–N(PMP)$SiMe_3$ (**19a-h**)

(PMP = *p*-methoxyphenyl)

R = Ph (a), 4-$MeOC_6H_4$ (b), 4-MeC_6H_4 (c), 4-ClC_6H_4 (d), 2-Py (e), 2-thienyl (f), C_6H_{13} (g), H (h).

Yield: 82% (**19a**), 99% (**19b**), 88% (**19c**), 99% (**19d**), 72% (**19e**), 63% (**19f**), 77% (**19g**), 47% (**19h**).

Metallic magnesium, which serves as a convenient electron source, proves useful for the C-F bond breaking process of trifluoromethylated aromatics. As a promising application for material science, we developed the C-F bond cleavage sequence for the synthesis of octafluoro[2.2]paracyclophane (AF4, **23**), which is an excellent precursor of the insulating parylene polymer (Scheme 7) (*55*). Mg(0)-promoted defluorinative silylation of 1,4-bis(trifluoromethyl)benzene (**20**) gave α,α,α',α',α'-pentafluoroxylene (**21**) in 48% yield. The subsequent conjugative 1,6-elimination from **21** induced by the catalytic use of CsF provided fluorocyclophane AF4 (**23**) in 53% yield (*56*).

Scheme 7

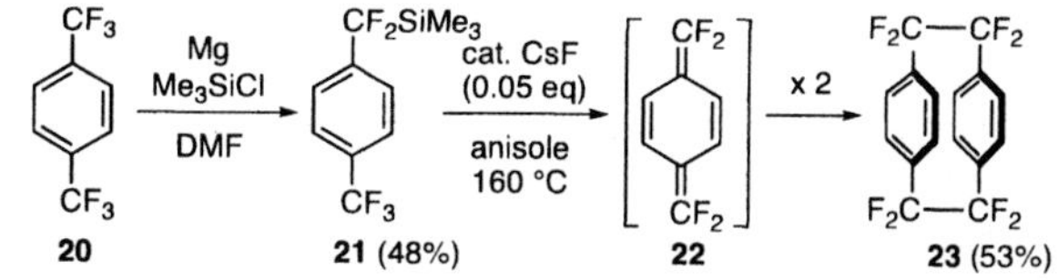

CF_3 Mg Me_3SiCl DMF CF_2SiMe_3 cat. CsF (0.05 eq) anisole 160 °C CF_2 x 2 F_2C—CF_2

CF_3 **20** CF_3 **21** (48%) CF_2 **22** F_2C—CF_2 **23** (53%)

Trifluoromethyl sulfoxides and sulfones, endowed with S-O double bonds, are considered to be good candidates for the reductive C-F bond cleavage reactions. Recently, Prakash and Olah *et al.* reported the interesting, general and efficient method for the preparation of tri- and difluoromethylsilanes by means of unusual Mg metal-mediated reductive tri- and difluoromethylation of chlorosilanes (Scheme 8) (*57*). The defluorination reactions of fluoromethylated sulfoxides and sulfones did not occur. The electron-transfer from magnesium metal to sulfoxide **24**, the subsequent reductive cleavage of the C-S bond to generate anionic CF_3-species was anticipated over the C-F bond fission, providing trifluoromethylsilanes **25** selectively.

Scheme 8

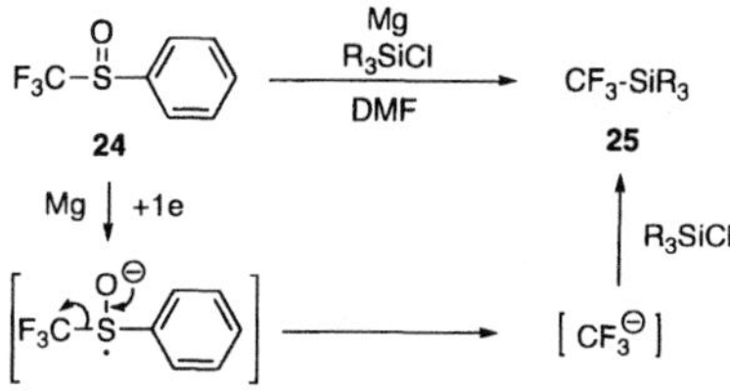

$F_3C-S(O)-C_6H_5$ **24** —Mg, R_3SiCl, DMF→ CF_3-SiR_3 **25**

Mg ↓ +1e

$[F_3C-S(O^{\ominus})\cdot-C_6H_5]$ ——→ $[CF_3^{\ominus}]$ —R_3SiCl→ **25**

Synthetic Application to Difluorinated Amino Acids. Among various organofluorine compounds, fluorinated amino acids have been studied as potential enzyme inhibitors and therapeutic agents (*58-60*). Recently, in the amino acid and peptide chemistry, amino acids possessing two fluorine atoms at the β-carbon have been paid much attention because they can act as potent inactivators of certain enzymes, in particular, highly selective inhibitors of pyridoxal phosphate-dependent enzymes *via* a suicide-type mechanism, and can block certain important metabolic pathways (*1c,61,62*).

A survey of our strategy to synthesize chiral β,β-difluorinated amino acids from trifluoroacetic acid (TFA) is shown in Scheme 9.

Scheme 9

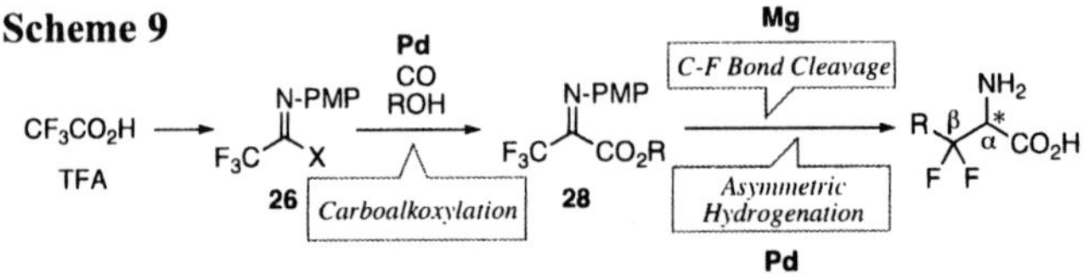

CF_3CO_2H (TFA) → F_3C–C(=N-PMP)–X **26** → [Pd, CO, ROH; *Carboalkoxylation*] → F_3C–C(=N-PMP)–CO_2R **28** → [Mg, *C-F Bond Cleavage*; Pd, *Asymmetric Hydrogenation*] → R–CF_2(β)–C*H(NH_2)(α)–CO_2H

Trifluoroacetimidoyl chloride **26a** is a stable compound and is easily prepared in high yield by refluxing a mixture of commercially available TFA, *p*-anisidine, PPh_3 and Et_3N in CCl_4 in one pot (Scheme 10) (*63*). Trifluoroacetimidoyl iodide **27** was obtained by halogen-exchange reaction of chloride **26a** with NaI in acetone quantitatively. We already established Pd-catalyzed carboalkoxylation of imidoyl iodide **27** to provide iminoesters **28** (*64*).

Scheme 10

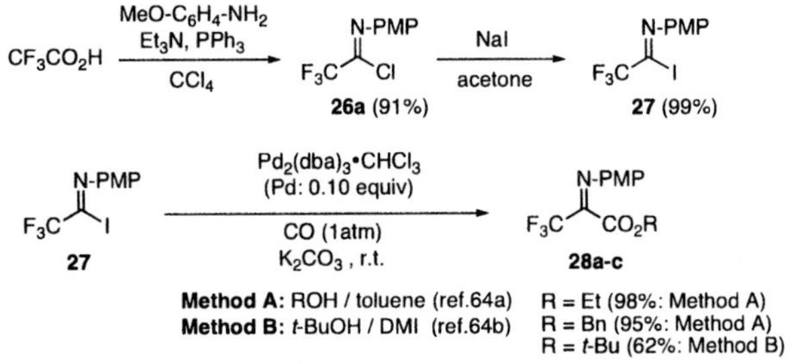

CF_3CO_2H → [MeO-C_6H_4-NH_2, Et_3N, PPh_3, CCl_4] → F_3C–C(=N-PMP)–Cl **26a** (91%) → [NaI, acetone] → F_3C–C(=N-PMP)–I **27** (99%)

F_3C–C(=N-PMP)–I **27** → [$Pd_2(dba)_3$•$CHCl_3$ (Pd: 0.10 equiv), CO (1atm), K_2CO_3, r.t.] → F_3C–C(=N-PMP)–CO_2R **28a-c**

Method A: ROH / toluene (ref.64a)
Method B: *t*-BuOH / DMI (ref.64b)

R = Et (98%: Method A)
R = Bn (95%: Method A)
R = *t*-Bu (62%: Method B)

Trifluoromethyl iminoester **28a** underwent reductive defluorination upon treatment with metallic magnesium and trimethylsilyl chloride, leading to enaminoester **29** (eq 20) (*54*).

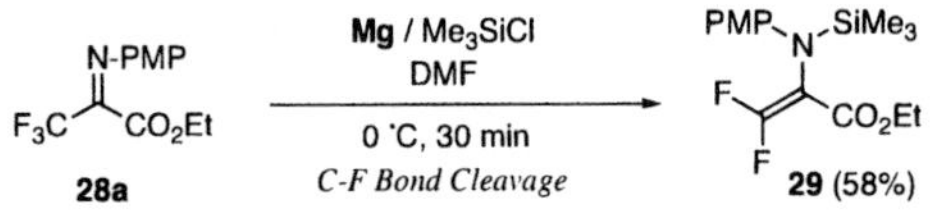

F_3C–C(=N-PMP)–CO_2Et **28a** → [Mg / Me_3SiCl, DMF, 0 °C, 30 min; *C-F Bond Cleavage*] → F_2C=C(N(PMP)$SiMe_3$)–CO_2Et **29** (58%) (20)

Aminodifluoroacrylate **29** is a very interesting compound possessing both *N*-silyl-difluoroenamine and difluoroacrylate moieties. Due to the unique structure, enamine **29** is useful precursor to a wide repertoire of difluorinated α-amino acids, since it can react with not only electrophiles, but also nucleophiles and radical species at the β-position, regioselectively (Scheme 11).

Scheme 11

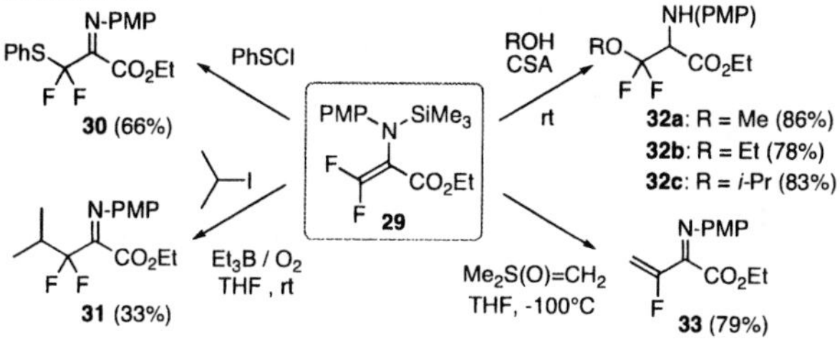

Electrophilic introduction of a thiophenyl group was accomplished by the exposure of **29** to the action of phenylsulfenyl chloride, affording difluorinated cysteine derivative (*54*). Performing the reaction of **29** with isopropyl iodide under radical conditions gave difluorinated leucine derivative **31** without accompanying defluorination (*33*). On the other hand, the difluoromethylene carbon of **29** is highly reactive even to a weak nucleophile like alcohols. Thus, regioselective nucleophilic addition of various alcohols to **29** proceeded smoothly in the presence of CSA to give the corresponding 3,3-difluoroserine derivatives **32** in good yields (*54*). Meanwhile, the reaction of difluoroenamine **29** with sulfur-ylide was accompanied with defluorination to afford α-fluoro-α,β-unsaturated imine **33** (*65*).

Also interesting and important is the asymmetric synthesis of difluorinated α-amino acid derivatives (Scheme 12).

Scheme 12

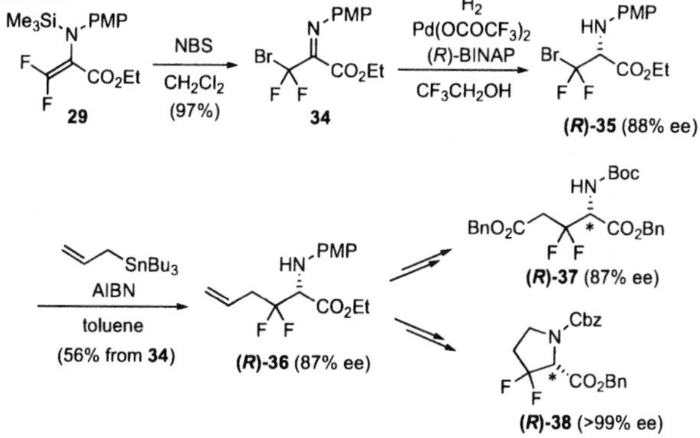

Enaminoester **29** was treated with NBS to provide the corresponding bromodifluoromethyl iminoester **34** in excellent yield. When iminoester **34** was subjected to a hydrogen pressure in the presence of small amount of

Pd($OCOCF_3$)$_2$ and (*R*)-BINAP in 2,2,2-trifluoroethanol (TFE) (*66*), the catalytic asymmetric hydrogenation of **34** proceeded smoothly at room temperature to yield aminoester (*R*)-**35** in 88% ee. Upon treatment with allyltributyltin and a catalytic amount of AIBN as a radical initiator, (*R*)-**35** underwent radical allylation to afford *C*-allylated aminoester (*R*)-**36**. By the use of chiral aminoester (*R*)-**36**, the syntheses of highly enantioenriched *N*- and *O*-protected β,β-difluoroglutamic acids (**37**) and β,β-difluoroprolines (**38**) were accomplished (*67*).

Development of Fluorinated Dianion Equivalents. A fluorinated synthon acting as a dianion equivalent allows a straightforward access to a large number of multifunctionalized organofluorine compounds by the reactions with various combinations of electrophiles. Among them, quite interesting are the dianion equivalents (**39-41**) endowed with acyl anion moieties (Fig. 2) (*11,12,20,68*).

Fig. 2

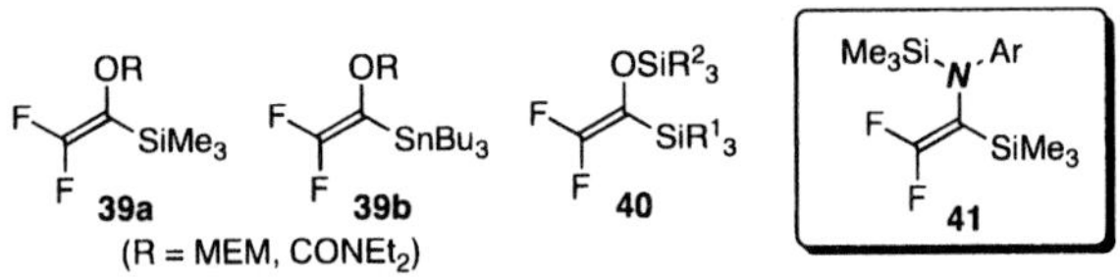

(R = MEM, CONEt$_2$)

Very recently, we have developed a new entry to the fluorinated bifunctional building blocks **41** serving as both an acyl anion synthon (*69*) and an enolate synthon. One-pot reaction sequence involving Mg(0)-promoted reductive C-F and C-Cl bond cleavage reactions of trifluoroacetimidoyl chlorides **26** afforded bis-silylated difluoroenamines **41** (Scheme 13) (*70*).

The preparation of the requisite dianion equivalents **41** is very simple. When imidoyl chlorides **26** were treated with Mg metal (8 mole equiv to **26**) and chlorotrimethylsilane (4 equiv) in distilled THF at 0 °C for 30 min, the dehalogenative double silylation reactions proceeded smoothly to afford bis-silylated difluoroenamines **41** in high yields. Both C-F and imidoyl C-Cl bonds in **26** were cleaved subsequently under mild conditions.

Scheme 13

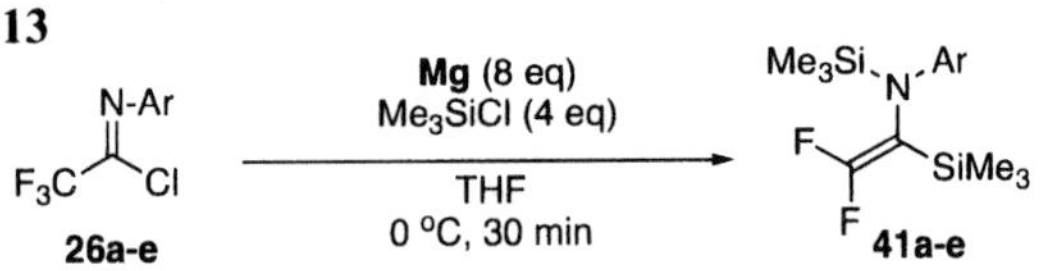

Ar = 4-MeOC_6H_4 (a), Ph (b), 4-MeC_6H_4 (c), 3-ClC_6H_4 (d), 4-ClC_6H_4.
Yield: 84% (**41a**), 85% (**41b**), 82% (**41c**), 80% (**41d**), 85% (**41e**).

By means of the successive double dehalogenation reactions as shown above, the resultant bis-silylated difluoroenamines **41** are very promising bifunctional synthetic blocks, which have not only *N*-silylenamine skeletons but also α-aminovinylsilane skeletons. Subsequent transformations of the bis(silyl)enamine **41** with two kinds of electrophiles gave a variety of difluorinated imines and enamines.

In Scheme 14, an application of **41** is represented by chemoselective sequential transformations with different electrophiles.

Scheme 14

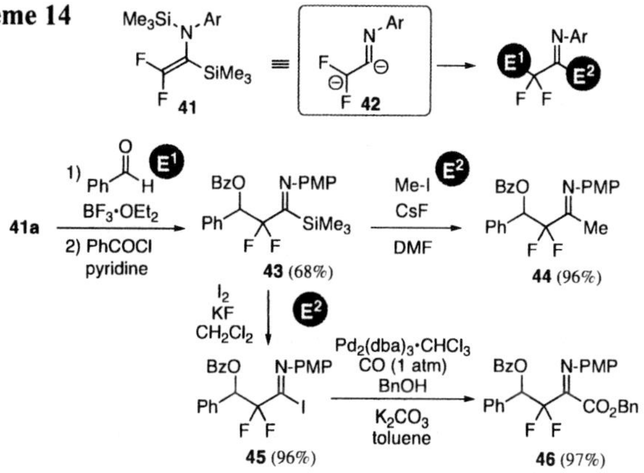

41 ≡ 42 → E1, E2

41a — 1) PhCHO (E1), $BF_3 \cdot OEt_2$; 2) PhCOCl, pyridine → **43** (68%) — Me-I (E2), CsF, DMF → **44** (96%)

43 — I_2, KF, CH_2Cl_2 (E2) → **45** (96%) — $Pd_2(dba)_3 \cdot CHCl_3$, CO (1 atm), BnOH, K_2CO_3, toluene → **46** (97%)

First, by utilizing their properties of difluoroenamines, **41** could react with electrophiles (E^1) to afford the difluorinated imidoyl silanes (*C*-alkylation products). Second, the resulting imidoyl silanes would undergo in turn the reactions with other electrophiles (E^2) by utilizing their properties of imidoylsilanes, giving a wide variety of difluorinated iminoacyl compounds. In general, an imidoylsilane is a stable compound, but it is amenable to a reaction as an acyl anion equivalent in the presence of fluoride ion (*71*). When *O*-protected difluoroimidoylsilane **43** was exposed to an excess amount of CH_3I in the presence of CsF, the corresponding methyl imine **44** was afforded in 96% yield. Furthermore, we have explored iodination of *O*-protected imidoylsilane **43**. In the presence of KF, the reaction of **32** with I_2 proceeded cleanly to afford imidoyl iodide **45** in high yield. The subsequent Pd-catalyzed carboalkoxylation reaction of imidoyl iodide **45** provided the iminoester **46**, a promising precursors of the highly functionalized α-amino acids.

Next, we show the alternative sequential double functionalization of bis(silyl)enamines **41** involving *N*-alkylation reactions (Scheme 15). Upon treatment with KF and CuI, the reaction of bis(silyl)enamines **41** with allyl bromide proceeded to afford the corresponding *N*-allyl difluoroenamine **48** in 92% yield. Then, the further transformations using the vinylsilane grouping in **48** were executed. In the presence of CsF, the reactions of *N*-allyl enamine **48** with benzaldehyde, cinnamyl aldehyde, and benzophenone proceeded smoothly to provide allylic alcohols **49**.

Scheme 15

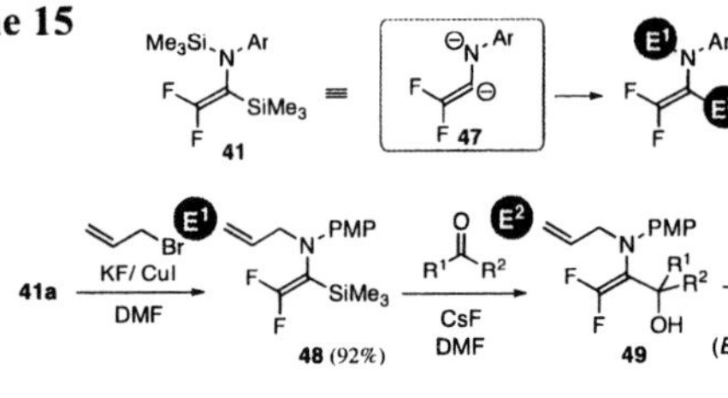

R^1	R^2	Yield (%)
Ph	H	92
(*E*)-PhCH=CH-	H	70
Ph	Ph	47

A particularly good use of *N*-allyl enamines **48** and **49** is made for thermal 3,3-sigmatropic rearrangement, so-called Claisen rearrangement. The transformations involving Claisen rearrangement of the fluorinated *allyl vinyl ethers* have received increasing attention (*72*). Actually, Percy, Taguchi, Shi and many other groups investigated Claisen rearrangement of fluorinated allyl vinyl ethers, to find the exclusive formation of the Claisen rearrangement products (*12e,16a,73*). In the thermal reactions of *N*-allylic difluoroenamine **48**, unusual reaction selectivities were observed as compared with that of allyl vinyl ethers (*74*).

What is the difference? Heating a solution of *N*-allyl difluoroenamine **48** to reflux in xylene led to complete consumption of the starting material. Against our expectations of the selective formation of the Claisen rearrangement product **50**, azabicyclo[2.1.1]hexane **51** was obtained as a major product (Scheme 16). Thermal [3,3]-sigmatropic rearrangement usually proceeds *via* a six-membered transition state A. The formation of bicyclo compound **51** could be explained *via* 1,5-crossed ring-closure (intramolecular [2+2]-cycloaddition) of allyl enamine **48**.

Scheme 16

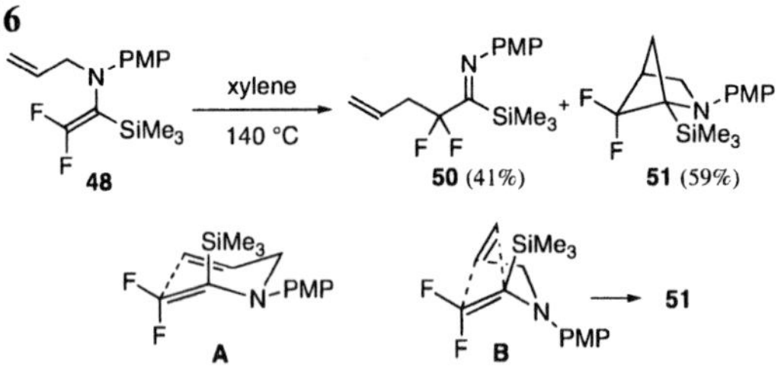

Furthermore, the thermal reaction of *N*-prenyl difluoroenamine **52** was examined. Unexpectedly, the intramolecular ene reaction proceeded smoothly to give isopropenyl pyrrolidine **53** as a sole product (Scheme 17). The exclusive formation of **53** (5-exo ring closure) can be explained *via* the transition state C, which leads to the *trans* relationship of the silyl and isopropenyl groups in **53**.

Scheme 17

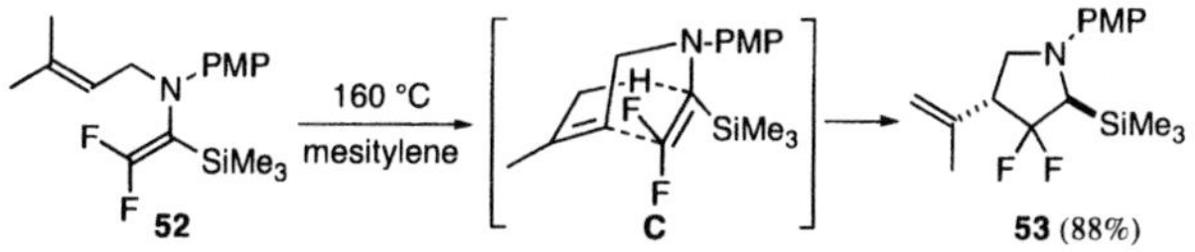

Conclusion. A property of a C-F bond, one of the strongest bonds to cleave, is quite attractive for the wide fields of science and technology. Due to their inertness and resistance to oxidative degradation, organofluorine compounds have found widespread use in practical applications. As a consequence, the progress in the development of selective C-F bond activation reactions has been slow despite their great synthetic potential.

The reactions described here present examples of practical synthetic transformations involving C-F bond cleavage. We have developed Mg-promoted defluorinative silylation of trifluoromethyl ketones, esters, imines, and aromatics in which each trifluoromethyl group is attached to π-electron system. The present reductive defluorination reactions give us a variety of useful fluorinated building blocks. Thus, the design for the reaction sequences involving C-F bond cleavage enables the development of various practical transformations, and expands the scope of organofluorine chemistry.

Acknowledgment. These works were supported by Grant-in-Aid from the Ministry of Education, Culture, Sports, Science and Technology of Japan. We also thank the SC-NMR laboratory of Okayama University for ^{19}F NMR analysis.

References

1. (a) Welch, J. T.; Eswarakrishnan, S. *Fluorine in Bioorganic Chemistry*; John Wiley & Sons, Inc.: New York, 1991. (b) Ojima, I; McCarthy, J. R.; Welch, J. T. ed., *Biomedical Frontiers of Fluorine Chemistry*, ACS Symposium Series 639, 1996. (c) Filler, R.; Kobayashi, Y. *Biomedicinal Aspects of Fluorine Chemistry*, Kodansha Ltd., Tokyo, 1982. (d) Ishikawa, N. Ed. *Synthesis and Speciality of Organofluorine Compounds*, CMC, Tokyo, 1987. (e) Ishikawa, N. Ed. *Biologically Active Organofluorine Compounds*, CMC, Tokyo, 1990.

2. (a) Welch, J. T. *Tetrahedron* **1987**, *43*, 3123–3197. (b) Resnati, G. *Tetrahedron* **1993**, *49*, 9385–9445.
3. (a) Gelb, M. H.; Svaren, J. P.; Abeles, R. H. *Biochemistry* **1985**, *24*, 1813–1817. (b) Gelb, M. H. *J. Am. Chem. Soc.* **1986**, *108*, 3146–3147.
4. Whitten, J. P.; Barney, C. L.; Huber, E. W.; Bey, P.; McCarthy, J. R. *Tetrahedron Lett.* **1989**, *30*, 3649–3652.
5. Shah, N. V.; Cama, L. D. *Heterocycles* **1987**, *25*, 221–227.
6. (a) Kitagawa, O.; Taguchi, T.; Kobayashi, Y. *Tetrahedron Lett.* **1988**, *29*, 1803–1806. (b) Taguchi, T.; Kitagawa, O.; Suda, Y.; Ohkawa, S.; Hashimoto, A.; Iitaka, Y.; Kobayashi, Y. *Tetrahedron Lett.* **1988**, *29*, 5291–5294. (c) Kitagawa, O.; Hashimoto, A.; Kobayashi, Y.; Taguchi, T. *Chem. Lett.* **1990**, 1307–1310.
7. Burton, D. J.; Easdon, J. C. *J. Fluorine Chem.* **1988**, *38*, 125–129.
8. (a) Iseki, K.; Kuroki, Y.; Asada, D.; Kobayashi, Y. *Tetrahedron Lett.* **1997**, *38*, 1447–1448. (b) Iseki, K.; Kuroki, Y.; Asada, D.; Takahashi, M.; Kishimoto, S.; Kobayashi, Y. *Tetrahedron* **1997**, *53*, 10271–10280. (c) Iseki, K. *Tetrahedron* **1998**, *54*, 13887–18914. (d) Iseki, K.; Asada, D.; Kuroki, Y. *J. Fluorine Chem.* **1999**, *97*, 85–89. (e) Kuroki, Y.; Asada, D.; Iseki, K. *Tetrahedron Lett.* **2000**, *41*, 9853–9858.
9. (a) Yamana, M.; Ishihara, T.; Ando, T. *Tetrahedron Lett.* **1983**, *24*, 507–510. (b) Ishihara, T.; Yamana, M.; Ando, T. *Chem. Lett.* **1984**, 1165–1168. (c) Kuroboshi, M.; Ishihara, T. *Bull. Chem. Soc. Jpn.* **1990**, *63*, 428–437.
10. (a) Nakai, T.; Tanaka, K.; Ishikawa, N. *Chem. Lett.* **1976**, 1263–1266. (b) Nakai, T.; Tanaka, K.; Ishikawa, N. *Chem. Lett.* **1977**, 1379–1382. (c) Tanaka, K.; Nakai, T.; Ishikawa, N. *Tetrahedron. Lett.* **1978**, *19*, 4809–4810.
11. (a) Ichikawa, J.; Sonoda, T.; Kobayashi, H. *Tetrahedron Lett.* **1989**, *30*, 1641–1644. (b) Ichikawa, J.; Sonoda, T.; Kobayashi, H. *Tetrahedron Lett.* **1989**, *30*, 5437–5438. (c) Ichikawa, J.; Sonoda, T.; Kobayashi, H. *Tetrahedron Lett.* **1989**, *30*, 6379–6382. (d) Ichikawa, J.; Moriya, T.; Sonoda, T.; Kobayashi, H. *Chem. Lett.* **1991**, 961. (e) Ichikawa, J.; Hamada, S.; Sonoda, T.; Kobayashi, H. *Tetrahedron Lett.* **1992**, *33*, 337–340. (f) Okada, Y.; Minami, T.; Yamamoto, T.; Ichikawa, J. *Chem. Lett.* **1992**, 547–550. (g) Ichikawa, J.; Minami, T.; Sonoda, T.; Kobayashi, H. *Tetrahedron Lett.* **1992**, *33*, 3779–3782. (h) Ichikawa, J.; Ikeura, C.; Minami, T. *Synlett* **1992**, 739–740. (i) Ichikawa, J.; Yonemaru, S.; Minami, T. *Synlett* **1992**, 833–834.
12. (a) Balnaves, A. S.; Gelbrich, T.; Hursthouse, M. B.; Light, M. E.; Palmer, M. J.; Percy, J. M. *J. Chem. Soc., Perkin Trans. 1* **1999**, 2525–2535. (b) DeBoos, G. A.; Fullbrook, J. J. ; Owton, W. M.; Percy, J. M. ; Thomas, A. C. *Synlett* **2000**, 963–966. (c) Broadhurst, M. J.; Brown, S. J.; Percy. J. M.; Prime, M. E. *J. Chem. Soc., Perkin Trans. 1* **2000**, 3217–3226. (d) DeBoos, G. A.; Fullbrook, J. J.; Percy, J. M. *Org. Lett.* **2001**, *3*, 2859–2861. (e) Garayt, M. R.; Percy, J. M. *Tetrahedron Lett.* **2001**, *42*, 6377–6380.

13. (a) Qian, C.-P.; Nakai, T. *Tetrahedron Lett.* **1988**, *29*, 4119–4122. (b) Qian, C.-P.; Nakai, T.; Dixon, D. A.; Smart, B. E. *J. Am. Chem. Soc.* **1990**, *112*, 4602–4604. (c) T. Narita, T.; Hagiwara, T.; Hamana, H.; Tomooka, K.; Liu, Y. -Z.; Nakai, T. *Tetrahedron Lett.* **1995**, *36*, 6091–6094.
14. (a) Funabiki, K.; Suzuki, C.; Takamoto, S.; Matsui, M.; Shibata, K. *J. Chem. Soc., Perkin Trans. 1* **1997**, 2679–2680. (b) Funabiki, K.; Ohtsuki, T.; Ishihara, T.; Yamanaka, H. *J. Chem. Soc., Perkin Trans. 1* **1998**, 2413–2423. (c) Funabiki, K.; Fukushima, Y.; Matsui, M.; Shibata, K. *J. Org. Chem.* **2000**, *65*, 606–609. (d) Funabiki, K. *J. Synth. Org. Chem. Jpn.* **2004**, *62*, 607–615 and references cited therein.
15. Osipov, S. N.; Golubev, A.; Sewald, N.; Kolomiets, A. F.; Fokin, A. V.; Burger, K. *Synlett* **1995**, 1269–1270.
16. (a) Shi, G.-Q.; Cai, W.-L. *J. Org. Chem.* **1995**, *60*, 6289–6295. (b) Shi, G.-Q.; Cao, Z.-Y; Zhang, X.-B. *J. Org. Chem.* **1995**, *60*, 6608–6611. (c) Jiang, B.; Zhang, X.; Shi, G.-Q. *Tetrahedron Lett.* **2002**, *43*, 6819–6821.
17. (a) Jin, F.; Jiang, B.; Xu, Y. *Tetrahedron Lett.* **1992**, *33*, 1221–1224. (b) Jin, F.; Hu, Y.; Huang, W. *J. Chem. Soc., Chem. Commun.* **1993**, 814–816. (c) Jin, F.; Xu, Y.; Huang, W. *J. Chem. Soc., Perkin Trans. 1* **1993**, 795–799.
18. (a) Portella, C.; Brigaud, T.; Lefebvre, O.; Plantier-Royon, R. *J. Fluorine Chem.* **2000**, *101*, 193–198. (b) Berber, H.; Brigand, T.; Lefebvre, O.; Plantier-Royon, R. Portella, C. *Chem. Eur. J.* **2001**, *7*, 903–909. (c) Lefebvre, O.; Brigaud, T.; Portella, C. *J. Org. Chem.* **2001**, *66*, 1941–1946. (d) Lefebvre, O.; Brigaud, T.; Portella, C. *J. Org. Chem.* **2001**, *66*, 4348–4351. (e) Saleur, D.; Bouillon, J. P.; Portella, C. *J. Org. Chem.* **2001**, *66*, 4543–4548.
19. Fleming, I.; Roberts, R. S.; Smith, S. C. *J. Chem. Soc., Perkin Trans. 1* **1998**, 1215–1228.
20. (a) Higashiya, S.; Lim, D. S.; Ngo, S. C.; Toscano, P. J.; Welch, J. T. Abstract of Paper, 222nd National Meeting of the American Chemical Society, Chicago, IL, **2001**: Washington, DC, ORGN 41. (b) Ngo, S. C.; Chung, W. J.; Lim, D. S.; Shigashiya, S.; Welch, J. T. *J. Fluorine Chem.* **2002**, *117*, 207–211. (c) Chung, W. J.; Higashiya, S.; Oba, Y.; Welch, J. T. *Tetrahedron* **2003**, *59*, 10031–10036. (d) Chung, W. J.; Welch, J. T. *J. Fluorine Chem.* **2004**, *125*, 543–548. (e) Chung, W. J.; Ngo, S. C.; Higashiya. S.; Welch, J. T. *Tetrahedron Lett.* **2004**, *45*, 5403–5406.
21. Uneyama, K.; Yan, F.; Hirama, S.; Katagiri, T. *Tetrahedron Lett.* **1996**, *37*, 2045–2048.
22. Portella, C.; Shermolovich, Y. *Tetrahedron Lett.* **1996**, *37*, 4063–4064.
23. There are several recent reviews on C-F activation by metal reagents, see: (a) Kiplinger, J. L.; Richmond, T. G.; Osterberg, C. E. *Chem. Rev.* **1994**, *94*, 373–431. (b) Burdeniuc, J.; Jedlicka, B.; Crabtree, R. H. *Chem. Ber./Recueil* **1997**, *130*, 145–154. (c) Richmond, T. G. In *Activating of Unreactive Bonds*

and Organic Synthesis (Topics in Organometallic Chemistry, Vol. 3); Murai, S., Ed.; Springer: New York, 1999; Vol. 3, pp 243–269.
24. (a) van der Boom, M. E.; Ben-David, Y.; Milstein, D. *J. Am. Chem. Soc.* **1999**, *121*, 6652–6656. (b) Kraft, B. M.; Lachicotte, R. J.; Jones, W. D. *J. Am. Chem. Soc.* **2001**, *123*, 10973–10979 and references cited therein.
25. Ooi, T.; Uraguchi, D.; Kagoshima, N.; Maruoka, K. *Tetrahedron Lett.* **1997**, *38*, 5679–5682.
26. Terao, J.; Ikumi, A.; Kuniyasu, H.; Kambe, N. *J. Am. Chem. Soc.* **2003**, *125*, 5646–5647.
27. Hirano, K.; Fujita, K.; Yorimitsu, H.; Shinokubo, H.; Oshima, K. *Tetrahedron Lett.* **2004**, *45*, 2555–2557.
28. Barma, D. K.; Kundu, A.; Zhang, H. M.; Mioskowski, C.; Falck, J. R. *J. Am. Chem. Soc.* **2003**, *125*, 3218–3219.
29. (a) Guijarro, D.; Yus, M. *Tetrahedron* **2000**, *56*, 1135–1138. (b) Guijarro, D; Yus, M. *J. Organomet. Chem.* **2001**, *624*, 53–57. (c) Herrera, R. P.; Guijarro, A.; Yus, M. *Tetrahedron Lett.* **2003**, *44*, 1309–1312. (d) Herrera, R. P.; Guijarro, A.; Yus, M. *Tetrahedron Lett.* **2003**, *44*, 1313–1316. (e) Guijarro, D.; Martinez, P.; Yus, M. *Tetrahedron* **2003**, *59*, 1237–1244. (f) Yus, M.; Herrera, R. P.; Guijarro, A. *Tetrahedron Lett.* **2003**, *44*, 5025–5027.
30. Stocker, J. H.; Jenevein, R. M. *J. Chem. Soc., Chem. Commun.* **1968**, 934–935.
31. Electroreductive defluorination of fluoromethylarenes: (a) Saboureau, C.; Troupel, M.; Sibille, S.; Périchon, J. *J. Chem. Soc., Chem. Commun.* **1989**, 1138–1139. (b) Andrieux, C. P.; Combellas, C.; Kanoufi, F.; Savéant, J.-M.; Thiébault, A. *J. Am. Chem. Soc.* **1997**, *119*, 9527–9540. (c) Clavel, P.; Léger-Lambert, M. P.; Biran, C.; Serein-Spirau, F.; Bordeau, M.; Roques, N.; Marzouk, H.; *Synthesis*, **1999**, 829–834. (d) Clavel, P.; Lessene, G.; Biran, C.; Bordeau, M.; Roques, N.; Trévin, S.; de Montauzon, D. *J. Fluorine Chem.* **2001**, *107*, 301–310.
32. Uneyama, K.; Maeda, K.; Kato, T.; Katagiri, T. *Tetrahedron Lett.* **1998**, *39*, 3741–3744.
33. Uneyama, K.; Kato, T. *Tetrahedron Lett.* **1998**, *39*, 587–590.
34. Uneyama, K.; Mizutani, G. *Chem. Commun.* **1999**, 613–614.
35. Uneyama, K.; Mizutani G.; Maeda, K.; Kato, T. *J. Org. Chem.* **1999**, *64*, 6717–6723.
36. A short account: Uneyama, K.; Amii, H. *J. Fluorine Chem.* **2002**, *114*, 127–131.
37. Amii, H.; Kobayashi, T.; Hatamoto, Y.; Uneyama, K. *Chem. Commun.* **1999**, 1323–1324.
38. (a) Ishino, Y.; Kita, Y.; Maekawa, H.; Ohno, T.; Yamasaki, Y.; Miyata, T.; Nishiguchi, I. *Tetrahedron Lett.* **1999**, *40*, 1349–1352. (b) Ohno, T.; Sakai, M.; Ishino, Y.; Shibata, T.; Maekawa, H.; Nishiguchi, I. *Org. Lett.* **2001**, *3*,

3439–3442. (c) Nishiguchi, I.; Sakai, M.; Maekawa, H.; Ohno, T.; Yamamoto, Y.; Ishino, Y. *Tetrahedron Lett.* **2002**, *43*, 635–637. (d) Kyoda, M.; Yokoyama, T.; Kuwahara, T.; Maekawa, H.; Nishiguchi, I. *Chem. Lett.* **2002**, 228–229. (e) Nishiguchi, I.; Yamamoto, Y.; Sakai, M.; Ohno, T.; Ishino, Y.; Maekawa, H. *Synlett* **2002**, 759–762. (f) Yamamoto, Y.; Kawano, S.; Maekawa, H.; Nishiguchi, I. *Synlett* **2004**, 30–36. (g) Maekawa, H.; Sakai, M.; Uchida, T.; Kita, Y.; Nishiguchi, I. *Tetrahedron Lett.* **2004**, *45*, 607–609.

39. (a) Chorki, F.; Crousse, B.; Ourevitch, M.; Bonnet-Delpon, D.; Bégué, J.-P.; Brigaud T.; Portella, C. *Tetrahedron Lett.* **2001**, *42*, 1487–1489. (b) Chorki, F.; Grellepois, F.; Crousse, B.; Ourevitch, M.; Bonnet-Delpon, D.; Bégué, J.-P. *J. Org. Chem.* **2001**, *66*, 7858–7867.
40. (a) Takeuchi, Y.; Asahina, M.; Hori, K.; Koizumi, T. *J. Chem. Soc., Perkin Trans. 1* **1988**, 1149–1153. (b) Qiu, Z.-M.; Burton, D. J. *J. Org. Chem.* **1995**, *60*, 5570–5578.
41. (a) Prakash, G. K. S.; Hu, J.; Alauddin, M. M. *J. Fluorine Chem.* **2003**, *121*, 239–243. (b) Prakash, G. K. S.; Alauddin, M. M.; Hu, J. B.; Conti, P. S.; Olah, G. A. *J. Labelled. Compd. Rad.* **2003**, *46*, 1087–1092.
42. Reviews: (a) Danishefsky, S. *Acc. Chem. Res.* **1981**, *14*, 400–406. (b) Danishefsky, S. J.; DeNinno, M. P. *Angew. Chem., Int. Ed. Engl.* **1987**, *26*, 15–23. (c) Danishefsky, S. *Chemtracts: Org. Chem.* **1989**, *2*, 273–297.
43. Amii, H.; Kobayashi, T.; Terasawa, H.; Uneyama, K. *Org. Lett.* **2001**, *3*, 3103–3105.
44. Hojo, M.; Masuda, R.; Kokuryo, Y.; Shioda, H.; Matsuo, S. *Chem. Lett.* **1976**, 499–502.
45. Prakash, G. K. S.; Hu, J.; Olah, G. A. *J. Fluorine Chem.* **2001**, *112*, 355–360.
46. Allmendinger, T.; Furet, P.; Hungerbühler, E. *Tetrahedron Lett.* **1990**, *31*, 7297–7300.
47. (a) Boros, L. G.;Corte, B. D.; Gimi, R. H.; Welch, J. T.; Wu, Y.; Handschumacher, R. E. *Tetrahedron Lett.* **1994**, *35*, 6033–6036. (b) Welch, J. T.; Lin, J. *Tetrahedron* **1996**, *52*, 291–304. (c) Lin, J.; Welch, J. T. *Tetrahedron Lett.* **1998**, *39*, 9613–9616. (d) Welch, J. T.; Allmendinger, T. Fluoroolefin Peptide Isosteres. In *Peptidomimetic Protocols*, Ed. Kazmierski, W. M. Humana Press, New Jersey, **1999**. pp 375–384. (e) Welch, J. T.; Lin, J.; Boros, L. G.; DeCorte, B.; Bergmann, K.; Gimi, R. Fluoro-olefin Isosteres as Peptidomimetics, In ref 1b, pp 129–142.
48. Abraham, R. J.; Ellison, S. L. R.; Schonholzer, P.; Thomas, W. A. *Tetrahedron* **1986**, *42*, 2101–2110.
49. (a) Otaka, A.; Mitsuyama, E.; Watanabe, H.; Tamamura, H.; Fujii, N. *Chem. Commun.* **2000**, 1081–1082. (b) Otaka, A.; Watanabe, H.; Mitsuyama, E.; Yukimasa, A.; Tamamura, H.; Fujii, N. *Tetrahedron Lett.* **2001**, *42*, 285–287. (c) Otaka, A.; Watanabe, H.; Yukimasa, A.; Oishi, S. Tamamura, H.; Fujii, N. *Tetrahedron Lett.* **2001**, *42*, 5443–5446. (d) Otaka, A.; Mitsuyama, E.;

Watanabe, J.; Watanabe, H.; Fujii, N. *Biopolymers* **2004**, *76*, 140–144. (e) Otaka, A.; Watanabe, J.; Yukimasa, A.; Sasaki, Y.; Watanabe, H.; Kinoshita, T.; Oishi, S.; Tamamura, H.; Fujii, N. *J. Org. Chem.* **2004**, *69*, 1634–1645.

50. (a) Sano, S.; Yokoyama, K.; Teranishi, R.; Shiro, M.; Nagao, Y. *Tetrahedron Lett.* **2002**, *43*, 281-284. (b) Sano, S.; Teranishi, R.; Nagao, Y. *Tetrahedron Lett.* **2002**, *43*, 9183–9186. (c) Sano, S.; Saito, K.; Nagao, Y. *Tetrahedron Lett.* **2003**, *44*, 3987–3990. (d) Sano, S.; Takemoto, Y.; Nagao, Y. *Tetrahedron Lett.* **2003**, *44*, 8853–8855.
51. (a) Okada, M.; Nakamura, Y.; Saito, A.; Sato, A.; Horikawa, H.; Taguchi, T. *Chem. Lett.* **2002**, 28–29. (b) Okada, M.; Nakamura, Y.; Saito, A.; Sato, A.; Horikawa, H.; Taguchi, T. *Tetrahedron Lett.* **2002**, *43*, 5845–5847. (c) Nakamura, Y.; Okada, M.; Horikawa, H.; Taguchi, T. *J. Fluorine Chem.* **2002**, *117*, 143–148.
52. Hata, H.; Kobayashi, T.; Amii, H.; Uneyama, K.; Welch, J. T. *Tetrahedron Lett.* **2002**, *43*, 6099–6102.
53. Amii, H.; Kobayashi, T.; Uneyama, K. *Synthesis* **2000**, 2001–2003.
54. Mae, M.; Amii, H.; Uneyama, K. *Tetrahedron Lett.* **2000**, *41*, 7893–7896.
55. (a) Moore, J. A.; Lang, C.-I. Vapor Deposition Polymerization as a Route to Fluorinated Polymers. In *Fluoropolymers 1: Synthesis*; Hougham, G.; Cassidy, P. E.; Johns, K.; Davidson, T. Eds.; Plenum Press: New York, 1999; pp 273–312. (c) Dolbier, W. R., Jr.; Duan, J.-X.; Roche, A. J. *Org. Lett.* **2000**, *2*, 1867–1869. (c) Dolbier, W. R., Jr.; Beach, W. F. *J. Fluorine Chem.* **2003**, *122*, 97–104 and references cited therein.
56. Amii, H.; Hatamoto, Y.; Seo, M.; Uneyama, K. *J. Org. Chem.* **2001**, *66*, 7216–7218.
57. Prakash, G. K. S.; Hu, J.; Olah, G. A. *J. Org. Chem.* **2003**, *68*, 4457–4463.
58. (a) Ondetti, M. A.; Cushman, D. W.; Rubin, B. *Science* **1977**, *196*, 441–444. (b) Cushman, D. W.; Cheung, H. S.; Sabo, E. F.; Ondetti, M. A. *Biochemistry* **1977**, *16*, 5484–5491. (c) Kollonitsch, J.; Perkins, L. M.; Patchett, A. A.; Doldouras, G. A.; Marburg, S.; Duggan, D.E.; Maycock, A. L.; Aster, S. D. *Nature* **1978**, *274*, 906–908. (d) Wyvratt, M. J.; Tristram, E. W.; Ikeler, T. J.; Lohr, N. S.; Joshua, H.; Springer, J. P.; Arison, B. H.; Patchett, A. A. *J. Org. Chem.* **1984**, *49*, 2816–2819.
59. Kukhar, V. P.; Soloshonok, V. A. *Fluorine-containing Amino Acids: Synthesis and Properties*, John Wiley & Sons, Inc.: New York, 1995.
60. (a) Abeles, R. H.; Maycock, A. L. *Acc. Chem. Res.* **1976**, *9*, 313–319. (b) Kumadaki, I. *J. Syn. Org. Chem. Jpn.* **1984**, *42*, 786–793. (c) Faraci, W. S.; Walsh, C. T. *Biochemistry* **1989**, *28*, 431–437.
61. (a) Walsh, C. *Tetrahedron* **1982**, *38*, 871–909. (b) Bey, P. *Ann. Chim. (Paris)* **1984**, *9*, 695–702. (c) Rando, R. R. *Pharmacol. Rev.* **1984**, *36*, 111–142. (d) Rathod, A. H.; Tashjian, A. H. Jr.; Abeles, R. H. *J. Biol. Chem.* **1986**, *261*,

6461–6469. (e) Ojima, I.; Inoue, T.; Chakravarty, S. *J. Fluorine Chem.* **1999**, *97*, 3–10.

62. (a) Uneyama, K. Asymmetric Synthesis of Fluoro-Amino Acids. In *Enantiocontrolled Synthesis of Fluoro-Organic Compounds*, Soloshonok, V. A. Ed.; John Wiley & Sons, Ltd.: Chichester, 1999; pp 391–418. (b) Uneyama, K.; Katagiri, T.; Amii, H. *J. Synth. Org. Chem. Jpn.* **2002**, *60*, 1069–1075.
63. (a) Tamura, K.; Mizukami, H.; Maeda, K.; Watanabe, H.; Uneyama, K. *J. Org. Chem.* **1993**, *58*, 32–35. (b) Uneyama, K. *J. Synth. Org. Chem. Jpn.* **1995**, *53*, 43–52. (c) Uneyama, K. *J. Fluorine Chem.* **1999**, *97*, 11–25.
64. (a) Watanabe, H.; Hashizume, Y.; Uneyama, K. *Tetrahedron Lett.* **1992**, *33*, 4333–4336. (b) Amii, H.; Kishikawa, Y.; Kageyama, K.; Uneyama, K. *J. Org. Chem.* **2000**, *65*, 3404–3408.
65. Mae, M.; Matsuura, M.; Amii, H.; Uneyama, K. *Tetrahedron Lett.* **2002**, *43*, 2069–2072.
66. Abe, H.; Amii, H.; Uneyama, K. *Org. Lett.* **2001**, *3*, 313–315.
67. Suzuki, A.; Mae, M.; Amii, H.; Uneyama, K. *J. Org. Chem.* **2004**, *69*, 5132–5134.
68. (a) Patel, S. T.; Percy, J. M.; Wilkes, R. D. *Tetrahedron* **1995**, *51*, 9201–9216. (b) Howarth, J. A.; Owton, W. M.; Percy, J. M. *J. Chem. Soc., Chem. Commun.* **1995**, 757–758. (c) Howarth, J. A.; Owton, W. M.; Percy, J. M.; Rock, M. H. *Tetrahedron* **1995**, *51*, 10289–10302. (d) Crowley, P. J.; Howarth, J. A.; Owton, W. M.; Percy, J. M.; Stansfield. K. *Tetrahedron Lett.* **1996**, *37*, 5975–5938.
69. Bordeau, M.; Clavel, P.; Barba, A.; Berlande, M.; Biran, C.; Roques, N. *Tetrahedron Lett.* **2003**, *44*, 3741–3744.
70. Kobayashi, T.; Nakagawa, T.; Amii, H.; Uneyama, K. *Org. Lett.* **2003**, *5*, 4297–4300.
71. Uneyama, K.; Noritake, C.; Sadamune, K. *J. Org. Chem.* **1996**, *61*, 6055–6057.
72. A review of rearrangement-based methods for the synthesis of organofluorine compounds; Percy, J. M.; Prime, M. E. *J. Fluorine Chem.* **1999**, *100*, 147–156.
73. (a) Ito, H.; Sato, A.; Kobayashi, T.; Taguchi, T. *Chem. Commun.* **1998**, 2441–2442. (b) Yuan, W.; Berman, R. J.; Gelb, M. H. *J. Am. Chem. Soc.* **1987**, *109*, 8071–8081. (c) Metcalf, B. W.; Jarvi, E. T.; Burkhart, J. P. *Tetrahedron Lett.* **1985**, *26*, 2861–2864.
74. Amii, H.; Ichihara, Y.; Nakagawa, T.; Kobayashi, T.; Uneyama, K. *Chem. Commun.* **2003**, 2902–2903.

Chapter 28

Reactions of Ethyl Bromodifluoroacetate in the Presence of Copper Powder

Kazuyuki Sato, Masaaki Omote, Akira Ando, and Itsumaro Kumadaki*

Faculty of Pharmaceutical Sciences, Setsunan University, 45–1, Nagaotoge-cho, Hirakata, Osaka 573–0101, Japan

Copper-mediated reactions of ethyl bromodifluoroacetate (**1**) as procedures for the synthesis of compounds containing a CF_2 group are presented. The complex formed in the reaction of **1** in the presence of copper powder reacted with vinyl or aryl iodides to give cross-coupling products, with Michael acceptors to give 1,4-addition products and with unactivated olefins to give radical addition products. Some applications of these reactions for synthesis of fluorine analogues of bioactive compounds are presented. A novel reaction mediated by the Wilkinson catalyst will be also presented.

Introduction

The efficient methodologies for the syntheses of organofluorine compounds have become quite important in recent organic synthesis, since many compounds with fluorine substituents are used in various fields, such as medicines, agricultural chemicals, liquid crystals, and so forth. (*1*). One of the authors has been engaged in developing new methodologies for the syntheses of organofluoine compounds, and reported copper-mediated trifluoromethylation of aromatic (*2a*) and aliphatic halides (*2b*) with trifluoromethyl iodide, the ene reaction of trifluoromethyl ketones (*3*) and imines (*4*), and new reactions using Halothane (*5*). Typical examples of these reactions are shown in Scheme 1. Some applications of these reactions for synthesis of fluorine analogues of biologically important compounds (*6,3b*) are also illustrated at the bottom of the scheme.

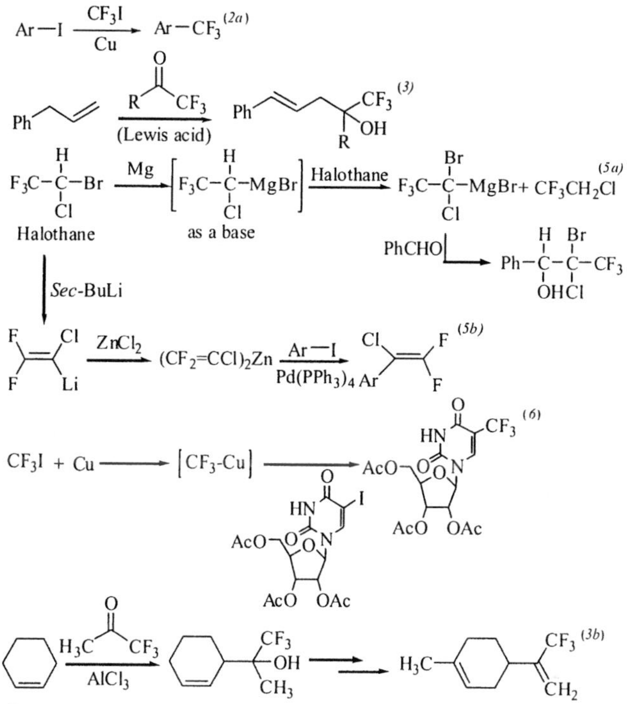

Ar—I —(CF_3I, Cu)→ Ar—CF_3 (*2a*)

Ph-CH=CH$_2$ + RC(O)CF_3 —(Lewis acid)→ Ph-CH=CH-CH$_2$-C(R)(OH)CF_3 (*3*)

$F_3C-CHBrCl$ (Halothane) —Mg→ [$F_3C-CHCl-MgBr$] (as a base) —Halothane→ $F_3C-CBrCl-MgBr$ + CF_3CH_2Cl (*5a*); —PhCHO→ Ph-CH(OH)-CBrCl-CF_3

Halothane —*Sec*-BuLi→ $F_2C=CCl(Li)$ —$ZnCl_2$→ $(CF_2{=}CCl)_2Zn$ —(Ar—I, $Pd(PPh_3)_4$)→ Ar(Cl)C=CF_2 (*5b*)

CF_3I + Cu ——→ [CF_3-Cu] —(5-iodouridine triacetate)→ 5-trifluoromethyluridine triacetate (*6*)

cyclohexene + H_3CC(O)CF_3 —$AlCl_3$→ cyclohexenyl-C(CF_3)(OH)CH_3 ——→ H_3C-cyclohexenyl-C(CF_3)=CH_2 (*3b*)

Scheme 1

Importance of compounds with a difluoromethylene (CF_2) unit are also increasing recently, since this unit is biologically more stable than a methylene unit and it is taken as an isostere of an ether oxgen. Once, we planned to synthesize a difluoro analogue of a biologically active compound. At the start of that research, methods for introducing a CF_2 unit into an organic compound had been investigated energetically (*7*). For this purpose, the following reactions were reported: Addition of CF_2X_2 to olefins (*8*), cyclopropanation of olefins by the reaction of difluorocarbene formed from $ClCF_2COOX$ (*9*), various reactions using metal difluorovinylides (*10*), and so on. Among them, Reformatsky reactions (*11*) using halodifluoroacetate or aldol reaction (*12*) of its enol form had been widely used to introduce a CF_2 unit. These reactions are very useful to give α,α-difluoro-β-hydroxyesters. Thus, ethyl bromodifluoro-acetate (**1**) is recognized as a good starting material for CF_2 compounds. However, it is difficult to deprive the hydroxyl group of the products owing to the high electronegativity of the α-fluorine atoms (*13*). These methodologies were not suitable for our purpose, since our target had no hydroxyl group at the β-position. Therefore, we planned to develop a new methodology for the synthesis of α,α-difluoroesters that do not have a β-hydroxyl group.

As mentioned above, one of the authors had reported copper-mediated trifluoromethylation of halogen compounds with trifluoromethyl iodide. Kobayashi and Taguchi applied this methodology to methyl difluoroiodoacetate (**2**) (*14*). However, **2** is not commercially available, is quite unstable and somewhat difficult to use. Therefore, we planned to reinvestigate the reactions of **1**. An important advantage using this compound as a synthon is that it is commercially available and that several types of reactions can be developed. For example, the bromine atom of **1** could be attacked by an electrophile or a radical (*15*). Furthermore, the products could be modified to a variety of other functional groups by reactions of nucleophiles on the ester moiety (*16*).

Here, we would like to discuss the results of our research using **1** in the presence of active copper powder (*17*) in the next section. Further, a novel reaction of **1** mediated by the Wilkinson catalyst will be presented in the following section.

Results and Discussion

Cross-coupling Reaction

Among synthetic methodologies for an α,α-difluoroester without a β-hydroxyl group, we paid attention to the cross-coupling reaction of methyl difluoroiodoacetate (**2**) with organic halides. In 1986, Kobayashi and co-workers reported a cross-coupling reaction of **2** with various halides in the presence of copper powder (*14d*). The reaction proceeded smoothly in good to excellent yields. However, **2** is not commercially available, is quite unstable and is somewhat difficult to use. Therefore, we examined a procedure similar to that of Kobayashi's group using three equivalents of **1** with (*E*)-1-iodo-1-hexene in HMPA (*18*). The cross-coupled product, ethyl (*E*)-2,2-difluoro-3-octenoate (**3a**), was obtained in a low yield of only 25% and a much larger amount of 1-bromo-1-hexene was formed. These results suggested that the reaction of **1** with copper powder gave a radical anion and Cu(I) ion. The radical anion reacted with another mole of copper to give ethoxycarbonyl-difluoromethylcopper and bromide ion, which reacted with the Cu(I) ion to give

CuBr. Both copper compounds reacted with the iodo compounds to give **3a** and 1-bromo-1-hexene, respectively. We thought that the reaction of the iodo compound with CuBr might be accelerated by the solvent, HMPA. When DMSO was used as a solvent, the yield of **3a** was improved to 42%, but still a considerable amount of the 1-bromo-1-hexene was formed, along with another byproduct, diethyl (*Z*)-4-butyl-2,5,5-trifluoro-2-hexenedioate (**4**). This appeared to be formed by the reaction of **3a** with another equivalent of the organocopper species. In other words, the species formed from **1** must have reacted with **3a** to give **4** through an S_N2' type reaction. To avoid the formation of the byproducts, we used one equivalent of **1**. The yield of **3a** was improved up to 64%, and the formation of the byproducts was suppressed (Scheme 2). These results suggest that the intermediate has an anionic property.

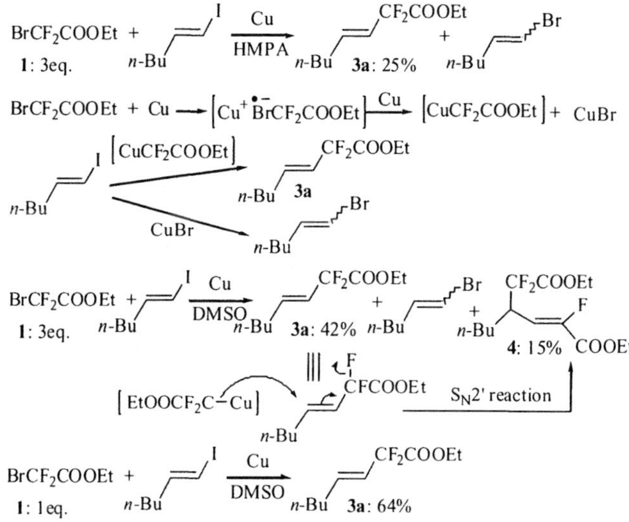

$BrCF_2COOEt$ + I / n-Bu — Cu, HMPA → CF_2COOEt / n-Bu **3a**: 25% + Br / n-Bu
1: 3eq.

$BrCF_2COOEt + Cu \rightarrow [Cu^+ \dot{Br}^-CF_2COOEt] \xrightarrow{Cu} [CuCF_2COOEt] + CuBr$

I / n-Bu — $[CuCF_2COOEt]$ → CF_2COOEt / n-Bu **3a**; — CuBr → Br / n-Bu

$BrCF_2COOEt$ + I / n-Bu — Cu, DMSO → CF_2COOEt / n-Bu **3a**: 42% + Br / n-Bu + CF_2COOEt, F, COOEt / n-Bu **4**: 15%
1: 3eq.

$[EtOOCF_2C-Cu]$ + F / $CFCOOEt$ / n-Bu — S_N2' reaction → **4**

$BrCF_2COOEt$ + I / n-Bu — Cu, DMSO → CF_2COOEt / n-Bu **3a**: 64%
1: 1eq.

Scheme 2

This cross-coupling reaction was applied to other halogen compounds using one equivalent of **1** in DMSO (Table I). Vinyl and aryl iodides gave the corresponding products stereospecifically in high yields, but alkyl or alkynyl iodides and aryl bromide did not give any coupling products.

Recently, Ashwood's group has applied our cross-coupling reaction to various heteroaryl bromides (*19*), and Sakamoto's group has reported synthesis of antifungals using this reaction with heteroaryl halides (*20*).

In this reaction, the reactive species seemed to be an organocopper complex derived from **1**. However, this reaction was different from the cross-

Table I. Reaction of 1 with Halogen Compounds

$$R{-}X + BrCF_2COOEt\ (\mathbf{1}) \xrightarrow[\text{DMSO, 55 °C}]{\text{Cu}} R{-}CF_2COOEt\ (\mathbf{3})$$

Entry	R—X	Time (h)	3		Yield (%)
1	(E)-n-Bu-CH=CH-I	12	(E)-n-Bu-CH=CH-CF_2COOEt	**3a**	64
2	(Z)-n-Bu-CH=CH-I	12	(Z)-n-Bu-CH=CH-CF_2COOEt	**3b**	69
3	n-Bu-C≡C-I	0	n-Bu-C≡C-CF_2COOEt	**3c**	0[a]
4	alkyl iodide	6	alkyl-CF_2COOEt	**3d**	0
5	Ph—I	5	Ph—CF_2COOEt	**3e**	53
6	Ph—Br	5	Ph—CF_2COOEt	**3e**	0
7	O_2N-C_6H_4-I	9	O_2N-C_6H_4-CF_2COOEt	**3f**	66
8	MeO-C_6H_4-I	5	MeO-C_6H_4-CF_2COOEt	**3g**	55
9	TBSO-alkenyl iodide	9	TBSO-alkenyl-CF_2COOEt	**3h**	62
10	BzO-alkenyl iodide	8	BzO-alkenyl-CF_2COOEt	**3i**	60
11	n-Bu-CH=C(TMS)-I	30	n-Bu-CH=C(TMS)-CF_2COOEt	**3j**	21

a) Homo coupling occurred at rt. *(Reproduced with permission from reference 18. Copyright 1999. Pharmaceutical Society of Japan.)*

coupling reaction using **2** by Kobayashi's group. They reported that a copper complex was fairly stable and that it was observed by ^{19}F-NMR (*14c*). To compare the property of our organocopper complex with that of Kobayashi's group, **1** was treated with copper powder, and then (*E*)-1-iodo-1-hexene was added to the mixture. At the first stage, **1** was completely consumed after 12 h, but after the second stage, we could not observe the formation of **3a** at all, and only 1-bromo-1-hexene was isolated. This suggested that our organocopper complex was unstable, or decomposition of the complex was accelerated by bromide ion. However, we could not exclude by the above experiment the possibility that (*E*)-1-iodo-1-hexene reacted with copper powder first and that this organocopper intermediate reacted with **1** to give **3a**. So, (*E*)-1-iodo-1-hexene was treated with copper powder first for 12 h. In this stage, no decrease of (*E*)-1-iodo-1-hexene was observed by GLC, and addition of **1** to the mixture resulted in formation of **3a** (Scheme 3). This result shows that **1** reacts with copper first and the organocopper complex reacts at once with (*E*)-1-iodo-1-hexene, as it is formed.

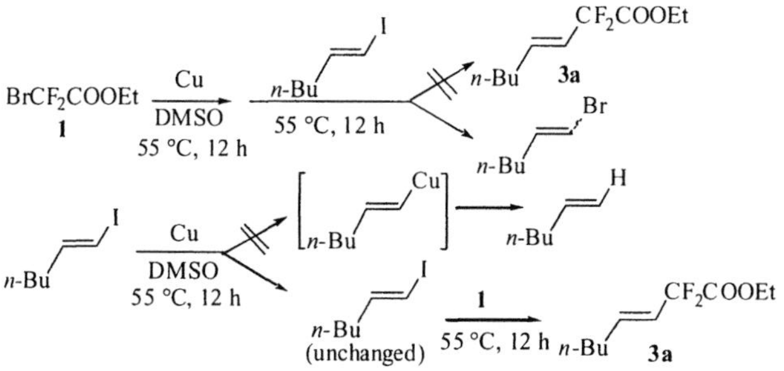

Scheme 3 (Reproduced with permission from reference 18. Copyright 1999. Pharmaceutical Society of Japan.)

Thus, our organocopper complex is not stable in our reaction condition, and it decomposes in the absence of (*E*)-1-iodo-1-hexene. If the olefin is added later, it reacts with copper bromide to give 1-bromo-1-hexene. All of our experiments are consistent with the suggestion that an unstable organocopper complex is formed and that it reacts with iodides as soon as it is formed. These results show that our organocopper complex from **1** is different from Kobayashi's copper complex from the iodide (**2**).

Michael-type Reaction (1,4-Addition Reaction).

There are many reports on the Michael type reaction (1,4-addition reaction) of organocopper reagents (*21*). Most of the organocopper reagents for these 1,4-addition reactions have been usually produced by the reaction of carbanion intermediates, such as organolithium or Grignard reagents with copper salts. As mentioned in the previous section, when an excess of **1** was used, compound **4** was formed probably through an S_N2' reaction. This result suggested that the active intermediate must behave as an anion. If this is true, it seemed likely that the intermediate might react with a Michael acceptor. Here, we describe the 1,4-addition reaction of the copper reagent obtained directly from a halogen compound, **1**, and copper powder.

We chose 2-cyclohexen-1-one as a Michael acceptor, and examined its reaction with **1** in the presence of copper powder under a similar condition as that used in the previous section. The Michael type adduct, ethyl 2,2-difluoro-2-(3-oxocyclohexyl)acetate (**5a**), was obtained in a fairly good yield (*22*).

When the reaction was carried out with other solvents, the 1,4-addition product could not be obtained in satisfactory yields. (See eq. 1)

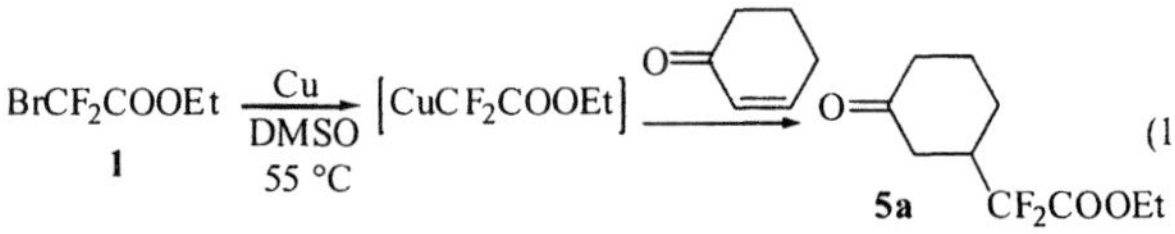

$$BrCF_2COOEt \xrightarrow[\text{DMSO, 55 °C}]{Cu} [CuCF_2COOEt] \longrightarrow \textbf{5a} \quad (1)$$

1; 5a: CF_2COOEt

In the earlier work related, Taguchi and coworkers had reported that the organocopper reagent derived from **2** did not react with α,β-unsaturated carbonyl compounds, although it did react with unconjugated olefins (*14b*). This supports the former hypothesis that their intermediate is different from that derived from **1**.

The results of the reaction of **1** with other Michael acceptors are shown in Table II.

Table II. Reaction of 1 with Michael Acceptors

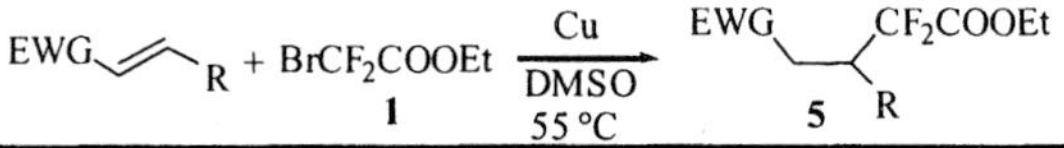

$$EWG\text{-}CH{=}CH\text{-}R + BrCF_2COOEt\ (\textbf{1}) \xrightarrow[\text{DMSO, 55 °C}]{Cu} EWG\text{-}CH_2\text{-}CH(R)\text{-}CF_2COOEt\ (\textbf{5})$$

Entry	EWG–CH=CH–R	Time (h)	5	Yield (%)
1	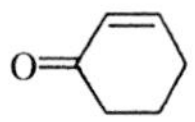 2-cyclohexen-1-one	3	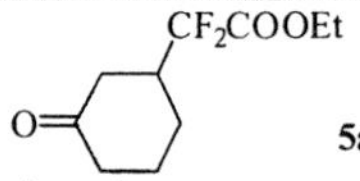 5a	45
2	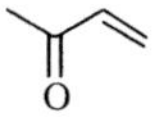 methyl vinyl ketone	5	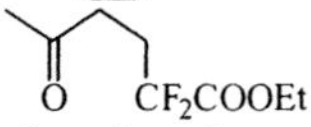 5b	54
3	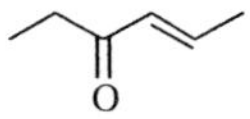 ethyl propenyl ketone	3	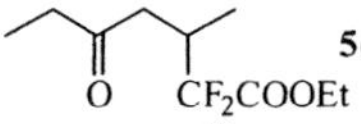 5c	49
4	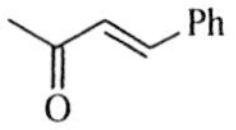 4-phenyl-3-buten-2-one (Ph)	8	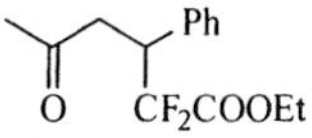 5d	8[a)]
5	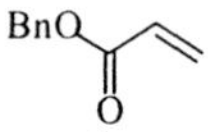 BnO acrylate	3	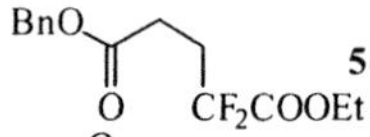 5e	16
6	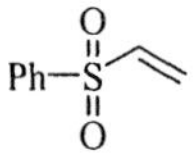 Ph–SO_2–CH=CH_2	7	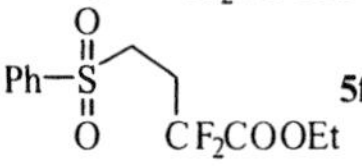 5f	40
7	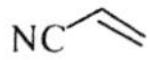 NC–CH=CH_2	5	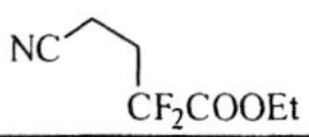 5g	42

a) (*Z*)-**6** and (*E*)-**6** were obtained as byproducts in 16% and 10%, respectively. *(Reproduced with permission from reference 22. Copyright 2000. Pharmaceutical Society of Japan.)*

This 1,4-addition reaction proceeded smoothly and regiospecifically with several Michael acceptors examined. However, the reaction with 4-phenyl-3-

buten-2-one gave a poor yield of the Michael adduct and produced large amounts of ethyl 3-acetyl-2,2-difluoro-4-phenyl-3-butenoates (**6**). Compounds **6** seemed to be formed via a radical mechanism. Reduction of **6** gave **7**, the regioisomer of compound **5d**.

These results show that the organocopper reagent derived from **1** has an anionic property and that it reacts regioselectively with many Michael acceptors. However, certain Michael acceptors stabilizing a radical intermediate can react with the radical anion intermediate with elimination of CuBr, as shown in Scheme 4.

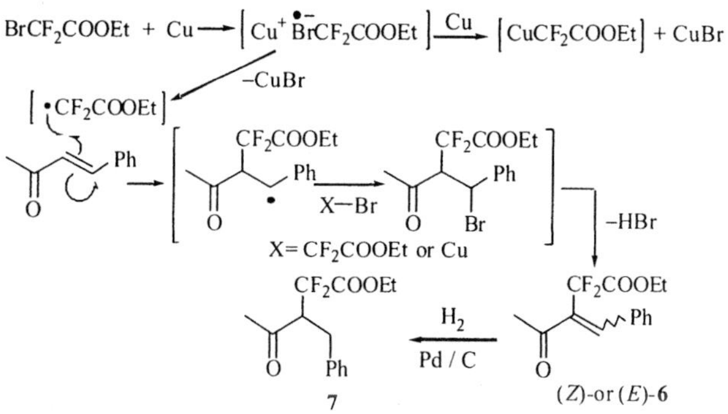

$BrCF_2COOEt + Cu \rightarrow [Cu^+ \overset{\bullet-}{Br}CF_2COOEt] \xrightarrow{Cu} [CuCF_2COOEt] + CuBr$

$-CuBr$

$[\bullet CF_2COOEt]$

Ph

CF_2COOEt Ph O

$X{-}Br$

CF_2COOEt Ph O Br

$X = CF_2COOEt$ or Cu

$-HBr$

CF_2COOEt O Ph **7**

H_2 Pd / C

CF_2COOEt Ph O (*Z*)-or (*E*)-**6**

Scheme 4

Taguchi and co-workers reported addition reaction of **2** to Michael acceptors using zinc powder (*14a*). In their reaction, 1,2-addition products were obtained along with the 1,4-addition products. Thus, the regioselectivity of their zinc reagent is lower than our organocopper reagent.

Improvement of the Michael Type Reaction Using TMEDA as an Additive

In the previous section, we described that the Michael type 1,4-addition reaction proceeded in good yields. However, in order to obtain the 1,4-adducts, DMSO must be used as the solvent. Furthermore, when a substrate had a substituent that stabilized a radical intermediate, considerable amounts of radical adducts were obtained at the expense of the 1,4-adduct. This seemed to restrict the scope of this reaction.

On the other hand, it is well known that some additives, such as PPh_3, TMEDA or HMPA, improve the yield of the 1,4-addition reaction (*23*). In attempt to overcome the problem shown above, we tried various additives that would help the 1,4-addition reaction proceed smoothly in low-boiling solvents without the formation of byproducts from a radical reaction. On this study, we found that the addition of TMEDA in THF led to the successful 1,4-addition reaction of **1** with Michael acceptors without formation of the byproducts (*24*).

The results of the application of the above result to some Michael acceptors are shown in Table III.

Table III. Reaction of 1 with Some Michael Acceptors: Effect of TMEDA

EWG–CH=CH–R + $BrCF_2COOEt$ (**1**) —(Cu, TMEDA, THF)→ EWG–CH_2–CH(R)–CF_2COOEt (**5**)

Entry	EWG–CH=CH–R	Temp.	Time (h)	5	Yield (%)
1	O= (cyclohexenone)	reflux	7	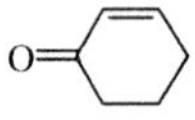 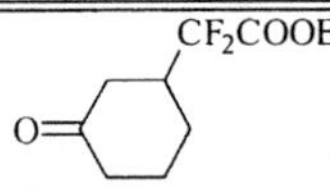 CF_2COOEt **5a**	73
2	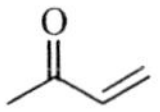 O	r.t.	5	 O, CF_2COOEt **5b**	52
3	O	reflux	2	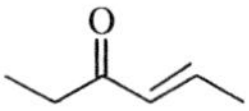 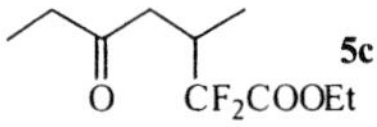 O, CF_2COOEt **5c**	68
4	O, Ph	reflux	2	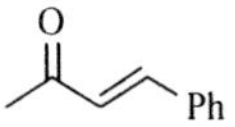 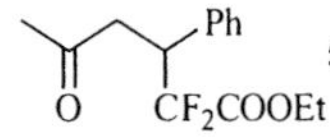 Ph, O, CF_2COOEt **5d**	21
5	O, Ph, Ph	reflux	1	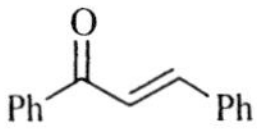 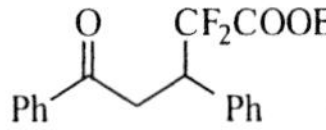 O, CF_2COOEt, Ph, Ph **5h**	23
6	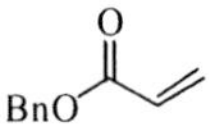 O, BnO	r.t.	5	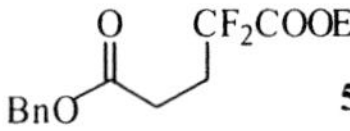 O, CF_2COOEt, BnO **5e**	60
7	O, H	reflux	3	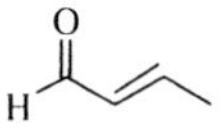 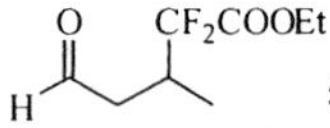 O, CF_2COOEt, H **5i**	23
8	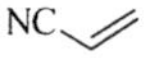 NC	r.t.	2	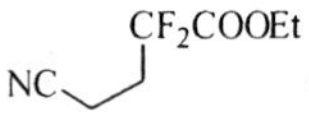 CF_2COOEt, NC **5g**	40
9	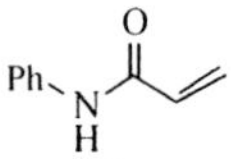 O, Ph–NH	r.t.	5	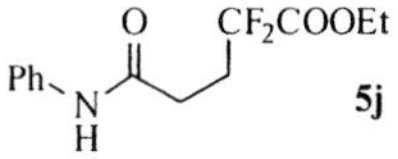 O, CF_2COOEt, Ph–NH **5j**	32
10	Ph–SO_2	reflux	1	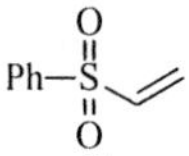 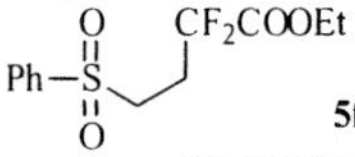 Ph–SO_2, CF_2COOEt **5f**	73
11	$(EtO)_2P(O)$	reflux	1	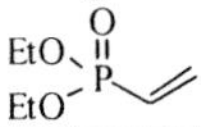 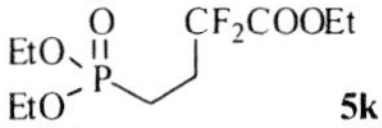 $(EtO)_2P(O)$, CF_2COOEt **5k**	69

(Reproduced with permission from reference 24. Copyright 2003. Elsevier.)

The 1,4-addition reaction proceeded with most of the Michael acceptors in good yields. Further, even the substrates stabilizing a radical, which gave products through radical reaction in the absence of TMEDA, afforded only 1,4-addition products as shown in entries 4 and 5, though the yields were not high.

Summarizing the above results, **1** reacted with various Michael acceptors to give 1,4-adducts in moderate to good yields, when TMEDA was used as an additive. The reaction proceeded regiospecifically; neither 1,2-addition products nor radical reaction products were obtained under this condition. THF was the best solvent, though the 1,4-addition reaction proceeded in many other low-boiling solvents. This improvement of the 1,4-addition reaction using TMEDA as an additive makes this reaction more useful, since the work-up is more convenient than the workp-up when DMSO is used as the solvent.

Radical Addition Reaction

Free radical reaction has become an important methodology for a C-C bond formation and a ring construction (*25*). The most general method for formation of new C-C bonds via radical intermediates involves addition of the radical to an alkene. The adduct itself is a new radical, which can propagate a radical chain reaction. These radicals for addition reaction can be generated by a halogen atom abstraction from halogen compounds by metals.

On the other hand, there are many reports concerning a radical addition reaction using trialkyltin hydride, triethyl borane or copper powder as a method to introduce a CF_2COOR group (*14b, 26*). In the former section, we described that 4-phenyl-3-buten-2-one gave much larger amounts of **6** that might be formed through a radical mechanism than that through the 1,4-addition. If this radical reaction proceeded commonly with unactivated olefins, this would offer another route to CF_2 compounds (Eq. 2) (*27*).

$$\text{R-CH=CH-R'} + \underset{\mathbf{1}}{BrCF_2COOEt} \longrightarrow \underset{\mathbf{8}}{\text{R-CH(Br)-CH(R')-}CF_2COOEt} \quad (2)$$

Among the various reaction conditions examined using allylbenzene as a substrate, the condition previously used for the cross-coupling reaction was found to give the best result for this radical reaction. Under this condition, the product from radical addition, ethyl 4-bromo-2,2-difluoro-5-phenylpentanoate (**8a**), was obtained along with smaller amounts of ethyl 2,2-difluoro-5-phenyl-pentanoate (**9**) and ethyl (*E*)-2,2-difluoro-5-phenyl-4-pentenoate (**10**). A speculative mechanism for the formation of these compounds is shown in Scheme 5. Thus, the primary complex from **1** and copper affords ethoxy-carbonyldifluoromethyl radical, which reacts with allylbenzene, an unactivated olefin, to give a methylene radical. This radical abstracts a bromine atom from **1** to give **8a** or disproportinates to afford **9** and **10**.

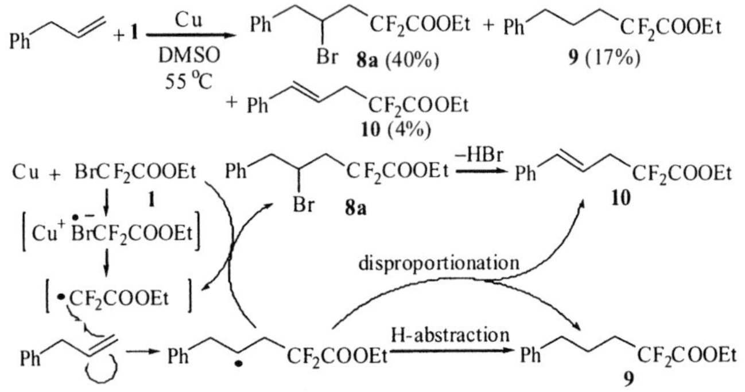

Scheme 5 (Reproduced with permission from reference 27. Copyright 2002. Collect. Czech. Chem. Commun.)

When α-methylstyrene was used as a substrate, none of the simple addition product was obtained. Instead, a dimeric product (**13**) was isolated in 60% yield along with traces of byproducts (Scheme 6). Apparently, the intermediate benzylic radical was so stable that it could not abstract bromine from **1**, and instead dimerized to give **13** as a diastereomeric mixture.

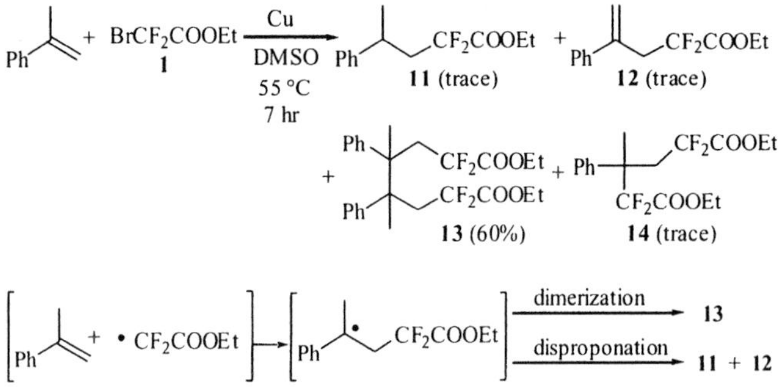

Scheme 6 (Reproduced with permission from reference 27. Copyright 2002. Collect. Czech. Chem. Commun.)

When the silyl enol ether of acetophenone was subjected to this reaction, ethyl (*E*)-2-fluoro-4-oxo-4-phenyl-2-butenoate (**15**) and ethyl (*Z*)-2-fluoro-4-oxo-4-phenyl-2-butenoate (**16**) were obtained in 4 and 14% yields, respectively (Scheme 7). These low yields must be due to the formation of acetophenone as a byproduct. An interesting point is that **15** was found by GLC analysis to isomerize during the reaction to **16**. Namely, though the ester **15** was observed initially as a major peak, this peak decreased with time, while that of **16** increased.

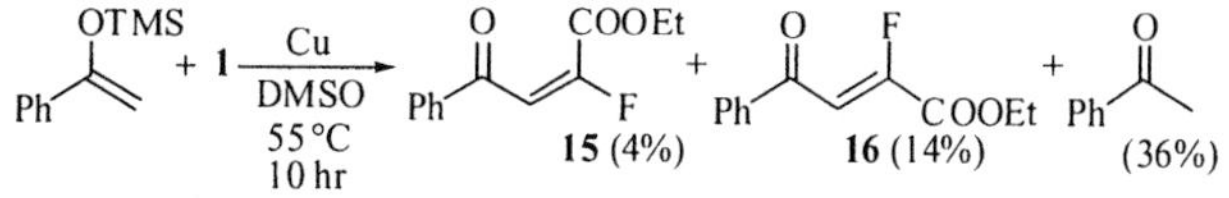

Scheme 7 *(Reproduced with permission from reference 27. Copyright 2002. Collect. Czech. Chem. Commun.)*

We rationalized these results as follows. Among the conformations of the proposed primary adduct (**17**), conformers *A* and *B* must be slightly more stable than conformer *C*. Conformers *A* and *B* would give **15** by anti-elimination, while *C* would form an isomer **16**. However, **16** was estimated to be slightly more stable than **15** by MOPAC calculation. Thus, **15** isomerized to **16**. This ready isomerization might occur because of the low double bond character due to +Iπ interaction of the fluorine atom and the double bond (Figure 1). In fact, Ichikawa et al. have also reported the same theorization using Nazarov cyclization of 2-butyl-1,1-difluoro-4-methyl-1,4-pentadien-3-one (*28*).

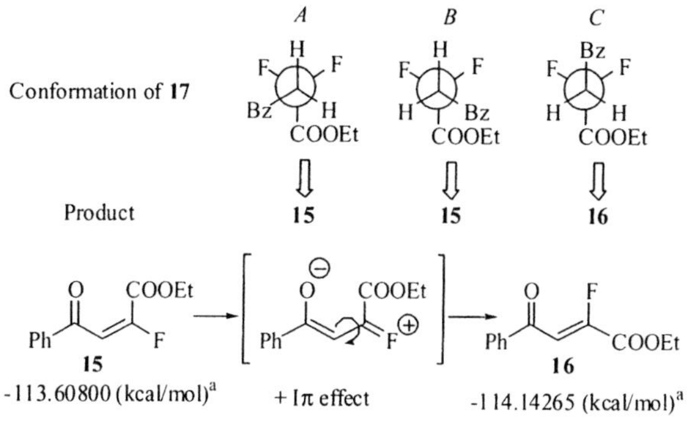

a) Heat of Formations of **15** and **16** calculated with AM1 on MOPAC.

Figure 1 (Reproduced with permission from reference 27. Copyright 2002. Collect. Czech. Chem. Commun.)

In the previous reaction, the products were formed in a low yield, since the TMS radical did not abstract a bromine atom from **1**, but abstracted another TMS from the silyl ether. On the other hand, there are many reports using the trialkyltin radical to abstract a halogen from halodifluoroacetate (*26*). Therefore, we used allyltributyltin as a substrate because this would produce tributyltin radical *in situ*. As expected, allyltributyltin smoothly reacted with **1** and gave ethyl 2,2-difluoro-4-pentenoate (**18**) as a single product in 48% yield by ^{1}H-NMR analysis (Scheme 8). Since the isolated yield of **18** was low (9%)

owing to its high volatility, crude **18** was treated with bromine. Ethyl 4,5-dibromo-2,2-difluoropentanoate (**19**) was obtained in 38% yield.

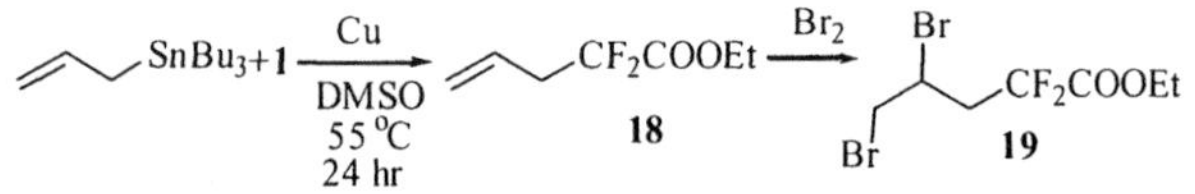

Scheme 8 (Reproduced with permission from reference 27. Copyright 2002. Collect. Czech. Chem. Commun.)

As described in this section, the radical reaction proceeded with olefins that do not possess electron-withdrawing groups. Thus, this reaction provides another route for CF_2 compounds, although the yield of the reaction must be improved.

Summary of Mechanisms of the Above Reactions

From the results in previous sections, we propose mechanisms of these reactions as shown in Scheme 9.

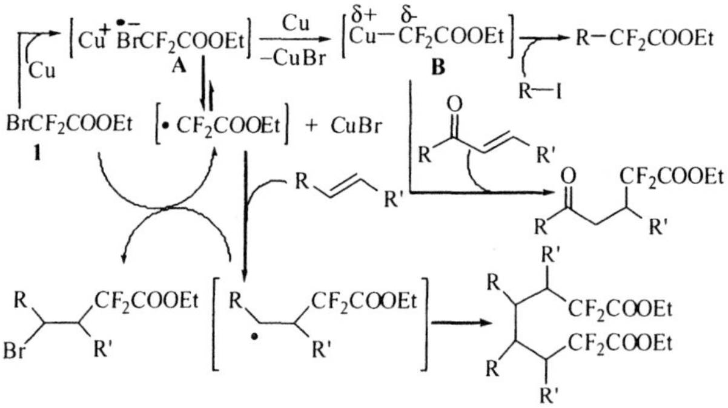

Scheme 9 (Reproduced with permission from reference 27. Copyright 2002. Collect. Czech. Chem. Commun.)

Ethyl bromodifluoroacetate (**1**) reacts with an equivalent of copper to give an organocopper complex (**A**). In the presence of an electron rich olefin, this complex (**A**) generates a radical species, and this reacts with the olefins to give the monomeric or dimeric products.

On the other hand, in the absence of electron rich olefins, the organocopper complex (**A**) reacts with another equivalent of copper to give another organocopper complex (**B**). The complex (**B**) reacts with vinyl or aryl iodides to give the cross-coupling products. The complex (**B**) possesses an anionic property, and reacts with various Michael acceptors to give 1,4-addition products. Actually, two equivalents of copper were necessary for the cross-coupling reaction or 1,4-addition reaction to proceed completely. These results support the mechanism mentioned above.

Application of the Reactions for Synthsesis of Fluorine Analogues of Biologically Important Compounds.

Synthesis of 4,4-Difluoro-α-tocopherol by the Cross-coupling Reaction

Vitamin E is a very important biologically active compound that has been used in various fields (*29*). It consists of tocopherols and tocotrienols, among which α-tocopherol is believed to be the most active component. The main biological effect of α-tocopherol is believed to be based an anti-oxidizing effect, but the clinical effect and the biological behaviors have not been solved clearly. So, to elucidate the biological effect more clearly, we have synthesized α-tocopherols with one or two CF_3 groups (F_3- or F_6-α-tocopherols) in the place of the methyl groups (Figure 2). The study of biological behavior of these fluorine analogues as probes on ^{19}F-NMR gave interesting results (*30*).

Now, we have a method to introduce a CF_2 group. So, we tried to synthesize α-tocopherol with a CF_2 group in the chroman ring (F_2-α-tocopherol) using our cross-coupling reaction (Figure 2) (*31*).

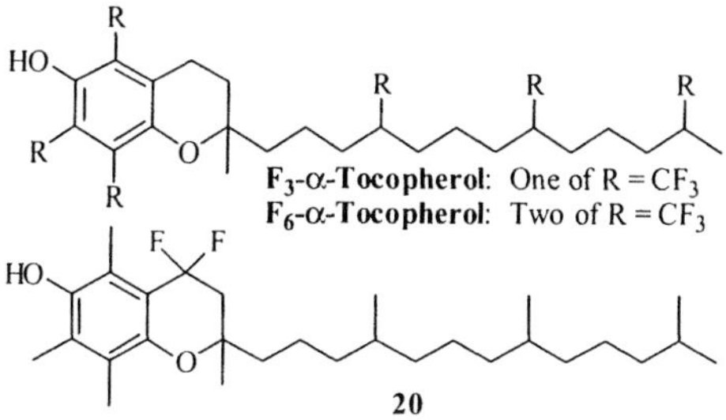

F_3-α-Tocopherol: One of R = CF_3
F_6-α-Tocopherol: Two of R = CF_3

20

Figure 2

First, an aryl iodide, 2-iodo-3,5,6-trimethylhydroquinone (**21**), was synthesized according to the literature (*32*). After protection of the hydroxyl groups with MOM, this was treated with **1** in the presence of copper powder to give the difluoro ester (**23**) in 96% yield. The ester was reduced to an aldehyde. The Horner-Emmons reaction of the aldehyde gave **24**. After catalytic reduction of the double bond, the ester group was reduced and dehydrated. The double bond was oxidized with *m*CPBA to an epoxide. Treatment of the epoxide with

CF_3COOH gave a chroman ring. The Swern oxidation of the primary alcohol gave the chroman aldehyde (**25**).

For introduction of the side chain, **25** was treated with a Wittig reagent (**26**) derived from 3,7,11-trimethyldodecyl bromide, which was obtained according to the literature (*33*). With a few functional group transformations, we completed the synthesis of 4,4-difluoro-α-tocopherol (**20**) (Scheme 10).

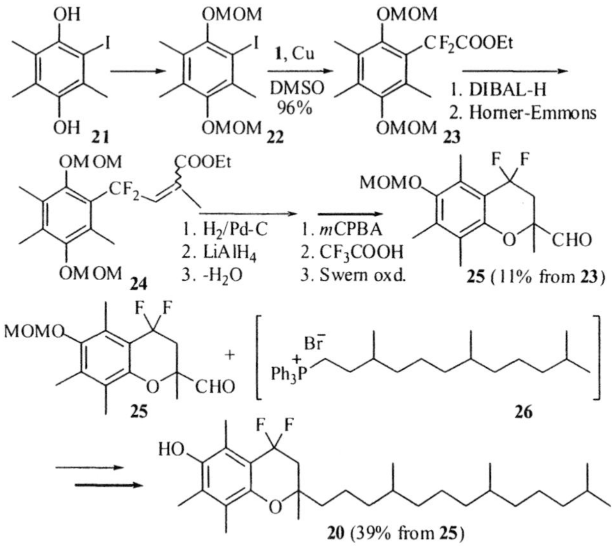

OH, OMOM, I, **21**, **22**, 1, Cu, DMSO, 96%, CF_2COOEt, **23**, 1. DIBAL-H, 2. Horner-Emmons, OMOM, CF_2, COOEt, **24**, 1. H_2/Pd-C, 2. $LiAlH_4$, 3. $-H_2O$, 1. *m*CPBA, 2. CF_3COOH, 3. Swern oxd., MOMO, F, F, O, CHO, **25** (11% from **23**), +, Br^-, Ph_3P^+, **26**, HO, **20** (39% from **25**)

Scheme 10

Our cross-coupling reaction of **1** with an aryl iodide was successfully applied for the synthesis of 4,4-difluoro-α-tocopherol (**20**). This compound may have clinical uses based on the electronic effect of CF_2, besides the use as a probe on ^{19}F-NMR study. The successful coupling reaction suggests that this method is widely applicable for the syntheses of aryldifluoromethylene compounds.

Simple Synthesis of α,α-Difluoro Analogue of Fatty Acid Using Radical Reaction

Fatty acids are biologically very important, and their fluorine analogues are used widely for investigation of fatty acids, since fluorine atoms are easily monitered by ^{19}F-NMR. We had reported investigation of Langmuir membrane of difluorostearic acid (*34*). Here, a simple synthesis of α,α-difluoro analogue of fatty acid using the radical reaction is illustrated in Scheme 11 (*27*).

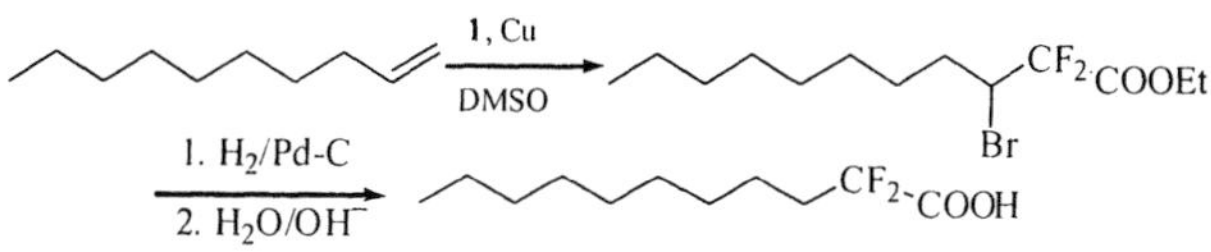

1, Cu / DMSO → CF_2COOEt, Br; 1. H_2/Pd-C, 2. H_2O/OH^- → CF_2-COOH

Scheme 11

Epilogue: Rhodium Catalyzed Reaction

Here, we would like to show our preliminary results of a rhodium catalyzed reaction, which gives quite interesting products. As discussed in the former section, copper mediated reaction of ethyl bromodifluoroacetate (**1**) was found to provide new methodologies for various compounds with a CF_2 unit. On the other hand, rhodium catalyzed reactions are investigated intensively, and 1,4-addition reaction with electron deficient olefins using organoboronic acids, organostannanes or organosilanes has also been reported (*35*). We expected that if this method could be applied for the reaction of **1**, a new methodology would be provided. Namely, we thought that **1** and α,β-unsaturated ketones would react with **1** in the presence of diethylzinc and a rhodium catalyst to give the corresponding 1,4-addition product (**5**). However, we did not obtain **5a** by the reaction with 2-cyclohexen-1-one. An unexpected product (**27**), where **1** added on the α-carbon of α,β-unsaturated ketones, was obtained with a small amount of a 1,2-addition product (**28**) (Scheme 12) (*36*).

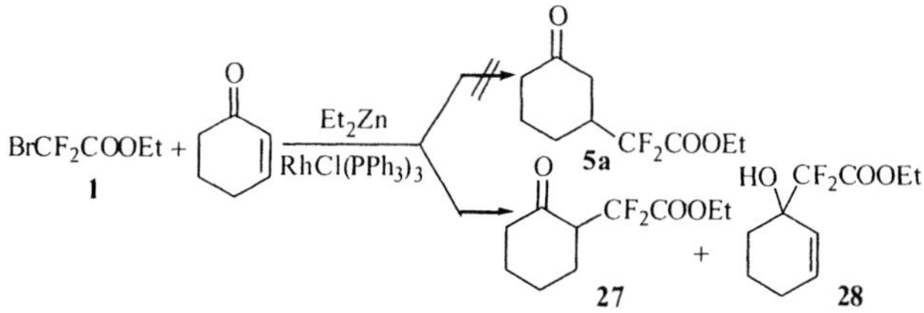

$BrCF_2COOEt$ + (2-cyclohexen-1-one) → Et_2Zn, $RhCl(PPh_3)_3$ → (not) **5a** CF_2COOEt; **27** CF_2COOEt + **28** HO CF_2COOEt

Scheme 12

Interestingly, when THF was used as the solvent, compound **27** was obtained in 77% as a major product, while compound **28** was obtained in 80% using CH_3CN as the solvent. The mechanism for the formation of **27** is not clear, but the ethyl group of diethylzinc must play an important role, since dimethylzinc or diphenylzinc are not effective for this reaction. Only the Wilkinson catalyst was effective among Rh complexes examined. Without the catalyst, the reaction was very slow, and only 1,2-adduct, which was expected from the reaction of zinc compound, was obtained.

This reaction seems to be applied for the synthesis of other perfluoroalkyl compounds. Thus, CF_3I was used in the place of **1**, α-trifluoromethylated product was obtained (Eq. 3) (*37*).

$$CF_3I + \text{cyclohexenone} \xrightarrow[RhCl(PPh_3)_3]{Et_2Zn} \text{2-}CF_3\text{-cyclohexanone} \quad (3)$$

This new reaction would work complementally with the reaction of **1** in the presence of copper powder for the syntheses of various CF_2 compounds and other fluoroalkyl compounds.

Concluding remark

The reaction of ethyl bromodifluoroacetate (**1**) in the presence of copper powder was found to be useful for the introduction of a CF_2 unit into various molecules. Thus, the reaction with aryl or alkenyl iodides gave cross-coupling products. Michael acceptors gave 1,4-addition products. TMEDA improved this reaction remarkably: THF can be used as a solvent, the radical reaction is surpressed, and regioselectivity was improved. The radical reaction of **1** with unactivated olefins in the presence of copper powder gave addition products. We believe that these reactions of commercially available **1** in the presence of copper powder provide useful methods for the syntheses of compounds with a CF_2 functional group. A preliminary result of the investigation of a rhodium catalyzed reaction showed that a quite interesting product, where CF_2COOEt group was introduced to α-position of α,β-unsaturated carbonyl compound, was formed. This new reaction seems to be applied to other perfluoroalkyl compounds. The detailed mechanism and scope of this reaction are under investigation.

Acknowledgement

Part of this work was presented at ACS symposium and published in the special issue for this symposium of Journal of Fluorine Chemistry, **2004**, *125*, 509-515.

References

1. (a) Smart, B. E.; *Chem. Rev.* **1996**, *96*, 1555-1823. (b) Resnati, G; Soloshonok, V. A.; *Tetrahedron* **1996**, *52*, 1-330. (c) "*Organofluorine*

Compounds in Medicinal Chemistry and Biomedical Applications", Filler, R.; Kobayashi, Y.; Yagupolskii, L. M., Eds.; Elsevier, Amsterdam, 1993. (d) Welch, J. T.; *Tetrahedron* **1987**, *43*, 3123-3197.

2. (a) Kobayashi, Y.; Kumadaki, I.; *Tetrahedron Lett.* **1969**, 4095-4096. (b) Kobayashi, Y.; Kumadaki, I.; Sato, S.; Hara, N.; Chikami, E.; *Chem. Pharm. Bull.* **1970**, *18*, 2334-2339. (c) Kobayashi, Y.; Yamamoto, K.; Kumadaki, I.; *Tetrahedron Lett.* **1979**, 4071-4072.
3. (a) Kobayashi, Y.; Kumadaki, I.; Nagai, T.; *Chem. Pharm. Bull.* **1984**, *32*, 5031-5035. (b) A review: Kumadaki, I.; *Reviews on Heteroatom Chemistry*, **1993**, *9*, 181-204, and references therein.
4. Shimada, T.; Ando, A.; Takagi, T.; Koyama, M.; Miki, T.; Kumadaki, I.; *Chem. Pharm. Bull.* **1992**, *40*, 1665-1666.
5. Takagi, T.; Takesue, A.; Koyama, M.; Ando, A.; Miki, T.; Kumadaki, I.; *J. Org. Chem.* **1992**, *57*, 3921-3923. M. Nishihara, M.; Nakamura, Y.; Maruyama, N.; Sato, K.; Omote, M.; Ando, A.; Kumadaki, I.; *J. Fluorine Chem.* **2003**, *122*, 247-249.
6. Kobayashi, Y.; Kumadaki, I.; Yamamoto, K.; *J. Chem. Soc., Chem. Commun.* **1977**, 536-537.
7. Tozer, M. J.; Herpin, T. F.; *Tetrahedron* **1996**, *52*, 8619-8683.
8. (a) Wu, F. -H.; Huang, W. -Y.; *J. Fluorine Chem.* **2001**, *110*, 59-61. (b) Asai, H.; Uneyama, K.; *Chem. Lett.* **1995**, *24*, 1123-1124. (c) Chen, J.; Hu, C. -M.; *J. Chem. Soc., Perkin Trans. 1* **1994**, 1111-1114.
9. (a) Kirihara, M.; Kawasaki, M.; Takuwa, T.; Kakuda, H.; Wakikawa, T.; Takeuchi, Y.; Kirk, K.; *Tetrahedron: Asymmetry* **2003**, *14*, 1753-1761. (b) Mitsukura, K.; Korekiyo, S; Itoh, T.; *Tetrahedron Lett.* **1999**, *40*, 5739-5742.
10. (a) Ando, A.; Takahashi, J.; Nakamura, Y.; Maruyama, N.; Nishihara, M.; Fukushima, K.; Moronaga, J.; Inoue, M.; Sato, K.; Omote, M.; Kumadaki, I.; *J. Fluorine Chem.* **2003**, *123*, 283-285. (b) Anilkumar, R.; Burton, D. J.; *Tetrahedron Lett.* **2002**, *43*, 6979-6982. (c) Dimartino, G.; Percy, J. M.; *Chem. Commun.* **2000**, 2339-2340. (d) Maruoka, K.; Shimada, I.; Akakura, M.; Yamamoto, H.; *Synlett* **1994**, 847-848.
11. (a) Braun, M.; Vonderhagen, A.; Waldmhller, D.; *Liebigs Ann.* **1995**, 1447-1450. (b) Lang, E. W.; Schaub, B.; *Tetrahedron Lett.* **1988**, *29*, 2943-2946. (c) Hallinan, E. A.; Fried, J.; *Tetrahedron Lett.* **1984**, *25*, 2301-2302.
12. (a) Iseki, K.; *Tetrahedron* **1998**, *54*, 13887-13914. (b) Iseki, K.; Kuroki, Y.; Asada, D.; Kobayashi, Y.; *Tetrahedron Lett.* **1997**, *38*, 1447-1448. (c) Kitagawa, O.; Taguchi, T.; Kobayashi, Y.; *Tetrahedron Lett.* **1988**, *29*, 1803-1806.
13. (a) Konas, D. W.; Pankuch, J. J.; Coward, J. K.; *Synthesis* **2002**, 2616-2626. (b) Konno, K.; Ojima, K.; Hayashi, T.; Takayama, H.; *Chem. Pharm.*

Bull. **1992**, *40*, 1120-1124. (c) Morikawa, T.; Nishiwaki, T.; Nakamura, K.; Kobayashi, Y.; *Chem. Pharm. Bull.* **1989**, *37*, 813-15.

14. (a) Kitagawa, O.; Hashimoto, A.; Kobayashi, Y.; Taguchi, T.; *Chem. Lett.* **1990**, 1307-1310. (b) Kitagawa, O.; Miura, A.; Kobayashi, Y.; Taguchi, T.; *Chem. Lett.* **1990**, 1011-1014. (c) Kitagawa, O.; Taguchi, T.; Kobayashi, Y.; *Chem. Lett.* **1989**, 389-392. (d) Taguchi, T.; Kitagawa, O.; Morikawa, T.; Nishiwaki, T.; Uehara, H.; Endo, H.; Kobayashi, Y.; *Tetrahedron Lett.*, **1986**, *27*, 6103-6106.
15. (a) Staas, D. D.; Savage, K. L.; Homnick, C. F.; Tsou, N. N.; Ball, R. G.; *J. Org. Chem.* **2002**, *67*, 8276-8279. (b) Zhao, G.; Sun, H. -L.; Qian, Z. -S.; Yin, W. -X.; *J. Fluorine Chem.* **2001**, *111*, 217-219. (c) Schwaebe, M. K.; McCarthy, J. R.; Whitten, J. P.; *Tetrahedron Lett.* **2000**, *41*, 791-794. (d) Yoshida, M.; Suzuki, D.; Iyoda, M.; *Chem. Lett.* **1996**, *25*, 1097-1098. (e) Yamamoto, T.; Ishibuchi, S.; Ishizuka, T.; Haratake, M.; Kunieda, T.; *J. Org. Chem.* **1993**, *58*, 1997-1998.
16. (a) Croxtall, B.; Fawcett, J.; Hope, E. G.; Stuart, A. M.; *J. Fluorine Chem.* **2003**, *119*, 65-73. (b) Kakino, R.; Shimizu, I.; Yamamoto, A.; *Bull. Chem. Soc. Jpn.* **2001**, *74*, 371-376. (c) Fukuda, H.; Tetsu, M.; Kitazume, T.; *Tetrahedron* **1996**, *52*, 157-164. (d) Lanier, M.; Pastor, R.; Riess, J. G.; *Tetrahedron Lett.* **1993**, *34*, 5093-5094. (e) Bravo, P.; Pregnolato, M.; Resnati, G.; *J. Org. Chem.* **1992**, *57*, 2726-2731. (f) Kitazume, T.; Ohnogi, T.; Lin, J. T.; Yamazaki, T.; Ito, K.; *J. Fluorine Chem.* **1989**, *42*, 17-29.
17. Brewster, R. Q.; Groening, T.; "*Org. Synthesis*", Coll. Vol. II, **1948**, 445-446.
18. Sato, K.; Kawata, R.; Ama, F.; Omote, M.; Ando, A.; Kumadaki, I.; *Chem. Pharm. Bull.* **1999**, *47*, 1013-1016.
19. Ashwood, M. S.; Cottrell, I. F.; Cowden, C. J.; Wallace, D. J.; Davies, A. J.; Kennedy, D. J.; Dolling, U. H.; *Tetrahedron Lett.* **2002**, *50*, 9271-9273.
20. Eto, H.; Kaneko, Y.; Sakamoto, T.; *Chem. Pharm. Bull.* **2000**, *48*, 982-990.
21. (a) A. Alexakis; C. Benhaim; *Eur. J. Org. Chem.* **2002**, 3221-3236. (b) E., Nakamura; S. Mori; *Angew. Chem., Int. Ed.* **2000**, *39*, 3750-3771. (c) P., Perlmutter; "*Conjugate Addition Reactions in Organic Synthesis*", Pergamon Press Ltd., **1992**.
22. Sato, K.; Tamura, M.; Tamoto, K.; Omote, M.; Ando, A.; Kumadaki, I.; *Chem. Pharm. Bull.* **2000**, *48*, 1023-1025.
23. "*Organocopper Reagents*", Taylor, R. J. K., Eds.; Oxford University Press, New York, **1994**.
24. Sato, K.; Nakazato, S.; Enko, H.; Tsujita, H.; Fujita, K.; Yamamoto, T.; Omote, M.; Ando, A.; Kumadaki, I.; *J. Fluorine Chem.* **2003**, *121*, 105-107.
25. (a) Curran, D. P.; Porter, N. A.; Giese, B.; "*Stereochemistry of Radical Reactions*", VCH, Weinheim, **1996**. (b) Giese, B; Kopping, B.; Gobel, T.;

Dickhaut, J.; Thoma, G.; Kulicke, K. J.; Trach, F.; *Org. React.* **1996**, *48*, 301-856. (c) Jasperse, C. P.; Curran, D. P.; Fevig, T. L.; *Chem. Rev.* **1991**, *91*, 1237-1286.
26. (a) Itoh, T.; Sakabe, K.; Kudo, K.; Zagatti, P.; Renou, M.; *Tetrahedron Lett.* **1998**, *39*, 4071-4074. (b) Itoh, T.; Ohara, H.; Emoto, S.; *Tetrahedron Lett.* **1995**, *36*, 3531-3534. (c) Morikawa, T.; Uejima, M.; Kobayashi, Y.; Taguchi, T.; *J. Fluorine Chem.* **1993**, *65*, 79-89. (d) Arnone, A.; Bravo, P.; Cavicchio, G.; Frigerio, M.; Viani, F.; *Tetrahedron* **1992**, *48*, 8523-8540. (e) Yang, Z. -Y.; Burton, D. J.; *J. Chem. Soc., Chem. Commun.* **1992**, 233-234. (f) Yang, Z. -Y.; Burton, D. J.; *J. Org. Chem.* **1991**, *56*, 5125-5132. (g) Yang, Z. -Y.; Burton, D. J.; *J. Fluorine Chem.* **1989**, *45*, 435-9.
27. Sato, K.; Ogawa, Y.; Tamura, M.; Harada, M.; Ohara, T.; Omote, M.; Ando, A.; Kumadaki, I.; *Coll. Czech. Chem. Commun.* **2002**, *67*, 1285-1295.
28. Ichikawa, J.; Miyazaki, S.; Fujiwara, M.; Minami, T.; *J. Org. Chem.* **1995**, *60*, 2320-2321.
29. (a) "*Handbook of Vitamins*, Third Edition", Rucker, R B.; Suttie, J. W.; McCormick, D B.; Machlin, L J., Eds. Dekker, New York, 2001. (b) "*Handbook of Vitamins, nutritional, biochemical and clinical aspects*", Machlin, L. J., Eds.; Marcel Dekker Inc., 1984.
30. (a) Yano, T.; Yajima, S.; Hasegawa, K.; Kumadaki, I.; Yano, Y.; Otani, S.; Uchida, M.; *Carcinogenesis* **2000**, *21*, 2129-2133. (b) Koyama, M.; Takagi, T.; Ando, A.; Kumadaki, I.; *Chem. Pharm. Bull.* **1995**, *43*, 1466-1474, and references therein.
31. Sato, K.; Nishimoto, T.; Tamoto, K.; Omote, M.; Ando, A.; Kumadaki, I.; *Heterocycles* **2002**, *56*, 403-412.
32. Cressman, H. W. J.; Thirtle, J. R.; *J. Org. Chem.* **1966**, *31*, 1279-1281.
33. (a) Cohen, N.; Lopresti, R J.; Saucy, G.; *J. Am. Chem. Soc.* **1979**, *101*, 6710-6716. (b) Scott, J. W.; Bizzarro, F. T.; Parrish, D. R.; Saucy, G.; *Helv. Chim. Acta* **1976**, *59*, 290-306. (c) Mayer, H.; Schudel, P.; Rüegg, R.; Isler, O.; *Helv. Chim. Acta* **1963**, *46*, 650-671.
34. O'Hagan, D.; Kumadaki, I.; Petty, M.; Takaya, H.; Pearson, C.; *J. Fluorine Chem.* **1998**, *90*, 133-138.
35. (a) Hayashi, T.; *Bull. Chem. Soc. Jpn.* **2004**, *77*, 13-21. (b) Fagnou, K.; Lautens, M.; *Chem. Rev.* **2003**, *103*, 169-196. (c) Hayashi, T.; *Synlett* **2001**, 879-887.
36. Sato, K.; Tarui, A.; Kita, T.; Ishida, Y.; Tamura, H.; Omote, M.; Ando, A.; Kumadaki, I.; *Tetrahedron Lett.* **2004**, *45*, 5735-5737.
37. Presented at *the 14th European Symposium on Fluorine Chemistry*, **2004**, Poznan, Poland.

Application of Fluorinated Synthons in Crystal Engineering

Chapter 29

Fluorine-Based Crystal Engineering

Taizo Ono and Yoshio Hayakawa

Research Institute of Instrumentation Frontier, National Institute of Advanced Industrial and Science Technology (AIST), 2266–98, Anagahora, Shimoshidami, Moriyama, Nagoya 463–8560, Japan

Hybrid compounds R_F-X-R_H consisting of perfluoro and hydro moieties, R_F and R_H, which are connected with X functionality, are described. New crystal engineering based on these hybrid compounds is proposed as a rational method for the design of the functional materials. Hexafluoropropene trimers are selected as the fluoro synthons. A systematic investigation on the crystal structures of the hybrid compounds revealed that there are five packing motifs in connection with the topology of the molecules. Topochemical requirements for the solid-state polymerization of diacetylene functionality were satisfied with the precise tuning by tailor-made hybrid synthons. As a result, a black crystalline material with a metallic luster was synthesized. This new fluorine-based crystal engineering will be one of the unique molecular design methods for the material science.

Highly ordered supramolecular structures are ubiquitous in nature from organic to inorganic realms. The most sophisticated structures seen in living organisms are the results of the cooperation of the intermolecular forces, after a bottom-up process based on the covalent architecture of the molecules. A crystal is also a highly ordered supramolecular bottom-up expression of molecules with enticing beauty and thus has attracted many scientists. In accordance with the surge of nanotechnology, the strong desire for the rational design of the highly ordered materials rekindled and gave an impetus to the research in crystal engineering as one of the most important tools in the supramolecular science (*1*). However, our understanding of the way of assemblage of molecules is yet in the embryonic stage (*2*). Thus, the apriori prediction of the crystal structures seems to be an awful challenge or impossible (*3*). According to Desiraju, a general solution to this problem would bring in a billion dollars to the pharmaceutical industry and a possible Nobel Prize to whoever cracks it (*4*). In this context, engineering the crystal packing, instead of the ab initio prediction of the crystal packing of any given molecules, could be a more realistic approach for the design of crystalline materials.

A variety of molecular interactions have been investigated for aligning the molecules by using so-called supramolecular synthons approach (*5*). Most of the attention has been paid to the hydrogen-bonding-based synthons due mainly to an impetus from the DNA base pair structures (*6*). A stronger ionic bond has been widely used to make the structures (*7*). Recent advancement in the metal-ligand coordination is also striking (*8*).

In contrast to these strong bonds, we will focus on the weak non-bonded interactions in this chapter. First, we introduce various types of inter-molecular synthons, which have been used in crystal engineering. Second, we survey the fluorine-based crystal engineering, and then explain our recently devised crystal engineering based on the fluorine-containing hybrid compounds, R_F-X-R_H.

Crystal Engineering based on the weak non-bonded intermolecular interactions

Also molecules for which no strong molecular interaction is expected crystallize in many cases. X-ray analysis of such crystals has revealed the existence of a variety of weak non-bonded interactions.

Charge-transfer interactions (*9*), weak hydrogen bonds, and halogen-halogen interactions are seen in the following combination of intermolecular synthons.

Charge-transfer interactions: π-π interactions (*10*), C-X (X=F, Cl, Br, I) ---π systems (*11*), halogen bondings between a halogen and a lone pair of electrons of hetero atoms (N, O, S etc.); π-π interaction and halogen bonding recently immerged as new tools for the crystal engineering in an interdisciplinary connection with fluorine chemistry.

Weak hydrogen bondings: C-H---X (X=O, N, Cl, F, and an aromatic π system); Weak hydrogen bondings are recent concerns in this field (*12*).

Halogen-halogen interactions: C-X---X-C (X=Cl, Br, I); The halogen-halogen interaction has been taken as a subject of crystal engineering due to a ubiquitous appearance of such interactions in the organic halides (*13*).

Fluorine-based crystal engineering

There has been reported some crystal engineering based on the fluoro organic compounds. One strategy is based on the fluorine's size, H-mimicry, to finely tune the crystal packing of the organic molecules with a subtle change of the molecular shape by partial substitutions of H(s) of the molecules with F(s) (*14*). Some other reports claim C-H---F-C hydrogen bonding synthons or π---F-C synthons for the crystal engineering (*15*) but the existence itself for such kinds of weak interactions are still controversial (*16*).

Another fluorine-based crystal engineering, which has been recently reported by a Resnati group, is very innovative. Their method is based on another fluorine's unique property, its extremely high electronegativity. The method opened the door in an interdisciplinary field between supramolecular and fluorine chemistries (*17*). The group showed the high potentiality of the approach by an elegant application for a chiral resolution of halofluorocarbons which is otherwise impossible (*18*). The unique system he devised is the halogen bonding strengthened in the perfluorocarbon-halogen synthons (*19*). Halogen bonding itself has been known for a long time (*20*), but its elegant use in the crystal engineering has been realized by his new idea to use the R_F-X---N, O (X=Br, I) bases interactions for aligning the molecules. This type of fluoro-strengthened halogen bonding is stronger than the hydrogen bonding (*21*). The details are to be referred to the next relevant chapter.

Another way of the fluorine's use in the molecular assemblage is based on a face-to-face stacking between perfluoro aromatics and non-fluorinated aromatics (*22*). However, the C-F---H-C interactions prevail over the above-mentioned π-π stackings in some cases (*23*). This strategy is now actively investigated for a specific material design (*24*). Recently proposed fluoro specific interactions of anions with perfluoro aromatic compounds (*25*) and fluoroaromatic-fluoroaromatic interactions are needed to be investigated further (*26*).

Crystal engineering by the fluorine-containing hybrid compounds

In contrast to the former fluorine-based crystal engineering methods, our fluorine-containing hybrid compound method is based on a different uniqueness of fluorine's property, *i.e.* its extremely low polarizability. This property gives the perfluoro compounds a repelling nature against both oil and water. This nature of the perfluoro group gives to the fluoro organic compounds a unique molecular recognizing capability when forming a supramolecular architecture (*27*). Conventional crystal engineering has focused on the attractive forces, but we focused on such a repelling nature (very weak dispersive interaction in genuine meaning) of fluorine to the non-fluorine elements. This lipophobic and hydrophobic fluorine's nature, called sometimes as "fluorophilic", has been a hot topic in the recent fluorous chemistry (*28*). The fluorous chemistry is a liquid technology characterized by the fluorous phase separation. Our crystal technology is based on the corresponding phenomenon seen in the solid state. Fluoro segregation, or fluoro bilayer structure in the crystalline phase is very ubiquitous in the X-ray structures of the organo fluorine compounds having a perfluoro subtitutent (*29*). This micro phase segregation is the result of the "very stern molecular recognition" caused by the very unique fluorine's fluorophilic nature. With this simple packing phenomenon in mind, we initiated a systematic structural investigation on the fluorine-containing hybrid synthons having the structure of R_F-X-R_H (R_F and R_H are perfluorinated and non-fluorinated groups and X is a junction group) to find the packing-structure relationship in the belief that such systematic investigation of the structure-packing relationship will give us a bottom-up process towards crystalline materials.

(a)

(b)

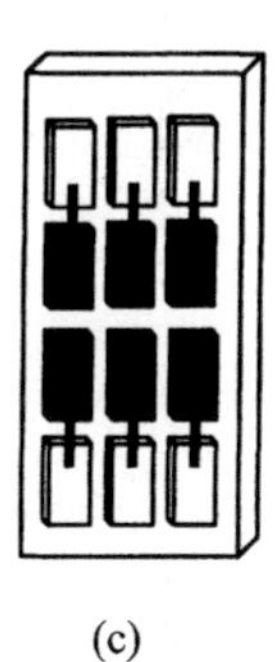

(c)

Scheme 1 (a) Perfluorinated blocks (filled boxes) and hydrocarbon blocks (open boxes) each form its own phase in a liquid state. (b) Hybrids of perfluorinated and hydrocarbon blocks form a homogeneous phase. (c) On crystallization, hybrids self-assemble to a fluoro- and hydro-segregated supramolecular architecture.

One of the most important points we would like to emphasize here is that the formation of the fluorous segregated supramolecular assemblage, the features commonly seen in the crystal packing, was realized by the molecular hybridization of fluoro- and hydro-blocks. This fluorous segregation is in sharp contrast with the simple separation of perfluorinated and non-fluorinated blocks if not hybridized (see the cartoon in scheme 1). It should be reminded that the work on the nickel dithiolene complex with CF_2 substitutents reported by Fournigue et al. is also based on control of the crystal packing by the fluorine's extremely low polarizability (*30*).

Requirement for the fluorinated synthons, R_F-X-R_H

The characters required for the fluorinated hybrid synthons are summarized as follows.

1. Hybrid compounds should be crystalline.
2. Enough size of fluoro segment for molecular recognition by fluorophilicity and fluorophobicity for the supramolecular alignment.
3. Easy preparation in high-yield and with a flexible design.
4. Fluoro segment should be commercially available, not expensive, desirably produced in an industrial scale.

Commercially available perfluoro olefins could be a choice for the purpose. Hexafluoropropene dimers are conceivable, but some attempts showed the difficulty in the crystallization and the size as the fluoro blocks seems to be not enough large for the solid state fluorous technology as is analogously seen in the liquid phase fluorous technology, *i.e.*, miscible nature of perfluorohexane and hexane. We chose hexafluoropropene trimers as the fluoro synthons, as it satisfies all the above requirements.

Synthesis of the fluorinated synthons, R_F-X-R_H

We fixed the R_F moiety and took the R_H moiety as a variable to get an insight into the crystal packing and structure relationship. An ether linkage was used for X due to the synthetic convenience and easiness of crystallization and the function's stability (*31*). The imine linkage might be also a choice for X, but it proved to be not suitable due mainly to the non-crystalline nature and due partly to the instability of the function and the poor yields (*32*). The variables R_H used are summarized with the yield data in Table 1.

Table 1. Structures of the hybrid compounds, R_F-O-R_H

R_H =

1 (97.5%) **2** (94.8%) **3** (41.2%) **4** (90%) **5** (99.2%) **6** (97.8%)

CHO

7 (98.3%) **8** (quantit.) **9** (99.8%) **10** (quantit.)

H_3CO

11 (quantit.) **12** (97.7%) **13** (98.9%) **14** (96.0%)

15 (quantit.) **16** (quantit.)

R_F = F_3C CF CF_3 C C CF CF_3 CF_3 CF_3

The reaction was conducted in DMF at room temperature with a slight excess of a hexafluoropropene trimer mixture consisting of *F*-4-methyl-3-isopropyl-2-pentene (**T-2**) and *F*-2-methyl-3-isopropyl-2-pentene (**T-3**) and the phenol derivatives (**T-3** reacts as **T-2** due to a fast equilibrium between **T-2** and **T-3** under the conditions used). The products are purified by using fluorous phase technique (extraction with FC-72) or by subjecting to a simple short-path silica gel column. Nearly quantitative yields were attained by these simple procedures in all cases. High quality crystals with a few millimeter sizes are obtained in almost all cases so that the potential application could be conceivable for the design of the non-linear optical materials.

Crystal structures of the fluoro synthons

All crystals are X-ray grade crystals and analyzed by the direct methods available in the teXan program package. All compounds have same common features of the packing, thus the hybrid molecules are aligned to form the segregated fluoro- and hydro-segments in the crystal packing. No exception appeared so that this micro phase separation is very robust. Therefore, the fluorine's effect of hexafluoropropene trimer was proved to be strong enough for the micro-fluorous phase formation. It is maybe that the shape of the globular R_F and non-globular R_H synergistically contribute to this micro-phase separation (*33*). One of the representative crystal packings is shown in Fig. 1.

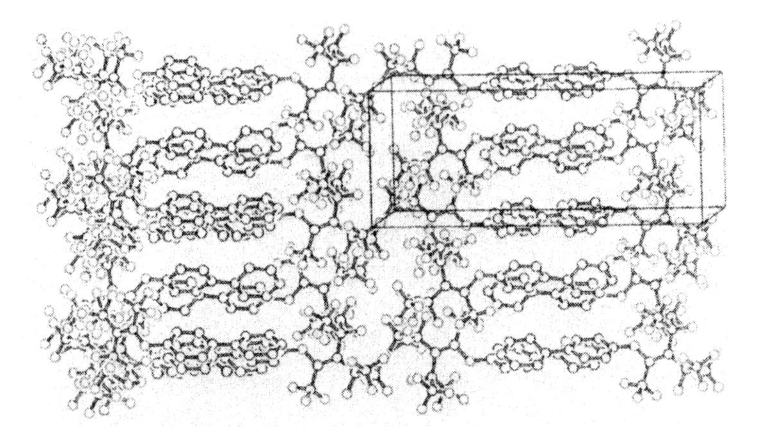

Fig. 1 A crystal packing of the hybrid molecule **12**: biphenyl moiety disordered in two conformations with 1 : 1 occupancy, one of which and hydrogen atoms are omit for clarity.

The hybrid molecules aligned in the anti-parallel manner with a biphenyl moiety overlapped. The formation of alternating fluoro- and hydro-layers is quite clear with segregated layer structures in the packing diagram. A main driving force should be the attraction between the hydrocarbon moieties of the hybrid molecules and at the same time the fluorine's repelling nature against the hydrocarbon, fluorophobic from the viewpoint of the hydrocarbon side, fluorophilic from the viewpoint of the fluorine's side. Recently, Row emphasized the F-F interactions found in the organic fluorine compounds (*34*). Such F-F interactions are claimed analogously to the known Cl-Cl interactions, but still quite obscure and remained to be discussed in the future. In accordance with the Row's view and contrary to our understanding on the driving force of the micro fluorous formation, it is noteworthy to point the disorder structure found in the crystal packing of this compound. The scaffold for the packing of this hybrid compound seems to be the fluoro moiety of the hybrid molecule judging from the circumstantial evidence of the disorder structure seen only in the biphenyl moiety but not in the fluoro moiety (one of the disorder structures was omitted for the sake of clarity). Disorder structures are frequently seen in the fluoro moiety in the fluoro-substituted molecules due to their weak interactions with the surrounding contacted molecules. The fluoro moieties lined together to form a fluorous segregated layer and the whole fluoro segment was buried in this micro fluorous phase. Taking into account the weak F-F interactions in addition to the stronger C-H---aromatic interactions, it is very intriguing to find the disorder structure only in the two aromatic rings but not in the fluoro part. This kind of reversed disorder structure seen here is the first reported case to the best of our knowledge. We analyzed all packings throughout the compounds and deduced the found packing modes into five packing motifs (Scheme 2).

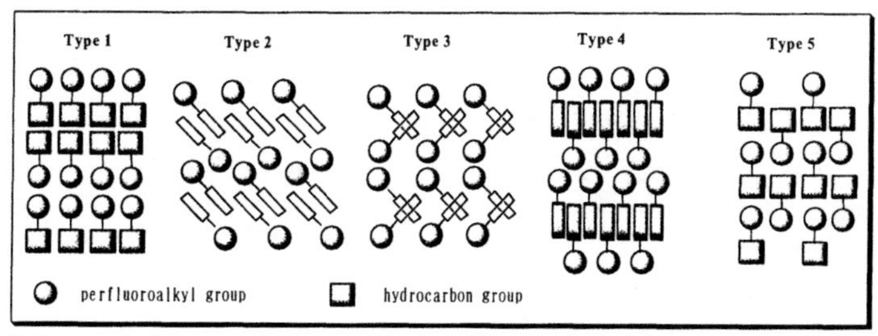

Scheme 2 Five packing motifs found in the hybrid molecules

The quantitative analysis of structure-packing relationship remained to be done, but even qualitative structure-packing relationship is useful to expect the packing of the designed molecules by the topological analogy. The application for the material science of our new method shown in the next paragraph will testify the potential.

Application for the design of electron-conducting polymer

To show the usefulness of this fluorine-based crystal engineering we chose motif **2** as an example for the molecular design of conjugated polymers, which are interesting materials with semiconducting and unique optical properties, applicable for LED and sensing devises (*35*). The structure and ORTEP figure of the fluoro synthon (**10**), which belongs to the motif **2**, is shown in Fig. 2. The stereo view of the crystal packing of **10** is shown in Fig. 3.

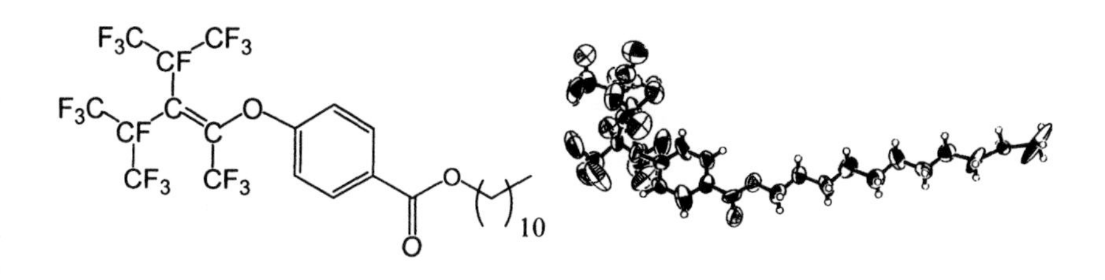

Fig. 2 The structural formula and ORTEP figure of the hybrid compound **10**

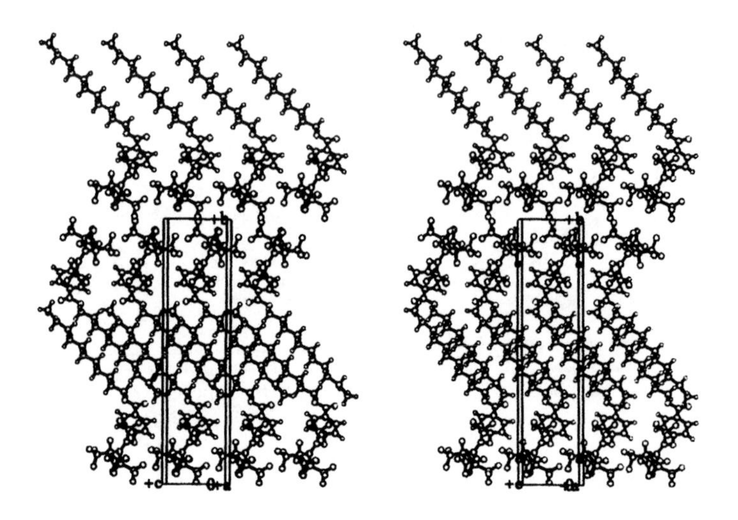

Fig. 3. A stereo view of the crystal packing of **10** which belongs to motif **2**

We designed the molecules, **DA-(n,m)**, having a diacetylene function in the middle of the long alkyl chain of **10**. The structure and ORTEP figure of the compound **DA-(9,12)** are shown in Fig. 4. As is expected from the molecular structure, the long alkyl chain bends at the both ends of the diacetylene moiety. This structural disturbance, however, did not change the packing mode (Fig. 5) so that the type **2** packing was proven to be rather robust and reserved intact all through. If the type **2** packing holds in the whole series of diacetylene derivatives, we can do the fine tuning of the mutual distance and orientation of diacetylene functions in the crystal packing by changing the carbon numbers m and n in the series, thus the topological requirement for the solid state polymerization could be realized in a certain combination of m and n. We therefore planned the synthesis of the compounds having several combinations of m and n.

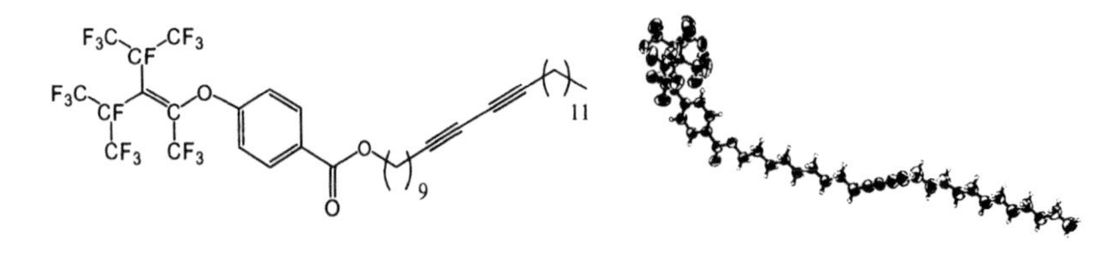

Fig. 4 The structural formula and ORTEP figure of the hybrid compound **DA-(9,12)**

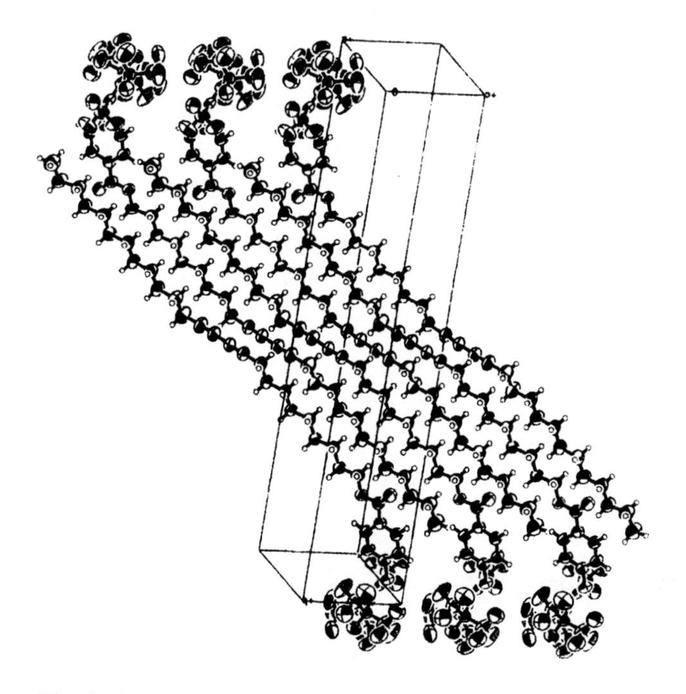

Fig. 5 A crystal packing of the hybrid compound **DA-(9,12)**

Synthesis of designed fluoro synthons with diacetylene group

The synthesis of the fluoro synthons **DA-(n,m)** having diacetylene functionality is outlined in scheme 3. First, lithiated terminal alkynes was brominated, then coupled with ω-hydroxyl terminal alkynes under the presence of cuprous chloride to give the desired *n*-alkanol with a diacetylene function in the middle of the long alkyl chain in 32-51% isolated yields. The obtained alcohols were condensed with benzoic acid having a fluoro moiety by using DCC to give the desired fluoro synthons in fairly good yields (82-95%). The benzoic acid with a fluoro moiety was easily prepared by the reaction of *p*-hydroxybenzoic acid with one of the isomer of hexafluoropropene trimers, *F*-2,4-dimethyl-3-ethyl-2-pentene, in high yield.

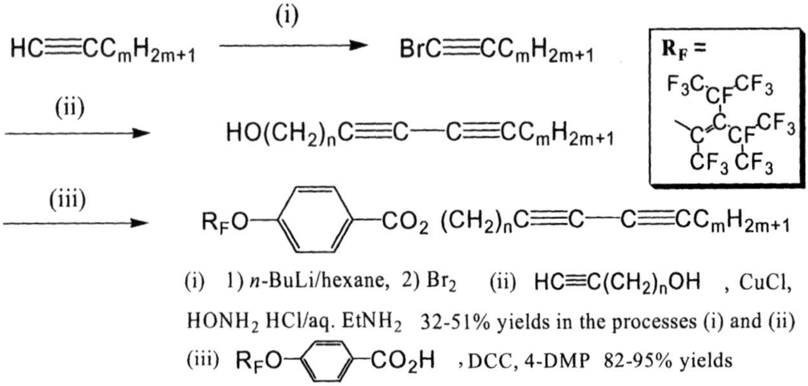

$$HC{\equiv}CC_mH_{2m+1} \xrightarrow{(i)} BrC{\equiv}CC_mH_{2m+1}$$

$$\xrightarrow{(ii)} HO(CH_2)_nC{\equiv}C{-}C{\equiv}CC_mH_{2m+1}$$

$$\xrightarrow{(iii)} R_FO{-}C_6H_4{-}CO_2\ (CH_2)_nC{\equiv}C{-}C{\equiv}CC_mH_{2m+1}$$

R_F = (perfluorinated branched alkenyl group: F_3C, CF, CF_3, C=C, CF, CF_3, CF_3, CF_3)

(i) 1) *n*-BuLi/hexane, 2) Br_2 (ii) $HC{\equiv}C(CH_2)_nOH$, CuCl, $HONH_2$ HCl/aq. $EtNH_2$ 32-51% yields in the processes (i) and (ii)
(iii) $R_FO{-}C_6H_4{-}CO_2H$, DCC, 4-DMP 82-95% yields

Scheme 3

Solid-state uv-initiated polymerization of fluoro synthons

The solid-state uv-initiated polymerization of **DA-(n,m)** was carried out by irradiating the crystals dispersed in perfluorodecalin under Ar atmosphere at ambient temperatures (*36*). The low-pressure 6W Hg lamp was used as the uv source. The photopolymerizability was summarized in Table 2.

Table 2 Photopolymerizability of the hybrid diacetylene derivatives

Diacetylene Derivatives	Photopolymerizability
DA-(3,3)	-
DA-(3,4)	-
DA-(4,3)	-
DA-(4,4)	+
DA-(9,8)	-
DA-(9,9)	+
DA-(9,10)	+
DA-(9,11)	-
DA-(9,12)	-

The clear transparent crystals of **DA-(9,9)**, and **DA-(9,10)** were colored to the blue soon after the irradiation and its color deepened within a minute and then to the black within next few minutes. The crystals of **DA-(4,4)** is also polymerizable but less reactive. These uv-initiated photopolymerizations occur only in a solid state but not at all in a liquid state even under the presence of radical initiator of AIBN at elevated temperatures. The surface of the black-colored crystals had a metal luster as is shown in Fig. 6.

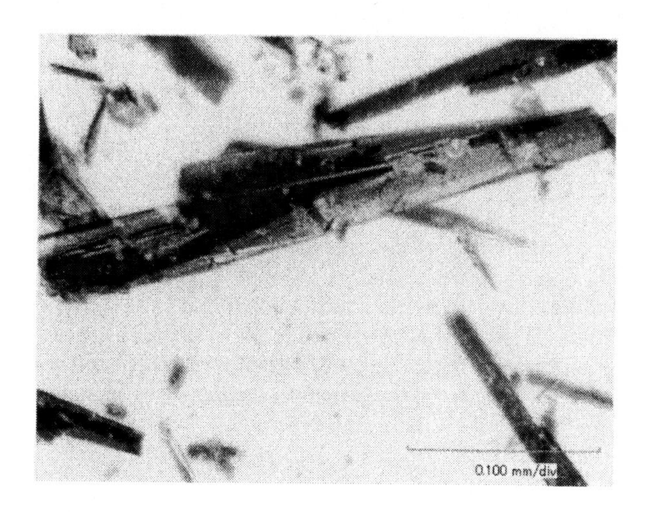

Fig. 6 The black-colored crystals of compound **DA-(9,9)** with a metallic luster which were obtained by the solid-state uv-initiated photopolymerization.

The crystal packing of the uv-stable **DA-(9,12)** showed the diacetylene moieties lined in an anti-parallel manner with the closest distance of 0.686 nm. As is expected from the distance, the crystals of **DA-(9,12)** remained unchanged after a prolonged time of uv-irradiation. On the other hand, the crystal packing of the uv-sensitive **DA-(9,9)**, which was X-ray analyzed at lower temperature (-120°C), revealed the corresponding distance between the diacetylene moieties is as close as 0.379 nm. The crystals of **DA-(9,10)** are too sensitive to X-ray irradiation and thus, we were not able to analyze the compound even at lower temperatures. Therefore, the distances between the diacetylene functions should be shorter. In contrast to our solid state photopolymerization, thermally initiated topochemical polymerization of diacetylene derivatives which were designed based on a logical host-guest approach has recently been reported (*37*). The rational design of the topochemical solid-state reaction is still very difficult, however, the fluorine-based crystal engineering including our hybrid fluoro synthon method will be continually exploited and revised in more sophisticated manners for the more reliable molecular design (*38*).

Summary

The crystalline compounds having the hybrid structures with fluorinated R_F and unfluorinated R_H parts connected by a junction group X, R_F-X-R_H, are systematically investigated by X-ray diffraction method. The packing modes found in a series of hybrid compounds are featured by the micro phase separation with segregated fluoro layers, which are grouped into 5 packing motifs. The new fluorine-based crystal engineering was devised based on the topological relationships between the packing motifs and the molecular structures. The usefulness of this methodology was proven by the molecular design for the photopolymerization. The crystals of the designed diacetylene derivatives successfully polymerized under uv irradiation to give black-colored crystalline materials with a metal luster. We believe that our new crystal engineering is still in its infancy stage but its potential is high and will be used in many ways, especially in the field of the functional material design.

References

1. a) *Solid-state Supramolecular Chemistry: Crystal Engineering;* MacNicol, D. D., Toda, F., Bishop, R., Eds.; *Comprehensive Supramolecular Chemistry Vol. 6.;* Atwood, J. L.; Davies J. E. D.; MacNicol, D. D.; Voegtle, F., Eds.; Lehn, J.-M., Chairman of the Editorial Board; Pergamon, 1996. b) *Crystals as Supermolecules;* Dunitz, J. D.; *Perspectives in Supramolecuar Chemistry, Vol.2. The Crystal as a Supramolecular Entity*; Desiraju, G. R., Eds.; Wiley: Chichester, 1995; p 1.
2. a) Hsing, W.-Y. *Mathematical Intelligencer* **1995**, *17*, 35-42. b) Lommerse, J. P. M.; Motherwell, W. D. S.; Ammon, H. L.; Dunitz, J. D.; Gavezzotti, A.; Hofmann, D. W. M.; Leusen, F. J. J.; Mooji, W. T. M.; Price, S. L.; Schweizer, B.; Schmidt, M. S.; van Eijck, B. P.; Verwer, P.; Williams, D. E. *Acta Cryst.* **2000**, *B56*, 697-714.
3. a) Duchamp, D. J., Marshi, R. E., *Acta Cryst.* **1969**, *B25*, 5-19. b) Dunitz, J. D. *Chem. Commun.* **2003**, 545-548.
4. a) Desiraju, G. R. *Nature Materials* **2002**, *1*, 77-79. b) Sarma, J. A. R. P.; Desiraju, G. R. *Crystal Growth & Design* **2002**, *2*, 93-100.

5. a) Desiraju, G. R. *Angew. Chem. Int. Ed. Engl.* **1995**, *34*, 2311-2327. b) *Supramolecular Organization and Materials Design;* Jones, W; Rao, C. N. R., Eds.; Cambridge University Press, **2002**, p 391.
6. a) Jeffrey, G. A. An Introduction to Hydrogen Bonding; Oxford University Press: Oxford, 1997. b) Desiraju, G. R. *Acc. Chem. Res.*. **2002**, *35*, 565-573.
7. a) Russell, V. A., Evans, C. C., Li, W, Ward, M. D. *Science* **1997**, *276*, 575-579. b) Holman, K. T., Pivovar, A. M., Ward, M. D. *Science* **2001**, *294*, 1907-1911. c) Holman, K. T., Pivovar, A. M., Swift, J. A., Ward, M., D. *Acc. Chem. Res.* **2001**, *34*, 107-118. d) Kosal, M. E., Chou, J.-H., Wilson, S. R., Suslick, K. S. *Natute Mater.* **2002**, *1*, 118-121.
8. a) Ziessel, R. *Synthesis* **1999**, *11*, 1839-1865. b) Leininger, S.; Olenyuk, B.; Stang, P. *J. Chem. Rev.* **2000**, *100*, 853-908. c) Oh, M.; Carpenter, G. B.; Sweigart, D. A. *Acc. Chem. Res.* **2004**, *1*, 1-11.
9. Prout, C. K.; Wright, J. D. *Angew. Chem. Int. Ed. Engl.* **2003**, *7*, 659.
10. Herbstein, F. H. *Perspect. Struct. Chem.* **1971**, *4*, 166-395.
11. a) Prasanna, M. D.; Row, T. N. G. *Crystal Eng.* **2000**, *3*, 135-154. b) Choudhury, A. R.; Urs, U. K.; Smith, P. S.; Goddard, R.; Howard, J. A. K.; Guru Row, T. N. *J. Mol. Struct.* **2002**, *641*, 225-232. c) Adams, H.; Cockroft, S. L.; Guardigli, C.; Hunter, C. A.; Lawson, K. R.; Perkins, J.; Spey, S. E.; Urch, C. J.; Ford, R. *ChemBioChem* **2004**, *5*, 657-665.
12. a) Desiraju, G. R.; Steiner, T. *The Weak Hydrogen Bond in Structural Chemitry and Biology*; Oxford University Press: Oxford, 1999. b) Sarma, J. A. R. P.; Desiraju, G. R. *Acc. Chem. Res.* **1986**, *19*, 222-228. c) Biradha, K; Fujita, M. *Chem. Lett.* **2000**, 350-351. d) Bach, A.; Lentz, D.; Luger, P.; Messerschmidt, M.; Olesch, C.; Patzschke, M. *Angew. Chem. Int. Ed. Engl.* **2002**, *41*, 296-299. e) Khan, M. S.; Al-Mandhary, M. R. A.; Al-Suti, M. K.; Corcoran, T. C.; Al-Mahrooqi, Y.; Attfield, J. P.; Feeder, N.; David, W. I. F.; Shankland, K.; Friend, R. H.; Koeler, A.; Marseglia, E. A.; Tedesco, E.; Tang, C. C.; Raithby, P. R.; Collings, J. C.; Roscoe, K. P.; Batsanov, A. S.; Stimson, L. M.; Marder, T. B. *New J. Chem.* **2003**, *27*, 140-149.
13. a) Desiraju, G. R. *Crystal Engineering, The Design of Organic Solids*, Material Science Monograph, 54, Elsevier, Amsterdam-Oxford-New York-Tokyo, 1989; p 175. b) Tanaka, K.; Fujimoto, D.; Oeser, T.; Irngartinger, H.; Toda, F. *Chem. Commun.* **2000**, 413-414. c) Choudhury, A. R.; Urs, U. K.; Guru Row, T. N.; Nagarajan, K. *J. Mol. Str.* **2002**, *605*, 71-77. d) Fan, J.; Sun, W.-Y.; Okamura, T.; Zheng, Y.-Q.; Sui, B.; Tang, W.-X.; Ueyama, N. *Cryst. Growth & Design* **2004**, *4*, 579-584. e) Zaman, B.; Udachin, K. A., Ripmeester, J. A. *Crystal Growth & Design* **2004**, *4*, 585-589.]
14. a) Guru Row, T. N.; *Coord. Chem. Rev.* **1999**, *183*, 81-100. b) Vishnumurthy, K.; Row, T. N. G., Venkatesan, K. *Tetrahedron* **1999**, *55*, 4095-4108.
15. Choudhury, A. R.; Row, T. N. G. *Crystal Growth & Design* **2004**, *4*, 47-52.

16. a) *Organic Fluorine Chemistry;* Sheppard, W. A.; Sharts, C. M., Eds.; W. A. Benjamin, Inc.: New York, 1969; p 2. b) Howard, J. A. K.; Hoy, V. J.; O'Hagan, E.; Smith, G. T. *Tetrahedron* **1996**, *52*, 12613-12622. c) Dunitz, J. D.; Taylor, R. *Chem. Eur. J.* **1997**, *3*, 89-98.
17. a) Messina, M. T.; Metrangolo, P.; Panzeri, W.; Ragg, E.; Resnati, G. *Tetrahedron Lett.* **1998**, *39*, 9069-9072. b) Amico, V.; Meille, S. V.; Corradi, E.; Messina, M. T.; Resnati, G. *J. Am. Chem. Soc.* **1998**, *120*, 8261-8262.
18. Farina, A.; Meille, S. V.; Messina, M. T.; Metrangolo, P.; Resnati, G.; Vecchio, G. *Angew. Chem. Int. Ed.* **1999**, *38*, 2433-2436.
19. a) Caronna, T.; Liantonio, R.; Logothetis, T. A.; Metrangolo, P.; Pilati, T.; Resnati, G. *J. Am. Chem. Soc.* **2004**, *126*, 4500-4501. b) Fox, D. B.; Liantonio, R.; Metrangolo, P.; Pilati, T.; Resnati, G. *J. Fluorine Chem.* **2004**, *125*, 271-281.
20. Haszeldine, R. N. *J. Chem. Soc.* **1953**, 2622-2626.
21. a) Valerio, G; Raos, G; Meille, S. V.; Metrangolo, P.; Resnati, G. *J. Phys. Chem.* **2000**, *104*, 1617-1620. b) Corradi, E.; Meille, S. V.; Messina, M. T.; Metrangolo, P.; Resnati, G. *Angew. Chem. Int. Ed. Engl.* **2000**, *39*, 1782-1786.
22. a) Patric, C. R.; Prosser, G. S. *Nature* **1960**, *187*, 1021. b) Haneline, M. R.; Tsunoda, M.; Gabbai, F. P. *J. Am. Chem. Soc.* **2002**, *124*, 3737-3742. c) Gdaniec, M.; Jankowski, W.; Milewska, M. J.; Polonski, T. *Angew. Chem. Int. Ed. Engl.* **2003**, *42*, 3903-3906. d) Collings, J. C.; Smith, P. S., Yufit, D. S.; Batsanov, A. S.; Howard, J. A. K.; Marder, T. B. Cryst. Eng. Comm. **2004**, *6*, 25-28.
23. a) Haffer, U.; Rotard, W.; Pickardt, J. *J. Fluorine Chem.* **1995**, *73*, 265-266. b) Thalladi, V. R.; Weiss, H.-C.; Blaeer, D.; Boese, R.; Nangia, A.; Desiraju, G. R. *J. Am. Chem. Soc.* **1998**, *120*, 8702-8710. c) Dai, C; Nguyen, P.; Marder, T. B.; Scott, A. J.; Clegg, W.; Viney, C. *Chem. Commun.* **1999**, 2493-2494.
24. a) Smith, C. E.; Smith, P. S.; Thomas, R. L.; Robins, E. G.; Collings, J. C.; Dai, C.; Scott, A. J.; Borwick, S.; Batsanov, A. S.; Watt, S. W.; Clark, S. J., Viney, C.; Howard, J. A. K., Clegg, W.; Marder, T. B. *J. Mater. Chem.* **2004**, *14*, 413-420.
25. a) Alkorta, I; Rozas, I; Elguero, J. *J. Am. Chem. Soc.* **2002**, *124*, 8593-8598. b) Kim, D; Tarakeshwar, P.; Kim, K. S. *J. Phys. Chem. A* **2004**, *108*, 1250-1258.
26. Kim, C.-Y.; Chandra, P. P.; Jain, A.; Christianson, D. W. *J. Am. Chem. Soc.* **2001**, *123*, 9620-9627.
27. a) Paleta, O.; Benes, M.; Koutnikova, J.; Kral, V. *Tetrahedron Lett.* **2002**, *43*, 5827-6831. b) Percec, V.; Glodde, M.; Johansson, G; Balagurusamy, V.

S. K.; Heinery, P. A. *Angew. Chem. Int. Ed. Engl.* **2003**, *42*, 4338-4342. c) Tomalia, D. A. *Nature Materials* **2003**, *2*, 711-712.

28. a) Horvath, I. T.; Rabai, J. *Science* **1994**, *266*, 72-75. b) Zhang, W. *Chem. Rev.* **2004**, *104*, 2531-2556.
29. a) Sato, T.; Yano, S. *Mol. Cryst. Liq. Cryst.* **1987**, *144*, 179-189. b) Kromm, P.; Bideau, J. –P.; Cotrait, M.; Destrade, C.; Nguyen, H. *Acta Cryst.* **1994**, *C50*, 112-115. c) Buchel, R.; Aebi, R.; Keese, R.; Venugopalan, P. *Acta. Cryst.* **1994**, *C50*, 1803-1805. d) Kromm, P.; Allouchi, H.; Bideau, J. –P.; Cotrait, M.; Nguyen, H. T. *Acta. Cryst.* **1995**, *C51*, 1229-1231. e) Jablonski, C. R.; Zhou, Z. *Can. J. Chem.* **1992**, *70*, 2544-2551. f) Schilling, M.; Bartmann, K.; Mootz, D. *J. Fluorine Chem.* **1995**, *73*, 225-228. g) Caronna, T.; Corradi, E; Meille, S. V.; Novo, B.; Resnati, G; Sidoti, G. *J. Fluorine Chem.* **1999**, *97*, 183-190.
30. a) Dautel, O. J.; Fourmigue, M. *J. Org. Chem.* **2000**, *65*, 6479-6486. b) Dautel O. J.; Fourmigue, M.; Canadell, E. *Chem. Eur. J.* **2001**, *7*, 2635-2643. c) Dautel, O. J.; Fourmigue, M. *Inorg. Chem.* **2001**, *40*, 2083-2087. d) Dautel, J. O.; Fourmigue, M.; Canadell, E.; Auban-Senzier, P. *Adv. Funct. Matr.* **2002**, *12*, 693-698.
31. Maruta, M; Ishikawa, N. *Nippon Kagaku Kaishi* **1978**, *2*, 253-258.
32. Del'tsova, D. P.; Gervits, L. L.; Kadyrov, A. A. *J. Fluorine Chem.* **1996**, *79*, 97-102.
33. Adams, M.; Dogic, Z.; Keller, S. L.; Fraden, S. *Nature* **1998**, *393*, 349-352.
34. a) Vishnumurthy, K.; Row, T. N. G.; Venkatesan, K. *J. Chem. Soc., Perkin Trans. 2* **1997**, 615-619. b) Prasanna, M. D.; Guru Row, T. N. *J. Mol. Struct.* **2001**, *562*, 55-61. c) Choudhury, A. R.; Nagarajan, K.; Guru Row, T. N. *Crystal Engineering* **2004**, *6*, 43.
35. a) Masley, D. W.; Sellmyer, M. A.; Daida, E. J.; Jacobson, J. M. J. Am. Chem. Soc. 2003, 125, 10532-10533. b) Song, J.; Cisar, J. S.; Bertozzi, C. R. *J. Am. Chem. Soc.* **2004**, *126*, 8459-8465. c) Samuel I. D. W. *Nature* **2004**, *429*, 709-711.
36. Hayakawa, Y.; Ono, T. Jpn. Pat. 55,038, 2003.
37. Ouyang, X.; Fowler, F. W.; Lauher, J. W. *J. Am. Chem. Soc.* **2003**, *125*, 12400-12401.
38. a) *Rectivity in Molecular Crystals*; Ohashi, Y., Ed.; Kodansha, Tokyo; VCH, Weinheim, New York, Basel, Cambridge, 1993; p 115. b) Nomura, S.; Itoh, T.; Nakasho, H.; Uno, T.; Kubo, M.; Sada, K.; Inoue, K.; Miyata, M. *J. Am. Chem. Soc.* **2004**, *126*, 2035-2041.

Chapter 30

Haloperfluorocarbons: Versatile Tectons in Halogen Bonding Based Crystal Engineering

Pierangelo Metrangolo*, Franck Meyer, Giuseppe Resnati*, and Maurizio Ursini

Department of Chemistry, Materials and Chemical Engineering, "G. Natta", Polytechnic of Milan, Via L. Mancinelli 7, I–20131 Milan, Italy

The physical and chemical properties of perfluorocarbons (PFCs), and of their functional derivatives, are dramatically different from those of the corresponding hydrocarbons (HCs). For instance, aromatic PFCs and HCs have quadrupolar moments of similar magnitude, but of opposite sign (*1*). Aliphatic PFCs are endowed with greater compressibilities and viscosities but lower internal pressures and refractive indexes than HC parents. Due to the high ionization energy and low polarizability of fluorine atoms, saturated PFCs have particularly weak intermolecular interactions and low surface energies. Heats of solution and cohesive pressures of PFCs are remarkably different from those of corresponding HCs and enthalpies of interaction between PFCs and HCs are smaller than those between HCs. Thus, the miscibility of perfuorinated alkanes, ethers and tertiary amines with many organic solvents is quite limited (*2-4*). With more polar compounds such as water and organic or inorganic salts (OSs or ISs, respectively) the miscibility is even less.

In general, PFC and HC derivatives have a low affinity for each other and this makes additional and non minor problems when the synthesis of hybrid PFC/HC materials is pursued. As a result, a few examples of molecular, macromolecular, or supramolecular systems are described in the literature where PFC and HC residues coexist. These mixed materials are endowed with quite unique and useful properties. As it is the case for many other hybrid materials, these properties are not the simple sum of the properties of the residues which are composed of.

The di- or triblock compounds, R_F-R_H and R_F-R_H-R_F, respectively (*5-7*), the heterophasic diblock copolymers, and the comb-shaped polymers with PFC side chains (*8-9*) are examples of the materials where PFC and HC moieties are connected through covalent bonds. The materials where PFC and HC moieties are held together through non-covalent interactions are a virtually unknown class of compounds. The unique combination of physical properties that PFCs show relative to HCs, requires that if these hybrid and non-covalent materials are pursued, specifically tailored intermolecular interactions have to be used.

Halogen bonding is an attractive, non-covalent interaction occurring between halogen atoms, which work as electron acceptors (Lewis acids, halogen bonding donors), and lone pair possessing atoms, which work as electron donors (Lewis bases, halogen bonding acceptors) (*10-16*). The term halogen bonding has been employed to stress the analogy existing between this interaction and the hydrogen bonding. Two of these similarities are the interactions strengths (10-200 kJ mol^{-1}) and the bonding directions along the axis of the lone-pair orbital on the electron donor.

In this chapter we will show how the halogen bonding is a new paradigm in supramolecular chemistry. It effectively overcomes the low reciprocal affinity between PFC and HC derivatives, enacts their intermolecular recognition, and drives successfully their self-assembly into supramolecular architectures. The halogen bonding is a sufficiently consistent interaction to allow a wide diversity of PFC and HC modules to interact after a predictable fashion. This predictability allows an halogen bonding based crystal engineering to be developed. The structure of the formed hybrid material can be anticipated with an uncommon accuracy. We will describe the preparation of both two-component heteromeric materials and three-component heteromeric materials, PFC/HC and PFC/HC/IS systems, respectively. We will also report on the specific contribution that different analytical techniques can give in identifying and characterising the halogen bonding formation. It will be also described how, reminiscent of the low reciprocal affinity between PFCs and HCs or ISs, strong module segregation is often present and gives rise to alternating layers of nanometric size. At the end of the chapter, some useful applications of the halogen bonding driven self-assembly of PFC and CH derivatives will be briefly outlined.

Self assembly with neutral electron donors

The first reports on the ability of halogen atoms involved in attractive interaction with lone pair possessing atoms dates back to two centuries ago (*17-18*). It is now well documented (*10-16*) that halogen bonded adducts form in the solid, liquid, and gas phases. The halogen bonding donors can be halogen atoms from both inorganic and organic compounds. Calculations and experimental data show that the strength of the interactions given by different halogens is opposite to their electronegativity, namely I > Br > Cl > F. It is of particular relevance to observe that only few reports are available where fluorine works as halogen bonding donor and the formed adducts have never been characterized in the solid state (*19-20*). The substituents near the halogen atoms have a marked influence on the strength of the formed interaction. Indeed, strong electron-withdrawing species, such as perfluorinated groups, decrease the electronic density of the halogen atom that becomes a stronger electron acceptor (*21-22*).

The electron donor species can be neutral electron rich heteroatoms (P, N, O, S, Se), or anions (I^-, Br^-, Cl^-....). Theoritical calculations and experimental measurements have been used to rank electron donors as a funtion of their strenght as halogen bonding acceptors. The obtained scales may vary depending on the electron acceptor used. The electron density on the electron donor site plays a key role in determining the strength of the formed halogen bonding. Its increase results in stronger interactions. In most cases anions are better halogen bonding acceptors than neutral species and phosphorous and nitrogen are better halogen bonding acceptors than oxygen and sulfur. The electron donor strength can also be tuned by controlling its hybridization states (N (sp^3) > N (sp^2) > N (sp) and O (sp^3) > O (sp^2)).

The first study on halogen bonding adducts involving fluoroorganic derivatives was reported by Haszeldine in 1953. He described dramatic changes of UV spectra of iodofluorocarbon when the solvent was changed from alkanes to alcohols or amines (*23*). Then, investigations were reported in the sixties and seventies where the formed adducts were studied in the gas and liquid phases by using IR and NMR techniques (*24-31*). It was shown how complexes formation between perfluoroalkyl iodides and amines was characterised, in IR spectra, by the appearance of an absorption around 100 cm^{-1} corresponding to the N···I bonding and, in ^{19}F NMR spectra, by an upfield chemical shift of the -CF_2X signal (X = I, Br). Then, it was suggested that this halogen bonding would enable supramolecular structures to be formed as solids at room temperature. However, when solid complexes were reported (*32-33*), their structures were unproperly assigned.

Our advancement in the field was the development of an halogen bonding based crystal engineering enabling the preparation of multicomponent heteromeric architectures. A wide diversity of iodo- and bromo-PFCs have been used as effective tectons in self-assembly processes and X-ray analyses has been proven as a powerful tool for the characterization of thus formed hybrid fluorous materials.

Our first studies of co-crystal formations were based on interactions of perfluoroalkyl iodides with a wide variety of aromatic and aliphatic diamines (Scheme 1). The self-assembly of bidendate acids 1 and bases 2 is initiated by their recognition in solution as shown by 19F NMR and IR spectra (*34-36*). The modules tend to behave as telechelic acids and bases, respectively, the halogen bonding is reiterated at either ends of the molecules, and the slow evaporation of

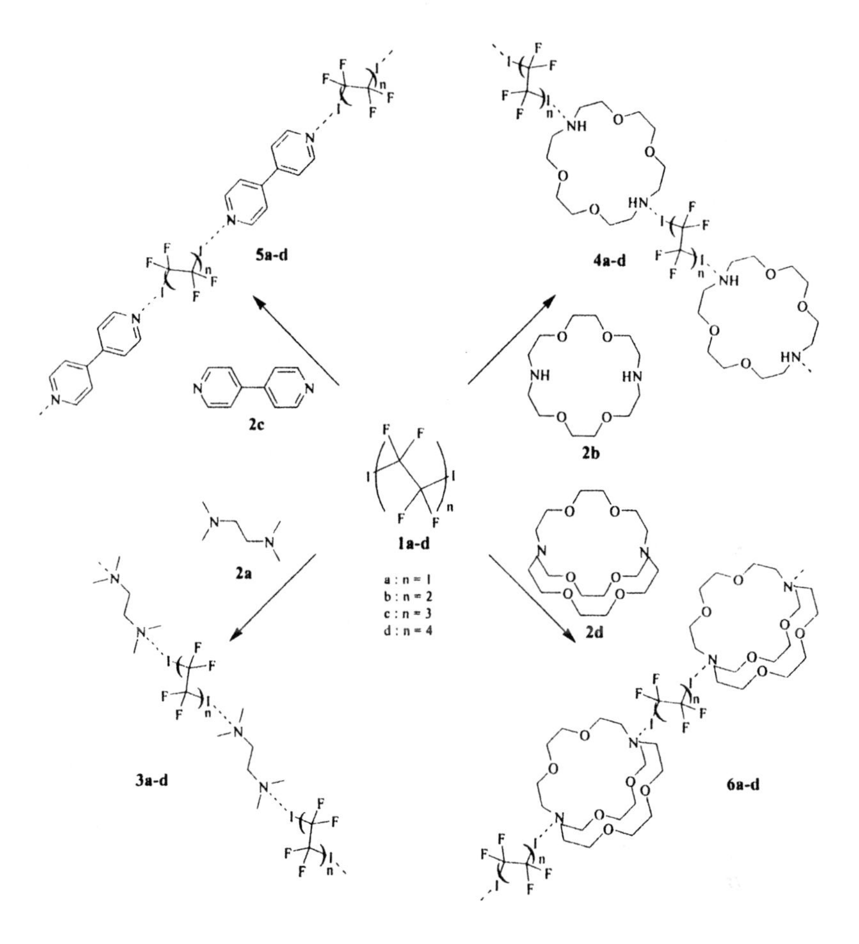

*Scheme 1. Infinite 1D chains **3-6** where N···I halogen bondings drives the self-assembly of diiodoperfluoroalkanes **1** with dinitrogen substituted derivatives **2**.*

the solvent leads to the formation of non-covalent co-polymers **3-6**. The physical chracteristics of these materials (IR; Raman; DSC; ^{19}F, ^{14}N, ^{15}N, ^{13}C NMRs, X-ray) demonstrate the presence of N···I halogen bondings.

For instance, an equimolar amount of 1,2-diiodotetrafluoroethane **1a** and *N,N,N',N'*-tetramethylethylenediamine (TMEDA) **2a** leads to the co-crystal **3a** the melting point of which is dramatically higher than the pure starting modules (m.p. **1a** = –27 °C, m.p. **2a** = –55 °C, and m.p. **3a** = 105 °C) (*37*). This melting point increase can be also observed with the co-crystal **6a** (m.p. = 119 °C) obtained from **1a** and the cryptand 4,7,13,16,21,24-hexaoxa-1,10-diazabicyclo[8,8,8]hexacosane (Kryptofix® 2.2.2., K.2.2.2.) **2d** (m.p. = 73 °C) (*38*). Diiodoperfluoroalkanes and dinitrogen substituted hydrocarbons form one dimensional (1D) infinite chains with a melting point invariably higher than the starting modules. This melting point increase is in correlation with the strength of the N···I-R_F interactions (R_F = perfluoroalkyl chain). It reveals the formation of well defined co-crystals rather than mechanical mixtures, and it can be used as a simple proof of the co-crystal formation.

The N···I halogen bonding is a general, effective, and reliable tool to overcome the low affinity between PFCs and HCs and self-assemble them into 1D chains. However, the reluctance of the two modules to mix survives in the formed adducts. The supramolecular packing of the halogen bonded infinite chains is governed by the low reciprocal affinity that minimizes the contacts between fluorocarbon and hydrocarbon domains. Figure 1 shows the packing

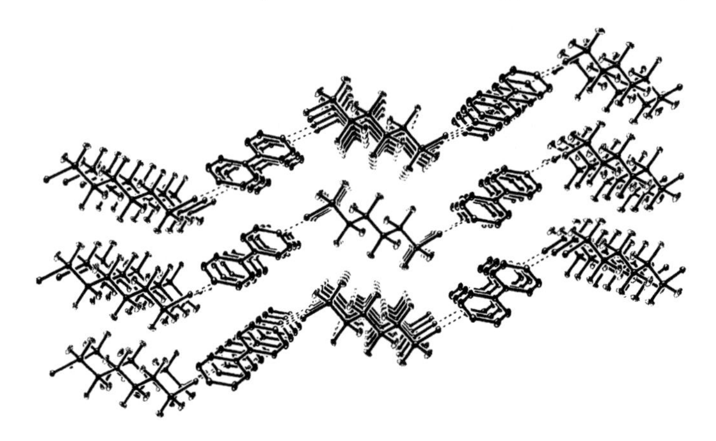

*Figure 1. ORTEP view of co-crystal **5d**. Segregation of dipyridyl **2c** and 1,8-diiodoperfluorooctane (**1d**) gives rise to parallel layers joined by N···I interactions.*

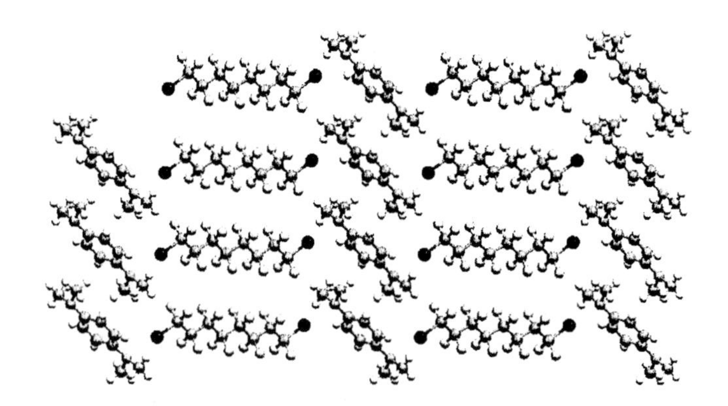

*Figure 2. Crystal packing of the co-crystal **7a**. N···Br interactions are responsible for the self-assembly of N,N,N',N'-tetramethyl-1,4-phenylenediamine (**2e**) and 1,8-dibromoperfluorooctane (**8a**). Module segregation is apparent.*

diagram of the co-crystal **5d** formed by dipyridyl (**2c**) and 1,8-diiodoperfluorooctane (**1d**). Modules segregation clearly occurs and alternating PFC and HC layers are formed. As clearly shown by Figure 2, layered structures are present also when bromo-PFCs are used and when dinitrogen substituted tectons other than **2a-d** are employed (*39*). Single crystal X-ray structures afford many useful structural informations, such as the N···I distance and the N···I-C angle. These properties are discussed in the co-crystal **9** formed on slow evaporation of a chloroform solution of TMEDA **2a** and (*E*)-1,2-diiodo-1,2-difluoroethene (**10**) (Scheme 2) (*40*).

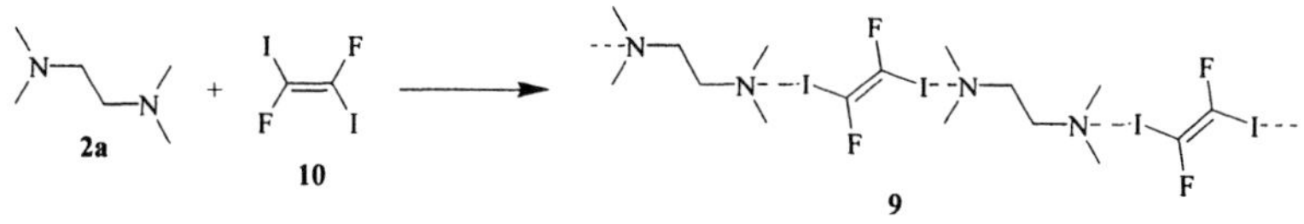

*Scheme 2. Formation of 1D infinite network **9** through halogen bonding driven self-assembly of starting modules **2a** and **10**.*

This co-crystal, where the two starting modules are present in the 1:1 ratio, is formed from solutions where the starting modules ratio spans in the range 0.5 - 2.5. A similar behaviour has been shown by several other co-crystals **3-6** thus proving the strength of the N···I halogen bonding and its ability to control self-assembly processes. In the co-crystal **9**, no disorder affects the HC module **2a** while the PFC module **10** is disordered over two positions simulating a cross (Figure 3). Once again, a similar behaviour has been observed in several other co-crystals **3-6** and it is consistent with the particularly weak intermolecular interactions that characterise PFCs compared to HCs. Two slightly different N···I distances are present, N···I_1= 2.805 Å and N···I_2 = 2.815 Å.

These values are greater than the average covalent N–I bond (2.07 Å) (*41*) but approximatly 0.8 times the sum of van der Waals radii (1.98 Å for iodine and 1.55 Å for nitrogen) (*42*). They are quite typical for N···I-R_F interactions. Consistent with an $n \rightarrow \sigma^*$ electron donation from the nitrogen to iodine, N···I-C angles in **10** are approximately linear, N···I_1-C is 175.9° and N···I_2-C is 175.6°. This high directionality is typical for any N···I-R_F halogen bonding. It is a particularly useful property in crystal engineering as it allows the geometry of the formed supramolecules to be anticipated. It has allowed useful applications of the interaction (*43*). The formation of co-crystals **4** and **6** starting from K.2.2. and K.2.2.2., **2b** and **2d**, respectively, is perfectly consistent with the general statement that nitrogen is a stronger electron donor than oxygen. However, a careful selection of the employed tectons allows also the opposite to occur. Specifically, if the electron density on the oxygen atom is conveniently increased, the electron donor behaviour of this atom can dramatically increase and become even stronger than that of a typical nitrogen atom. This is the case when pyridyl derivatives and their *N*-oxide analogues are compared (*44*).

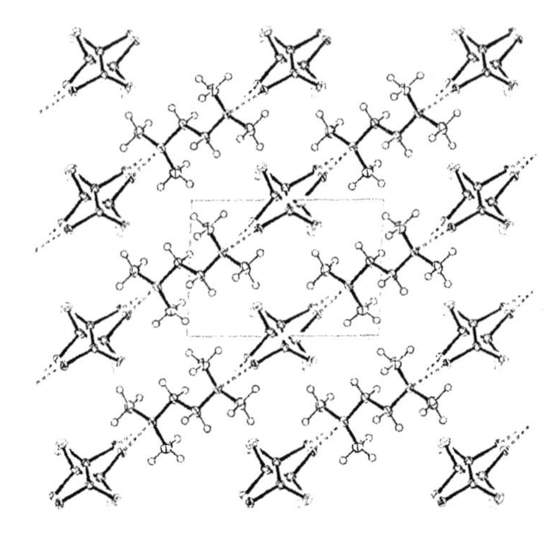

*Figure 3. Packing of the co-crystal **10**.*

In solution, the N···I interactions that 1,4-diiodotetrafluorobenzene (**11**) gives with pyridine and its 4-methyl analogue are weaker than the O···I interactions formed with the corresponding *N*-oxides. The same holds also in the solid. Both 4,4'-dipyridyl **2c** and 4,4'-dipyridyl-N,N'-dioxide **13** self-assemble with 1,4-diiodotetrafluorobenzene **11**. On crystallization of two component solutions, infinite chains **12** and **14** are formed, respectively (Scheme 3a and 3b). Interestingly, the melting point of **14**, where O···I halogen bondings are present, is definitively higher than that of **12**, where N···I halogen bondings are present (> 300 and 180-185 °C, respectively). Slow evaporation of the solvent (chloroform/methanol) from mixtures containing equimolar amounts of **2c**, **13**, and **11** leads to the preferential formation of the co-crystal **14** that has been isolated as a pure product. **2c** was recovered in the liquid phase (Scheme 3c).

Clearly, the stronger O···I halogen bondings, between **13** and **11**, prevail over the weaker N···I halogen bondings, between **2c** and **11**, in identifying the modules to be involved in self-assembly processes. This is perfectly consistent with the expectation that the high electron density on oxygen atoms of *N*-oxide derivatives boosts their ability to work as particularly strong halogen bonding acceptors.

Mixtures of hydroquinone, 1,4-diiodotetrafluorobenzene, and 1,2-bis(4-pyridyl)ethane were used in similar competitive formations of co-crystals. Through this experiments, it was shown that the N···I-R_F halogen bonding can prevail over N···H-O hydrogen bonding in driving self-assembly processes (*45*).

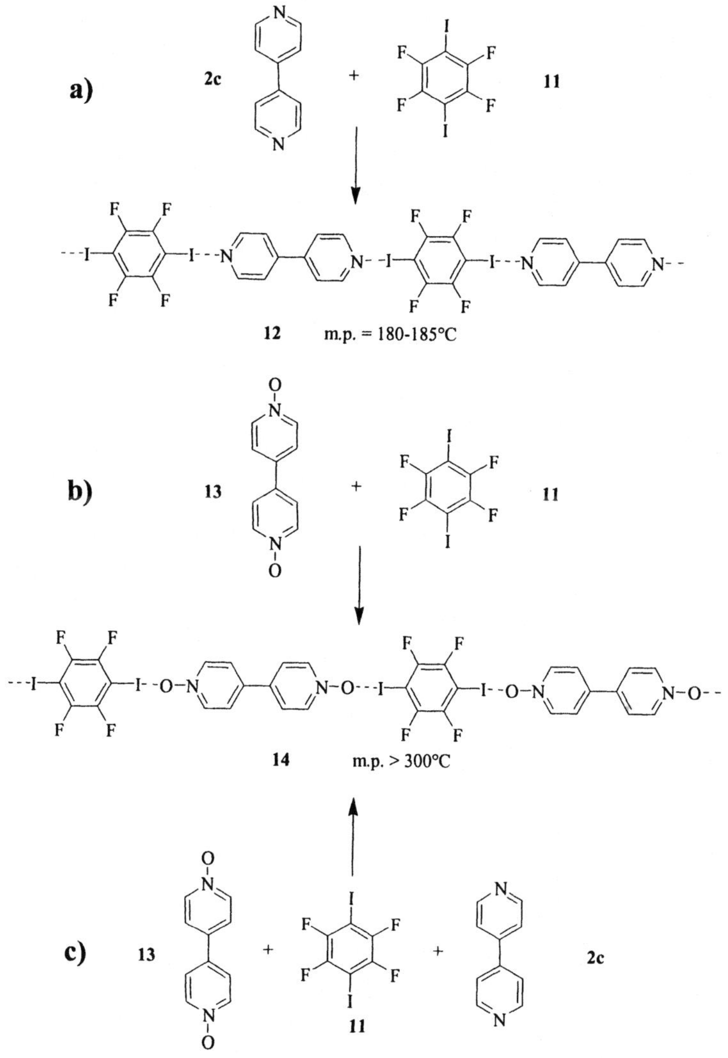

a) 2c + 11

12 m.p. = 180-185°C

b) 13 + 11

14 m.p. > 300°C

c) 13 + 11 + 2c

*Scheme 3. Selective supramolecular synthesis where the stronger O···I halogen bondings, between **13** and **11**, prevail over the weaker N···I halogen bondings, between **2c** and **11**, and lead to the preferential formation of co-crystal **14**.*

Other competitive co-crystals formations allowed us to prove that also the O···I-R_F halogen bondings can prevail over N···H-O hydrogen bondings.

Occasional reports describe how also S···I-R_F interactions can play a key role in molecular recognition processes. 2-Mercapto-1-methylimidazole (**15**) interacts with the diiodotetrafluorobenzene **11** to give the 1D infinite chain **16** in which the sulfur atom is engaged in two S···I halogen bondings. The presence of one S···H-N hydrogen bonding also contributes in driving the self-assembly (Scheme 4) (*46*).

The co-crystals described above are all obtained starting from bidendate electron donor and acceptor species. Several other similar supramolecular architectures have been obtained (*47-56*), but we have also shown that when mononitrogen- or monooxygen-HCs interact with diiodo-PFCs (*41*, *57*), or when dinitrogen-HCs interact with monoiodo-PFCs, well defined trimeric adducts are formed in the co-crystallization. For instance, interactions between the 1-iodoperfluoroheptane **17a** and TMEDA **2c** afford the co-crystal **18a** where the two starting modules are present in 2:1 ratio in order to pair the binding sites (Scheme 5) (*58*).

Several properties of these discrete adducts recall those of the material where infinite chains are present. The melting point of the formed product **18a** is 53 °C, dramatically higher than the melting point of the pure starting compounds (m.p. **2c** = –55 °C and m.p. **17a** = –8 °C). The structural organisation was established through single crystal X-ray analysis and it shows a strong segregation between the PFC and HC modules (Figure 4). Well-defined $CF_3(CF_2)_6$-I···TMEDA···I-$(CF_2)_6CF_3$ units are present where two N···I interactions bind one diamine molecule to two different 1-iodoperfluoroheptane molecules.

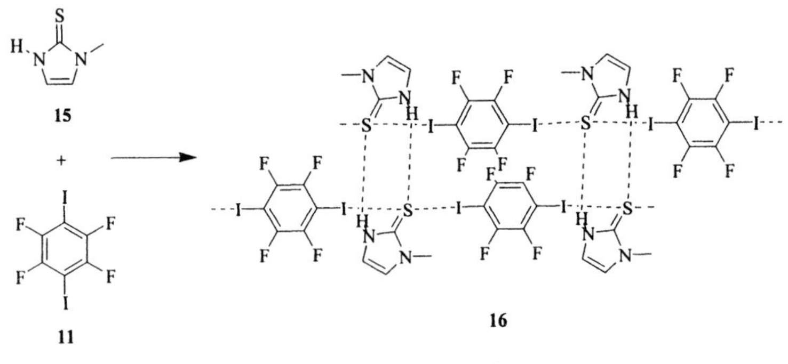

*Scheme 4. Formation of 1D infinite chain **16** through S···I halogen bonding and S···H hydrogen bonding.*

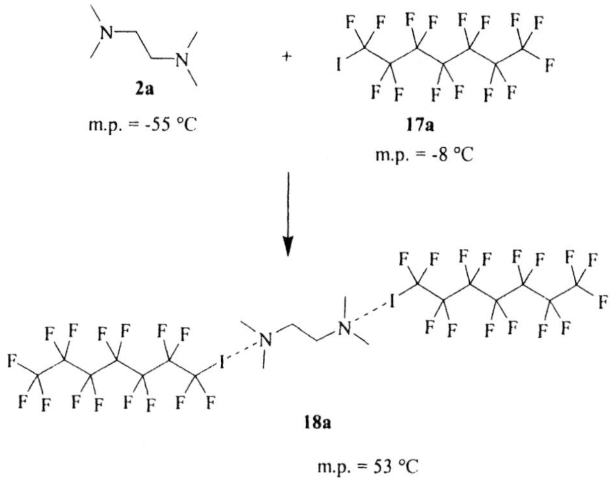

2a

m.p. = -55 °C

17a

m.p. = -8 °C

18a

m.p. = 53 °C

*Scheme 5. Formation of trimeric aggregate **18a** through N···I interactions between TMEDA **2a** and iodoperfluoroheptane **17a**.*

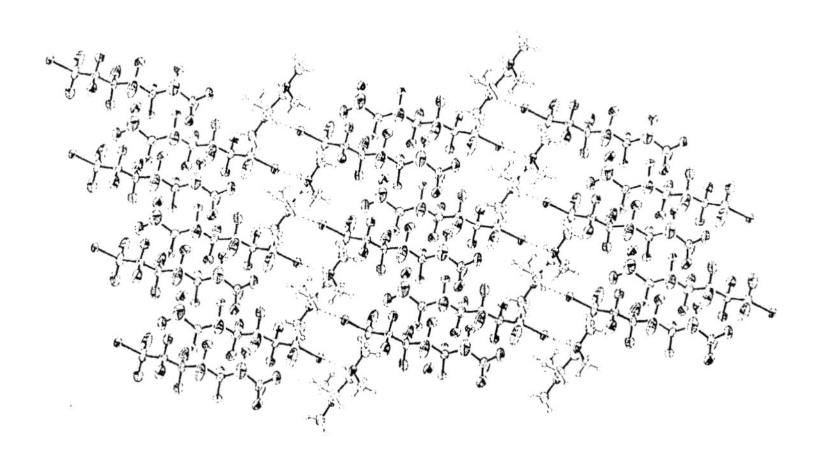

*Figure 4. Packing of the co-crystal **18a** formed by TMEDA **2a** and iodoperfluoroheptane **17a** evidencing how interdigitation of trimeric adducts occurs in the layered structure.*

In solution, TMEDA adopts a *gauche* conformation while in the co-crystal **18a**, nitrogen atoms adopt a *trans* arrangement, as already described in other adducts. A dramatic rotational disorder affects the perfluoroheptyl chain and it has been modelled by splitting atoms over two equally populated locations. As is the case in other perfluorinated chains, the difluoromethylene groups adopt a distorted *trans*-planar arrangement, probably to minimise the intramolecular repulsion between 1,3-positioned difluoromethylenes. The crystallographic analysis established the N···I-C angle is 172.4(1)°, a value slightly smaller than in structurally related infinite chains given by TMEDA and other diamines with various α,ω-diiodoperfluoroalkanes. The N···I length is 2.762(3) Å, quite similar with the N···I distance shown in other co-crystals.

In order to demonstrate that a specific D···X-R_F (D = electron donor atom, X = I, Br, Cl) interaction occurs also in the liquid phase, investigations have been carried out by NMR (*59*). The approach proved a simple, reliable, and effective technique to establish also the relative strength of different interactions.

First of all, an equimolar mixture of quinuclidine **19** and 1-iodo-perfluoropropane **17b** in deuterated chloroform has been studied by ^{1}H NMR and ^{13}C NMR (*60-61*). Spectra show that an attractive intermolecular interaction occurs between the two species and the chemical shift variations are consistent with an n→σ* electron donation from the nitrogen to the iodine (Scheme 6).

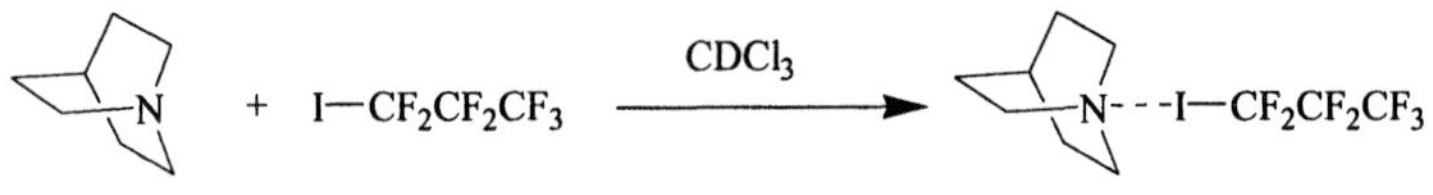

$$\mathbf{19} + \text{I—CF}_2\text{CF}_2\text{CF}_3\ (\mathbf{17b}) \xrightarrow{\text{CDCl}_3} \text{N}\cdots\text{I—CF}_2\text{CF}_2\text{CF}_3\ (\mathbf{18b})$$

*Scheme 6. N···I interaction between quinuclidine **19** and 1-iodo-perfluoropropane **17b**.*

But a definite evidence of the N···I interaction is given by the ^{19}F and ^{14}N NMR spectra. Indeed, a 7 ppm chemical shift was observed in the ^{14}N NMR spectrum and the signal width passed from 960 Hz to 2160 Hz. These results establish clearly the increase of the quadrupolar relaxation related to a specific N···I intermolecular association. The ^{19}F NMR spectrum shows that the chemical shift of the -CF_2I group changes. On increasing the iodide/amine ratio the signal moves from -61.13 ppm (δ_{free}) to -77.80 ppm (δ_{max}).

The ^{19}F NMR signals of 1,2-diiodo- and 1,2-dibromotetrafluoroethanes in numerous heteroatom substituted HCs used as solvents were also studied. Many different D···X-R_F interactions could thus be ranked as a function of their relative strength.

A comparison of the observed chemical shift changes with the data from some other analytical techniques (IR, Raman, DSC) and from experimental results (competitive co-crystals formation) reveals that the stronger the halogen bonding is, the larger the induced ^{19}F NMR chemical shift change is. *n*-Pentane (unable to work as an electron donor) was used as reference solvent and Table I lists a selection of the obtained $\Delta\delta_{CF2X} = \delta_{n\text{-pentane}} - \delta_{\text{used solvent}}$ values.

Table I. ^{19}F NMR chemical shift differences of 1,2-dihalotetrafluoroethanes in different solvents.[a]

Solvents	ICF_2CF_2I $\Delta\delta_{CF_2I}$ (ppm)[b,c]	$BrCF_2CF_2Br$ $\Delta\delta_{CF_2Br}$ (ppm)[b,d]
Piperidine	11.23	2.40
Morpholine	10.59	1.95
Thiomorpholine	9.94	1.62
Hexamethylphosphoramide (HMPA)	8.23	2.76
Pyridine	7.32	1.12
Dimethyl sulfoxide	7.22	2.03
Tetrahydrothiophene	5.38	0.76
Tetrahydrofuran	4.13	0.95
Acetone	3.63	0.90
Acetonitrile	2.74	0.45
Thiophene	1.41	0.26
Furan	1.11	0.13

[a] $CFCl_3$ was used as internal standard. [b] $\Delta\delta_{CF2X} = \delta_{n\text{-pentane}} - \delta_{\text{used solvent}}$. [c] $\delta_{ICF2CF2I\text{ in }n\text{-pentane}}$ = -52.42 ppm. [d] $\delta_{BrCF2CF2Br\text{ in }n\text{-pentane}}$ = -63.32 ppm.

The $\Delta\delta_{CF2I}$ values span over the range of 1.11 to 11.23 ppm while the $\Delta\delta_{CF2Br}$ values vary from 0.13 to 2.76 ppm. These values imply that the iodine atom gives stronger interactions than the bromine atom independent from the type of the electron donor. The nitrogen atom is confirmed a very good electron donor. Surprisingly, the HMPA is the best electron donor toward the bromide acid while it is not with the iodide acid.

Also IR and Raman spectroscopies are useful tools to detect the formation of halogen bonded adducts and to quantify the relative strength of different interactions. The vibration shifts and the intensity variations are of particular relevance to the formation of intermolecular interactions in the solid. Thus, frequencies of perfluoroalkyl iodides **1a-c** and diazabicyclooctane (DABCO) **2f** have been compared with the formed adducts **20a-c** (Scheme 7) (*62*).

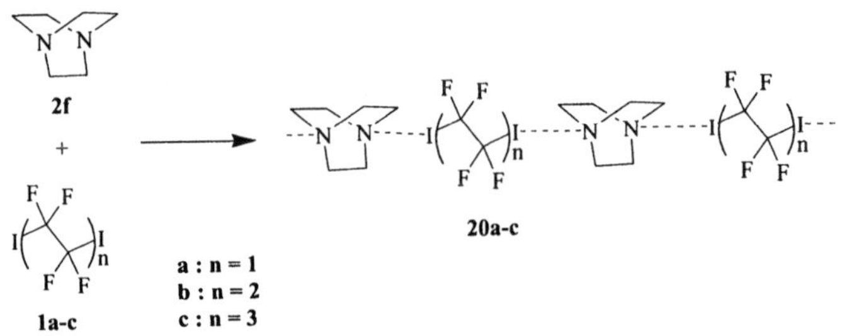

a : n = 1
b : n = 2
c : n = 3

*Scheme 7. Formation of 1D infinite chains by self-assembly of perfluoroalkyl iodides **1a-c** and diazabicyclooctane (DABCO) **2f**.*

Particularly relevant absorptions are the strong C-F stretching motions (1200–1050 cm^{-1} range) in IR spectra of PFCs **1a-c** and the C-H bonds stretchings in IR and Raman spectra of DABCO **2f** (3000-2800 cm^{-1} region). In the IR spectra of co-polymers **20a-c**, the $\nu_{C\text{-}F}$ values shift to lower frequencies than in pure **1a-c**. Interestingly, shifts of **20b,c** are smaller than those observed in co-polymer **20a**, consistent with the fact that the internal difluoromethylene groups are less sensitive to the N···I interaction formation. As far as the electron donor module is concerned, the $\nu_{C\text{-}H}$ absorptions of **20a-c** are at higher frequencies in IR and Raman spectra as compared to the pure DABCO and their intensities markedly decrease. As anticipated by a general rule, the higher the $\nu_{C\text{-}H}$ frequency, the lower its vibration intensity. This behavior can be correlated with a higher positive charge on, namely a more acidic character for, the H atoms in co-polymers **20a-c** than in the pure amine **2f**. It is therefore perfectly consistent with the electron donation from nitrogen to iodine on halogen bonding formation. The C-I stretching at 268 cm^{-1} in the Raman spectrum of pure **1a** is red-shifted to 254 cm^{-1} in the co-crystal **20a**. This lower frequency shift implies a weaker C-I bond force constant, consistent with an $n \rightarrow \sigma^*$ electron donation from the nitrogen of DABCO to the iodine atoms of diiodoperfluoroalkanes **1**.

Self-assembly with anionic electron donors

The comparison of pyridine derivatives and the corresponding *N*-oxides (Scheme 3) demonstrated that the greater the electron density on the halogen bonding acceptor is, the stronger the formed interactions are. The anion···X-R_F halogen bonding was thus anticipated to be stronger than the N···X-R_F halogen bonding and robust enough to induce the self-assembly of modules with a particularly low reciprocal affinity (*63-64*). The stability of the I_3^- anion confirms the anticipation. Calculation (density functional methods, B3LYP/LACVP basis set) estimated an interaction dissociation energy of 41.6 kcal/mol in the $I^- \cdots I_2$ complex and of 6.02 kcal/mol in the 4,4'-dipyridylethylene·**11** infinite chain (*65-66*).

To confirm this hypothesis, the relative halogen bonding acceptor abilities of piperidine and different tetrabutylammonium salts have been established by considering the change they induce on the ^{19}F NMR signal of 1,2-diiodotetrafluoroethane (**1a**) (*67*). Table II reports the chemical shift changes shown by **1a** in deuterochloroform ($\Delta\delta_{CF2I} = \delta_{CF2I} - \delta_{CF2I + donor}$).

Halide anions are particularly good electron donors and the interaction strength decrease according to the scale $I^- > Br^- > Cl^-$. Moreover, TBA-I gives the highest chemical shift variation while piperidine, a very good nitrogen electron donor, gives nearly the smallest change. We thus focussed our attention on iodide salts. Its cation was chosen in order to have a good salt solubility in polar organic solvents and a high ion pair dissociation.

Table II. ^{19}F NMR Chemical shift differences of 1,2-diiodotetrafluoroethane (1a) in the presence of different donors in $CDCl_3$.[a-c]

Donors	ICF_2CF_2I $\Delta\delta_{CF2I}$ (ppm)[d]
Piperidine	0.97
TBA-I	3.54
TBA-Br	3.10
TBA-Cl	2.17
TBA-Succinimide	2.00
TBA-CN	1.78
TBA-SCN	1.38
TBA-Br_3	0.99
TBA-NO_3	0.40

[a] $CFCl_3$ was used as internal standard. [b] Donor/acceptor molar ratio = 2.8. [c] TBA = tetrabutylammonium. [d] $\Delta\delta_{CF2I} = \delta_{pure\ ICF2CF2I} - \delta_{complexed\ ICF2CF2I}$.

In this way the electron donor ability of the anion is maximized. Specifically, cryptated potassium iodide was investigated. Previous studies have shown that a mixture of **2d** and **1a-d** form infinite chains **6a-d** thanks to N···I halogen bondings (Scheme 8).

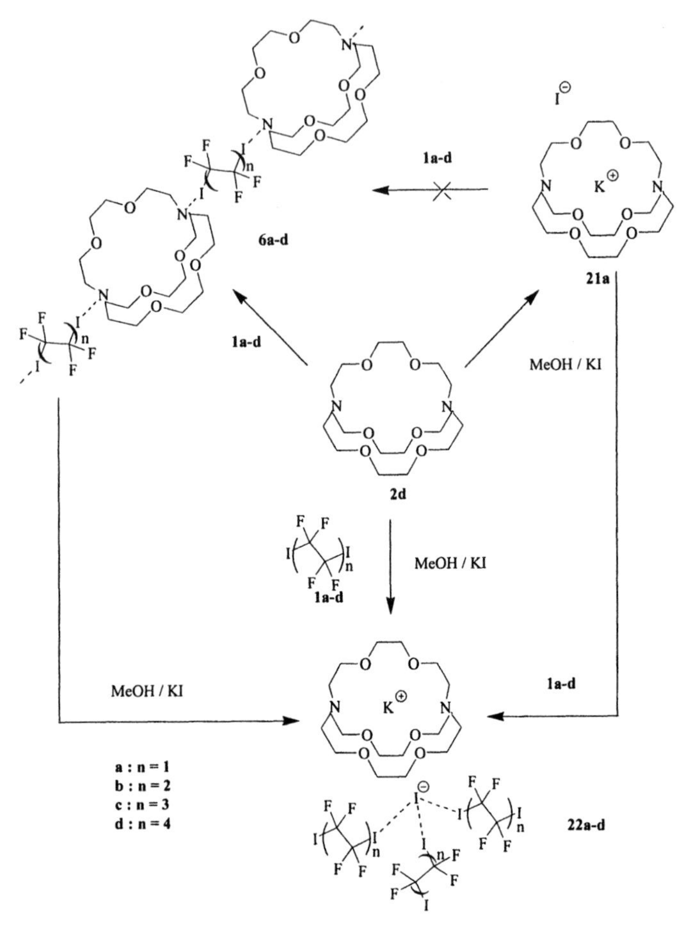

*Scheme 8. Self-assembly of three component systems. One-pot/two-steps or two-pot/two-steps supramolecular syntheses both afford the thermodynamically more stable complexes **22a-d**.*

When **2d** is crystallised in the presence of potassium iodide, the solid cryptate K.2.2.2.⊂ KI **21a** is isolated thanks to the well-known ability of **2d** to work as an alkaline metal receptor (Scheme 8). When the three components **1a-d**, **2d**, and potassium iodide are mixed, the solid cryptate K.2.2.2.⊂ KI is formed and the naked iodide functions as electron donor towards the iodine atoms of **1a–d**. The colourless products **22a-d** are obtained. In these three-component heteromeric architectures the PFC, HC, and IS elements are present in 1.5:1:1 ratio.

The three products behave as a system of two Lewis acid motifs (potassium cations and iodine atoms of **1a–d**) and two Lewis base motifs (iodide anions and heteroatom array of **2d**). Although the two-component complexes **6a-d** and **21a** (acid-base moieties) are stable co-crystals, the pairing affording three-component complexes **22a–d** prevails. The preferred pairing is that predicted by the HSAB (hard and soft acid-base) theory (*68*). Multi-step supramolecular syntheses occur where two intermolecular interactions (potassium cryptation and $I^{-}\cdots$I-PFC halogen bonding) prevail over other possible interactions, and work in a hierarchical fashion. The formation of the three-component complexes **22b–d** is the thermodynamically more stable pathway in which the supramolecular architectures are obtained through a one-pot/two-steps or two-pot/two-steps synthesis.

^{19}F NMR spectra of these systems have been registered. The shifts of the -CF_2I signals give further evidence that the $I^{-}\cdots$I halogen bonding is stronger than the N···I one. Compared to pure **1b**, the co-crystal **22b** gives an upfield shift of 2.90 ppm while an equimolar solution of **6b** affords a variation of 0.19 ppm for the same signal.

The X-ray analyses of complexes **22b,d** show the two co-crystals present numerous analogies from the structural point of view. In both cases the iodide ions chelate three different iodine atoms of three different iodo-PFCs **1b,d** while the diiodo-PFCs **1b,d** behave as bidentate and telechelic electron acceptors. The $I^{-}\cdots$I-PFC distances are 3.475 Å in **22b** and 3.474 Å in **22d**. Once again, these values are longer than the covalent I-I bond (2.666 Å) but approximately 0.8 times the sum of van der Waals radii (1.98 Å for iodine atom and 2.20 Å for iodide ion) (*69*). The $I\cdots I^{-}\cdots I$ angles are 73.0° in **22b** and 73.1° in **22d**.

The low affinity existing between PFCs and HCs dominates the overall crystal packing. A strong segregation between K.2.2.2.⊂ KI **21a** and iodo-PFCs **1b,d** occurs and results in the formation of subnanometric layers held together by the $I^{-}\cdots$I halogen bonding (Figure 5). Potassium cations are embedded in the HC layer. Iodide anions are located at the interface and hold together the alternating PFC and HC layers through $I^{-}\cdots$I-PFC halogen bondings. The general organisation forms infinite 2D honeycomb-like structures topologically equivalent to (6,3) nets (*70*). In **2d** these nets interpenetrate. Interpenetration is common in highly structured and ordered systems, while PFC derivatives usually form disordered and amorphous materials. This is the first report on interpenetrated networks involving PFC derivatives.

The nature of the cryptated metal cation affects the stoechiometry of the formed co-crystal but leaves unchanged the preference for the self-assembly of three-component heteromeric architectures with respect to the alternative self-assembly processes. For instance, when K.2.2.2. (**2d**) and 1,2-diiodotetrafluoroethane (**1a**) are crystallized in the presence of barium iodide, cryptation affords K.2.2.2.⊂ BaI_2 **21b** and the naked iodide anions work as strong electron donors towards the iodine atoms of **1a**. The three-component complex $[Ba(K.2.2.2.)^{2+}][(C_2F_4I_4)^{2-}]$ **22e** is formed as colourless crystals where the PFC, HC, and IS elements are present in 1:1:1 ratio (Scheme 9) (*71*).

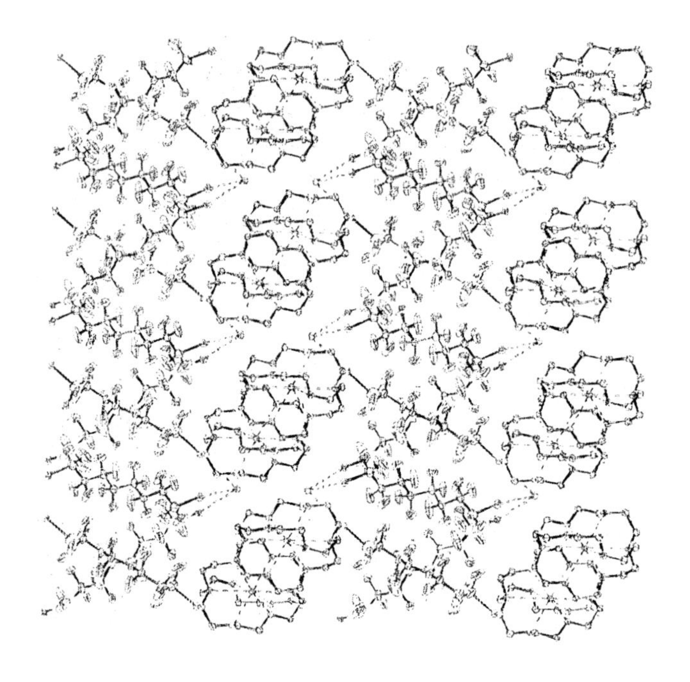

*Figure 5. ORTEP view of the crystal packing of **22d**. $I^{-}\cdots I$ Halogen bondings are dashed lines. The cation cryptate atoms (C, N, O, and K) as well as iodide ions and iodine atoms are not disordered, while the carbon and fluorine atoms of the fluorinated chains show dramatic rotational disorder (not shown) despite that the data were collected at 127K.*

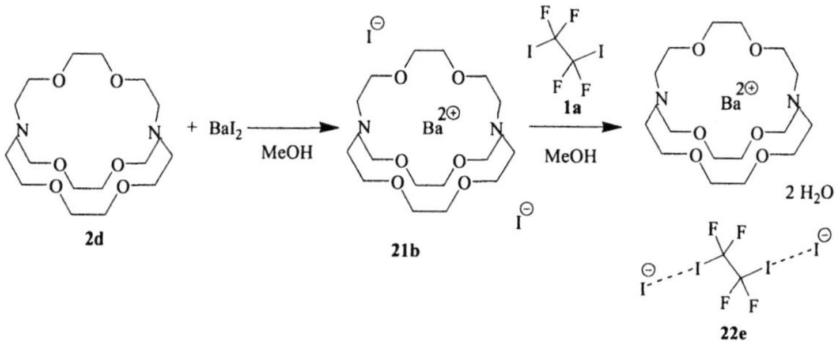

*Scheme 9. Self-assembly of K.2.2.2. (**2d**), BaI_2, and 1,2-diiodotetrafluoroethane (**1a**) providing the three-component derivative **22e**.*

Structural details obtained from the X-ray analysis (Figure 6) show that two molecules of water are also present to complete the coordination sphere of the barium cations that are embedded in the cryptand. The I^-···I-PFC distance is 3.391 Å. Once again it is approximately 0.8 times the sum of van der Waals radii of iodine atom and iodide ion. The I^-···I-C angle is 177.3°. Contrary to the previous case, the iodide anions function as monodendate electron donors probably due to the fact the PFC chains of **1a** are too short to bridge two different anions. As a result, discrete and fluorinated polyiodide addutcs $(C_2F_4I_4)^{2-}$ are present instead of infinite 2D honeycomb like networks.

The anion driven self-assembly of three-component heteromeric adducts may give rise to new interesting applications. For instance, it may allow iodo-PFCs to drive cryptated inorganic or organic salts from aqueous and solid phases into fluorous phase. In this way iodo-PFCs will behave as a new class of effective phase transfer catalysts tailored to fluorous solvents. In fact, at 293 K, K.2.2.2.⊂ KI has a solubility >200 mg/mL in water and ~1 mg/mL in benzene (a standard organic solvent). On the other hand, it is insoluble in perfluorohexane (<0.05 mg/mL) while it has a solubility of ~0.5 mg/mL in 1-bromoperfluorooctane (a frequently used fluorous solvent) and has a solubility >500 mg/mL in 1-iodoperfluorohexane. The greater solubility of the cryptate in iodo-PFCs than in bromo-PFCs is due to the greater ability of the former compounds to work as halogen bonding donors. The partition coefficient between 1-iodoperfluorohexane and water is 7.2 (established through ^{1}H NMR). This proves the greater affinity of the cryptate K.2.2.2.⊂ KI for iodo-PFCs than for water.

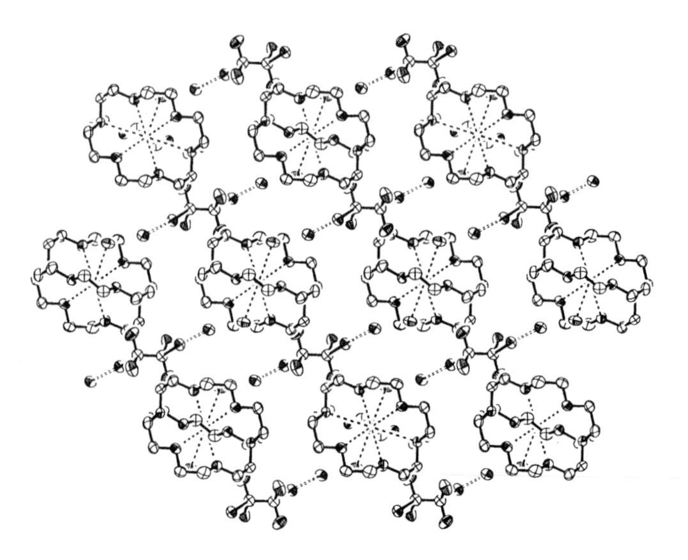

*Figure 6. ORTEP view of the crystal packing of **22e** formed by K.2.2.2. **(2d)**, barium iodide, and 1,2-diiodotetrafluoroethane **(1a)**. For the sake of clarity hydrogen atoms have been omitted. The $I^-\cdots I$ halogen bondings and the coordination to barium are dashed lines.*

Applications of the self-assembly

Applications of non-covalent hybrid systems obtained through the halogen bonding driven self-assembly of haloperfluorocarbons are just starting to emerge. The covalent approach to hybrid materials containing PFC residues encounters non-minor synthetic problem related to the connecting of the HC and PFC residues in a single molecule. The non-covalent approach to hybrid materials overcomes these problems. Moreover, the final properties of a material depend on the properties of its components at a molecular level and on the organisation of the molecules in the bulk material. The directionality and strength of the halogen bonding allow the supramolecular architecture of the formed two-component and three-component assemblies to be anticipated. This adds an extra advantage for the approach here described if the obtaining of materials with pre-established properties is pursued.

The resolution of racemic mixtures via diastereoisomeric adducts formation with enantiopure derivatives is a standard and economical protocol for the obtaining of chiral and enantiopure HCs. A halogen bonding based co-crystal formation allowed the first resolution of chiral and racemic halo-PFCs to be successfully performed. It is worth nothing that PFC-halides are a class of compounds of high technological relevance (*72*) and that enantiopure PFC derivatives are a nearly unknown class of compounds. Moreover, co-crystal formation followed by X-ray analyses might also be a reliable strategy to go over difficulties encountered for determining the absolute configuration of highly fluorinated or perfluorinated derivatives (*73-74*).

Preliminary investigations with racemic 1,2-dibromohexafluoropropane (**8b**) and various chiral and enantiopure diamines showed that the N···Br interactions are too weak to give rise to solid adducts when volatile dibromo-PFCs are used (*75-77*). Halide anions are stronger halogen bonding acceptors than nitrogen atoms and Br^-···Br halogen bondings proved effective in the resolution of **8b**.

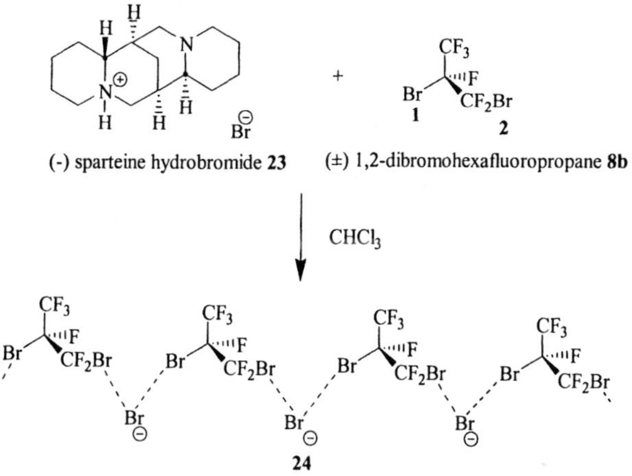

(-) sparteine hydrobromide **23** (±) 1,2-dibromohexafluoropropane **8b**

$CHCl_3$

24

*Scheme 10. Formation of the co-crystal **24** by self-assembly of (-) sparteine hydrobromide **23** and (S) 1,2-dibromohexafluoropropane **8b**.*

Specifically, slow evaporation of a chloroform solution of enantiopure (-)-sparteine hydrobromide **23** and of the racemic dibromo-PFC **8b** affords the co-crystal **24** where only (*S*)-**8b** is present (Scheme 10) (*78*). The cohesion of the co-crystal and the strength of the Br^-···Br halogen bonding are revealed by the co-crystal melting point (105 °C) that is definitively higher than the pure racemic dibromo-PFC module **8b** (-95 °C).

Details of the supramolecular architecture of chiral and enaniopure co-crystal **24** show all the characteristics typical for an halogen bonded system. PFC and HC modules segregate. Two distinct Br^-···Br interactions (3.369 Å and 3.260 Å) are present. They are longer than the Br-Br covalent bond distance and approximately 0.8 times the sum of van der Waals radii. Consistent with an n→σ* electron donation from bromide anions to bromine atoms, the two different $Br^-\cdots Br$-C angles are approximately linear (173.6° and 178.4°). Each bromine anion connects a primary and a secondary bromine of two distinct PFC units of **8b** and gives raise to enantiopure and infinite twofold helices in which the resolution process of **8b** results from a highly specific inclusion in enantiopure helixes (Figure 7). This arrangement maximizes the information transfer from the optically pure HC module to the PFC module that is resolved.

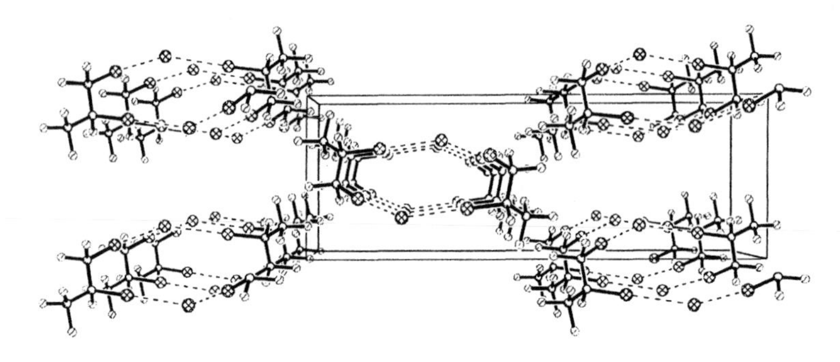

*Figure 7. Perspective view of the co-crystal **24** along the crystallographic b direction. Sparteinium ions are omitted to better evidence the halogen-bonded PFC helices.*

Another interesting application of hybrid PFC-HC materials is in the field of solid-state synthesis. This topic is receiving growing interest due to its inherently green profile and to the high selectiveties it can give rise to. A non-stereoselective UV-induced [2+2] cycloaddition reaction occurs on irradiation of concentrated solutions of (*E*)-1,2-bis(4-pyridyl)ethylene (4,4'-bpe, **2g**). Differently, no reaction is observed when the 4,4'-bpe is in the solid state as the distance and the orientation of the double bonds are not suitable for dimer formation (*79*). An iodo-PFC halogen bonding donor has been used in the context of a template/co-crystal system to induce a stereoselective photocyclization of **2g**.

N···I interactions drive the self-assembly of the pentaerytritol tetrakis-4-(iodotetrafluorophenyl)ether (**25**), a tetratopic electron acceptors, with 4,4'-bpe **2g**, a ditopic electron donors. The 1D infinite ribbons **26** are formed (Scheme 11) (*24-31*).

The co-crystal structure, determined through single-crystal X-ray analysis (Figure 8), reveals that intramolecular $\pi\cdots\pi$ interactions between couples of fluorophenyl rings preorganize the template arms at distances ranging from 3.524 to 3.897 Å.

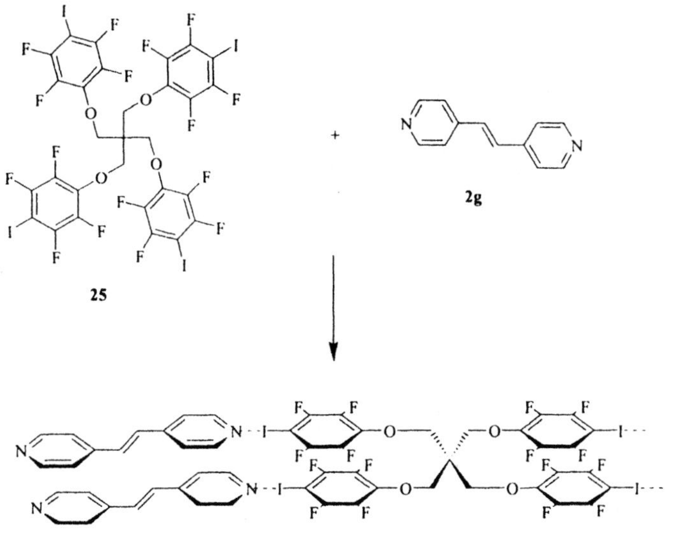

Scheme 11. *Formation of the co-crystal **26** by self-assembly of pentaerythritol ether **25** and 4,4'-bpe **2g**.*

Intermolecular N···I halogen bondings assembles the template **25** and the olefin carrier **2g** in ribbons. Thanks to the high directionality and the remarkable strength of the halogen bondings, the template arms alignment is

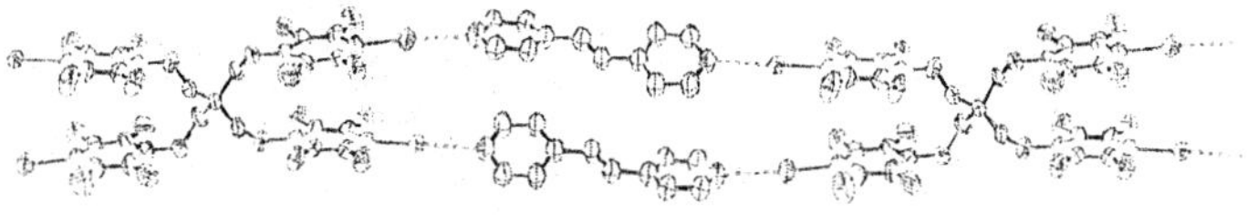

*Figure 8. 1D Infinite ribbons **26** formed through N···I halogen bonding driven self-assembly of pentaerythritol ether **25** and 4,4'-bpe **2g**.*

translated into olefins alignment at the distance required for the phodimerization. In fact, when the co-crystal **26** is irradiated using a Rayonet at 300nm, a photochemical cycloaddition readily occurs under topochemical control and produces tetrakis(4-pyridyl)cyclobutane **27** (Scheme 12).

^{1}H NMR and mass spectroscopies of the crude reaction mixture confirm that the reaction gives quantitatively and stereospecifically the *rctt* isomer as a pure product. Interestingly, cross-plane photoreactions are prevented by the absence of any short contacts between the olefins belonging to adjacent ribbons in the crystal stacking (inter-ribbons distance > 7 Å).

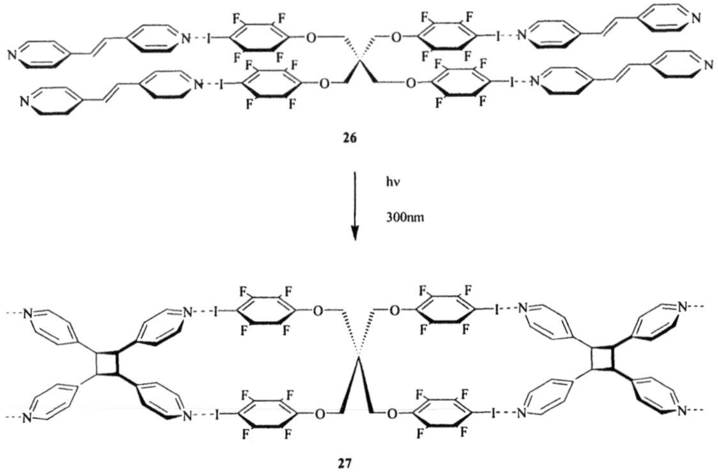

26

hν

300nm

27

*Scheme 12. Solid-state photochemical cycloaddition of ribbons **26** to give cyclobutanes **27** under topochemical control.*

Another recent application of halogen bonded material containing halo-PFCs has been reported in the liquid crystals field. It is well known that the hydrogen bonding is capable to induce liquid crystallinity on self-assembly of non-mesomorphic species. The halogen bonding has shown the same ability when alkoxystilbazoles **28** and iodopentafluorobenzene (**17c**) are used as starting materials (*80*).

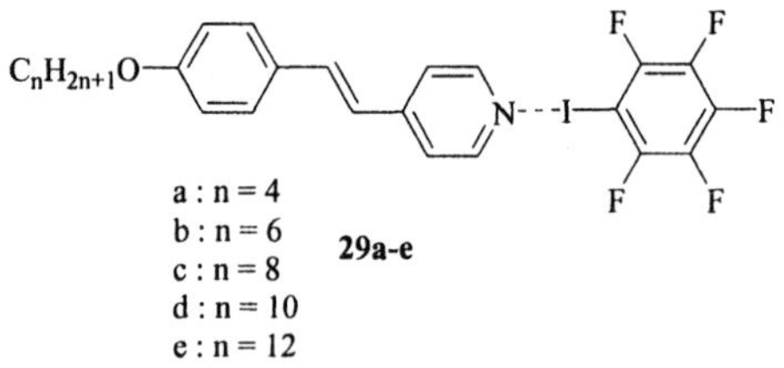

$C_nH_{2n+1}O$

a : n = 4
b : n = 6
c : n = 8
d : n = 10
e : n = 12

29a-e

*Scheme 13. N···I halogen bonding driven self-assembly of alkoxystilbazoles **28** and iodopentafluorobenzene (**17c**).*

When equimolar amounts of alkoxystilbazoles **28** and iodopentafluorobenzene are mixed in THF, the slow evaporation of the solvent gives rise to pale yellow products. This change of colour is consistent with an electron shifting from the nitrogen atom toward the iodine atom due to the halogen bonding interaction. A unequivocal evidence of the formation of dimeric complexes **29** has been obtained through the X-ray analysis of **29c**. Similar to the other adducts described above, the N···I distance is 2.811 Å and the N···I-C angle is 168.41°. All the non covalent dimers **29a-e** present liquid crystal properties despite pure stilbazole components **28** have a nonmesomorphic nature. **29a,b** show a monotropic nematic phase while enantiotropic SmA phases are observed with complexes **29c-e**. Further investigations have been carried out by replacing iodopentafluorobenzene with bromopentafluorobenzene, but no conclusive proof of complex formation was obtained by the thermal behaviour of the formed product. The N···Br interaction is probably too weak for allowing the co-crystal to exist at temperatures high enough to observe the liquid crystalline properties.

An interesting application of the halogen bonding has been reported in the engineering of fluorinated comb-like polymers. These products have recently received interest due to their unique surface behaviour (*81-83*). Indeed, the strength and the specificity of the halogen bonding has driven the self-assembly of iodo-PFCs and HC polymers. Non covalent polymers bearing perfluoroalkyl side chains have been obtained. PFC and HC components segregation provides the surfaces coating with fluorinated materials. In the detail, when the poly(4-vinylpyridine) (P4VP) **30** is in the presence of α,ω-diiodoperfluoroalkanes **1a-d**, the recognition between the nitrogen and the iodine atoms gives rise to complexes **31a-d** that are isolated as dense oils in the case of **31a-c** and as paste-like product in the case of **31d** (Scheme 14) (*84*).

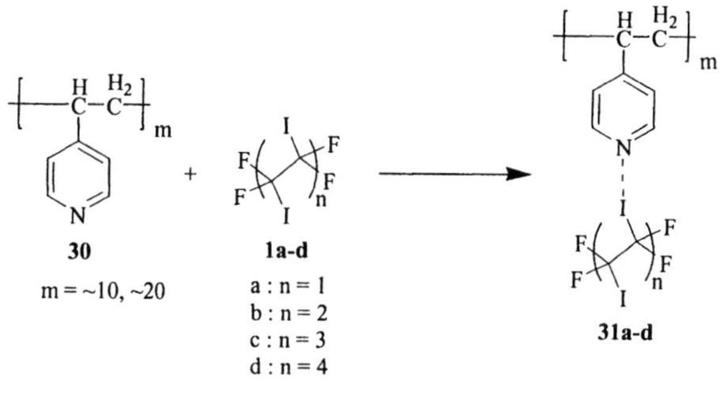

30

m = ~10, ~20

1a-d

a : n = 1
b : n = 2
c : n = 3
d : n = 4

31a-d

*Scheme 14. Formation of fluorous com-like polymeric complexes **31a-d** by self-assembly of P4VP **30** and diiodoperfluoroalkanes **1a-d**.*

The formation of new materials is confirmed by IR spectroscopy. For instance, the P4VP bands in the 3100-2900 cm^{-1} region for the pure components **1a-d**, show blue shift and intensity decrease in adducts **31a-d**. Similarly, the pyridine band at 1000-900 cm^{-1} is shifted to higher frequencies in the complexes. The features typical for smectic-type liquid crystallinity are evidenced in complex **31c** with polarized light. The macroscopic organisation remained unchanged also upon heating to 70°C, thus suggesting the lyotropic character of this hybrid material. Moreover, the study of the molten complex by applying shear stress showed behaviour consistent with a lamellar organization, strictly similar to related literature system (*85*). The segregation occurring between the PFCs and HCs contributes to the liquid crystalline properties of such materials and demonstrates the great efficiency of the halogen bonding for the preparation of fluorinated coatings. This new methodology offers an alternative way to the covalent synthesis of fluorinated graft copolymers.

Conclusion

Through the description of numerous examples, we have shown how iodo- and bromo-PFCs work as effective electron accepting species when interacting with a wide variety of lone pair possessing atoms, either neutral or anionic. Halo-PFCs are therefore versatile tectons for the elaboration of supramolecular architectures by halogen bonding. Theoretical calculations and experimental results prove that iodo-PFCs and anionic species give stronger halogen bondings than bromo-PFCs and neutral donors, respectively. The interaction is strong enough to overcome the low affinity PFC compounds show for any non-fluorinated compound, either organic or inorganic. Two-component and three-component materials have thus been obtained where the low reciprocal affinity among the starting modules survives in the marked segregation. Subnanometric layers of different compositions are often present. The overall architectures of these materials can be anticipated from the geometry of the complementary modules and the directionality of the halogen bonding. This is emblematically shown in the structure of the first discrete and fluorous polyiodide networks **22**.

The usefulness of several analytical techniques in the characterization of the obtained hybrid materials has also been presented. In particular, the pinning of halo-PFCs in halogen bonded heteromeric systems allowed X-ray crystallography to be routinely applied for the analyses of perfluoroalkyl chains as long as 1-iodioperfluorodecane. This is a particularly noteworthy result if we remember the high conformational disorder typical for perfluoroalkyl chains and their tendency to give rise to waxy materials.

Recent studies have provided the first applications of these hybrid materials in the fields of liquid crystals, polymers coating, solid phase synthesis, and resolution of racemic PFCs. Thanks to its ease of use, the approach may have a great potential in the preparation of well-ordered nanostructures that could be applied, to biotechnologies, electronics…

References

1. Adams, H.; Blanco, J. L. J.; Chessari, G.; Hunter, C. A.; Low, C. M. R.; Sanderson, J. M. *Chem. Eur. J.* **2001**, *7*, 3494-3503.
2. Smart, B. E. In: Banks, R. E., Smart, B. E.; Tatlov, J. C., editors. *Organofluorine chemistry: principles and commercial applications*. New York: Plenum Press, **1994**.
3. Hidelbrand, J.; Cochran, D. R. F. *J. Am. Chem. Soc.* **1949**, *71*, 22-25.
4. Dorset, D. L. *Macromolecules* **1990**, *23*, 894-901.
5. Ku, C. Y.; Lo Nostro, P.; Chen, S. H. *J. Phys. Chem. B* **1997**, *101*, 908-914.
6. Ren, Y.; Imamura, K. I.; Ogawa, A.; Kato, T. *J. Phys. Chem. B* **2001**, *105*, 4305-4312.
7. Lo Nostro, P.; Chen, S. H. *J. Phys. Chem.* **1993**, *97*, 6535-6540.
8. Katano, Y.; Tomono, H.; Nakajima, T. *Macromolecules* **1994**, *27*, 2342-2344.
9. Guan, Z.; DeSimone, J. M. *Macromolecules* **1994**, *27*, 5527-5532.
10. Metrangolo, P.; Resnati, G. *Chem. Eur. J.* **2001**, *7*, 2511-2519.
11. Metrangolo, P.; Resnati, G. Halogen Bonding. in *Encyclopedia of Supramolecular Chemistry*; Atwood, J. L.; Steed J. W.; Eds.; Marcel Dekker Inc, New York, **2004**, 628-635.
12. Metrangolo, P.; Pilati, T.; Resnati, G.; Stevenazzi, A. *Current Opin. Colloid. Interface Sci.* **2003**, *8*, 215-222.
13. Metrangolo, P.; Pilati, T.; Resnati, G. In *Self-assembly of hybrid fluorous materials*; Handbook of fluorous chemistry; Gladysz, J.A.; Curran, D.P.; Horváth. I.T.; Eds.; Wiley-VCH, Weinheim, **2004**, 507-520.
14. Hassel, O. *Science* **1970**, *170*, 497-502.
15. Dumas, J.M.; Gomel, L.; Guerin, M. in: S. Patai, Z. Rappoport (Eds.), *The Chemistry of Functional Groups*, Suppl. D, Wiley, New York, **1983**, 985-1020.
16. Bent, H.A. *Chem. Rev.* **1968**, *68*, 587-648.
17. Guthrie, F. *J. Chem. Soc.* **1863**, *16*, 239-244.
18. Remsen, I.; Norris, J. F. *Am. Chem. J.* **1896**, *18*, 90-95.
19. Burdeniuc, J.; Sanford, M.; Crabtree, R. H. *J. Fluorine Chem.* **1998**, *91*, 49-54.
20. Burdeniuc, J.; Crabtree, R. H. *Organometallics* **1998**, *17*, 1582-1586.
21. Lommerse, J. P. M.; Stone, A. J.; Taylor, R.; Allen, F. H. *J. Am. Chem. Soc.* *118*, **1996**, 3108-3116.
22. Weiss, R.; Schwab, O.; Hampel, F. *Chem. Eur. J.* **1999**, *5*, 968-974.
23. Hazeldine, R.N. *J. Chem. Soc.* **1953**, 2622-2626.
24. Cheetham, N. F.; Pullin, A. D. E. *Chem. Commun.* **1965**, 418-419.
25. Larsen, D. W.; Allred, A. L. *J. Phys. Chem.* **1965**, *69*, 2400-2401.

26. Cheetham, N. F.; Pullin, A. D. E. *Aust. J. Chem.* **1971**, *24*, 479-487.
27. Cheetham, N. F.; McNaught, I. J.; Pullin, A. D. E. *Aust. J. Chem.* **1971**, *24*, 973-985.
28. Mishra, A.; Pullin, A. D. E. *Aust. J. Chem.* **1971**, *24*, 2493-2507.
29. Cheetham, N. F.; McNaught, I. J.; Pullin, A. D. E. *Aust. J. Chem.* **1974**, *27*, 987-1007.
30. McNaught, J.; Pullin, A. D. E. *Aust. J. Chem.* **1974**, *27*, 1009-1015.
31. Legon, A. C.; Millen, D. J.; Rogers, S.C. *Chem. Commun.* **1975**, 580-581.
32. Chen, Q. Y.; Qiu, Z. M. *J. Fluorine Chem.* **1987**, *35*, 79.
33. Chen, Q. Y.; Li, Z. T.; Zhou, C. M. *J. Chem. Soc. Perkin Trans. 1* **1993**, 2457-2462.
34. Amico, V.; Meille, S. V.; Corradi, E.; Messina, M. T.; Resnati, G. *J. Am. Chem. Soc.* **1998**, *120*, 8261-8262.
35. Lunghi, A.; Cardillo, P.; Messina, T.; Metrangolo, P.; Panzeri, W.; Resnati, G. *J. Fluorine Chem.* **1998**, *91*, 191-194.
36. Navarrini, W.; M. T.; Metrangolo, P.; Pilati, T.; Resnati, G. *New J. Chem.* **2000**, *24*, 777-780.
37. Cardillo, P.; Corradi, E.; Lunghi, A.; Meille, S. V.; Messina, M. T.; Metrangolo, P.; Resnati, G. *Tetrahedron* **2000**, 56, 5535-5550.
38. Corradi, E.; Meille, S. V.; Messina, M. T.; Metrangolo, P.; Resnati, G. *Tetrahedron Lett.* **1999**, *40*, 7519-7523.
39. Liantonio, R.; Metrangolo, P.; Pilati, T.; Resnati, G.; Stevenazzi, A. *Cryst. Growth Des.* **2003**, *3*, 799-803.
40. Burton, D. D.; Fontana, F.; P. Metrangolo, P.; Panzeri, W.; Pilati, T. ; P.; Resnati, G. *Tetrahedron Lett.* **2003**, *44*, 645-648.
41. Padmanabhan, K.; Paul, I. E.; Curtin, D.Y. *Acta Crystallogr., Sect. C* **1990**, *46*, 88-92.
42. Bondi, A. *J. Phys. Chem.* **1964**, *68*, 441-451.
43. Caronna, T.; Liantonio, R.; Logothetis T. A.; Metrangolo, P.; Pilati, T.; Resnati, G. *J. Am. Chem. Soc.* **2004**, *126*, 4500-4501.
44. Messina, M. T.; Metrangolo, P.; Panzeri, W. ; Pilati, T. ; P.; Resnati, G. *Tetrahedron* **2001**, 57, 8543-8550.
45. Corradi, E.; Meille, S.V.; Messina, M.T.; Metrangolo, P.; Resnati, G. *Angew. Chem. Int. Ed.* **2000**, *39*, 1782-1786.
46. Jay, J. I.; Padgett, C. W.; Walsh, R. D. B.; Hanks, T. W.; Pennington, W. T. *Cryst. Growth Des.* **2001**, *1*, 501-507.
47. Valerio, G.; Raos, G.; Meille, S. V.; Metrangolo, P.; Resnati, G. *J. Phys. Chem. A* **1999**, *104*, 1617-1620.
48. Messina, M. T.; Metrangolo, P.; Pappalardo, S.; Parisi, M. F.; Pilati, T.; Resnati. G. *Chem. Eur. J.* **2000**, *6*, 3495-3500.
49. Messina, M. T.; Metrangolo, P.; Quici, S.; Manfredi, A.; Pilati, T.; Resnati, G. *Supramol. Chem.* **2001**, *12*, 405-410.

50. Liantonio, R.; Luzzati, S.; Metrangolo, P.; Pilati, T.; Resnati, G. *Tetrahedron* **2002**, *58*, 4023-4029.
51. Liantonio, R.; Metrangolo, P.; Pilati, T.; Resnati, G. *Acta Crystallog.* **2002**, *E58*, 575-577.
52. Liantonio, R.; Mele, M. L.; Metrangolo, P.; Resnati, G. *Supramol. Chem.* **2003**, *15*, 177-188.
53. De Santis, A.; Forni, A.; Liantonio, R.; Metrangolo, P.; Pilati, T.; Resnati, G. *Chem. Eur. J.* **2003**, *9*, 3974-3983.
54. Amati, M.; Lelj, F.; Liantonio, R.; Metrangolo, P.; Luzzati, S.; Pilati, T. Resnati, G. *J. Fluorine Chem.* **2004**, *125*, 629-640.
55. Logothetis, T. A.; Meyer, F.; Metrangolo, P. Pilati, T. Resnati, G. *New. J. Chem.* **2004**, *28*, 760-763.
56. Forni, A.; Metrangolo, P.; Pilati, T.; Resnati, G. *Cryst. Growth Des.* **2004**, *4*, 291-295.
57. Liantonio, R.; Logothetis, T. A.; Messina, M. T.; Metrangolo, P.; Resnati, G.; Pilati, T. *Coll. Czech. Chem. Commun.* **2002**, *67*, 1373-1382.
58. Fontana, F.; Forni, A.; Metrangolo, P.; Panzeri, W.; Pilati, T. ; Resnati, G. *Supramol. Chem.* **2002**, *14*, 47-55.
59. For a ^{13}C NMR study of halogen bonding, see : Glaser, R.; Chen, N.; Wu, H.; Knotts, N.; Kaupp, M. *J. Am. Chem. Soc.* **2004**, *126*, 4412-4419.
60. Messina, M. T.; Metrangolo, P.; Panzeri, W.; Ragg, E.; Resnati, G. *Tetrahedron Lett.* **1998**, *39*, 9069-9072.
61. Metrangolo, P.; Panzeri, W.; Recupero, F.; Resnati, G. *J. Fluorine Chem.* **2002**, *114*, 27-33.
62. Messina, M. T.; Metrangolo, P.; Navarrini W.; Radice, S.; Resnati, G.; Zerbic, G. *J. Mol. Struct.* **2000**, *524*, 87-94.
63. Lehn, J.-M. *Supramolecular Chemistry-Concepts and Perspectives*, VCH, Weinheim, **1995**.
64. Vilar, R.; Mingos, D. M. P.; White, A. J. P.; Williams, D. J. *Angew. Chem. Int. Ed.* **1998**, *37*, 1258-1261.
65. Walsh, R. B.; Padgett, C. W.; Metrangolo, P.; Resnati, G.; Hanks, T.W.; Pennington, W. T. *Cryst. Growth Des.* **2001**, *1*, 165-175.
66. Landrum G. A.; Goldberg, N.; Hoffmann, R. *J. Chem. Soc., Dalton Trans.* **1997**, 3605-3613.
67. Liantonio, R.; Metrangolo, P.; Pilati, T. ; P.; Resnati, G. *Cryst. Growth Des.* **2003**, *3*, 355-361.
68. Tse-Lok, H. *Hard and Soft Acids and Bases Principle in Organic Chemistry*, Academic Press: New York, **1977**.
69. van der Waals radii are from Winter, M. J. *WebElements* (http://www.webelements.com); WebElements: Sheffield, **2001**.
70. Batten, S.R.; Robson, R. *Angew. Chem. Int. Ed.* **1998**, *37*, 1460-1494.

71. Fox, D. B.; Liantonio, R.; Metrangolo, P.; Pilati, T.; Resnati, G. *J. Fluorine Chem.* **2004**, *125*, 271-281.
72. Wakselman, C.; Lantz, A. *Organofluorine Chemistry: Principles and Commercial Applications* (Eds.: R. E. Banks, B. E. Smart, J. C. E. Tatlow), Plenum, New York, **1994**.
73. Polavarapu, P. L. ; Cholli, A. L.; Vernice, G. *J. Pharm. Sci.* **1993**, *82*, 791-793.
74. Schurig, V.; Juza, M.; Green, B. E.; Horakh, J.; Simon, A. *Angew. Chem. Int. Ed. Engl.* **1996**, *35*, 1680-1682.
75. Skell, P. S.; Pavlis, R. R.; Lewis, D. C.; Shea, K. S. *J. Am. Chem. Soc.* **1973**, *95*, 6735-6745.
76. Skell, P. S.; Pavlis, R. R.; Lewis, D. C.; Shea, K. S. *Science* **1970**, *170*, 497-502.
77. Wilen, S. H.; Bunding, K. A.; Kascheres, C. M.; Weider, M. J. *J. Am. Chem. Soc.* **1985**, *107*, 6997-6998.
78. Farina, A.; Meille, S.V.; Messina, M.T.; Metrangolo, P.; Resnati, G. *Angew. Chem. Int. Ed.* **1999**, *38*, 2433-2436.
79. Vansant, J.; Toppets, S.; Smets, G.; Declercq, J. P.; Germain, G.; Van Meerssche M. *J. Org. Chem.* **1980**, *45*, 1565-1573.
80. Nguyen, H. L.; Horton, P. N.; Hursthouse, M. B.; Legon, A. C.; Bruce, D. W. *J. Am. Chem. Soc.* **2004**, *126*, 16-17.
81. Ishiwari, K.; Ohmori, A.; Koizumi, S. *Nippon Kagaku Gakkaishi* **1985**, *10*, 1924-1928.
82. Schineider, J.; Erdelen, C.; Ringsdorf, H.; Rabolt, J. F. *Macromolecules* **1989**, *22*, 3475-3480.
83. Katano, Y.; Nakajima, T.; Tomono, H. *Macromolecules* **1994**, *27*, 2342-2344.
84. Bertani, R.; Metrangolo, P.; Moiana, A.; Perez, E.; Pilati, T.; Resnati, G.; Rico-Lattes, I.; Sassi, A. *Adv. Mater.* **2002**, *14*, 1197-1201.
85. Kimura, T.; Sakurai, K.; Takahashi, T. *Polymer* **1999**, *40*, 5939-5945.

Fluorine-Containing Amino Acids and Peptides: Fluorinated Synthons for Life Sciences

Chapter 31

Fluoro-β-lactams as Useful Building Blocks for the Synthesis of Fluorinated Amino Acids, Dipeptides, and Taxoids

Iwao Ojima, Larisa Kuznetsova, Ioana Maria Ungureanu, Antonela Pepe, Ilaria Zanardi, and Jin Chen

Department of Chemistry, State University of New York at Stony Brook, Stony Brook, NY 11794–3400

Enantiopure 1-acyl-3-hydroxyl-4-R_f-azetidin-2-ones serve as versatile intermediates for the syntheses of fluorine-containing α-hydroxy-β-amino acids, dipeptides, and taxoid anticancer agents. 3-Hydroxy-4-CF_2H-β-lactams with >99% ee can be obtained in high yields through the DAST reaction of the corresponding enantiopure 4-formyl-β-lactam obtained through ketene-imine [2+2] cycloaddition, followed by enzymatic optical resolution. An efficient method has been developed for the synthesis of 3-hydroxy-4-CF_3-β-lactams (>99% ee) through [2+2] cycloaddition of a CF_3-imine with acetoxyketene and subsequent enzymatic optical resolution. Practical processes for the preparations of these enantiopure 3-hydroxy-4-R_f-β-lactams (R_f = CF_2H and CF_3) as well as their synthetic applications are described.

Introduction

β-Amino acids have been attracting considerable interests because of their inherent biological activities and their useful characteristics as building blocks for the design and development of potential therapeutic drugs and "β-peptides", possessing unique properties *(1,2)*. β-Amino acids are also useful for the studies of enzymatic reaction mechanisms *(1,2)*. Among various types in the β-amino acids family, α-hydroxy-β-amino acids (isoserines) are one of the most important members because many of them act as potent enzyme inhibitors and they also serve as crucial building blocks for the compounds of biological and medicinal importance *(2,3)*.

For example, α-hydroxy-β-amino acid moieties are found in paclitaxel, (antitumor agent) *(4-6)*, bestatin *(7,8)* (inhibitor of aminopeptidases, immunological response modifier), microginin *(9)* (an ACE inhibitor, KNI inhibitors), the kinostatins (HIV-1 protease inhibitors) *(10,11)*.

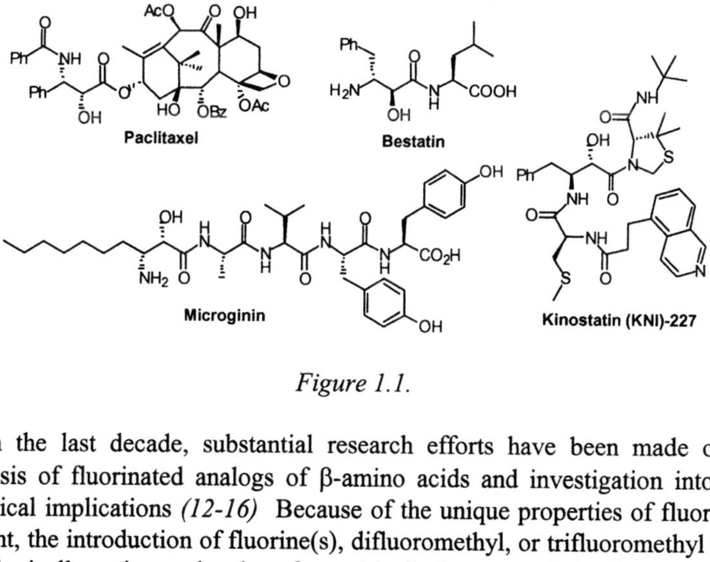

Paclitaxel

Bestatin

Microginin

Kinostatin (KNI)-227

Figure 1.1.

In the last decade, substantial research efforts have been made on the synthesis of fluorinated analogs of β-amino acids and investigation into their biological implications *(12-16)* Because of the unique properties of fluorine as element, the introduction of fluorine(s), difluoromethyl, or trifluoromethyl group to biologically active molecules often critically improves their pharmacological properties (e.g., membrane permeability, hydrophobic binding stability against metabolic oxidation, etc.) *(12,17)*. Moreover, the sensitivity of ^{19}F NMR spectroscopy along with large ^{19}F-^{1}H coupling constants and the virtual absence of ^{19}F in the living tissue render fluorine incorporation a particularly powerful tool for the investigation of biological processes *(18-20)*. Therefore, fluorine-containing α-hydroxy-β-amino acids are expected to serve as useful bioactive compounds with a wide range of potential applications in medicinal chemistry and chemical biology. However, only a few methods have been reported to date for the synthesis of fluorine-containing α-hydroxy-β-amino acids.

One of the literature known methods for the stereoselective synthesis of α-hydroxy-β-trifluoromethyl-β-amino acid is the use of highly regio- and stereoselective hydrolysis of aziridine **1.2(+)** at the C-2 position (Scheme 1.2). The aziridine **1.2(+)** with >98% ee can be obtained via enzymatic resolution of racemic **1.2** with *Candida antarctica* lipase (CAL) (35%, 50% in theory) *(21)*. This method gives *anti*-isomer exclusively. The corresponding *syn*-isomer can be obtained by applying the Mitsunobu reaction.

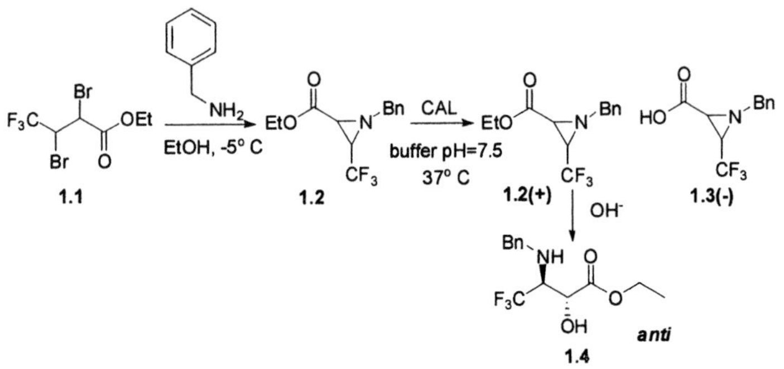

Scheme 1.2

Racemic *anti*-α-hydroxy-β-fluoroalkyl-β-amino esters can also be synthesized *via* base-catalyzed intramolecular rearrangement of imino ethers and subsequent reduction of the resulting α-hydroxy-β-imino esters **1.7** (Scheme 1.3) *(22)*. The diastereoselective reduction of **1.7** with $NaBH_4$-$ZnCl_2$ gives *anti*-ester **1.8** in 86% yield accompanied by a small amount (4%) of *syn*-ester **1.9**.

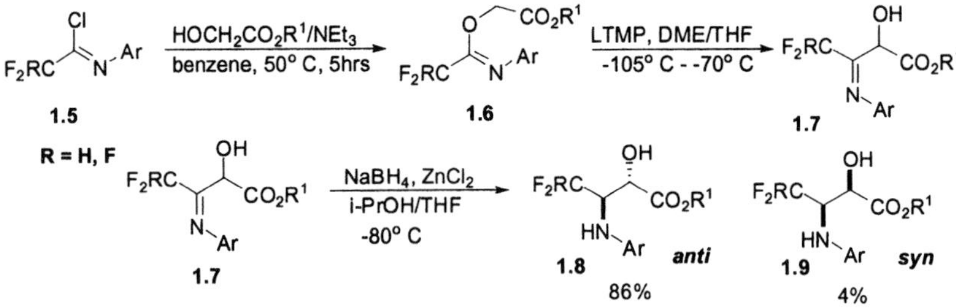

Scheme 1.3

Both enantiomers of *syn*-α-hydroxy-β-trifluoromethyl-β-amino acids (**1.12** and **1.13**) can be obtained through a ketene-imine cycloaddition reaction, followed by acidic methanolysis of the resulting β-lactam intermediates **1.11**

(Scheme 1.4) *(23)*. The Staudinger reaction of chiral imine **1.10** with benzyloxyketene gives a diastereomeric mixture of *cis* β-lactams **1.11** with low diastereoselectivity (29%(+)/41%(-)). Two diastereomers are separated by chromatography and crystallization. Each diastereomer is subjected to methanolysis and hydrogenolysis to give the corresponding enantiopure *syn*-ester **1.12** or **1.13**.

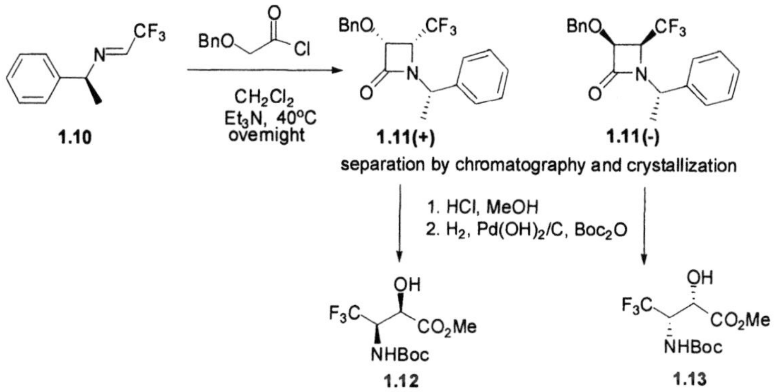

BnOCH$_2$COCl, CH_2Cl_2, Et_3N, 40°C, overnight

1.10 → 1.11(+) + 1.11(-)

separation by chromatography and crystallization

1. HCl, MeOH
2. H_2, $Pd(OH)_2/C$, Boc_2O

1.12, 1.13

Scheme 1.4

The major drawback of these methods is the limitation in scope and/or rather tedious access to both enantiomers. Therefore, new and efficient approaches to enantiopure fluorine-containing α-hydroxy-β-amino acids need to be developed. To this end, the *β-Lactam Synthon Method* (β-LSM) developed in these laboratories offers an attractive protocol for the synthesis of enantiopure CF_2H- and CF_3-containing α-hydroxy-β-amino acids and their congeners.

We describe in this chapter a concise account of our work on the efficient synthesis of enantiopure 1-acyl-3-hydroxy-4-R_f-azetidin-2-ones ($R_f = CF_2H$, CF_3) and their applications to the synthesis of R_f-containing α-hydroxy-β-amino acids, dipeptides and taxoids. Scheme 1.5 illustrates representative transformations of *N*-*t*-Boc-3-PO-4-R_f-β-lactams (P = hydroxyl protecting group).

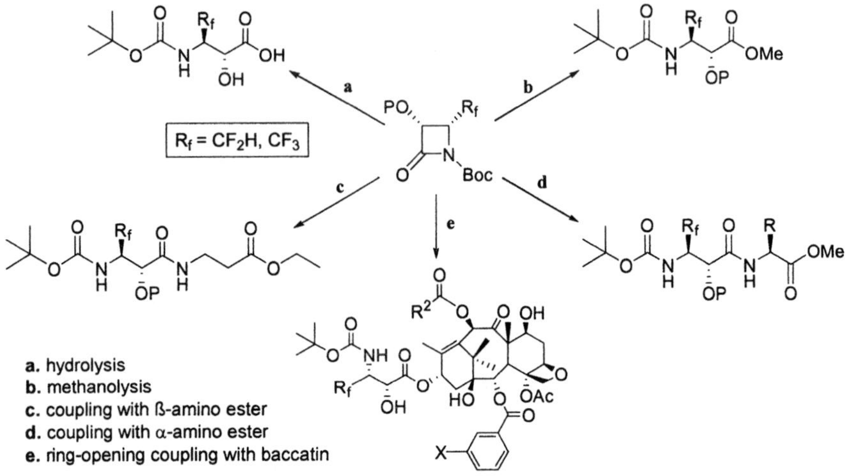

a. hydrolysis
b. methanolysis
c. coupling with ß-amino ester
d. coupling with α-amino ester
e. ring-opening coupling with baccatin

$R_f = CF_2H, CF_3$

Scheme 1.5

2. Synthesis of Enantiopure β-Lactams

2.1. Synthesis of enantiopure 3-hydroxy-4-CF_2H-β-lactams

Enantioselective synthesis of (3*R*,4*S*)-3-TIPSO-4-(2-methyl-1-propenyl)-azetidin-2-one was previously reported by our laboratory using chiral enolate – imine cyclocondensation *(15,16,24)*. Although this procedure cleanly provided the desired enantiopure β-lactam, it had some inconvenience for scale up. Accordingly, we have recently developed another method, which allows for a large-scale synthesis, using enzymatic optical resolution of racemic *cis*-3-AcO-4-(2-methyl-1-propenyl)azetidin-2-ones.

Racemic *cis*-1-PMP-3-AcO-4-(2-methyl-1-propenyl)azetidin-2-one (**2.2**) (PMP = *p*-methoxyphenyl) was prepared through [2+2] ketene-imine cycloaddition. Acetoxyketene generated *in situ* from acetoxyacetyl chloride and triethylamine was reacted with *N*-PMP-3-methyl-2-butenaldimine (**2.1**) to give cis-β-lactam **2.2(±)** in good yield (Scheme 2.1). Then, the enzymatic optical resolution of β-lactam **2.2(±)** was carried out using the "PS-Amano" lipase *(25)* in a buffer solution (pH 7) at 50°C. This enzyme selectively hydrolyzes the acetate moiety of (-)-β-lactam **2.2(-)** to afford kinetically resolved (3*R*,4*S*)-3-

AcO-β-lactam **2.2(+)** (>99% ee) and (3*S*,4*R*)-3-hydroxy-β-lactam **2.3(-)** (96-99% ee) with extremely high enantiopurity in high recovery yields *(26)*.

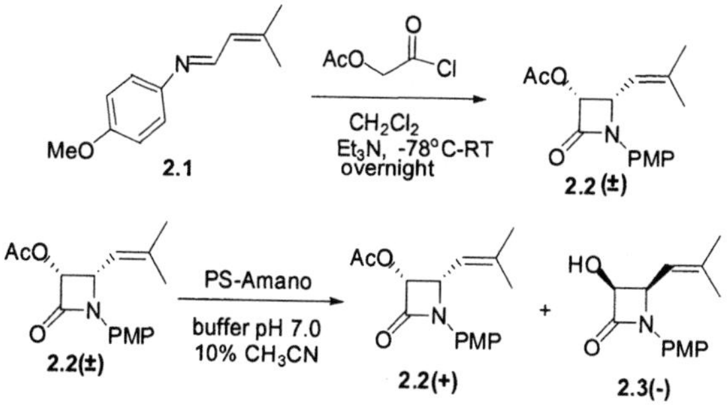

Scheme 2.1

Since the acetyl group was not tolerated in the diethylaminosulfur trifluoride (DAST) reaction, the protecting group of the 3-hydroxyl moiety of β-lactam **2.2(+)** was changed to triisopropylsilyl (TIPS). The resulting (3*R*,4*S*)-1-PMP-3-TIPSO-4-(2-methyl-1-propenyl)azetidin-2-one (**2.4(+)**) was subjected to ozonolysis to give (3*R*,4*S*)-1-PMP-3-TIPSO-4-formylazetidin-2-one which was immediately reacted with DAST to afford the corresponding 4-difluoromethylazetidin-2-one (**2.5(+)**) in high yield. Finally the PMP group was removed using cerium ammonium nitrate (CAN) to give enantiopure (3*R*,4*R*)-3-TIPSO-4-difluoromethylazetidin-2-one (**2.6(+)**) (Scheme 2.2). In a similar manner, (3*S*,4*S*)-3-hydroxy-β-lactam **2.3(-)** was converted to enantiopure (3*S*,4*S*)-3-TIPSO-4-difluoromethylazetidin-2-one (**2.6(-)**) (Scheme 2.2).

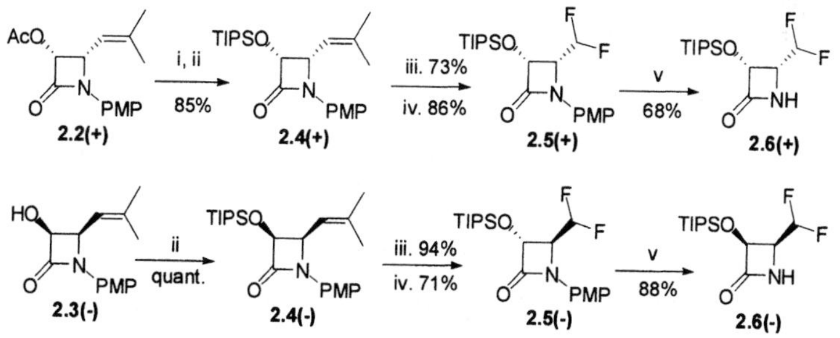

i) KOH, THF, 0 °C; ii) TIPSCl, Et_3N, DMAP; iii) O_3, $MeOH/CH_2Cl_2$, -78 °C; iv)DAST,CH_2Cl_2; v) CAN, H_2O/CH_3CN, -15 °C.

Scheme 2.2

2.2. Synthesis of enantiopure 3-hydroxy-4-CF_3-β-lactams

A different strategy was employed for the synthesis of enantiopure trifluoromethyl-β-lactams. This time the trifluoromethyl moiety was introduced from the very beginning at the imine stage of the [2+2] ketene-imine cycloaddition. *N*-PMP-trifluoroacetaldimine **2.7** was reacted with the ketene generated *in situ* from benzyloxyacetyl chloride to afford racemic cis-3-benzyloxy-4-trifluoromethylazetidin-2-one (**2.8(±)**) in moderate yield, as reported by us and others *(27-29)*. Because of the reduced nucleophilicity of the nitrogen of the imine, it was necessary to run the reaction at 40 °C to achieve a high conversion, a temperature much higher than that used for the reactions of usual aldimines. Hydrogenolysis of β-lactam **2.8(±)**, followed by acetylation gave the corresponding racemic *cis*-3-acetoxy-β-lactam **2.10(±)** in good overall yield (Scheme 2.3).

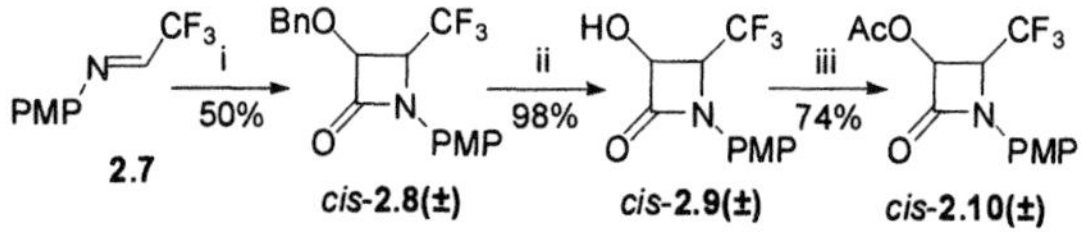

2.7 → (i, 50%) → *cis*-2.8(±) → (ii, 98%) → *cis*-2.9(±) → (iii, 74%) → *cis*-2.10(±)

i) $PhCH_2OCH_2COCl$, Et_3N, CH_2Cl_2, 40 °C; ii) H_2, Pd/C, MeOH, 45°C; iii) Ac_2O, DMAP, Py, CH_2Cl_2

Scheme 2.3

The kinetic optical resolution of β-lactam **2.10(±)** was performed under the same conditions as those described for the resolution of β-lactam **2.2(±)** (PS-Amano, 50 °C, pH 7). It was found that the enzymatic resolution gave (3*R*,4*R*)-3-AcO-4-CF_3-β-lactam **2.10(+)** in high yield (42%, 50% in theory). However, the corresponding kinetically hydrolyzed product, (3*S*,4*S*)-3-hydroxy-4-CF_3-β-lactam **2.9(-)**, was not isolated at all, presumably due to the further hydrolysis of the β-lactam ring of this product under these conditions. Changing the pH to 6 (slower reaction) or 8 (faster reaction) did not solve this problem. Only by decreasing the temperature to 0-5 °C at pH 7, it was possible to control the over-reaction and isolate (3*S*,4*S*)-3-hydroxy-4-CF_3-β-lactam **2.9(-)**with 97% ee in good yield (36%, 50% in theory) in addition to **2.10(+)**. The results are summarized in Table 2.1.

Table 2.1. Enzymatic optical resolution of racemic *cis*-3-AcO-4-CF_3-β-lactam 2.10(±)

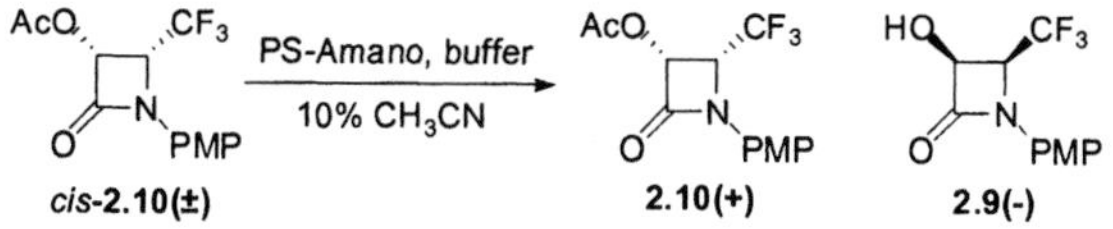

β-lactam	pH	Temp.	Time	2. 10(+)		2.9(-)	
				yield (%)	ee (%)	yield (%)	ee (%)
cis-2.10(±)	7	50 °C	4 days	42	100	-	-
cis-2.10(±)	7	0-5 °C	12 h	45	99.9	36	97

Source: Reproduced with permission from reference 26. Copyright 2004 Elsevier.

We also investigated the efficiency of the enzymatic resolution when another key β-lactam, i.e., NH-free cis-3-acetoxy-4-CF_3-azetidin-2-one (**2.11(±)**), was employed as the substrate. This β-lactam **2.11(±)** was readily obtained by removal of the *N*-PMP group from **2.10(±)**. It was found that the enzymatic resolution proceeded much faster than that of **2.10(±)**, i.e., 50% conversion was reached in 8 h at 3 °C to give (+)-3-AcO-4-CF_3-β-lactam (**2.11(+)**) as well as (-)-3-hydroxy-4-CF_3-azetidin-2-one (**2.12(-)**) without the ring-opening over-reaction. Moreover, (-)-β-lactam **2.12(-)** turned out to be stable (i.e., no over-reaction by the lipase) even at 25 °C under the reaction conditions. Thus, the kinetic resolution of β-lactam **2.11(±)** using the "PS-Amano" lipase at 25 °C and pH 7 reached the 50% conversion in 3 h to afford 3-AcO-β-lactam **2.11(+)** and 3-hydroxy-β-lactam **2.12(-)** with excellent enantiopurity in excellent yields (Scheme 2.4).

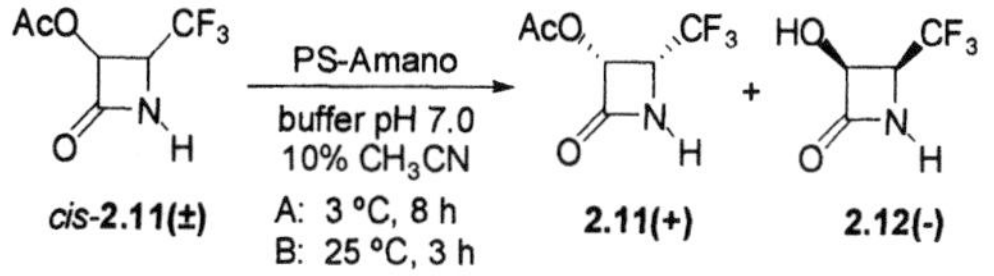

Scheme 2.4

The enantiopure 3-AcO-4-CF_3-β-lactam **2.10(+)** and 3-hydroxy-4-CF_3-β-lactam **2.9(-)** thus obtained were converted to the corresponding (3*R*,4*R*)-3-TIPSO-4-CF_3-β-lactam **2.14(+)** and (3*S*,4*S*)-3-TIPSO-4-CF_3-β-lactam **2.14(-)** using the same protocol as that described for the preparation of 3-TIPSO-4-CF_2H-β-lactams, **2.6(+)** and **2.6(-)** (*vide supra*) (Scheme 2.5).

2.3. Synthesis of enantiopure 1-acyl-3-hydroxy-4-R_f-β-lactams

The *β-Lactam Synthon Method* (β-LSM) has been developed by exploiting the unique nature of this strained four-membered skeleton for its facile ring-opening reactions with a variety of nucleophiles *(30,31)*. When the nitrogen of this strained cyclic amide is acylated (including carbalkoxy, carbamoyl, thiocarbamoyl, and sulfonyl groups besides the standard acyl groups), the resulting *N*-acyl-β-lactam becomes exceptionally reactive for nucleophilic attacks, leading to facile ring-opening coupling. This unique feature of *N*-acyl-β-lactams has been successfully utilized in organic synthesis and medicinal chemistry *(30)*. Scheme 2.6 illustrates readily available *N*-acyl-β-lactams through *N*-acylation of NH-free 3-PO-4-R_f-β-lactams (P = hydroxyl protecting group).

Table 2.2 shows a series of *N*-acyl-3-TIPSO-3-R_f-β-lactams **2.15** ($R_f = CF_2H$) and **2.16** ($R_f = CF_3$) prepared through acylation (including carbalkoxylation and sulfonylation) of 3-TIPSO-4-CF_2H-azetidin-2-one (**2.6(-)**) and 3-TIPSO-4-CF_3-azetidin-2-one (**2.14(+)**) as examples (*t*-Boc = *t*-butoxycarbonyl; Cbz = carbobenzoxy; Ts = *p*-toluenesulfonyl; yields are not optimized) *(3,32-34)*.

Table 2.2. Synthesis of *N*-acyl-3-TIPSO-4-R_f-β-lactams

TIPSO, R_f, N–H → (i or ii) → TIPSO, R_f, N–R

2.6(-) ($R_f = CF_2H$) **2.15(-)** ($R_f = CF_2H$)
2.14(+) ($R_f = CF_3$) **2.16(+)** ($R_f = CF_3$)

i) RCl or R_2O, Et_3N, DMAP, CH_2Cl_2, 25 °C; ii) LiHMDS 1eq, RCl (1.2 equiv.), THF, -78°C

entry	NH-β-lactam	configuration	R_f	conditions	*N*-acyl-β-lactam	R	yield (%)
1	**2.6(-)**	3*S*,4*S*	CF_2H	i	**2.15a(-)**	*t*-Boc	80
2	**2.6(-)**	3*S*,4*S*	CF_2H	i	**2.15b(-)**	Cbz	60
3	**2.6(-)**	3*S*,4*S*	CF_2H	i	**2.15c(-)**	Tosyl	66
4	**2.6(-)**	3*S*,4*S*	CF_2H	i	**2.15d(-)**	4-F-Bz	54
5	**2.14(+)**	3*R*,4*R*	CF_3	i	**2.16a(+)**	*t*-Boc	87
6	**2.14(+)**	3*R*,4*R*	CF_3	ii	**2.16b(+)**	Cbz	60

Source: Reproduced with permission from reference 26. Copyright 2004 Elsevier.

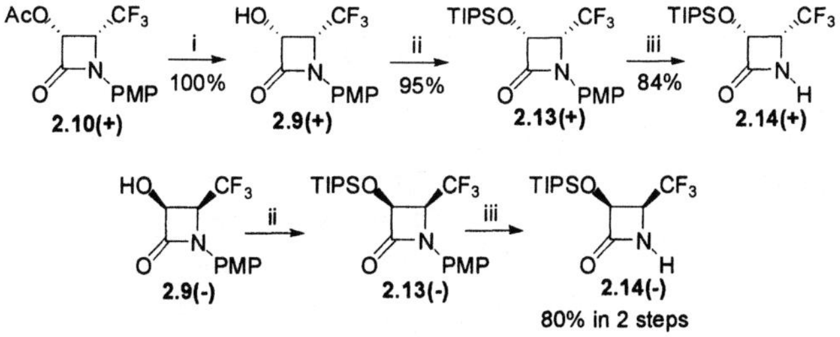

i) KOH, THF, -5°C; ii) TIPSCl, Et_3N, CH_2Cl_2; iii) CAN, CH_3CN/H_2O, -10 °C.

Scheme 2.5

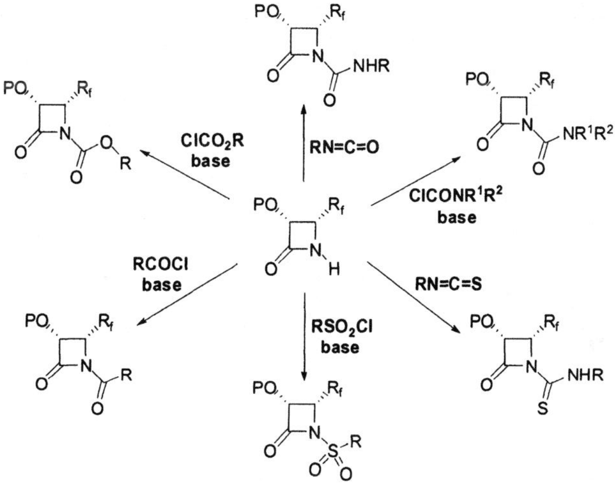

Scheme 2.6

3. Fluorinated β-lactams as versatile synthetic building blocks

3.1. Synthesis of enantiopure β-CF_2H- and β-CF_3- α-hydroxy-β-amino acids and esters via facile hydrolysis/alcoholysis of *N*-acyl-β-lactams

Enantiopure β-R_f-α-hydroxy-β-amino acids (R_f = CF_2H or CF_3) can be obtained through facile ring-opening hydrolysis of *N*-acyl-3-TIPSO-4-R_f-β-lactams (*vide supra*). Thus, the reaction of β-lactams, **2.15c(-)** and **2.16a(-)**, with potassium hydroxide in H_2O/THF at ambient temperature gave the corresponding *O*-protected acids, **3.1(-)** and **3.2(-)** in high isolated yields (Table 3.1). *O*-TIPS protecting group can be easily removed by HF/Pyr as needed.

Table 3.1. Synthesis of β-CF_2H- and CF_3-α-TIPSO-β-amino acids

KOH, THF, H_2O, 25 °C, 12 h

2.15c(-)(R_f = CF_2H)
2.16a(-)(R_f = CF_3)

3.1(-)(R_f = CF_2H)
3.2(-)(R_f = CF_3)

entry	β-lactam	configuration	R_f	R	β-Rf-α-TIPSO-β-amino acid	configuration	yield (%)
1	**2.15c(-)**	3S,4S	CF_2H	Tosyl	**3.1(-)**	2*S*,3*S*	83
2	**2.16a(-)**	3S,4S	CF_3	*t*-Boc	**3.2(-)**	2*S*,3*S*	80

Source: Reproduced with permission from reference 26. Copyright 2004 Elsevier.

In a similar manner enantiopure *O*-TIPS-β-R_f-α-hydroxy-β-amino acid methyl esters, **3.3(-)** (R_f = CF_2H) and **3.4(+)** (R_f = CF_3), were synthesized through a facile methanolysis of R_f-β-lactams, **2.15(-)** and **2.16(+)**, in the presence of triethylamine and a catalytic amount of 4-(*N,N*-dimethylamino)pyridine (DMAP) at ambient temperature in good to quantitative yields. Results are summarized in Table 3.2 (yields are not optimized). *O*-TIPS protecting group can be easily removed by HF/Pyr as needed.

3.2. Synthesis of β-CF_3-α-hydroxy-α-methyl-β-amino acid via extremely stereoselective methylation of *N*-acyl-β-lactam

Enantiopure β-CF_3-α-hydroxy-α-methyl-β-amino acid **3.7** can be synthesized via extremely stereoselective methylation of β-lactam **3.5** at the C-3

Table 3.2. Synthesis of β-CF_2H- and CF_3-α-TIPSO-β-amino esters

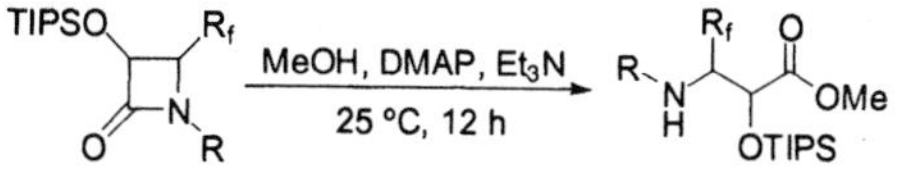

TIPSO, R_f β-lactam, N–R; MeOH, DMAP, Et_3N, 25 °C, 12 h → R–NH–CH(R_f)–CH(OTIPS)–C(=O)OMe

2.15(-) (R_f = CF_2H) → **3.3(-)** (R_f = CF_2H)
2.16(+) (R_f = CF_3) → **3.4(+)** (R_f = CF_3)

entry	β-lactam	configuration	R_f	R	β-R_f-α-TIPSO-β-amino ester	configuration	yield (%)
1	**2.15a(-)**	3S,4S	CF_2H	*t*-Boc	**3.3a(-)**	2*S*,3*S*	98
2	**2.15b(-)**	3S,4S	CF_2H	Cbz	**3.3b(-)**	2*S*,3*S*	78
3	**2.15c(-)**	3S,4S	CF_2H	Tosyl	**3.3c(-)**	2*S*,3*S*	87
4	**2.16a(+)**	3R,4R	CF_3	*t*-Boc	**3.4a(+)**	2*R*,3*R*	70
5	**2.16b(+)**	3R,4R	CF_3	Cbz	**3.4b(+)**	2*R*,3*R*	58
6	**2.16c(+)**	3R,4R	CF_3	Tosyl	**3.4c(+)**	2*R*,3*R*	100

Source: Reproduced with permission from reference 26. Copyright 2004 Elsevier.

position, followed by deprotection and hydrolysis (Scheme 3.1). The β-lactam enolate was generated by treating **3.5** with LDA at –30 °C, where dimethylphenylsilyl (DMPS) was used as the *O*-protecting group instead of very bulky TIPS since this reaction is sensitive to the bulkiness of the C-3 substituent. This enolate was reacted with methyl iodide at –78 °C to give *O*-DMPS-3-methyl-β-lactam **3.6** as single diastereomer in 65% yield. After the removal of *N*-PMP group with CAN, acidic hydrolysis gives β-CF_3-α-hydroxy-α-methyl-β-amino acid **3.7**. 3-Methyl-β-lactam **3.6** can also be converted to *N*-*t*-Boc-3-methyl-β-lactam **3.8**, which would serve as another useful synthetic intermediate

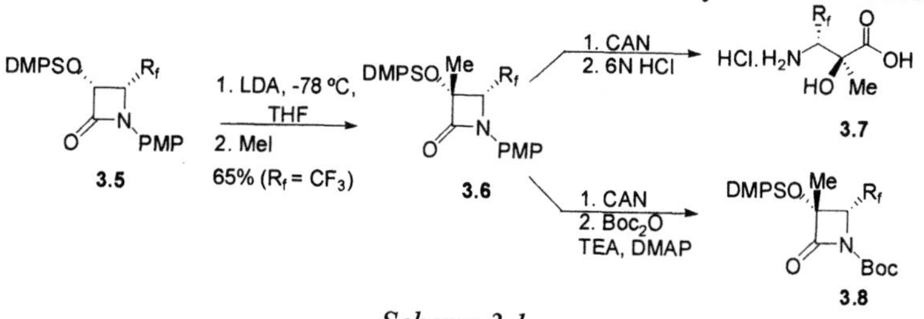

DMPSO, R_f, N-PMP β-lactam **3.5** — 1. LDA, -78 °C, THF; 2. MeI; 65% (R_f = CF_3) → DMPSO, Me, R_f, N-PMP **3.6**

3.6 — 1. CAN; 2. 6N HCl → HCl.H_2N–CH(R_f)–C(HO)(Me)–C(=O)OH **3.7**

3.6 — 1. CAN; 2. Boc_2O, TEA, DMAP → DMPSO, Me, R_f, N-Boc **3.8**

Scheme 3.1

for the introduction of a conformationally restricted CF_3-containing α-hydroxy-β-amino acid residue to fluoro-peptides and peptidomimetics. The same procedure is applicable to other R_f groups including difluoromethyl group.

3.3. Synthesis of R_f-containing isoserine dipeptides through efficient ring-opening coupling *N*-acyl-β-lactams with amino esters

The ring-opening coupling of *N*-acyl-β-lactams, **2.15** and **2.16** with various α- and β-amino acid esters provides a very easy access to a library of R_f-containing isoserine dipeptides. Since these coupling reactions do not need any peptide coupling reagents such as *N.N*-dicyclohexylcarbodiimide (DCC) and *N,N*-diisopropylcarbodiimide (DIC), the "atom economy" *(35)* is extremely

Table 3.3. Ring-opening coupling of *N*-acyl-β-lactams with amino esters

α–AA-OMe.HCl, NMM, CH_2Cl_2, 25 °C → **3.9** (R_f = CF_2H), **3.10** (R_f = CF_3)

2.15(-) (R_f = CF_2H), **2.16(+)** (R_f = CF_3)

β-AlaOEt.HCl, NMM, CH_2Cl_2, 25 °C → **3.11** (R_f = CF_2H), **3.12** (R_f = CF_3)

entry	β-lactam	configuration	R_f	R	amino ester	reaction time (h)	dipeptide	isolated yield (%)
1	**2.15a(-)**	3*S*,4*S*	CF_2H	*t*-Boc	(*S*)-Phe-OMe	24	**3.9a**	no rxn
2	**2.15a(-)**	3*S*,4*S*	CF_2H	*t*-Boc	Gly-OMe	18	**3.9b**	62
3	**2.15a(-)**	3*S*,4*S*	CF_2H	*t*-Boc	β-Ala-OEt	18	**3.11a**	50
4	**2.15c(-)**	3*S*,4*S*	CF_2H	Tosyl	(*S*)-Val-OMe	12	**3.9c**	82
5	**2.15c(-)**	3*S*,4*S*	CF_2H	Tosyl	(*S*)-Leu-OMe	12	**3.9d**	94
6	**2.15c(-)**	3*S*,4*S*	CF_2H	Tosyl	(*S*)-Met-OMe	12	**3.9e**	89
7	**2.15c(-)**	3*S*,4*S*	CF_2H	Tosyl	(*S*)-Phe-OMe	12	**3.9f**	83
8	**2.15d(-)**	3*S*,4*S*	CF_2H	4-F-Bz	(*S*)-Phe-OMe	12	**3.9g**	100
9	**2.16b(+)**	3*R*,4*R*	CF_3	Cbz	(*S*)-Phe-OMe	96	**3.10a**	65
10	**2.16b(+)**	3*R*,4*R*	CF_3	Cbz	β-Ala-OEt	48	**3.12b**	65

Source: Reproduced with permission from reference 26. Copyright 2004 Elsevier.

high. The coupling reactions gave the corresponding R_f-containing isoserine dipeptides, **19-22**, in good to quantitative yields. Results are summarized in Table 3.3.

The results described above as well as our previous works *(36,37)* allow us to envision the versatile utility of the *N*-acyl-3-PO-4-R_f-lactams (P = hydroxyl protecting group) for the synthesis of R_f-containing isoserine dipeptides, depsipeptides, peptidomimetics, and key synthetic building blocks for R_f-containing hydroxyethylene, dihyroxylethylene, and hydroxyethylamine dipeptide isosteres. Scheme 3.2 illustrates the possible transformations of *N*-t-Boc-3-PO-4-R_f-lactams.

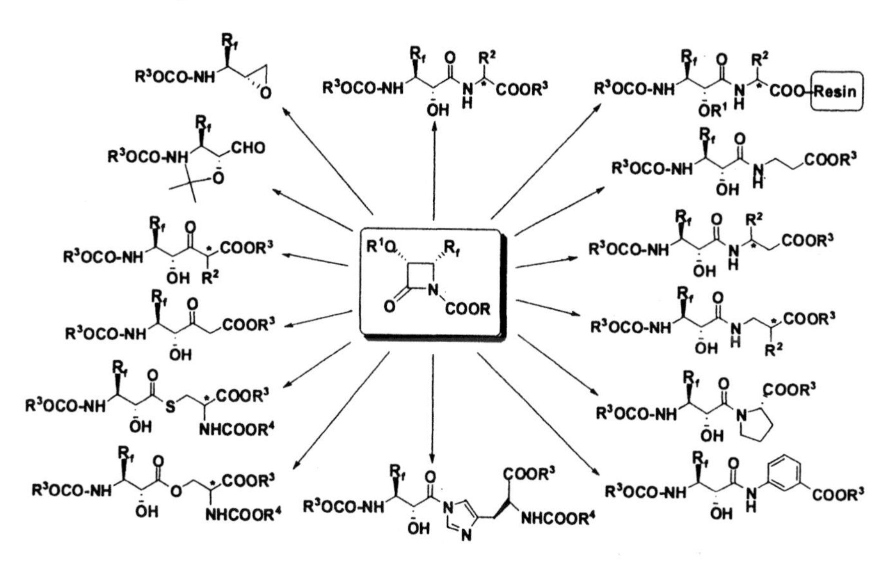

Scheme 3.2

3.4. Synthesis of CF_2H- and CF_3-containing taxoid anticancer agents through ring-opening coupling of *N*-acyl-β-lactams with baccatins

As a part of our continuing work on the synthesis and structure-activity relationship studies of fluoro-taxoid anticancer agents *(38,39)*, we applied the ring-opening coupling reaction of (3*R*,4*R*)-*N*-*t*-Boc-3-TIPSO-4-R_f-β-lactams, **2.15a** and **2.16a**, with baccatins for the synthesis of two new fluoro-taxoids.

Thus, the reactions of (3*R*,4*R*)-N-*t*-Boc-3-TIPSO-4-CF_2H-β-lactam **2.15a** and (3*R*,4*R*)-N-*t*-Boc-3-TIPSO-4-CF_3-β-lactam **2.16a** with 2-debenzoyl-2-(3-chlorobenzoyl)-10-deacetyl-10-propanyl-baccatin (**3.13**) and with 2-debenzoyl-

2-(3-azidobenzoyl)-10-deacetyl-10-propanylbaccatin (**3.14**), respectively, were carried out in the presence of LiHMDS in THF at –40 °C. The subsequent removal of the silyl protecting groups by HF/pyridine gave the corresponding new fluoro-taxoids **3.15** and **3.16** in fairly good overall yields (Scheme 3.3).

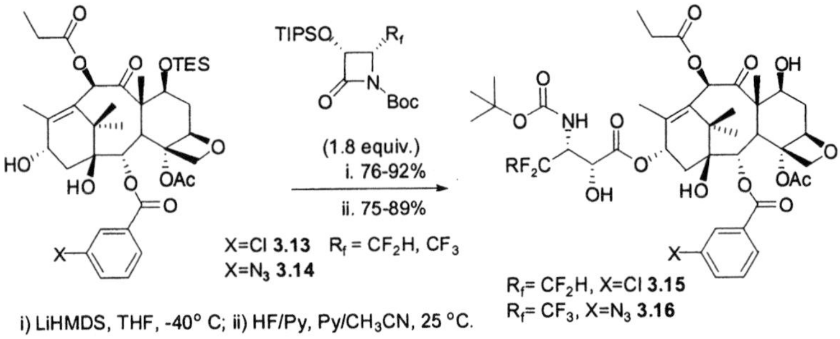

X=Cl **3.13** $R_f = CF_2H, CF_3$
X=N_3 **3.14**

(1.8 equiv.) i. 76-92% ii. 75-89%

$R_f = CF_2H$, X=Cl **3.15**
$R_f = CF_3$, X=N_3 **3.16**

i) LiHMDS, THF, -40° C; ii) HF/Py, Py/CH_3CN, 25 °C.

Scheme 3.3

Cytotoxicity of the two new fluoro-taxoids was evaluated *in vitro* against both types of human breast cancer cell lines, sensitive and drug resistant. The IC_{50} values were determined through 72 h exposure of the fluoro-taxoids to the cancer cells according to protocol developed by Skehan et al *(40)*. As Table 3.4 shows, the new fluoro-taxoids, **3.15** and **3.16**, possess more than two orders of magnitude higher cytotoxicity than paclitaxel against the drug-resistant cell lines, MCF7-R and LCC6-MDR, and several times higher potency than paclitaxel against the drug-sensitive cell lines, MCF7-S and LCC6-WT. Accordingly, these two fluoro-taxoids are highly promising additions to the arsenal of the "second-generation taxoid anticancer agents".

Table 3.4. *In vitro* cytotoxicity (IC_{50} nM)[a] of fluoro-taxoids

taxoid	MCF7-S (breast)	MCF7-R (breast)	R/S[b]	LCC6-WT (breast)	LCC6-MDR (breast)	R/S[b]	H460 (ovarian)	HT-29 (colon)
Paclitaxel	1.8	484	269	3.4	216	64	5.5	3.6
3.11	0.6	6.4	11	0.6	3.1	5.2	0.3	0.5
3.12	0.4	2.6	6.5	1.2	1.6	1.3	0.2	0.4

[a]The concentration of compound which inhibits 50% (IC_{50}) of the growth of the human tumor cell line after 72 h drug exposure. [b]R/S = drug-resistance factor = IC_{50}(drug-resistant cell line)/IC_{50}(drug-sensitive cell line)

Source: Reproduced with permission from reference 26.

4. Conclusion

Efficient methods have been developed for the synthesis of 3-hydroxy-4-R_f-β-lactams (>99% ee) (R_f = CF_2H and CF_3) through ketene-imine cycloaddition, followed by enzymatic optical resolution. Enantiopure 1-acyl-3-hydroxyl-4-R_f-β-lactams, readily obtained from the 3-hydroxy-4-R_f-β-lactams via *N*-acylation, serve as versatile intermediates for the synthesis of a variety of fluorine-containing α-hydroxy-β-amino acids, isoserine dipeptides, and taxoid anticancer agents.

Acknowledgment

This research was supported by grants from the National Institutes of Health (NIGMS and NCI). Generous gift of 10-deacetylbaccatin III from Indena, SpA is gratefully acknowledged.

References

1. Juaristi, E. *Enantioselective Synthesis of β-Amino Acids*, 1997.
2. Cole, D. C. *Tetrahedron* **1994,** *50*, 9517-9582.
3. Ojima, I.; Delaloge, F. In *Peptidomimetics Protocols;* Kazmierski, W. M. Ed.; Humana Press: Totowa, New Jersey, **1998**; 137-160.
4. Kingston, D. G. I. *Chem. Comm.* **2001,** *10*, 867-880.
5. Ojima, I.; Lin, S.; Wang, T. *Curr. Med. Chem.* **1999,** *6*, 927-954.
6. Georg, G. I., Chen T. T., Ojima I., Wyas D. *Taxane Anticancer Agents: Basic Science and Current Status*; American Chemical Society: Washington D. C., 1995.
7. Umezawa, H., Aoyagi, T., Suda, H., Hamada, M., Takeuchi, T. *J. Antibiot.* **1976,** *29*, 97-99.
8. Pearson, W. H. H. *J.Org. Chem.* **1989,** *54*, 4235-4237.
9. Okino, T. M., H.; Murakami, M.; Yamaguchi, K. *Tetrahedron Lett.* **1993,** *34*, 501-504.
10. Mimoto, T. H., N.; Takaku, H.; Kisanuki, S.; Fukazawa, T.; Terashima, K.; Kato, R.; Nojima, S.; Misawa, S.; Ueno, T.; Imai, J.; Enomoto, H.; Tanaka,

S.; Sakikawa, H.; Shintani, M.; Hayashi, H.; Kiso, Y. *Chem. Pharm.Bul.* **2000,** *48*, 1310-1326.

11. Kiso, Y. Y., S.; Matsumoto, H.; Mimoto, T.; Kato, R.; Nojima, S.; Takaku, H.; Fukazawa, T.; Kimura, T.; Akaji. *Arch. Pharm.* **1998,** *331*, 87-89.
12. Ojima, I.; McCarthy, J. M.; Welch, J. T. *Biomedical Frontiers of Fluorine Chemistry*; American Chemical Society: Washington, D. C., 1996.
13. Ojima, I. *ChemBioChem* **2004,** *5*, 628-635.
14. Hook, D. F.; Gessier, F.; Noti, C.; Kast, P.; Seebach, D. *ChemBioChem* **2004,** *5*, 691-706.
15. Ojima, I.; Inoue, T.; Chakravarty, S. *J. Fluorine Chem.* **1999,** *97*, 3-10.
16. Ojima, I.; Inoue, T.; Slater, J. C.; Lin, S.; Kuduk, S. C.; Chakravarty, S.; Walsh, J. J.; Gilchrist, L.; McDermott, A. E.; Cresteil, T.; Monsarrat, B.; Pera, P.; Bernacki, R. J. In *"Asymmetric Fluoroorganic Chemistry: Synthesis, Application, and Future Directions"; ACS Symp. Ser. 746;* Ramachandran Ed.; American Chemical Society: Washington, D. C., **1999**; Chapter 12, 158-181.
17. Kukhar, V. P., Soloshonok, V. A., *Fluorine-containing Amino Acids: Synthesis and Properties*; Wiley: Chichester, 1994.
18. O'Hagan D., S. C., Cobb S.L., Hamilton J. T. G., Cormac D. Murphy C. D. *Nature* **2002,** *416*, 279-280.
19. Gerhard U, T. S., Mortishire-Smith R. *J. Pharm. Biol. Analysis* **2003**, 531-538.
20. Martino R, M.-M. M., Gilard V. *Curr. Drug Metab.* **2000**, 271-303.
21. Davoli, P.; Forni, A.; Franciosi, C.; Moretti, I.; Prati, F. *Tetrahedron: Asymmetry* **1999,** *10*, 2361-2371.
22. Uneyama, K. H., J.; Amii, H. *Tetrahedron Lett.* **1998,** *39*, 4079-4082.
23. Abouabdellah A., B. J. P., Bonnet-Delpon D., Nga T.T.T. *J.Org. Chem.* **1997,** *62*, 8826-8833.
24. Ojima, I.; Habus, I.; Zhao, M.; Zucco, M.; Park, Y. H.; Sun, C. M.; Brigaud, T. *Tetrahedron* **1992,** *48*, 6985-7012.
25. Brieva, R.; Crich, J. Z.; Sih, C. J. *J. Org. Chem.* **1993,** *58*, 1068-1075.
26. Kuznetsova, L.; Ungureanu, I. M.; Pepe, A.; Zanardi, I.; Wu, X.; Ojima, I. *J. of Fluorine Chem.* **2004,** *125*, 487-500.
27. Abouabdellah, A.; Bégué, J.-P.; Bonnet-Delpon, D. *Synlett* **1996**, 399-400.
28. Ojima, I.; Slater, J. C. *Chirality* **1997,** *9*, 487-494.
29. Lin, S., Geng X.D., Qu CX, Tynebor R., Gallagher D.J., Pollina E., Rutter J., Ojima I. *Chirality* **2000,** *12*, 431-441.
30. Ojima, I.; Delaloge, F. *Chem. Soc. Rev.* **1997,** *26*, 377-386.
31. Ojima, I. *Acc. Chem. Res.* **1995,** *28*, 383-389 and references cited therein.
32. Gerard, S.; Dive, G.; Clamot, B.; Touillaux, R.; Marchand-Brynaert, J. *Tetrahedron* **2002,** *58*, 2423-2433.
33. Ojima, I.; Chen, H. J. C. *J. Chem. Soc., Chem. Comm.* **1987,** *8*, 625-626.

34. Arimoto, M.; Hayano, T.; Soga, T.; Yoshioka, T.; Tagawa, H.; Furukawa, M. *J. Antibiot.* **1986,** *39*, 1243-1256.
35. Trost, B. M. *Acc. Chem. Res.* **2002,** *35*, 695-705.
36. Ojima, I.; Sun, C. M.; Park, Y. H. *J. Org. Chem.* **1994,** *59*, 1249-1250.
37. Ojima, I.; Wang, H.; Wang, T.; Ng, E. W. *Tetrahedron Lett.* **1998,** *39*, 923-926.
38. Ojima, I.; Lin, S.; Slater, J. C.; Wang, T.; Pera, P.; Bernacki, R. J.; Ferlini, C.; G., S. *Bioorg. Med. Chem.* **2000,** *8*, 1576-1585.
39. Ojima, I.; Slater, J. C.; Pera, P.; Veith, J. M.; Abouabdellah, A.; Bégué, J.-P.; Bernacki, R. J. *Bioorg. Med. Chem. Lett.* **1997,** *7*, 209-214.
40. Skehan, P.; Streng, R.; Scudierok, D.; Monks, A.; McMahon, J.; Vistica, D.; Warren, J. T.; Bokesch, H.; Kenney, S.; Boyd, M. R. *J. Nat. Cancer Inst.* **1990,** *82*, 1107-1112.

Chapter 32

Synthesis of Fluorinated Prolines and Pyroglutamic Acids

Feng-Ling Qing[1,2] and Xiao-Long Qiu[1]

[1]Key Laboratory of Organofluorine Chemistry, Shanghai Institute of Organic Chemistry, Chinese Academy of Sciences, 354 Fenglin Lu, Shanghai 200032, China
[2]College of Chemistry and Chemistry Engineering , Donghua University, 1882 West Yanan Lu, Shanghai 200051, China

Recent achievements in the synthesis of fluorinated prolines and pyroglutamic acids are described. 4(3)-Fluoroprolines and 4,4(3,3)-difluoroprolines were prepared by the fluorination of 4(3)-hydroxyprolines and 4(3)-ketoprolines respectively. 4-Trifluoromethyl- and 4-difluoromethylprolines were obtained starting from the fluorine-containing building blocks. The oxidation of fluorinated prolines provided fluorinated pyroglutamic acids.

It is well recognized that the introduction of fluorine atom(s) into many biologically active molecules can bring remarkable and profound changes in physical, chemical and biological properties (*1*). Of these fluorinated compounds, the area of fluorinated amino acids (FAAs) is rapidly expanding, taking an important place (*2*). Besides being ultilized as a powerful tool for illumination of the principles of protein design (*3*), FAAs has also found their application in medicinal, agricultural and material sciences (*4*). Cyclic FFAs recently received increased attentions because their incorporation of medium size ring into key positions in peptide chains plays an important role, which constitutes the most prominent pathway to conformationally constrained peptide-mimetics (5). Thus, synthetic routes towards cyclic conformationally constrained fluorinated amino acids have been intensively developed, although there are many syntheses of acyclic fluorinated amino acids (*2*),(*6*).

Prolines and pyroglutamic acids are two types of very important conformationally constrained amino acids and have been extensively used in the pharmaceutical industry, protein design and engineering (*7*). In connection with the importance of fluorinated amino acids and the rigidity of the two types of amino acids, a series of fluorinated prolines and pyroglutamic acids were recently synthesized.

Syntheses of fluorinated prolines

Prolines substituted at 4-position have shown to enhance the thermal stability of collagenmimetic triple helices, with *trans*-4-fluoroproline yielding the most striking results (*7*). In general, 4-fluoroproline derivatives **3** and 4,4-difluoroproline derivatives **4** were prepared from the corresponding 4-hydroxy proline derivatives **1** and 4-keto proline derivatives **2** via direct fluorination with DAST (*8*), morph-DAST (*9*), 2-chloro-1,1,2-trifluorotriethylamine (*8a*), and $C_4H_9SO_2{}^{18}F$ (*10*) (Scheme 1). In addition, 4-fluoroproline derivatives were also prepared via nucleophilic substitution of 4-tosyloxyproline derivatives (*11*).

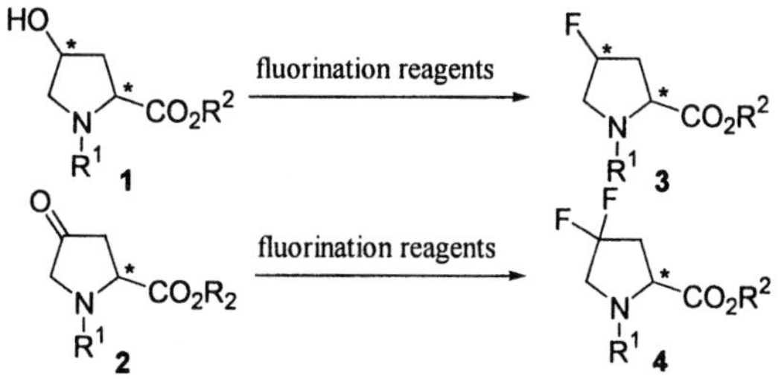

HO, * , * CO_2R^2, N, R^1 **1** — fluorination reagents → F, * , * CO_2R^2, N, R^1 **3**

O, * CO_2R_2, N, R^1 **2** — fluorination reagents → F, F, * CO_2R^2, N, R^1 **4**

fluorination reagent: DAST, Morph-DAST, 2-Chloro-1,1,2-trifluorotriethylamine,$C_4H_9SO_2F$

Scheme 1. General strategies for 4-fluorinated prolinates.

In 1976, Wakseman et al. (*12*) reported the synthesis of racemic methyl *N-tert*-4-trifluoromethylproline by means of a [3+2] cycloaddition. Recently, Boc-protected *cis*- and *trans*-4-trifluoromethyl-D-prolines were also synthesized starting from Garner's aldehyde **5** (Scheme 2) (*13*). The key step is construction of proline ring skeleton via the cyclization reaction. The compound **6** was prepared from **5** in two steps according to the reported procedures. Initial hydrogenation of the double bond with Raney Ni in MeOH followed by reduction of the ester group with $LiAlH_4$ afforded the alcohol **7** in 94% yield. Benzylation of **7** followed by hydrolysis of the hemiaminal moiety with 80% AcOH afforded **8** in 75% yield over two steps. Protection of **8** with a TBDMS group gave **9** and the two diastereoisomers could be separated by flash chromatography. Hydrogenation of **9a** and **9b** gave the alcohol **10a** and **10b** in 99 and 91% yield, respectively. Mesylation of **10a** and **10b** followed by treatment with KHMDS furnished the desired cyclisation products **11a** and **11b** in 83 and 80% yield, respectively. yield. Removal of the protecting groups followed by oxidation with the Jones reagent gave the desired Boc-protected *trans*- and *cis*-4-trifluoromethyl-D-prolines **12a** and **12b.**

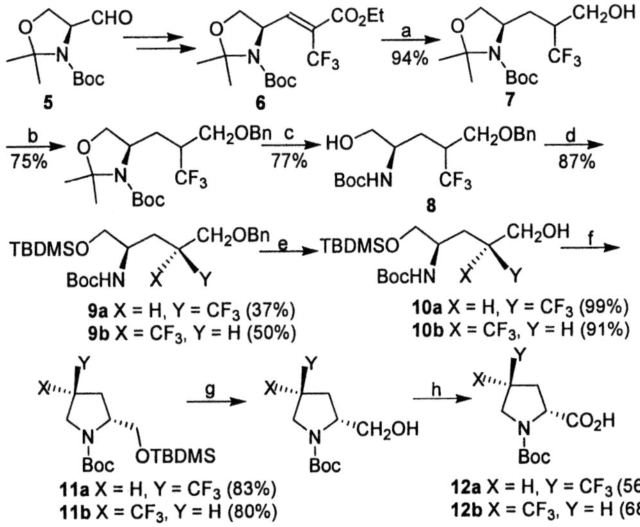

9a X = H, Y = CF_3 (37%)
9b X = CF_3, Y = H (50%)
10a X = H, Y = CF_3 (99%)
10b X = CF_3, Y = H (91%)
11a X = H, Y = CF_3 (83%)
11b X = CF_3, Y = H (80%)
12a X = H, Y = CF_3 (56%)
12b X = CF_3, Y = H (68%)

Scheme 2. Reagents and conditions: (a) i. H_2, Raney Ni, MeOH, rt, overnight; ii. $LiAlH_4$, Et_2O, 0°C; (b) BnBr, NaH, TBAI, THF, rt, 5 h; (c) 80% aq. AcOH, 50°C, overnight; (d) TBDMSCl, imidazole, CH_2Cl_2, rt, 1 h; (e) 10% Pd/C, H_2, EtOH, rt, overnight; (f) i. MsCl, Et_3N, CH_2Cl_2, rt, overnight; ii. KHMDS, THF, 0°C, 24 h; (g) TBAF, THF, rt, 2 h; (h) Jones reagent.

Goodman and Del Valle (*14*) stereoselectively synthesised Boc-protected *cis*- and *trans*-4-trifluoromethyl-L-prolines by asymmetric hydrogenation reactions as the key steps (Scheme 3). Treatment of the ketone **2** with trimethyl(trifluoromethyl)silane (CF_3TMS) gave the tertiary alcohol **13** in 56% yield. Reduction of the ester group followed by selective protection of the resulting primary alcohol afforded **14** in 84% yield over two steps. Tosylation of the tertiary alkoxide of **14** followed by direct treatment with *t*-BuOK furnished the key pyrroline intermediate **15** in 76% yield. The heterogenous hydrogenation of **15** with Pd/C as a catalyst resulted in almost complete removal of the TBDMS protecting group. The best facial selectivity (15:1, *cis/trans*) was observed by using 5% Pd/C as a catalyst in EtOAc. Oxidation of **16** gave the desired Boc-protected *cis*-4-trifluoromethyl-L-proline **17**. On the other hand, removal of the protecting group of compound **15** followed by hydroxy-directed asymmetric hydrogenation of the compound **18** gave the fluorinated alcohol **19** in 77% yieid over two steps. The best diastereoselectivity (158:1, *trans/cis*) was obtained with 2 mol% [Ir(cod)(py)PCy_3] as a catalyst. Oxidation of **19** gave the Boc-protected *anti*-4-trifluoromethyl-L-proline **20** in 96% yield.

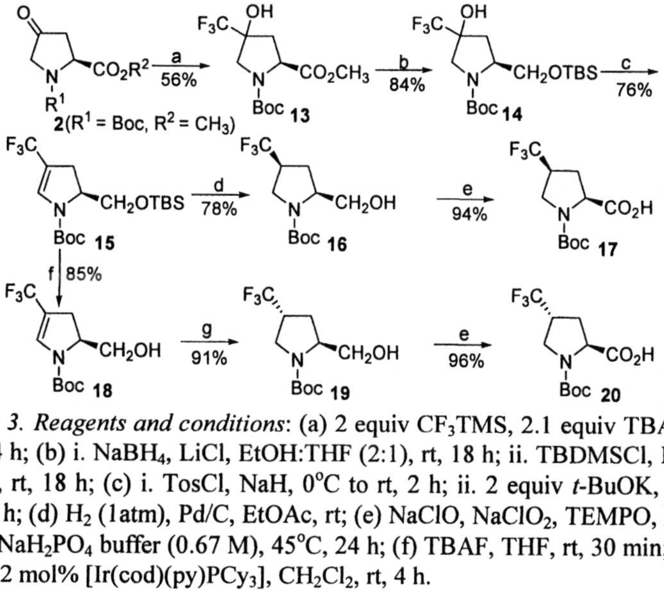

Scheme 3. Reagents and conditions: (a) 2 equiv CF_3TMS, 2.1 equiv TBAF, 0°C to rt, 24 h; (b) i. $NaBH_4$, LiCl, EtOH:THF (2:1), rt, 18 h; ii. TBDMSCl, DMAP, CH_2Cl_2, rt, 18 h; (c) i. TosCl, NaH, 0°C to rt, 2 h; ii. 2 equiv *t*-BuOK, THF, -40°C, 2 h; (d) H_2 (1atm), Pd/C, EtOAc, rt; (e) NaClO, $NaClO_2$, TEMPO, MeCN, pH 6.7 NaH_2PO_4 buffer (0.67 M), 45°C, 24 h; (f) TBAF, THF, rt, 30 min; (g) H_2 (1atm), 2 mol% [Ir(cod)(py)PCy_3], CH_2Cl_2, rt, 4 h.

In addition, Boc-protected *cis*-4-trifluoromethyl- and *cis*-4-difluoromethyl-L-prolines were also been synthesized via another short and efficient route from the key intermediate **2** (*15*). Trifluoromethylation of the carbonyl group of **2** with CF_3TMS gave an alcohol **21** in 81% yield (Scheme 4). The alcohol **21** was treated with $SOCl_2$/pyridine under reflux conditions to give the olefin **22** in 78% yield. Hydrogenation and deprotection of **22** with Pd/C as a catalyst in EtOH yielded a single diastereoisomer, *N*-Boc-*cis*-4-trifluoromethyl-L-proline **23**. On the other hand, treatment of **2** with CF_2Br_2/Zn/HMPT in THF gave the olefin **24** in 48% yield. Hydrogenation of **24** stereoselectively afforded the fluorinated amino acid, *N*-Boc-*cis*-4-difluoromethyl-L-proline **25**.

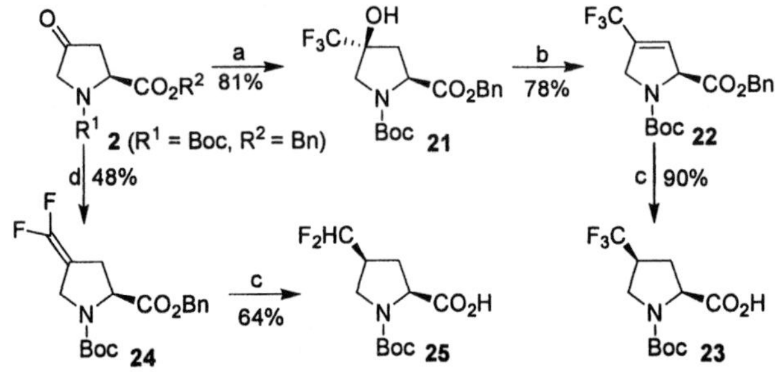

Scheme 4. Reagents and conditions: (a) i. CF_3SiMe_3, TBAF (cat.), rt, overnight; ii. satd. aq. NH_4Cl, rt, 15 min, then TBAF, rt, 1 h; (b) $SOCl_2$, pyridine, reflux, 20 min; (c) Pd/C, H_2, EtOH; (d) CF_2Br_2, Zn, HMPT, THF, reflux, 3.5 h.

In 2001, Dugave et. al (*16*) reported the synthesis of optically pure *N*-Boc-protected 3-fluoroprolines **28** and **31** (Scheme 5). The key steps were introduction of fluorine atom into 3-position of proline derivatives via fluorination of compounds **26** and **29** with DAST. After hydrogenation of the resulting compounds **27** and **30**, the target molecules **28** and **31** were obtained in 26% (four steps) and 13% (seven steps) yield, respectively.

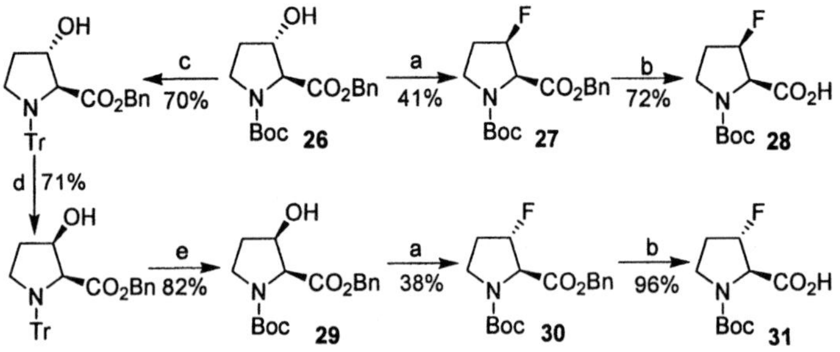

Scheme 5. Reagents and conditions: (a) DAST, CH_2Cl_2; (b) H_2, Pd/C, MeOH; (c) TFA, CH_2Cl_2; (d) PPh_3, DEAD, $PhCO_2H$, toluene, then KOH in MeOH; (e) HCO_2H, 1,2-dichloroethane, then Boc_2O, Et_3N, CH_2Cl_2.

Synthesis of 3,3-difluoroproline was first realized by Coward et al in 1993 (*17*). Compound **32** was obtained in 73% yield by oxidation of 6-hydroxy-1-aza-3-oxabicyclo[3.3.0]octan-2-one under Swern conditions. Treatment of **32** with DAST furnished the difluorinated compound **33** in moderate yield (Scheme 6). Hydrolysis of bicyclic compound **33** to the prolinol derivative followed by protection of amino group, oxidation of hydroxymethyl and esterification provided the 3,3-difluoroproline derivative **34**.

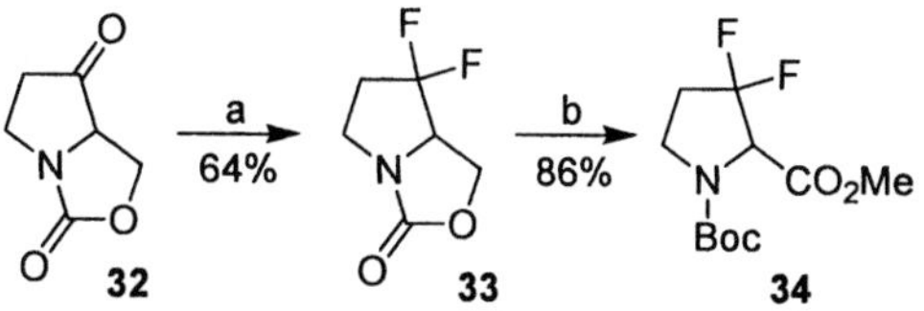

Scheme 6. *Reagents and conditions:* (a) DAST, CH_2Cl_2, -78°C to rt; (b) i. 6N HCl, heat; ii. Boc_2O, $NaHCO_3$, $CHCl_3$, H_2O; iii. $RuO_2{\cdot}xH_2O$, 10% $NaIO_4$, EtOAc; iv. CH_2N_2, Et_2O.

Later, Shi et al (*18*) also synthesized racemic 3,3-difluoroproline **39** starting from δ,ε-unsaturated β,β-difluoro-α-keto ester **35**. Treatment of the ester **35** with H_2NOCH_3 followed by reduction with Zn powder and protection of the resulting amino group afforded the amino ester **36** (Scheme 7). Ozonolysis of the ester **36** followed by treatment with Me_2S provided the cyclic hemiaminal **37** in high yield. After removal of the protecting group by hydrogenation, the resulting cyclic imine was further concomitantly reduced under the hydrogenation condition to yield the hydrochloride salt **38**. The compound **38** was subjected to acidic hydrolysis to furnish the desired 3,3-difluoroproline **39.**

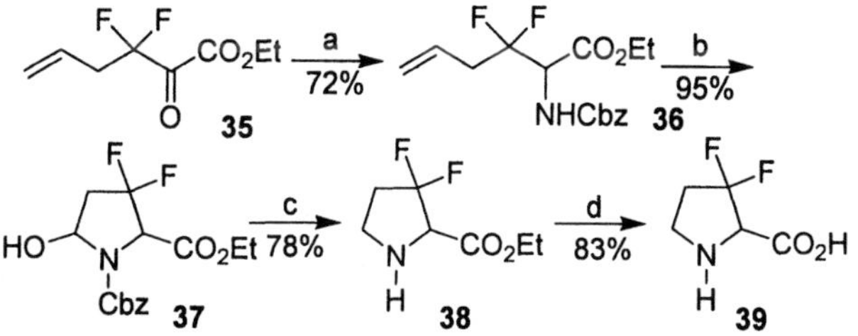

Scheme 7. *Reagents and conditions:* (a) i. H_2NOCH_3, EtOH, 40°C; ii. Zn, aq. HCO_2H; iii. Boc_2O; (b) i. O_3, CH_2Cl_2; ii. Me_2S; (c) H_2, Pd/C, HCl, MeOH; (d) 6N HCl, reflux.

Syntheses of fluorinated pyroglutamic acids

Pyroglutamic acid and its derivatives are important amino acids in many bioactive compounds and recently, there have appeared many reports about the synthesis and charterization of substituted pyroglutamic acids and peptides containing them (*19*). 4-Substituted groups of pyroglutamic acids play an important role for conformation and activity of pyroglutamic acid derivatives (*20*).

The RuO_4 oxidation of fluorinated proline derivatives is the general strategy for the synthesis of fluorinated pyroglutamic acid derivatives. 3,3-Difluoro-,[17] 4-fluoro- (*8a*), (*8b*), (*21*), 4-trifluoromethyl- (*22*), 4-difluoromethyl (*22*)-

pyroglutamic acid derivatives were conveniently synthesized by means of this strategy (Scheme 8) although it cannot be applied to the synthesis of 4,4-difluoropyroglutamate due to the strong electron-withdrawing CF_2 group at C-4. In addition, *cis*-4-fluoro-L-pyroglutamate could also be prepared via direct fluorination of 4-hydroxy-L-pyroglutamate with DAST (*23*).

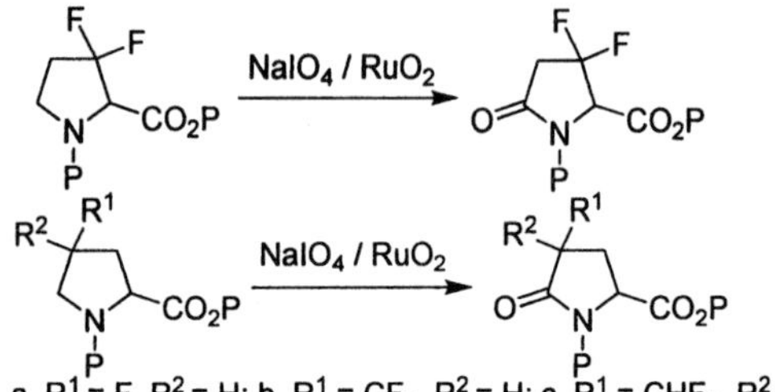

a. R^1 = F, R^2 = H; b. R^1 = CF_3, R^2 = H; c. R^1 = CHF_2, R^2 = H
P = protecting groups

Scheme 8. General strategies for fluorinated pyroglutamates.

A 4,4-difluoropyroglutamic acid derivative was prepared starting from enantiomerically pure bicyclic lactam **40** (*24*), derivated from pyroglutamic acid (Scheme 9). Electrophilic fluorination of the compound **40** with *N*-fluorobenzenesulfonimide (NFSi) gave the difluorinated lactam **41** in 42% yield, which was further converted to 4,4-difluoropyroglutamate **42** via acidic treatment followed by oxidation and esterification. Similarly, 4-fluoropyroglutamate **44** could be synthesized from 2-pyrrolinone **43** (*25*). In addition, Tolman et al. (*26*) also synthesized 4-fluoropyroglutamic acid by means of cyclization of the corresponding 4-fluoroglutamic acid.

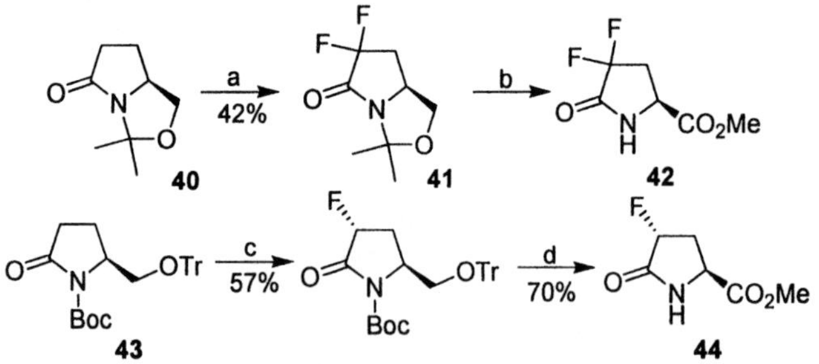

Scheme 9. Reagents and conditions: (a) i. LDA, NFSi, THF, -78°C; ii. LDA, NFSi, THF, -78°C; (b) i. AcOH, CH_3CN, H_2O, 90°C; ii. H_2CrO_4, acetone; iii. CH_2N_2; (c) i. LDA, THF, -78°C; ii. NFSi, THF, -78 to 55°C; (d) i. TFA, Et_3SiH, CH_2Cl_2, 0°C or HCl, MeOH; ii. Oxidation; iii. CH_2N_2, Et_2O.

Besides the aforementioned strategy, 3-fluoro-pyroglutamic acid **46** was also synthesized via direct treatment of the D,L-erythro-3-hydroxyglutamic acid **45** with SF_4 in liquid hydrogen fluoride at -78°C (Scheme 10) (*27*).

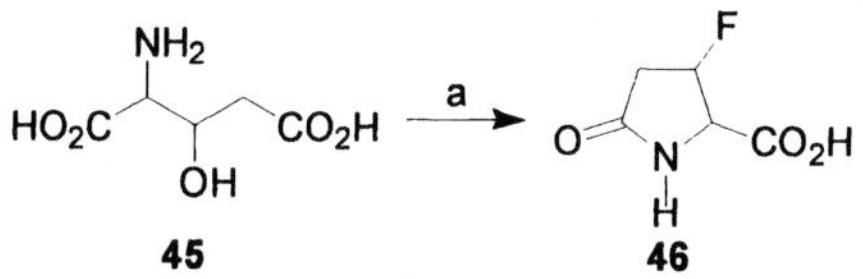

Scheme 10. Reagents and conditions: SF_4, liquid HF, -78°C.

Recently, 4-monofluoromethylenyl- and *cis*-4-monofluoromethyl-L-pyroglutamic acids **49** and **51** were prepared in an accidentially found dehydrogenation reaction (Scheme 11) (*28*). Treatment of the *cis*-4-difluoromethyl-L-pyroglutamate **47** with organic amino base (Et_3N being the best) yielded the defluorinated compound **48** in high yield. Following hydrogenation of resulting defluorinated compound **48** afforded the compound **50** in good diastereoselectivity and good yield. One-step removal of protective groups with TFA successfully provided 4-monofluoromethylenyl- and *cis*-4-monofluoromethyl-L-pyroglutamic acids **49** and **51**.

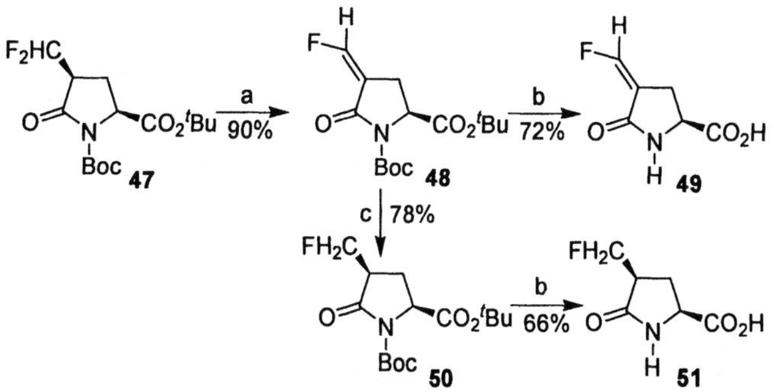

Scheme 11. Reagents and conditions: (a) organic amino base, rt. (b) TFA, CH_2Cl_2, rt; (c) Pd-$BaSO_4$, H_2, EtOAC or dioxane.

Acknowledgements

We thank the thank the National Natural Science Foundation of China and and the Ministry of Education of China for financial support.

References

1. (a) *Organnofluorine Chemistry: Principles and Commerical Applications:* Banks, R. E.; Smart, B. E.; Tatlow, J. C., Eds; Plenum: New Yoek, 1994. (b) Welch, J. T.; Eswaraksrishnan, S. *Fluorine in Bioorganic Chemistry*; Wiley: New York, 1991. (c) *Biomedical Frontiers of Fluorine Chemistry*; Ojima, I.; McCarthy, J. R.; Welch, J. T., Eds.; ACS Symposium Series 639; American Chemical Society: Washington, DC, 1996. (d) *Organoflurine Chemicals and Their Industrial Applications*; Banks, R. E., Eds.; Ellis. Harwood: New York, 1979. (e) Peters, R. *Carbon-fluorine Compounds Chemistry, Biochemistry and Biological Activities.* A Ciba foundation symposium. Amsterdam: Elsevier; 1972.
2. Kukhar, V. P.; Soloshonok, V. A. *Fluorine-Containing Amino Acids: Synthesis and Properties*, John Wiley and Sons Ltd., New York, 1995.
3. (a) Borman, S. *C&EN* **2001**, *79*, 41-42. (b) Dagani, R. *C&EN* **2003**, *81*, 40-44.
4. (a) *Biomedical Frontiers of Fluorine Chemistry*; Ojima, I.; McCarthy, J. R., Welch, J. T., Eds.; ACS Symposium Series 639; American Chemical Society: Washington, DC, 1996. (b) Yoder, N. C.; Kumar, K. *Chem. Soc. Rev.* **2002**, *31*, 335-341. (c) Marsh, E. N. G. *Chemistry & Biology* **2000**, 7, 153-157.
5. (a) Giannis, A.; Kolter, T. *Angew. Chem. Int. Ed. Engl.* **1993**, *32*, 1244-1267. (b) Gante, J. *Angew. Chem. Int. Ed. Engl.* **1994**, *33*, 1699-1720.
6. (a) Qiu, X.-L.; Meng, W.-D.; Qing, F.-L. *Tetrahedron* **2004**, 60, 6711-6745. (b) Sutherland, A.; Willis, C. L. *Nat. Prod. Rep.* **2000**, *17*, 621-631.
7. (a) Holmgren, S. K.; Taylor, K. M.; Bretscher, L. E.; Raines, R. T. *Nature* **1998**, *392*, 666-667. (b) Holmgren, S. K.; Bretscher, L. E.; Taylor, K. M.; Raines, R. T. *Chem. Biol.* **1999**, *6*, 63-70. (c) Eberhardt, E. S.; Panasik, N. Jr.; Raines, R. T. *J. Am. Chem. Soc.* **1996**, *118*, 12261-12266.
8. (a) Hudlicky, M. *J. Fluorine Chem.* **1993**, *60*, 193-210. (b) Hudlicky, M.; Merola, J. S. *Tetrahedron Lett.* **1990**, *31*, 7403-7406. (c) Smith, E. M.; Swiss, G. F.; Neustadi, B. R.; Gold, E. H.; Sommer, J. A. et al. *J. Med. Chem.* **1988**, *31*, 875-885. (d) Kronenthal, D. R.; Mueller, R. H.; Kuester, P. L.; Kissick, T. P.; Johnson, E. J. *Tetrahedron Lett.* **1990**, *31*, 1241-1244. (e) Tran, T. T.; Patino, N.; Condom, R.; Frogier, T.; Guedj, R. *J. Fluorine Chem.* **1997**, *82*, 125-130. (f) O'Neil, L. A.; Thompson, S.; Kalindjian, S. B.; Jenkins, T. C. *Tetrahedron Lett.* **2003**, *44*, 7809-7812. (g) Demange, L.; Menez, A.; Dugave, C. *Tetrahedron Lett.* **1998**, *39*, 1169-1172. (h) Burger, K.; Rudolph, M., Fehn, S.; Sewald, N. *J. Fluorine Chem.* **1994**, *66*, 87-90. (i) Karanewsky, D. S.; Badia, M. C.; Cushman, D. W.; DeForrest, J. M.; Dejneka, T. et al. *J. Med. Chem.* **1990**, *33*, 1459-1469. (j) Hamacher, K. *J. Labelled Compd. Radiopharm.* **1999**, *42*, 1135-1144.

9. (a) Bretscher, L. E.; Jenkins, C. L.; Taylor, K. M.; DeRider, M. L.; Raines, R. T.; *J. Amer. Chem. Soc.* **2001**, *123*, 777-778. (b) Hodges, J. A.; Raines, R. T. *J. Amer. Chem. Soc.* **2003**, *125*, 9262-9263.
10. Jelinski, M.; Hamacher, K.; Coenen, H. H. *J. Labelled Compd. Radiopharm.* **2001**, *44*, S151-S153.
11. (a) Mazza, S. M. *J. Labelled Compd. Radiopharm.* **2000**, *43*, 1047-1058. (b) Hamacher, K. *J. Labelled Compd. Radiopharm.* **1999**, *42*, 1135-1144.
12. Leroy, J.; Wakselman, C. *Can. J. Chem.* **1976**, *54*, 218-225.
13. Qiu, X.-l.; Qing, F.-l. *J. Chem. Soc., Perkin Trans. 1* **2002**, 2052-2057.
14. Dell Valle, J. R.; Goodman, M. *Angew. Chem. Int. Ed. Engl.* **2002**, *41*, 1600-1602.
15. Qiu, X.-L.; Qing, F.-L. *J. Org. Chem.* **2002**, *67*, 7162-7164.
16. Demange, L.; Cluzeau, J.; Ménez, A.; Dugave, C. *Tetrahedron Lett.* **2001**, *42*, 651-653.
17. (a) Hart, B. P.; Coward, J. K. *Tetrahedron Lett.* **1993**, *34*, 4917-4920. (b) Hart, B. P.; Haile, W. H.; Licato, N. J.; Bolanowska, W. E.; McGuire, J. J.; Coward, J. K. *J. Med. Chem.* **1996**, *39*, 56-65.
18. Shi, G.-q.; Cai, W.-l. *J. Org. Chem.* **1995**, *60*, 6289-6295.
19. Bernardi, F.; Garavelli, M.; Scatizzi, M.; Tomasini, C.; Trigari, V.; Crisma, M.; Formaggio, F.; Peggion, C.; Toniolo, C. *Chem. Eur. J.* **2002**, *8*, 2516-2525.
20. Hon, Y.-S.; Chang, Y.-C.; Gong, M.-I. *Heterocycles* **1990**, *31*, 191-195.
21. Hudlicky, M. *J. Fluorine Chem.* **1991**, *54*, 195.
22. Qiu, X.-L.; Qing, F.-L. *J. Org. Chem.* **2003**, *68*, 3614-3617.
23. Avent, A. G.; Bowler, A. N.; Doyle, P. M.; Marchand, C. M.; Young, D. W. *Tetrahedron Lett.* **1992**, *33*, 1509-1512.
24. (a) Konas, D. W.; Coward, J. K. *Org. Lett.* **1999**, *1*, 2105-2107. (b) Doyle, M. P.; Hu, W.; Phillips, I. M.; Moody, C. J.; Pepper, A. G.; Slawin, A. M. Z. *Adv. Synth. Catal.* **2001**, *343*, 112-117.
25. Konas, D. W.; Coward, J. K. *J. Org. Chem.* **2001**, *66*, 8831-8842.
26. (a) Tolman, V.; Vlasáková, V.; Źivný, K. *J. Fluorine Chem.* **1985**, *29*, 138. (b) Tolman, V.; Vlasáková, V.; Nemecek, J. *J. Fluorine Chem.* **1993**, *60*, 185-191. (c) Tolman, V.; Sedmera, P. *J. Fluorine Chem.* **2000**, *101*, 5-10.
27. Vidal-Cros, A.; Gaudry, M.; Marquet, A. *J. Org. Chem.* **1985**, *50*, 3163-3167.
28. Qiu, X.-L.; Meng, W.-D.; Qing, F.-L. *Tetrahedron* **2004**, *60*, 5201-5206.

Chapter 33

Synthesis and Properties of New Fluorinated Peptidomimetics

Monica Sani, Marco Molteni, Luca Bruché, Alessandro Volonterio*, and Matteo Zanda*

C.N.R.-Istituto di Chimica del Riconoscimento Molecolare, sezione "A. Quilico", and Dipartimento C.M.I.C. "G. Natta", Politecnico di Milano, via Mancinelli 7, I–20131 Milano, Italy

Stereocontrolled protocols, both in solution and in solid-phase, for the synthesis of many different fluoroalkyl peptidomimetics are surveyed in this paper: (a) Tfm-malic peptidomimetics as micromolar inhibitors of some matrix metalloproteinases; (b) partially modified retro (PMR) and retro-inverso (PMRI)-ψ[CH(CF_3)NH]-peptides with a strong proclivity to assume turn-like conformations; (c) ψ[CH(CF_3)NH]-peptide mimics holding a great potential as hybrids between natural peptides and hydrolytic transition state analogs; (d) the first PMR-peptides incorporating a trifluoroalanine surrogate.

Introduction

In bioorganic and medicinal chemistry, judicious introduction of fluorine atoms or appropriate fluorinated functions into a molecule has become a first choice in order to modify and tune its biological properties (*1-3*). Thus, for example, a fluorine atom has been used with great success as a replacement for a hydrogen atom or a hydroxy group, a CF_2 has been used as a mimic for an oxygen atom, the CF_3 as an isostere of methyl, isopropyl, phenyl (*3b*), and so on.

In the realm of peptides, incorporation of tailored fluorinated functions has been used to replace and/or mimic critical peptide bonds. For example, a fluorovinyl group has been demonstrated to be a hydrolytically stable replacement for a peptide bond and a much better electronic mimic than the unfluorinated vinyl function (*4*). Another important application consists in the incorporation of a CF_3CO or $COCF_2CO$ function in a key backbone position in order to achieve inhibition of proteolytic enzymes, in particular serine proteases, thanks to the ability of the fluoroalkyl group to stabilize the *gem*-diolic form, that is able to mimic the hydrolitic transition state (*5,6*). Recently, fluorinated amino-acids (trifluoroleucine, trifluorovaline, trifluoromethionine, etc.) have been incorporated into protein structures, resulting in a deep modification of properties such as enhanced thermal and chemical stability, affinity for lipid bilayer membranes, stronger self-association, and so on (*7-10*). In another interesting application, 4-fluoroproline has been demonstrated to bring about peculiar conformational features to collagen-like structures, depending on the stereochemistry of the fluorinated stereogenic center (*11*).

However, very little is known about the synthesis and the properties of backbone-modified peptides with fluoroalkyl functions incorporated in critical positions. With this in mind, about five years ago we undertook a study aimed at a better understanding and rationalization of the "fluorine-effect" in peptidomimetic structures.

Trifluoro-analogs of Peptidomimetic Metalloproteinase Inhibitors

Matrix metalloproteinases (MMPs) are zinc (II)-dependent proteolytic enzymes involved in the degradation of the extracellular matrix (*12-15*). More than 25 human MMPs have been identified so far. Loss in the regulation of their activity can result in the pathological destruction of connective tissue, a process associated with a number of severe diseases, such as cancer and arthritis. The inhibition of various MMPs has been envisaged as a strategy for the therapeutic intervention against such pathologies. To date, however, a number of drawbacks have hampered the successful exploitation of MMPs as pharmacological targets. In particular, the toxicity demonstrated by many MMPs' inhibitors in clinical trials has been ascribed to nonspecific inhibition.

Recently, Jacobson and coworkers described a new family of potent peptidomimetic hydroxamate inhibitors **A** (Fig. 1) of MMP-1, -3 and –9, bearing a quaternary α-methyl-alcoholic moiety at P1 position, and several different R^1 groups at P1' (*16*). Interestingly, the other stereoisomers, including the epimers at the quaternary carbinol function, showed much lower activity, as the authors demonstrated that the hydroxamic binding function was moved away from the catalytic Zn^{2+} center.

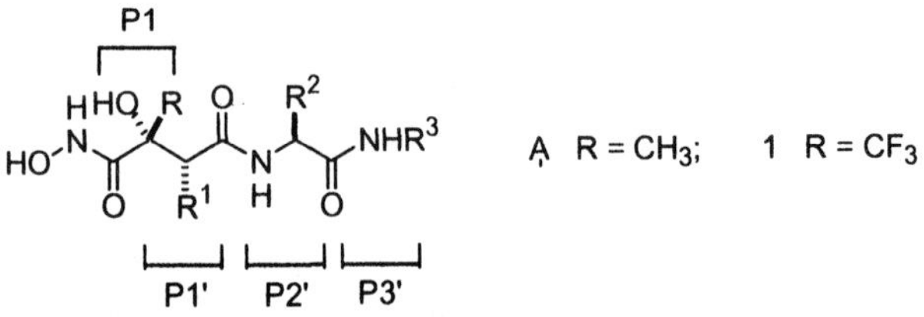

P1
HHO R O R2
HO N N NHR3
O R1 H O
P1' P2' P3'

A R = CH_3; 1 R = CF_3

Figure 1. The DuPont Merck's MMP inhibitors (A) and their CF_3-analogues (1).

We hypothesized that incorporation of fluoroalkyl substituents on the P1 quaternary position of the inhibitors **A** could be an effective strategy for changing and tuning their binding properties. In fact, albeit the α-CF_3 group was expected to diminish to some extent the chelating properties of the hydroxamate function, this could lead to increased selectivity toward the different MMPs. We therefore accomplished the synthesis of the Tfm-analogs **1** (Fig. 1) of **A**, and studied the effect of the replacement of the α-CH_3 group with a CF_3 on the inhibition of MMPs (*17*).

We decided to concentrate our efforts on the substrates **1** having R^1 = $(CH_2)_3Ph$, since their analogs **A** were reported to be very active. First attempts to synthesize the α-Tfm-malic unit of **1** *via* titanium (IV) catalyzed aldol reaction of trifluoropyruvic esters with enantiopure *N*-acyl oxazolidin-2-ones gave disappointing results (*18*). Although this reaction was *per se* satisfactory (61-87% yields, d.r. up to 8:1 depending on the *N*-acyl group), the subsequent exocyclic cleavage of the oxazolidin-2-one auxiliary could not be performed, despite intensive efforts. We therefore turned our attention to oxazolidin-2-thiones (*19*), whose cleavage was reported to occur much more smoothly (*20*).

The $TiCl_4$ catalyzed reaction of the N-acyl-oxazolidin-2-thione **2** (Figure 2) with ethyl trifluoropyruvate **3** afforded the two diastereomeric adducts **4** and **5**, out of four possible, in low diastereomeric ratio. The reaction featured a favourable scale-up effect, affording ca. 70% yield on a hundreds-milligrams scale, and 90% on a ten-grams scale. A number of alternative conditions were explored, but neither significant improvement nor switch of the diastereocontrol could be achieved.

The synthesis of the major diastereomer derivatives **11a-c** (Fig. 3) was developed first. Under the standard conditions reported in the literature (BnOH, cat. DMAP, DCM, r.t.) the exocyclic cleavage of the oxazolidinone on **4** was very slow, affording modest conversion to the corresponding Bn-ester and partial epimerization of the secondary stereocenter.

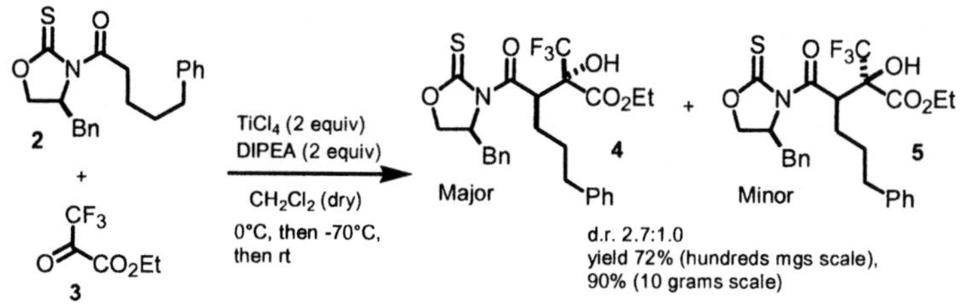

Figure 2. Stereocontrolled synthesis of the Tfm-malic framework.

However, we found that solid K_2CO_3 in moist dioxane (rt, 10-12 h), was able to produce directly the key carboxylic acid intermediate **6** in satisfactory yields and with very low α-epimerization (2%).

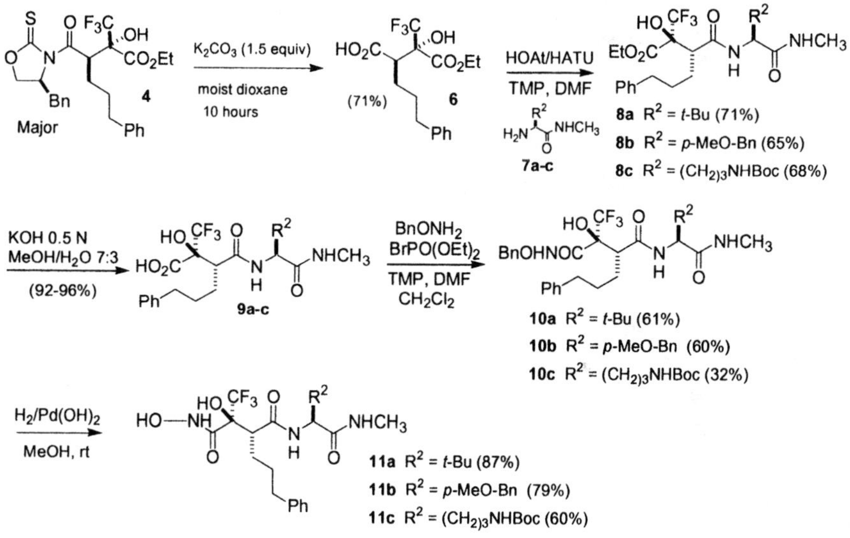

Figure 3. Total synthesis of the hydroxamates derived from the major diastereomer.

Coupling of the acid **6** with α-amino acid amides **7a-c** was achieved in good yields with the HOAt/HATU system (*21*). The resulting peptidomimetic esters **8a-c** were submitted to saponification, affording the acids **9a-c** in high yields.

The subsequent coupling of **9a-c** with *O*-Bn hydroxylamine proved to be extremely challenging, owing to the low reactivity and high steric hindrance of the carboxylic group bound to the quaternary α-Tfm carbinolic center. A number of "conventional" coupling agents for peptides were unsuccessfully tested, but finally we found that freshly prepared $BrPO(OEt)_2$ was able to promote the coupling in reasonable yields (32-61%). With **10a-c** in hand we addressed the final O-Bn cleavage by hydrogenolysis that provided the targeted hydroxamates **11a-c** in good yields.

Since **11a-c** are the "wrong" diastereomers with respect to **A**, we deemed it necessary to synthesize at least one analog having the correct stereochemistry, in order to have a complete set of biological data on the effect of the introduction of the Tfm group. However, a tailored synthetic protocol had to be developed *ex-novo*, because the minor diastereomer **5** (Fig. 4) featured a dramatically different reactivity in the key steps of the synthesis. Since we noticed that the coupling of **5** and **7a** with HATU/HOAt gave rise to relevant amounts of the β-lactone **13** (which had to be processed separately) besides the expected coupling product **14**, we decided to prepare first the intermediate **13**.

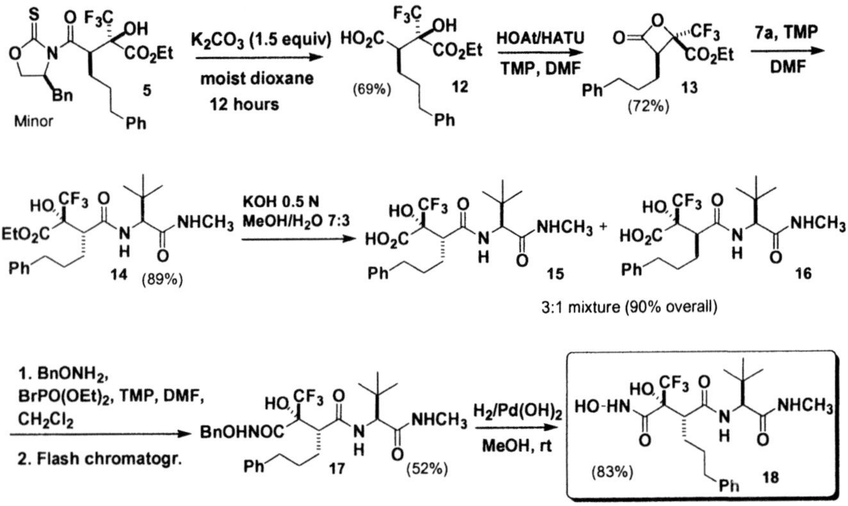

Figure 4. Total synthesis of the hydroxamate derived from the minor diastereomer.

The latter was reacted with free **7a**, affording the desired molecule **14** in high yields. Basic hydrolysis of the ester **14** occurred effectively, although a partial epimerization of the $[Ph(CH_2)_3]$-stereocenter occurred, affording a 3:1 mixture of diastereomers **15** and **16** which were subjected together to coupling with $BnONH_2$. The resulting diastereomeric O-Bn hydroxamates could be separated by FC which afforded pure **17** that was hydrogenated to the target free hydroxamate **18**.

The hydroxamates **11a-c** and **18** were tested for their ability to inhibit MMP-2, MMP-3 and MMP-9 activity using zymographic analysis. The IC_{50} values (μM) portrayed in Table 1 show that diastereomers **11a-c** displayed low inhibitory activity, in line with the parent CH_3-compounds. Disappointingly, **18** showed a much lower activity than the exact CH_3-analog **A** that was reported to be a low nanomolar inhibitor of MMP-3 and -9 (K_i = 13 nM toward MMP-3 and < 1 nM toward MMP-9). It is also worth noting that **11a** and **18** showed little selectivity, whereas **11b** and **11c** showed a fairly better affinity for MMP-9, in comparison with MMP-2 and -3. A possible explanation for the dramatic drop of activity upon replacement of the quaternary methyl with a CF_3, is that **18**, biased by the bulky CF_3-group, is unable to assume the crucial binding conformation of the CH_3-analog **A**. Alternatively, one can hypothesize that the bulky and highly electron-rich CF_3 group is unable to fit the S1 pocket of the hitherto tested MMPs.

Table 1. IC_{50} (μM) of the target Tfm-hydroxamates.

Compound	*MMP-2*	*MMP-3*	*MMP-9*
11a	156	> 1000	121
11b	407	> 1000	84
11c	722	> 1000	23
18	23	43	15

Current work is actively pursuing the synthesis of novel fluorinated analogs of MMPs inhibitors having improved pharmacological properties.

Partially-Modified Retro and Retro-inverso ψ[NHCH(CF_3)]Gly-Peptides

The need of drug-like molecules retaining both activity and potency of the parent peptides, while being much more stable and orally active (*22*), has been a major driving-force for the development of a variety of peptide mimics. Two very attractive ways of generating bio-active peptide mimics with improved bio-stability are: (a) to replace a peptide bond with a surrogate unit X, which is usually symbolized as ψ(X) (*23-25*); (b) to reverse all or some of the peptide bonds (NH-CO instead of CO-NH) giving rise to the so called retro- or partially-modified-retro (PMR) peptides, respectively (*26-28*). When the stereochemistry of one or more amino acids of the reversed segment is inverted, the resulting pseudo-peptide is termed as retro-inverso. A malonic unit is classically incorporated to provide partially-modified retro-peptides, while the direction can be restored incorporating an additional *gem*-diaminoalkyl unit. Recently, we proposed a new strategy for generating peptidomimetic sequences, based on the idea of combining both the "surrogate-unit" and the "direction-reversal" strategies in a novel class of pseudo-peptides having a ψ[NHCH(CF_3)] instead of the ψ(NHCO) unit featured by retro-peptides (Fig. 5) (*29,30*). A first intuitive consequence of this structural modification is that the symmetry of conventional malonyl PM-retropeptides is broken, resulting in the generation of two families of regioisomeric PMR-ψ[NHCH(CF_3)] and -ψ[CH(CF_3)NH]-peptides, each one existing in two epimeric forms at the CF_3-substituted carbon.

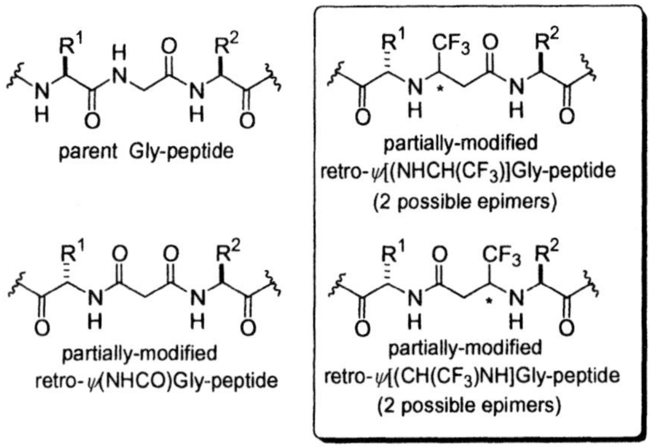

Figure 5. Different isomers of ψ*[NHCH(CF_3)] Gly- and* ψ*[CH(CF_3)NH] Gly-retropeptides.*

In this way a great diversity of structurally related, potentially bio-active molecules can be created. Further important biological and physico-chemical features are expected to arise from the presence of the [CH(CF_3)NH] group. For example, the ψ[NHCH(CF_3)]-surrogate should be very stable toward proteolytic degradation. Furthermore, the stereo-electronically demanding CF_3 group is expected to constraint the molecule, limiting the number of accessible conformational states and leading to peptidomimetics displaying well-defined conformational motifs. Last but not least, it should modify the binding properties of the parent peptide, changing the hydrogen bonding or coordinative features of the ligand in the putative receptor sites.

Our synthetic approach to the ψ[NHCH(CF_3)]-containing dipeptide units involved an aza-Michael reaction of a series of structurally varied L-α-amino acid esters **19** (Fig. 6) with enantiomerically and geometrically pure Michael acceptors (*S*)-(*E*)-**20**.

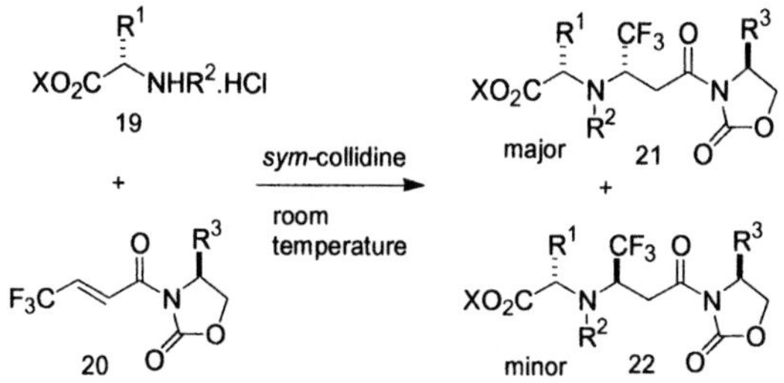

*Figure 6. Aza-Michael reaction with amino-ester nucleophiles **19**.*

Diastereomeric pseudo-dipeptides **21** (major) and **22** (minor) were formed in excellent yields at room temperature in 16-88 hours, under kinetic control. The solvent has a strong influence on stereoselectivity, and the best results were obtained in DCM (50% d.e.), while a substantial drop of d.e. was observed with more polar solvents such as ethanol, acetonitrile, THF, DMF or mixtures. The base also plays an active role, as demonstrated by the fact that the use of DABCO, instead of TMP, accelerates the reaction but slightly lowers the d.e. of the products. In the absence of a base, namely pre-generating the free α-amino-ester from the hydrochloride by treatment with $NaHCO_3$, the diastereoselectivity was similar to that achieved with TMP. In the light of these results, DCM and TMP were used as, respectively, the solvent and the base of choice for the preparation of a number of pseudo-dipeptides **21**. We could also demonstrate that the facial diastereoselectivity of these reactions is mainly controlled by the nucleophiles **19** rather than by the acceptors **20**. The degree of diastereoselectivity, followed the trend R^1 = *iso*-Pr > *iso*-Bu > Me > Bn > H (d.e. up to 78%). Modest d.e. was obtained with the cyclic α-amino ester L-Pro-OBn, but the reaction occurred also in this case with very good yield. On the other hand, the R^3 substituent on the oxazolidinone stereocenter had a lower effect on the stereoselectivity. No meaningful effect was exerted by the X group of the nucleophile. These results can be rationalized if one considers that in the absence of chelating agents *N*-(*E*)-enoyl-oxazolidin-2-ones **20** exist in *transoid* conformation (as portrayed in Fig. 6 and 7), with the R^3 substituent pointed away from the C=C bond, thus exerting little control of the facial selectivity. In contrast, the R^1 side-chain of α-amino esters **19** should be spatially close to the forming stereogenic center in the transition state, therefore its influence is much more important. Chelation with Lewis acids for biasing the Tfm-oxazolidin-2-ones **20** in *cisoid* conformation was tried with little success, probably due to their poor basic character.

The method is viable for preparing longer PMR and PMRI ψ[NHCH(CF_3)]Gly-peptidyl-oxazolidinones by using unprotected *N*-terminal peptides as nucleophiles in the aza-Michael reactions with **20** that occur with excellent yields and good stereocontrol.

The chemoselective cleavage of the oxazolidinone auxiliary was achieved in 55-82% yields upon treatment of the oxazolidinone pseudopeptides **21, 22** and **25** (Fig. 7) with LiOH/H_2O_2 (*31*). The resulting pseudo-peptides having a terminal CO_2H group were coupled with another α-amino ester. The final diastereomeric PMR tripeptides **23, 24** and PMRI tripeptides **26**, orthogonally protected at the carboxy end-groups, and therefore suitable for further selective elongation, were obtained in quantitative yields, often as solid materials.

A very interesting feature of PMR ψ[NHCH(CF_3)]Gly-peptides like **23** is their weakly basic nature, which is due to the presence of the strongly electron-withdrawing CF_3 in the α-position to the amino group. These compounds do not form stable salts with trifluoroacetic acid or 2N hydrochloric acid. One can therefore say that, in terms of basicity, the [NHCH(CF_3)] moiety resembles more closely a retropeptide unit [NHCO] than a conventional secondary amine moiety.

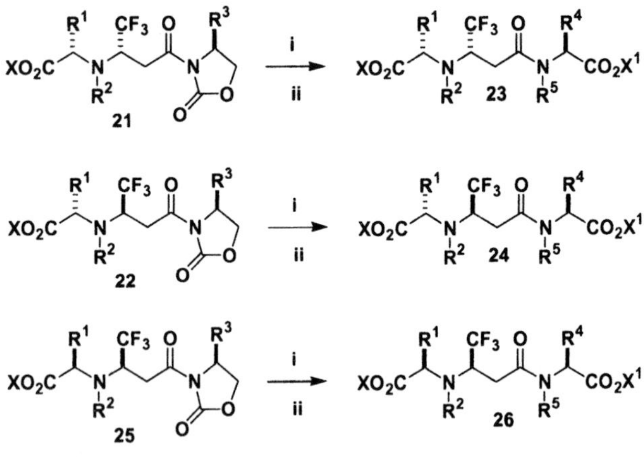

Figure 7. i) LiOH, H_2O_2; ii) HATU/HOAt, sym-*collidine, DMF, α-amino ester.*

The parallel solid-phase synthesis of small libraries of PMR-Ψ[NHCH(CF_3)]Gly tri-, tetra-, and pentapeptides, which in perspective should be applicable in the preparation of wider libraries of PMR-Ψ[NHCH(CF_3)] polypeptides for high-throughput biological screening, was also developed (*32,33*). As an example, in Fig. 8 the synthesis of PMR-Ψ[NHCH(CF_3)]Gly pentapeptides is described. The tripeptide resin H-Val-Val-Ala-Owang **27** was prepared by Fmoc-chemistry and subjected to the conjugate addition with **20a**, taking place very effectively in 3 days at r.t. A 10:1 mixture of diastereoisomers **28** was formed. In general these aza-Michael reactions in solid-phase occurred with diastereocontrols comparable with those in solution (see above). Interestingly, in the case of tripeptides as polymer supported nucleophiles, we observed the highest diastereoselection as compared with the conjugate additions of H-Val-OWang and H-Val-Gly-OWang (7:1 and 4.5:1 mixtures of diastereoisomers, respectively). This shows that additional stereocenters, even in remote positions of the nucleophile, can have a strong influence on the stereochemical outcome of the conjugate additions.

The resin **28** was chemoselectively hydrolysed at the C-terminus by treatment with lithium hydroperoxide generated *in situ*. The resulting pseudotetrapeptide resin **29** was coupled (HOAt/DIC) with different α-amino acid esters generating, after release from the resin, the PMR-Ψ[NHCH(CF_3)]Gly pentapeptides **30** with very good overall yields and purity.

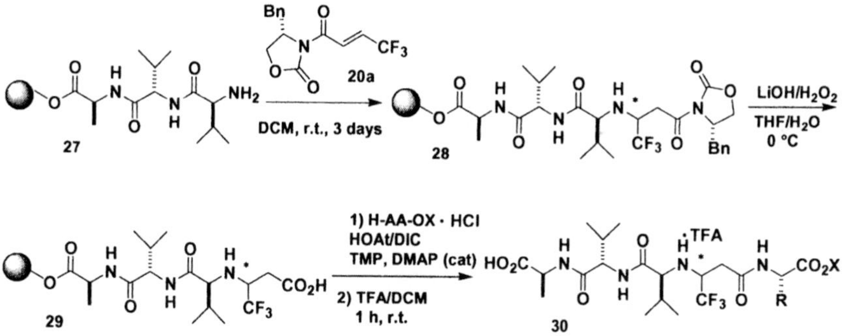

Figure 8. Solid-phase synthesis of PMR ψ[NHCH(CF_3)] Gly-peptides.

This method was also adapted to the solution and solid-phase synthesis of PMR- and PMRI-ψ[NHCH(CF_3)]-Gly peptidyl hydroxamates (*34,35*), which are of interest since the HONHCO– end group is very effective in coordinating the Zn^{2+} cofactor of MMP (see above). In the solid-phase approach, the hydroxylamine resin **31** (Fig. 9) was prepared in two steps from commercial

Wang resin, according to the method of Floyd (*36*), and coupled to an excess of L-Fmoc-Ala to give the protected alanine polymer **32**, from which the Fmoc group was cleaved with 20% piperidine in DMF. The resulting resin-bound α-amino hydroxamate was submitted to 1,4-conjugate addition with the achiral oxazolidin-2-one **33**. As a general trend, low diastereoselectivity was observed in these aza-Michael reactions involving hydroxamates as nucleophiles. Thus, in analogy with the synthesis in solution, **34** were formed as nearly equimolar mixtures of epimers at the Tfm-substituted centre. Treatment of **34** with lithium hydroperoxide cleaved the oxazolidin-2-one with excellent chemoselectivity. Coupling of **35** to α-amino esters **36** afforded the tripeptidyl resins **37**, from which the retro- and retro-inverso hydroxamates **38** were released in good yields and purity upon treatment with TFA.

Tetrapeptidyl hydroxamates and a tripeptidyl hydroxamate having a methylamide terminus, which is often encountered in MMPs inhibitors, were prepared as well by the same method, thus demonstrating its wide scope.

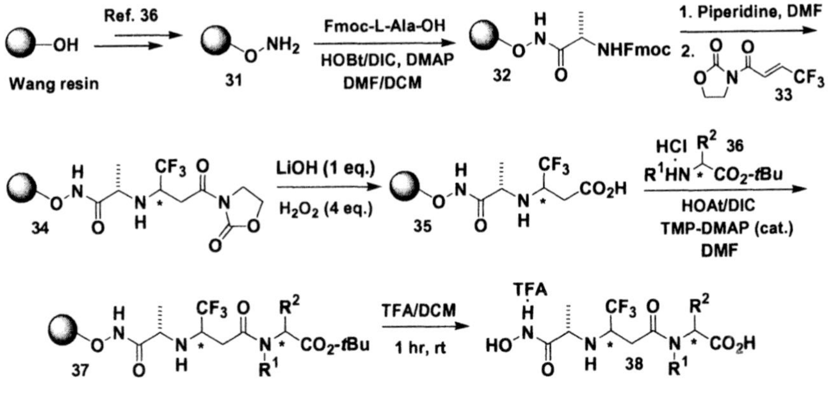

Figure 9. Solid-phase synthesis of PMR ψ[NHCH(CF_3)]Gly-peptidyl hydroxamates.

The conformational features of PMR *ψ*[NHCH(CF_3)]Gly-peptides were studied in detail. It has been shown that very simple PMR-peptides, such as the triamide **39** (Fig. 10), adopt turn-like nine-membered folding patterns with an intramolecular N-H···O=C hydrogen bond in the solid state, as well as in low-polarity solvent solutions (*37*). Racemic *ψ*[NHCH(CF_3)]-diamide **40** (Fig. 10) was synthesized in good overall yield by means of the standard protocol (see Fig. 6 and 7). ^{1}H NMR spectroscopy supported by MD calculations showed the stability of turn-like conformations for **40** in a weakly hydrogen bonding solvent, such as $CDCl_3$, which are comparable to that of parent malonyl-based retropeptides **39**. Similar turn-like conformations were found in the solid-state, as demonstrated by X-ray diffraction studies of several ψ[NHCH(CF_3)]Gly-peptides, such as **41**.

This very interesting conformational behaviour is a likely consequence of two main factors: (1) severe torsional restrictions about sp^3 bonds in the [CO–CH_2–CH(CF_3)–NH–CH(R)–CO] module, which is biased by the stereoelectronically demanding CF_3 group and the R side-chain; (2) formation of nine-membered intramolecularly hydrogen-bonded rings, which have been clearly detected both in $CHCl_3$ solution and in some crystal structures.

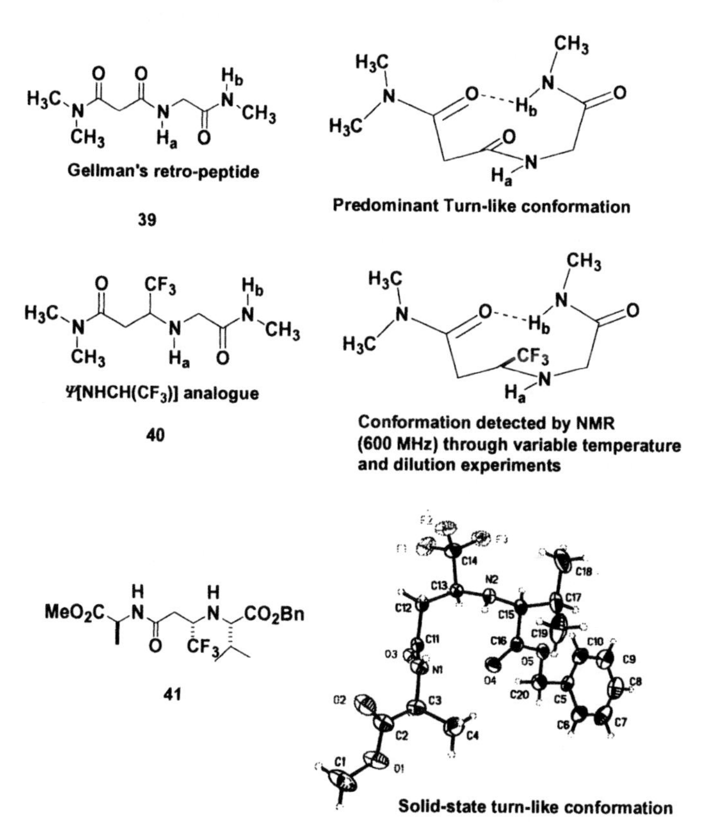

Figure 10. Solution and solid-state conformation of some PMR ψ[NHCH(CF_3)]Gly-peptides.

The first factor seems to be more important, as turn-like conformations were found in the solid-state even in the absence of intramolecular hydrogen bonding. The relative configuration of the –C*H(CF_3)NHC*H(R)– stereogenic centers was found to have a major effect on the stability of the turn-like conformation, which seems to require a *syn* stereochemistry, such as **23,41**, whereas the diastereomers **24** investigated so far did not show the same conformational properties.

This could represent a new general concept for the rational design of linear peptidomimetics incorporating a turn-like secondary structure.

X-Ray diffraction and *ab initio* computational studies showed that the [–CH(CF_3)NH–] group can be seen as a sort of hybrid between a peptide bond mimic and a proteolytic transition state analog, as it combines some of the properties of a peptidyl -CONH- group (low NH basicity, CH(CF_3)–NH–CH backbone angle close to 120 degrees, C–CF_3 bond substantially isopolar with the C=O) with some others of the tetrahedral intermediate [–C(OX)(O^-)NH–] involved in the protease-mediated hydrolysis reaction of a peptide bond (high electron density on the CF_3 group, tetrahedral backbone carbon).

ψ[CH(CF$_3$)NH]Gly-Peptides

A significant advancement in the development of backbone-modified peptidomimetics having a [CH(CF_3)NH] module as a replacement of a peptide bond [CONH], is represented by ψ[CH(CF_3)NH]Gly-Peptides **42** (Fig. 11) (*38*), which are much closer to natural peptides than PMR-ψ[NHCH(CF_3)]Gly-peptides.

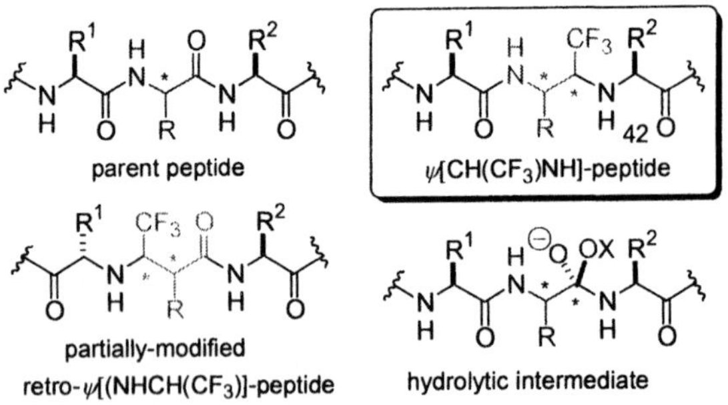

Figure 11. Structure of natural and ψ[CH(CF$_3$)NH]Gly-Peptides.

In analogy with the latter compounds, the [CH(CF_3)NH] unit is expected to behave as a sort of hybrid between a [CONH] mimic and a proteolytic transition-state analogue.

The stereocontrolled synthesis of these brand new peptidomimetics is based on another key aza-Michael reaction involving 3,3,3-trifluoro-1-nitropropene **43** (Fig. 12) and an array of α-amino esters, generated *in situ* from the hydrochlorides **44** with a base. The reactions took place almost instantaneously at r.t., affording the diastereomeric α'-Tfm-β'-nitro α-amino esters **45** (major) and **46** (minor) under kinetic control.

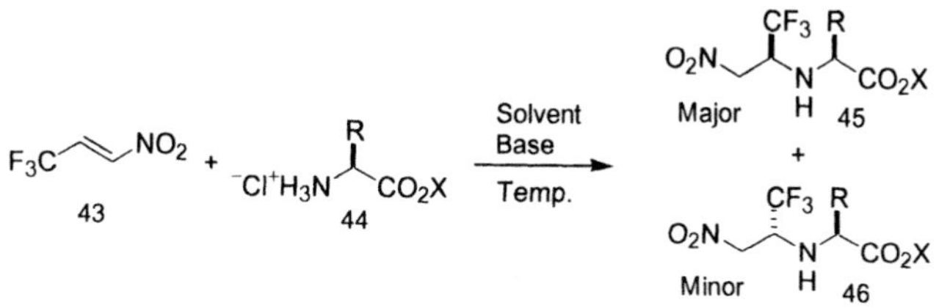

Figure 12. The aza-Michael reaction to give the ψ[NHCH(CF_3)]Gly-peptide backbone.

The diastereoselectivity of the process was studied in detail. We found that it depends mainly on four reaction parameters: (1) base, (2) solvent, (3) stoichiometry of the base, (4) R side chain of **44**. Concerning the base, the best stereocontrol (63% de using L-Val esters as nucleophile) was achieved with DIPEA, whereas $NaHCO_3$, TMP and DABCO gave modest results. As observed for other aza-Michael reactions (see above), low-polarity solvents provided remarkably higher diastereocontrol. Thus, toluene afforded 84% d.e., whereas DCM and THF afforded modest d.e.s. Quite surprisingly, intermediate results were observed using apolar CCl_4. Even more surprisingly, also the stoichiometry of DIPEA was found to have a profound effect on the stereocontrol. The optimum amount was found to be 1.1 equiv. (as used in the experiments cited above). In the absence of free DIPEA the d.e. dropped dramatically. Accordingly, a progressive decrease of stereoselectivity was observed by increasing the amount of DIPEA from 1.1 to 1.7 equiv., whereas little variation occurred beyond this quantity. Other bases, such as TMP and $NaHCO_3$, did not feature the same "stoichiometry-effect" affording comparable d.e.s upon changing the number of equivalents used. The effect of the R side chain of **44** was in line with the expectations. In fact, the highest d.e.s were observed with bulky R groups (iso-Pr, sec-Bu) whereas lower degrees of stereocontrol where observed with R = Me, Bn, etc.

Room temperature was found to be essential in order to achieve high yields of **45** and **46**, whereas surprisingly at lower temperatures (for example –40 or –70°C) very complex mixtures of products were obtained. All the experimental evidence above suggests that these aza-Michael reactions occur through a tight, polar, termolecular transition state (TS), involving **43**, **44** and DIPEA, which appears to play a fundamental catalytic role. Polar solvents, as well as the presence of more than one molecule of DIPEA, may disrupt this TS, thus lowering the stereocontrol.

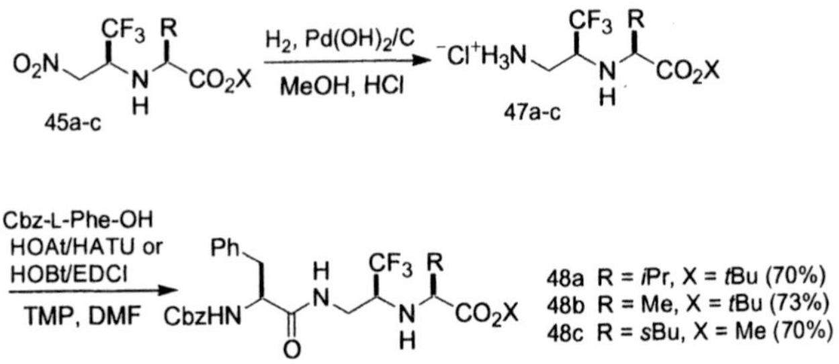

45a-c
H_2, $Pd(OH)_2/C$
MeOH, HCl
47a-c
Cbz-L-Phe-OH
HOAt/HATU or
HOBt/EDCI
TMP, DMF
48a R = *i*Pr, X = *t*Bu (70%)
48b R = Me, X = *t*Bu (73%)
48c R = *s*Bu, X = Me (70%)

*Figure 13. Elaboration of the aza-Michael adducts **45** into ψ[CH(CF_3)NH]Gly-Peptides **48**.*

Elaboration of the major adducts **45a-c** into the target ψ[CH(CF_3)NH]Gly-peptides **48a-c** is shown in Fig. 13. The nitro group of **45a-c** was hydrogenated to amino group using the Pearlman's catalyst, and the resulting diamino compounds were trapped as hydrochlorides **47a-c** and submitted without purification to coupling with Cbz-L-Phe-OH affording the ψ[CH(CF_3)NH]Gly-tripeptides **48a-c** in good overall yields.

In order to proof the value of this synthetic methodology in the preparation of more complex ψ[CH(CF_3)NH]Gly-peptides, we have recently completed the synthesis of all of the four diastereomers of tetrapeptide **49** (Fig. 14) (*39*). The latter is an interesting model for conformational studies, because some of the analogues incorporating a natural Gly have been shown to assume very stable and highly populated β-hairpin conformations, depending on the configuration of the Pro residue (*40*). A detailed study of the conformational features of the stereoisomers **49** is currently in progress.

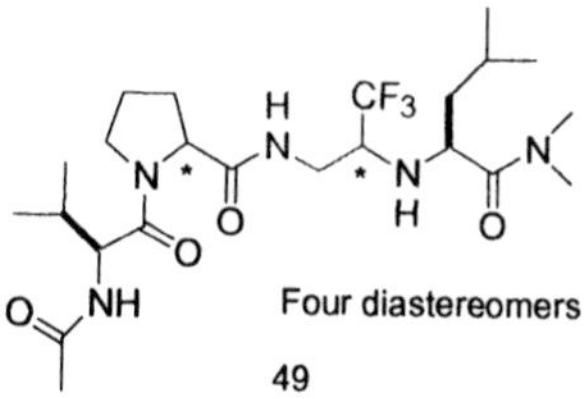

Four diastereomers

49

Figure 14. Complex model ψ[CH(CF₃)NH]Gly-tetrapeptides.

Partially Modified Retro-Peptide Mimics Incorporating a 3,3,3-Trifluoroalanine (α-Trifluoromethyl Glycine) Surrogate

Incorporation of 3,3,3-trifluoroalanine (TF-Ala), also called α-trifluoromethyl-glycine, into a peptide sequence **B** (Fig. 15) is a considerably challenging endeavour, owing to its low chemical and configurational stability at pH > 6 (*41*). On the other hand, no mimics of TF-Ala-containing peptides had been described in the literature. For this reason, we undertook the synthesis of PMR ψ[$NHCH_2$]-peptide mimics **50** incorporating a chemically stable and stereo-defined [$CH_2CH(CF_3)CO$] surrogate of TF-Ala (*42*).

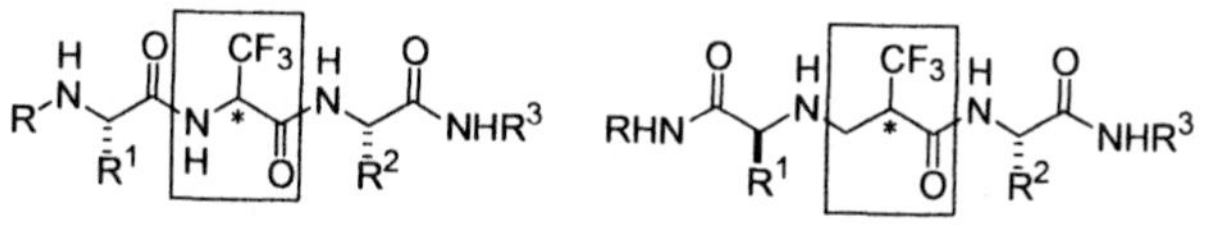

B

50

*Figure 15. A peptide incorporating Trifluoroalanine (**B**) and a PMR-ψ[$NHCH_2$]-peptide (**50**) incorporating a Trifluoroalanine mimic.*

Retrosynthetic analysis of the problem suggested that asymmetric aza-Michael addition of α-amino esters to *N*-(α-trifluoromethyl)acryloyl-α-amino esters could represent a viable entry to the the target structures **50**. Michael acceptors **51** (Fig. 16) (obtained upon reacting the appropriate α-amino ester H-AA-OX^1 with α-Tfm-acryloyl chloride) were reacted with α-amino-esters (generated *in situ* from **44** with a tertiary amine) affording an array of PMR-tripeptides **53** in 75-98% yields. Also in this case the diastereoselectivity of the process was studied in detail. We found that it depends mainly on four reaction parameters: (1) base, (2) solvent, (3) R and R^1 side chains of **44** and **51**, respectively, (4) the relative stereochemistry of **44** and **51**. Concerning the base, the best diastereocontrol was achieved with DABCO, whereas lower d.e.s were achieved with TMA, $NaHCO_3$, DIPEA, TEA, DMAP, TMP (in the order). Interestingly, chiral bases such as quinidine and cinchonine gave the worst diastereocontrols. The solvent is another key factor in this reaction, with low-polarity or apolar solvents providing much higher diastereocontrol. Thus, the best stereocontrol was achieved with CCl_4, followed in the order by toluene, THF, CH_2Cl_2 and acetonitrile.

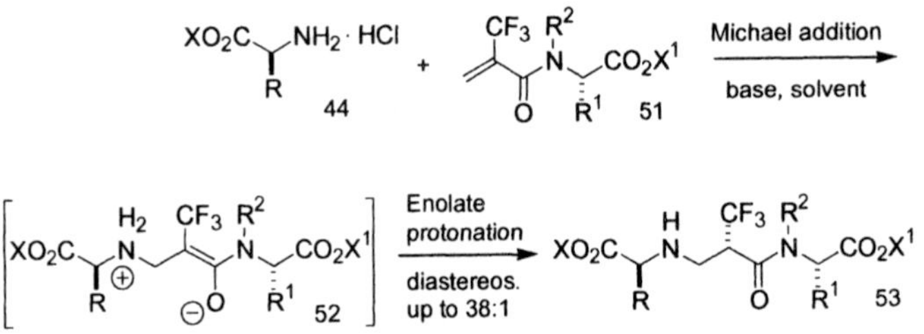

*Figure 16. Tandem reaction producing PMR-ψ[NHCH$_2$]-tripeptides **53**.*

The R side-chain of **44** was proven to have a strong influence: the de of the products **53** increased in the sense R = H < *i*Bu < Bn < Me < *s*Bu < *i*Pr, therefore the best stereocontrols were achieved with the bulkiest R groups. Also the R^1 side-chain belonging to the acceptors **51** had a strong influence, and its effect on the de substantially followed the same trend observed for R. The configuration of the reaction partners **44** and **51** had also a profound effect, as demonstrated by the use of matched/mismatched pairs of reactants. In general, the *like* combination (D/D or L/L) provided remarkably higher stereoselectivity. In the best case, when L-Val esters **44** were reacted with L-Val derived acceptors **51** in CCl_4, with DABCO as base, a d.e. as high as 95% was achieved.

Although detailed kinetic and computational studies will be necessary in order to draw a reliable mechanistic picture of this process, the body of experimental evidence discussed above suggests that amino-ester **44**, Michael acceptor **51** and DABCO might form a tight termolecular ion-pair TS, which is energetically favored and not disrupted in non-polar solvents, during the critical stereogenic intramolecular proton transfer from the intermediate zwitterion **52** (Fig. 16) to the final product **53**.

An efficient stereocontrolled synthesis of PMR $\psi[NHCH_2]$-tripeptide mimics was developed on solid-phase as well (*43*). *N*-Fmoc α-amino acids loaded on Wang resin **54** (Fig. 17) were deprotected to **55** upon treatment with piperidine. Next, the resins **55** were reacted with an excess of 2-trifluoromethyl-propenoyl chloride. This process provided the trifluoromethylated resins **56**, functionalized as chiral Michael acceptors. The crucial aza-Michael reactions were performed by addition of 3 equiv. of the appropriate α-amino ester to a suspension of resin **56** in the appropriate solvent, in the presence of 6 equiv. of base, producing the desired resins **57** in a very effective manner. Release of the PMR peptides **58** from the solid support was achieved upon treatment of **57** with TFA in DCM. The target compounds **58** were invariably obtained with good to excellent chemical purity. In general, one can say that the main features of the solution-phase process were retained in the solid-phase version. In fact, the stereocontrol could be strongly improved (up to 15:1) by using apolar solvents like carbon tetrachloride and DABCO as base. The main difference was that the solid-phase reactions were less diastereoselective than those in solution. This could be due to the fact that the polymeric support biases the reaction partners in a different transition state with respect to that assumed in the highly stereoselective solution-phase process. However, this drawback is counterbalanced by the much greater potential of this solid-phase version for a fast, automated generation of large arrays of PMR- $\psi[NHCH_2]$-peptides **58** for high throughput assays, and for the synthesis of polypeptide mimetics as well.

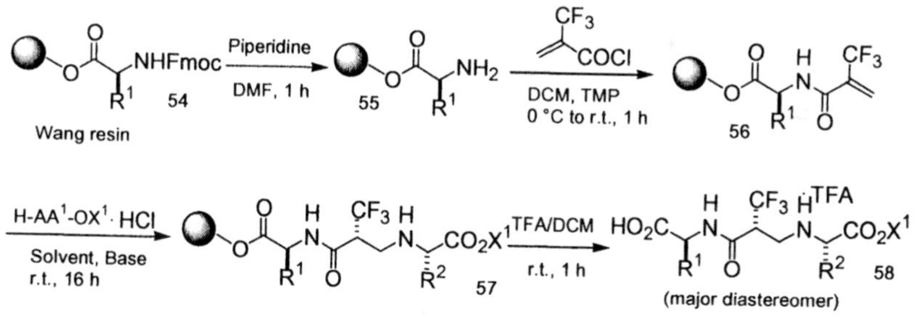

NHFmoc Piperidine DMF, 1 h 54 Wang resin 55 NH_2 CF_3 COCl DCM, TMP 0 °C to r.t., 1 h 56 CF_3

$H\text{-}AA^1\text{-}OX^1 \cdot HCl$ Solvent, Base r.t., 16 h 57 CO_2X^1 TFA/DCM r.t., 1 h HO_2C CO_2X^1 TFA 58 (major diastereomer)

Figure 17. Solid-phase synthesis of PMR ψ[NHCH₂]-peptides 72 incorporating a 3,3,3-trifluoroalanine mimic.

Conclusions

In conclusion, we have shown that a wide range of enantiomerically pure Tfm-containing peptides and pseudopeptides can be synthesized in a stereocontrolled manner both in solution and in solid-phase. The work carried out so far is expected to open up the route to further classes and combinatorial libraries of fluorinated peptidomimetics, allowing for a systematic study of their hitherto largely unknown biological, conformational and structural properties, which are likely to be extremely interesting and peculiar owing to the presence of fluorine. Further research will hopefully contribute to shed light on the chemistry and the biology of backbone-modified fluorinated peptidomimetics, which have typically been difficult to address owing to the complexity of their synthesis in stereodefined manner. It is easy to predict that the field of fluorine-containing peptides and mimics will see important and exciting developments in the next future (*44*).

Acknowledgements

Special thanks are due to Dr. Fiorenza Viani (CNR, Milan), Massimo Frigerio (Politecnico di Milano), Dr. Stefano Bellosta (University of Milan, Italy), Gabriele Candiani and Dr. Francoise Pecker (Creteille, France), Dr. Dorina Belotti and Dr. Raffaella Giavazzi (Istituto Mario Negri, Bergamo, Italy) for their contribution to these projects. We are also grateful to the CNR, Politecnico di Milano, MIUR (Cofin 2002, Project "Peptidi Sintetici Bioattivi"), and European Commission (IHP Network grant "FLUOR MMPI" HPRN-CT-2002-00181) for financial support.

References

(1) Ismail, F. M. D. *J. Fluorine Chem.* **2002**, *118*, 27-33.
(2) Smart, B. E. *J. Fluorine Chem.* **2001**, *109*, 3-11.
(3) (a) Ojima, I.; McCarthy, J. R.; Welch, J. T. *Biomedical Frontiers of Fluorine Chemistry*, Eds. ACS Books, American Chemical Society, Washington, D.C., 1996. (b) Ojima, I. *ChemBioChem* **2004**, *5*, 628-635.
(4) See for example: Bartlett, P. A.; Otake, A. *J. Org. Chem.* **1995**, *60*, 3107-3111 and references therein.
(5) For $COCF_3$ terminal peptides, see for example: Garrett, G. S.; McPhail, S. J.; Tornheim, K.; Correa, P. E.; McIver, J. M. *Bioorg. Med. Chem. Lett.* **1999**, *9*, 301-306.

(6) For peptides incorporating a $COCF_2CO$ function, see for example: Eda, M.; Asimori, A.; Akahoshi, F.; Yoshimura, T.; Inoue, Y.; Fukaya, C.; Nakajima, M.; Fukuyama, H.; Imada, T.; Nakamura, N. *Bioorg. Med. Chem. Lett.* **1998**, *8*, 913-918.
(7) Bilgicer, B.; Kumar, K. *Tetrahedron* **2002**, *58*, 4105-4112.
(8) Wang, P.; Tang, Y.; Tirrell, D. A. *J. Am. Chem. Soc.* **2003**, *125*, 6900-6906.
(9) Horng, J.-C.; Raleigh, D. P. *J. Am. Chem. Soc.* **2003**, *125*, 9286-9287.
(10) Duewel, H. S.; Daub, E.; Robinson, V.; Honek, J. F. *Biochemistry* **2001**, *40*, 13167-13176.
(11) Barth, D.; Milbradt, A. G.; Renner, C.; Moroder, L. *Chem. Biochem.* **2004**, *5*, 79-86.
(12) Coussens, L. M.; Fingleton, B.; Matrisian, L. M. *Science* **2002**, *295*, 2387-2392.
(13) Whittaker, M.; Floyd, C. D.; Brown, P.; Gearing, A. J. H. *Chem. Rev.* **1999**, *99*, 2735-2776.
(14) Bode, W.; Huber, R. *Biochim. Biophys. Acta,* 2000, **1477**, 241-252.
(15) Gavazzi, R.; Taraboletti, G. *Crit. Rev. Oncol. Hematol.* **2001**, *37*, 53-60.
(16) (a) Jacobson, I. C.; Reddy, P. G.; Wasserman, Z. R.; Hardman, K. D.; Covington, M. B.; Arner, E. C.; Copeland, R. A.; Decicco, C. P.; Magolda, R. L. *Bioorg. Med. Chem. Lett.* **1998**, *8*, 837-842. (b) For the standard nomenclature of substrate residues and their corresponding binding sites on the enzyme, see: Babine, R. E.; Bender, L. E. *Chem. Rev.* **1997**, *97*, 1359-1472.
(17) Sani, M.; Belotti, D.; Giavazzi, R.; Panzeri, W.; Volontario, A.; Zanda, M. *Tetrahedron Lett.* **2004**, *45*, 1611-1615.
(18) Zucca, C.; Bravo, P.; Malpezzi, L.; Volontario, A.; Zanda, M. *J. Fluorine Chem.* **2002**, *114/2*, 215-223.
(19) Crimmins, M. T.; McDougall, P. J. *Org. Lett.* **2003**, *5*, 591-594 and references therein.
(20) Su, D.-W.; Wang, Y.-C.; Yan, T.-H. *Tetrahedron Lett.* **1999**, *40*, 4197-4198.
(21) Carpino, L. A.; El Faham, A. *J. Org. Chem.* **1995**, *60*, 3561-3564 and references therein.
(22) Loffet, A. *J. Peptide Sci.* **2002**, *8*, 1-7.
(23) Olson, G. L.; Bolin, D. R.; Bonner, M. P.; Bös, M.; Cook, C. M.; Fry, D. C.; Graves, B. J.; Hatada, M.; Hill, D. E.; Kahn, M.; Madison, V. S.; Rusiecki, V. K.; Sarabu, R.; Sepinwall, J.; Vincent, G. P.; Voss, M. E. *J. Med. Chem.* **1993**, *36*, 3039-3049.
(24) Gante, J. *Angew. Chem. Int. Ed. Engl.* **1994**, *33*, 1699-1720.
(25) Leung, D.; Abbenante, G.; Fairlie, D. P. *J. Med. Chem.* **2000**, *43*, 305-341.
(26) Goodman, M.; Chorev, M. *Acc. Chem. Res.* **1979**, *12*, 1-7.
(27) Chorev, M.; Goodman, M. *Acc. Chem. Res.* **1993**, *26*, 266-273.
(28) Fletcher, M. D.; Campbell, M. M. *Chem. Rev.* **1998**, *98*, 763-795.
(29) Volonterio, A.; Bravo, P.; Zanda, M. *Org. Lett.* **2000**, *2*, 1827-1830.

(30) Volonterio, A.; Bellosta, S.; Bravin, F.; Bellucci, M. C.; Bruché, L.; Colombo, G.; Malpezzi, L.; Mazzini, S.; Meille, S. V.; Meli, M.; Ramírez de Arellano, C.; Zanda, M. *Chem. Eur. J.* **2003**, *9*, 4510-4522.
(31) Evans, D. A.; Britton, T. C.; Ellman, J. A.; Dorow, R. L. *J. Am. Chem. Soc.* **1990**, *112*, 4011-4030.
(32) Volonterio, A.; Bravo, P.; Moussier, N.; Zanda, M. *Tetrahedron Lett.* **2000**, *41*, 6517-6521.
(33) Sani, M.; Bravo, P.; Volontario, A.; Zanda, M. *Collect. Czech. Chem. Commun.* **2002**, *67*, 1305-1319.
(34) Volonterio, A.; Bravo, P.; Zanda, M. *Tetrahedron Lett.* **2001**, *42*, 3141-3144.
(35) Volonterio, A.; Bellosta, S.; Bravo, P.; Canavesi, M.; Corradi, E.; Meille, S. V.; Monetti, M.; Moussier, N.; Zanda, M. *Eur. J. Org. Chem.* **2002**, 428-438.
(36) Floyd, C. D.; Lewis, C. N.; Patel, S. R.; Whittaker, M. *Tetrahedron Lett.* **1996**, *37*, 8045-8048.
(37) Dado, G. P.; Gellman, S. H. *J. Am. Chem. Soc.* **1993**, *115*, 4228-4245 and references therein.
(38) Molteni, M.; Volonterio, A.; Zanda, M. *Org. Lett.* **2003**, *5*, 3887-3890.
(39) Molteni, M.; M. Zanda, unpublished results.
(40) Haque, T. S.; Little, J. C.; Gellman, S. H. *J. Am. Chem. Soc.* **1996**, *118*, 6975-6985.
(41) (a) Bordusa, F.; Dahl, C.; Jakubke, H.-D.; Burger, K.; Koksch, B. *Tetrahedron: Asymmetry* **1999**, *10*, 307-313 and references therein. (b) Koksch, B.; Sewald, N.; Jakubke, H.-D.; Burger, K. *Synthesis and incorporation of α-trifluoromethyl-substituted amino acids into peptides.* In: *Biomedical frontiers of fluorine chemistry;* Ojima, I.; McCarthy, J. R.; Welch, J. T., Eds.; ACS Symposium Series 639; American Chemical Society: Washington, DC, 1996; pp. 42-58.
(42) Sani, M.; Bruché, L.; Chiva, G.; Fustero, S.; Piera, J.; Volonterio, A.; Zanda, M. *Angew. Chem. Int. Ed.* **2003**, *42*, 2060-2063.
(43) Volonterio, A.; Fustero, S.; Piera, J.; Chiva, G.; Sanchez Rosello, M.; Sani, M.; Zanda, M. *Tetrahedron Lett.* **2003**, *44*, 7019-7022.
(44) Key to the abbreviations and acronyms used in this paper: DCM = dichloromethane; TMP = *sym*-collidine (2,4,6-trimethylpyridine); DCC = dicyclohexylcarbodiimide; DMAP = 4-(N,N-dimethylamino)pyridine; HATU = O-(7-azabenzotriazol-1-yl)-*N,N,N',N'*-tetramethyluronium hexafluorophosphate; HOAt = 1-hydroxy-7-azabenzotriazole; PyBroP = Bromotripyrrolidinophosphonium hexafluorophosphate; DIPEA = diisopropylethylamine; DMF = N,N-dimethylformamide; DABCO = 1,4-diazabicyclo[2,2,2]octane; TFA = trifluoroacetic acid; TEA = triethylamine; TMA = trimethylamine.

Chapter 34

Fluorinated β-Enamino Esters as Versatile Synthetic Intermediates: Synthesis of Fluorinated β-Amino Acids and Uracils

Santos Fustero*, Juan F. Sanz-Cervera*, Julio Piera, María Sánchez-Roselló, Diego Jiménez, and Gema Chiva

Departamento de Química Orgánica, Universidad de Valencia, Avenida Vicente Andrés Estellés s/n, E–46100 Burjassot, Spain

β-Enamino esters are versatile synthetic intermediates that can be prepared with a diversity of methods. The reactivity of these compounds both in the asymmetric synthesis of fluorinated β-amino acids and in the preparation of heterocyclic systems such as fluorinated uracils and thiouracils has been studied.

Introduction

In Nature, the great majority of molecules, including proteins and nucleic acids as well as most biologically active compounds, contain nitrogen. Therefore, developing new synthetic methods for the construction of nitrogenous molecules has defined the frontiers of organic synthesis since its very beginning. In this context, the past few decades have seen a concerted effort to develop building blocks for this purpose. One such group of building blocks is comprised of the β-enamino esters (*1*,*2*), an important class of molecules which have proven their utility as valuable and versatile intermediates for the synthesis of biologically active compounds such as α and β-amino acids, conformationally restricted peptide analogs, heterocyclic derivatives, and alkaloids, these last generally through aza-annulation processes.

One of the keys to the versatility of these compounds is the various reactions in which they can be prepared. Thus, in addition to the classic condensation reaction between β-keto esters and amines, other routes, including the nucleophilic addition of acid derivative enolates to nitriles or imidoyl halides, the Michael addition of amines to alkynoates, and the reaction of imines with activated carbonic acid derivatives, have all been undertaken with more or less success (*1*, *2*).

In the course of our ongoing study of the synthesis and reactivity of 1,3-difunctionalized derivatives (*3*), we became interested in the development of new strategies for the synthesis of fluorine-containing nitrogen derivatives such as acyclic and cyclic β-amino acids as well as biologically active heterocyclic systems like fluorinated uracils and thiouracils, all starting from a common intermediate, namely fluorinated β-enamino esters (Figure 1).

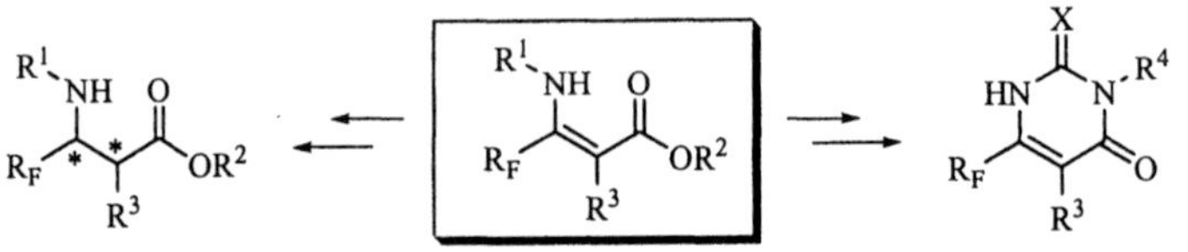

Figure 1. Fluorinated β-enamino esters as starting materials for the preparation of β-amino acids and (thio)uracils.

In contrast to their non-fluorinated counterparts, γ-fluorinated β-enamino esters have received far less attention in the past. A survey of the literature reveals that relatively few methodologies have been developed for the synthesis of these derivatives. Some of the most recent and useful procedures described include:

Synthesis of fluorinated β-enamino esters.

Condensation of fluorinated β-keto esters and amines.

In this context, Soloshonok has very recently described an improved synthesis of fluorinated β-enamino esters **2** and **3** which takes into account the chemo- and regioselectivity in the reactions of highly electrophilic fluorine-containing β-keto esters **1** and aliphatic (*4*) and aromatic amines (*5*) (Figure 2). The process is highly dependent on the reaction conditions and can be applied on a large-scale. These systems are useful key intermediates for the preparation both of enantiomerically pure fluorinated β-amino acids **4** by biomimetic transamination processes (*6*) and of biologically interesting heterocyclic systems such as 2-trifluoromethyl 4-quinolinones **5** by intramolecular cyclizations (*5*).

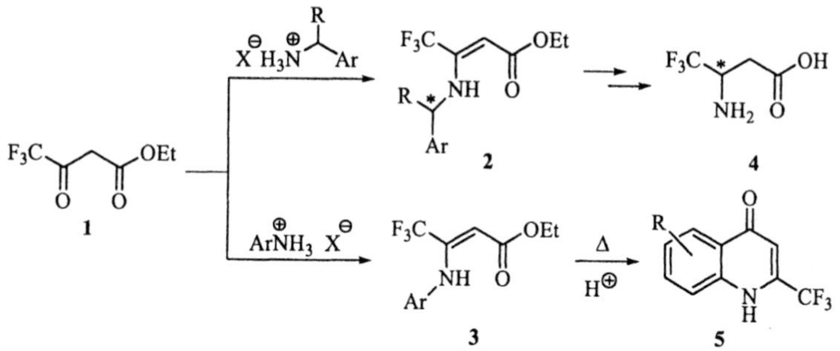

F_3C OEt
O O
1
$X^{\ominus}\ H_3N^{\oplus}$ R Ar
F_3C OEt
R * NH O
Ar
2
F_3C * OH
NH_2 O
4
$Ar\overset{\oplus}{N}H_3\ X^{\ominus}$
F_3C OEt
NH O
Ar
3
Δ
$H^{\oplus}$
O
R
N H CF_3
5

Figure 2. Condensation reaction of fluorinated β-keto esters and amines.

Michael addition of amines to fluorinated alkynoates.

Abarbri *et al.* have recently used this well-known reaction for the preparation of optically active perfluoroalkyl-oxazepin-7-ones **6** by reaction of perfluoro-2-alkynoates **7** with bifunctional heteronucleophiles such as optically active amino alcohols **8** (*7*) (Figure 3). The global two-step process involves an intermolecular Michael addition that generates **9**, followed by lactone formation.

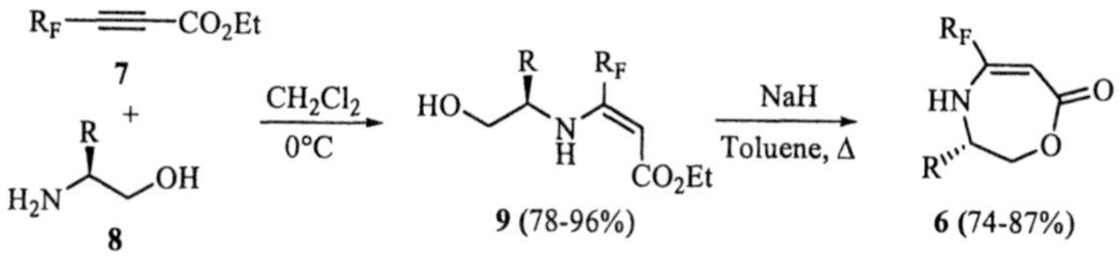

R_F—≡—CO_2Et
7
+
R
H_2N OH
8
CH_2Cl_2
0°C
HO R R_F
N H
CO_2Et
9 (78-96%)
NaH
Toluene, Δ
R_F
HN =O
R O
6 (74-87%)

Figure 3. Michael addition of amines to fluorinated alkynoates.

In a similar fashion, (*E*)- or (*Z*)-perfluoroalkyl β-enamino esters had been previously obtained quantitatively through direct addition of primary or secondary aliphatic amines to ethyl perfluoroalkynoates without the need for a catalyst (*8*).

Wittig reaction of fluorinated amides.

N-Aryl fluorinated β-imino/enamino esters **10** can be obtained easily and in excellent yields by reacting fluorinated *N*-aryl substituted amides **11** with phosphoranes **12** (*9*) (Figure 4). These enamines are precursors of synthetically and biologically important heterocycles such as indole **13** and quinolone **14** derivatives, both of which can be obtained from a common intermediate. Thus, an intramolecular Heck reaction in N_2 atmosphere provided indole derivatives in moderate yields. Quinolones were prepared under similar reaction conditions except that the reactions were performed under CO atmosphere. The overall transformation implies a sequential Wittig-Heck reaction.

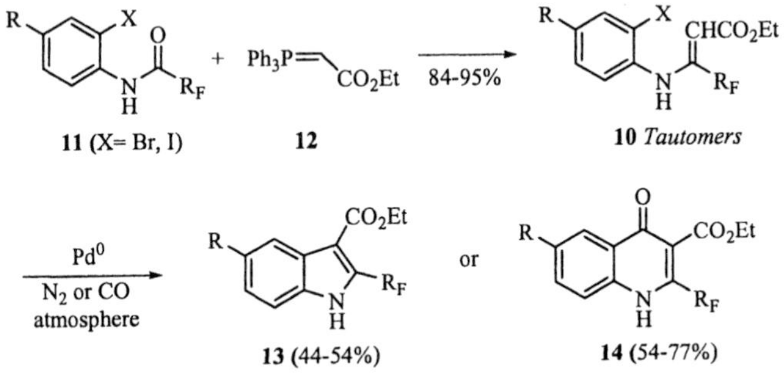

11 (X= Br, I) + $Ph_3P{=}CHCO_2Et$ **12** → (84-95%) **10** *Tautomers*

→ Pd^0, N_2 or CO atmosphere → **13** (44-54%) or **14** (54-77%)

Figure 4. Wittig reaction of fluorinated amides.

Intramolecular Wittig type rearrangement of imino (thio)ethers.

This strategy, described in 1998 by Uneyama for iminoethers **15**, provided α-hydroxy-β-imino-γ-fluorinated esters **16** in good yields (>80%) (Figure 5). Compounds **15** were converted into **16** after treatment with a base at low temperature *via* Wittig rearrangment (*10*). The best results were obtained when lithium 2,2,6,6-tetramethylpiperidide (LTMP) was used as base and when the reaction temperature was kept between –105°C and –70°C for 1 h. Compounds **16** are precursors of racemic α-hydroxy-β-amino-γ-fluorinated acids **17** through stereoselective reduction processes.

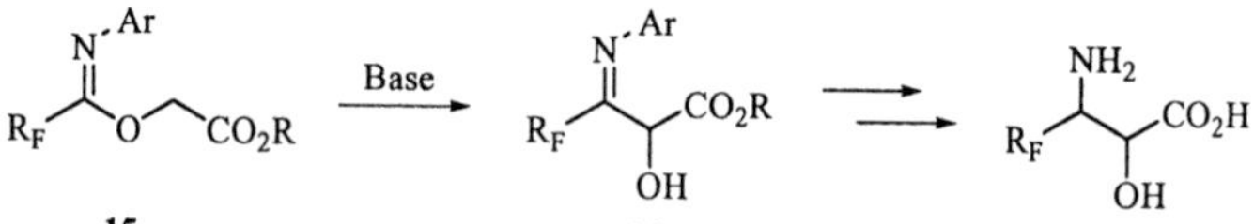

Figure 5. *Intramolecular Wittig type rearrangement of imino (thio)ethers.*

The same author studied an extension of the *O*-Wittig rearrangement to the corresponding thio-analogues with unexpected results (*11*) (Figure 6). The starting thioglycolates **18** were prepared in good yields from fluorinated imidoyl chlorides. Compounds **18** were next subjected to the base-catalyzed (LDA) Wittig type rearrangement. The reaction proceeded at a higher temperature (-40°C) and with a longer reaction time (9 h) than those of oxygen analogues. Surprisingly, a facile desulfurization was observed, affording β-enamino esters **19** bearing no sulfur moiety.

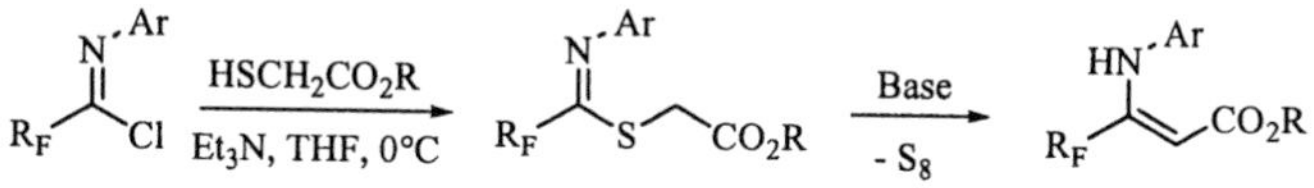

Figure 6. Preparation of fluorinated thioglycolates and their desulfurization to β-enamino esters.

Ester enolate condensation with fluorinated imidoyl halides or nitriles.

This strategy represents one of the simplest as well as one of the most efficient and general routes to fluorinated β-enamino esters **20**. In 1997, we described the condensation reaction between lithium ester enolates and imidoyl chlorides **21** (*12,13*). Subsequent treatment of alkyl esters with 2.0 equivalents of lithium diisopropylamide (LDA) in THF, followed by addition at –78°C of a variety of fluorinated *N*-alkyl or *N*-aryl imidoyl chlorides provided, after standard workup, the corresponding fluorinated β-enamino esters **20**, which were isolated as a mixture of imino and enamino tautomers (Figure 7). In general, the process works well, with good yields (64-95%) being obtained regardless of the nature of the starting materials. No excess of starting material is necessary with the exception of LDA, for which a two-fold excess should be used in order to ensure the presence of the intermediate **22**, which results in a significant improvement of the chemical yield.

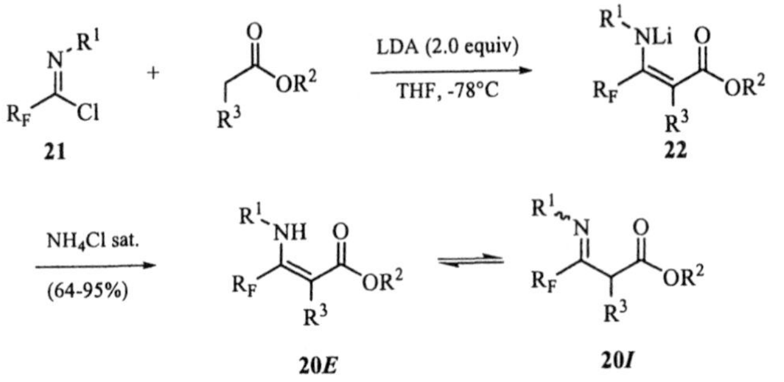

Figure 7. *Ester enolate condensation with fluorinated imidoyl halides.*

In order to demonstrate the scope of this approach, we extended the process to other β-enamino acid derivatives, such as fluorinated β-enamino amides derived from chiral nonracemic acyclic and cyclic amides. Thus, compounds **21** reacted with a range of chiral amides to afford the enamino tautomers exclusively and in good yields. Some representative examples are shown in Figure 8.

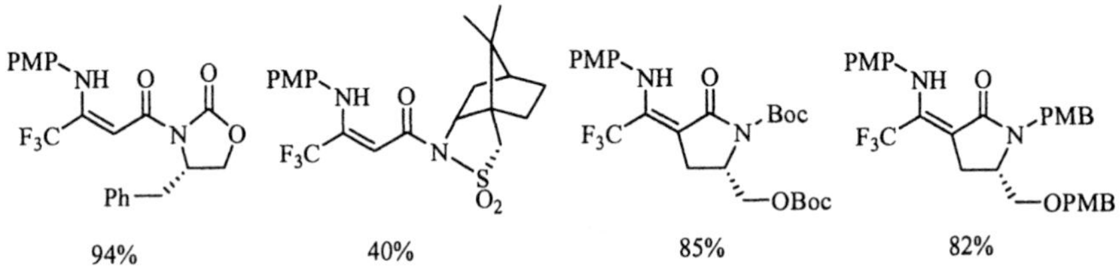

Figure 8. Fluorinated β-enamino amides derived from chiral nonracemic acyclic and cyclic amides.

In the same vein, we have also reported a simple route to *N*-substituted C-protected β-enamino acid derivatives **23** by reacting 2-alkyl-Δ^2-oxazolines **24** (X=O) and 2-alkyl-Δ^2-thiazolines **24** (X=S) with imidoyl chlorides **21** (Figure 9). Thus, azaenolates derived from 2-alkyl-Δ^2-oxa(thia)zolines react with compounds **21** under the same conditions as described above to provide the enamino tautomers **23** in good yields (*14,15*).

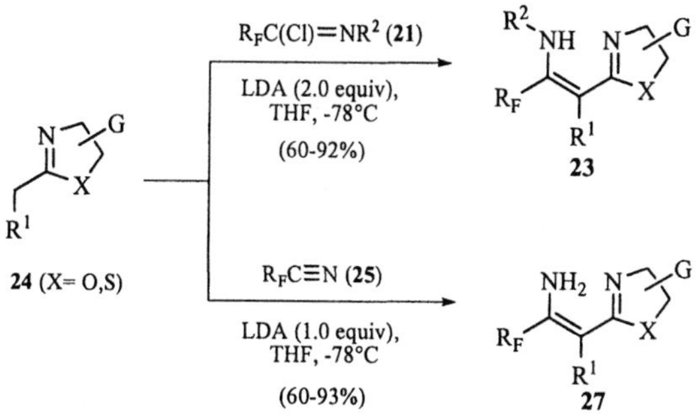

$R_FC(Cl)=NR^2$ (**21**)
LDA (2.0 equiv), THF, -78°C
(60-92%)
23
24 (X= O,S)
$R_FC{\equiv}N$ (**25**)
LDA (1.0 equiv), THF, -78°C
(60-93%)
27

Figure 9. Preparation of Δ^2-oxazoline protected β-enamino esters.

The reactivity of fluorinated nitriles in this kind of process was also studied. We found that lithium ester enolates and lithium azaenolates derived from 2-alkyl-Δ^2-oxazolines condensed smoothly with fluorinated aromatic and aliphatic nitriles **25** to afford excellent yields of fluorinated β-enamino acid derivatives **26** (*16*) (Figure 10) and **27** (*17*) (Figure 9). In this case, only one equivalent of LDA was necessary to ensure the success of the process.

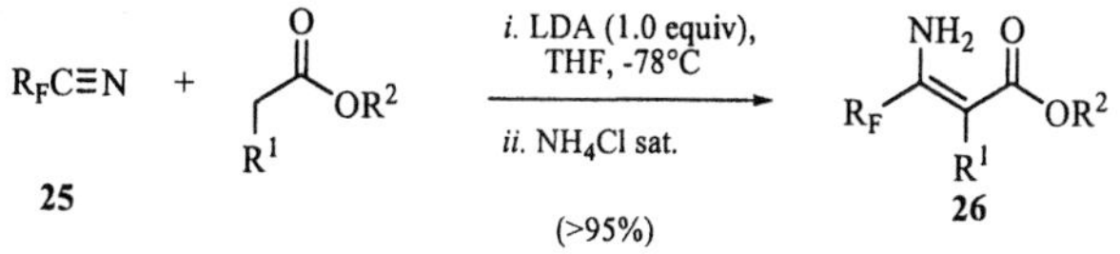

$R_FC{\equiv}N$ + ...
25
i. LDA (1.0 equiv), THF, -78°C
ii. NH_4Cl sat.
(>95%)
26

Figure 10. Ester enolate condensation with fluorinated nitriles.

In turn, compounds **26** and **27** have been used as starting materials for the preparation of racemic and chiral nonracemic fluorinated β-amino acid derivatives (*15*) and heterocycles such as uracils and thiouracils (see below).

Reactivity and applications of fluorinated β-enamino esters.

Synthesis of fluorinated β-amino acids through reduction of fluorinated β-enamino esters.

Although the chemo- and stereoselective reduction of chiral non-racemic β-enamino ester derivatives constitutes a simple and attractive route to enantiopure β-amino acids, very few examples of this strategy being used for preparing enantiopure fluorinated β-amino acids have been reported (*6*). In 1999, our group described a new and efficient two-step procedure for the diastereoselective synthesis of racemic (*18*) and chiral non-racemic *syn*-α-alkyl-β-fluoroalkyl-β-amino esters **29** (*13*). In this protocol, a chiral auxiliary group was used in the alcohol moiety of the ester. The condensation reaction yielded optically active fluorinated enamino esters **28**, which were then reduced with $NaBH_4/ZnI_2$ in an aprotic, non-chelating solvent (CH_2Cl_2) to give the corresponding fluorinated β-amino esters **29**. The best asymmetric induction was achieved with (−)-8-phenylmenthol as the chiral auxiliary. While the yields were high, a mixture of both diastereomers was obtained in all cases. These could, however, be separated by means of column chromatography. With few exceptions, the *syn* diastereomer was predominant, with the diastereomeric excesses ranging between moderate and good (*d.e.* up to 96%). In addition, the researchers found that the amino group could be easily deprotected with CAN. Thus, the use of (−)-8-phenylmenthol as a chiral auxiliary allowed for the preparation of optically active β-amino acids **30** in good yields (Figure 11).

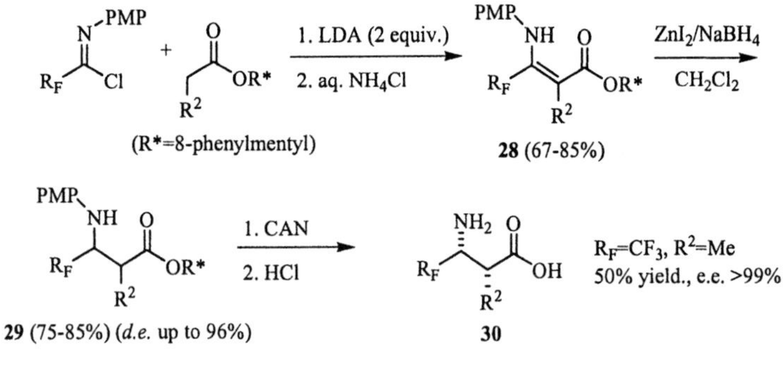

Figure 11. Chemo- and stereoselective reduction of chiral non-racemic β-enamino ester derivatives for the preparation of enantiopure β-amino acids.

An explanation for the stereochemical outcome of the reduction of chiral β-enamino esters **28** (R_F = CF_3, R^2 = Me) could involve the participation of two diastereomeric chelate models, in which the hydride attack is conditioned by the presence of the 8-phenyl group of the chiral auxiliary (1,5-asymmetric induction, Figure 12).

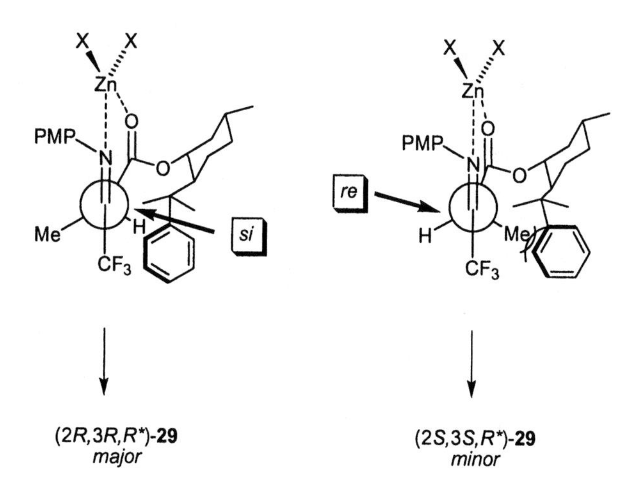

*Figure 12. A possible explanation for the stereochemical outcome of the reduction of chiral β-enamino esters **28**.*

Considering that several well-known amino acid drugs (*i.e.* Captopril and its trifluoromethylated analogue) contain a thiol group and that isoserine derivatives display an important biological activity, Uneyama *et al.* focused on the synthesis of fluorinated isoserine analogs. In the course of their work, they were able to achieve the diastereoselective synthesis of both diastereomers of racemic *S*-tert-butyl-β-(trifluoromethyl)isocysteine (*19*) by coupling fluorinated imidoyl chloride **31** with the enolate **32** to provide β-imino esters **33** in good yields. To suppress the desulfurization during the Wittig-type rearrangement (see Figure 6 above), *S-tert*-butyl protection of the sulfur moiety was necessary (*11*) (Figure 13).

The conversion of compounds **33** into the corresponding fluorinated β-amino acid diastereomers *syn*-**34** and *anti*-**34** was achieved by means of sterocontrolled reduction with hydride via either the chelated intermediate or the non-chelated Felkin-Ahn intermediate, respectively. For this purpose, $NaBH_4$ in CH_2Cl_2 and in the presence of $ZnBr_2$ led to the preferential reduction of the imino moiety, affording the *syn* product exclusively. It was deemed convenient to stop the reaction when *ca.* 20% of the starting material was still present, as prolonged reaction times led to the appearance of alcohol as a result of the ester group reduction. In contrast, reduction of compounds **33** with $NaBH_4$ in a solvent that traps sodium ions to generate naked borohydride [*e.g.* THF/di(ethyleneglycol)dimethylether] gave a mixture of both diastereoisomers, with the *anti* being predominant, in a proportion of 11:89 *syn:anti*.

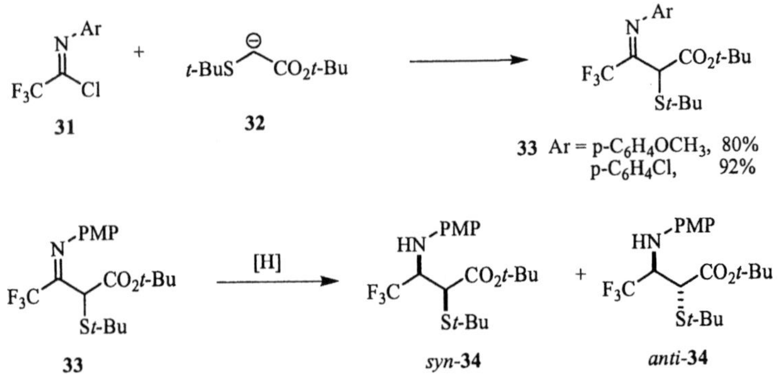

Figure 13. Uneyama's synthesis of isoserine fluorinated analogs.

Both diastereomers **34** were easily deprotected by first removing the PMP group by means of CAN oxidation, followed by *t*-butyl group removal through acid-catalyzed hydrolysis (Figure 14). Additionally, the deprotection of the *S*-*tert*-butyl group was achieved in two steps. First, the C-S bond was cleaved with *o*-nitrobenzenesulfenyl chloride to give the unsymmetrical disulfide **35** in 88% yield. This compound was then reduced with $NaBH_4$ to afford the thiol **36** in 64% yield.

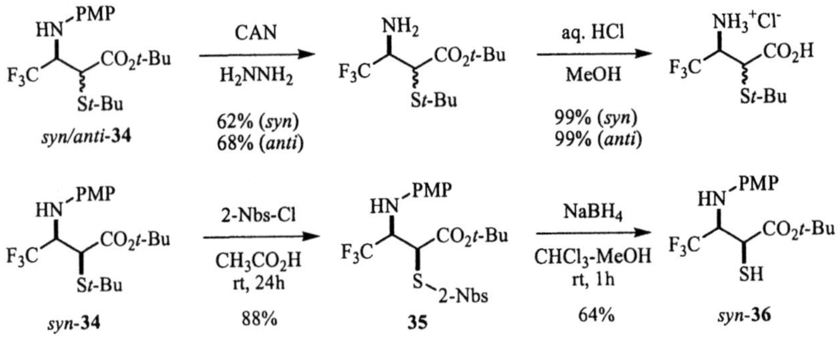

Figure 14. Deprotection reactions of isoserine fluorinated analogs.

Cyclic β-amino acids represent an interesting class of compounds because of their potential as therapeutic agents. Cyclic β-amino acids are also useful intermediates in the synthesis of natural products, β-peptides, and peptidomimetics. While the chemistry of their non-fluorinated derivatives has received a great deal of attention in the past few years (*20*), very little is known about their fluorinated counterparts. Although the literature includes several examples of β-amino acids with seven-membered rings, no fluorinated seven-membered β-amino acids have been described until recently.

Our group described the first diastereoselective preparation of fluorinated seven-membered β-amino acid derivatives through Ring-Closing Metathesis (RCM) (*21*, *22*) (Figure 15). The first step of the synthesis consisted of the condensation between the imidoyl chloride **37** and the enolate of ester **38**. This yielded compound **39**, which appeared as a mixture of enamino and imino tautomers. Although a variety of reagents were used to reduce compound **39**, no stereoselectivity was achieved. Thus, column chromatography was used to separate the *syn*-**40** and *anti*-**40** diastereomers (in 1:1 ratio), which were then submitted to an RCM reaction with second-generation ruthenium catalyst (IHMes)(PCy_3)Cl_2Ru=CHPh (*23*) to yield the respective seven-membered β-amino esters **41** in low yields.

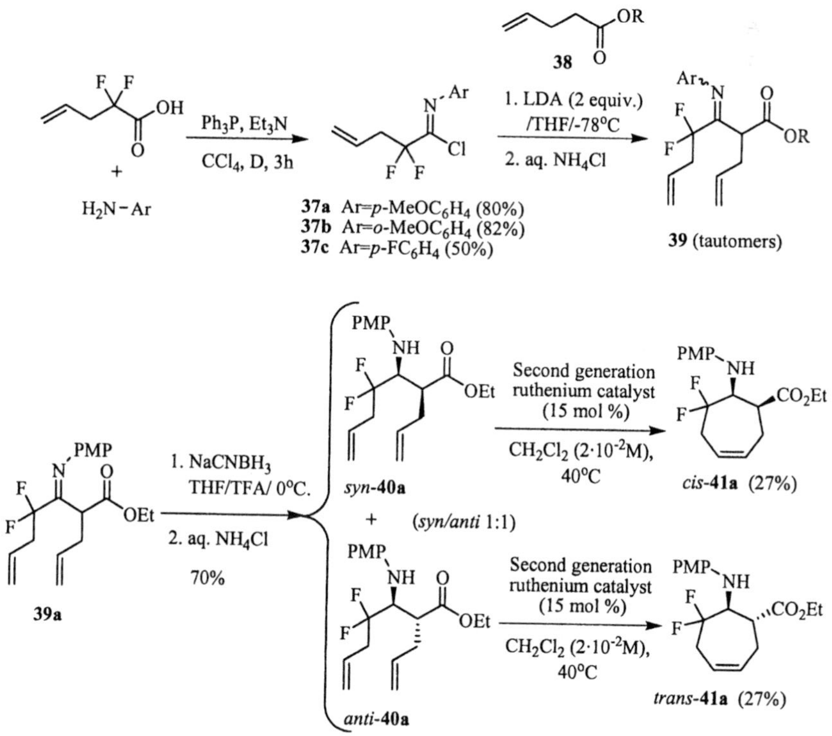

*Figure 15. Diastereoselective preparation of seven-membered β-amino acid derivatives **41** by means of Ring-Closing Metathesis (RCM).*

Since neither the diastereoselectivity nor the yields for this approach were satisfactory, the reactions were next performed in inverse order. Thus, compounds **39** were cyclized by reaction with either first-generation ruthenium catalyst catalysts $(PCy_3)_2Cl_2Ru{=}CHPh$ (*24*) or second-generation ruthenium catalyst $(IHMes)(PCy_3)Cl_2Ru{=}CHPh$ (the latter gave slightly better yields under milder conditions) to furnish the cyclized imino esters **42** in 60-90% yield, depending on the substituents (Figure 16). These compounds were then reduced with $NaCNBH_3$/TFA in a completely stereoselective fashion to give the *cis* diastereomer **41** in 72-90% yield.

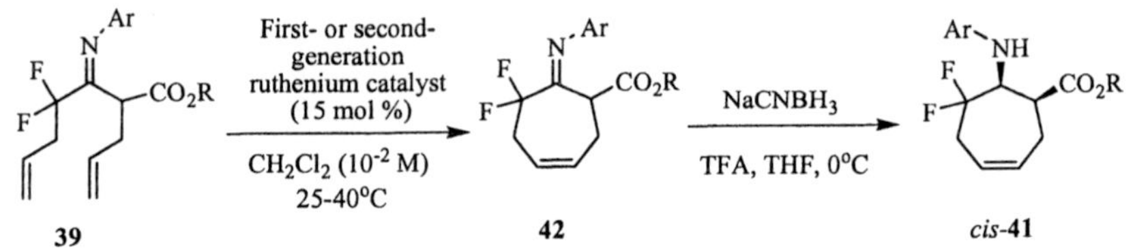

First- or second-generation ruthenium catalyst (15 mol %)
CH_2Cl_2 (10^{-2} M) 25-40°C
$NaCNBH_3$
TFA, THF, 0°C
39 **42** *cis*-**41**

*Figure 16. RCM and diastereoselective reduction of compound **39**.*

Finally, it was possible to either deprotect the amino group by means of PMP removal with CAN, or hydrogenate the double bond with hydrogen and a palladium on charcoal catalyst, with both procedures producing very high yields of **43** or **44**, respectively (Figure 17).

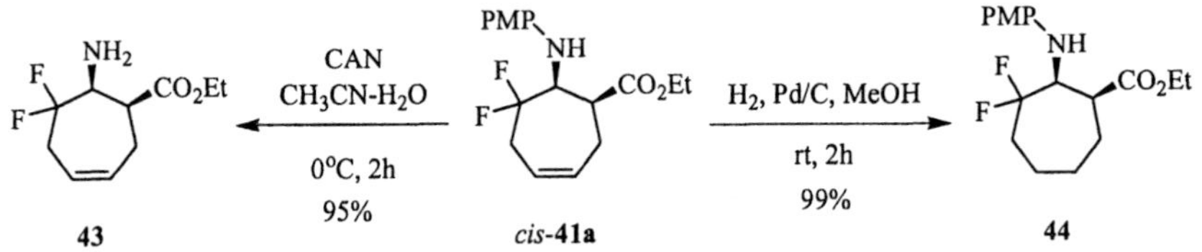

CAN CH_3CN-H_2O 0°C, 2h 95%
H_2, Pd/C, MeOH rt, 2h 99%
43 *cis*-**41a** **44**

*Figure 17. Deprotection reactions on compound **41a**.*

Synthesis of fluorinated uracils from β-enamino esters and iso(thio)cyanates.

Our research group has developed several strategies for using β-enamino esters as starting materials in the synthesis of fluorinated uracils. In some cases, the synthesis is applicable for obtaining non-fluorinated compounds as well, but quite often it is precisely the differential reactivity of fluorinated organic compounds that makes the synthesis possible. The two main strategies involved are the reaction of ester enolates with fluorinated nitriles and the reaction of Δ^2-oxazolin-*C*-protected β-enamino ester enolates with fluorinated nitriles.

In the first uracil synthesis developed by our group from β-enamino esters (*25*), compounds **26** are condensed with iso(thio)cyanates **45** to furnish the corresponding (thio)uracils after treatment with NaH in DMF. This condensation leads to the *N*-acylation of the enamine, which then undergoes a cyclization to yield the (thio)uracils **46**. When isocyanates were used, the corresponding uracils were obtained in good yields (64-90%), with the yields for the thiouracils being slightly lower (64-70%). In this way, 20 new fluorinated (thio)uracils **46** were easily prepared in only two steps and in high yields from fluorinated nitriles, esters, and iso(thio)cyanates (Figure 18).

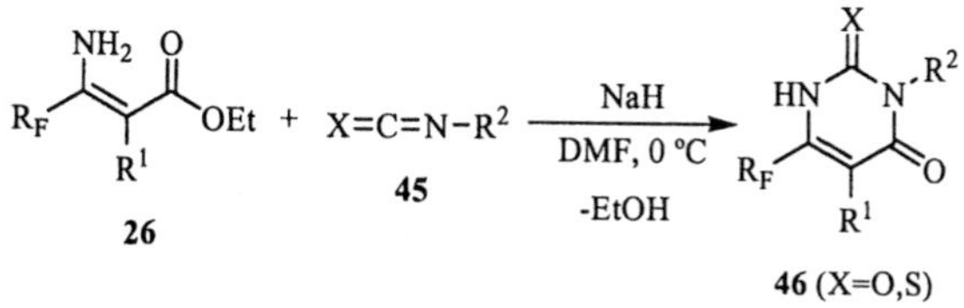

NH_2 O R_F OEt R^1 **26** + $X{=}C{=}N{-}R^2$ **45** → NaH, DMF, 0 °C, -EtOH → X HN N R^2 R_F O R^1 **46** (X=O,S)

*Figure 18. Reaction of compounds **26** with iso(thio)cyanates **45**.*

Interestingly, the direct halogenation of uracils at C-5 when R^1 = H does not seem to be possible. For this reason, the halogenation with either N-chlorosuccinimide (NCS) or N-bromosuccinimide (NBS) of the corresponding β-enamino esters is a useful approach to C-5 chlorinated or brominated uracils. Unfortunately, we have not been successful in our efforts at direct fluorination or iodination with a variety of reagents. Thus far, then, we have only been able to introduce a Cl or a Br atom into this position easily and in good yields (*26*).

Although the procedure for the uracil synthesis outlined above appears to be general, there is one notable exception: the reaction of β-enamino esters **26** with 2-chloroethyl isocyanate **45a** did not provide the desired uracils **46**, and only an ill-defined mixture of products were formed (Figure 19). This type of compound remains, however, an interesting target, since the chlorine atom would allow the introduction of functionality in that position through nucleophilic substitution reactions. The failure of this particular reaction may reside in the strongly basic medium, which could cause an HCl elimination in the isocyanate. In the next section, we will discuss the strategy we developed for the synthesis of this particular kind of uracils.

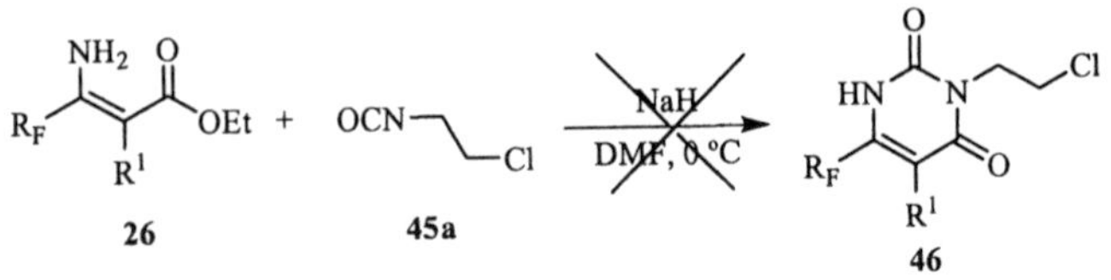

Figure 19. *β-Enamino esters **26** fail to condense with 2-chloroethylisocyanate **45a**.*

Our original two-step uracil synthesis, however, seemed a good candidate for solid-phase methodology not only due to its high yields, but also for the ease with which diversity can be introduced into the molecule. For instance, diversity can be easily introduced into groups R^1 (from different esters), R^2 (from different iso(thio)cyanates), R_F (from different fluorinated nitriles), and X (using either an isocyanate or an isothiocyanate), thereby facilitating the preparation of small libraries of fluorinated uracils for their subsequent biological evaluation. The literature contains a single precedent of the synthesis of uracils in solid-phase, but in that example the diversity introduced was reduced and only difficult to separate mixtures of isomeric, non-fluorinated uracils were obtained (*27*). By the same token, there are also very few examples of solid-phase parallel syntheses of fluorinated compounds (*28*), a fact that makes the search for new solid-phase syntheses for organofluorinated compounds even more appealing.

Our solid-phase synthesis of fluorinated uracils as outlined in Figure 20 is directly adapted from that described above (*29*). Thus, Wang resin **47** was first acetylated to afford its acetylated counterpart **48**. Next, the ester enolate of this acetylated resin was formed with an excess of LDA in THF at -50°C, which was then treated with difluorophenylacetonitrile at -78°C for 3 h to afford resin **49**, which was finally treated with an excess of NaH in DMF at 0°C, followed by reaction with several different iso(thio)cyanates **45**. This sequence led to the formation of the uracils with concomitant cleavage from the resin, which precluded the need for a specific cleavage step (*30*). With this procedure, C-6-difluorobenzylated uracils **46** (X=O) were obtained in good yields (67-89%) and with high purity (65-99%). In contrast, the corresponding C-6-difluorobenzylated thiouracils **46** (X=S) prepared with thioisocyanates instead of isocyanates were obtained in lower yields (55-63%) and with lower purity (61-73%), as had been the case in the previous solution synthesis. Although we only used one ester and one nitrile in our study, it should be possible to use many different esters and nitriles, which in turn would facilitate the preparation of small libraries of (thio)uracils.

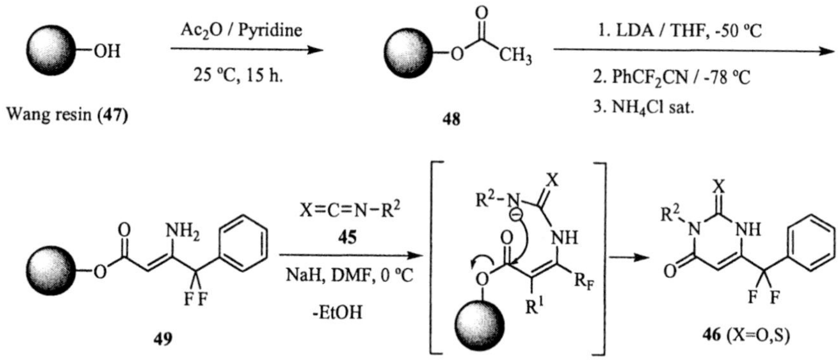

Figure 20. Solid phase synthesis of fluorinated uracils via β-enamino esters.

Thus, our first uracil synthesis from β-enamino esters allows for the simple and efficient preparation of fluorinated (thio)uracils with a methodology that can be easily adapted to solid-phase methodology.

Synthesis of fluorinated uracils from Δ^2-oxazolin-*C*-protected β-enamino esters and triphosgene.

Our second uracil synthesis (*31*) allows for the preparation of C-6 fluoroalkylated N-3 alkylated pyrimidin-2,4-diones **50** (Figures 21 and 22) from 2-alquil-Δ^2-oxazolines and fluorinated nitriles.

In this synthesis, the Δ^2-oxazolin-C-protected fluorinated enamino esters **27** were reacted with triphosgene (*32*) to give a mixture of isomeric oxazolopyrimidinones **51** and **52** in yields ranging from 70 to 95%. (Figure 21).

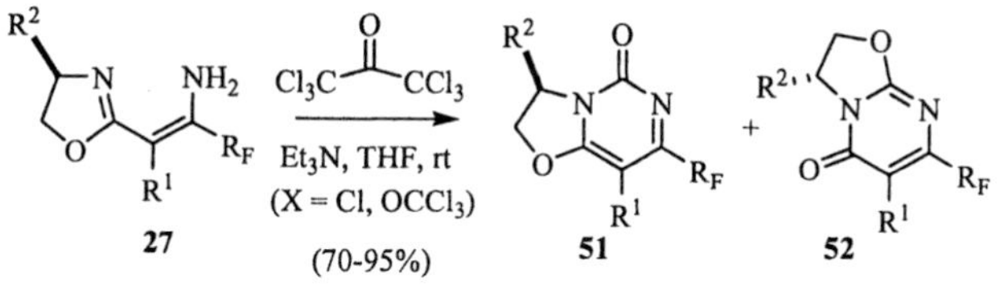

Figure 21. Reaction of compounds 27 with triphosgene.

While the pyrimidinone derivatives **51** were the predominant products of the condensation reaction, their isomers **52** were formed in all instances as well. In several cases, both compounds **51** and **52** were isolated and purified, but in some instances only **51** was isolated. Because of the similarities between isomeric compounds **51** and **52**, their structural elucidation was only possible through X-ray diffraction analysis.

It is worth noting that, unlike our first uracil synthesis, this alternative method is also useful for the preparation of non-fluorinated uracils, as the oxazoline aza-enolates are reactive enough to furnish the corresponding protected β-amino esters through reaction with non-fluorinated nitriles (*33*). In contrast, ester enolates will not react with non-fluorinated nitriles.

The final step in this synthesis consisted of an oxazoline ring-opening reaction by a nucleophile (Figure 22), a reaction for which there are several precedents in the literature (*34,35*). We studied the reactivity of pyrimidinones **51** and **52** with nucleophiles such as MeOH, EtOH, H_2O, AcOH, and HCl, under basic or acidic conditions, to give compounds **50** (Figure 22).

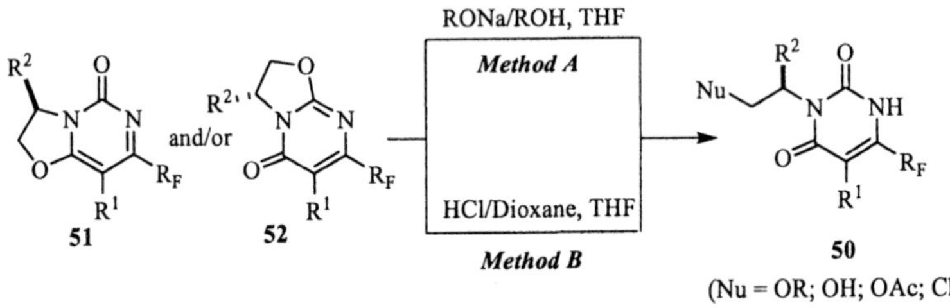

*Figure 22. Ring-opening reaction of pyrimidinones **51** and/or **52** with nucleophiles. Synthesis of uracils **50**.*

Compounds **51** and **52** underwent oxazoline ring opening under basic conditions in refluxing THF; subsequent hydrolysis with aq. NH_4Cl solution furnished uracils **50** (Method A, Figure 22). The ring-opening reaction can be carried out under milder acidic conditions as well. Thus, when compounds **51** and/or **52** were dissolved in THF and treated with 4M HCl in dioxane at room temperature, subsequent hydrolysis with aq. NH_4Cl solution also afforded uracils **50** (Nu = Cl; Method B, Figure 22). It is remarkable that the same uracil **50** was obtained as a reaction product regardless of whether compound **51**, **52**, or a mixture of **51**+**52** was used. In all cases it was observed that the ring-opening reaction in acidic medium proceeded faster (0.5-2 h) and with better yields (80-98%) than the corresponding reactions under basic conditions (5-7 h; 72-80%). In resume, this method provides a straightforward synthesis of fluorinated and non-fluorinated uracil derivatives **50** from Δ^2-oxazolines and fluorinated and non-fluorinated nitriles in only three steps with satisfactory chemical yields. The ring-opening reaction of intermediate oxazolopyrimidinones **51** and **52** by a number of different nucleophiles allows the preparation of a variety of potentially interesting analogues.

Acknowledgements

We thank the Ministerio de Ciencia y Tecnología and the Generalitat Valenciana of Spain for financial support (BQU2003-01610 and GRUPOS03/193, respectively).

References

1. See for example: Fustero, S.; García de la Torre, M.; Jofré, V.; Pérez Carlón, R.; Navarro, A.; Simón Fuentes, A. *J. Org. Chem.* **1998**, *63*, 8825-8836.
2. Bartoli, G.; Bosco, M.; Locatelli, M.; Marcantoni, E.; Melchiorre, P.; Sambri, L. *Synlett* **2004**, 239-242 and literature therein.
3. Fustero, S.; García de la Torre, M.; Pina, B.; Simón Fuentes, A. *J. Org. Chem.* **1999**, *64*, 5551-5556 and literature therein.
4. Ohkura, H.; Berbasov, D.O.; Soloshonok, V. *Tetrahedron* **2003**, *59*, 1647-1656.
5. Berbasov, D.O.; Soloshonok, V. *Synthesis* **2003**, 2005-2010.
6. Soloshonok, V.; Soloshonok, I.V.; Kukhar, V.; Svedas, V. K. *J. Org. Chem.* **1998**, *63*, 1878-1884.
7. Prié, G.; Richard, S.; Guignard, A.; Thibonnet, J.; Parrain, J.; Duchêne, A.; Abarbri, M. *Helv. Chim. Acta* **2003**, *86*, 726-732.
8. Richard, S.; Prié, G.; Parrain, J.; Duchêne, A.; Abarbri, M. *J. Fluorine Chem.* **2002**, *117*, 35-41.
9. Stanforth, S.P. *Tetrahedron* **2001**, *57*, 1833-1836 and literature therein.
10. Uneyama, K.; Hao, J.; Amii, H. *Tetrahedron Lett.* **1998**, *39*, 4079-4082.
11. Uneyama, K.; Ohkura, H.; Hao, J.; Amii, H. *J. Org. Chem.* **2001**, *66*, 1026-1029.
12. Fustero, S.; Pina, B.; Simón-Fuentes, A. *Tetrahedron Lett.* **1997**, *38*, 6771-6774.
13. Fustero, S.; Pina, B.; Salavert, E.; Navarro, A.; Ramírez de Arellano, C.; Simón Fuentes, A. *J. Org. Chem.* **2002**, *67*, 4667-4679.
14. Fustero, S.; Navarro, A.; Díaz, D.; García de la Torre, M.; Asensio, A.; Sanz, F.; Liu, M. *J. Org. Chem.* **1996**, *61*, 8849-8859.

15. Fustero, S.; Salavert, E.; Pina, B.; Ramírez de Arellano, C.; Asensio, A. *Tetrahedron* **2001**, *57*, 6475-6486.
16. Piera, J. Ph.D. Dissertation, University of Valencia, Valencia, Spain, 2004.
17. Salavert, E. Ph.D. Dissertation, University of Valencia, Valencia, Spain, 2002.
18. Fustero, S.; Pina, B.; García de la Torre, M.; Navarro, A.; Ramírez de Arellano, C.; Simón, A. *Org. Lett.* **1999**, *1*, 977-980.
19. Ohkura, H.; Handa, M.; Katagiri, T.; Uneyama, K. *J. Org. Chem.* **2002**, *67*, 2692-2695.
20. Fülöp, F. *Chem. Rev.* **2001**, *101*, 2181-2204.
21. Fustero, S.; Bartolomé, A.; Sanz-Cervera, J.F.; Sánchez-Roselló, M.; García Soler, J.; Ramírez de Arellano, C.; Simón, A. *Org. Lett.* **2003**, *5*, 2523-2526.
22. Abell *et al.* have very recently described the synthesis of cyclic non-fluorinated β-amino acid esters from methionine, allylglycine, and serine: Gardiner, J.; Anderson, K.H.; Downard, A.; Abell, A.D. *J. Org. Chem.* **2004**, *69*, 3375-3382.
23. Scholl, M.; Ding, S.; Lee, C.W.; Grubbs, R.H. *Org. Lett.* **1999**, *1*, 953-956.
24. Schwab, P.; France, M. B.; Ziller, J.W.; Grubbs, R.H. *Angew. Chem. Int. Ed. Engl.* **1995**, *34*, 2039-2041.
25. Fustero, S.; Piera, J.; Sanz-Cervera, J.F.; Catalán, S.; Ramírez de Arellano, C. *Org. Lett.* **2004**, *6*, 1417-1420.
26. Fustero, S.; Salavert, E.; Sanz-Cervera, J.F.; Román, R.; Fernández-Gutiérrez, B.; Asensio, A. *Lett. Org. Chem.* **2004**, *1*, 163-167.
27. Wahhab, A.; Leban, J. *Tetrahedron Lett.* **2000**, *41*, 1487-1490.
28. Vidal, A.; Nefzi, A.; Houghten, R.A. *J. Org. Chem.* **2001**, *66*, 8268-8272.
29. Volonterio, A.; Chiva, G.; Fustero, S.; Piera, J.; Sánchez Roselló, M.; Sani, M.; Zanda, M. *Tetrahedron Lett.* **2003**, *44*, 7019-7022.
30. Bräse, S.; Dahmen, S. in *Handbook of Combinatorial Chemistry*; Nicolau, K.C., Hanko, R., Hartwig, W. Eds; Wiley-VCH: Weinheim, 2002; Vol. 1, pp 59-169.
31. Fustero, S.; Salavert, E.; Sanz-Cervera, J.F.; Piera, J.; Asensio, A. *Chem. Commun.* **2003**, 844-845.
32. For a review see: Cotarca, L.; Delogu, P.; Nardelli, A.; Sunji, V. *Synthesis* **1996**, 553-576.
33. Díaz-Hernández, D., Ph.D. Dissertation, University of Valencia, Valencia, Spain, 1997.
34. Lis, R.; Morgan, T.K.; Marisca, A.J.; Gómez, R.P.; Lind, J.M.; Davey, D.D.; Philips, G.B.; Sullivan, M.E. *J. Med. Chem.* **1990**, *33*, 2883-2891.
35. Agami, C.; Dechoux, L.; Hamon, L.; Melaimi, M. *J. Org. Chem.* **2000**, *65*, 6666-6669, and references cited therein.

Chapter 35

Using the Potential of Fluorine for Peptide and Protein Modification

Christian Jäckel and Beate Koksch*

Freie Universität Berlin, Institut für Chemie, Takustrasse 3, 14195 Berlin, Germany

Methods for the enzyme-catalyzed resolution of a broad variety of fluorinated amino acids as well as for their incorporation into peptides and proteins using commercially available proteases have been developed. The synthetic strategies described here extend the scope of methods available for site-specific peptide and protein modification by fluorinated amino acids using simple and environmentally attractive routes. Moreover, a new screening system which has been developed for a systematic investigation of the molecular interactions of fluoro-substituted amino acids with native polypeptides is introduced.

Extending the spectra of building blocks which can be used for peptide and protein engineering beyond the natural amino acids broadens the scope of peptide and proteins. (*1-3*) Highly functionalized amino acid residues can serve as valuable tools to be used as biophysical probes for detailed studies of structure-function relationships or for the construction of tailor-made biomolecules which will expand the repertoire of protein functions. Many functional groups like e.g. halides are rarely found in the natural amino acid pool. Incorporation of the unique electronic properties of fluorine into amino acids creates, therefore, a new and exciting class of building blocks for peptide and protein modification. (*4*) Beta-fluorinated amino acids themselves have

gained prominence as mechanism based inhibitors of amino acid decarboxylases and transaminases. (*5*) C^{α}-fluoroalkyl substituted amino acids bearing a fluorinated substituent instead of the α-proton are known to be able to increase metabolic stability (*6*) of peptides as well as to stabilize peptide secondary structure. (*7*) The incorporation of fluorine usually shows dramatic effects on protein stability, protein-protein interactions, and the physical properties of protein based materials. (*8-11*) Furthermore, the ^{19}F atom as well as the C-F bond serves as a highly specific and powerful label for spectroscopic investigations of pathways, metabolisms, and structure-activity relationships using NMR or Raman spectroscopy, respectively. (*12-14*)
The impact of the fluorine substitution on all of the above-mentioned peptide and protein properties strongly depends on the position as well as the content of fluorine substitution within a special amino acid while the main discrimination has to be made between side chain fluorination and C^{α}-fluoroalkyl substitution. This difference not only dictates the properties exerted by the fluorinated building block but also the methods which have to be used for amino acid synthesis as well as for their incorporation into peptides.
It is part of our research program to develop routine methods for the synthesis of a broad variety of fluorinated amino acids as well as for their incorporation into peptides and proteins. Moreover, we study the properties of fluorinated amino acids and their interaction pattern with native amino acids within a native polypeptide environment. The first part of this review, therefore, will summarize new methods for resolution of racemic C^{α}-fluoroalkyl substituted amino acids as well as the protease-catalyzed incorporation of these sterically demanding amino acids into peptides. The second part introduces a new screening system which has been developed for a systematic investigation of the molecular interactions of fluoro-substituted amino acids with native polypeptides.

Part 1. Methods for Enzymatic Resolution of Racemic C^{α}-Fluoroalkyl Substituted Amino Acids and for their Protease-Catalyzed Incorporation into Peptides

1.1. Enzymatic Resolution of Racemic C^{α}-Fluoroalkyl Substituted Amino Acids

Several routes towards racemic C^{α}-fluoroalkyl substituted amino acids in which the α-proton is substituted by a fluoroalkyl group have been developed. The most convenient synthesis of this class of fluorinated amino acids is based on an amidoalkylation of carbon nucleophiles with highly electrophilic acylimines of 3,3,3-trifluoropyruvate (*5*, *15-22*; Scheme 1).

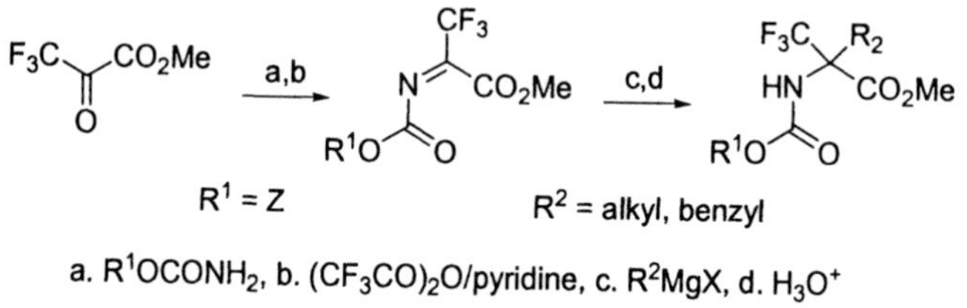

F_3C–C(=O)–CO_2Me $\xrightarrow{a,b}$ CF_3–C(=N–C(=O)OR^1)–CO_2Me $\xrightarrow{c,d}$ F_3C–C(R_2)(NH–C(=O)OR^1)–CO_2Me

R^1 = Z R^2 = alkyl, benzyl

a. R^1OCONH_2, b. $(CF_3CO)_2O$/pyridine, c. R^2MgX, d. H_3O^+

Scheme 1. *Synthesis route towards fully protected C^α-fluoroalkyl substituted amino acids.*

This route enables the synthesis of α-(trifluoromethyl) amino acids (αTfm amino acids) with orthogonal protective groups. (*23-26*) Analogously, α-(difluoromethyl) α-amino acids (αDfm amino acids), the virtually unknown α-(chlorodifluoromethyl) and α-(bromodifluoromethyl) α-amino acids can be obtained *via* addition of C nucleophiles to acylimines of corresponding partially fluorinated pyruvates. (*27*)

Due to an increasing interest in fluoroalkyl amino acids for peptide and protein modification and considering the divergent biological activities of the enantiomers of C^α-fluoroalkyl substituted amino acids and their diastereomeric peptide derivatives, the availability of these compounds in enantiomerically pure form is highly desirable. Important effords have been made in the development of new methods for the enantioselective preparation of β-fluorinated α-amino acids during the last years. *(28-30)* However, several synthetic routes to optically pure C^α-fluoroalkyl substituted amino acids rely on chemical (*31-33*) and enzymatic resolution. (*34*) A promising strategy for a diastereoselective synthesis of aTfm amino acids proceeds via amidoalkylation of carbon nucleophiles with *in situ* formed homochiral cyclic acyl imines (*35*,*36*). The dioxopiperazines (DOP) obtained with good stereoselectivity can be transformed into homochiral dipeptide esters by regioselective acidolysis in methanol. (*35*) However, the majority of fluorinated $C^{\alpha,\alpha}$-dialkylated amino acids is prepared chemically followed by enzymatic resolution of the enantiomers. The separation of the optical isomers of H-(αTfm)Ala-OH by partial hydrolysis of the racemic N-trifluoroacetyl derivative with hog kidney aminoacylase has been reported by Keller et al. (*34*) It was reported by our group that proteases like subtilisin, α-chymotrypsin or papain accept C^α-fluoroalkyl substituted amino acid esters as substrates only to a very limited extent. (*37*) Therefore, the application of these proteases for the resolution of enantiomeric C^{α}-fluoroalkyl substituted amino acid derivatives is excluded except for Z-(αTfm)Gly-OMe. (*38*)

At DSM Pharma Chemicals (Geleen, The Netherlands) several methods for the preparation of enantiomerically pure $C^{\alpha,\alpha}$-dialkylated amino acids via enzymatic resolution of racemic amino acid amides have been developed. Amidases from *Mycobacterium neoaurum* (ATCC 25795) and *Ochrobactrum anthropi* (NCIMB 40321), both exhibiting a high L-stereoselectivity, have been applied for large-scale preparation of many different optically active $C^{\alpha,\alpha}$-dialkylated amino acids (*39,40*; Scheme 2). It was, therefore, interesting to study the influence of the electronically modified amino acid derivatives on individual enzyme-substrate interactions and, thus, the catalytic efficiency and enantioselectivity of the amidases (*41*; Table I).

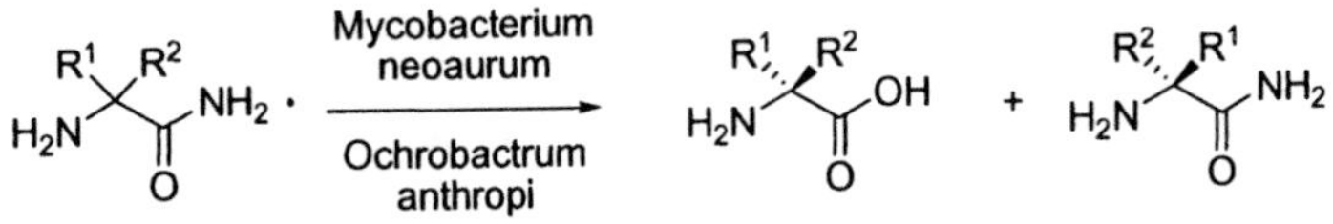

Mycobacterium neoaurum / Ochrobactrum anthropi

Scheme 2. *Enzymatic resolution of racemic C^{α}-fluoroalkyl amino acid amides.*

Amidase from *Mycobacterium neoaurum* (ATCC 25795) hydrolyzes *R,S*-H-(αTfm)Ala-NH_2, *R,S*-H-(αCF_2Cl)Ala-NH_2 and *R,S*-H-(αCF_2Br)Ala-NH_2 with high enantioselectivity (E > 200) to give the pure *R*-amino acids. In case of *R,S*-H-(αTfm)Ala-NH_2 the reaction was carried out in preparative scale and both the amino acid and the unconverted amide were isolated. As a proof of the enantioselectivity of the amidase reaction the amino acid was coupled to H-Ala-NH_2 *via* standard chemical peptide synthesis and the amino acid amide was likewise coupled to Z-Ala-OH. The resulting dipeptide products were shown to be diastereoisomerically pure by ^{19}F NMR as well as ^{1}H NMR analysis.

Mycobacterium neoaurum usually expresses a high enantioselectivity for the *L*-form of a racemic mixture. The absolute configuration of the converted H-(αTfm)Ala-NH_2 could be assigned to be *R* which corresponds to *D*-Ala if comparing the positions of the methyl groups. Apparently, the Tfm group as the larger of the two substituents at the C^{α}-atom is bound by the enzyme in the pocket which usually binds the methyl group of *L*-Ala. To what extent the electronic properties of the fluorine substituents add to the size effect would need to be investigated further. The growing size of the substituents at the halogenated group does not seem to influence the enantioselectivity of the enzyme (E > 200) for reaction with *R,S*-H-(αCF_2Cl)Ala-NH_2 and *R,S*-H-(αCF_2Br)Ala-NH_2. Merely the reaction rate is significantly lower for these substrates than for the aTfm congener.

Ochrobactrum anthropi (NCIMB 40321) accepts *R,S*-H-(αDfm)Phe-NH_2 as substrate and allows, for the first time, a successful enantioselective enzymatic hydrolysis of a C^{α}-fluoroalkyl substituted Phe derivative. In contrast, Phe derivatives which bear more than two halogen residues at the C^{α}-alkyl group are not accepted by the two enzymes. No significant hydrolysis was observed in these cases. These results indicate that the steric constraint exhibited by a trifluoromethyl or difluorochloromethyl group combined with the presence of a benzyl group, both at the α-carbon atom of an amino acid, is too high even for enzymes which provide a wide substrate specificity such as *Ochrobactrum anthropi* amidase. However, this technology can now be applied for the preparation of a variety of enantiopure C^{α}-fluoroalkyl substituted amino acids in preparative scale. (*41*)

Table I. Enzymatic Resolutions of Racemic C^{α}-Fluoroalkyl Amino Acid Amides using Amidases from Mycobacterium Neoaurum and Ochrobactrum Anthropi

	R¹	*R²*	*Enzyme*	*Conversion (%)*	*Amide ee (%)*	*Amino Acid ee (%)*	*E value*[a]
1	CF_3 (Tfm)	CH_3	*Mycobacterium neoaurum*	47	98.0	96.0	>200
2	CF_2Cl	CH_3	*Mycobacterium neoaurum*	48	99.5	94.7	>200
3	CF_2Br	CH_3	*Mycobacterium neoaurum*	49	99.5	94.8	>200
6	CH_2Ph[b]	CF_2H (Dfm)	*Ochrobactrum anthropi*	58	98.7	67.9	25

[a)] E values were calculated on the basis of the experimentally determined; e.e. values of the amide and the acid. [b)] This assignment is tentative.

1.2. Protease Catalyzed Peptide Synthesis for the Site-Specific Incorporation of α-Fluoroalkyl Amino Acids into Peptides

In our attempt to provide a broad variety of methods for the fast and simple incorporation of fluorinated amino acids into peptides, methods have been developed using commercially available proteases that make the direct enzymatic coupling of these sterically demanding and electronically modified amino acids possible. The synthetic strategy introduced here extends the scope of methods available for site-specific peptide and protein modification by fluorinated amino acids using simple and environmentally attractive routes.

1.2.1. Incorporation of α-Fluoroalkyl Amino Acids into Peptides using Trypsin and α-Chymotrypsin through Substrate and Medium Engineering

Effective site-specific incorporation of a wide variety of nonnatural amino acids into peptides and proteins remains a topic of high interest as it provides the opportunity for a more detailed understanding of protein structure and function. The combination of chemical synthetic methods with enzymatic peptide bond formation for the site-specific incorporation of nonnatural amino acids into peptides and proteins represents an attractive alternative to classical peptide chemistry. Enzymes work generally racemization free, highly regio- and stereoselectively, under mild reaction conditions, and require only minimal side chain protection. (*42,43*) However, the substrate specificity of available proteases usually restricts the number of residues between which a peptide bond can be synthesized. While di- and tripeptide methyl esters containing *N*-terminal αTfm amino acids are accepted as substrates by subtilisin, α-chymotrypsin, trypsin, and clostripain (*37,44-46*), direct enzymatic coupling of α-fluoroalkyl amino acids have been unsuccessful. Even in the case of Z-(αTfm)Gly-OMe which was shown to be a very specific substrate for subtilisin, protease-catalyzed peptide synthesis failed. (*38*) Recently, a powerful concept was established which overcomes these limitations of the classical enzymatic approach. (*47-52*) This concept is based on the binding site specific 4-guanidinophenyl ester (OGp) functionality to mediate acceptance of nonspecific amino acid moieties in the specificity-determining S_1 position of the enzyme (notation according to Schechter and Berger; *53*). Applying the advantages of the substrate mimetic concept to fluoroalkyl amino acids, we have succeeded for the first time in incorporating these sterically demanding $C^{\alpha,\alpha}$-dialkyl amino acids into the P_1 position of peptides enzymatically. (*54*)
4-Guanidinophenylester of C^{α}-fluoroalkyl Ala derivatives or Aib, respectively, can be easily prepared by reaction of the *N*-protected amino acids with 4-[*N'*,*N''*-bis(*tert*.-butyloxycarbonyl)guanidino]phenol using TBTU as coupling reagent. The sterically higher demanding C^{α}-fluoroalkyl or methyl Leu and Phe derivatives, respectively, can be synthesized in high yields as well, but have to be activated with DIC/HOAt and reacted with the lithium salt of the guanidinophenol (Scheme 3).
αDfm and αTfm substituted Ala, Leu, and Phe derivatives can be coupled directly to various nucleophiles - different in sequence and length - by trypsin (see Table II). Remarkably, in all cases the efficiency of peptide synthesis is much higher for αDfm-substituted amino acids compared to αTfm- as well as α-methyl-substituted derivatives. Remarkable differences in product yields between the (αDfm)Ala derivative and both Aib and (αTfm)Ala substrate

mimetics indicate a significant influence of the second α-substituent at the acyl donor on individual enzyme-substrate interactions. The efficiency of peptide synthesis using Z-(αDfm)Ala-OGp as the acyl component is at least twice as high in most cases as for the corresponding Aib and (αTfm)Ala derivatives, respectively, while product yields for both of the latter were found within the same range.

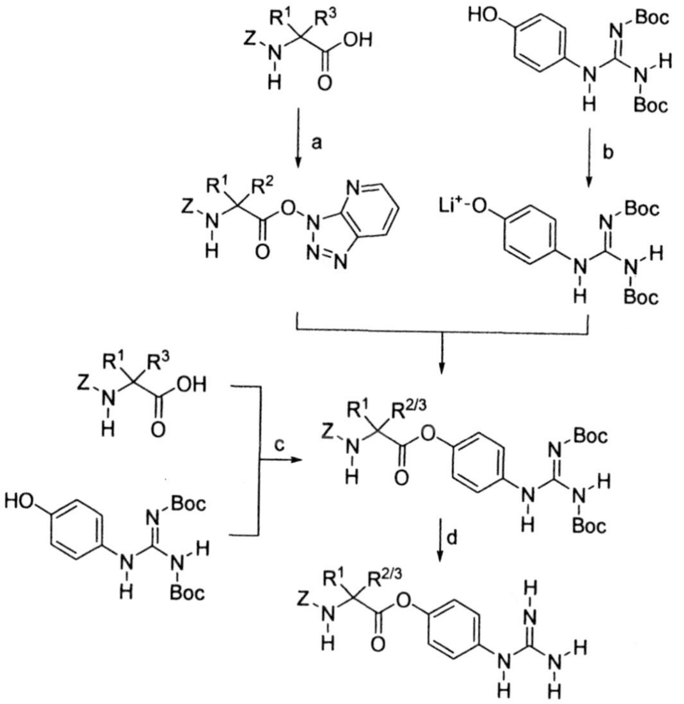

Scheme 3. *Synthesis of $C^{\alpha\alpha}$-dialkyl amino acid-4-guanidinophenyl esters. R^1: CH_3, CF_2H, CF_3; R^2: $CH_2C_6H_5$, $CH_2CH(CH_3)_2$; R^3: CH_3; a: DIC, HOAt, THF; b: n-butyl lithium, THF; c: TBTU, DIEA, DMF; d: TFA, ultra sound.*

Interpreting the differences in product yields between (αDfm)Ala derivatives and Aib and (αTfm)Ala substrate mimetics, respectively, the use of racemic mixtures in case of Z-protected (αDfm)Ala and (αTfm)Ala esters unlike for the Aib derivative has to be taken into account. A significant influence of the absolute configuration of the dialkyl amino acids on enzyme-substrate interactions within the active site of trypsin seems to be a more likely reason for this difference than the steric demand of the second substituent (alkyl or fluoroalkyl, respectively) at the C^{α}-atom. To prove this assumption, enzymatic peptide synthesis of Z-(αDfm)Ala-Met-NH_2 and Z-(αTfm)Ala-Met-NH_2 was carried out in a semi-preparative scale. Diastereomers were separated by HPLC and characterized by ^{19}F NMR. In case of the αTfm-substituted peptide a diastereomer-ratio of 1:3 was found, while for the corresponding αDfm substituted peptide the ratio was 1:1. Obviously, both enantiomers of the Z-(αDfm)Ala-enzyme complexes are deacylated by the nucleophile at the same rate which gives identical product yields. In contrast, one enantiomer of the Z-(αTfm)Ala-enzyme complex appears to be hydrolyzed faster by water than aminolyzed by the nucleophile, resulting in the formation of the amino acid instead of peptide bond formation. Moreover, the ability of the Dfm group to function as a hydrogen bond donor compared to the Tfm functionality certainly contributes to the observed difference in the interaction pattern between the fluoroalkyl amino acids and the enzyme. (*55,56*) Unlike the Tfm group, the αDfm amino acid could interact with the S′ region of the enzyme stabilizing the acyl enzyme intermediate. This would lead to a delayed reaction of at least one of the enantiomers with any kind of nucleophile which would result in a simultaneous aminolysis.

Table II. Yields (%) of Trypsin-Catalyzed Peptide Synthesis using Substrate Mimetics of $C^{\alpha,\alpha}$-Dialkylated Amino Acids.

Acyl donor	*Z-X_{C^α}Ala-OGp*			*Z-X_{C^α}Phe-OGp*				*Z-X_{C^α}Leu-OGp*	
Acyl acceptor *X_{C^α}:*	*Me*	*Dfm*	*Tfm*	*L-Me*	*D-Me*	*Dfm*	*Tfm*	*Dfm*	*Tfm*
H-Gly-NH_2	18	35	14	16	50	47	40	58	13
H-Leu-NH_2	19	51	18	23	53	62	46	64	27
H-Met-NH_2	35	70	27	45	82	72	53	88	36
H-Ala-Ala-OH	20	65	22	41	68	87	63	90	40
H-Ala-Met-OH	25	62	25	28	56	82	47	88	48
H-Ala-Arg-OH	50	83	33	63	88	92	71	94	54
H-Ala-Ala-Lys-OH	22	51	17	47	77	79	53	83	37
H-Ala-Ala-Ala-OH	22	54	19	43	79	86	61	88	45
H-Ala-Ala-Pro-OH	37	64	28	54	86	87	61	90	41

The influence of the stereochemistry in case of αTfm substituted Ala derivatives for substrate-enzyme interactions were shown to be of even higher importance in the frozen state. Using the advantage of substrate mimetics in combination with reactions in the frozen aqueous system had been already described as a powerful concept for an irreversible and efficient protease-catalyzed peptide synthesis independently of the primary protease specificity as well as without the risk of proteolytic side reactions. (*57*) Application of freezing reaction conditions on reaction of Z-R,S-(αTfm)Ala-OGp with various nucleophiles succeeded in a further increase of peptide yield up to 72% (Figure 1).

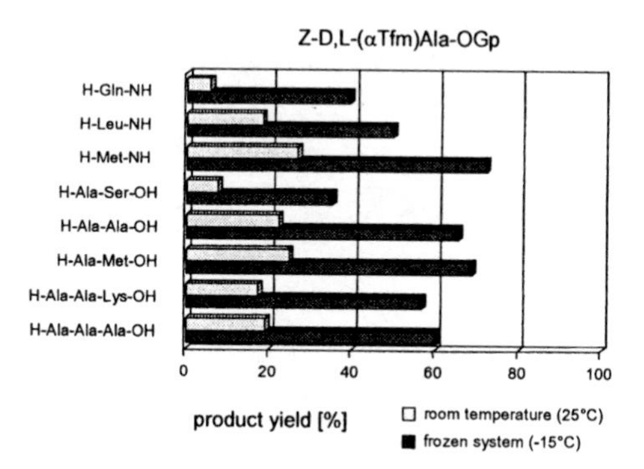

Figure 1. *Product yields of trypsin-catalyzed substrate mimetica mediated peptide synthesis at room temperature and in the frozen aqueous system.*

Motivated by the efficient application of α-chymotrypsin for linking nonnatural as well as non-peptidic acyl residues with amino acid and dipeptide amides this protease has also been tested for the incorporation of C^{α}-fluoroalkyl amino acids into the P_1 position of peptides. It was found that, in contrast to trypsin, α-chymotrypsin obviously does not accept C^{α}-dialkyl amino acids which carry a sterically higher demanding amino acid side chain at the α-carbon atom than a methyl group. A phenyl or isopropyl residue in combination with an α-fluoroalkyl or α-methyl substituent creates a sterical demand which seems to be too high. Therefore, only the Ala derived substrate mimetics Z-Aib-OGp, Z-(αDfm)Ala-OGp and Z-(αTfm)Ala-OGp were reacted with a small library of amino acid amides, di- and tripeptides catalyzed by α-chymotrypsin (results not shown here). The results convincingly show that the αDfm substituted derivative acts as the more efficient acyl donor again. α-Chymotrypsin efficiently catalyzes the peptide bond formation for both enantiomers of (αDfm)Ala with the same rate while in case of (αTfm)Ala one of the enantiomers is preferably incorporated into peptides. This finding is in close agreement with the results found for trypsin catalyzed reactions.

1.2.2. Incorporation of α-Fluoroalkyl Amino Acids into Peptides using Carboxypeptidase Y

Carboxypeptidase Y (CPY) is known to possess high catalytic activity and enantioselectivity for kinetic resolutions of many α-tertiary substituted carboxylic acid esters. (*58*) In contrast to the above-mentioned proteases, CPY has been shown to be provided with a broad S_1 and S_1' specificity while hydrophobic enzyme-substrate interactions are especially important. (*59-61*)

The suitability of C^{α}-fluoroalkyl alanine methyl ester to serve as acyl donors for CPY-catalyzed peptide bond formation was studied (*62*; Scheme 4). The results are summarized in Figure 2.

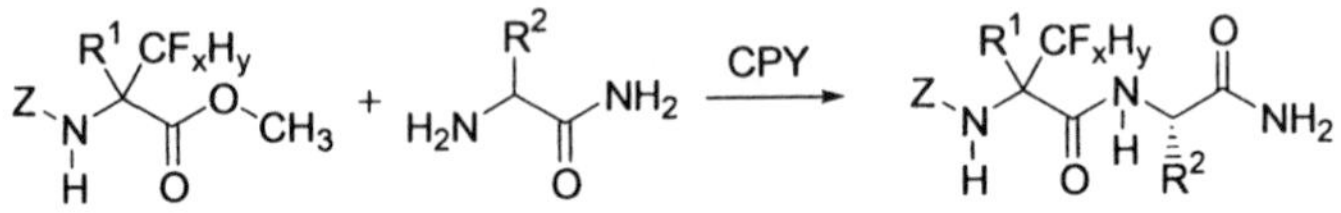

Scheme 4. *General scheme for CPY-catalyzed peptide synthesis.*

CPY catalyzes the acyl transfer to all of the four amino acid amides to give the desired *N*-protected dipeptide amides. However, the extent of peptide bond formation was considerably influenced by the second substituent at the C^{α} atom. In general, CPY was found to be considerably more efficient in catalyzing peptide bond formation in the case of the non-fluorinated amino acid (Aib). Highest peptide yields (80% and 70%, respectively) were obtained for reactions using nucleophiles with bulky residues (Leu and Val, respectively) which are in agreement with the known specificity of the protease for hydrophobic residues within both binding sites. The same tendency was observed for both of the corresponding C^{α}-fluoroalkyl amino acid esters. Here, the highest peptide yields (50%) could be obtained for reaction of *Z-R,S*-(αDfm)Ala-OMe with H-Leu-NH_2 and reaction of *Z-R,S*-(αTfm)Ala-OMe with H-Leu-NH_2 (45%), respectively. The influence of the absolute configuration of the α-fluoroalkyl amino acid on the efficiency of peptide bond formation by CPY was studied for the Tfm derivative in more detail. It was found that, in general, CPY accepts both enantiomers of (αTfm)Ala while certainly always preferring the *S*-configuration in P_1 position. (Figure 3).

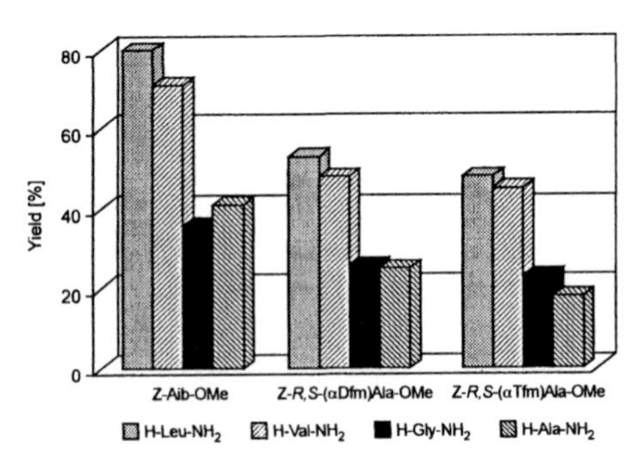

Figure 2. *Results of CPY- catalyzed peptide synthesis using C^{α}-fluoroalkylated and methylated amino acids, respectively, and four different nucleophiles.*

The strongest discrimination between both diastereomeric substrates was observed for the synthesis of Z-Phe-(αTfm)Ala-Ala-NH_2. Reaction of the diastereomeric peptide containing the *S*-enantiomer of the fluorinated amino acid in P_1 position gave 55% yield while product formation dropped to 20% when *S*-(αTfm)Ala was replaced by the *R*-enantiomer. Therefore, product formation increased up to 40% by applying the pure diastereomer of Z-Phe-*S*-(αTfm)Ala-OMe and is now even slightly higher than in the case of the corresponding Aib peptide. These results imply that substitution of a methyl group for a Tfm group at the C^{α}-atom of Aib can result, depending on the absolute configuration, in an improved binding of the substrate within the active site of CPY. Remarkably, in the case of the αTfm analogues, reaction of the diastereomer possessing the fluorinated, bulkier side chain in the same position as it would be for the side chain in the case of a natural amino acid results in a lower peptide yield. Surprisingly, CPY obviously accepts the bulky Tfm group within the binding site for the α-proton. The assumption can be made that the αDfm substituted Ala derivative will show a similar reaction behavior, however, this has not been explicitly investigated yet.

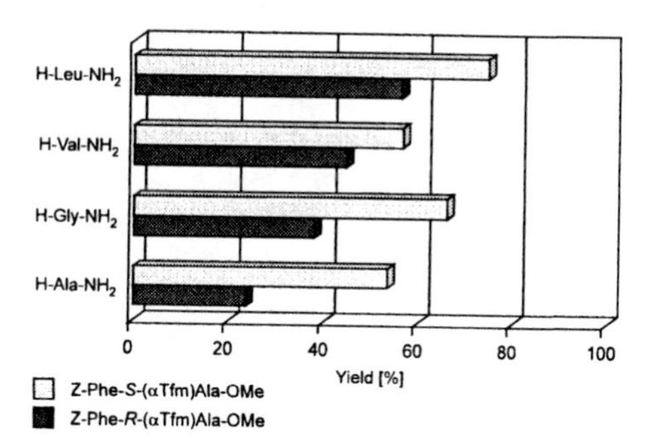

Figure 3. *Influence of the absolute configuration of (αTfm)Ala on the efficiency of CPY-catalyzed peptide synthesis.*

These results of CPY-catalyzed peptide synthesis represent the first example of a direct enzymatic coupling of two different, sterically demanding C^{α}-fluoroalkyl amino acids to amino acid nucleophiles without any further activation of the electrophilic substrate and without any medium or enzyme engineering. In this study, amino acid amides were used as nucleophiles. The application of amino acids (*63*) as nucleophiles will open up the possibility of incorporating a variety of fluorinated residues into biologically relevant peptides by enzymatic fragment condensation.

Part 2. Systematic Evaluation of the Molecular Interactions of Fluorinated Amino Acids with Native Polypeptides

The role of fluorine in bioorganic molecules, its impact on protein properties and biological activity, and the design of highly specific protein-protein interfaces has already been the subject of several studies. (*74*) Consideration of their results points out that fluorinated amino acids are extremely interesting for peptide and protein design because of the unique properties of the fluorine atom. However, some characteristics of fluorocarbon groups such as space filling, hydrophobicity, and the ability to participate in hydrogen bonding are still controversially discussed. However, knowing these characteristics is essential for the directed application of fluorinated amino acids for the *de novo* design of peptide and proteins, the directed manipulation of side chain interactions as well as folding stabilities. Therefore, a screening system has been developed which enables the systematic investigation of fluorinated amino acids regarding these properties and thus, can contribute to a better understanding of how their characteristics evolve in the context of a protein environment.

An essential requirement for investigations of specific amino acid side chain interactions as well as the influence of new building blocks on structural aspects of peptides and proteins like conformation and stability is a model system providing the following characteristics. First, the residue position to be substituted by the investigated nonnatural amino acid has to have one particular amino acid residue defined as its interaction partner. Second, a stable secondary structure of this model system is required which tolerates the substitutions to be made. Third, the model should allow a sensitive detection of the influence of the nonnatural building blocks on peptide/protein conformation and stability. To accomplish these requirements, α-helical coiled coil peptides can perfectly serve as such model systems. (*57,58*) Approximately 3% of all protein residues form coiled coils (*59*), which shows that this secondary structure is a very common motif in biological systems as found in transcription factors (*60*) and tropomyosin (*61*), for example. The structural principles of the coiled coil motif were analyzed in detail and therefore, have become very well understood.

The Coiled Coil Structure Motif (*64,65, 69-71* and references therein)

Coiled coils typically consist of two to five right-handed α-helices which are wrapped around each other to form a superhelical twist in a left-handed manner. The primary structure of each helix is characterized by a periodicity of seven residues, the so-called 4-3 heptad repeat which is commonly denoted (a-b-c-d-e-f-g)$_n$. The positions **a** and **d** are typically occupied by nonpolar residues (Leu, Ile, Val, Met) and form a special interaction surface at the interface of the helices by hydrophobic core packing ("knobs-into-holes"). In contrast, the positions **e** and **g** are solvent exposed and often occupied by charged amino acids (most common ones are Glu and Lys) forming inter-helical ionic interactions. In the remaining heptad repeat positions **b**, **c**, and **f**, located solvent exposed at the outer sides of the motif, polar residues are mostly found which can stabilize the conformation by intra-helical salt bridges. The hydrophobic core provides the major contribution to the structural stability of the coiled coil while the allocation of the two positions **a** and **d** of this interface by the different types of hydrophobic residues, β- and γ-branched, controls the order of aggregate formation. In contrast, the inter-helical ionic pairing positions **e** and **g** mainly dictate the orientation specificity (parallel versus antiparallel) as well as the preference for homo- or heterotypic coiled coil formation. Amino acids in these positions provide an abated contribution to the overall stability of the structure motif and the determination of the oligomerization state compared to that of the hydrophobic core residues.

The Coiled Coil Based Screening System

The screening system we use for our investigations of the properties of fluorinated amino acids is based on an antiparallel homodimeric coiled coil peptide containing 41 residues per helix (Figure 4). The relative orientation of the helices is determined by the design of the charged **e** and **g** positions. This results in a maximum number of attractive inter-helical ionic interactions (which would be all repulsive and therefore destabilizing in the the case of parallel folding) and a homogenous hydrophobic core consisting of Leu residues leading to a preference of the antiparallel over the parallel orientation (*72*) and a dimerization over higher oligomers in general. (*73*) Two residues in this screening model, K8 in a **g**-position of the charged interaction side and L9 in an **a**-position of the hydrophobic core, serve as substitution positions for the fluorinated amino acids to be investigated. Determined by the antiparallel orientation of the coiled coil, the side chain of nonnatural building blocks take part in heterogenous interactions with their natural partners, which are residues E29 for K8 and L33 for L9 of the opposite helix strand.
Therefore, this model peptide system provides perfect conditions to systematically analyzing properties of fluorinated amino acid side chains in a hydrophobic as well as a polar environment. The development of two simple screens facilitates the detection of the influences of the incorporated fluorinated amino acids in the substitution positions on inter-helical side chain – side chain interactions and conformational stability. Because such a structure motif has been shown to respond very sensitively to even single variations within either of the recognition domains in regard to dimmer stability (*74*), one screen of the interactions between a broad range of fluorinated building blocks and native amino acids measures the stability of such modified α-helical coiled coil dimers.

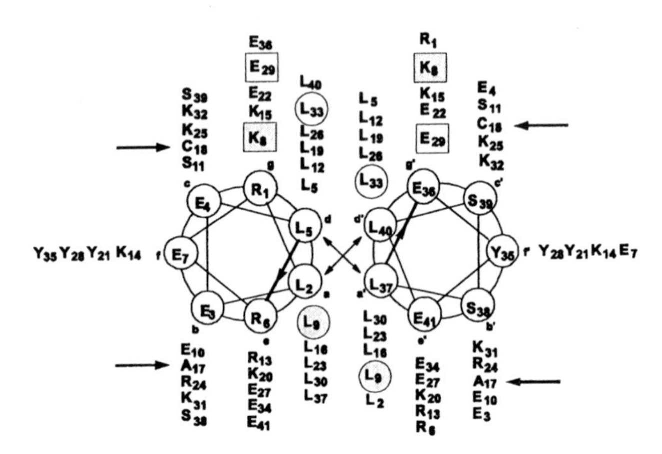

peptide sequences:

nucleophilic fragment	CLKYELRKLEYELKKLEYELSSLE

electrophilic fragments of native amino acids:

control sequence	Ac–RLEELREKLESLRKKLA
L9E	Ac–RLEELREK**E**ESLRKKLA
L9A	Ac–RLEELREK**A**ESLRKKLA
K8A	Ac–RLEELRE**A**LESLRKKLA

electrophilic fragments of side-chain modified amino acids

L9Abu	Ac–RLEELREK**Abu**ESLRKKLA
K8Abu	Ac–RLEELRE**Abu**LESLRKKLA
L9DfeGly	Ac–RLEELREK**DfeGly**ESLRKKLA
K8DfeGly	Ac–RLEELRE**DfeGly**LESLRKKLA
L9TfeGly	Ac–RLEELREK**TfeGly**ESLRKKLA
K8TfeGly	Ac–RLEELRE**DfeGly**LESLRKKLA

Figure 4. *Helical-wheel representation of the α-helical coiled coil peptides and amino acid sequences of all synthesized fragments. The substitution positions g,g'-K8 and a,a'-L9 (shaded in grey) and their interaction partners g',g-E29 and d',d-L33 (unshaded) are highlighted with squares (charged interface) and circles (hydrophobic interface). The ligation sites of the electrophilic and nucleophilic fragments are marked with arrows.*

The recording of thermal unfolding profiles using CD-spectroscopy and the proximate determination of peptide melting points of the substituted coiled coils afford conclusions about the interaction profiles within the recognition sites as well as intra-helical conformational tensions and, therefore, about the properties of fluorinated and backbone modified residues. A second screen is based on the property of peptides with hydrophobic heptad repeats to self-replicate (replicase reaction cycle; *75*). Consequently, we designed and synthesized the α-helical coiled coil strands of the screening system in two parts, a 24-residues long C-terminal nucleophilic fragment (Nu) and 17-residues long N-terminal electrophilic fragments bearing the substitutions in positions 8 and 9 (E_X; while the X denotes the respective substitution). These electrophilic fragments were synthesized as C-terminal benzylthioesters (*76-78*) which allows them to undergo a native chemical peptide ligation with Nu bearing an N-terminal Cys-residue. (*79,80*) The efficiency of template assisted peptide-bond formation strongly depends on the degree of complement of the interacting amino acid side chains of both recognition domains. Therefore, the recording of ligation reaction rates of the different substituted electrophilic fragments with Nu provides information about the impact of fluorinated amino acid side chains on dimer-interaction as well and will primarily give evidence of the sensitivity of the chosen system.

The Replicase Reaction Cycle

The replicase reaction cycle is a template-assisted auto-catalyzed self-replication process. (*75*) The initial step is an un-catalyzed peptide bond formation between both ligation fragments forming a full-length coiled coil strand. In the first cycle step this monomer acts as a template for the attachment of new ligation fragments in a coiled coil like manner. This non-covalent aggregation provides a close proximity of the functional groups taking part in the ligation process. Therefore, the template catalyzes its own formation, which results in a coiled coil homodimer. The dissociation/association equilibrium of this aggregate provides two new templates for further replicase cycles.

Investigations of Amino Acids carrying Fluorinated Side Chains

One controversial discussion about fluoroalkyl groups is held about the ability of participating in hydrogen bonding. The C-H-F-C interaction is described to be one of the weakest hydrogen bonds. (*55,56,82*) This interaction appears to be an important tool for protein engeneering. The hydrophobicity as well as the space filling requirement of fluoroalkyl groups remain controversial as well. Studies of the steric bulk of a trifluoromethyl group came to different results. Tirrell et al. consider substitution of hydrogen by fluorine in an amino acid to be isosteric. (*83*) In contrast, Kumar et al. assume that a trifluoromethyl group has twice the molar volume of a methyl group. (*11*,*84*) Both research groups substituted native leucine residues by 5,5,5-trifluoro- (*83*,*11*,*8*) and 5,5,5,5',5',5'-hexafluoroleucines (*84*,*10*,*9*) in the hydrophobic core of a parallel coiled coil dimer to analyze their influences on thermal and chemical structure stability. However, the most accepted opinion about this issue describes the space filling properties of fluorine to lie approximately halfway between hydrogen and methyl. (*85*) Therefore, a CF_3-group is supposed to be at least as large as isopropyl. (*86*,*87*) Accordingly, trifluoroethylglycine (TfeGly) should be sterically demanding similar to native leucine in the unpolar recognition domain of a coiled coil peptide. Thus, we decided to use fluorinated ethylglycines as starting points for investigating space filling and electronic properties of fluoroalkyl substituents using the screening system described above (Scheme 5).

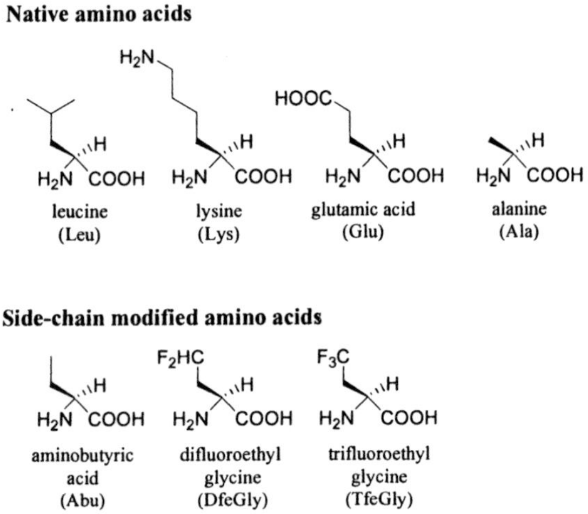

Native amino acids

leucine (Leu) | lysine (Lys) | glutamic acid (Glu) | alanine (Ala)

Side-chain modified amino acids

aminobutyric acid (Abu) | difluoroethyl glycine (DfeGly) | trifluoroethyl glycine (TfeGly)

Scheme 5. *Structure of all investigated amino acids within the substitution positions 8 and 9 of the electrophilic coiled coil fragments.*

In addition to the mentioned Nu and the nonsubstituted electrophilic fragment E_o we synthesized more E-variants with the amino acids alanin, ethylglycine (aminoisobutyric acid, Abu), difluoroethylglycine and trifluoroethylglycine in the substitution position L9 of the hydrophobic core ($\mathbf{E_{L9A}}$, $\mathbf{E_{L9Abu}}$, $\mathbf{E_{L9DfeGly}}$ and $\mathbf{E_{L9TfeGly}}$) as well as variants with these residues substituted into the ionic interface position K8 ($\mathbf{E_{K8A}}$, $\mathbf{E_{K8Abu}}$, $\mathbf{E_{K8DfeGly}}$ and $\mathbf{E_{K8TfeGly}}$). In a further peptide variant, the charged amino acid, Glu, was incorporated into the apolar surface

(E_{L9E}) in order to achieve information about the response of the screening system to strong destabilizing factors within the hydrophobic core. This helps to interpret and rank the results obtained for the fluorinated residues. To prove that the control sequence has self-replicating properties, E_o and Nu were incubated with and without added full-length peptide. (*88*) Addition of increasing amounts of template results in acceleration of product formation which indicates an autocatalytic system (results not shown). In further experiments, preparative ligation experiments with Nu and the above mentioned electrophilic fragments were performed, and the purified reaction products were analyzed in term of thermal stability by CD-spectroscopy in 5M guanidine hydrochloride. The use of urea as denaturizing reagent did not lead to a sufficient destabilization of the coiled coil peptides even at high concentrations (8 M). The results of these studies demonstrate the much stronger impact of substitutions within the hydrophobic core compared to those located in the charged recognition domain. All things considered, both the rates of product formation and the thermal stabilities of all variants differ from the control sequence peptide. This clearly proves our screening system to be sufficiently sensitive to detect even minor differences between the incorporated side chains, best demonstrated for the residues DfeGly and TfeGly differing only in the content of one fluorine atom. The incorporation of the charged residue Glu into the hydrophobic core instead of the native Leu leads to a dramatically slower reaction rate of the replicase system. This result is consistent with the L9G-variant unique characteristic of not forming a dimer at room temperature in 5M guanidine hydrochloride. This points out an extreme example of the destabilization of the hydrophobic core of a coiled coil by substituting one single Leu residue. All L9-variants show both strong retardation of the replication rates and significant decrease in thermal stability but represent differences in the order of influence within the set of incorporated analogues between both screens. While L9Abu is the most destabilized coiled coil within the series of ethylglycine, followed by L9DfeGly and L9TfeGly, the order of decrease in replicase reaction rate of these variants is inverse. As mentioned in the general description of the replicase cycle, the rate of production formation is affected by both the provision of free monomeric template, which means the dissociation rate of the dimeric ligation product, and the association efficiency of template and nucleophilic and electrophilic coiled coil fragments. All things considered, a faster turnover rate indicates a substituted electrophile to be a better catalyst in the system. Thus, the antipodal effects of stabilization, increasing the reaction rate by faster association and decreasing it caused by slower dissociation (vice versa for the impact of destabilization) yield in the overall rate of product formation which is determined by the contribution of both processes to the replication cycle. In contrast, the thermal unfolding experiments allow straight conclusions about the influence of the observed substituent on secondary structure stability and thus give direct information on the interaction profile and steric, as well as, electronic properties of this amino acid in the respective coiled coil recognition domain. However, due to the content of information from both association and dissociation processes, the replicase experiments may provide a more detailed insight into the character of fluoroalkyl groups compared to the pure stability screens. L9A is the most destabilized variant within this L9-series which is caused by a lack of hydrophobicity in the core packing (Figure 5).

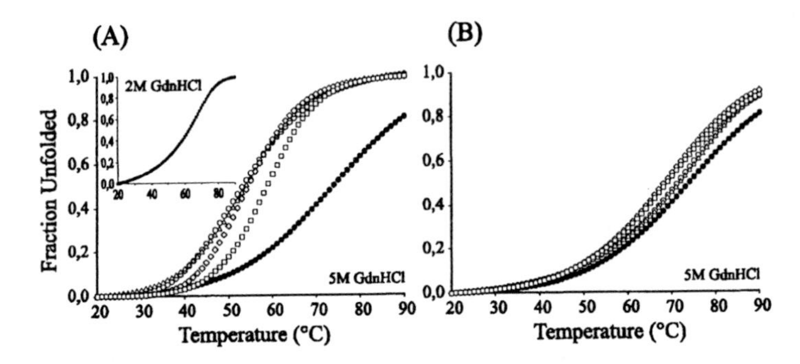

Figure 5. *Thermal unfolding profiles in 5M guanidine hydrochloride of the ligated coiled coil peptides of the control sequence (closed circles) and with substitutions in position 9* ***(A)*** *and position 8* ***(B)*** *with Ala (open circles), Abu (open triangles), DfeGly (open diamonds,) and TfeGly (open squares).* ***Inset diagram:*** *thermal unfolding profile of the L9E variant in 2M guanidine hydrochloride which was already fully unfolded at 20°C in 5M GdnHCl.*

A side-chain elongation of one methyl group (L9Abu) has only a minor stabilizing effect, indicated by the increase of the melting point of less than 1 K. While a further increase in side-chain volume and hydrophobicity by substituting two hydrogens with fluorine (L9DfeGly) leads to an additional stabilization to a similar extent, the incorporation of a third fluorine into the methyl group of the ethylglycine (L9TfeGly) results in a much stronger increase in secondary structure stability and a melting point enhancement of 5 K compared to the L9DfeGly variant. Considering these results, the fluorination of a methyl group has a much higher impact on hydrophobic core stability than the addition of a methyl group. This proves either the described hydrophobic or space filling or both of these properties of fluoroalkylation. The stabilizing effect shows no linear correlation with the order of fluorine substitution. This fact is based on the electron-withdrawing effect of the two fluorine substituents in the Dfe group causing an acidity of the proton attached. Thus, the acidic proton seems to disturb the interaction within the hydrophobic core, which is characterized by a lowered extent of stabilization effect. However, even the most stabile variant of this substitution series shows serious destabilization compared to the original sequence which bears a leucine in position 9 (Table III).

Table III. Melting temperatures of the α-helical coiled-coil peptides

Peptide	T_m *(°C)* [a)]
control sequence	73.9
L9Ala	53.2
L9Abu	54.0
L9DfeGly	54.4
L9TfeGly	59.0
K8Ala	71.3
K8Abu	71.9
K8DfeGly	68.3
K8TfeGly	68.9

[a)] T_m is defined as the temperature at which 50% of the peptide is unfolded.

Considering the hypothesis that the steric bulk of trifluoroethylglycin is similar to that of leucine and the postulated high lipophilicity of fluoroalkyl groups, which is due to the low polarizibility of the fluorine atoms (*89*), the opposite result should have been expected. Tirrell et al. and Kumar et al. could show that the incorporation of both trifluoroleucine (*11,8*) as well as hexafluoroleucine (*9*) into the hydrophobic core of a parallel coiled coil results in enhanced thermal as well as structural stability of the resulting peptide or protein assembly. The magnitude of the observed effect was correlated with the content of fluoroalkylated (*83*) amino acid. Considering the parallel orientation of the dimers used in these experiments, the fluoroleucines generally interacted with fluoroleucines. In further studies, disulfide-exchange assays with parallel self-sorting coiled coil peptides with hydrophobic cores composed entirely of leucine or hexafluoroleucine resulted in less than 3% heterodimeric assemblies with interactions between native and fluorinated leucines. (*84,10*) These heterodimers showed only a minor increase in stability compared to the homodimers containing native leucine residues. Nishino et al. could show in studies with amphiphilic helices containing either leucines or 2-(2,2,2-trifluoroethyl)glycines) as hydrophobic residues that these fluoroalkylated side chains tend to gather with each other. (*90*) These results indicate that fluorocarbon-fluorocarbon interactions are much stronger than both mixed fluorocarbon-hydrocarbon as well as hydrocarbon-hydrocarbon interactions. Bringing these studies in relation to our investigations on fluorocarbon-hydrocarbon interactions with fluorinated ethylglycines, we conclude that either the space filling of fluoroalkyl groups is much overestimated and/or the high electronegativity of fluorine disturbs a throughout formation of a hydrophobic mixed fluorocarbon-hydrocarbon core packing. The observed destabilization effect of trifluoroethylglycine within the apolar recognition site is accompanied by a strong retardation of the product formation rate in the ligation reaction due to a negative effect on the association of template and modified electrophile (Figure 6).

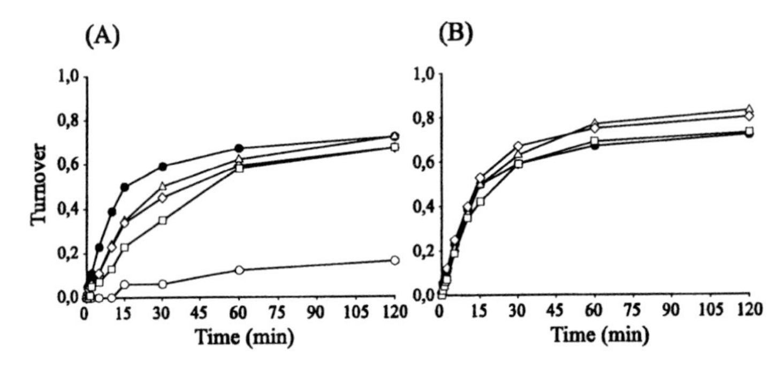

Figure 6. *Replicase turnover rates of electrophilic peptide fragments with the control sequence (closed circles) and with substitutions in position 9* ***(A)*** *and position 8* ***(B)*** *with Abu (open triangles), DfeGly (open diamonds) and TfeGly (open squares). Open circles: substitution of Leu in position 9 with Glu.*

Further destabilization of the coiled coil by decreasing the content of fluorine leads to a stepwise acceleration of the replicase reaction. This effect is based on a positive effect on the dissociation rate of the dimer. We conclude that the reduction of side-chain length within the hydrophobic core decreases the rate of coiled coil formation (assembling process), while the substitution of hydrogen by fluorine slows down the dissociation of the α-helical coiled coil strands. The reaction rate of the background reaction, which is the template-independent chemical peptide ligation, was shown to be uninfluenced by the type of substitution in position 9 of the coiled coil peptide. (*88*) Substitution of Lys in position 8 within the charged surface by the ethylglycine fluorination series had less effect on coiled coil stability and the product formation rate of the replicase reaction compared to the L9-variants. This result conforms with the present knowledge on the charged interface which provides a smaller contribution to coiled coil stability compared to the apolar core packing. (*65*) In general, breaking the K8-E29 saltbridge destabilizes the structure motif and results in a lowered melting point by ca. 2 K (K8A). The role of interhelical saltbridges in the aspect of stability of coiled coil peptides is a controversially studied issue as well. (*65*) While most attractive interhelical ionic pairings seem to stabilize the assembly (*91,92*) and general rules for optimization of intra- and interhelical saltbridges to achieve most coiled coil stability could be drawn up recently (*93*), examples of an opposite impact were observed. (*94*) In general, the contribution of such ion pairings is influenced by both energetically favorable coulombic attractions and unfavorable desolvation of the charges of involved residues (*93*). The elongation of the Ala side chain by one methyl group within the charged interface (K8Abu) does not lead to any detectable further destabilization, although the hydrophobic volume exposed to the polar solvent is doubled. In contrast, the fluorination of the ethyl side chain shows a significant impact on coiled coil stability. The dimer melting points are decreased by about 3 K for both variants, K8DfeGly and K8TfeGly, which shows no detectable differences between them. This proves the results for the set of ethylglycine fluorination in the substitution position L9 within the hydrophobic core. Again, the exchange of methyl hydrogens by fluorine show a stronger impact on lipophilicity and space filling property, respectively, than the addition of one methyl to the side chain. The elimination of the K8 from the attractive salt bridge (K8Abu) leads to an acceleration of the replicase reaction rate due to a faster dissociation of the coiled coil assembly, while a destabilization of the hydrophobic core in this manner resulted in a slower association process. This finding is in agreement with the recently published work of Matthews et al. suggesting that stabilizing ion pairings do not urgently accelerate the coiled coil folding process. (*95*) A stepwise fluorination of ethylglycine within the charged interface correlates, comparable to the substitution series of the hydrophobic core, with a decelaration of the ligation reaction, while in this case a destabilizing effect of fluorination has been observed. Interestingly, in contrast to the incorporation of fluoroalkyl groups into lipophilic interaction surfaces, fluorination of solvent exposed residues retards the association process of peptide domains, possibly caused by the formation of fluorine-fluorine interactions in the unfolded state which have to be disrupted during coiled coil assembly.

In summary, the screening system introduced here has been proven to be sufficiently sensitive to even detect the difference of one single fluorine atom within an alkyl group. First results gave already interesting information about the interplay between space-filling and hydrophobicity of fluorinated amino acid side chains within the context of a polypeptide environment. On this basis, a general protocol is being developed for studying systematically the steric, electronic, and hydrophobic effects of a variety of fluorinated amino acids, which differ in the content of fluorine as well as the position of the fluorination.

Acknowledgement

This work was funded by Deutsche Forschungsgemeinschaft (Innovationskolleg "Chemical signal – Biological answer" and KO1976/2-1). We are thankful to Mrs. H. Boettcher for technical assistance and Dr. Pamela Winchester for proofreading of the manuscript.

References

1. van Hest, J.C.M.; Tirrell, D.A. *Chem. Commun.* **2001**, 1897-1904.
2. Yoshikawa, E; Fournier, M.J.; Mason, T.L.; Tirrell, D.A. *Macromolecules* **1994**, *27*, 5471-5475.
3. Muir, T.W.; Kent, S.B. *Curr. Opin. Biotechnol.* **1993**, *4*, 420-427.
4. Koksch, B.; Sewald, N.; Jakubke, H.-D.; Burger, K. In *Biomedical frontiers of fluorine chemistry;* McCarthy, J.R.; Welch, J.T., Eds.; ACS Symposium Series 639; American Chemical Society: Washington, DC, **1996**; pp 42-58.
5. Burger, K.; Hoess, E.; Gaa, K.; Sewald, N.; Schierlinger, C.Z. *Naturforsch.* **1991**, *46b*, 361-384.
6. Koksch, B.; Sewald, N.; Hofmann, H.-J.; Burger, K.; Jakubke, H.-D. *J. Peptide Sci.* **1997**, *3,* 157-167.
7. Koksch, B.; Jakubke, H.-D.; Wenschuh, H.; Dietmeier, K.; Starostin, A.; Woolley, A.; Dathe, M.; Müller, G.; Gussmann, M.; Hofmann, H.-J.; Michel, T.; Burger, K. In *Peptides: Proceedings of the Twenty-Fifth European Peptide Symposium;* Bajusz, S.; Hudecz, F., Eds.; Akadémiai Kiadó: Budapest, **1999**; pp 670-671.
8. Tang, Y.; Ghirlanda, G.; Vaidehi, N.; Kua, J.; Mainz, D.T.; Goddard III, W. A.; DeGrado, W. F.; Tirrell, D. A. *Biochemistry*, **2001**, *40*, 2790-2796.
9. Tang, Y.; Tirrell, D.A. *J. Am. Chem. Soc.* **2001**, *123*, 11089-11090.
10. Bilgicer, B.; Xing, X.; Kumar, K. *J. Am. Chem. Soc.* **2001**, *123*, 11815-11816.
11. Bilgicer, B.; Fichera, A.; Kumar, K. *J. Am. Chem. Soc.* **2001**, *123*, 4393-4399.
12. Entress, R.M.H.; Dancer, R.J.; O'Brien, D.P.; Try, A.C.; Cooper, M.A.; Williams, D.H. *Chem. Biol.* **1998**, *5*, 329-337.
13. Ulrich, A.S. In *Encyclopedia of Spectroscopy and Spectrometry;* Lindon, J.; Tranter, G.; Holmes, J., Eds.; Academic Press, **2000**; pp 813-825.
14. Shart, C.M.; Gorelik, V.S. U.S. patent 6,307,625 B1, **2001**.
15. Soloshonok, V.A.; Gerus, I.I.: Yagupolskii, Y.L. *Zh. Org. Khim.* **1986**, *22*, 1335-1337.
16. Yagupolskii, Y.L.; Soloshonok, V.A.; Kukhar, V.P. *Zh. Org. Khim.* **1986**, *22*, 517-521.
17. Soloshonok, V.A.; Gerus, I.I.; Yagupolskii, Y.L.; Kukhar, V.P. *Zh. Org. Khim.* **1987**, *23*, 2308-2313.
18. Soloshonok, V.A.; Yagupolskii, Y.L.; Kukhar, V.P. *Zh. Org. Khim.* **1988**, *24*, 1638-1644.
19. Burger, K.; Gaa K. *Chem. Ztg.* **1991**, *114*, 101-104.
20. Burger K.; Sewald N. *Synthesis* **1990**, *2*, 115-118.
21. Osipov S.N.; Chkanikov N.D.; Kolomiets A.F.; Fokin A.V. *Bull. Acad. Sci. USSR, Chem. Sect. (Engl.)* **1986**, , 1256.

22. Osipov S.N.; Kolomiets A.F.; Fokin A.V. *Russ. Chem. Rev.* **1992**, *61*, 798.
23. Burger K.; Hollweck W.; *Synlett* **1994**, *9*, 751-753.
24. Sewald N.; Hollweck W.; Mütze K.; Schierlinger C.; Seymour L.C.; Gaa K.; Burger K.; Koksch B.; Jakubke H.-D. *Amino Acids* **1995,** *8*, 187-194.
25. Hollweck W.; Burger K. *J. Prakt. Chem.* **1995,** *337,* 391-396.
26. Hollweck W.; Sewald N.; Michel T.; K. Burger K. *Liebigs Ann. Chem.* **1997**, *12*, 2549-2551.
27. Osipov S.N.; Golubev A.S.; Sewald N.; Michel T.; Kolomiets A.F.; Fokin, A.V.; Burger K. *J. Org. Chem.* **1996**, *61*, 7521-7528.
28. Lazzaro, F.; Crucianelli, M.; De Angelis, F.; Frigerio, M.; Malpezzi, L.; Volonterio, A.; Zanda, M. *Tetrahedron: Asymmetry* **2004**, *15,* 889-893.
29. Lebouvier, N.; Laroche, C.; Huguenot, F.; Brigaud, T. *Tetrahedron Letters* **2002**, *43*, 2827-2830.
30. Asensio, A.; Bravo, P.; Crucianelli, M.; Farina, A.; Fustero, S.; Soler, J. G.; Meille, S. V.; Panzeri, W.; Viani, F.; Volonterio, A.; Zanda, M.. *Eur. J. Org. Chem.* **2001**, *8*, 1449-1458.
31. Galushko, S.V.; Shiskina, I.P.; Kobzev, S.P.; Soloshonok, V.A.; Yagupol`skii, Yu.L.; Kukhar`, V.P. *Zh. Anal. Khim.* **1988**, *43*, 2067-2069; *C.A.* **1989**, *110*, 111035z.
32. Keller, J.W.; Dick, K.O. *J. Chromatogr.* **1986**, *367*, 187-190.
33. Bravo, P.; Crucianelli, M.; Vergani, B.; Zanda, M. *Tetrahedron Lett.* **1998**, *39*, 7771-7774.
34. Keller, J.W.; Hamilton, B.J. *Tetrahedron Lett.* **1986**, 1249-1250.
35. Sewald N.; Seymour L.C.; Burger K.; Osipov S.N.; Kolomiets A.F.; Fokin A.V. *Tetrahedron: Asymmetry* **1994**, *5*, 1051-1060.
36. Bravo P.; Capelli S.; Meille S.V.; Viani F.; Zanda M.; Kukhar′ V.P.; Soloshonok V.A. *Tetrahedron: Asymmetry* **1994**, *5*, 2009-2018.
37. Burger, K.; Mütze, K.; Hollweck, W.; Koksch, B.; Kuhl, P.; Jakubke, H.-D.; Riede, J.; Schier, A. *J. prakt. Chem.* **1993**, *335*, 321-331.
38. Koksch, B.; Ueberham, U.; Jakubke, H.-D. *Pharmazie* **1997**, *52*, 74-75.
39. Schoemaker, H.E.; Boesten, W.H.J.; Kaptein, B.; Roos, E.C.; Broxterman, Q.B.; van den Tweel, W.J.J.; Kamphuis, J. *Acta Chem. Scand.* **1996**, *50*, 225-233.
40. Kaptein, B.; Boesten, W.H.J.; Broxterman. Q.B.; Peters, P.J.H.; Schoemaker, H.E.; Kamphuis, J. *Tetrahedron Asymm.* **1993**, *4*, 1113-1116.
41. Koksch, B.; Michel, T.; Kaptein, P.; Quadflieg, P.; Broxterman, Q.B. *Tetrahedron Asymm.* **2004**, *15*, 1401-1407.
42. Schellenberger, V.; Jakubke, H.-D. *Angew. Chem. Int. Ed. Engl.* **1991**, *30*, 1437-1449.
43. Moree, W.J.; Sears, P.; Kawashiro, K.; Witte, K.; Wong, C.-H. *J. Am. Chem. Soc.* **1997**, *119*, 3942-3947.

44. Bordusa, F.; Dahl, C.; Jakubke, H.-D.; Burger, K.; Koksch, B. *Tetrahedron: Asymmetry* **1999**, *10*, 307-313.
45. Koksch, B.; Sewald, N.; Burger, K.; Jakubke, H.-D. *Amino Acids* **1996**, *11*, 425-434.
46. Jakubke, H.-D. *J. Chin. Chem. Soc.* **1994**, *41*, 355-370.
47. Mitin, Y.V.; Schellenberger, V.; Schellenberger, U.; Jakubke, H.-D.; Zapevalova, N.P. In *Peptides*; Giralt, E.; Andreu, D., Eds.; ESCOM: Leiden, **1990**; pp 287-288.
48. Schellenberger, V.; Jakubke, H.-D.; Zapevalova, N.P.; Mitin, Y.V. *Biotechnol. Bioeng.* **1991**, *38*, 104-108.
49. Schellenberger, V.; Schellenberger, U.; Jakubke, H.-D.; Zapevalova, N.P.; Mitin, Y.V. *Biotechnol. Bioeng.* **1991**, *38*, 319-321.
50. Thormann, M.; Thust, S.; Hofmann, H.-J.; Bordusa, F. *Biochemistry* **1999,** *38*, 6056-6062.
51. Cerovsky, V.; Bordusa, F. *Journal of Peptide Research* **2000**, *55*, 325-329.
52. Cerovsky V.; Kockskamper J.; Glitsch H.G.; Bordusa, F. *Chembiochem* **2000**, *1*, 126-129.
53. Schechter, I.; Berger, A.C. *Biochem. Biophys. Res. Commun.* **1967**, *27*, 157-162.
54. S. Thust; B. Koksch. *J. Org. Chem.* **2003**, *68*, 2290-2296.
55. Erickson, J.A.; McLoughlin, J.I. *J. Org. Chem.* **1995**, *60*, 1626-1631.
56. Caminati, W.; Melandri, S.; Moreschini, P.; Favero, P.G. *Angew. Chem.* **1999**, *111*, 3105-3107.
57. Wehofsky, N.; Kirbach, S.W.; Haensler, M.; Wissmann, J.-D.; Bordusa, F. *Org. Lett.* **2000**, *2*, 2027-2030.
58. Kallwass, H.K.W.; Yee, C.; Blythe, T.A.; McNabb, T.J.; Rogers, E.E.;. Shames, S.L. *Bioorg. Med. Chem.* **1994**, *2*, 557-566.
59. Bai, Y.; Hayashi, R.; Hata, T. *J. Biochem. (Tokyo)* **1975**, *78*, 617-626.
60. Breddam, K. *Carlsberg Res. Commun.* **1984**, *49*, 535-554.
61. Endrizzi, J.A.; Breddam, K.; Remington, S.J. *Biochemistry* **1994**, *33*, 11106-11120.
62. Thust, S.; Koksch, B. Tetrahedron Lett. **2004**, *45*,
63. Eckstein, H.; Renner, H.-J. In *Peptides; Proceedings of the 24th European Peptide Symposium;* Edinburgh, **1996**, Mayflower Scientific, Kingswinford, UK, **1998**, pp 355-356.
64. Kohn, W.D.; Hodges, R.S. *Trends in Biotechnology* **1998**, *16*, 379-389.
65. Yu, Y.B. *Advanced Drug Delivery Reviews* **2002**, *54*, 1113-1129.
66. Wolf, E.; Kim, P.S.; Berger, B. *Protein Science* **1997**, *6*, 1179-1189.
67. O'Shea, E.K.; Klemm, J.D.; Kim, P.S.; Alber, T. *Science* **1991**, *254*, 539-544.
68. Sodek, J.; Hodges, R.S.; Smillie, L.B.; Jurasek, L. *Proc. Natl. Acad. Sci. USA* **1972**, *69*, 3800-3804.

69. Adamson, J.G.; Zhou, N.E.; Hodges, R.S. *Current Biology* **1993**, *4*, 428-437.
70. Tripet, B.; Wagschal, K.; Lavigne, P.; Mant, C.T.; Hodges, R.S. *J. Mol. Biol.* **2000**, *300*, 377-402.
71. Mason, J.; Arndt, K.M. *Chembiochem* **2004**, *5*, 170-176.
72. Monera, O.D.; Kay, C.M.; Hodges, R.S. *Biochemistry* **1994**, *33*, 3862-3871.
73. Moitra, J.; Szilák, L.; Krylov, D.; Vinson, C. *Biochemistry* **1997**, *36*, 12567-12573.
74. Saghatelian, A.; Yokobayashi, Y.; Soltani, K.; Ghadiri, M.K. *Nature* **2001**, *409*, 797-801.
75. Severin, K.; Lee, D.H.; Kennan, A.J.; Ghadiri, M.R. *Nature* **1997**, *389*, 706-709.
76. Ingenito, R.; Bianchi, E.; Fattori, D.; Pessi, A. *J. Am. Chem. Soc.* **1999**, *121*, 11369-11374.
77. Shin, Y.; Winans, K.A.; Backes, B.J.; Kent, S.B.H.; Ellman, J.A.; Bertozzi, C.R. *J. Am. Chem. Soc.* **1999**, *121*, 11684-11689.
78. Swinnen, D.; Hilvert, D. *Org. Lett.* **2000**, *2*, 2439-2442.
79. Tam, J.P.; Yu, Q.; Miao, Z. *Biopolymers* **1999**, *51*, 311-332.
80. Tam, J.P.; Xu, J.; Eom, K.D. *Biopolymers* **2001**, *60*, 194-205.
81. Kumar, K.; Yoder, N.C. *Chem. Soc. Rev.* **2002**, *31*, 335-341.
82. Kui, S.C.F.; Zhu, N.; Chan, M.C.W. *Angew. Chem.* **2003**, *115*, 1666-1670; *Angew. Chem. Int. Ed.* **2003**, *42*, 1628-1632.
83. Tang, Y.; Ghirlanda, G.; Petka, W.A.; Nakajima, T.; DeGrado, W.F.; Tirrell, D.A. *Angew. Chem. Int. Ed.* **2001**, *40*, 1494-1496.
84. Bilgiçer, B.; Kumar, K. *Tetrahedron* **2002**, *58*, 4105-4112.
85. Schlosser, M.; Michel, D. *Tetrahedron* **1996**, *52*, 99-108.
86. Nagai, T.; Nishioka, G.; Koyama, M.; Ando, A.; Miki, T.; Kumadaki, I. *J. Fluorine Chem.* **1992**, *57*, 229-237.
87. Mikami, K.; Itoh, Y.; Yamanaka, M. *Chemical Reviews* **2003**, *104*, 1-16.
88. Jäckel, C.; Seufert, W.; Thust S.; Koksch, B. *Chembiochem* **2004**, *5*, 717-720.
89. Resnati, G. *Tetrahedron* **1993**, *49*, 9385-9445.
90. Arai, T.; Imachi, T.; Kato, T.; Nishino, N. *Bull. Chem. Soc. Jpn.* **2000**, *73*, 439-445.
91. Zhou, N.E.; Kay, C.M.; Hodges, R.S. *J. Mol. Biol.* **1994**, *237*, 500-512.
92. Meier, M.; Lustig, A.; Aebi, U.; Burkhard, P. *J. Struct. Biol.* **2002**, *137*, 65-72.
93. Burkhard, P.; Ivaninski, S.; Lustig, A. *J. Mol. Biol.* **2002**, *318*, 901-910.
94. Phelan, P.; Gorfe, A.A.; Jelesarov, I.; Marti, D.N.; Warwicker, J.; Bossard, H.R. *Biochemistry* **2002**, *41*, 2998-3008.
95. Molero, B.I.; Zitzewitz, J.A.; Matthews, C.R. *J. Mol. Biol.* **2003**, *336*, 989-996.

Indexes

Author Index

Subject Index

A

B

D

E

F

G

H

I

K

L

M

N

O

P

R

S

T

U

V

W